INTRODUCTION TO CHEMICAL REACTION ENGINEERING AND KINETICS

INTRODUCTION TO CHEMICAL REACTION ENGINEERING AND KINETICS

Ronald W. Missen
Charles A. Mims
Bradley A. Saville

Department of Chemical Engineering and Applied Chemistry
University of Toronto

John Wiley & Sons, Inc.
New York • Chichester • Weinheim • Brisbane • Singapore • Toronto

Acquisitions Editor	Wayne Anderson
Marketing Manager	Katherine Hepburn
Freelance Production Manager	Jeanine Furino
Designer	Laura Boucher
Illustration Editor	Gene Aiello
Outside Production Management	Hermitage Publishing Services
Cover Design	Keithley Associates

This book was set in Times Ten by Publication Services and printed and bound by Hamilton Printing. The cover was printed by Lehigh Press.

This book is printed on acid-free paper. ∞

The paper in this book was manufactured by a mill whose forest management programs include sustained yield harvesting of its timberlands. Sustained yield harvesting principles ensure that the number of trees cut each year does not exceed the amount of new growth.

Copyright 1999 © John Wiley & Sons, Inc. All rights reserved.

No part of this publication may be reproduced, stored in a retrieval system or transmitted in any form or by any means, electronic, mechanical, photocopying, recording, scanning or otherwise, except as permitted under Sections 107 and 108 of the 1976 United States Copyright Act, without either the prior written permission of the Publisher, or authorization through payment of the appropriate per-copy fee to the Copyright Clearance Center, 222 Rosewood Drive, Danvers, MA 01923, (508) 750-8400, fax (508) 750-4470. Requests to the Publisher for permission should be addressed to the Permissions Department, John Wiley & Sons, Inc., 605 Third Avenue, New York, NY 10158-0012, (212) 850-6011, fax (212) 850-6008, E-Mail: PERMREQ@WILEY.COM.

Library of Congress Cataloging-in-Publication Data:

Missen, Ronald W. (Ronald William), 1928–
    Introduction to chemical reaction engineering and kinetics /
Ronald W. Missen, Charles A. Mims, Bradley A. Saville.
      p.   cm.
    Includes bibliographical references and index.
    ISBN 0-471-16339-2 (cloth : alk. paper)
    1. Chemical reactors. 2. Chemical kinetics. I. Mims, Charles A.
II. Saville, Bradley A. III. Title.
TP157.M538 1999
660′.2832—dc21   98-27267
                                                                         CIP

Printed in the United States of America

10 9 8 7 6 5 4 3 2

To our children

Preface

Introduction to Chemical Reaction Engineering and Kinetics is written primarily for a first course in chemical reaction engineering (CRE) for undergraduate students in chemical engineering. The purpose of the work is to provide students with a thorough introduction to the fundamental aspects of chemical reactor analysis and design. For this purpose, it is necessary to develop a knowledge of chemical kinetics, and therefore the work has been divided into two inter-related parts: chemical kinetics and CRE. Included with this book is a CD-ROM containing computer software that can be used for numerical solutions to many of the examples and problems within the book. The work is primarily based on material given to undergraduate students in the Department of Chemical Engineering and Applied Chemistry at the University of Toronto.

Scope and Organization of Material

The material in this book deals with kinetics and reactors. We realize that students in many institutions have an introduction to chemical kinetics in a course on physical chemistry. However, we strongly believe that for chemical engineering students, kinetics should be fully developed within the context of, and from the point of view of, CRE. Thus, the development given here differs in several important respects from that given in physical chemistry. Ideal-flow reactor models are introduced early in the book (Chapter 2) because of their use in kinetics investigations, and to get students accustomed to the concepts early. Furthermore, there is the additional purpose of drawing a distinction between a *reaction* model (network) or kinetics scheme, on the one hand, and a *reactor* model that incorporates a kinetics scheme, on the other. By a reaction model, we mean the development in chemical engineering kinetics of an appropriate (local or point) rate law, including, in the case of a multiphase system, the effects of rate processes other than chemical reaction itself. By contrast, a reactor model uses the rate law, together with considerations of residence-time and (if necessary) particle-size distributions, heat, mass, and momentum transfer, and fluid mixing and flow patterns, to establish the global behavior of a reacting system in a vessel.

We deliberately separate the treatment of characterization of ideal flow (Chapter 13) and of nonideal flow (Chapter 19) from the treatment of reactors involving such flow. This is because (1) the characterization can be applied to situations other than those involving chemical reactors; and (2) it is useful to have the characterization complete in the two locations so that it can be drawn on for whatever reactor application ensues in Chapters 14–18 and 20–24. We also incorporate nonisothermal behavior in the discussion of each reactor type as it is introduced, rather than treat this behavior separately for various reactor types.

Our treatment of chemical kinetics in Chapters 2–10 is such that no previous knowledge on the part of the student is assumed. Following the introduction of simple reactor models, mass-balance equations and interpretation of rate of reaction in Chapter 2, and measurement of rate in Chapter 3, we consider the development of rate laws for single-phase simple systems in Chapter 4, and for complex systems in Chapter 5. This is followed by a discussion of theories of reaction and reaction mechanisms in Chapters 6 and 7. Chapter 8 is devoted to catalysis of various types. Chapter 9 is devoted to reactions in multiphase systems. The treatment of chemical kinetics concludes in Chapter 10 with a discussion of enzyme kinetics in biochemical reactions.

Our treatment of Chemical Reaction Engineering begins in Chapters 1 and 2 and continues in Chapters 11–24. After an introduction (Chapter 11) surveying the field, the next five Chapters (12–16) are devoted to performance and design characteristics of four ideal reactor models (batch, CSTR, plug-flow, and laminar-flow), and to the characteristics of various types of ideal flow involved in continuous-flow reactors. Chapter 17 deals with comparisons and combinations of ideal reactors. Chapter 18 deals with ideal reactors for complex (multireaction) systems. Chapters 19 and 20 treat nonideal flow and reactor considerations taking this into account. Chapters 21–24 provide an introduction to reactors for multiphase systems, including fixed-bed catalytic reactors, fluidized-bed reactors, and reactors for gas-solid and gas-liquid reactions.

Ways to Use This Book in CRE Courses

One way in which the material can be used is illustrated by the practice at the University of Toronto. Chapters 1–8 (sections 8.1–8.4) on chemical kinetics are used for a 40-lecture (3 per week) course in the fall term of the third year of a four-year program; the lectures are accompanied by weekly 2-hour tutorial (problem-solving) sessions. Chapters on CRE (11–15, 17, 18, and 21) together with particle-transport kinetics from section 8.5 are used for a similarly organized course in the spring term. There is more material than can be adequately treated in the two terms. In particular, it is not the practice to deal with all the aspects of nonideal flow and multiphase systems that are described. This approach allows both flexibility in choice of topics from year to year, and material for an elective fourth-year course (in support of our plant design course), drawn primarily from Chapters 9, 19, 20, and 22–24.

At another institution, the use of this material depends on the time available, the requirements of the students, and the interests of the instructor. The possibilities include:

   (1) a basic one-semester course in CRE primarily for simple, homogeneous systems, using Chapters 1–4 (for kinetics, if required) and Chapters 11–17;
   (2) an extension of (1) to include complex, homogeneous systems, using Chapters 5 (for kinetics) and 18 in addition;
   (3) a further extension of (1) and (2) to include heterogeneous systems using Chapters 8 and 9 (for kinetics), and selected parts of Chapters 21–24;
   (4) a final extension to nonideal flow, using Chapters 19 and 20.

In addition, Chapters 6 and 7 could be reserved for the enrichment of the treatment of kinetics, and Chapter 10 can be used for an introduction to enzyme kinetics dealing with some of the problems in the reactor design chapters.

Reviewers have suggested that this book may be used both at the undergraduate level and at the beginning of a graduate course. The latter is not our intention or our practice, but we leave this to the discretion and judgement of individual instructors.

Problem Solving and Computer Tools

We place primary emphasis on developing the students' abilities to establish the working equations of an appropriate model for a particular reactor situation, and of course to interpret and appreciate the significance of quantitative results. In an introductory text in a field such as CRE, it is important to emphasize the development of principles, and to illustrate their application by means of relatively simple and idealized problem situations that can be solved with a calculator. However, with the availability of computer-based solution techniques, it is desirable to go beyond this approach for several reasons:

(1) Computer software allows the solution of more complex problems that require numerical, as opposed to analytical, techniques. Thus, a student can explore situations that more closely approximate real reactor designs and operating conditions. This includes studying the sensitivity of a calculated result to changing operating conditions.
(2) The limitations of analytical solutions may also interfere with the illustration of important features of reactions and of reactors. The consequences of linear behavior, such as first-order kinetics, may be readily demonstrated in most cases by analytical techniques, but those of nonlinear behavior, such as second-order or Langmuir-Hinshelwood kinetics, generally require numerical techniques.
(3) The development of mechanistic rate laws also benefits from computer simulations. All relevant elementary steps can be included, whereas, with analytical techniques, such an exploration is usually impossible.
(4) Computer-aided visual demonstrations in lectures and tutorials are desirable for topics that involve spatial and/or time-dependent aspects.

For these reasons, we include examples and problems that require numerical techniques for their solution together with suitable computer software (described below).

Computer Software: E-Z Solve: The Engineer's Equation Solving and Analysis Tool

Accompanying this book is a CD-ROM containing the computer software E-Z Solve, developed by IntelliPro, Inc and distributed by John Wiley & Sons, Inc. It can be used for parameter estimation and equation solving, including solution of sets of both nonlinear algebraic equations and differential equations. It is extremely easy to learn and use. We have found that a single 2-hour tutorial is sufficient to instruct students in its application. We have also used it in research problems, such as modeling of transient behavior in kinetics investigations. Other computer software programs may be used, if appropriate, to solve most of the examples and problems in the text that are solved with the aid of E-Z Solve (indicated in the text by a computer icon shown in the margin above). The successful use of the text is not restricted to the use of E-Z Solve for software support, although we encourage its use because of its capabilities for nonlinear parameter estimation and solution of coupled differential and algebraic equations. Appendix D provides examples illustrating the use of the software for these types of problems, along with the required syntax.

Web Site

A web site at www.wiley.com/college/missen is available for ongoing support of this book. It includes resources to assist students and instructors with the subject matter, such as sample files, demonstrations, and a description of the E-Z Solve software appearing on the CD-ROM that accompanies this book.

Acknowledgments

We acknowledge our indebtedness to those who have contributed to the literature on the topics presented here, and on whose work we have drawn. We are grateful for the contributions of S.T. Balke, W.H. Burgess, and M.J. Phillips, who have participated in the undergraduate courses, and for discussions with W.R. Smith. We very much appreciate the comments on the manuscript received from reviewers. CAM credits, in addition to his academic colleagues, his former coworkers in industry for a deep and continuing education into the subject matter.

We are also grateful for the assistance given by Esther Oostdyk, who entered the manuscript; by Lanny Partaatmadja, who entered material for the "Instructor Resources"; and by Mark Eichhorn, Nick Palozzi, Chris Ho, Winnie Chiu and Lanny Partaatmadja, who worked on graphics and on problems for the various chapters. We also thank Nigel Waithe, who produced copies of draft material for the students. We thank our students for their forbearance and comments, both written and oral, during the development of this book.

The development of the computer tools and their integration with the subject matter required strong support from Wayne Anderson and the late Cliff Robichaud at Wiley, and Philippe Marchal and his staff at Intellipro. Their assistance is gratefully acknowledged. We also thank the staff at Wiley and Larry Meyer and his staff at Hermitage Publishing Services for their fine work during the production phase.

Support for the development of the manuscript has been provided by the Department of Chemical Engineering and Applied Chemistry, the Faculty of Applied Science and Engineering, and the Office of the Provost, University of Toronto.

Ronald W. Missen
Charles A. Mims
Bradley A. Saville
Toronto, Ontario. May, 1998

Contents

1 • INTRODUCTION 1
    1.1 Nature and Scope of Chemical Kinetics 1
    1.2 Nature and Scope of Chemical Reaction Engineering 1
    1.3 Kinetics and Chemical Reaction Engineering 2
    1.4 Aspects of Kinetics 3
        1.4.1 Rate of Reaction—Definition 3
        1.4.2 Parameters Affecting Rate of Reaction: The Rate Law 4
        1.4.3 Measurement of Rate of Reaction—Preliminary 5
        1.4.4 Kinetics and Chemical Reaction Stoichiometry 6
        1.4.5 Kinetics and Thermodynamics/Equilibrium 14
        1.4.6 Kinetics and Transport Processes 15
    1.5 Aspects of Chemical Reaction Engineering 15
        1.5.1 Reactor Design and Analysis of Performance 15
        1.5.2 Parameters Affecting Reactor Performance 16
        1.5.3 Balance Equations 16
        1.5.4 An Example of an Industrial Reactor 18
    1.6 Dimensions and Units 19
    1.7 Plan of Treatment in Following Chapters 21
        1.7.1 Organization of Topics 21
        1.7.2 Use of Computer Software for Problem Solving 21
    1.8 Problems for Chapter 1 22

2 • KINETICS AND IDEAL REACTOR MODELS 25
    2.1 Time Quantities 25
    2.2 Batch Reactor (BR) 26
        2.2.1 General Features 26
        2.2.2 Material Balance; Interpretation of r_i 27
    2.3 Continuous Stirred-Tank Reactor (CSTR) 29
        2.3.1 General Features 29
        2.3.2 Material Balance; Interpretation of r_i 31
    2.4 Plug-Flow Reactor (PFR) 33
        2.4.1 General Features 33
        2.4.2 Material Balance; Interpretation of r_i 34
    2.5 Laminar-Flow Reactor (LFR) 36
    2.6 Summary of Results for Ideal Reactor Models 38
    2.7 Stoichiometric Table 39
    2.8 Problems for Chapter 2 40

3 • EXPERIMENTAL METHODS IN KINETICS: MEASUREMENT OF RATE OF REACTION 42
    3.1 Features of a Rate Law: Introduction 42
        3.1.1 Separation of Effects 42
        3.1.2 Effect of Concentration: Order of Reaction 42
        3.1.3 Effect of Temperature: Arrhenius Equation; Activation Energy 44

3.2 **Experimental Measurements: General Considerations** 45
3.3 **Experimental Methods to Follow the Extent of Reaction** 46
    3.3.1 *Ex-situ* and *In-situ* Measurement Techniques 46
    3.3.2 Chemical Methods 46
    3.3.3 Physical Methods 47
    3.3.4 Other Measured Quantities 48
3.4 **Experimental Strategies for Determining Rate Parameters** 48
    3.4.1 Concentration-Related Parameters: Order of Reaction 49
    3.4.2 Experimental Aspects of Measurement of Arrhenius Parameters A and E_A 57
3.5 **Notes on Methodology for Parameter Estimation** 57
3.6 **Problems for Chapter 3** 61

4 • DEVELOPMENT OF THE RATE LAW FOR A SIMPLE SYSTEM 64

4.1 **The Rate Law** 64
    4.1.1 Form of Rate Law Used 64
    4.1.2 Empirical versus Fundamental Rate Laws 65
    4.1.3 Separability versus Nonseparability of Effects 66
4.2 **Gas-Phase Reactions: Choice of Concentration Units** 66
    4.2.1 Use of Partial Pressure 66
    4.2.2 Rate and Rate Constant in Terms of Partial Pressure 67
    4.2.3 Arrhenius Parameters in Terms of Partial Pressure 68
4.3 **Dependence of Rate on Concentration** 69
    4.3.1 First-Order Reactions 69
    4.3.2 Second-Order Reactions 71
    4.3.3 Third-Order Reactions 72
    4.3.4 Other Orders of Reaction 75
    4.3.5 Comparison of Orders of Reaction 75
    4.3.6 Product Species in the Rate Law 78
4.4 **Dependence of Rate on Temperature** 79
    4.4.1 Determination of Arrhenius Parameters 79
    4.4.2 Arrhenius Parameters and Choice of Concentration Units for Gas-Phase Reactions 80
4.5 **Problems for Chapter 4** 80

5 • COMPLEX SYSTEMS 87

5.1 **Types and Examples of Complex Systems** 87
    5.1.1 Reversible (Opposing) Reactions 87
    5.1.2 Reactions in Parallel 88
    5.1.3 Reactions in Series 88
    5.1.4 Combinations of Complexities 88
    5.1.5 Compartmental or Box Representation of Reaction Network 89
5.2 **Measures of Reaction Extent and Selectivity** 90
    5.2.1 Reaction Stoichiometry and Its Significance 90
    5.2.2 Fractional Conversion of a Reactant 91
    5.2.3 Yield of a Product 91
    5.2.4 Overall and Instantaneous Fractional Yield 92
    5.2.5 Extent of Reaction 93
    5.2.6 Stoichiometric Table for Complex System 93
5.3 **Reversible Reactions** 94
    5.3.1 Net Rate and Forms of Rate Law 94
    5.3.2 Thermodynamic Restrictions on Rate and on Rate Laws 95
    5.3.3 Determination of Rate Constants 97
    5.3.4 Optimal T for Exothermic Reversible Reaction 99
5.4 **Parallel Reactions** 100
5.5 **Series Reactions** 103

5.6 Complexities Combined 106
   5.6.1 Concept of Rate-Determining Step (*rds*) 106
   5.6.2 Determination of Reaction Network 106
5.7 Problems for Chapter 5 108

6 • FUNDAMENTALS OF REACTION RATES 115
6.1 Preliminary Considerations 115
   6.1.1 Relating to Reaction-Rate Theories 115
   6.1.2 Relating to Reaction Mechanisms and Elementary Reactions 116
6.2 Description of Elementary Chemical Reactions 117
   6.2.1 Types of Elementary Reactions 117
   6.2.2 General Requirements for Elementary Chemical Reactions 120
6.3 Energy in Molecules 120
   6.3.1 Potential Energy in Molecules—Requirements for Reaction 120
   6.3.2 Kinetic Energy in Molecules 126
6.4 Simple Collision Theory of Reaction Rates 128
   6.4.1 Simple Collision Theory (SCT) of Bimolecular Gas-Phase Reactions 129
   6.4.2 Collision Theory of Unimolecular Reactions 134
   6.4.3 Collision Theory of Bimolecular Combination Reactions; Termolecular Reactions 137
6.5 Transition State Theory (TST) 139
   6.5.1 General Features of the TST 139
   6.5.2 Thermodynamic Formulation 141
   6.5.3 Quantitative Estimates of Rate Constants Using TST with Statistical Mechanics 143
   6.5.4 Comparison of TST with SCT 145
6.6 Elementary Reactions Involving Other Than Gas-Phase Neutral Species 146
   6.6.1 Reactions in Condensed Phases 146
   6.6.2 Surface Phenomena 147
   6.6.3 Photochemical Elementary Reactions 149
   6.6.4 Reactions in Plasmas 150
6.7 Summary 151
6.8 Problems for Chapter 6 152

7 • HOMOGENEOUS REACTION MECHANISMS AND RATE LAWS 154
7.1 Simple Homogeneous Reactions 155
   7.1.1 Types of Mechanisms 155
   7.1.2 Open-Sequence Mechanisms: Derivation of Rate Law from Mechanism 155
   7.1.3 Closed-Sequence Mechanisms; Chain Reactions 157
   7.1.4 Photochemical Reactions 163
7.2 Complex Reactions 164
   7.2.1 Derivation of Rate Laws 164
   7.2.2 Computer Modeling of Complex Reaction Kinetics 165
7.3 Polymerization Reactions 165
   7.3.1 Chain-Reaction Polymerization 166
   7.3.2 Step-Change Polymerization 168
7.4 Problems for Chapter 7 170

8 • CATALYSIS AND CATALYTIC REACTIONS 176
8.1 Catalysis and Catalysts 176
   8.1.1 Nature and Concept 176
   8.1.2 Types of Catalysis 178
   8.1.3 General Aspects of Catalysis 179
8.2 Molecular Catalysis 182
   8.2.1 Gas-Phase Reactions 182
   8.2.2 Acid-Base Catalysis 183

8.2.3 Other Liquid-Phase Reactions   186
8.2.4 Organometallic Catalysis   186
8.3 Autocatalysis   187
8.4 Surface Catalysis: Intrinsic Kinetics   191
8.4.1 Surface-Reaction Steps   191
8.4.2 Adsorption Without Reaction: Langmuir Adsorption Isotherm   192
8.4.3 Langmuir-Hinshelwood (LH) Kinetics   195
8.4.4 Beyond Langmuir-Hinshelwood Kinetics   197
8.5 Heterogeneous Catalysis: Kinetics in Porous Catalyst Particles   198
8.5.1 General Considerations   198
8.5.2 Particle Density and Voidage (Porosity)   199
8.5.3 Modes of Diffusion; Effective Diffusivity   199
8.5.4 Particle Effectiveness Factor η 201
8.5.5 Dependence of η on Temperature   210
8.5.6 Overall Effectiveness Factor η_o 212
8.6 Catalyst Deactivation and Regeneration   214
8.6.1 Fouling   214
8.6.2 Poisoning   215
8.6.3 Sintering   215
8.6.4 How Deactivation Affects Performance   216
8.6.5 Methods for Catalyst Regeneration   216
8.7 Problems for Chapter 8   218

9 • MULTIPHASE REACTING SYSTEMS   224
9.1 Gas-Solid (Reactant) Systems   224
9.1.1 Examples of Systems   224
9.1.2 Constant-Size Particle   225
9.1.3 Shrinking Particle   237
9.2 Gas-Liquid Systems   239
9.2.1 Examples of Systems   239
9.2.2 Two-Film Mass-Transfer Model for Gas-Liquid Systems   240
9.2.3 Kinetics Regimes for Two-Film Model   242
9.3 Intrinsic Kinetics of Heterogeneous Reactions Involving Solids   255
9.4 Problems for Chapter 9   257

10 • BIOCHEMICAL REACTIONS: ENZYME KINETICS   261
10.1 Enzyme Catalysis   261
10.1.1 Nature and Examples of Enzyme Catalysis   261
10.1.2 Experimental Aspects   263
10.2 Models of Enzyme Kinetics   264
10.2.1 Michaelis-Menten Model   264
10.2.2 Briggs-Haldane Model   266
10.3 Estimation of K_m and V_{max} 267
10.3.1 Linearized Form of the Michaelis-Menten Equation   267
10.3.2 Linearized Form of the Integrated Michaelis-Menten Equation   269
10.3.3 Nonlinear Treatment   269
10.4 Inhibition and Activation in Enzyme Reactions   269
10.4.1 Substrate Effects   270
10.4.2 External Inhibitors and Activators   272
10.5 Problems for Chapter 10   276

11 • PRELIMINARY CONSIDERATIONS IN CHEMICAL REACTION ENGINEERING   279
11.1 Process Design and Mechanical Design   279
11.1.1 Process Design   279
11.1.2 Mechanical Design   283

11.2 Examples of Reactors for Illustration of Process Design Considerations   283
    11.2.1 Batch Reactors   283
    11.2.2 Stirred-Tank Flow Reactors   284
    11.2.3 Tubular Flow Reactors   284
    11.2.4 Fluidized-Bed Reactors   290
    11.2.5 Other Types of Reactors   291
11.3 Problems for Chapter 11   292

12 • BATCH REACTORS (BR)   294

12.1 Uses of Batch Reactors   294
12.2 Batch Versus Continuous Operation   295
12.3 Design Equations for a Batch Reactor   296
    12.3.1 General Considerations   296
    12.3.2 Isothermal Operation   300
    12.3.3 Nonisothermal Operation   304
    12.3.4 Optimal Performance for Maximum Production Rate   307
12.4 Semibatch and Semicontinuous Reactors   309
    12.4.1 Modes of Operation: Semibatch and Semicontinuous Reactors   309
    12.4.2 Advantages and Disadvantages (Semibatch Reactor)   310
    12.4.3 Design Aspects   311
12.5 Problems for Chapter 12   313

13 • IDEAL FLOW   317

13.1 Terminology   317
13.2 Types of Ideal Flow; Closed and Open Vessels   318
    13.2.1 Backmix Flow (BMF)   318
    13.2.2 Plug Flow (PF)   318
    13.2.3 Laminar Flow (LF)   318
    13.2.4 Closed and Open Vessels   318
13.3 Characterization of Flow By Age-Distribution Functions   319
    13.3.1 Exit-Age Distribution Function E 319
    13.3.2 Cumulative Residence-Time Distribution Function F 321
    13.3.3 Washout Residence-Time Distribution Function W 322
    13.3.4 Internal-Age Distribution Function I 322
    13.3.5 Holdback H 322
    13.3.6 Summary of Relationships Among Age-Distribution Functions   322
    13.3.7 Moments of Distribution Functions   323
13.4 Age-Distribution Functions for Ideal Flow   325
    13.4.1 Backmix Flow (BMF)   325
    13.4.2 Plug Flow (PF)   327
    13.4.3 Laminar Flow (LF)   330
    13.4.4 Summary of Results for Ideal Flow   332
13.5 Segregated Flow   332
13.6 Problems for Chapter 13   333

14 • CONTINUOUS STIRRED-TANK REACTORS (CSTR)   335

14.1 Uses of a CSTR   336
14.2 Advantages and Disadvantages of a CSTR   336
14.3 Design Equations for a Single-Stage CSTR   336
    14.3.1 General Considerations; Material and Energy Balances   336
    14.3.2 Constant-Density System   339
    14.3.3 Variable-Density System   344
    14.3.4 Existence of Multiple Stationary States   347
14.4 Multistage CSTR   355
    14.4.1 Constant-Density System; Isothermal Operation   357
    14.4.2 Optimal Operation   358
14.5 Problems for Chapter 14   361

15 • PLUG FLOW REACTORS (PFR)   365
15.1 Uses of a PFR   365
15.2 Design Equations for a PFR   366
  15.2.1 General Considerations; Material, Energy and Momentum Balances   366
  15.2.2 Constant-Density System   370
  15.2.3 Variable-Density System   376
15.3 Recycle Operation of a PFR   380
  15.3.1 Constant-Density System   381
  15.3.2 Variable-Density System   386
15.4 Combinations of PFRs: Configurational Effects   387
15.5 Problems for Chapter 15   389

16 • LAMINAR FLOW REACTORS (LFR)   393
16.1 Uses of an LFR   393
16.2 Design Equations for an LFR   394
  16.2.1 General Considerations and Material Balance   394
  16.2.2 Fractional Conversion and Concentration (Profiles)   395
  16.2.3 Size of Reactor   397
  16.2.4 Results for Specific Rate Laws   397
  16.2.5 Summary of Results for LFR   399
  16.2.6 LFR Performance in Relation to SFM   400
16.3 Problems for Chapter 16   400

17 • COMPARISONS AND COMBINATIONS OF IDEAL REACTORS   402
17.1 Single-Vessel Comparisons   402
  17.1.1 BR and CSTR   402
  17.1.2 BR and PFR   404
  17.1.3 CSTR and PFR   405
  17.1.4 PFR, LFR, and CSTR   406
17.2 Multiple-Vessel Configurations   408
  17.2.1 CSTRs in Parallel   409
  17.2.2 CSTRs in Series: RTD   410
  17.2.3 PFR and CSTR Combinations in Series   413
17.3 Problems for Chapter 17   418

18 • COMPLEX REACTIONS IN IDEAL REACTORS   422
18.1 Reversible Reactions   422
18.2 Parallel Reactions   426
18.3 Series Reactions   429
  18.3.1 Series Reactions in a BR or PFR   429
  18.3.2 Series Reactions in a CSTR   430
18.4 Choice of Reactor and Design Considerations   432
  18.4.1 Reactors for Reversible Reactions   433
  18.4.2 Reactors for Parallel-Reaction Networks   435
  18.4.3 Reactors for Series-Reaction Networks   437
  18.4.4 Reactors for Series-Parallel Reaction Networks   441
18.5 Problems for Chapter 18   445

19 • NONIDEAL FLOW   453
19.1 General Features of Nonideal Flow   453
19.2 Mixing: Macromixing and Micromixing   454
19.3 Characterization of Nonideal Flow in Terms of RTD   455
  19.3.1 Applications of RTD Measurements   455
  19.3.2 Experimental Measurement of RTD   455

Contents xvii

19.4 One-Parameter Models for Nonideal Flow 471
    19.4.1 Tanks-in-Series (TIS) Model 471
    19.4.2 Axial Dispersion or Dispersed Plug Flow (DPF) Model 483
    19.4.3 Comparison of DPF and TIS Models 490
19.5 Problems for Chapter 19 490

20 • REACTOR PERFORMANCE WITH NONIDEAL FLOW 495
20.1 Tanks-in-Series (TIS) Reactor Model 495
20.2 Axial Dispersion Reactor Model 499
20.3 Segregated-Flow Reactor Model (SFM) 501
20.4 Maximum-Mixedness Reactor Model (MMM) 502
20.5 Performance Characteristics for Micromixing Models 504
20.6 Problems for Chapter 20 508

21 • FIXED-BED CATALYTIC REACTORS FOR FLUID-SOLID REACTIONS 512
21.1 Examples of Reactions 512
21.2 Types of Reactors and Modes of Operation 514
    21.2.1 Reactors for Two-Phase Reactions 514
    21.2.2 Flow Arrangement 514
    21.2.3 Thermal and Bed Arrangement 514
21.3 Design Considerations 516
    21.3.1 Considerations of Particle and Bed Characteristics 516
    21.3.2 Fluid-Particle Interaction; Pressure Drop $(-\Delta P)$ 517
    21.3.3 Considerations Relating to a Reversible Reaction 519
21.4 A Classification of Reactor Models 523
21.5 Pseudohomogeneous, One-Dimensional, Plug-Flow Model 527
    21.5.1 Continuity Equation 527
    21.5.2 Optimal Single-Stage Operation 528
    21.5.3 Adiabatic Operation 529
    21.5.4 Nonadiabatic Operation 542
21.6 Heterogeneous, One-Dimensional, Plug-Flow Model 544
21.7 One-Dimensional Versus Two-Dimensional Models 546
21.8 Problems for Chapter 21 546

22 • REACTORS FOR FLUID-SOLID (NONCATALYTIC) REACTIONS 552
22.1 Reactions and Reaction Kinetics Models 552
22.2 Reactor Models 553
    22.2.1 Factors Affecting Reactor Performance 553
    22.2.2 Semicontinuous Reactors 553
    22.2.3 Continuous Reactors 554
    22.2.4 Examples of Continuous Reactor Models 556
    22.2.5 Extension to More Complex Cases 563
22.3 Problems for Chapter 22 566

23 • FLUIDIZED-BED AND OTHER MOVING-PARTICLE REACTORS FOR FLUID-SOLID REACTIONS 569
23.1 Moving-Particle Reactors 570
    23.1.1 Some Types 570
    23.1.2 Examples of Reactions 572
    23.1.3 Advantages and Disadvantages 573
    23.1.4 Design Considerations 574
23.2 Fluid-Particle Interactions 574
    23.2.1 Upward Flow of Fluid Through Solid Particles: $(-\Delta P)$ Regimes 575
    23.2.2 Minimum Fluidization Velocity (u_{mf}) 575

23.2.3 Elutriation and Terminal Velocity (u_t) 577
23.2.4 Comparison of u_{mf} and u_t 578
23.3 Hydrodynamic Models of Fluidization 579
23.3.1 Two-Region Model (Class (1)) 579
23.3.2 Kunii-Levenspiel (KL) Bubbling-Bed Model (Class (2)) 580
23.4 Fluidized-Bed Reactor Models 584
23.4.1 KL Model for Fine Particles 584
23.4.2 KL Model for Intermediate-Size Particles 592
23.4.3 Model for Large Particles 595
23.4.4 Reaction in Freeboard and Distributor Regions 595
23.5 Problems for Chapter 23 596

24 • REACTORS FOR FLUID-FLUID REACTIONS 599
24.1 Types of Reactions 599
24.1.1 Separation-Process Point of View 599
24.1.2 Reaction-Process Point of View 599
24.2 Types of Reactors 600
24.2.1 Tower or Column Reactors 600
24.2.2 Tank Reactors 602
24.3 Choice of Tower or Tank Reactor 602
24.4 Tower Reactors 603
24.4.1 Packed-Tower Reactors 603
24.4.2 Bubble-Column Reactors 608
24.5 Tank Reactors 614
24.5.1 Continuity Equations for Tank Reactors 614
24.5.2 Correlations for Design Parameters for Tank Reactors 615
24.6 Trickle-Bed Reactor: Three-Phase Reactions 618
24.7 Problems for Chapter 24 619

APPENDIX A 623
A.1 Common Conversion Factors for Non-SI Units to SI Units 623
A.2 Values of Physicochemical Constants 623
A.3 Standard SI Prefixes 624

APPENDIX B: BIBLIOGRAPHY 625
B.1 Books on Chemical Reactors 625
B.2 Books on Chemical Kinetics and Catalysis 626

APPENDIX C: ANSWERS TO SELECTED PROBLEMS 627

APPENDIX D: USE OF E-Z SOLVE FOR EQUATION SOLVING AND PARAMETER ESTIMATION 635

NOMENCLATURE 643

REFERENCES 652

INDEXES 657

Chapter 1

Introduction

In this introductory chapter, we first consider what chemical kinetics and chemical reaction engineering (CRE) are about, and how they are interrelated. We then introduce some important aspects of kinetics and CRE, including the involvement of chemical stoichiometry, thermodynamics and equilibrium, and various other rate processes. Since the rate of reaction is of primary importance, we must pay attention to how it is defined, measured, and represented, and to the parameters that affect it. We also introduce some of the main considerations in reactor design, and parameters affecting reactor performance. These considerations lead to a plan of treatment for the following chapters.

Of the two themes in this book, kinetics and CRE, the latter is the main objective, and we consider kinetics primarily as it contributes to, and is a part of, CRE.

1.1 NATURE AND SCOPE OF CHEMICAL KINETICS

Chemical kinetics is concerned with the rates of chemical reactions, that is, with the quantitative description of how fast chemical reactions occur, and the factors affecting these rates. The chemist uses kinetics as a tool to understand fundamental aspects of reaction pathways, a subject that continues to evolve with ongoing research. The applied chemist uses this understanding to devise new and/or better ways of achieving desired chemical reactions. This may involve improving the yield of desired products or developing a better catalyst. The chemical engineer uses kinetics for reactor design in chemical reaction or process engineering.

A legitimate objective of chemical kinetics is to enable us to predict beforehand the rate at which given chemical substances react, and to control the rate in some desirable fashion; alternatively, it is to enable us to "tailor" chemical reactions so as to produce substances with desirable chemical characteristics in a controllable manner, including choice of an appropriate catalyst. Quantum mechanical calculations theoretically provide the tools for such predictions. Even with today's powerful computers, however, we are far from being in a position to do this in general, and we must study experimentally each reacting system of interest in order to obtain a quantitative kinetics description of it.

1.2 NATURE AND SCOPE OF CHEMICAL REACTION ENGINEERING

Chemical reaction engineering (CRE) is concerned with the rational design and/or analysis of performance of chemical reactors. What is a chemical reactor, and what does its rational design involve? A chemical reactor is a device in which change in composition of matter occurs by chemical reaction. The chemical reaction is normally the most important change, and the device is designed to accomplish that change. A reactor is usually the "heart" of an overall chemical or biochemical process. Most industrial chemical processes are operated for the purpose of producing chemical products such as ammonia and petrochemicals. Reactors are also involved in energy production, as in engines (internal-combustion, jet, rocket, etc.) and in certain electrochemical cells (lead-acid, fuel). In animate objects (e.g., the human body), both are involved. The rational design of this last is rather beyond our capabilities but, otherwise, in general, design includes determining the type, size, configuration, cost, and operating conditions of the device.

A legitimate objective of CRE is to enable us to predict, in the sense of rational design, the performance of a reactor created in response to specified requirements and in accordance with a certain body of information. Although great strides have been taken in the past few decades toward fulfilling this objective, in many cases the best guide is to base it, to some extent, on the performance of "the last one built."

1.3 KINETICS AND CHEMICAL REACTION ENGINEERING

In chemical kinetics, the chemical reactor used to carry out the reaction is a tool for determining something about the reacting system: rate of reaction, and dependence of rate on various factors, such as concentration of species i (c_i) and temperature (T). In chemical reaction engineering (CRE), the information obtained from kinetics is a means to determine something about the reactor: size, flow and thermal configuration, product distribution, etc. Kinetics, however, does not provide all the information required for this purpose, and other rate processes are involved in this most difficult of all chemical engineering design problems: fluid mechanics and mixing, heat transfer, and diffusion and mass transfer. These are all constrained by mass (stoichiometric) and energy balances, and by chemical equilibrium in certain cases.

We may consider three levels of system size to compare further the nature of kinetics and of CRE. In order of increasing scale, these levels are as follows:

(1) *Microscopic or molecular*—a collection of reacting molecules sufficiently large to constitute a point in space, characterized, at any given instant, by a single value for each of c_i, T, pressure (P), and density (ρ); for a fluid, the term "element of fluid" is used to describe the collection;

(2) *Local macroscopic*—for example, one solid particle reacting with a fluid, in which there may be gradients of c_i, T, etc. within the particle; and

(3) *Global macroscopic*—for example, a collection or bed of solid particles reacting with a fluid, in which, in addition to local gradients within each particle, there may be global gradients throughout a containing vessel, from particle to particle and from point to point within the fluid.

These levels are illustrated in Figure 1.1. Levels (1) and (2) are domains of *kinetics* in the sense that attention is focused on reaction (rate, mechanism, etc.), perhaps in conjunction with other rate processes, subject to stoichiometric and equilibrium constraints. At the other extreme, level (3) is the domain of CRE, because, in general, it is at this level that sufficient information about overall behavior is required to make decisions about reactors for, say, commercial production. Notwithstanding these comments, it is possible under certain ideal conditions at level (3) to make the required decisions based on information available only at level (1), or at levels (1) and (2) combined. The concepts relating to these ideal conditions are introduced in Chapter 2, and are used in subsequent chapters dealing with CRE.

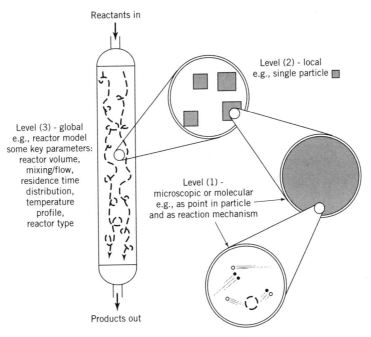

Figure 1.1 Levels for consideration of system size

1.4 ASPECTS OF KINETICS

1.4.1 Rate of Reaction—Definition

We define the rate of reaction *verbally* for a species involved in a reacting system either as a reactant or as a product. The system may be single-phase or multiphase, may have fixed density or variable density as reaction proceeds, and may have uniform or varying properties (e.g., ρ, c_A, T, P) with respect to position at any given time. The *extensive rate of reaction* with respect to a species A, R_A, is the observed rate of formation of A:

$$R_A = \frac{moles\ A\ formed}{unit\ time}, \text{ e.g., } \frac{mol}{s} \qquad (1.4\text{-}1)$$

The *intensive rate of reaction*, r_A, is the rate referred to a specified normalizing quantity (NQ), or rate basis, such as volume of reacting system or mass of catalyst:

$$r_A = \frac{moles\ A\ formed}{(unit\ time)(unit\ NQ)}, \text{ e.g., } \frac{mol}{(s)(m^3)} \qquad (1.4\text{-}2)$$

The rate, R_A or r_A, as defined is <u>negative if A is consumed, and is positive if A is</u> produced. One may also define a species-independent rate of reaction for a single reaction or step in a mechanism, but this requires further consideration of stoichiometry (Section 1.4.4).

The rate r_A is independent of the size of the reacting system and of the physical circumstances of the system, whereas R_A is not. Thus, r_A may be considered to be the

4 Chapter 1: Introduction

"point" or "intrinsic" rate at the molecular level and is the more useful quantity. The two rates are related as follows, with volume V as NQ:

For a uniform system, as in a well-stirred tank,

$$R_A = r_A V \qquad (1.4\text{-}3)$$

For a nonuniform system,

$$R_A = \int_V r_A \, dV \qquad (1.4\text{-}4)$$

The operational interpretation of r_A, as opposed to this verbal definition, *does* depend on the circumstances of the reaction.[1] This is considered further in Chapter 2 as a consequence of the application of the conservation of mass to particular situations. Furthermore, r_A depends on several parameters, and these are considered in Section 1.4.2. The rate with respect to any other species involved in the reacting system may be related to r_A directly through reaction stoichiometry for a simple, single-phase system, or it may require additional kinetics information for a complex system. This aspect is considered in Section 1.4.4, following a preliminary discussion of the measurement of rate of reaction in Section 1.4.3.

1.4.2 Parameters Affecting Rate of Reaction: The Rate Law

Rate of reaction depends on a number of parameters, the most important of which are usually

(1) The nature of the species involved in the reaction;
(2) Concentrations of species;
(3) Temperature;
(4) Catalytic activity;
(5) Nature of contact of reactants; and
(6) Wave-length of incident radiation.

These are considered briefly in turn.

(1) Many examples of types of very fast reactions involve ions in solution, such as the neutralization of a strong acid by a strong base, and explosions. In the former case, the rate of change may be dictated by the rate at which the reactants can be brought into intimate contact. At the other extreme, very slow reactions may involve heterogeneous reactions, such as the oxidation of carbon at room temperature. The reaction between hydrogen and oxygen to form water can be used to illustrate both extremes. Subjected to a spark, a mixture of hydrogen and oxygen can produce an explosion, but in the absence of this, or of a catalyst such as finely divided platinum, the reaction is extremely

---

[1] Attempts to define operationally the rate of reaction in terms of certain derivatives with respect to time (t) are generally unnecessarily restrictive, since they relate primarily to closed static systems, and some relate to reacting systems for which the stoichiometry must be explicitly known in the form of *one* chemical equation in each case. For example, a IUPAC Commission (Mills, 1988) recommends that a species-independent rate of reaction be defined by $r = (1/\nu_i V)(dn_i/dt)$, where ν_i and n_i are, respectively, the stoichiometric coefficient in the chemical equation corresponding to the reaction, and the number of moles of species i in volume V. However, for a flow system at steady-state, this definition is inappropriate, and a corresponding expression requires a particular application of the mass-balance equation (see Chapter 2). Similar points of view about rate have been expressed by Dixon (1970) and by Cassano (1980).

slow. In such a case, it may be wrongly supposed that the system is at equilibrium, since there may be no detectable change even after a very long time.

(2) Rate of reaction usually depends on concentration of reactants (and sometimes of products), and usually increases as concentration of reactants increases. Thus, many combustion reactions occur faster in pure oxygen than in air at the same total pressure.

(3) Rate of reaction depends on temperature and usually increases nearly exponentially as temperature increases. An important exception is the oxidation of nitric oxide, which is involved in the manufacture of nitric acid; in this case, the rate decreases as T increases.

(4) Many reactions proceed much faster in the presence of a substance which is itself not a product of the reaction. This is the phenomenon of catalysis, and many life processes and industrial processes depend on it. Thus, the oxidation of SO_2 to SO_3 is greatly accelerated in the presence of V_2O_5 as a catalyst, and the commercial manufacture of sulfuric acid depends on this fact.

(5) The nature or intimacy of contact of reactants can greatly affect the rate of reaction. Thus, finely divided coal burns much faster than lump coal. The titration of an acid with a base occurs much faster if the acid and base are stirred together than if the base is simply allowed to "dribble" into the acid solution. For a heterogeneous, catalytic reaction, the effect may show up in a more subtle way as the dependence of rate on the size of catalyst particle used.

(6) Some reactions occur much faster if the reacting system is exposed to incident radiation of an appropriate frequency. Thus, a mixture of hydrogen and chlorine can be kept in the dark, and the reaction to form hydrogen chloride is very slow; however, if the mixture is exposed to ordinary light, reaction occurs with explosive rapidity. Such reactions are generally called photochemical reactions.

The way in which the rate of reaction depends on these parameters is expressed mathematically in the form of a *rate law*; that is, for species A in a given reaction, the rate law takes the general form

$$r_A = r_A(conc., temp., cat.\ activity, etc.) \qquad (1.4\text{-}5)$$

The form of the rate law must be established by experiment, and the complete expression may be very complex and, in many cases, very difficult, if not impossible, to formulate explicitly.

1.4.3 Measurement of Rate of Reaction—Preliminary

The rate of chemical reaction must be measured and cannot be predicted from properties of chemical species. A thorough discussion of experimental methods cannot be given at this point, since it requires knowledge of types of chemical reactors that can be used, and the ways in which rate of reaction can be represented. However, it is useful to consider the problem of experimental determination even in a preliminary way, since it provides a better understanding of the methods of chemical kinetics from the outset.

We require a means to follow the progress of reaction, most commonly with respect to changing composition at fixed values of other parameters, such as T and catalytic activity. The method may involve intermittent removal of a sample for analysis or continuous monitoring of an appropriate variable measuring the extent of reaction, without removal of a sample. The rate itself may or may not be measured directly, depending on the type of reactor used. This may be a nonflow reactor, or a continuous-flow reactor, or one combining both of these characteristics.

6 Chapter 1: Introduction

A common laboratory device is a batch reactor, a nonflow type of reactor. As such, it is a closed vessel, and may be rigid (i.e., of constant volume) as well. Sample-taking or continuous monitoring may be used; an alternative to the former is to divide the reacting system into several portions (aliquots), and then to analyze the aliquots at different times. Regardless of which of these sampling methods is used, the rate is determined indirectly from the property measured as a function of time. In Chapter 3, various ways of converting these direct measurements of a property into measures of rate are discussed in connection with the development of the rate law.

EXAMPLE 1-1

To illustrate a method that can be used for continuous monitoring of the composition of a reacting system, consider a gas-phase reaction carried out in a constant-volume batch reactor at a given temperature. If there is a change in moles of gas as reaction takes place, the measured total pressure of the system changes continuously with elapsed time. For example, suppose the reaction is A $\rightarrow$ B + C, where A, B, and C are all gases. In such a case, the rate of reaction, r_A, is related to the rate of decrease in the partial pressure of A, p_A, which is a measure of the concentration of A. However, it is the total pressure (P) that is measured, and it is then necessary to relate P to p_A. This requires use of an appropriate equation of state. For example, if the reacting system can be assumed to be a mixture of ideal gases, and if only A is present initially at pressure p_{Ao}, $p_A = 2p_{Ao} - P$ at any instant. Thus, the reaction can be followed noninvasively by monitoring P with respect to time (t). However, r_A must be obtained indirectly as a function of P (i.e., of p_A) by determining, in effect, the slope of the P (or p_A)–t relation, or by using an integrated form resulting from this (Chapter 3).

Other properties may be used in place of pressure for various kinds of systems: for example, color, electrical conductivity, IR spectroscopy, and NMR.

Other methods involve the use of continuous-flow reactors, and in certain cases, the rate is measured directly rather than indirectly. One advantage of a flow method is that a steady-state can usually be established, and this is an advantage for relatively fast reactions, and for continuous monitoring of properties. A disadvantage is that it may require relatively large quantities of materials. Furthermore, the flow rate must be accurately measured, and the flow pattern properly characterized.

One such laboratory flow reactor for a gas-phase reaction catalyzed by a solid (particles indicated) is shown schematically in Figure 1.2. In this device, the flowing gas mixture (inlet and outlet indicated) is well mixed by internal recirculation by the rotating impeller, so that, everywhere the gas contacting the exterior catalyst surface is at the same composition and temperature. In this way, a "point" rate of reaction is obtained.

Experimental methods for the measurement of reaction rate are discussed further in Chapter 3, and are implicitly introduced in many problems at the ends of other chapters. By these means, we emphasize that chemical kinetics is an experimental science, and we attempt to develop the ability to devise appropriate methods for particular cases.

1.4.4 Kinetics and Chemical Reaction Stoichiometry

All chemical change is subject to the law of conservation of mass, including the conservation of the chemical elements making up the species involved, which is called chemical stoichiometry (from Greek relating to measurement (*-metry*) of an element (*stoichion*)). For each element in a closed reacting system, there is a conservation equa-

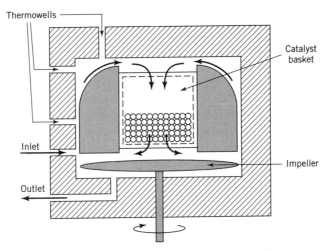

Figure 1.2 Laboratory flow reactor for solid-catalyzed gas-phase reaction (schematic adapted from Mahoney, 1974)

tion stating that the amount of that element is fixed, no matter how combined or recombined, and regardless of rate of reaction or whether equilibrium is attained.

Alternatively, the conservation of atomic species is commonly expressed in the form of chemical equations, corresponding to chemical reactions. We refer to the stoichiometric constraints expressed this way as *chemical reaction stoichiometry*. A simple system is represented by one chemical equation, and a complex system by a set of chemical equations. Determining the number and a proper set of chemical equations for a specified list of species (reactants and products) is the role of chemical reaction stoichiometry.

EXAMPLE 1-2

The oxidation of sulfur dioxide to sulfur trioxide in the manufacture of sulfuric acid is an example of a simple system. It involves 3 species (SO_2, O_2 and SO_3) with 2 elements (S and O). The stoichiometry of the reaction can be represented by one, and only one, chemical equation (apart from a multiplicative factor):

$$2\,SO_2 + O_2 = 2\,SO_3 \tag{A}$$

or

$$-2\,SO_2 - O_2 + 2\,SO_3 = 0 \tag{B}$$

Equation (A) or (B) stems from the fact that the *two* element balances involve *three* quantities related to amounts of the species. These balances may be written as follows:

For S:

$$1\Delta n_{SO_2} + 0\Delta n_{O_2} + 1\Delta n_{SO_3} = 0 \tag{C}$$

For O:

$$2\Delta n_{SO_2} + 2\Delta n_{O_2} + 3\Delta n_{SO_3} = 0 \tag{D}$$

where Δn_{SO_2} = the change in moles of SO_2 by reaction, and similarly for Δn_{O_2} and Δn_{SO_3}. The coefficients in equations (C) and (D) form a matrix $\mathbf{A}$ in which each column represents a species and each row an element:

$$\mathbf{A} = \begin{pmatrix} 1 & 0 & 1 \\ 2 & 2 & 3 \end{pmatrix} \tag{E}$$

The entries in $\mathbf{A}$ are the subscripts to the elements in the molecular formulas of the substances (in an arbitrary order). Each column is a vector of the subscripts for a substance, and $\mathbf{A}$ is called a *formula matrix*.

In this case, $\mathbf{A}$ can be transformed by elementary row operations (multiply the second row by 1/2 and subtract the first row from the result) to the unit-matrix or reduced row-echelon form:

$$\mathbf{A}^* = \begin{pmatrix} 1 & 0 & 1 \\ 0 & 1 & 1/2 \end{pmatrix} \tag{F}$$

The form in (F) provides a solution for Δn_{SO_2} and Δn_{O_2} in equations (C) and (D) in terms of Δn_{SO_3}. This is

$$\Delta n_{SO_2} = -\Delta n_{SO_3}; \text{ and } \Delta n_{O_2} = -(1/2)\Delta n_{SO_3} \tag{G}$$

which may be written as

$$\frac{\Delta n_{SO_2}}{-1} = \frac{\Delta n_{O_2}}{-1/2} = \frac{\Delta n_{SO_3}}{1} \tag{G'}$$

The numbers -1, $-1/2$, and 1 in (G') are in proportion to the stoichiometric coefficients in equation (B), which provides the same interpretation as in (G) or (G'). The last column in (F) gives the values of the stoichiometric coefficients of SO_2 and O_2 (on the left side) in a chemical equation involving one mole of SO_3 (on the right side):

$$+1SO_2 + \frac{1}{2}O_2 = 1SO_3 \tag{H}$$

or, in conventional form, on elimination of the fraction:

$$2SO_2 + O_2 = 2SO_3 \tag{H'}$$

SO_2 and O_2 are said to be component species, and SO_3 is a noncomponent species. The number of components C is the rank of the matrix $\mathbf{A}$ (in this case, 2):

$$\text{rank}(\mathbf{A}) = C \tag{1.4-6}$$

Usually, but not always, <u>C is the same as the number of elements,</u> M. In this sense, C is the smallest number of chemical "building blocks" (ultimately the elements) required to form a system of specified species.

More generally, a simple system is represented by

$$\sum_{i=1}^{N} \nu_i \mathbf{A}_i = 0 \tag{1.4-7}$$

where N is the number of reacting species in the system, ν_i is the stoichiometric coefficient for species i [negative ($-$) for a species written on the left side of $=$ and positive ($+$) for a species written on the right side], and A_i is the molecular formula for species i. For a simple system, if we know the rate of reaction for one species, then we know the rate for any other species from the chemical equation, which gives the ratios in which species are reacted and formed; furthermore, it is sometimes convenient to define a species-independent rate of reaction r for a simple system or single step in a mechanism (Chapter 6). Thus, in Example 1-2, incorporating both of these considerations, we have

$$\boxed{r = \frac{r_{SO_2}}{-2} = \frac{r_{O_2}}{-1} = \frac{r_{SO_3}}{2}}$$

where the signs correspond to consumption ($-$) and formation ($+$); r is positive.

More generally, for a *simple* system, the rates r and r_i are related by

$$\boxed{r = r_i/\nu_i; \qquad i = 1, 2, \ldots, N} \qquad (1.4\text{-}8)$$

We emphasize that equation 1.4-7 represents only reaction stoichiometry, and has no necessary implications for reaction mechanism or reaction equilibrium.[2] In many cases of simple systems, the equation can be written by inspection, if the reacting species and their molecular formulas are known.

A *complex* reacting system is defined as one that requires more than one chemical equation to express the stoichiometric constraints contained in element balances. In such a case, the number of species usually exceeds the number of elements by more than 1. Although in some cases a proper set of chemical equations can be written by inspection, it is useful to have a universal, systematic method of generating a set for a system of any complexity, including a simple system. Such a method also ensures the correct number of equations (R), determines the number (C) and a permissible set of components, and, for convenience for a very large number of species (to avoid the tedium of hand manipulation), can be programmed for use by a computer.

A procedure for writing or generating chemical equations has been described by Smith and Missen (1979; 1991, Chapter 2; see also Missen and Smith, 1989). It is an extension of the procedure used in Example 1-2, and requires a list of all the species

---

[2] We use various symbols to denote different interpretations of chemical statements as follows (with SO_2 oxidation as an example):

$$2SO_2 + O_2 = 2SO_3, \qquad (1)$$

as above, is a *chemical equation* expressing only conservation of elements S and O;

$$2SO_2 + O_2 \rightarrow 2SO_3 \qquad (2)$$

(also expresses conservation and) indicates *chemical reaction* occurring in the *one direction* shown at some finite rate;

$$2SO_2 + O_2 \rightleftarrows 2SO_3 \qquad (3)$$

(also expresses conservation and) indicates *chemical reaction* is to be considered to occur simultaneously *in both directions* shown, each at some finite rate;

$$2SO_2 + O_2 \rightleftharpoons 2SO_3 \qquad (4)$$

(also expresses conservation and) indicates the system is at *chemical equilibrium*; this implies that (net rate) $r = r_i = 0$.

involved, their molecular formulas, and a method of solving the linear algebraic equations for the atom balances, which is achieved by reduction of the **A** matrix to **A***. We illustrate the procedure in the following two examples, as implemented by the computer algebra software *Mathematica*[3] (Smith and Missen, 1997).[4] (The systems in these examples are small enough that the matrix reduction can alternatively be done readily by hand manipulation.) As shown in these examples, and also in Example 1-2, the maximum number of linearly independent chemical equations required is[5]

$$R = N - \text{rank}(\mathbf{A}) = N - C \qquad (1.4\text{-}9)$$

A *proper set* of chemical equations for a system is made up of R linearly independent equations.

EXAMPLE 1-3

The dehydrogenation of ethane (C_2H_6) is used to produce ethylene (C_2H_4), along with H_2, but other species, such as methane (CH_4) and acetylene (C_2H_2), may also be present in the product stream. Using *Mathematica*, determine C and a permissible set of components, and construct a set of chemical equations to represent a reacting system involving these five species.

SOLUTION

The system is formally represented by a list of species, followed by a list of elements, both in arbitrary order:

$$\{(C_2H_6, H_2, C_2H_4, CH_4, C_2H_2), (C, H)\}$$

The procedure is in four main steps:

(1) The entry for each species (in the order listed) of the formula vector formed by the subscripts to the elements (in the order listed):

$$C2H6 = \{2, 6\}$$
$$H2 = \{0, 2\}$$
$$C2H4 = \{2, 4\}$$
$$CH4 = \{1, 4\}$$
$$C2H2 = \{2, 2\}$$

---

[3] *Mathematica* is a registered trademark of Wolfram Research, Inc.
[4] Any software that includes matrix reduction can be used similarly. For example, with *Maple* (Waterloo Maple, Inc.), the first three steps in Example 1-3 are initiated by (1) with (linalg: ; (2) transpose (array ([list of species as in (1)])); (3) rref ("). In many cases, the matrix reduction can be done conveniently by hand manipulation.
[5] Chemical reaction stoichiometry is described more fully on a Web site located at http://www.chemical-stoichiometry.net. The site includes a tutorial and a Java applet to implement the matrix reduction method used in the examples here.

(2) The construction of the formula matrix **A** by the statement:

$$\text{MatrixForm}[\text{Transpose}[A = \{C2H6, H2, C2H4, CH4, C2H2\}]]$$

which is followed by the response:

$$\begin{matrix} 2 & 0 & 2 & 1 & 2 \\ 6 & 2 & 4 & 4 & 2 \end{matrix}$$

(3) The reduction of **A** to the unit-matrix form **A*** by the statement:

$$\text{RowReduce}[\%]$$

which is followed by the response:

$$\begin{matrix} 1 & 0 & 1 & 1/2 & 1 \\ 0 & 1 & -1 & 1/2 & -2 \end{matrix}$$

(4) Obtaining the chemical equation(s):

$$C = \text{rank}(\mathbf{A}) = 2$$

(the number of 1's in the unit submatrix on the left). The columns in the unit submatrix represent the components, C_2H_6 and H_2 (in that order) in this case. Each of the remaining three columns gives the values of the stoichiometric coefficients of the *components* (on the left side) in a chemical equation involving 1 mole of each of the *noncomponents* (on the right side) in the order in the list above. Thus, the maximum number of linearly independent chemical equations is

$$R = N - C = 5 - 2 = 3$$

The set of three equations is

$$+1C_2H_6 - 1H_2 = 1C_2H_4$$

$$+\frac{1}{2}C_2H_6 + \frac{1}{2}H_2 = 1CH_4$$

$$+1C_2H_6 - 2H_2 = 1C_2H_2$$

This is referred to as a canonical form of the set, since each equation involves exclusively 1 mole of one noncomponent, together with the components as required. However, we conventionally write the equations without minus signs and fractions as:

$$C_2H_6 = H_2 + C_2H_4 \tag{A}$$

$$C_2H_6 + H_2 = 2CH_4 \tag{B}$$

$$C_2H_6 = 2H_2 + C_2H_2 \tag{C}$$

This set is not unique and does not necessarily imply anything about the way in which reaction occurs. Thus, from a stoichiometric point of view, (A), (B), and (C) are properly called *equations* and not *reactions*. The nonuniqueness is illustrated by the fact that any one of these three linearly independent equations can be replaced by a combination of equations (A), (B), and (C). For example, (A) could be replaced by 2(B) − (A):

$$2H_2 + C_2H_4 = 2CH_4, \quad \text{(D)}$$

so that the set could consist of (B), (C), and (D). However, this latter set is not a canonical set if C_2H_6 and H_2 are components, since two noncomponents appear in (D).

There is a disadvantage in using *Mathematica* in this way. This stems from the arbitrary ordering of species and of elements, that is, of the columns and rows in **A**. Since columns are not interchanged to obtain **A*** in the commands used, the unit submatrix does not necessarily occur as the first C columns as in Example 1-3. The column interchange can readily be done by inspection, but the species designation remains with the column. The following example illustrates this. (Alternatively, the columns may be left as generated, and **A*** interpreted accordingly.)

EXAMPLE 1-4

Using *Mathematica*, obtain a set of chemical equations in canonical and in conventional form for the system

$$\{(CO_2, H_2O, H_2, CH_4, CO), (H, C, O)\}$$

which could refer to the steam-reforming of natural gas, primarily to produce H_2.

SOLUTION

Following the first two steps in the procedure in Example 1-3, we obtain

$$\mathbf{A} = \begin{pmatrix} (1) & (2) & (3) & (4) & (5) \\ 0 & 2 & 2 & 4 & 0 \\ 1 & 0 & 0 & 1 & 1 \\ 2 & 1 & 0 & 0 & 1 \end{pmatrix}$$

Here the numbers at the tops of the columns correspond to the species in the order given, and the rows are in the order of the elements given. After row reduction, *Mathematica* provides the following:

$$\mathbf{A}^* = \begin{pmatrix} (1) & (2) & (3) & (4) & (5) \\ 0 & 0 & 1 & 4 & 1 \\ 1 & 0 & 0 & 1 & 1 \\ 0 & 1 & 0 & -2 & -1 \end{pmatrix}$$

This matrix can be rearranged by column interchange so that it is in the usual form for **A***; the order of species changes accordingly. The resulting matrix is

$$\mathbf{A}^* = \begin{pmatrix} (3) & (1) & (2) & (4) & (5) \\ 1 & 0 & 0 & 4 & 1 \\ 0 & 1 & 0 & 1 & 1 \\ 0 & 0 & 1 & -2 & -1 \end{pmatrix}$$

From this matrix, $C = \text{rank}(\mathbf{A}^*) = \text{rank}(\mathbf{A}) = 3$; the three components are H_2, CO_2, and H_2O in order. The two noncomponents are CH_4 and CO. Also, $R = N - C = 5 - 3 = 2$. Therefore, a proper set of equations, indicated by the entries in the last two columns, is:

$$+4H_2 + 1CO_2 - 2H_2O = 1CH_4$$
$$+1H_2 + 1CO_2 - 1H_2O = 1CO$$

in canonical form, or, in conventional canonical form,

$$4H_2 + CO_2 = 2H_2O + CH_4$$
$$H_2 + CO_2 = H_2O + CO$$

In general, corresponding to equation 1.4-7 for a simple system, we may write a set of chemical equations for a complex system as

$$\sum_{i=1}^{N} \nu_{ij} A_i = 0; \quad j = 1, 2, \ldots, R \tag{1.4-10}$$

where ν_{ij} is the stoichiometric coefficient of species i in equation j, with a sign convention as given for equation 1.4-7.

These considerations of stoichiometry raise the question: Why do we write chemical equations in kinetics if they don't necessarily represent reactions, as noted in Example 1-3? There are three points to consider:

(1) A proper set of chemical equations provides an aid in chemical "book-keeping" to determine composition as reaction proceeds. This is the role of chemical stoichiometry. On the one hand, it prescribes elemental balances that must be obeyed as constraints on reaction; on the other hand, in prescribing these constraints, it reduces the amount of other information required (e.g., from kinetics) to determine the composition.

(2) For a given system, one particular set of chemical *equations* may in fact correspond to a set of chemical *reactions* or steps in a kinetics scheme that *does* represent overall reaction (as opposed to a kinetics mechanism that represents details of reaction as a reaction path). The important consequence is that the maximum number of steps in a kinetics scheme is the same as the number (R) of chemical equations (the number of steps in a kinetics mechanism is usually greater), and hence stoichiometry tells us the maximum number of independent rate laws that we must obtain experimentally (one for each step in the scheme) to describe completely the macroscopic behavior of the system.

(3) The canonical form of equation 1.4-10, or its corresponding conventional form, is convenient for relating rates of reaction of substances in a complex system, corresponding to equation 1.4-8 for a simple system. This convenience arises because the rate of reaction of each noncomponent is independent. Then the net rate of reaction of each component can be related to a combination of the rates for the noncomponents.

EXAMPLE 1-5

For the system in Example 1-3, relate the rates of reaction of each of the two components, $r_{C_2H_6}$ and r_{H_2}, to the rates of reaction of the noncomponents.

14 Chapter 1: Introduction

SOLUTION

From equation (A) in Example 1-3,

$$\frac{r_{C_2H_6}(A)}{-1} = \frac{r_{C_2H_4}}{1}$$

Similarly from (B) and (C),

$$\frac{r_{C_2H_6}(B)}{-1} = \frac{r_{CH_4}}{2}$$

and

$$\frac{r_{C_2H_6}(C)}{-1} = \frac{r_{C_2H_2}}{1}$$

Since $r_{C_2H_6} = r_{C_2H_6}(A) + r_{C_2H_6}(B) + r_{C_2H_6}(C)$,

$$(-r_{C_2H_6}) = r_{C_2H_4} + \frac{1}{2}r_{CH_4} + r_{C_2H_2}$$

Similarly,

$$r_{H_2} = r_{C_2H_4} - \frac{1}{2}r_{CH_4} + 2r_{C_2H_2}$$

If we measure or know *any* 3 of the 5 rates, then the other 2 can be obtained from these 2 equations, which come entirely from stoichiometry.

For a system involving N species, R equations, and C components, the results of Example 1-5 may be expressed more generally as

$$r_i = \sum_{k=C+1}^{N} \frac{\nu_{ij}}{\nu_{kj}} r_k; \qquad i = 1, 2, \ldots, C; j = 1, 2, \ldots, R \qquad \textbf{(1.4-11)}$$

corresponding to equation 1.4-8. Equations 1.4-11 tell us that we require a maximum of $R = N - C$ (from equation 1.4-9) independent rate laws, from experiment (e.g., one for each noncomponent). These together with element-balance equations enable complete determination of the time-course of events for the N species. Note that the rate of reaction r defined in equation 1.4-8 refers only to an individual reaction in a kinetics scheme involving, for example, equations (A), (B), and (C) as reactions in Example 1-3 (that is, to $r_{(A)}$, $r_{(B)}$, and $r_{(C)}$), and not to an "overall" reaction.

1.4.5 Kinetics and Thermodynamics/Equilibrium

Kinetics and thermodynamics address different kinds of questions about a reacting system. The methods of thermodynamics, together with certain experimental information, are used to answer questions such as (1) what is the maximum possible conversion of a reactant, and the resulting equilibrium composition of the reacting system at given conditions of T and P, and (2) at given T and P, how "far" is a particular reacting system from equilibrium, in terms of the "distance" or affinity measured by the Gibbs-energy driving force (ΔG)? Another type of question, which cannot be answered by thermodynamic methods, is: If a given reacting system is not at equilibrium, at what rate, with respect to time, is it approaching equilibrium? This is the domain of kinetics.

These questions point up the main differences between chemical kinetics and chemical thermodynamics, as follows:

(1) Time is a variable in kinetics but not in thermodynamics; rates dealt with in the latter are with respect to temperature, pressure, etc., but not with respect to time; equilibrium is a time-independent state.
(2) We may be able to infer information about the mechanism of chemical change from kinetics but not from thermodynamics; the rate of chemical change is dependent on the path of reaction, as exemplified by the existence of catalysis; thermodynamics, on the other hand, is not concerned with the path of chemical change, but only with "state" and change of state of a system.
(3) The ΔG of reaction is a measure of the affinity or tendency for reaction to occur, but it tells us nothing about how fast reaction occurs; a very large, negative ΔG, as for the reaction $C + O_2 \to CO_2$, at ambient conditions, although favorable for high equilibrium conversion, does not mean that the reaction is necessarily fast, and in fact this reaction is very slow; we need not be concerned about the disappearance of diamonds at ambient conditions.
(4) Chemical kinetics is concerned with the rate of reaction and factors affecting the rate, and chemical thermodynamics is concerned with the position of equilibrium and factors affecting equilibrium.

Nevertheless, equilibrium can be an important aspect of kinetics, because it imposes limits on the extent of chemical change, and considerable use is made of thermodynamics as we proceed.

1.4.6 Kinetics and Transport Processes

At the molecular or microscopic level (Figure 1.1), chemical change involves only chemical reaction. At the local and global macroscopic levels, other processes may be involved in change of composition. These are diffusion and mass transfer of species as a result of differences in chemical potential between points or regions, either within a phase or between phases. The term "chemical engineering kinetics" includes all of these processes, as may be required for the purpose of describing the overall rate of reaction. Yet another process that may lead to change in composition at the global level is the mixing of fluid elements as a consequence of irregularities of flow (nonideal flow) or forced convection.

Still other rate processes occur that are not necessarily associated with change in composition: heat transfer and fluid flow. Consideration of heat transfer introduces contributions to the energy of a system that are not associated with material flow, and helps to determine T. Consideration of fluid flow for our purpose is mainly confined to the need to take frictional pressure drop into account in reactor performance.

Further details for quantitative descriptions of these processes are introduced as required.

1.5 ASPECTS OF CHEMICAL REACTION ENGINEERING

1.5.1 Reactor Design and Analysis of Performance

Reactor design embodies many different facets and disciplines, the details of some of which are outside our scope. In this book, we focus on process design as opposed to

16 Chapter 1: Introduction

mechanical design of equipment (see Chapter 11 for elaboration of these terms). Other aspects are implicit, but are not treated explicitly: instrumentation and process control, economic, and socioeconomic (environmental and safe-operation). Reactor design is a term we may apply to a new installation or modification; otherwise, we may speak of the analysis of performance of an existing reactor.

1.5.2 Parameters Affecting Reactor Performance

The term "reactor performance" usually refers to the operating results achieved by a reactor, particularly with respect to fraction of reactant converted or product distribution for a given size and configuration; alternatively, it may refer to size and configuration for a given conversion or distribution. In any case, it depends on two main types of behavior: (1) rates of processes involved, including reaction and heat and mass transfer, sometimes influenced by equilibrium limitations; and (2) motion and relative-motion of elements of fluid (both single-phase and multiphase situations) and solid particles (where involved), whether in a flow system or not.

At this stage, type (1) is more apparent than type (2), and we provide some preliminary discussion of (2) here. Flow characteristics include relative times taken by elements of fluid to pass through the reactor (residence-time distribution), and mixing characteristics for elements of fluid of different ages: point(s) in the reactor at which mixing takes place, and the level of segregation at which it takes place (as a molecular dispersion or on a macroscopic scale). Lack of sufficient information on one or both of these types is a major impediment to a completely rational reactor design.

1.5.3 Balance Equations

One of the most useful tools for design and analysis of performance is the balance equation. This type of equation is used to account for a conserved quantity, such as mass or energy, as changes occur in a specified system; element balances and stoichiometry, as discussed in Section 1.4.4, constitute one form of *mass* balance.

The balance is made with respect to a "control volume" which may be of finite (V) or of differential (dV) size, as illustrated in Figure 1.3(a) and (b). The control volume is bounded by a "control surface." In Figure 1.3, $\dot{m}$, F, and q are mass (kg), molar (mol), and volumetric (m^3) rates of flow, respectively, across specified parts of the control surface,[6] and $\dot{Q}$ is the rate of heat transfer to or from the control volume. In (a), the control volume could be the contents of a tank, and in (b), it could be a thin slice of a cylindrical tube.

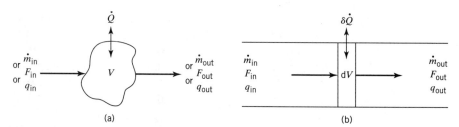

Figure 1.3 Control volumes of finite (V) size (a) and of differential (dV) size (b) with material inlet and outlet streams and heat transfer ($\dot{Q}$, $\delta\dot{Q}$)

---

[6]The "dot" in $\dot{m}$ is used to distinguish flow rate of mass from static mass, m. It is not required for F and q, since these symbols are not used for corresponding static quantities. However, it is also used for rate of heat transfer, $\dot{Q}$, to distinguish it from another quantity.

The balance equation, whether for mass or energy (the two most common uses for our purpose), is of the form:

$$\begin{pmatrix} \text{rate of input} \\ \text{of mass (or} \\ \text{energy) to} \\ \text{control volume} \end{pmatrix} - \begin{pmatrix} \text{rate of output} \\ \text{of mass (or} \\ \text{energy) from} \\ \text{control volume} \end{pmatrix} = \begin{pmatrix} \text{rate of} \\ \text{accumulation of} \\ \text{mass (or energy)} \\ \text{within control} \\ \text{volume} \end{pmatrix} \quad (1.5\text{-}1)$$

Equation 1.5-1 used as a mass balance is normally applied to a chemical species. For a simple system (Section 1.4.4), only one equation is required, and it is a matter of convenience which substance is chosen. For a complex system, the maximum number of independent mass balance equations is equal to R, the number of chemical equations or noncomponent species. Here also it is largely a matter of convenience which species are chosen. Whether the system is simple or complex, there is usually only one energy balance.

The input and output terms of equation 1.5-1 may each have more than one contribution. The input of a species may be by convective (bulk) flow, by diffusion of some kind across the entry point(s), and by formation by chemical reaction(s) within the control volume. The output of a species may include consumption by reaction(s) within the control volume. There are also corresponding terms in the energy balance (e.g., generation or consumption of enthalpy by reaction), and in addition there is heat transfer ($\dot{Q}$), which does not involve material flow. The accumulation term on the right side of equation 1.5-1 is the net result of the inputs and outputs; for steady-state operation, it is zero, and for unsteady-state operation, it is nonzero.

The control volume depicted in Figure 1.3 is for one fixed in position (i.e., fixed observation point) and of fixed size but allowing for variable mass within it; this is often referred to as the Eulerian point of view. The alternative is the Lagrangian point of view, which focuses on a specified mass of fluid moving at the average velocity of the system; the volume of this mass may change.

In further considering the implications and uses of these two points of view, we may find it useful to distinguish between the control volume as a region of space and the system of interest within that control volume. In doing this, we consider two ways of describing a system. The first way is with respect to flow of material:

(F1) *Continuous-flow system:* There is at least one input stream *and* one output stream of material; the mass inside the control volume *may* vary.
(F2) *Semicontinuous-flow or semibatch system:* There is at least one input stream *or* one output stream of material; the mass inside the control volume *does* vary for the latter.
(F3) *Nonflow or static system:* There are no input or output streams of material; the mass inside the control volume *does not* vary.

A second way of describing a system is with respect to *both* material and energy flows:

(S1) An open system can exchange both material and energy with its surroundings.
(S2) A closed system can exchange energy but not material with its surroundings.
(S3) An isolated system can exchange neither material nor energy with its surroundings.

In addition,

(S4) An adiabatic system is one for which $\dot{Q} = 0$.

These two ways of classification are not mutually exclusive: S1 may be associated with F1 or F2; S2 with F1 or F3; S3 only with F3; and S4 with F1 or F2 or F3.

1.5.4 An Example of an Industrial Reactor

One of the most important industrial chemical processes is the manufacture of sulfuric acid. A major step in this process is the oxidation of SO_2 with air or oxygen-enriched air in the reversible, exothermic reaction corresponding to equation (A) in Example 1-2:

$$SO_2 + \frac{1}{2}O_2 \rightleftarrows SO_3$$

This is carried out in a continuous-flow reactor ("SO_2 converter") in several stages, each stage containing a bed of particles of catalyst (promoted V_2O_5).

Figure 1.4 shows a schematic diagram of a Chemetics SO_2 converter. The reactor is constructed of stainless steel and consists of two vertical concentric cylinders. The inner cylinder contains a heat exchanger. The outer cylinder contains four stationary beds of catalyst, indicated by the rectangular shaded areas and numbered 1, 2, 3, and 4. The direction of flow of gas through the reactor is indicated by the arrows; the flow is downward through each bed, beginning with bed 1. Between the beds, which are separated by the inverted-dish-shaped surfaces, the gas flows from the reactor to heat exchangers for adjustment of T and energy recovery. Between beds 3 and 4, there is

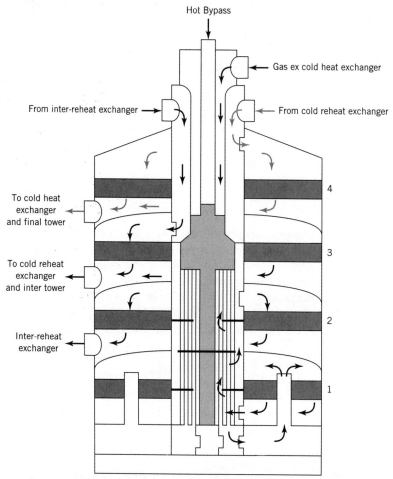

Figure 1.4 Schematic diagram of a four-stage Chemetics SO_2 converter (courtesy Kvaerner-Chemetics Inc.)

also flow through an "inter tower" for partial absorption of SO_3 (to form acid). The gas from bed 4 flows to a "final tower" for complete absorption of SO_3. During passage of reacting gas through the beds, the reaction occurs adiabatically, and hence T rises. The operating temperature range for the catalyst is about 400°C to 600°C. The catalyst particles contain a few percent of the active ingredients, and are either cylindrical or ringlike in shape, with dimensions of a few mm. From economic and environmental (low SO_2-emission) considerations, the fractional conversion of SO_2 should be as high as possible, and can be greater than 99%.

Some important process design and operating questions for this reactor are:

(1) Why is the catalyst arranged in four shallow beds rather than in one deeper bed?
(2) What determines the amount of catalyst required in each bed (for a given plant capacity)? How is the amount calculated?
(3) What determines the depth and diameter of each bed? How are they calculated?
(4) What determines the temperature of the gas entering and leaving each stage?

The answers to these questions are contained in part in the reversible, exothermic nature of the reaction, in the adiabatic mode of operation, and in the characteristics of the catalyst. We explore these issues further in Chapters 5 and 21.

1.6 DIMENSIONS AND UNITS

For the most part, in this book we use SI dimensions and units (SI stands for *le système international d'unités*). A dimension is a name given to a measurable quantity (e.g., length), and a unit is a standard measure of a dimension (e.g., meter (for length)). SI specifies certain quantities as primary dimensions, together with their units. A primary dimension is one of a set, the members of which, in an *absolute* system, cannot be related to each other by definitions or laws. All other dimensions are secondary, and each can be related to the primary dimensions by a dimensional formula. The choice of primary dimensions is, to a certain extent, arbitrary, but their minimum number, determined as a matter of experience, is not. The number of primary dimensions chosen may be increased above the minimum number, but for each one added, a dimensional constant is required to relate two (or more) of them.

The SI primary dimensions and their units are given in Table 1.1, together with their dimensional formulas, denoted by square brackets, and symbols of the units. The number of primary dimensions (7) is one more than required for an absolute system, since

Table 1.1 SI primary dimensions and their units

Dimension (quantity)	Dimensional formula	Unit	Symbol of unit
length	[L]	meter	m
mass	[M]	kilogram	kg
amount of substance	[M_m]	mole	mol
time	[t]	second	s
temperature	[T]	kelvin	K
electric current	[I]	ampere	A
luminous intensity	(not used here)	candela	cd
dimensional constant molar mass	[M][M_m]$^{-1}$	kg mol^{-1}	symbol M^a

[a] The value is specific to a species.

Table 1.2 Important SI secondary dimensions and their units

Dimension (quantity)	Dimensional formula	Unit	Symbol of unit
area	$[L]^2$	square meter	m^2
volume	$[L]^3$	cubic meter	m^3
force	$[M][L][t]^{-2}$	newton	N
pressure	$[M][L]^{-1}[t]^{-2}$	pascal	$Pa(\equiv N\,m^{-2})$
energy	$[M][L]^2[t]^{-2}$	joule	$J(\equiv N\,m)$
molar heat capacity	$[M][L]^2[t]^{-2}[M_m]^{-1}[T]^{-1}$	(no name)	$J\,mol^{-1}\,K^{-1}$

there are two (mass and amount of substance) that relate to the same quantity. Thus, a dimensional constant is required, and this is the molar mass, denoted by M, which is specific to the species in question.

Table 1.2 gives some important SI secondary dimensions and their units, together with their dimensional formulas and symbols of the units. The dimensional formulas may be confirmed from definitions or laws.

Table 1.3 gives some commonly used non-SI units for certain quantities, together with conversion factors relating them to SI units. We use these in some examples and problems, except for the calorie unit of energy. This last, however, is frequently encountered.

Still other units encountered in the literature and workplace come from various other systems (absolute and otherwise). These include "metric" systems (c.g.s. and MKS), some of whose units overlap with SI units, and those (FPS) based on English units. The Fahrenheit and Rankine temperature scales correspond to the Celsius and Kelvin, respectively. We do not use these other units, but some conversion factors are given in Appendix A. Regardless of the units specified initially, our approach is to convert the input to SI units where necessary, to do the calculations in SI units, and to convert the output to whatever units are desired.

In associating numerical values in specified units with symbols for physical quantities, we use the method of notation called "quantity calculus" (Guggenheim, 1967, p. 1). Thus, we may write $V = 4 \times 10^{-2}$ m^3, or $V/m^3 = 4 \times 10^{-2}$, or $10^2\, V/m^3 = 4$. This is useful in headings for columns of tables or labeling axes of graphs unambiguously. For example, if a column entry or graph reading is the number 6.7, and the column heading or axis label is $10^3 r_B/mol\,L^{-1}s^{-1}$, the interpretation is $r_B = 6.7 \times 10^{-3}$ $mol\,L^{-1}s^{-1}$.

Table 1.3 Commonly used non-SI units

Quantity	Unit	Symbol of unit	Relation to SI unit
volume	liter	L	$10^3\,cm^3 = 1\,dm^3$
			$= 10^{-3}\,m^3$
pressure	bar	bar	$10^5\,Pa = 100\,kPa$
			$= 10^{-1}\,MPa$
energy	calorie	cal	4.1840 J
temperature	degree Celsius	°C	$T/K = T/°C + 273.15$
time	minute	min	60 s
	hour	h	3600 s

1.7 PLAN OF TREATMENT IN FOLLOWING CHAPTERS

1.7.1 Organization of Topics

This book is divided into two main parts, one part dealing with reactions and chemical kinetics (Chapters 2 to 10), and the other dealing with reactors and chemical reaction engineering (Chapters 2 and 11 to 24). Each chapter is provided with problems for further study, and answers to selected problems are given at the end of the book.

Although the focus in the first part is on kinetics, certain ideal reactor models are introduced early, in Chapter 2, to illustrate establishing balance equations and interpretations of rate (r_i), and as a prelude to describing experimental methods used in measuring rate of reaction, the subject of Chapter 3. The development of rate laws for single-phase simple systems from experimental data is considered in Chapter 4, with respect to both concentration and temperature effects. The development of rate laws is extended to single-phase complex systems in Chapter 5, with emphasis on reaction networks in the form of kinetics schemes, involving opposing, parallel, and series reactions. Chapters 6 and 7 provide a fundamental basis for rate-law development and understanding for both simple and complex systems. Chapter 8 is devoted to catalysis of various types, and includes the kinetics of reaction in porous catalyst particles. A treatment of noncatalytic multiphase kinetics is given in Chapter 9; here, models for gas-solid (reactant) and gas-liquid systems are described. Chapter 10 deals with enzyme kinetics in biochemical reactions.

The second part of the book, on chemical reaction engineering (CRE), also begins in Chapter 2 with the first introduction of ideal reactor models, and then continues in Chapter 11 with further discussion of the nature of CRE and additional examples of various types of reactors, their modes of operation, and types of flow (ideal and nonideal). Chapter 12 develops design aspects of batch reactors, including optimal and semibatch operation. In Chapter 13, we return to the topic of ideal flow, and introduce the characterization of flow by age-distribution functions, including residence-time distribution (RTD) functions, developing the exact results for several types of ideal flow. Chapters 14 to 16 develop the performance (design) equations for three types of reactors based on ideal flow. In Chapter 17, performance characteristics of batch reactors and ideal-flow reactors are compared; various configurations and combinations of flow reactors are explored. In Chapter 18, the performance of ideal reactor models is developed for complex kinetics systems in which the very important matter of product distribution needs to be taken into account. Chapter 19 deals with the characterization of nonideal flow by RTD measurements and the use of flow models, quite apart from reactor considerations; an introduction to mixing behavior is also given. In Chapter 20, nonideal flow models are used to assess the effects of nonideal flow on reactor performance for single-phase systems. Chapters 21 to 24 provide an introduction to reactors for multiphase systems: fixed-bed catalytic reactors (Chapter 21); reactors for gas-solid (noncatalytic) reactions (Chapter 22); fluidized-bed reactors (Chapter 23); and bubble-column and stirred-tank reactors for gas-liquid reactions (Chapter 24).

1.7.2 Use of Computer Software for Problem Solving

The solution of problems in chemical reactor design and kinetics often requires the use of computer software. In chemical kinetics, a typical objective is to determine kinetics rate parameters from a set of experimental data. In such a case, software capable of parameter estimation by regression analysis is extremely useful. In chemical reactor design, or in the analysis of reactor performance, solution of sets of algebraic or differential equations may be required. In some cases, these equations can be solved analytically. However, as more realistic features of chemical reactor design are explored, analytical solutions are often not possible, and the investigator must rely on software packages capable of numerically solving the equations involved. Within this book, we present both analytical and numerical techniques for solving problems in reactor design and kinetics. The software used with this book is E-Z Solve. The icon shown in the margin here is used similarly throughout the book to indicate where the software is mentioned, or is employed in the solution of examples, or can be employed to advantage in the solution of end-of-chapter problems. The software has several features essential to solving problems in kinetics and reactor design. Thus, one can obtain

(1) Linear and nonlinear regressions of data for estimation of rate parameters;
(2) Solution of systems of nonlinear algebraic equations; and
(3) Numerical integration of systems of ordinary differential equations, including "stiff" equations.

The E-Z Solve software also has a "sweep" feature that allows the user to perform sensitivity analyses and examine a variety of design outcomes for a specified range of parameter values. Consequently, it is also a powerful design and optimization tool.

Many of the examples throughout the book are solved with the E-Z Solve software. In such cases, the computer file containing the program code and solution is cited. These file names are of the form exa-b.msp, where "ex" designates an example problem, "a" the chapter number, and "b" the example number within that chapter. These computer files are included with the software package, and can be readily viewed by anyone who has obtained the E-Z Solve software accompanying this text. Furthermore, these example files can be manipulated so that end-of-chapter problems can be solved using the software.

1.8 PROBLEMS FOR CHAPTER 1

1-1 For the ammonia-synthesis reaction, $N_2 + 3H_2 \rightarrow 2NH_3$, if the rate of reaction with respect to N_2 is $(-r_{N_2})$, what is the rate with respect to (a) H_2 and (b) NH_3 in terms of $(-r_{N_2})$?

1-2 The rate law for the reaction $C_2H_4Br_2 + 3KI \rightarrow C_2H_4 + 2KBr + KI_3$ in an inert solvent, which can be written as $A + 3B \rightarrow$ products, has been found to be $(-r_A) = k_A c_A c_B$, with the rate constant $k_A = 1.34$ L mol^{-1} h^{-1} at 74.9°C (Dillon, 1932).

(a) For the rate of disappearance of KI, $(-r_B)$, what is the value of the rate constant k_B?

(b) At what rate is KI being used up when the concentrations are $c_A = 0.022$ and $c_B = 0.22$ mol L^{-1}?

(c) Do these values depend on the nature of the reactor in which the reaction is carried out? (They were obtained by means of a constant-volume batch reactor.)

1-3 (a) In Example 1-4, of the 5 rate quantities r_i (one for each species), how many are independent (i.e., would need to be determined by experiment)?

(b) Choose a set of these to exclude r_{H_2}, and relate r_{H_2} to them.

1-4 For each of the following systems, determine C (number of components), a permissible set of components, R (maximum number of independent chemical equations), and a proper set of chemical equations to represent the stoichiometry. In each case, the system is represented by a list of species followed by a list of elements.

(a) {(NH_4ClO_4, Cl_2, N_2O, $NOCl$, HCl, H_2O, N_2, O_2, ClO_2), (N, H, Cl, O)} relating to explosion of NH_4ClO_4 (cf. Segraves and Wickersham, 1991, equation (10)).

(b) {($C(gr)$, $CO(g)$, $CO_2(g)$, $Zn(g)$, $Zn(\ell)$, $ZnO(s)$), (C, O, Zn)} relating to the production of zinc metal (Denbigh, 1981, pp 191–193). ($Zn(g)$ and $Zn(\ell)$ are two different species of the same substance Zn.)

(c) {(Cl_2, NO, NO_2, HCl, N_2O, H_2O, HNO_3, NH_4ClO_4, $HClO_4 \bullet 2H_2O$), (Cl, N, O, H)} relating to the production of perchloric acid (Jensen, 1987).

(d) $\{(H^+, OH^-, NO^+, Tl^+, H_2O, NO_2^-, N_2O_3, HNO_2, TlNO_2), (H, O, N, Tl, p)\}$ relating to the complexation of Tl^+ by NO_2^- in aqueous solution (Cobranchi and Eyring, 1991). (Charge p is treated as an element.)

(e) $\{(C_2H_4, C_3H_6, C_4H_8, C_5H_{10}, C_6H_{12}), (C, H)\}$ relating to the oligomerization of C_2H_4.

(f) $\{(ClO_2^-, H_3O^+, Cl_2, H_2O, ClO_3^-, ClO_2), (H, Cl, O, p)\}$ (Porter, 1985).

1-5 The hydrolysis of a disaccharide, $C_{12}H_{22}O_{11}(A) + H_2O \rightarrow 2C_6H_{12}O_6$, takes place in a constant-volume 20-L container. Results of the analysis of the concentration of disaccharide as a function of time are:

t/s	0	5	10	15	20	30	40	50
c_A/mol L^{-1}	1.02	0.819	0.670	0.549	0.449	0.301	0.202	0.135

(a) What is the relationship between the rate of disappearance of disaccharide (A) and the rate of appearance of monosaccharide ($C_6H_{12}O_6$)?

(b) For each time interval, calculate the rate of disappearance of disaccharide (A), in terms of both the total or extensive rate, $(-R_A)$, and the volumetric or intensive rate, $(-r_A)$.

(c) Plot the rates calculated in (b) as functions of c_A. What conclusions can you draw from this plot?

1-6 In a catalytic flow reactor, CO and H_2 are converted to CH_3OH.

(a) If 1000 kg h^{-1} of CO is fed to the reactor, containing 1200 kg of catalyst, and 14% of CO reacts, what is the rate of methanol production per gram of catalyst?

(b) If the catalyst has 55 m^2 g^{-1} surface area, calculate the rate per m^2 of catalyst.

(c) If each m^2 of catalyst has 10^{19} catalytic sites, calculate the number of molecules of methanol produced per catalytic site per second. This is called the turnover frequency, a measure of the activity of a catalyst (Chapter 8).

1-7 The electrode (or half-cell) reactions in a H_2-O_2 fuel cell are:

$$H_2(g) \rightarrow 2H^+(aq) + 2e \quad \text{(anode)}$$

$$O_2(g) + 4H^+(aq) + 4e \rightarrow 2H_2O(\ell) \quad \text{(cathode)}$$

If a battery of cells generates 220 kW at 1.1 V, and has 10 m^2 of Pt electrode surface,

(a) What is the rate of consumption of H_2, in mol m^{-2} s^{-1}?

(b) What is the rate of consumption of O_2?

($N_{Av} = 6.022 \times 10^{23}$ mol^{-1}; electronic charge is 1.6022×10^{-19} C)

1-8 At 7 A.M., the ozone (O_3) content in the atmosphere over a major city is 0.002 ppmv (parts per million by volume). By noon, the measurement is 0.13 ppmv and a health alert is issued. The reason for the severity is that the region acts as a batch reactor—the air is trapped horizontally (by mountains) and vertically (by a temperature inversion at 1000 m). Assume the area of the region is 10,000 km^2 and is home to 10 million people. Calculate the following:

(a) The average intensive rate, $\bar{r}_{O_3}$, during this period, in mol m^{-3} s^{-1};

(b) The average extensive rate, $\bar{R}_{O_3}$, during this period; and

(c) $\bar{r}_{O_3}$ in mol person^{-1} h^{-1}.

(d) If the reaction is $3O_2 \rightarrow 2O_3$ (A), calculate $(-\bar{r}_{O_2})$ and $\bar{r}_{(A)}$.

1-9 The destruction of 2-chlorophenol (CP, $M = 128.5$ g mol^{-1}), a toxic organochlorine compound, by radiative treatment was investigated by Evans et al. (1995). The following data were measured as a function of time in a 50 cm^3 closed cell:

t/h	0	6	12	18	24
c_{CP}/mg L^{-1}	0.340	0.294	0.257	0.228	0.204

(a) What is $(-r_{CP})$ in mol L^{-1} s^{-1} during the first time interval?

(b) What is $(-r_{CP})$ in mol L^{-1} s^{-1} during the last time interval?

(c) What is $(-\bar{R}_{CP})$, the average total rate of chlorophenol destruction over the whole interval?
(d) What would $(-\bar{R}_{CP})$ be for a 10^4 m^3 holding pond under the same conditions?
(e) If the concentration must be reduced to 0.06 mg L^{-1} to meet environmental standards, how long would it take to treat a 0.34 mg L^{-1} solution?
  (i) 2 days;   (ii) less than 2 days;   (iii) more than 2 days. State any assumptions made.

Chapter 2

Kinetics and Ideal Reactor Models

In this chapter, we describe several ideal types of reactors based on two modes of operation (batch and continuous), and ideal flow patterns (backmix and tubular) for the continuous mode. From a kinetics point of view, these reactor types illustrate different ways in which rate of reaction can be measured experimentally and interpreted operationally. From a reactor point of view, the treatment also serves to introduce important concepts and terminology of CRE (developed further in Chapters 12 to 18). Such ideal reactor models serve as points of departure or first approximations for actual reactors. For illustration at this stage, we use only simple systems.

Ideal flow, unlike nonideal flow, can be described exactly mathematically (Chapter 13). Backmix flow (BMF) and tubular flow (TF) are the two extremes representing mixing. In backmix flow, there is complete mixing; it is most closely approached by flow through a vessel equipped with an efficient stirrer. In tubular flow, there is no mixing in the direction of flow; it is most closely approached by flow through an open tube. We consider two types of tubular flow and reactors based on them: plug flow (PF) characterized by a flat velocity profile at relatively high Reynolds number (Re), and laminar flow (LF) characterized by a parabolic velocity profile at relatively low Re.

In this chapter, we thus focus on four types of ideal reactors:

(1) Batch reactor (BR), based on complete mixing;
(2) Continuous-flow stirred tank reactor (CSTR), based on backmix flow;
(3) Plug-flow reactor (PFR), based on plug flow; and
(4) Laminar-flow reactor (LFR), based on laminar flow.

We describe each of these in more detail in turn, with particular emphasis on the material-balance equation in each of the first three cases, since this provides an interpretation of rate of reaction; for the last case, LFR, we consider only the general features at this stage. Before doing this, we first consider various ways in which time is represented.

2.1 TIME QUANTITIES

Time is an important variable in kinetics, and its measurement, whether direct or indirect, is a primary consideration. Several time quantities can be defined.

(1) Residence time (t) of an element of fluid is the time spent by the element of fluid in a vessel. In some situations, it is the same for all elements of fluid, and in others there is a spread or distribution of residence times. Residence-time distribution (RTD) is described in Chapter 13 for ideal-flow patterns, and its experimental measurement and use for nonideal flow are discussed in Chapter 19.

(2) Mean residence time ($\bar{t}$) is the average residence time of all elements of fluid in a vessel.

(3) Space time (τ) is usually applied only to flow situations, and is the time required to process one reactor volume of inlet material (feed) measured at *inlet* conditions. That is, τ is the time required for a volume of feed equal to the volume of the vessel (V) to flow through the vessel. The volume V is the volume of the vessel accessible to the fluid. τ can be used as a scaling quantity for reactor performance, but the reaction conditions must be the same, point-by-point, in the scaling.

(4) Space velocity (S_v) is the reciprocal of space time, and as such is a frequency (time^{-1}): the number of reactor volumes of feed, measured at inlet conditions, processed per unit time.

2.2 BATCH REACTOR (BR)

2.2.1 General Features

A batch reactor (BR) is sometimes used for investigation of the kinetics of a chemical reaction in the laboratory, and also for larger-scale (commercial) operations in which a number of different products are made by different reactions on an intermittent basis.

A batch reactor, shown schematically in Figure 2.1, has the following characteristics:

(1) Each batch is a closed system.
(2) The total mass of each batch is fixed.
(3) The volume or density of each batch may vary (as reaction proceeds).
(4) The energy of each batch may vary (as reaction proceeds); for example, a heat exchanger may be provided to control temperature, as indicated in Figure 2.1.
(5) The reaction (residence) time t for all elements of fluid is the same.
(6) The operation of the reactor is inherently unsteady-state; for example, batch composition changes with respect to time.
(7) Point (6) notwithstanding, it is assumed that, at any time, the batch is uniform (e.g., in composition, temperature, etc.), because of efficient stirring.

As an elaboration of point (3), if a batch reactor is used for a liquid-phase reaction, as indicated in Figure 2.1, we may usually assume that the volume per unit mass of material is constant (i.e., constant density), but if it is used for a gas-phase reaction, this may not be the case.

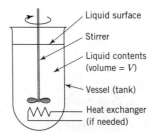

Figure 2.1 Batch reactor (schematic, liquid-phase reaction)

2.2.2 Material Balance; Interpretation of r_i

Consider a reaction represented by A + ... → products taking place in a batch reactor, and focus on reactant A. The general balance equation, 1.5-1, may then be written as a material balance for A with reference to a specified control volume (in Figure 2.1, this is the volume of the liquid).

For a batch reactor, the only possible input and output terms are by reaction, since there is no flow in or out. For the reactant A in this case, there is output but not input. Equation 1.5-1 then reduces to

rate of formation of A by reaction = rate of accumulation of A

or, in mol s^{-1}, say[1],

$$(-r_A)V = -dn_A/dt, \quad (2.2\text{-}1)$$

where V is the volume of the reacting system (not necessarily constant), and n_A is the number of moles of A at time t. Hence the interpretation of r_A for a batch reactor in terms of amount n_A is

$$(-r_A) = -(1/V)(dn_A/dt) \quad (2.2\text{-}2)$$

Equation 2.2-2 may appear in various forms, if n_A is related to other quantities (by normalization), as follows:

(1) If A is the limiting reactant, it may be convenient to normalize n_A in terms of f_A, the <u>fractional conversion of A</u>, defined by

$$\boxed{f_A = (n_{Ao} - n_A)/n_{Ao}} \quad \text{(BR)} \quad (2.2\text{-}3)$$

where n_{Ao} is the initial amount of A; f_A may vary between 0 and 1. Then equation 2.2-2 becomes

$$(-r_A) = (n_{Ao}/V)(df_A/dt) \quad (2.2\text{-}4)$$

(2) Whether A is the limiting reactant or not, it may be convenient to normalize by means of the extent of reaction, ξ, defined for any species involved in the reaction by

$$d\xi = dn_i/\nu_i; \quad i = 1, 2, \ldots, N \quad (2.2\text{-}5)$$

---

[1] Note that the rate of formation of A is r_A, as defined in section 1.4; for a reactant, this is a negative quantity. The rate of disappearance of A is $(-r_A)$, a positive quantity. It is this quantity that is used subsequently in balance equations and rate laws for a reactant. For a product, the rate of formation, a positive quantity, is used. The symbol r_A may be used generically in the text to stand for "rate of reaction of A" where the sign is irrelevant and correspondingly for any other substance, whether reactant or product.

28 Chapter 2: Kinetics and Ideal Reactor Models

Then equation 2.2-2 becomes, for $i \equiv A$,

$$(-r_A) = -(\nu_A/V)(d\xi/dt) \qquad (2.2\text{-}6)$$

(3) Normalization may be by means of the system volume V. This converts n_A into a volumetric molar concentration (molarity) of A, c_A, defined by

$$c_A = n_A/V \qquad (2.2\text{-}7)$$

If we replace n_A in equation 2.2-2 by $c_A V$ and allow V to vary, then we have

$$(-r_A) = -\frac{dc_A}{dt} - \frac{c_A}{V}\frac{dV}{dt} \qquad (2.2\text{-}8)$$

Since $(-r_A)$ is now related to two quantities, c_A and V, we require additional information connecting c_A (or n_A) and V. This is provided by an equation of state of the general form

$$\boxed{V = V(n_A, T, P)} \qquad (2.2\text{-}9)$$

(3a) A special case of equation 2.2-8 results if the reacting system has constant volume (i.e., is of constant density). Then $dV/dt = 0$, and

$$(-r_A) = -dc_A/dt \text{ (constant density)} \qquad (2.2\text{-}10)$$

Thus, for a constant-density reaction in a BR, r_A may be interpreted as the slope of the c_A–t relation. This is illustrated in Figure 2.2, which also shows that r_A itself depends on t, usually decreasing in magnitude as the reaction proceeds, with increasing t.

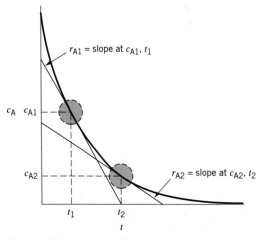

Figure 2.2 Interpretation of r_A for an isothermal, constant-density batch system

EXAMPLE 2-1

For a reaction represented by A → products, in which the rate, $(-r_A)$, is proportional to c_A, with a proportionality constant k_A, show that the time (t) required to achieve a specified fractional conversion of A (f_A) is independent of the initial concentration of reactant c_{Ao}. Assume reaction occurs in a constant-volume batch reactor.

SOLUTION

The rate law is of the form

$$(-r_A) = k_A c_A$$

If we combine this with the material-balance equation 2.2-10 for a constant-density reaction,

$$-dc_A/dt = k_A c_A$$

From this, on integration between c_{Ao} at $t = 0$ and c_A at t,

$$t = (1/k_A)\ln(c_{Ao}/c_A) = (1/k_A)\ln[1/(1 - f_A)]$$

from equation 2.2-3. Thus, the time t required to achieve any specified value of f_A under these circumstances is independent of c_{Ao}. This is a characteristic of a reaction with this form of rate law, but is not a general result for other forms.

2.3 CONTINUOUS STIRRED-TANK REACTOR (CSTR)

2.3.1 General Features

A continuous stirred-tank reactor (CSTR) is normally used for liquid-phase reactions, both in a laboratory and on a large scale. It may also be used, however, for the laboratory investigation of gas-phase reactions, particularly when solid catalysts are involved, in which case the operation is batchwise for the catalyst (see Figure 1.2). Stirred tanks may also be used in a series arrangement (e.g., for the continuous copolymerization of styrene and butadiene to make synthetic rubber).

A CSTR, shown schematically in Figure 2.3(a) as a single vessel and (b) as two vessels in series, has the following characteristics:

(1) The flow through the vessel(s), both input and output streams, is continuous but not necessarily at a constant rate.
(2) The system mass inside each vessel is not necessarily fixed.
(3) The fluid inside each vessel is perfectly mixed (backmix flow, BMF), and hence its properties are uniform at any time, because of efficient stirring.
(4) The density of the flowing system is not necessarily constant; that is, the density of the output stream may differ from that of the input stream.
(5) The system may operate at steady-state or at unsteady-state.
(6) A heat exchanger may be provided in each vessel to control temperature (not shown in Figure 2.3, but comparable to the situation shown in Figure 2.1).

There are several important consequences of the model described in the six points above, as shown partly in the property profiles in Figure 2.3:

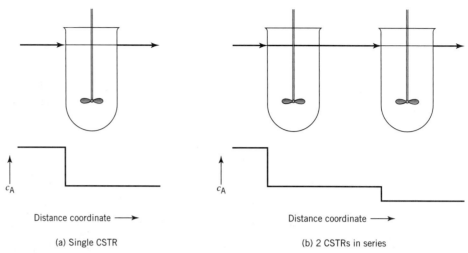

Figure 2.3 Property profile (e.g., c_A for A + ... → products) in a CSTR

[1] Since the fluid inside the vessel is uniformly mixed (and hence elements of fluid are uniformly distributed), all fluid elements have equal probability of leaving the vessel in the output stream at any time.

[2] As a consequence of [1], the output stream has the same properties as the fluid inside the vessel.

[3] As a consequence of [2], there is a step-change across the inlet in any property of the system that changes from inlet to outlet; this is illustrated in Figure 2.3(a) and (b) for c_A.

[4] There is a continuous distribution (spread) of residence times (t) of fluid elements; the spread can be appreciated intuitively by considering two extremes: (i) fluid moving directly from inlet to outlet (short t), and (ii) fluid being caught up in a recycling motion by the stirring action (long t); this distribution can be expressed exactly mathematically (Chapter 13).

[5] The mean residence time, $\bar{t}$, of fluid inside the vessel for steady-state flow is

$$\bar{t} = V/q \quad \text{(CSTR)} \tag{2.3-1}$$

where q is the steady-state flow rate (e.g., m³ s⁻¹) of fluid *leaving* the reactor; this is a consequence of [2] above.

[6] The space time, τ for steady-state flow is

$$\tau = V/q_o \tag{2.3-2}$$

where q_o is the steady-state flow rate of feed at inlet conditions; note that for constant-density flow, $q_o = q$, and $\tau = \bar{t}$. Equation 2.3-2 applies whether or not density is constant, since the definition of τ takes no account of this.

[7] In steady-state operation, each stage of a CSTR is in a stationary state (uniform c_A, T, etc.), which is independent of time.

It is important to understand the distinction between the implications of points [3] and [5]. Point [3] implies that there is instantaneous mixing at the point of entry between the input stream and the contents of the vessel; that is, the input stream instantaneously blends with what is already in the vessel. *This does not mean that any reaction taking place in the fluid inside the vessel occurs instantaneously.* The time required for the change in composition from input to output stream is $\bar{t}$, point [5], which may be small or large.

2.3.2 Material Balance; Interpretation of r_i

Consider again a reaction represented by A + ... → products taking place in a single-stage CSTR (Figure 2.3(a)). The general balance equation, 1.5-1, written for A with a control volume defined by the volume of fluid in the reactor, becomes

$$\begin{pmatrix} \text{rate of input} \\ \text{of A by} \\ \text{flow} \end{pmatrix} - \begin{pmatrix} \text{rate of output} \\ \text{of A by} \\ \text{flow} \end{pmatrix} + \begin{pmatrix} \text{rate of formation} \\ \text{of A by reaction} \\ \text{within control volume} \end{pmatrix}$$

$$= \begin{pmatrix} \text{rate of accumulation} \\ \text{of A within} \\ \text{control volume} \end{pmatrix} \quad (1.5\text{-}1a)$$

or, on a molar basis,

$$F_{Ao} - F_A + r_A V = dn_A/dt \quad (2.3\text{-}3)$$

(for unsteady-state operation)

$$\left[\begin{array}{c} F_{Ao} - F_A + r_A V = 0 \\ \text{(for steady-state operation)} \end{array} \right] \quad (2.3\text{-}4)$$

where F_{Ao} and F_A are the molar flow rates, mol s^{-1}, say, of A entering and leaving the vessel, respectively, and V is the volume occupied by the fluid inside the vessel. Since a CSTR is normally only operated at steady-state for kinetics investigations, we focus on equation 2.3-4 in this chapter.

As in the case of a batch reactor, the balance equation 2.3-3 or 2.3-4 may appear in various forms with other measures of flow and amounts. For a flow system, the fractional conversion of A (f_A), extent of reaction (ξ), and molarity of A (c_A) are defined in terms of F_A rather than n_A:

$$\left\{ \begin{array}{l} f_A = (F_{Ao} - F_A)/F_{Ao} \\ \xi = \Delta F_A/\nu_A = (F_A - F_{Ao})/\nu_A \\ \boxed{c_A = F_A/q} \end{array} \right\} \text{Flow system} \quad \begin{array}{l} (2.3\text{-}5) \\ (2.3\text{-}6) \\ (2.3\text{-}7) \end{array}$$

(cf. equations 2.2-3, -5, and -7, respectively).

From equations 2.3-4 to -7, r_A may be interpreted in various ways as[2]

$$(-r_A) = (F_{Ao} - F_A)/V = -\Delta F_A/V = -\Delta F_A/q\bar{t} \qquad (2.3\text{-}8)$$
$$= F_{Ao} f_A/V \qquad (2.3\text{-}9)$$
$$= -\nu_A \xi/V \qquad (2.3\text{-}10)$$
$$= (c_{Ao} q_o - c_A q)/V \qquad (2.3\text{-}11)$$

where subscript o in each case refers to inlet (feed) conditions. These forms are all applicable whether the density of the fluid is constant or varies, but apply only to steady-state operation.

If density *is* constant, which is usually assumed for a liquid-phase reaction (but is usually *not* the case for a gas-phase reaction), equation 2.3-11 takes a simpler form, since $q_o = q$. Then

$$(-r_A) = (c_{Ao} - c_A)/(V/q)$$
$$= -\Delta c_A/\bar{t} \quad (constant\ density) \qquad (2.3\text{-}12)$$

from equation 2.3-1. If we compare equation 2.2-10 for a BR and equation 2.3-12 for a CSTR, we note a similarity and an important difference in the interpretation of r_A. Both involve the ratio of a concentration change and time, but for a BR this is a derivative, and for a CSTR it is a finite-difference ratio. Furthermore, in a BR, r_A changes with t as reaction proceeds (Figure 2.2), but for steady-state operation of a CSTR, r_A is constant for the stationary-state conditions (c_A, T, etc.) prevailing in the vessel.

EXAMPLE 2-2

For a liquid-phase reaction of the type A + ... → products, an experimental CSTR of volume 1.5 L is used to measure the rate of reaction at a given temperature. If the steady-state feed rate is 0.015 L s^{-1}, the feed concentration (c_{Ao}) is 0.8 mol L^{-1}, and A is 15% converted on flow through the reactor, what is the value of $(-r_A)$?

SOLUTION

The reactor is of the type illustrated in Figure 2.3(a). From the material balance for this situation in the form of equation 2.3-9, together with equation 2.3-7, we obtain

$$(-r_A) = F_{Ao} f_A/V = c_{Ao} q_o f_A/V = 0.8(0.015)0.15/1.5 = 1.2 \times 10^{-3} \text{mol L}^{-1}\text{s}^{-1}$$

---

[2]For comparison with the "definition" of the species-independent rate, r, in footnote 1 of Chapter 1 (which corresponds to equation 2.2-2 for a BR),

$$r(\text{CSTR}) = r_i/\nu_i = (1/\nu_i)(\Delta F_i/V) = (1/\nu_i q)(\Delta F_i/\bar{t}) \qquad (2.3\text{-}8a)$$

2.4 PLUG-FLOW REACTOR (PFR)

2.4.1 General Features

A plug-flow reactor (PFR) may be used for both liquid-phase and gas-phase reactions, and for both laboratory-scale investigations of kinetics and large-scale production. The reactor itself may consist of an empty tube or vessel, or it may contain packing or a fixed bed of particles (e.g., catalyst particles). The former is illustrated in Figure 2.4, in which concentration profiles are also shown with respect to position in the vessel.

A PFR is similar to a CSTR in being a flow reactor, but is different in its mixing characteristics. It is different from a BR in being a flow reactor, but is similar in the progressive change of properties, with position replacing time. These features are explored further in this section, but first we elaborate the characteristics of a PFR, as follows:

(1) The flow through the vessel, both input and output streams, is continuous, but not necessarily at constant rate; the flow in the vessel is PF.
(2) The system mass inside the vessel is not necessarily fixed.
(3) There is no axial mixing of fluid inside the vessel (i.e., in the direction of flow).
(4) There is complete radial mixing of fluid inside the vessel (i.e., in the plane perpendicular to the direction of flow); thus, the properties of the fluid, including its velocity, are uniform in this plane.
(5) The density of the flowing system may vary in the direction of flow.
(6) The system may operate at steady-state or at unsteady-state.
(7) There may be heat transfer through the walls of the vessel between the system and the surroundings.

Some consequences of the model described in the seven points above are as follows:

[1] Each element of fluid has the same residence time t as any other; that is, there is no spread in t.

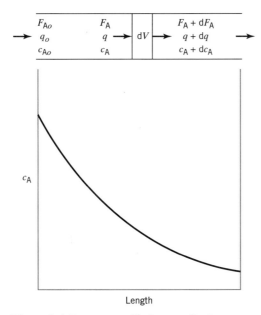

Figure 2.4 Property profile (e.g., c_A for A+... → products) in a PFR (at steady-state)

[2] Properties may change continuously in the direction of flow, as illustrated for c_A in Figure 2.4.

[3] In the axial direction, each portion of fluid, no matter how large, acts as a closed system in motion, not exchanging material with the portion ahead of it or behind it.

[4] The volume of an element of fluid does not necessarily remain constant through the vessel; it may change because of changes in T, P and n_t, the total number of moles.

2.4.2 Material Balance; Interpretation of r_i

Consider a reaction represented by A + ... → products taking place in a PFR. Since conditions may change continuously in the direction of flow, we choose a differential element of volume, dV, as a control volume, as shown at the top of Figure 2.4. Then the material balance for A around dV is, from equation 1.5-1a (preceding equation 2.3-3):

$$F_A - (F_A + dF_A) + r_A dV = dn_A/dt \qquad (2.4\text{-}1)$$

(for unsteady-state operation)

$$F_A - (F_A + dF_A) + r_A dV = 0 \qquad (2.4\text{-}2)$$

(for steady-state operation)

From equation 2.4-2 for steady-state operation, together with the definitions provided by equations 2.3-5 to -7, the interpretations of r_A in terms of F_A, f_A, ξ, and c_A, corresponding to equations 2.3-8 to -11, are[3]

$$\begin{aligned}
(-r_A) &= -dF_A/dV = -dF_A/q\,dt & (2.4\text{-}3) \\
&= F_{A_o} df_A/dV & (2.4\text{-}4) \\
&= -\nu_A d\xi/dV & (2.4\text{-}5) \\
&= -d(c_A q)/dV & (2.4\text{-}6)
\end{aligned}$$

These forms are all applicable whether or not the density of the fluid is constant (through the vessel).

If density *is* constant, equation 2.4-6 takes the form of equation 2.2-10 for constant density in a BR. Then, since q is constant,

$$\begin{aligned}
(-r_A) &= -dc_A/(dV/q) & (2.4\text{-}7) \\
&= -dc_A/dt \quad (\textit{constant density}) & (2.2\text{-}10)
\end{aligned}$$

where t is the time required for fluid to flow from the vessel inlet to the point at which the concentration is c_A (i.e., the residence time to that point). As already implied in equations 2.4-7 and 2.2-10, this time is given by

---

[3] For comparison with the "definition" of the species-independent rate, r, in footnote 1 of Chapter 1, we have the similar result:

$$r(\text{PFR}) = r_i/\nu_i = (1/\nu_i)(dF_i/dV) = (1/\nu_i q)(dF_i/dt) \qquad (2.4\text{-}3a)$$

2.4 Plug-Flow Reactor (PFR) 35

$$t = V/q_o \quad \text{(constant density)} \quad (2.4\text{-}8)$$

whether V represents the total volume of the vessel, in which case t is the residence time of fluid in the vessel ($\equiv \bar{t}$ for a CSTR in equation 2.3-1), or part of the volume from the inlet ($V = 0$). Equation 2.2-10 is the same for both a BR and a PFR for constant density with this interpretation of t for a PFR.

EXAMPLE 2-3

Calculate (a) the residence time, t, and (b) the space time, τ, and (c) explain any difference between the two, for the gas-phase production of C_2H_4 from C_2H_6 in a cylindrical PFR of constant diameter, based on the following data and assumptions:

(1) The feed is pure C_2H_6 (A) at 1 kg s^{-1}, 1000 K and 2 bar.
(2) The reaction rate is proportional to c_A at any point, with a proportionality constant of $k_A = 0.254$ s^{-1} at 1000 K (Froment and Bischoff, 1990, p. 351); that is, the rate law is $(-r_A) = k_A c_A$.
(3) The reactor operates isothermally and at constant pressure.
(4) $f_A = 0.20$ at the outlet.
(5) Only C_2H_4 and H_2 are formed as products.
(6) The flowing system behaves as an ideal-gas mixture.

SOLUTION

(a) In Figure 2.4, the gas flowing at a volumetric rate q at any point generates the control volume dV in time dt. That is,

$$dV = q\,dt \quad \text{or} \quad dt = dV/q$$

The total residence time, t, is obtained by integrating from inlet to outlet. For this, it is necessary to relate V and q to one quantity such as f_A, which is zero at the inlet and 0.2 at the outlet. Thus,

$$t = \int dV/q \quad (2.4\text{-}9)$$

$$= \int F_{Ao}\,df_A/q(-r_A) \quad \text{from equation 2.4-4}$$

$$= \int F_{Ao}\,df_A/q k_A c_A \quad \text{from rate law given}$$

$$= (F_{Ao}/k_A)\int df_A/F_A \quad \text{from equation 2.3-7}$$

$$= (F_{Ao}/k_A)\int df_A/F_{Ao}(1 - f_A) \quad \text{from equation 2.3-5}$$

$$= (1/k_A)\int_0^{0.2} df_A/(1 - f_A)$$

$$= (1/0.254)[-\ln(0.8)] = 0.89 \text{ s}$$

(b) From the definition of space time given in Section 2.1, as in equation 2.3-2,

$$\tau = V/q_o \quad (2.3-2)$$

This is the same result as for residence time t in *constant-density* flow, equation 2.4-8. However, in this case, density is not constant through the PFR, and the result for τ is different from that for t obtained in (a).

Using equation 2.4-4 in integrated form, $V = \int F_{Ao} df_A/(-r_A)$, together with the stoichiometry of the reaction, from which the total molar flow rate at any point is

$$\begin{aligned} F_t &= F_A + F_{C_2H_4} + F_{H_2} \\ &= F_{Ao}(1 - f_A) + F_{Ao}f_A + F_{Ao}f_A \\ &= F_{Ao}(1 + f_A) \end{aligned}$$

and the ideal-gas equation of state, from which the volumetric flow rate at any point is

$$q = F_t RT/P$$

where R is the gas constant, and the inlet flow rate is

$$q_o = F_{to}RT/P = F_{Ao}RT/P$$

we obtain, on substitution into equation 2.3-2,

$$\begin{aligned} \tau &= \left[\int F_{Ao} df_A/(-r_A)\right]/(F_{Ao}RT/P) \\ &= (1/k_A)\int_0^{0.2} (1 + f_A)\,df_A/(1 - f_A) \\ &= (1/k_A)[-2\ln(0.8) - 0.2] \\ &= 0.99 \text{ s} \end{aligned}$$

(c) $\tau > t$, because τ, based on inlet conditions, does not take the acceleration of the flowing gas stream into account. The acceleration, which affects t, is due to the continuous increase in moles on reaction.

2.5 LAMINAR-FLOW REACTOR (LFR)

A laminar-flow reactor (LFR) is rarely used for kinetic studies, since it involves a flow pattern that is relatively difficult to attain experimentally. However, the *model* based on laminar flow, a type of tubular flow, may be useful in certain situations, both in the laboratory and on a large scale, in which flow approaches this extreme (at low Re). Such a situation would involve low fluid flow rate, small tube size, and high fluid viscosity, either separately or in combination, as, for example, in the extrusion of high-molecular-weight polymers. Nevertheless, we consider the general features of an LFR at this stage for comparison with features of the other models introduced above. We defer more detailed discussion, including applications of the material balance, to Chapter 16.

The general characteristics of the simplest model of a continuous LFR, illustrated schematically in Figure 2.5, are as follows:

(1) The flow through the vessel is laminar (LF) and continuous, but not necessarily at constant rate.

2.5 Laminar-Flow Reactor (LFR) 37

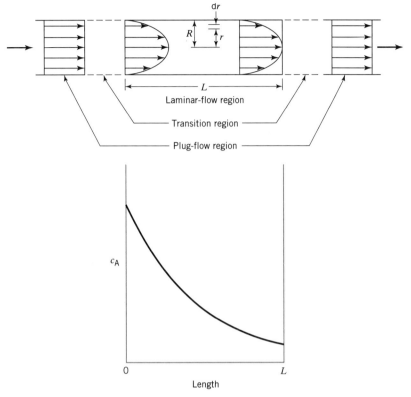

Figure 2.5 LFR: velocity and concentration (for A + ... → products) profiles (at steady-state)

(2) The system mass inside the vessel is not necessarily fixed.
(3) There is no axial mixing of fluid inside the vessel.
(4) There is no radial mixing of fluid inside the vessel.
(5) The density of the flowing system is not necessarily constant.
(6) The system may operate at steady-state or at unsteady-state.
(7) There may be heat transfer through the walls of the vessel between the system and the surroundings.

These seven points correspond to those posed for a PFR in Section 2.4.1. However, there are important differences in points (1) and (4) relating to the type of flow and to mixing in the radial direction in a cylindrical tube. These are illustrated in Figure 2.5 (for a cylindrical vessel). In Figure 2.5, we focus on the laminar-flow region of length L and radius R; fluid is shown entering at left by PF and leaving at right by PF, with a transition region between PF and LF; in other words, regardless of how fluid enters and leaves, we assume that there is a region in which LF is fully established and maintained; r is the (variable) radius between the center line ($r = 0$) and the wall ($r = R$). For simplicity in this case, we consider only steady-state behavior, in spite of the more general situation allowed in points (1), (2), and (6).

Some consequences of the model described in the seven points above are as follows:

[1] From point (1), the velocity profile is parabolic; that is, the linear (axial) velocity u depends quadratically on radial position r, as described by fluid mechanics (see, e.g., Kay and Nedderman, 1974, pp. 69–71):

$$u(r) = u_o[1 - (r/R)^2] \tag{2.5-1}$$

where u_o is the (maximum) velocity at the center of the vessel, and the mean velocity $\bar{u}$ is

$$\bar{u} = u_o/2 \tag{2.5-2}$$

[2] Points (3) and (4) above imply no molecular diffusion in the axial and radial directions, respectively.

[3] A cylindrical LFR can be pictured physically as consisting of a large number of thin cylindrical shells (each of thickness dr) of increasing radius (from center to wall) moving or slipping past each other with decreasing velocity (from center to wall); the residence time of a thin cylindrical shell at radius r is

$$t(r) = L/u(r) \tag{2.5-3}$$

and the mean residence time of all fluid in the vessel is

$$\bar{t} = L/\bar{u} \tag{2.5-4}$$
$$= 2t(r)[1 - (r/R)^2] \tag{2.5-5}$$

from equations 2.5-1 to -3.

2.6 SUMMARY OF RESULTS FOR IDEAL REACTOR MODELS

The most important results obtained in this chapter for ideal reactor models, except the LFR, are summarized in Table 2.1. The relationships for the items listed in the first

Table 2.1 Summary of results for ideal reactor models[a,b]

Item	BR	CSTR	PFR
(1) definitions			
f_A	$(n_{Ao} - n_A)/n_{Ao}$ (2.2-3)[c]	$(F_{Ao} - F_A)/F_{Ao}$ (2.3-5)	
c_A	n_A/V (2.2-7)	F_A/q (2.3-7)	
(2) $(-r_A)$	$(n_{Ao}/V)\, df_A/dt$ (2.2-4)	$F_{Ao}f_A/V$ (2.3-9)	$F_{Ao}\, df_A/dV$ (2.4-4)
(3) time quantities			
τ	(N/A)	V/q_o (2.3-2)	
t	$t = \bar{t}$	[d]	$t = \bar{t}$
$\bar{t}$	$= n_{Ao}\int df_A/V(-r_A)$	V/q (2.3-1)	$= \int dV/q$ (2.4-9)
(4) special case of constant-density system			
f_A		$(c_{Ao} - c_A)/c_{Ao}$	
$(-r_A)$	$-dc_A/dt$ (2.2-10)	$(c_{Ao} - c_A)q_o/V$ (2.3-12)	$-dc_A/dt$ (2.2-10)
t	$t = \bar{t}$	[d]	$t = \bar{t}$
$\bar{t}$	$= -\int dc_A/(-r_A)$ (from 2.2-10)	$V/q_o = \tau$	$= V/q_o = \tau$ (2.4-8)

[a] Excluding LFR.
[b] For reaction A + ... → products with A as limiting reactant.
[c] Equation number in text.
[d] There is a distribution of residence time (t); see Chapter 13.

column are given in the next three columns for a BR, CSTR, and PFR in turn. The equation number in the text is given in each case. The results for items (1), (2), and (3) in the first column apply to either variable or constant density. Those under item (4) apply only to the special case of a constant-density system.

2.7 STOICHIOMETRIC TABLE

A useful tool for dealing with reaction stoichiometry in chemical kinetics is a "stoichiometric table." This is a spreadsheet device to account for changes in the amounts of species reacted for a basis amount of a closed system. It is also a systematic method of expressing the moles, or molar concentrations, or (in some cases) partial pressures of reactants and products, for a given reaction (or set of reactions) at any time or position, in terms of initial concentrations and fractional conversion. Its use is illustrated for a simple system in the following example.

EXAMPLE 2-4

For the gas-phase oxidation of ethylene to ethylene oxide, construct a stoichiometric table in terms of moles on the basis that only the reactants are present initially, and ethylene is the limiting reactant.

SOLUTION

The stoichiometry of the reaction is represented by the equation

$$C_2H_4(g) + \frac{1}{2}O_2(g) = C_2H_4O(g)$$

A stoichiometric table is constructed as follows:

species	initial state	change	final state
C_2H_4 (A)	n_{Ao}	$-f_A n_{Ao}$	$n_{Ao}(1 - f_A)$
O_2 (B)	n_{Bo}	$-\frac{1}{2}f_A n_{Ao}$	$n_{Bo} - \frac{1}{2}f_A n_{Ao}$
C_2H_4O (C)	0	$+f_A n_{Ao}$	$f_A n_{Ao}$
Total:	$n_{to} = n_{Ao} + n_{Bo}$	$-\frac{1}{2}f_A n_{Ao}$	$n_t = n_{to} - \frac{1}{2}f_A n_{Ao}$

As indicated, it is suggested that the table be constructed in symbolic form first, and numerical values substituted afterwards. If molar amounts are used, as in the table above, the results are valid whether the density is constant or not. If density is constant, molar concentrations, c_i, may be used in a similar manner. If both density and temperature are constant, partial pressure, p_i, may be used in a similar manner.

The first column lists all the species involved (including inert species, if present). The second column lists the basis amount of each substance (in the feed, say); this is an arbitrary choice. The third column lists the change in the amount of each species from the basis or initial state to some final state in which the fractional conversion is f_A. Each change is in terms of f_A, based on the definition in equation 2.2-3, and takes the stoichiometry into account. The last column lists the amounts in the final state as the sum of the second and third columns. The total amount is given at the bottom of each column.

2.8 PROBLEMS FOR CHAPTER 2

2-1 The half-life ($t_{1/2}$) of a reactant is the time required for its concentration to decrease to one-half its initial value. The rate of hydration of ethylene oxide (A) to ethylene glycol ($C_2H_4O + H_2O \rightarrow C_2H_6O_2$) in *dilute* aqueous solution is proportional to the concentration of A, with a proportionality constant $k_A = 4.11 \times 10^{-5}$ s^{-1} at 20°C for a certain catalyst (HClO$_4$) concentration (constant). Determine the half-life ($t_{1/2}$), or equivalent space-time ($\tau_{1/2}$), in s, of the oxide (A) at 20°C, if the reaction is carried out
(a) In a batch reactor,
(b) In a CSTR operating at steady-state.
(c) Explain briefly any difference between the two time quantities in (a) and (b).

2-2 Calculate the mean residence time ($\bar{t}$) and space time (τ) for reaction in a CSTR for each of the following cases, and explain any difference between ($\bar{t}$) and τ:
(a) Homogeneous liquid-phase reaction, volume of CSTR (V) = 100 L, feed flow rate (q_o) = 10 L min^{-1};
(b) Homogeneous gas-phase reaction, V = 100 L, q_o = 200 L min^{-1} at 300 K (T_o); stoichiometry: A(g) = B(g) + C(g); reactor outlet temperature (T) = 350 K; reactor inlet and outlet pressures essentially the same and relatively low; conversion of A, 40%.

2-3 For the experimental investigation of a homogeneous gas-phase reaction occurring in a CSTR, explain briefly, but quantitatively, under what circumstances $\bar{t}/\tau > 1$. Consider separately each factor affecting this ratio. Assume steady-state operation, ideal-gas behavior, and equal inlet and outlet flow areas.

2-4 For a homogeneous gas-phase reaction occurring in a plug-flow reactor, explain briefly under what circumstances $\bar{t}/\tau < 1$. Consider each factor affecting this ratio separately. Give an example (chemical reaction + circumstance(s)) for illustration. Assume steady-state operation and constant cross-sectional area.

2-5 The decomposition of phosphine (PH$_3$) to phosphorus vapor (P$_4$) and hydrogen is to take place in a plug-flow reactor at a constant temperature of 925 K. The feed rate of PH$_3$ and the pressure are constant. For a conversion of 50% of the phosphine, calculate the residence time ($\bar{t}$) in the reactor and the space time (τ); briefly explain any difference. Assume the rate of decomposition is proportional to the concentration of PH$_3$ at any point, with a proportionality constant $k = 3.6 \times 10^{-3}$ s^{-1} at 925 K.

2-6 An aqueous solution of ethyl acetate (A), with a concentration of 0.3 mol L^{-1} and flowing at 0.5 L s^{-1}, mixes with an aqueous solution of sodium hydroxide (B), of concentration 0.45 mol L^{-1} and flowing at 1.0 L s^{-1}, and the combined stream enters a CSTR of volume 500 L. If the reactor operates at steady-state, and the fractional conversion of ethyl acetate in the exit stream is 0.807, what is the rate of reaction ($-r_A$)?

2-7 An experimental "gradientless" reactor (similar to that in Figure 1.2), which acts as a CSTR operating *adiabatically*, was used to measure the rate of oxidation of SO$_2$, to SO$_3$ with a V$_2$O$_5$ catalyst (Thurier, 1977). The catalyst is present as a *fixed bed* (200 g) of solid particles within the reactor, with a bulk density (mass of catalyst/volume of bed) of 500 g L^{-1} and a bed voidage (m^3 void space m^{-3} bed) of 0.40; a rotor within the reactor serves to promote BMF of *gas*. Based on this information and that given below for a particular run at steady-state, calculate the following:
(a) The fraction of SO$_2$ converted (f_{SO_2}) in the exit stream;
(b) The rate of reaction, $-r_{SO_2}$, mol SO$_2$ reacted (g cat)$^{-1}$ s^{-1}; at what T does this apply?
(c) The mean residence time of gas ($\bar{t}$) in the catalyst bed, s;
(d) The space time, τ, for the gas in the catalyst bed, if the feed temperature T_o is 548 K.
Additional information:
  Feed rate (total F_{to}): 1.2 mol min^{-1}
  Feed composition: 25 mole % SO$_2$, 25% O$_2$, 50% N$_2$ (inert)
  T (in reactor): 800 K; P (inlet and outlet): 1.013 bar
  Concentration of SO$_3$ in *exit* stream: 10.5 mole %

2-8 Repeat Example 2-4 for the case with O_2 as the limiting reactant.

2-9 (a) Construct a stoichiometric table in terms of partial pressures (p_i) for the gas-phase decomposition of nitrosyl chloride (NOCl) to nitric oxide (NO) and chlorine (Cl_2) in a constant-volume batch reactor based on the following initial conditions: $p_{NOCl,o} = 0.5$ bar, $p_{Cl_2,o} = 0.1$ bar, and (inert) $p_{N_2,o} = 0.4$ bar.

(b) If the reaction proceeds to 50% completion at a constant temperature, what is the total pressure (P) in the vessel?

(c) If the temperature changes as the reaction proceeds, can the table be constructed in terms of moles? molar concentrations? partial pressures? Explain.

2-10 For the system in problem 1-3, and the equations obtained for part (b), construct an appropriate stoichiometric table. Note the significance of there being more than one chemical equation (in comparison with the situation in problems 2-8 and 2-9).

Chapter 3

Experimental Methods in Kinetics: Measurement of Rate of Reaction

The primary use of chemical kinetics in CRE is the development of a rate law (for a simple system), or a set of rate laws (for a kinetics scheme in a complex system). This requires experimental measurement of rate of reaction and its dependence on concentration, temperature, etc. In this chapter, we focus on experimental methods themselves, including various strategies for obtaining appropriate data by means of both batch and flow reactors, and on methods to determine values of rate parameters. (For the most part, we defer to Chapter 4 the use of experimental data to obtain values of parameters in particular forms of rate laws.) We restrict attention to single-phase, simple systems, and the dependence of rate on concentration and temperature. It is useful at this stage, however, to consider some features of a rate law and introduce some terminology to illustrate the experimental methods.

3.1 FEATURES OF A RATE LAW: INTRODUCTION

3.1.1 Separation of Effects

In the general form of equation 1.4-5 (for species A in a reaction), we first assume that the effects of various factors can be separated as:

$$r_A = r'_A(conc.)r''_A(temp.)r'''_A(cat.\ activity)\ldots \qquad (3.1\text{-}1)$$

This separation is not always possible or necessary, but here it means that we can focus on individual factors explicitly in turn. In this chapter, we consider only the first two factors (concentration and temperature), and introduce others in subsequent chapters.

3.1.2 Effect of Concentration: Order of Reaction

For the effect of concentration on r_A, we introduce the concept of "order of reaction." The origin of this lies in early investigations in which it was recognized that, in many cases, the rate at a given temperature is proportional to the concentration of a reactant raised to a simple power, such as 1 or 2. This power or exponent is the order of reaction with respect to that reactant.

Thus, for a reaction represented by

$$|\nu_A|A + |\nu_B|B + |\nu_C|C \rightarrow products \tag{A}$$

the rate of disappearance of A may be found to be of the form:

$$(-r_A) = k_A c_A^\alpha c_B^\beta c_C^\gamma \tag{3.1-2}$$

where α is the order of reaction with respect to reactant A, β is the order with respect to B, and γ is the order with respect to C. The overall order of reaction, n, is the sum of these exponents:

$$n = \alpha + \beta + \gamma \tag{3.1-3}$$

and we may refer to an nth-order reaction in this sense. *There is no necessary connection between a stoichiometric coefficient such as ν_A in reaction (A) and the corresponding exponent α in the rate law.*

The proportionality "constant" k_A in equation 3.1-2 is called the "rate constant," but it actually includes the effects of all the parameters in equation 3.1-1 other than concentration. Thus, its value usually depends on temperature, and we consider this in the next section.

For reaction (A), the rate may be written in terms of $(-r_B)$ or $(-r_C)$ instead of $(-r_A)$. These rates are related to each other through the stoichiometry, as described in Section 1.4.4. Corresponding rate constants k_B or k_C may be introduced instead of k_A, and these rate constants are similarly related through the stoichiometry. Such changes do not alter the form of equation 3.1-2 or values of α, β, and γ; it is a matter of convenience which species is chosen. In any case, it should clearly be specified. Establishing the form of equation 3.1-2, including the values of the various parameters, is a matter for experiment.

EXAMPLE 3-1

Repeat problem 1-2(a) in light of the above discussion.

SOLUTION

The reaction in problem 1-2(a) is represented by A + 3B → products. The rate law in terms of A is $(-r_A) = k_A c_A c_B$, and in terms of B is $(-r_B) = k_B c_A c_B$. We wish to determine the value of k_B given the value of k_A. From equation 1.4-8,

$$(-r_A)/(-1) = (-r_B)/(-3), \text{ or } (-r_B) = 3(-r_A)$$

Thus,

$$k_B c_A c_B = 3 k_A c_A c_B$$

and

$$k_B = 3 k_A = 3(1.34) = 4.02 \text{ L mol}^{-1} \text{ h}^{-1}$$

3.1.3 Effect of Temperature: Arrhenius Equation; Activation Energy

A rate of reaction usually depends more strongly on temperature than on concentration. Thus, in a first-order ($n = 1$) reaction, the rate doubles if the concentration is doubled. However, a rate may double if the temperature is raised by only 10 K, in the range, say, from 290 to 300 K. This essentially exponential behavior is analogous to the temperature-dependence of the vapor pressure of a liquid, p^*, or the equilibrium constant of a reaction, K_{eq}. In the former case, this is represented approximately by the Clausius-Clapeyron equation,

$$\frac{d \ln p^*}{dT} = \frac{\Delta H^{vap}(T)}{RT^2} \tag{3.1-4}$$

where ΔH^{vap} is the enthalpy of vaporization. The behavior of K_{eq} is represented (exactly) by the van't Hoff equation (Denbigh, 1981, p. 144),

$$\frac{d \ln K_{eq}}{dT} = \frac{\Delta H^\circ(T)}{RT^2} \tag{3.1-5}$$

where ΔH° is the standard enthalpy of reaction.

Influenced by the form of the van't Hoff equation, Arrhenius (1889) proposed a similar expression for the rate constant k_A in equation 3.1-2, to represent the dependence of $(-r_A)$ on T through the second factor on the right in equation 3.1-1:

$$\boxed{\frac{d \ln k_A}{dT} = \frac{E_A}{RT^2}} \tag{3.1-6}$$

where E_A is a characteristic (molar) energy, called the *energy of activation*. Since $(-r_A)$ (hence k_A) increases with increasing T in almost every case, E_A is a positive quantity (the same as ΔH^{vap} in equation 3.1-4, but different from ΔH° in equation 3.1-5, which may be positive or negative).

Integration of equation 3.1-6 on the assumption that E_A is independent of T leads to

$$\ln k_A = \ln A - E_A/RT \tag{3.1-7}$$

or

$$k_A = A \exp(-E_A/RT) \tag{3.1-8}$$

where A is a constant referred to as the pre-exponential factor. Together, E_A and A are called the Arrhenius parameters.

Equations 3.1-6 to -8 are all forms of the Arrhenius equation. The usefulness of this equation to represent experimental results for the dependence of k_A on T and the numerical determination of the Arrhenius parameters are explored in Chapter 4. The interpretations of A and E_A are considered in Chapter 6 in connection with theories of reaction rates.

EXAMPLE 3-2

It is sometimes stated as a rule of thumb that the rate of a chemical reaction doubles for a 10 K increase in T. Is this in accordance with the Arrhenius equation? Determine the value of the energy of activation, E_A, if this rule is applied for an increase from (a) 300 to 310 K, and (b) 800 to 810 K.

SOLUTION

From equations 3.1-1 and -2, we write

$$(-r_A) = k_A(T)r'_A(conc.)$$

and assume that $k_A(T)$ is given by equation 3.1-8, and that $r'_A(conc.)$, although unknown, is the same form at all values of T. If we let subscript 1 refer to the lower T and subscript 2 to the higher $T(T_2 = T_1 + 10)$, then, since $r_2 = 2r_1$,

$$A\exp(-E_A/RT_2)r'_A(conc.) = A\exp(-E_A/RT_1)2r'_A(conc.)$$

From this,

$$E_A = RT_1T_2 \ln 2/(T_2 - T_1)$$

(a) $\quad E_A = 8.314(300)310(\ln 2)/10 = 53{,}600 \text{ J mol}^{-1}$

(b) $\quad E_A = 8.314(800)810(\ln 2)/10 = 373{,}400 \text{ J mol}^{-1}$

These are very different values, which shows that the rule is valid for a given reaction only over a limited temperature range.

3.2 EXPERIMENTAL MEASUREMENTS: GENERAL CONSIDERATIONS

Establishing the form of a rate law experimentally for a particular reaction involves determining values of the reaction rate parameters, such as α, β, and γ in equation 3.1-2, and A and E_A in equation 3.1-8. The general approach for a simple system would normally require the following choices, not necessarily in the order listed:

(1) Choice of a species (reactant or product) to follow the extent of reaction (e.g., by chemical analysis) and/or for specification of the rate; if the reaction stoichiometry is not known, it may be necessary to establish this experimentally, and to verify that the system is a simple one.
(2) Choice of type of reactor to be used and certain features relating to its mode of operation (e.g., a BR operated at constant volume); these establish the numerical interpretation of the rate from the appropriate material balance equation (Chapter 2).
(3) Choice of method to follow the extent of reaction with respect to time or a time-related quantity (e.g., by chemical analysis).
(4) Choice of experimental strategy to follow in light of points (1) to (3) (i.e., how to perform the experiments and the number and type required).
(5) Choice of method to determine numerically the values of the parameters, and hence to establish the actual form of the rate law.

We consider these points in more detail in the remaining sections of this chapter. Points (1) and (3) are treated together in Section 3.3, and points (2) and (4) are treated together in Section 3.4.1. Unless otherwise indicated, it is assumed that experiments are carried out at fixed T. The effect of T is considered separately in Section 3.4.2. Some comments on point (5) are given in Section 3.5.

3.3 EXPERIMENTAL METHODS TO FOLLOW THE EXTENT OF REACTION

For a simple system, it is only necessary to follow the extent (progress) of reaction by means of *one* type of measurement. This may be the concentration of one species or one other property dependent on concentration. The former would normally involve a "chemical" method of analysis with intermittent sampling, and the latter a "physical" method with an instrument that could continuously monitor the chosen characteristic of the system. We first consider *ex-situ* and *in-situ* measurements.

3.3.1 *Ex-situ* and *In-situ* Measurement Techniques

A large variety of tools, utilizing both chemical and physical methods, are available to the experimentalist for rate measurements. Some can be classified as *ex-situ* techniques, requiring the removal and analysis of an aliquot of the reacting mixture. Other, *in-situ*, methods rely on instantaneous measurements of the state of the reacting system without disturbance by sample collection.

Of the *ex-situ* techniques, chromatographic analysis, with a wide variety of columns and detection schemes available, is probably the most popular and general method for composition analysis. Others include more traditional wet chemical methods involving volumetric and gravimetric techniques. A large array of physical analytical methods (e.g., NMR, mass spectroscopy, neutron activation, and infrared spectroscopy) are also available, and the experimenter's choice depends on the specific system (and availability of the instrument). For *ex-situ* analysis, the reaction must be "quenched" as the sample is taken so that no further reaction occurs during the analysis. Often, removal from the reactor operating at a high temperature or containing a catalyst is sufficient; however, additional and prompt intervention is sometimes necessary (e.g., immersion in an ice bath or adjustment of pH).

In-situ methods allow the measurement to be made directly on the reacting system. Many spectroscopic techniques, ranging from colorimetric measurements at one wavelength to infrared spectroscopy, are capable (with appropriate windows) of "seeing" into a reactor. System pressure (constant volume) is one of the simplest such measurements of reaction progress for a gas-phase reaction in which there is a change in the number of moles (Example 1-1). For a reactor with known heat transfer, the reactor temperature, along with thermal properties, also provides an *in-situ* diagnostic.

Figure 3.1 shows a typical laboratory flow reactor for the study of catalytic kinetics. A gas chromatograph (GC, lower shelf) and a flow meter allow the complete analysis of samples of product gas (analysis time is typically several minutes), and the determination of the molar flow rate of various species out of the reactor (R) contained in a furnace. A mass spectrometer (MS, upper shelf) allows real-time analysis of the product gas sampled just below the catalyst charge and can follow rapid changes in rate. Automated versions of such reactor assemblies are commercially available.

3.3.2 Chemical Methods

The titration of an acid with a base, or vice versa, and the precipitation of an ion in an insoluble compound are examples of chemical methods of analysis used to determine the concentration of a species in a liquid sample removed from a reactor. Such methods are often suitable for relatively slow reactions. This is because of the length of time that may be required for the analysis; the mere collection of a sample does not stop further reaction from taking place, and a method of "quenching" the reaction may be required. For a BR, there is the associated difficulty of establishing the time t at which the concentration is actually measured. This is not a problem for steady-state operation of a flow reactor (CSTR or PFR).

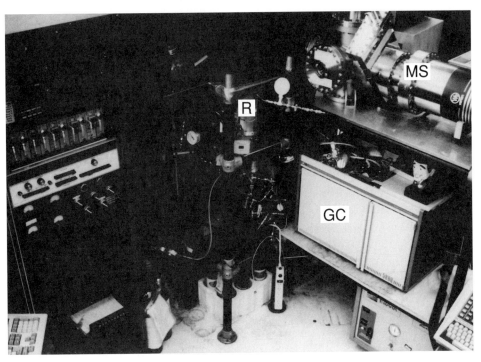

Figure 3.1 Example of a laboratory catalytic flow reactor

An alternative mode of operation for the use of a BR is to divide the reacting system into a number of portions, allowing each portion to react in a separate vessel (BR), and analysing the various portions at a series of increasing times to simulate the passage of time in a single BR. It may be more convenient to stop the reaction at a particular time in a single portion, as a sample, followed by analysis, than to remove a sample from a complete batch, followed by quenching and analysis.

3.3.3 Physical Methods

As chemical reaction proceeds in a system, physical properties of the system change because of the change in chemical composition. If an appropriate property changes in a measurable way that can be related to composition, then the rate of change of the property is a measure of the rate of reaction. The relation between the physical property and composition may be known beforehand by a simple or approximate model, or it may have to be established by a calibration procedure. An advantage of a physical method is that it may be possible to monitor continuously the system property using an instrument without disturbing the system by taking samples.

Examples of physical-property changes that can be used for this purpose are as follows:

(1) Change of pressure in a gas-phase reaction involving change of total moles of gas in a constant-volume BR (see Example 1-1); in this case, the total pressure (P) is measured and must be related to concentration of a particular species. The instrument used is a pressure gauge of some type.

(2) Change of volume in a liquid-phase reaction; the density of a reacting system may change very slightly, and the effect can be translated into a volume change magnified considerably by means of a capillary tube mounted on the reactor, which, for other purposes, is a constant-volume reactor (the change in volume is a very small percentage of the total volume). The reactor so constructed is called

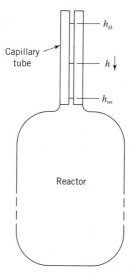

Figure 3.2 A BR in the form of a dilatometer

a dilatometer, and is illustrated in Figure 3.2. The change in volume is related to the change in the liquid level in the capillary, which can be followed by a traveling microscope.

(3) Change of optical rotation in a reacting system involving optically active isomers (e.g., the inversion of sucrose); the instrument used is a polarimeter to measure the angle of rotation of polarized light passing through the system.

(4) Change of electrical conductance in a reacting system involving ionic species (e.g., the hydrolysis of ethyl acetate); the reaction is carried out in a conductivity cell in an electrical circuit for measuring resistance.

(5) Change of refractive index involving use of a refractometer (for a liquid system) or an interferometer (for a gas system).

(6) Change of color—use of a cell in a spectrophotometer.

(7) Single-ion electrodes for measurement of concentration of individual species.

(8) Continuous mass measurement for solid reactant, or absorbent for capture of product(s).

3.3.4 Other Measured Quantities

In addition to chemical composition (concentration of a species) and properties in lieu of composition, other quantities requiring measurement in kinetics studies, some of which have been included above, are:

(1) Temperature, T; not only the measurement, but also the control of T is important, because of the relatively strong dependence of rate on T;
(2) Pressure, P;
(3) Geometric quantities: length, L, as in the use of a dilatometer described above; area, A, as in characterizing the extent of surface in a solid catalyst (Chapter 8); and volume, V, as in describing the size of a vessel;
(4) Time, t; and
(5) Rate of flow of a fluid, q (in a CSTR or PFR).

3.4 EXPERIMENTAL STRATEGIES FOR DETERMINING RATE PARAMETERS

In this section, we combine discussion of choice of reactor type and of experimental methods so as to develop the basis for the methodology of experimentation. We focus first on approaches to determine concentration-related parameters in the rate law, and then on temperature-related parameters. The objective of experiments is to obtain a set of point rates (Section 1.4.1) at various conditions so that best values of the parameters may be determined.

Methods of analyzing experimental data depend on the type of reactor used, and, in some types, on the way in which it is used. For a BR or a PFR, the methods can be divided into "differential" or "integral." In a differential method, a point rate is measured while a small or "differential" amount of reaction occurs, during which the relevant reaction parameters (c_i, T, etc.) change very little, and can be considered constant. In an integral method, measurements are made while a large or "integral" amount of reaction occurs. Extraction of rate-law parameters (order, A, E_A) from such integral data involves comparison with predictions from an assumed rate law. This can be done with simple techniques described in this and the next chapter, or with more sophisticated computer-based optimization routines (e.g., E-Z Solve). A CSTR generates point rates directly for parameter estimation in an assumed form of rate law, whether the amount of reaction taking place is small or large.

3.4.1 Concentration-Related Parameters: Order of Reaction

3.4.1.1 *Use of Constant-Volume BR*

For simplicity, we consider the use of a *constant-volume* BR to determine the kinetics of a system represented by reaction (A) in Section 3.1.2 with one reactant (A), or two reactants (A and B), or more (A, B, C, ...). In every case, we use the rate with respect to species A, which is then given by

$$(-r_A) = -dc_A/dt \quad (constant\ density) \qquad (2.2\text{-}10)$$

We further assume that the rate law is of the form $(-r_A) = k_A c_A^\alpha c_B^\beta c_C^\gamma$, and that the experiments are conducted at fixed T so that k_A is constant. An experimental procedure is used to generate values of c_A as a function of t, as shown in Figure 2.2. The values so generated may then be treated by a differential method or by an integral method.

3.4.1.1.1 *Differential methods*

Differentiation of concentration-time data. Suppose there is only one reactant A, and the rate law is

$$(-r_A) = k_A c_A^n \qquad (3.4\text{-}1)$$

From equation 2.2-10 and differentiation of the $c_A(t)$ data (numerically or graphically), values of $(-r_A)$ can be generated as a function of c_A. Then, on taking logarithms in equation 3.4-1, we have

$$\ln(-r_A) = \ln k_A + n \ln c_A \qquad (3.4\text{-}2)$$

from which linear relation ($\ln(-r_A)$ versus $\ln c_A$), values of the order n, and the rate constant k_A can be obtained, by linear regression. Alternatively, k_A and n can be obtained directly from equation 3.4-1 by nonlinear regression using E-Z Solve.

EXAMPLE 3-3

If there were two reactants A and B in reaction (A), Section 3.1.2, and the rate law were of the form

$$(-r_A) = k_A c_A^\alpha c_B^\beta \tag{3.4-3}$$

how would values of α, β, and k_A be obtained using the differentiation procedure?

SOLUTION

The procedure is similar to that for one reactant, although there is an additional constant to determine. From equation 3.4-3,

$$\ln(-r_A) = \ln k_A + \alpha \ln c_A + \beta \ln c_B \tag{3.4-4}$$

Like equation 3.4-2, this is a linear relation, although in three-dimensional $\ln(-r_A)$–$\ln c_A$–$\ln c_B$ space. It is also linear with respect to the constants $\ln k_A$, α, and β, and hence their values can be obtained by linear regression from an experiment which measures c_A as a function of t. Values of $(-r_A)$ can be generated from these as a function of c_A by differentiation, as described above for the case of a single reactant. The concentrations c_A and c_B are not independent but are linked by the reaction stoichiometry:

$$\frac{c_A - c_{Ao}}{\nu_A} = \frac{c_B - c_{Bo}}{\nu_B} \tag{3.4-5}$$

where c_{Ao} and c_{Bo} are the initial (known) concentrations. Values of c_B can thus be calculated from measured values of c_A. Alternatively, k_A, α, and β can be obtained directly from equation 3.4-3 by nonlinear regression using E-Z Solve.

Initial-rate method. This method is similar to the previous one, but only uses values of rates measured at $t = 0$, obtained by extrapolation from concentrations measured for a relatively short period, as indicated schematically in Figure 3.3.

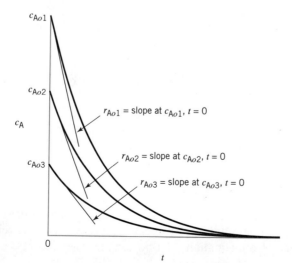

Figure 3.3 Initial-rate method

In Figure 3.3, $c_A(t)$ plots are shown for three different values of c_{Ao}. For each value, the *initial* slope is obtained in some manner, numerically or graphically, and this corresponds to a value of the *initial* rate $(-r_A)_o$ at $t = 0$. Then, if the rate law is given by equation 3.4-1,

$$(-r_A)_o = k_A c_{Ao}^n \tag{3.4-6}$$

and

$$\ln(-r_A)_o = \ln k_A + n \ln c_{Ao} \tag{3.4-7}$$

By varying c_{Ao} in a series of experiments and measuring $(-r_A)_o$ for each value of c_{Ao}, one can determine values of k_A and n, either by linear regression using equation 3.4-7, or by nonlinear regression using equation 3.4-6.

If more than one species is involved in the rate law, as in Example 3-3, the same technique of varying initial concentrations in a series of experiments is used, and equation 3.4-7 becomes analogous to equation 3.4-4.

3.4.1.1.2 Integral methods

Test of integrated form of rate law. Traditionally, the most common method of determining values of kinetics parameters from experimental data obtained isothermally in a constant-volume BR is by testing the integrated form of an assumed rate law. Thus, for a reaction involving a single reactant A with a rate law given by equation 3.4-1, we obtain, using the material balance result of equation 2.2-10,

$$-dc_A/c_A^n = k_A dt \tag{3.4-8}$$

Integration of this between the limits of c_{Ao} at $t = 0$, and c_A at t results in

$$\frac{1}{n-1}(c_A^{1-n} - c_{Ao}^{1-n}) = k_A t \quad (n \neq 1) \tag{3.4-9}$$

(the significance of $n = 1$ is explored in Example 3-4 below). Equation 3.4-9 implies that a plot of c_A^{1-n} versus t is a straight line with slope and intercept indicated in Figure 3.4. Since such a linear relation is readily identified, this method is commonly used to determine values of both n and k_A; however, since n is unknown initially, a value must

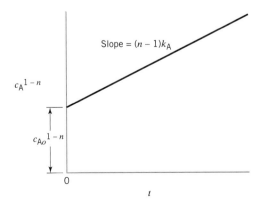

Figure 3.4 Linear integrated form of nth-order rate law $(-r_A) = k_A c_A^n$ for constant-volume BR $(n \neq 1)$

first be assumed to calculate values of the ordinate. (A nonlinear method of determining values of the parameters from experimental data may be used instead, but we focus on linear methods that can be demonstrated graphically in this section.)

EXAMPLE 3-4

As noted in equation 3.4-9, the form given there is not applicable to a first-order rate law (why not?). For $n = 1$, what is the form corresponding to equation 3.4-9?

SOLUTION

If $n = 1$, equation 3.4-9 becomes indeterminate ($k_A t = 0/0$). In this case, we return to equation 3.4-8, which then integrates to

$$c_A = c_{Ao} \exp(-k_A t) \quad (n = 1) \quad (3.4\text{-}10)$$

or, on linearization,

$$\ln c_A = \ln c_{Ao} - k_A t \quad (n = 1) \quad (3.4\text{-}11)$$

As illustrated in Figure 3.5, a linear relation for a first-order reaction is obtained from a plot of $\ln c_A$ versus t. (The result given by equation 3.4-10 or -11 can also be obtained directly from equation 3.4-9 by taking limits in an application of L'Hôpital's rule; see problem 3-8.)

If the rate law involves more than one species, as in equation 3.4-3, the same general test procedure may be used, but the integrated result depends on the form of the rate law.

EXAMPLE 3-5

What is the integrated form of the rate law $(-r_A) = k_A c_A c_B$ for the reaction $|\nu_A|A + |\nu_B|B \to$ products carried out in a constant-volume BR?

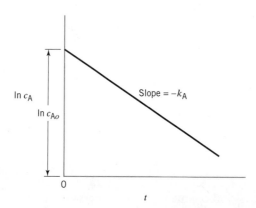

Figure 3.5 Linear integrated form of first-order rate law for constant-volume BR

3.4 Experimental Strategies for Determining Rate Parameters 53

SOLUTION

From the rate law and the material-balance equation 2.2-10, the equation to be integrated is

$$-\frac{dc_A}{c_A c_B} = k_A dt \qquad (3.4\text{-}12)$$

As in Example 3-3, c_B is not independent of c_A, but is related to it through equation 3.4-5, to which we add the extent of reaction to emphasize that there is only one composition variable:

$$\frac{c_A - c_{Ao}}{\nu_A} = \frac{c_B - c_{Bo}}{\nu_B} = \frac{\xi}{V} \qquad (3.4\text{-}5a)$$

where ξ is the extent of reaction introduced in equation 2.2-5, and equation 2.2-7 has been used to eliminate the mole numbers from 2.2-5. Equation 3.4-5a may then be used to eliminate both c_A and c_B from equation 3.4-12, which becomes:

$$\frac{d\xi}{(c_{Ao} + a\xi)(c_{Bo} + b\xi)} = -\frac{k_A}{a} dt \qquad (3.4\text{-}12a)$$

where $a = \nu_A/V$ and $b = \nu_B/V$. Integration by the method of partial fractions followed by reversion from ξ to c_A and c_B results in

$$\ln\left(\frac{c_A}{c_B}\right) = \ln\left(\frac{c_{Ao}}{c_{Bo}}\right) + \frac{k_A}{\nu_A}(\nu_B c_{Ao} - \nu_A c_{Bo})t \qquad (3.4\text{-}13)$$

Thus, $\ln(c_A/c_B)$ is a linear function of t, with the intercept and slope as indicated, for this form of rate law. The slope of this line gives the value of k_A, if the other quantities are known.

Equations 3.4-9, -10 or -11, and -13 are only three examples of integrated forms of the rate law for a constant-volume BR. These and other forms are used numerically in Chapter 4.

Fractional lifetime method. The half-life, $t_{1/2}$, of a reactant is the time required for its concentration to decrease to one-half its initial value. Measurement of $t_{1/2}$ can be used to determine kinetics parameters, although, in general, any fractional life, t_{f_A}, can be similarly used.

In Example 2-1, it is shown that t_{f_A} is independent of c_{Ao} for a first-order reaction carried out in a constant-volume BR. This can also be seen from equation 3.4-10 or -11. Thus, for example, for the half-life,

$$t_{1/2} = (\ln 2)/k_A \quad (n = 1) \qquad (3.4\text{-}14)$$

and is independent of c_{Ao}. A series of experiments carried out with different values of c_{Ao} would thus all give the same value of $t_{1/2}$, if the reaction were first-order.

More generally, for an nth-order reaction, the half-life is given (from equation 3.4-9) by

$$t_{1/2} = \frac{2^{n-1} - 1}{k_A(n-1)c_{Ao}^{n-1}} \quad (n \neq 1) \quad (3.4\text{-}15)$$

Both equations 3.4-14 and -15 lead to the same conclusion:

$$t_{1/2} c_{Ao}^{n-1} = a \text{ constant} \quad (\text{all } n) \quad (3.4\text{-}16)$$

This may be used as a test to establish the value of n, by trial, from a series of experiments carried out to measure $t_{1/2}$ for different values of c_{Ao}. The value of k_A can then be calculated from the value of n obtained, from equation 3.4-14 or -15. Alternatively, equation 3.4-15 can be used in linear form ($\ln t_{1/2}$ versus $\ln c_{Ao}$) for testing similar to that described in the previous section.

3.4.1.2 Use of a CSTR

Consider a constant-density reaction with one reactant, A $\rightarrow$ products, as illustrated for a liquid-phase reaction in a CSTR in Figure 3.6. One experiment at steady-state generates one point value of $(-r_A)$ for the conditions (c_A, q, T) chosen. This value is given by the material balance obtained in Section 2.3.2:

$$(-r_A) = (c_{Ao} - c_A)q/V \quad (2.3\text{-}12)$$

To determine the form of the rate law, values of $(-r_A)$ as a function of c_A may be obtained from a series of such experiments operated at various conditions. For a given reactor (V) operated at a given T, conditions are changed by varying either c_{Ao} or q. For a rate law given by $(-r_A) = k_A c_A^n$, the parameter-estimation procedure is the same as that in the differential method for a BR in the use of equation 3.4-2 (linearized form of the rate law) to determine k_A and n. The use of a CSTR generates point $(-r_A)$ data directly without the need to differentiate c_A data (unlike the differential method with a BR).

If there is more than one reactant, as in Examples 3-3 or 3-5, with a rate law given by $(-r_A) = k_A c_A^\alpha c_B^\beta$, the procedure to determine $(-r_A)$ is similar to that for one reactant, and the kinetics parameters are obtained by use of equation 3.4-4, the linearized form of the rate law.

EXAMPLE 3-6

How would the procedure described above have to be modified if density were not constant?

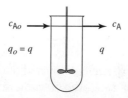

Figure 3.6 Steady-state operation of a CSTR for measurement of $(-r_A)$; constant density

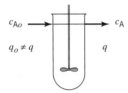

Figure 3.7 Steady-state operation of a CSTR for measurement of $(-r_A)$; variable density

SOLUTION

If density is not constant, the volumetric inlet and outlet flow rates, q_o and q, respectively, are not the same, as indicated in Figure 3.7. As a consequence, $(-r_A)$, for each experiment at steady-state conditions, is calculated from the material balance in the form

$$(-r_A) = (c_{Ao}q_o - c_A q)/V \qquad (2.3\text{-}11)$$

Apart from this, the procedure is the same as described above for cases of one or more than one reactant.

3.4.1.3 Use of a PFR

As in the case of a BR, a PFR can be operated in both a differential and an integral way to obtain kinetics data.

3.4.1.3.1 PFR as differential reactor. As illustrated in Figure 3.8, a PFR can be regarded as divided into a large number of thin strips in series, each thin strip constituting a differential reactor in which a relatively small but measurable change in composition occurs. One such differential reactor, of volume δV, is shown in the lower part of Figure 3.8; it would normally be a self-contained, separate vessel, and not actually part of a large reactor. By measuring the small change from inlet to outlet, at sampling points S_1 and S_2, respectively, we obtain a "point" value of the rate at the average conditions (concentration, temperature) in the thin section.

Consider steady-state operation for a system reacting according to A $\rightarrow$ products. The system is not necessarily of constant density, and to emphasize this, we write the material balance for calculating $(-r_A)$ in the form[1]

$$(-r_A) = F_{Ao}\delta f_A/\delta V \qquad (2.4\text{-}4a)$$

where δf_A is the small increase in fraction of A converted on passing through the small volume δV, and F_{Ao} is the *initial* flow rate of A (i.e., that corresponding to $f_A = 0$).

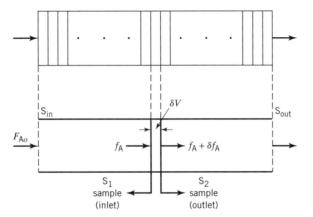

Figure 3.8 PFR as differential or integral reactor

---

[1] The ratio of $F_{Ao}\,\delta f_A/\delta V$ is an approximation to the instantaneous or point rate $F_{Ao}\,df_A/dV$.

Depending on the method of analysis for species A, f_A may be calculated from c_A, together with the flow rates, q and F_A, by equations 2.3-5 and -7. By varying c_{Ao} at the inlet, and/or by varying flow rate, in a series of experiments, each at steady-state at the same T, one can measure $(-r_A)$ as a function of c_A at the given T to obtain values of k_A and n in the rate law, in the same manner as described for a BR.

If there were more than one reactant, the procedure would be similar, in conjunction with the use of equations such as 3.4-4 and -5.

3.4.1.3.2 PFR as integral reactor. In Figure 3.8, the entire vessel indicated from sampling points S_{in} to S_{out}, over which a considerable change in f_A or c_A would normally occur, could be called an *integral* PFR. It is possible to obtain values of kinetics parameters by means of such a reactor from the material balance equation 2.4-4 rearranged as

$$\frac{V}{F_{Ao}} = \int \frac{df_A}{(-r_A)} \qquad (2.4\text{-}4b)$$

If the rate law (for $(-r_A)$) is such that the integral can be evaluated analytically, then it is only necessary to make measurements (of c_A or f_A) at the inlet and outlet, S_{in} and S_{out}, respectively, of the reactor. Thus, if the rate law is given by equation 3.4-1, integration of the right side of equation 2.4-4b results in an expression of the form $g(f_A)/k_A$, where $g(f_A)$ is in terms of the order n, values of which can be assumed by trial, and k_A is unknown. The left side of equation 2.4-4b for a given reactor (V) can be varied by changing F_{Ao}, and $g(f_A)$ is a linear function of V/F_{Ao} with slope k_A, if the correct value of n is used.

If the rate law is such that the integral in equation 2.4-4b cannot be evaluated analytically, it is necessary to make measurements from samples at several points along the length of the reactor, and use these in a numerical or graphical procedure with equation 2.4-4b.

EXAMPLE 3-7

If the gas-phase reaction A $\rightarrow$ B + C is first-order with respect to A, show how the value of the rate constant k_A can be obtained from measurements of c_A (or f_A) at the inlet and outlet of a PFR operated isothermally at T, and at (essentially) constant P.

SOLUTION

The rate law is

$$(-r_A) = k_A c_A$$

and c_A and f_A are related by, from equations 2.3-5 and -7,

$$c_A = F_{Ao}(1 - f_A)/q(f_A)$$

where it is emphasized by $q(f_A)$ that the volumetric flow rate q depends on f_A. If we assume ideal-gas behavior, and that only A is present in the feed, the dependence is given in this particular case by (with the aid of a stoichiometric table):

$$q = q_o(1 + f_A)$$

Substitution of the above equations for $(-r_A)$, c_A, and q in equation 2.4-4b results in

$$\frac{V}{q_o} = \frac{1}{k_A}\int_0^{f_A} \frac{(1+f_A)\,df_A}{1-f_A} = \frac{-[f_A + 2\ln(1-f_A)]}{k_A}$$

Thus, for given V, T, and P, if q_o is varied to obtain several values of f_A at the outlet, the expression $-[f_A + 2\ln(1-f_A)]$ is a linear function of V/q_o with slope k_A, from which the latter can be obtained. (The integration above can be done by the substitution $x = 1 - f_A$.)

3.4.2 Experimental Aspects of Measurement of Arrhenius Parameters A and E_A

So far, we have been considering the effect of concentration on the rate of reaction, on the assumption that temperature is maintained constant during the time of reaction in a batch reactor or throughout the reactor in a flow reactor. This has led to the idea of order of a reaction and the associated rate "constant." The rate of a chemical reaction usually depends more strongly on temperature, and measuring and describing the effect of temperature is very important, both for theories of reaction rates and for reactor performance. Experimentally, it may be possible to investigate the kinetics of a reacting system at a given temperature, and then to repeat the work at several other temperatures. If this is done, it is found that the rate constant depends on temperature, and it is through the rate constant that we examine the dependence of rate on temperature, as provided by the Arrhenius equation 3.1-6, -7, or -8. If this equation appropriately represents the effect of temperature on rate, it becomes a matter of conducting experiments at several temperatures to determine values of A and E_A, the Arrhenius parameters.

Taking T into account implies the ability to operate the reactor at a particular T, and hence to measure and control T. A thermostat is a device in which T is controlled within specified and measurable limits; an example is a constant-T water bath.

In the case of a BR, the entire reactor vessel may be immersed in such a device. However, maintaining constant T in the environment surrounding a reactor may be more easily achieved than maintaining constant temperature throughout the reacting system inside the reactor. Significant temperature gradients may be established within the system, particularly for very exothermic or endothermic reactions, unless steps are taken to eliminate them, such as by efficient stirring and heat transfer.

In the case of a CSTR, external control of T is usually not necessary because the reactor naturally operates internally at a stationary value of T, if internal mixing is efficiently accomplished. If may be necessary, however, to provide heat transfer (heating or cooling) through the walls of the reactor, to maintain relatively high or low temperatures. Another means of controlling or varying the operating T is by controlling or varying the feed conditions (T_o, q_o, c_{Ao}).

In the case of a PFR, it is usually easier to vary T in a controllable and measurable way if it is operated as a differential reactor rather than as an integral reactor. In the latter case, it may be difficult to eliminate an axial gradient in T over the entire length of the reactor.

3.5 NOTES ON METHODOLOGY FOR PARAMETER ESTIMATION

In Section 3.4, traditional methods of obtaining values of rate parameters from experimental data are described. These mostly involve identification of linear forms of the rate expressions (combinations of material balances and rate laws). Such methods are often useful for relatively easy identification of reaction order and Arrhenius parameters, but may not provide the best parameter estimates. In this section, we note methods that do not require linearization.

Generally, the primary objective of parameter estimation is to generate estimates of rate parameters that accurately predict the experimental data. Therefore, once estimates of the parameters are obtained, it is *essential* that these parameters be used to predict (recalculate) the experimental data. Comparison of the predicted and experimental data (whether in graphical or tabular form) allows the "goodness of fit" to be assessed. Furthermore, it is a general premise that differences between predicted and experimental concentrations be randomly distributed. If the differences do not appear to be random, it suggests that the assumed rate law is incorrect, or that some other feature of the system has been overlooked.

At this stage, we consider a reaction of the form of (A) in section 3.1.2:

$$|\nu_A|A + |\nu_B|B + |\nu_C|C \rightarrow \text{products} \qquad \textbf{(A)}$$

and that the rate law is of the form of equations 3.1-2 and 3.1-8 combined:

$$(-r_A) = k_A c_A^\alpha c_B^\beta \ldots = A \exp(-E_A/RT) c_A^\alpha c_B^\beta \ldots \qquad \textbf{(3.4-17)}$$

(In subsequent chapters, we may have to consider forms other than this straightforward power-law form; the effects of T and composition may not be separable, and, for complex systems, two or more rate laws are simultaneously involved. Nevertheless, the same general approaches described here apply.)

Equation 3.4-17 includes three (or more) rate parameters in the first part: $k_A, \alpha, \beta, \ldots$, and four (or more) in the second part: $A, E_A, \alpha, \beta, \ldots$. The former applies to data obtained at one T, and the latter to data obtained at more than one T. We assume that none of these parameters is known *a priori*.

In general, parameter estimation by statistical methods from experimental data in which the number of measurements exceeds the number of parameters falls into one of two categories, depending on whether the function to be fitted to the data is linear or nonlinear with respect to the parameters. A function is linear with respect to the parameters, if for, say, a doubling of the values of all the parameters, the value of the function doubles; otherwise, it is nonlinear. The right side of equation 3.4-17 is nonlinear. We can put it into linear form by taking logarithms of both sides, as in equation 3.4-4:

$$\ln(-r_A) = \ln A - (E_A/RT) + \alpha \ln c_A + \beta \ln c_B + \ldots \qquad \textbf{(3.4-18)}$$

The function is now $\ln(-r_A)$, and the parameters are $\ln A, E_A, \alpha, \beta, \ldots$.

Statistical methods can be applied to obtain values of parameters in both linear and nonlinear forms (i.e., by linear and nonlinear regression, respectively). Linearity with respect to the parameters should be distinguished from, and need not necessarily be associated with, linearity with respect to the variables:

(1) In equation 3.4-17, the right side is nonlinear with respect to both the parameters $(A, E_A, \alpha, \beta, \ldots)$ and the variables $(T, c_A, c_B, \ldots)$.

(2) In equation 3.4-18, the right side is linear with respect to both the parameters and the variables, *if* the variables are interpreted as $1/T, \ln c_A, \ln c_B, \ldots$. However, the transformation of the function from a nonlinear to a linear form may result in a poorer fit. For example, in the Arrhenius equation, it is usually better to estimate A and E_A by nonlinear regression applied to $k = A \exp(-E_A/RT)$, equation 3.1-8, than by linear regression applied to $\ln k = \ln A - E_A/RT$, equation 3.1-7. This is because the linearization is statistically valid only if the experimental data are subject to constant *relative* errors (i.e., measurements are subject to fixed percentage errors); if, as is more often the case, constant *absolute* errors are observed, linearization misrepresents the error distribution, and leads to incorrect parameter estimates.

(3) The function $y = a + bx + cx^2 + dx^3$ is linear with respect to the parameters a, b, c, d (which may be determined by linear regression), but not with respect to the variable x.

The reaction orders obtained from nonlinear analysis are usually nonintegers. It is customary to round the values to nearest integers, half-integers, tenths of integers, *etc.* as may be appropriate. The regression is then repeated with order(s) specified to obtain a revised value of the rate constant, or revised values of the Arrhenius parameters.

A number of statistics and spreadsheet software packages are available for linear regression, and also for nonlinear regression of algebraic expressions (e.g., the Arrhenius equation). However, few software packages are designed for parameter estimation involving numerical integration of a differential equation containing the parameters (e.g., equation 3.4-8). The E-Z Solve software is one package that can carry out this more difficult type of nonlinear regression.

EXAMPLE 3-8

Estimate the rate constant for the reaction A → products, given the following data for reaction in a constant-volume BR:

t/arb. units	0	1	2	3	4	6	8
c_A/arb. units	1	0.95	0.91	0.87	0.83	0.76	0.72

Assume that the reaction follows either first-order or second-order kinetics.

SOLUTION

This problem may be solved by linear regression using equations 3.4-11 ($n = 1$) and 3.4-9 (with $n = 2$), which correspond to the relationships developed for first-order and second-order kinetics, respectively. However, here we illustrate the use of nonlinear regression applied directly to the differential equation 3.4-8 so as to avoid use of particular linearized integrated forms. The method employs user-defined functions within the E-Z Solve software. The rate constants estimated for the first-order and second-order cases are 0.0441 and 0.0504 (in appropriate units), respectively (file ex3-8.msp shows how this is done in E-Z Solve). As indicated in Figure 3.9, there is little difference between the experimental data and the predictions from either the first- or second-order rate expression. This lack of sensitivity to reaction order is common when $f_A < 0.5$ (here, $f_A = 0.28$).

Although we cannot clearly determine the reaction order from Figure 3.9, we can gain some insight from a residual plot, which depicts the difference between the predicted and experimental values of c_A using the rate constants calculated from the regression analysis. Figure 3.10 shows a random distribution of residuals for a second-order reaction, but a nonrandom distribution of residuals for a first-order reaction (consistent overprediction of concentration for the first five datapoints). Consequently, based upon this analysis, it is apparent that the reaction is second-order rather than first-order, and the reaction rate constant is 0.050. Furthermore, the sum of squared residuals is much smaller for second-order kinetics than for first-order kinetics (1.28×10^{-4} versus 5.39×10^{-4}).

We summarize some guidelines for choice of regression method in the chart in Figure 3.11. The initial focus is on the type of reactor used to generate the experimental data

60 Chapter 3: Experimental Methods in Kinetics: Measurement of Rate of Reaction

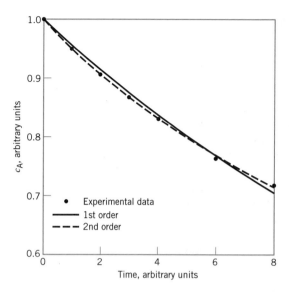

Figure 3.9 Comparison of first- and second-order fits of data in Example 3-8

(for a simple system and rate law considered in this section). Then the choice depends on determining whether the expression being fitted is linear or nonlinear (with respect to the parameters), and, in the case of a BR or integral PFR, on whether an analytical solution to the differential equation involved is available. The equations cited by number are in some cases only representative of the type of equation encountered.

In Figure 3.11, we exclude the use of differential methods with a BR, as described in Section 3.4.1.1.1. This is because such methods require differentiation of experimental $c_i(t)$ data, either graphically or numerically, and differentiation, as opposed to integration, of data can magnify the errors.

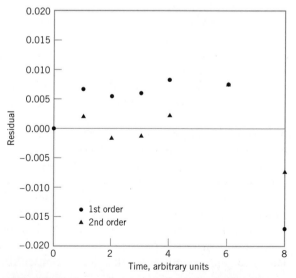

Figure 3.10 Comparison of residual values, $c_{A,calc} - c_{A,exp}$ for first- and second-order fits of data in Example 3-8

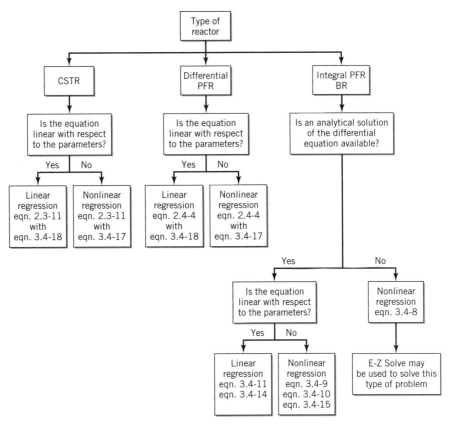

Figure 3.11 Techniques for parameter estimation

3.6 PROBLEMS FOR CHAPTER 3

3-1 For each of the following cases, what method could be used to follow the course of reaction in kinetics experiments conducted isothermally in a constant-volume BR?
  (a) The gas-phase reaction between NO and H_2 (with N_2 and H_2O as products) at relatively high temperature;
  (b) The liquid-phase decomposition of N_2O_5 in an inert solvent to N_2O_4 (soluble) and O_2;
  (c) The liquid-phase saponification of ethyl acetate with NaOH;
  (d) The liquid-phase hydration of ethylene oxide to ethylene glycol in dilute aqueous solution;
  (e) The hydrolysis of methyl bromide in dilute aqueous solution.

3-2 For the irreversible, gas-phase reaction $2A \rightarrow D$ studied manometrically in a rigid vessel at a certain (constant) T, suppose the measured (total) pressure P is 180 kPa after 20 min and 100 kPa after a long time (reaction complete). If only A is present initially, what is the partial pressure of D, p_D, after 20 min? State any assumptions made.

3-3 For the irreversible, gas-phase decomposition of dimethyl ether (CH_3OCH_3) to CH_4, H_2, and CO in a rigid vessel at a certain (constant) T, suppose the *increase* in measured (total) pressure, ΔP, is 20.8 kPa after 665 s. If only ether is present initially, and the *increase* in pressure after a long time (reaction complete) is 82.5 kPa, what is the partial pressure of ether, p_E, after 665 s? State any assumptions made.

3-4 If $c_{A o}/c_{B o} = \nu_A/\nu_B$, equation 3.4-13 cannot be used (show why). What approach would be used in this case in Example 3-5 to test the validity of the proposed rate law?

3-5 Sketch a plot of the rate constant, k (*not* $\ln k$), of a reaction against temperature (T), according to the Arrhenius equation, from relatively low to relatively high temperature, clearly indicating the limiting values of k and slopes. At what temperature (in terms of E_A) does this curve have an inflection point? Based on typical values for E_A (say 40,000 to 300,000 J mol^{-1}), would this temperature lie within the usual "chemical" range? Hence, indicate what part (shape) of the curve would typify chemical behavior.

3-6 Suppose the liquid-phase reaction $A + |\nu_B|B \rightarrow$ products was studied in a batch reactor at two temperatures and the following results were obtained:

$T/°C$	f_A	t/min
20	0.75	20
30	0.75	9

Stating all assumptions made, calculate E_A, the Arrhenius energy of activation, for the reaction. Note that the order of reaction is not known.

3-7 What is the expression corresponding to equation 3.4-13 for the same type of reaction ($|\nu_A|A + |\nu_B|B \rightarrow$ products, constant density) occurring in a CSTR of volume V with a steady-state flow rate of q?

3-8 By applying L'Hôpital's rule for indeterminate forms, show that equation 3.4-11 results from equation 3.4-9.

3-9 The reaction between ethylene bromide and potassium iodide in 99% methanol (inert) has been found to be first-order with respect to each reactant (second-order overall) (Dillon, 1932). The reaction can be represented by $C_2H_4Br_2 + 3KI \rightarrow C_2H_4 + 2KBr + KI_3$ or $A + 3B \rightarrow$ products.

(a) Derive an expression for calculating the second-order rate constant k_A (the equivalent of equation 3.4-13).

(b) At 59.7°C in one set of experiments, for which $c_{Ao} = 0.0266$ and $c_{Bo} = 0.2237$ mol L^{-1}, the bromide (A) was 59.1% reacted at the end of 15.25 h. Calculate the value of k_A and specify its units.

3-10 A general rate expression for the irreversible reaction $A + B \rightarrow C$ can be written as:

$$r_C = k c_A^\alpha c_B^\beta c_C^\gamma$$

Use a spreadsheet or equivalent computer program to calculate the concentration of product C as the reaction proceeds with time (t) in a constant-volume batch reactor (try the parameter values supplied below). You may use a simple numerical integration scheme such as $\Delta c_C = r_C \Delta t$.

set 1: simple rate laws:

	c_{Ao}	c_{Bo}	c_{Co}	k	α	β	γ
(a)	1	1	0	0.05	1	0	0
(b)	1	1	0	0.05	1	1	0
(c)	1	1	0	0.025	1	1	0
(d)	1	2	0	0.025	1	1	0

set 2: more complicated rate laws:

	c_{Ao}	c_{Bo}	c_{Co}	k	α	β	γ
(e)	1	1	0.0001	0.05	1	0	1
(f)	1	1	0.0001	0.005	1	0	-1

Observe what is happening by plotting c_C versus t for each case and answer the following:

(i) Qualitatively state the similarities among the different cases. Is component B involved in the reaction in all cases?

(ii) By graphical means, find the time required to reach 20%, 50%, and 90% of the ultimate concentration for each case.
(iii) Compare results of (a) and (b), (b) and (c), (c) and (d), (a) and (e), and (a) and (f). Explain any differences.

3-11 Diazobenzenechloride decomposes in solution to chlorobenzene and nitrogen (evolved):

$$C_6H_5N_2Cl(\text{solution}) \rightarrow C_6H_5Cl(\text{solution}) + N_2(g)$$

One liter of solution containing 150 g of diazobenzenechloride is allowed to react at 70°C. The cumulative volume of $N_2(g)$ collected at 1 bar and 70°C as a function of time is given in the following table:

t/min	0	1	2	3	4	5	6	7
volume of N_2/L	0	1.66	3.15	4.49	5.71	6.81	7.82	8.74

(a) Calculate the concentration of diazobenzenechloride at each time, and hence calculate the rate of reaction by a difference method for each interval.
(b) What reaction order fits the data?
(c) What is the value (and units) of the rate constant for the reaction order obtained in (b)?

Chapter 4

Development of the Rate Law for a Simple System

In this chapter, we describe how experimental rate data, obtained as described in Chapter 3, can be developed into a quantitative rate law for a simple, single-phase system. We first recapitulate the form of the rate law, and, as in Chapter 3, we consider only the effects of concentration and temperature; we assume that these effects are separable into reaction order and Arrhenius parameters. We point out the choice of units for concentration in gas-phase reactions and some consequences of this choice for the Arrhenius parameters. We then proceed, mainly by examples, to illustrate various reaction orders and compare the consequences of the use of different types of reactors. Finally, we illustrate the determination of Arrhenius parameters for the effect of temperature on rate.

4.1 THE RATE LAW

4.1.1 Form of Rate Law Used

Throughout this chapter, we refer to a single-phase, irreversible reaction corresponding to the stoichiometric equation 1.4-7:

$$\sum_{i=1}^{N} \nu_i A_i = 0 \quad \text{(4.1-1)}$$

where N is the number of reacting species, both "reactants" and "products"; for a reactant, ν_i is negative, and for a product, it is positive, by convention.

The corresponding reaction is written in the manner of reaction (A) in Section 3.1.2:

$$|\nu_A|A + |\nu_B|B + \ldots \rightarrow \nu_D D + \nu_E E + \ldots \quad \text{(4.1-2)}$$

We assume that the rate law for this reaction has the form, from equations 3.1-2 and 1.4-8,

$$r = k \prod_{i=1}^{N} c_i^{\alpha_i} = \frac{r_i}{\nu_i} = \frac{k_i}{|\nu_i|} \prod_{i=1}^{N} c_i^{\alpha_i} \quad \text{(4.1-3)}$$

where r and k are the species-independent rate and rate constant, respectively, and r_i and k_i refer to species i. Since k_i is positive for all species, the absolute value of ν_i is used in the last part of 4.1-3. In this equation, $\prod$ indicates a continued product ($c_1^{\alpha_1} c_2^{\alpha_2} \ldots$), and α_i is the order of reaction with respect to species i. In many cases, only reactants appear in the rate law, but equation 4.1-3 allows for the more general case involving products as well.

We also assume that the various rate constants depend on T in accordance with the Arrhenius equation. Thus, from equations 3.1-8 and 4.1-3,

$$k = A \exp(-E_A/RT) = \frac{k_i}{|\nu_i|} = \frac{A_i}{|\nu_i|} \exp(-E_A/RT) \qquad (4.1\text{-}4)$$

Note that, included in equations 4.1-3 and -4, and corresponding to equation 1.4-8 ($r = r_i/\nu_i$), are the relations

$$k = k_i/|\nu_i| \; ; \; i = 1, 2, \ldots, N \qquad (4.1\text{-}3a)$$

$$A = A_i/|\nu_i| \; ; \; i = 1, 2, \ldots, N \qquad (4.1\text{-}4a)$$

As a consequence of these various defined quantities, care must be taken in assigning values of rate constants and corresponding pre-exponential factors in the analysis and modeling of experimental data. This also applies to the interpretation of values given in the literature. On the other hand, the function $\prod c_i^{\alpha_i}$ and the activation energy E_A are characteristics *only* of the reaction, and are not specific to any one species.

The values of α_i, A, and E_A must be determined from experimental data to establish the form of the rate law for a particular reaction. As far as possible, it is conventional to assign small, integral values to α_1, α_2, etc., giving rise to expressions like first-order, second-order, etc. reactions. However, it may be necessary to assign zero, fractional and even negative values. For a zero-order reaction with respect to a particular substance, the rate is independent of the concentration of that substance. A negative order for a particular substance signifies that the rate decreases (is inhibited) as the concentration of that substance increases.

The rate constant k_i in equation 4.1-3 is sometimes more fully referred to as the specific reaction rate constant, since $|r_i| = k_i$ when $c_i = 1$ ($i = 1, 2, \ldots, N$). The units of k_i (and of A) depend on the overall order of reaction, n, rewritten from equation 3.1-3 as

$$n = \sum_{i=1}^{N} \alpha_i \qquad (4.1\text{-}5)$$

From equations 4.1-3 and -5, these units are (concentration)$^{1-n}$ (time)$^{-1}$.

4.1.2 Empirical versus Fundamental Rate Laws

Any mathematical function that adequately represents experimental rate data can be used in the rate law. Such a rate law is called an *empirical* or *phenomenological* rate law. In a broader sense, a rate law may be constructed based, in addition, on concepts of reaction mechanism, that is, on how reaction is inferred to take place at the molecular level (Chapter 7). Such a rate law is called a *fundamental* rate law. It may be more correct in functional form, and hence more useful for achieving process improvements.

66 Chapter 4: Development of the Rate Law for a Simple System

Furthermore, extrapolations of the rate law outside the range of conditions used to generate it can be made with more confidence, if it is based on mechanistic considerations. We are not yet in a position to consider fundamental rate laws, and in this chapter we focus on empirical rate laws given by equation 4.1-3.

4.1.3 Separability versus Nonseparability of Effects

In equation 4.1-3, the effects of the various reaction parameters (c_i, T) are separable. When mechanistic considerations are taken into account, the resulting rate law often involves a complex function of these parameters that cannot be separated in this manner. As an illustration of nonseparability, a rate law derived from reaction mechanisms for the catalyzed oxidation of CO is

$$(-r_{CO}) = k(T)c_{CO}c_{O_2}^{1/2}/[1 + K(T)c_{CO} + K'(T)c_{O_2}^{1/2}].$$

In this case, the effects of c_{CO}, c_{O_2}, and T cannot be separated. However, the simplifying assumption of a separable form is often made: the coupling between parameters may be weak, and even where it is strong, the simpler form may be an adequate representation over a narrow range of operating conditions.

4.2 GAS-PHASE REACTIONS: CHOICE OF CONCENTRATION UNITS

4.2.1 Use of Partial Pressure

The concentration c_i in equation 4.1-3, the rate law, is usually expressed as a molar volumetric concentration, equation 2.2-7, for any fluid, gas or liquid. For a substance in a gas phase, however, concentration may be expressed alternatively as partial pressure, defined by

$$p_i = x_i P; \quad i = 1, 2, \ldots, N_g \quad \textbf{(4.2-1)}$$

where N_g is the number of substances in the gas phase, and x_i is the mole fraction of i in the *gas phase*, defined by

$$x_i = n_i/n_t; \quad i = 1, 2, \ldots, N_g \quad \textbf{(4.2-2)}$$

where n_t is the total number of moles in the *gas phase*.

The partial pressure p_i is related to c_i by an equation of state, such as

$$p_i = z(n_i/V)RT = zRTc_i; \quad i = 1, 2, \ldots, N_g \quad \textbf{(4.2-3)}$$

where z is the compressibility factor for the gas mixture, and depends on T, P, and composition. At relatively low density, $z \simeq 1$, and for simplicity we frequently use the form for an ideal-gas mixture:

$$p_i = RTc_i; \quad i = 1, 2, \ldots, N_g \quad \textbf{(4.2-3a)}$$

EXAMPLE 4-1

For the gas-phase reaction 2A + 2B → C + 2D taking place in a rigid vessel at a certain T, suppose the measured (total) pressure P decreases initially at a rate of 7.2 kPa min^{-1}. At what rate is the partial pressure of A, p_A, changing? State any assumptions made.

4.2 Gas-Phase Reactions: Choice of Concentration Units

SOLUTION

Assume ideal-gas behavior (T, V constant). Then,

$$PV = n_t RT \quad \text{and} \quad p_A V = n_A RT$$

At any instant,

$$n_t = n_A + n_B + n_C + n_D$$

$$\begin{aligned} dn_t &= dn_A + dn_B + dn_C + dn_D \\ &= dn_A + dn_A - (1/2)\,dn_A - dn_A \\ &= (1/2)dn_A \end{aligned}$$

at $t = 0$
$$dn_{to} = (1/2)\,dn_{Ao}$$

Thus, from the equation of state and stoichiometry

$$\left(\frac{dP}{dt}\right)_o = \frac{RT}{V}\left(\frac{dn_t}{dt}\right)_o = \frac{RT}{2V}\left(\frac{dn_A}{dt}\right)_o = \frac{1}{2}\left(\frac{dp_A}{dt}\right)_o$$

and
$$(dp_A/dt)_o = 2(dP/dt)_o = 2(-7.2) = -14.4 \text{ kPa min}^{-1}$$

4.2.2 Rate and Rate Constant in Terms of Partial Pressure

If p_i is used in the rate law instead of c_i, there are two ways of interpreting r_i and hence k_i. In the first of these, the definition of r_i given in equation 1.4-2 is retained, and in the second, the definition is in terms of rate of change of p_i. Care must be taken to identify which one is being used in a particular case. The first is relatively uncommon, and the second is limited to constant-density situations. The consequences of these two ways are explored further in this and the next section, first for the rate constant, and second for the Arrhenius parameters.

4.2.2.1 Rate Defined by Equation 1.4-2

The first method of interpreting rate of reaction in terms of partial pressure uses the verbal definition given by equation 1.4-2 for r_i. By analogy with equation 4.1-3, we write the rate law (for a reactant i) as

$$(-r_i) = k'_{ip} \prod_{i=1}^{N} p_i^{\alpha_i} \tag{4.2-4}$$

where the additional subscript in k'_{ip} denotes a partial-pressure basis, and the prime distinguishes it from a similar but more common form in the next section. From equations 4.1-3 and -5, and 4.2-3a and -4, it follows that k_i and k'_{ip} are related by

$$k_i = (RT)^n k'_{ip} \tag{4.2-5}$$

The units of k'_{ip} are (concentration)(pressure)$^{-n}$(time)$^{-1}$.

4.2.2.2 Rate Defined by $-dp_i/dt$

Alternatively, we may redefine the rate of reaction in terms of the rate of change of the partial pressure of a substance. If density is constant, this is analogous to the use of $-dc_i/dt$ (equation 2.2-10), and hence is restricted to this case, usually for a constant-volume BR.

In this case, we write the rate law as

$$(-r_{ip}) = -dp_i/dt = k_{ip} \prod_{i=1}^{N} p_i^{\alpha_i} \quad \text{(constant density)} \quad (4.2\text{-}6)$$

where r_{ip} is in units of (pressure)(time)$^{-1}$. From equations 2.2-10 and 4.2-3a, and the first part of equation 4.2-3, r_{ip} is related to r_i by

$$\frac{r_{ip}}{r_i} = \frac{dp_i}{dc_i} = RT \quad \text{(constant density)} \quad (4.2\text{-}7)$$

regardless of the order of reaction.

From equations 4.1-2 and -5, and 4.2-3a, -6, and -7, k_i and k_{ip} are related by

$$k_i = (RT)^{n-1} k_{ip} \quad \text{(constant density)} \quad (4.2\text{-}8)$$

The units of k_{ip} are (pressure)$^{1-n}$(time)$^{-1}$.

EXAMPLE 4-2

For the gas-phase decomposition of acetaldehyde (A, CH$_3$CHO) to methane and carbon monoxide, if the rate constant k_A at 791 K is 0.335 L mol^{-1}s^{-1},

(a) What is the order of reaction, and hence the form of the rate law?
(b) What is the value of k_{Ap}, in Pa^{-1} s^{-1}, for the reaction carried out in a constant-volume BR?

SOLUTION

(a) Since, from equations 4.1-3 and -5, the units of k_A are (concentration)$^{1-n}$(time)$^{-1}$, $1 - n = -1$, and $n = 2$; that is, the reaction is second-order, and the rate law is of the form $(-r_A) = k_A c_A^2$.
(b) From equation 4.2-8,

$$k_{Ap} = k_A(RT)^{1-n} = 0.335/8.314(1000)791 = 5.09 \times 10^{-8} \text{ Pa}^{-1} \text{ s}^{-1}$$

4.2.3 Arrhenius Parameters in Terms of Partial Pressure

4.2.3.1 Rate Defined by Equation 1.4-2

We apply the definition of the characteristic energy in equation 3.1-6 to both k_i and k'_{ip} in equation 4.2-5 to relate E_A, corresponding to k_i, and E'_{Ap}, corresponding to k'_{ip}. From equation 4.2-5, on taking logarithms and differentiating with respect to T, we have

$$\frac{d \ln k_i}{dT} = \frac{n}{T} + \frac{d \ln k'_{ip}}{dT}$$

and using equation 3.1-6, we convert this to

$$E_A = E'_{Ap} + nRT \qquad (4.2\text{-}9)$$

For the relation between the corresponding pre-exponential factors A and A'_p, we use equations 3.1-8, and 4.2-5 and -9 to obtain

$$A = A'_p(RTe)^n \qquad (4.2\text{-}10)$$

where $e = 2.71828$, the base of natural logarithms.

If A and E_A in the original form of the Arrhenius equation are postulated to be independent of T, then their analogues A'_p and E'_{Ap} are not independent of T, except for a zero-order reaction.

4.2.3.2 Rate Defined by $-dp_i/dt$

Applying the treatment used in the previous section to relate E_A and E_{Ap}, corresponding to k_{ip}, and A and A_p, corresponding to k_{ip}, with equation 4.2-5 replaced by equation 4.2-8, we obtain

$$E_A = E_{Ap} + (n-1)RT \qquad (4.2\text{-}11)$$

and

$$A = A_p(RTe)^{n-1} \qquad (4.2\text{-}12)$$

These results are similar to those in the previous section, with $n - 1$ replacing n, and similar conclusions about temperature dependence can be drawn, except that for a first-order reaction, $E_A = E_{Ap}$ and $A = A_p$. The relationships of these differing Arrhenius parameters for a third-order reaction are explored in problem 4-12.

4.3 DEPENDENCE OF RATE ON CONCENTRATION

Assessing the dependence of rate on concentration from the point of view of the rate law involves determining values, from experimental data, of the concentration parameters in equation 4.1-3: the order of reaction with respect to each reactant and the rate constant at a particular temperature. Some experimental methods have been described in Chapter 3, along with some consequences for various orders. In this section, we consider these determinations further, treating different orders in turn to obtain numerical values, as illustrated by examples.

4.3.1 First-Order Reactions

Some characteristics and applications of first-order reactions (for A → products, $(-r_A) = k_A c_A$) are noted in Chapters 2 and 3, and in Section 4.2.3. These are summarized as follows:

(1) The time required to achieve a specified value of f_A is independent of c_{Ao} (Example 2-1; see also equation 3.4-16).

Chapter 4: Development of the Rate Law for a Simple System

(2) The calculation of time quantities: half-life ($t_{1/2}$) in a BR and a CSTR (constant density), problem 2-1; calculation of residence time t for variable density in a PFR (Example 2-3 and problem 2-5).
(3) The integrated form for constant density (Example 3-4), applicable to both a BR and a PFR, showing the exponential decay of c_A with respect to t (equation 3.4-10), or, alternatively, the linearity of $\ln c_A$ with respect to t (equation 3.4-11).
(4) The determination of k_A in an isothermal integral PFR (Example 3-7).
(5) The identity of Arrhenius parameters E_A and E_{Ap}, and A and A_p, based on c_A and p_A, respectively, for constant density (Section 4.2.3).

EXAMPLE 4-3

The rate of hydration of ethylene oxide (A) to ethylene glycol ($C_2H_4O + H_2O \rightarrow C_2H_6O_2$) in dilute aqueous solution can be determined dilatometrically, that is, by following the small change in volume of the reacting system by observing the height of liquid (h) in a capillary tube attached to the reaction vessel (a BR, Figure 3.1). Some results at 20°C, in which the catalyst ($HClO_4$) concentration was 0.00757 mol L^{-1}, are as follows (Brönsted et al., 1929):

t/min	h/cm	t/min	h/cm
0	18.48 (h_o)	270	15.47
30	18.05	300	15.22
60	17.62	330	15.00
90	17.25	360	14.80
120	16.89	390	14.62
240	15.70	1830	12.29 (h_∞)

Determine the order of this reaction with respect to ethylene oxide at 20°C, and the value of the rate constant. The reaction goes virtually to completion, and the initial concentration of ethylene oxide (c_{Ao}) was 0.12 mol L^{-1}.

SOLUTION

We make the following assumptions:

(1) The density of the system is constant.
(2) The concentration of water remains constant.
(3) The reaction is first-order with respect to A.
(4) The change in concentration of A ($c_{Ao} - c_A$) is proportional to the change in height ($h_o - h$).

To justify (1), Brönsted et al., in a separate experiment, determined that the total change in height for a 1-mm capillary was 10 cm for 50 cm³ of solution with $c_{Ao} = 0.2$ mol L^{-1}; this corresponds to a change in volume of only 0.16%.

The combination of (2) and (3) is referred to as a pseudo-first-order situation. H_2O is present in great excess, but if it were not, its concentration change would likely affect the rate. We then use the integral method of Section 3.4.1.1.2 in conjunction with equation 3.4-11 to test assumption (3).

4.3 Dependence of Rate on Concentration   71

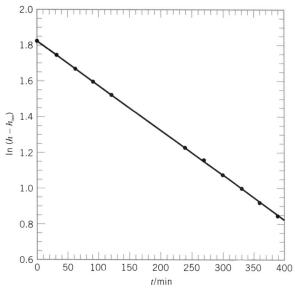

Figure 4.1 First-order plot for $C_2H_4O + H_2O \rightarrow C_2H_6O_2$; data of Brönsted et al. (1929)

Assumption (4) means that $c_{Ao} \propto h_o - h_\infty$ and $c_A \propto h - h_\infty$. Equation 3.4-11 then becomes

$$\ln(h - h_\infty) = \ln(h_o - h_\infty) - k_A t$$
$$= 1.823 - k_A t$$

Some of the data of Brönsted et al. are plotted in Figure 4.1, and confirm that the relation *is* linear, and hence that the reaction is first-order with respect to A. The value of k_A obtained by Brönsted et al. is 2.464×10^{-3} min^{-1} at 20°C.

4.3.2 Second-Order Reactions

A second-order reaction may typically involve one reactant (A $\rightarrow$ products, $(-r_A) = k_A c_A^2$) or two reactants ($|\nu_A|A + |\nu_B|B \rightarrow$ products, $(-r_A) = k_A c_A c_B$). For one reactant, the integrated form for constant density, applicable to a BR or a PFR, is contained in equation 3.4-9, with $n = 2$. In contrast to a first-order reaction, the half-life of a reactant, $t_{1/2}$ from equation 3.4-16, is proportional to c_{Ao}^{-1} (if there are two reactants, both $t_{1/2}$ and fractional conversion refer to the limiting reactant). For two reactants, the integrated form for constant density, applicable to a BR and a PFR, is given by equation 3.4-13 (see Example 3-5). In this case, the reaction stoichiometry must be taken into account in relating concentrations, or in switching rate or rate constant from one reactant to the other.

EXAMPLE 4-4

At 518°C, acetaldehyde vapor decomposes into methane and carbon monoxide according to $CH_3CHO \rightarrow CH_4 + CO$. In a particular experiment carried out in a constant-volume BR (Hinshelwood and Hutchison, 1926), the initial pressure of acetaldehyde was 48.4 kPa,

72 Chapter 4: Development of the Rate Law for a Simple System

and the following increases of pressure (ΔP) were noted (in part) with increasing time:

t/s	42	105	242	480	840	1440
ΔP/kPa	4.5	9.9	17.9	25.9	32.5	37.9

From these results, determine the order of reaction, and calculate the value of the rate constant in pressure units (kPa) and in concentration units (mol L^{-1}).

SOLUTION

It can be shown that the experimental data given do not conform to the hypothesis of a first-order reaction, by the test corresponding to that in Example 4-3. We then consider the possibility of a second-order reaction. From equation 4.2-6, we write the combined assumed form of the rate law and the material balance equation (for constant volume), in terms of CH_3CHO (A), as

$$(-r_{Ap}) = -dp_A/dt = k_{Ap} p_A^2 \tag{1}$$

The integrated form is

$$\frac{1}{p_A} = \frac{1}{p_{Ao}} + k_{Ap} t \tag{2}$$

so that $1/p_A$ is a linear function of t. Values of p_A can be calculated from each value of ΔP, since $P_o = p_{Ao}$, and

$$\Delta P = P - P_o = p_A + p_{CH_4} + p_{CO} - p_{Ao}$$
$$= p_A + 2(p_{Ao} - p_A) - p_{Ao} = p_{Ao} - p_A = 48.4 - p_A \tag{3}$$

Values of p_A calculated from equation (3) are:

t/s	42	105	242	480	840	1440
p_A/kPa	43.9	38.6	30.6	22.6	15.9	10.5

These values are plotted in Figure 4.2 and confirm a linear relation (i.e., $n = 2$). The value of k_{Ap} calculated from the slope of the line in Figure 4.2 is

$$k_{Ap} = 5.07 \times 10^{-5} \text{ kPa}^{-1} \text{ s}^{-1}$$

and, from equation 4.2-8 for k_A in $(-r_A) = k_A c_A^2$,

$$k_A = RT k_{Ap} = 8.314(791)5.07 \times 10^{-5} = 0.334 \text{ L mol}^{-1} \text{ s}^{-1}$$

4.3.3 Third-Order Reactions

The number of reactions that can be accurately described as third-order is relatively small, and they can be grouped according to:

(1) Gas-phase reactions in which one reactant is nitric oxide, the other being oxygen or hydrogen or chlorine or bromine; these are discussed further below.
(2) Gas-phase recombination of two atoms or free radicals in which a third body is required, in each molecular act of recombination, to remove the energy of

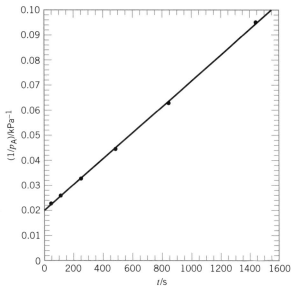

Figure 4.2 Linear second-order plot for Example 4-4

recombination; since consideration of these reactions requires ideas of reaction mechanism, they are considered further in Chapter 6.

(3) Certain aqueous-phase reactions, including some in which acid-base catalysis is involved; for this reason, they are considered further in Chapter 8.

Gas-phase reactions involving nitric oxide which appear to be third-order are:

$$2NO + O_2 \rightarrow 2NO_2$$
$$2NO + 2H_2 \rightarrow N_2 + 2H_2O$$
$$2NO + Cl_2 \rightarrow 2NOCl \text{ (nitrosyl chloride)}$$
$$2NO + Br_2 \rightarrow 2NOBr \text{ (nitrosyl bromide)}$$

In each case, the rate is found to be second-order with respect to NO(A) and first-order with respect to the other reactant (B). That is, as a special form of equation 4.1-3,

$$(-r_A) = k_A c_A^2 c_B \qquad (4.3\text{-}1)$$

(In each case, we are considering only the direction of reaction indicated. The reverse reaction may well be of a different order; for example, the decomposition of NO_2 is second-order.)

The first of these reactions, the oxidation of NO, is an important step in the manufacture of nitric acid, and is very unusual in that its rate *decreases* as T increases (see problem 4-12).

The consequences of using equation 4.3-1 depend on the context: constant or variable density and type of reactor.

EXAMPLE 4-5

Obtain the integrated form of equation 4.3-1 for the reaction $|\nu_A|A + |\nu_B|B \rightarrow$ products occurring in a constant-volume BR.

74 Chapter 4: Development of the Rate Law for a Simple System

SOLUTION

From the rate law and the material balance equation 2.2-10, the equation to be integrated is

$$-\frac{dc_A}{c_A^2 c_B} = k_A dt \tag{4.3-2}$$

The result is rather tedious to obtain, but the method can be the same as that in Example 3-5: use of the stoichiometric relationship and the introduction of ξ, followed by integration by partial fractions and reversion to c_A and c_B to give

$$\nu_B \ln\left(\frac{c_A}{c_B}\right) - \frac{M}{c_A} = \nu_B \ln\left(\frac{c_{Ao}}{c_{Bo}}\right) - \frac{M}{c_{Ao}} + \frac{M^2 k_A}{\nu_A} t \tag{4.3-3}$$

where $M = \nu_B c_{Ao} - \nu_A c_{Bo}$. The left side is a linear function of t; k_A can be determined from the slope of this function.

EXAMPLE 4-6

Suppose the following data were obtained for the homogeneous gas-phase reaction $2A + 2B \rightarrow C + 2D$ carried out in a rigid 2-L vessel at 800°C.

P_o/ kPa	x_{Ao}	$(dP/dt)_o$/ (kPa)min^{-1}
46	0.261	-0.8
70	0.514	-7.2
80	0.150	-1.6

Assuming that at time zero no C or D is present, obtain the rate law for this reaction, stating the value and units of the rate constant in terms of L, mol, s.

SOLUTION

From equation 4.2-6, in terms of A and initial rates and conditions, and an assumed form of the rate law, we write

$$(-r_{Ap})_o = -(dp_A/dt)_o = k_{Ap} p_{Ao}^\alpha p_{Bo}^\beta \tag{1}$$

Values of $(dp_A/dt)_o$ can be calculated from the measured values of $(dP/dt)_o$, as shown in Example 4-1. Values of p_{Ao} and p_{Bo} can be calculated from the given values of P_o and x_{Ao} (from equation 4.2-1). The results for the three experiments are as follows:

p_{rAo}/ kPa	p_{Bo}/ kPa	$(dp_A/dt)_o$/ kPa min^{-1}
12	34	-1.6
36	34	-14.4
12	68	-3.2

We take advantage of the fact that p_{B_o} is constant for the first two experiments, and p_{A_o} is constant for the first and third. Thus, from the first two and equation (1),

$$\frac{-1.6}{-14.4} = \frac{k_{Ap}(12)^\alpha(34)^\beta}{k_{Ap}(36)^\alpha(34)^\beta} = \left(\frac{1}{3}\right)^\alpha$$

from which

$$\alpha = 2$$

Similarly, from the first and third experiments,

$$\beta = 1$$

(The overall order, n, is therefore 3.) Substitution of these results into equation (1) for any one of the three experiments gives

$$k_{Ap} = 3.27 \times 10^{-4} \text{ kPa}^{-2} \text{ min}^{-1}$$

From equation 4.2-8,

$$k_A = (RT)^2 k_{Ap} = (8.314)^2(1073)^2 3.27 \times 10^{-4}/60 = 434 \text{ L}^2 \text{ mol}^{-2} \text{ s}^{-1}$$

4.3.4 Other Orders of Reaction

From the point of view of obtaining the "best" values of kinetics parameters in the rate law, equation 4.1-3, the value of the order can be whatever is obtained as a "best fit" of experimental data, and hence need not be integral. There is theoretical justification (Chapter 6) for the choice of integral values, but experiment sometimes indicates that half-integral values are appropriate. For example, under certain conditions, the decomposition of acetaldehyde is (3/2)-order. Similarly, the reaction between CO and Cl_2 to form phosgene ($COCl_2$) is (3/2)-order with respect to Cl_2 and first-order with respect to CO. A zero-order reaction in which the rate is independent of concentration is not observed for reaction in a single-phase fluid, but may occur in enzyme reactions, and in the case of a gas reacting with a solid, possibly when the solid is a catalyst. The basis for these is considered in Chapters 8 and 10.

4.3.5 Comparison of Orders of Reaction

In this section, we compare the effect of order of reaction n on $c_A/c_{Ao} = 1 - f_A$ for various conditions of reaction, using the model reaction

$$A \rightarrow \text{products} \tag{A}$$

with rate law

$$(-r_A) = k_A c_A^n \tag{3.4-1}$$

We do this for isothermal constant-density conditions first in a BR or PFR, and then in a CSTR. The reaction conditions are normalized by means of a dimensionless reaction number M_{An} defined by

76 Chapter 4: Development of the Rate Law for a Simple System

$$M_{An} = k_A c_{Ao}^{n-1} \bar{t} \qquad (4.3\text{-}4)$$

where $\bar{t}$ is the reaction time in a BR or PFR, or the mean residence time in a CSTR.

4.3.5.1 BR or PFR (Isothermal, Constant Density)

For an nth-order isothermal, constant-density reaction in a BR or PFR ($n \neq 1$), equation 3.4-9 can be rearranged to obtain c_A/c_{Ao} explicitly:

$$c_A^{1-n} - c_{Ao}^{1-n} = (n-1)k_A t \qquad (n \neq 1) \qquad (3.4\text{-}9)$$

$$= (n-1)M_{An}/c_{Ao}^{n-1} \qquad (3.4\text{-}9a)$$

(note that $\bar{t} \equiv t$ here). From equation 3.4-9a,

$$c_A/c_{Ao} = [1 + (n-1)M_{An}]^{1/(1-n)} \qquad (n \neq 1) \qquad (4.3\text{-}5)$$

For a first-order reaction ($n = 1$), from equation 3.4-10,

$$c_A/c_{Ao} = \exp(-k_A t) = \exp(-M_{A1}) \qquad (n = 1) \qquad (4.3\text{-}6)$$

The resulting expressions for c_A/c_{Ao} for several values of n are given in the second column in Table 4.1. Results are given for $n = 0$ and $n = 3$, although single-phase reactions of the type (A) are not known for these orders.

In Figure 4.3, c_A/c_{Ao} is plotted as a function of M_{An} for the values of n given in Table 4.1. For these values of n, Figure 4.3 summarizes how c_A depends on the parameters k_A, c_{Ao}, and $\bar{t}$ for any reaction of type (A). From the value of c_A/c_{Ao} obtained from the figure, c_A can be calculated for specified values of the parameters. For a given n, c_A/c_{Ao} decreases as M_{An} increases; if k_A and c_{Ao} are fixed, increasing M_{An} corresponds

Table 4.1 Comparison of expressions[a] for $c_A/c_{Ao} = 1 - f_A$

Order(n)	BR or PFR $c_A/c_{Ao} = 1 - f_A$	CSTR $c_A/c_{Ao} = 1 - f_A$
0	$= 1 - M_{A0}$; $M_{A0} \leq 1$ $= 0$; $M_{A0} \geq 1$	$= 1 - M_{A0}$; $M_{A0} \leq 1$ $= 0$; $M_{A0} \geq 1$
1/2	$= (1 - M_{A1/2}/2)^2$; $M_{A1/2} \leq 2$ $= 0$; $M_{A1/2} \geq 2$	$= 1 - \dfrac{M_{A1/2}^2}{2}\left[\left(1 + \dfrac{4}{M_{A1/2}^2}\right)^{1/2} - 1\right]$
1	$= \exp(-M_{A1})$	$= (1 + M_{A1})^{-1}$
3/2	$= (1 + M_{A3/2}/2)^{-2}$	from solution of cubic equation [in $(c_A/c_{Ao})^{1/2}$]: $M_{A3/2}(c_A/c_{Ao})^{3/2} + (c_A/c_{Ao}) - 1 = 0$
2	$= (1 + M_{A2})^{-1}$	$= \dfrac{(1 + 4M_{A2})^{1/2} - 1}{2M_{A2}}$
3	$= (1 + 2M_{A3})^{-1/2}$	from solution of cubic equation: $M_{A3}(c_A/c_{Ao})^3 + (c_A/c_{Ao}) - 1 = 0$

[a] For reaction A → products; $(-r_A) = k_A c_A^n$; $M_{An} = k_A c_{Ao}^{n-1} \bar{t}$; isothermal, constant-density conditions; from equations 4.3-5, -6, and -9.

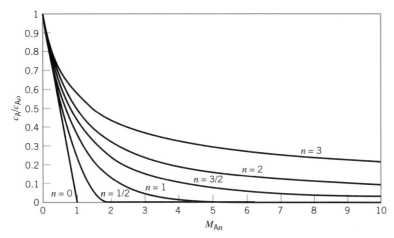

Figure 4.3 Comparison of c_A/c_{Ao} for various orders of reaction in a BR or PFR (for conditions, see footnote to Table 4.1)

to increasing reaction time, $\bar{t}$. For a given M_{An}, c_A/c_{Ao} increases with increasing order, n. We note that for $n = 0$ and $1/2$, c_A/c_{Ao} decreases to 0 at $M_{A0} = 1$ and $M_{A1/2} = 2$, respectively, whereas for the other values of n, c_A/c_{Ao} approaches 0 asymptotically. The former behavior is characteristic for $n < 1$; in such cases, the value of M_{An} for the conditions noted in Figure 4.3 is given from equation 4.3-5 by

$$M_{An}(c_A/c_{Ao} = 0) = 1/(1 - n); \qquad n < 1 \tag{4.3-7}$$

We also note that the slope s of the curves in Figure 4.3 is not the rate of reaction $(-r_A)$, but is related to it by $(-r_A) = -s(-r_A)_o$, where $(-r_A)_o$ is the initial rate at $M_{An} = 0$ $(-(r_A)_o = k_A c_{Ao}^n)$. The limiting slope at $M_{An} = 0$ is $s = -1$ in every case, as is evident graphically for $n = 0$, and can be shown in general from equations 4.3-5 and -6.

4.3.5.2 CSTR (Constant Density)

For an nth-order, constant-density reaction in a CSTR, the combination of equations 2.3-12 and 3.4-1 can be rearranged to give a polynomial equation in c_A/c_{Ao}:

$$(-r_A) = k_A c_A^n = (c_{Ao} - c_A)/\bar{t} \tag{4.3-8}$$

from which, using equation 4.3-4 for M_{An}, we obtain (for *all* values of n):

$$M_{An}\left(\frac{c_A}{c_{Ao}}\right)^n + \left(\frac{c_A}{c_{Ao}}\right) - 1 = 0 \tag{4.3-9}$$

Solutions for c_A/c_{Ao} from equation 4.3-9 are given in the third column in Table 4.1. For $n = 3/2$ and 3, the result is a cubic equation in $(c_A/c_{Ao})^{1/2}$ and c_A/c_{Ao}, respectively. The analytical solutions for these are cumbersome expressions, and the equations can be solved numerically to obtain the curves in Figure 4.4.

In Figure 4.4, similar to Figure 4.3, c_A/c_{Ao} is plotted as a function of M_{An}. The behavior is similar in both figures, but the values of c_A/c_{Ao} for a CSTR are higher than those for a BR or PFR (except for $n = 0$, where they are the same). This is an important characteristic in comparing these types of reactors (Chapter 17). Another difference is that c_A/c_{Ao} approaches 0 asymptotically for all values of $n > 0$, and not just for $n \geq 1$, as in Figure 4.3.

78 Chapter 4: Development of the Rate Law for a Simple System

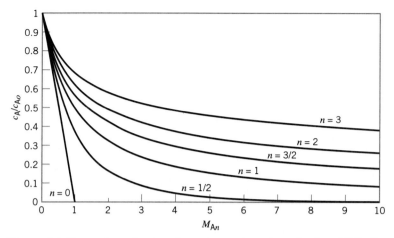

Figure 4.4 Comparison of c_A/c_{Ao} for various orders of reaction in a CSTR (for conditions, see footnote to Table 4.1)

4.3.6 Product Species in the Rate Law

The rate of reaction may also depend on the concentration of a product, which is included in equation 4.1-3. If α_i for a product is negative, the effect is called product inhibition, and is not uncommon in catalytic reactions (Chapter 8). If α_i for a product is positive, the reaction accelerates with increasing conversion, and the effect is called autocatalysis (Chapter 8). The possible involvement of product species in the rate law should be considered in the experimental investigation. This can be tested by measuring the rate at low conversions. Since reactant concentrations vary little in such cases, any relatively large changes in rate arise from the large percentage changes in product concentration, which increases from zero to a finite value.

EXAMPLE 4-7

Suppose the following rate data are obtained at the same T from a 400-cm³ CSTR in a kinetics investigation of the vapor-phase dehydration of ethyl alcohol to form ethyl ether:

$$2C_2H_5OH(A) \rightarrow (C_2H_5)_2O(B) + H_2O(C)$$

The values of $(-r_A)$ are calculated from the measured concentrations of A by means of equation 2.3-12 (constant density assumed).

expt.	q_o/ cm³ s⁻¹	c_{Ao}/ mol L⁻¹	c_A/ mol L⁻¹	$c_B = c_C$/ mol L⁻¹	f_A	$10^4(-r_A)$/ mol L⁻¹ s⁻¹
1	20	0.05	0.0476	0.00120	0.048	1.20
2	20	0.10	0.0966	0.00170	0.034	1.70
3	10	0.05	0.0467	0.00167	0.066	0.83
4	10	0.10	0.0952	0.00239	0.048	1.20

Propose a rate law for this reaction.

Table 4.2 Values of the Arrhenius parameters

Reaction	Order n	A/ (L mol^{-1})$^{n-1}$ s^{-1}	E_A/ kJ mol^{-1}	Reference*
$H_2 + I_2 \to 2HI$	2	1.3×10^{11}	163.2	(1)
$2HI \to H_2 + I_2$	2	7.9×10^{10}	184.1	(1)
$2C_4H_6 \to c\text{-}C_8H_{12}$	2	1.3×10^8	112.1	(1)
$CH_3 + CH_3 \to C_2H_6$	2	2.0×10^{10}	0	(1)
$Cl + H_2 \to HCl + H$	2	7.9×10^{10}	23	(1)
$NO + O_3 \to NO_2 + O_2$	2	6.3×10^8	10.5	(1)
$HOCl + I^- \to HOI + Cl^-$	2	1.6×10^9	3.8	(2)
$OCl^- + I^- \to OI^- + Cl^-$	1	4.9×10^{10}	50	(2)
$C_2H_5Cl \to C_2H_4 + HCl$	1	4.0×10^{14}	254	(3)
$c\text{-}C_4H_8 \to 2C_2H_4$	1	4.0×10^{15}	262	(3)

*(1) Bamford and Tipper (1969).
(2) Lister and Rosenblum (1963).
(3) Moore (1972, p. 395).

SOLUTION

We note that in experiments 1 and 3 c_A is approximately the same, but that $(-r_A)$ decreases as c_B or c_C increases, approximately in inverse ratio. Experiments 2 and 4 similarly show the same behavior. In experiments 2 and 3, c_B or c_C is approximately constant, and $(-r_A)$ doubles as c_A doubles. These results suggest that the rate is first-order (+1) with respect to A, and −1 with respect to B or C, or (less likely) B and C together. From the data given, we can't tell which of these three possibilities correctly accounts for the inhibition by product(s). However, if, for example, B is the inhibitor, the rate law is

$$(-r_A) = k_A c_A c_B^{-1}$$

and k_A can be calculated from the data given.

4.4 DEPENDENCE OF RATE ON TEMPERATURE

4.4.1 Determination of Arrhenius Parameters

As introduced in sections 3.1.3 and 4.2.3, the Arrhenius equation is the normal means of representing the effect of T on rate of reaction, through the dependence of the rate constant k on T. This equation contains two parameters, A and E_A, which are usually stipulated to be independent of T. Values of A and E_A can be established from a minimum of two measurements of k at two temperatures. However, more than two results are required to establish the validity of the equation, and the values of A and E_A are then obtained by parameter estimation from several results. The linear form of equation 3.1-7 may be used for this purpose, either graphically or (better) by linear regression. Alternatively, the exponential form of equation 3.1-8 may be used in conjunction with nonlinear regression (Section 3.5). Some values are given in Table 4.2.

EXAMPLE 4-8

Determine the Arrhenius parameters for the reaction $C_2H_4 + C_4H_6 \to C_6H_{10}$ from the following data (Rowley and Steiner, 1951):

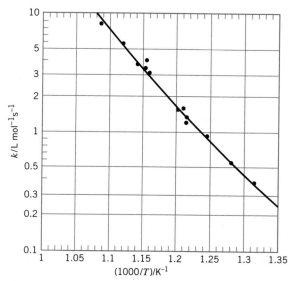

Figure 4.5 Arrhenius plot for $C_2H_4 + C_4H_6 \rightarrow C_6H_{10}$ (data of Rowley and Steiner, 1951)

T/K	k/L mol^{-1} s^{-1}	T/K	k/L mol^{-1} s^{-1}
760	0.384	863	3.12
780	0.560	866	4.05
803	0.938	867	3.47
832	1.565	876	3.74
822	1.34	894	5.62
823	1.23	921	8.20
826	1.59		

SOLUTION

The data of Rowley and Steiner are shown graphically in Figure 4.5, with k plotted on a logarithmic scale (equivalent to $\ln k$ on a linear scale) against $1000/T$. According to equation 3.1-7, the result should be a linear relation, with a slope of $-E_A/R$ and an intercept (not indicated in Figure 4.5) of $\ln A$. The values of E_A and A obtained by Rowley and Steiner in this way are 115,000 J mol^{-1} and 3.0×10^7 L mol^{-1} s^{-1}, respectively.

4.4.2 Arrhenius Parameters and Choice of Concentration Units for Gas-Phase Reactions

The consequences for the effect of different choices of concentration units developed in Section 4.2.3 are explored in problem 4-12 for the third-order NO oxidation reaction.

4.5 PROBLEMS FOR CHAPTER 4

4-1 The kinetics of the pyrolysis of mixtures of 2-butyne (A, C_4H_6) and vinylacetylene (B, C_4H_4) have been investigated by Harper and Heicklen (1988). Pyrolysis is a factor in soot formation, which involves polymerization at one stage. Although the major product in this case was a polymer, o-xylene (C, C_8H_{10}) was also produced, and this was chosen as the species of interest. Reaction was carried out in a constant-volume BR, and analysis was by mass spectrometry.

Initial rates of formation of C for various initial concentrations of A and B at 400°C are as follows:

$10^4\, c_{Ao}$ mol L^{-1}	$10^4\, c_{Bo}$ mol L^{-1}	$10^9\, r_{Co}$ mol L^{-1} s^{-1}
9.41	9.58	12.5
4.72	4.79	3.63
2.38	2.45	0.763
1.45	1.47	0.242
4.69	14.3	12.6
2.28	6.96	3.34
1.18	3.60	0.546
0.622	1.90	0.343
13.9	4.91	6.62
6.98	2.48	1.67
3.55	1.25	0.570
1.90	0.67	0.0796

(a) Test the hypothesis that the initial rate of formation of o-xylene is first-order with respect to each of A and B.

(b) For a rate law of the form $r_C = k_A c_A^\alpha c_B^\beta$, determine values of k_A, α, and β by nonlinear regression.

(c) From the following values of the rate constant, given by the authors, at five temperatures, determine the values of the Arrhenius parameters A and E_A, and specify their units.

$T/°C$	350	375	400	425	450
$10^3\, k/$L mol^{-1} s^{-1}	4.66	6.23	14.5	20.0	37.9

4-2 The rate of decomposition of dimethyl ether (CH$_3$OCH$_3$) in the gas phase has been determined by Hinshelwood and Askey (1927) by measuring the increase in pressure (ΔP) accompanying decomposition in a constant-volume batch reactor at a given temperature. The reaction is complicated somewhat by the appearance of formaldehyde as an intermediate product at the conditions studied, but we assume here that the reaction goes to completion according to CH$_3$OCH$_3$ → CH$_4$ + H$_2$ + CO, or A → M + H + C. In one experiment at 504°C, in which the initial pressure ($P_o = p_{Ao}$, pure ether being present initially) was 41.6 kPa, the following values of ΔP were obtained:

t/s	$\Delta P = (P - p_{Ao})/$kPa	t/s	$\Delta P = (P - p_{Ao})/$kPa
0	0	916	26.7
207	7.5	1195	33.3
390	12.8	1587	41.6
481	15.5	2240	53.6
665	20.8	2660	58.3
777	23.5	3155	62.3
		∞	82.5

Test the hypothesis that the reaction is first-order with respect to ether.

4-3 The hydrolysis of methyl bromide (CH$_3$Br) in dilute aqueous solution may be followed by titrating samples with AgNO$_3$. The volumes of AgNO$_3$ solution (V) required for 10 cm^3 samples at 330 K in a particular experiment in a constant-volume batch reactor were as follows

(Millard 1953, p. 453):

t/min	0	88	300	412	reaction complete
V/cm^3	0	5.9	17.3	22.1	49.5

(a) Write the equations for the reactions occurring during hydrolysis and analysis.
(b) If the reaction is first-order with respect to $CH_3Br(A)$, show that the rate constant may be calculated from $k_A = (1/t) \ln[V_\infty/(V_\infty - V)]$, where t is time, V_∞ is the volume of $AgNO_3$ required for titration when the reaction is complete, and V is the volume required at any time during the course of the reaction.
(c) Calculate values of k_A to show whether the reaction is first-order with respect to CH_3Br.

4-4 Ethyl acetate reacts with sodium hydroxide in aqueous solution to produce sodium acetate and ethyl alcohol:

$$CH_3COOC_2H_5(A) + NaOH \rightarrow CH_3COONa + C_2H_5OH$$

This saponification reaction can be followed by withdrawing samples from a BR at various times, adding excess standard acid to "quench" the reaction by neutralizing the unreacted hydroxide, and titrating the excess acid with base. In a particular experiment at 16°C, samples of 100 cm^3 were withdrawn at various times; the concentration of acid used (HCl) was 0.0416 mol L^{-1}. The following results were obtained (V_t is the volume of acid solution required to neutralize unreacted NaOH at time t) (Glasstone, 1946, p. 1058).

t/s	0	224	377	629	816	∞
V_t/cm^3	62.09	54.33	50.60	46.28	43.87	33.06

Using this information, obtain the rate law for the reaction.

4-5 The rate of decomposition of gaseous ethylene oxide (C_2H_4O), to CH_4 and CO, has been studied by Mueller and Walters (1951) by determination of the fraction (f_A) of oxide (A) reacted after a definite time interval (t) in a constant-volume batch reactor. In a series of experiments, the initial pressure of the oxide (P_{Ao}) was varied. Some of the results are as follows:

P_{Ao}/kPa	27.1	37.2	40.4	55.3	58.6
t/s	2664	606	2664	2664	1206
f_A	0.268	0.084	0.274	0.286	0.139

From these results, determine the order of reaction and the value of the rate constant (specify its units).

4-6 The rate of reaction between hydrocyanic acid (HCN) and acetaldehyde (CH_3CHO) to give acetaldehyde cyanohydrin has been studied in a constant-volume batch reactor at 25°C in dilute aqueous solution, buffered to keep the pH constant (Svirbely and Roth, 1953). The reaction is

$$HCN + CH_3CHO \rightarrow CH_3CH(OH)CN$$

A typical set of results is given below, where the concentrations are in mol L^{-1}

t/min	3.28	11.12	24.43	40.35	67.22	∞
$c_{HCN} \times 10^2$	6.57	6.19	5.69	5.15	4.63	2.73
$c_{CH_3CHO} \times 10^2$	3.84	3.46	2.96	2.42	1.90	0.00

Determine the rate law for this reaction at 25°C, and calculate the rate constant, and the initial concentrations of HCN(c_{Ao}) and CH_3CHO(c_{Bo}).

4-7 The rate of acetylation of benzyl chloride in dilute aqueous solution at 102°C has been studied by Huang and Dauerman (1969). The reaction is

$$CH_3COONa + C_6H_5CH_2Cl \rightarrow CH_3COOC_6H_5CH_2 + Na^+ + Cl^-$$

or A + B → products

Some of the data they obtained for a solution equimolar in reactants ($c_{Ao} = 0.757$ mol L^{-1}) in a constant-volume batch reactor are as follows (f'_B is the fraction of B *unconverted* at time t):

$10^{-3}t/s$	24.5	54.7	88.6	126.7
f'_B	0.912	0.809	0.730	0.638

Determine the form of the rate law and the value of the rate constant at 102°C based on these data.

4-8 The rate of decomposition of nitrogen pentoxide (N_2O_5) in the inert solvent CCl_4 can be followed by measuring the volume of oxygen evolved at a given temperature and pressure, since the unreacted N_2O_5 and the other products of decomposition remain in solution. Some results at 45°C from a BR are as follows (Eyring and Daniels, 1930):

t/s	162	409	1721	3400	∞
O_2 evolved/cm^3	3.41	7.78	23.00	29.33	32.60

What is the order of the decomposition reaction (which for this purpose can be written as $N_2O_5 \rightarrow N_2O_4 + \frac{1}{2}O_2$)? Assume the reaction goes to completion.

4-9 Rate constants for the first-order decomposition of nitrogen pentoxide (N_2O_5) at various temperatures are as follows (Alberty and Silbey, 1992, p. 635):

T/K	273	298	308	318	328	338
$10^5 k/s^{-1}$	0.0787	3.46	13.5	49.8	150	487

Show that the data obey the Arrhenius relationship, and determine the values of the Arrhenius parameters.

4-10 Rate constants for the liquid-phase, second-order, aromatic substitution reaction of 2-chloroquinoxaline (2CQ) with aniline in ethanol (inert solvent) were determined at several temperatures by Patel (1992). The reaction rate was followed by means of a conductance cell (as a BR). Results are as follows:

$T/°C$	20	25	30	35	40
$10^5 k/dm^3$ mol^{-1} s^{-1}	2.7	4.0	5.8	8.6	13.0

Calculate the Arrhenius parameters A and E_A for this reaction, and state the units of each.

4-11 Suppose the liquid-phase reaction A → B + C was studied in a 3-L CSTR at steady-state, and the following results were obtained:

expt.	q/cm^3 s^{-1}	$T/°C$	c_A/mol L^{-1}
1	0.50	25	0.025
2	6.00	25	0.100
3	1.50	35	0.025

Assuming that the rate law is of the form $(-r_A) = k_A c_A^n = A \exp(-E_A/RT)c_A^n$, determine A, E_A, and n, and hence k_A at 25°C and at 35°C. c_{Ao} in all three runs was 0.250 mol L^{-1}.

4-12 The oxidation of nitric oxide, NO(A) + $\frac{1}{2}O_2$ → NO_2, is a third-order gas-phase reaction (second-order with respect to NO). Data of Ashmore et al. (1962) for values of the rate constant at various temperatures are as follows:

T/K	377	473	633	633	692	799
$10^{-3} k_A/L^2$ mol^{-2} s^{-1}	9.91	7.07	5.83	5.73	5.93	5.71

(a) Calculate the corresponding values of k_{Ap} in $kPa^{-2}s^{-1}$.
(b) Determine the values of the Arrhenius parameters based on the values of k_A given above.
(c) Repeat (b) using the values calculated in (a) to obtain E_{Ap} and A_p.
(d) Compare the difference $E_A - E_{Ap}$ as calculated in (b) and (c) with the expected result.
(e) Which is the better representation, (b) or (c), of the experimental data in this case?
(See also data of Bodenstein et al. (1918, 1922), and of Greig and Hall (1967) for additional data for the range 273 to 622 K.)

4-13 The chlorination of dichlorotetramethylbenzene (A) in acetic acid at 30°C has been studied by Baciocchi et al. (1965). The reaction may be represented by

$$A + B \rightarrow products,$$

where B is chlorine. In one experiment in a batch reactor, the initial concentrations were $c_{Ao} = 0.0347$ mol L^{-1}, and $c_{Bo} = 0.0192$ mol L^{-1}, and the fraction of chlorine reacted (f_B) at various times was as follows:

t/min	0	807	1418	2255	2855	3715	4290
f_B	0	0.2133	0.3225	0.4426	0.5195	0.5955	0.6365

Investigate whether the rate law is of the form $(-r_A) = (-r_B) = kc_Ac_B$, and state your conclusion, including, if appropriate, the value of k and its units.

4-14 The reaction $2NO + 2H_2 \rightarrow N_2 + 2H_2O$ was studied in a constant-volume BR with equimolar quantities of NO and H_2 at various initial pressures:

P_o/kPa	47.2	45.5	50.0	38.4	33.5	32.4	26.9
$t_{1/2}$/s	81	102	95	140	180	176	224

Calculate the overall order of the reaction (Moore, 1972, p. 416).

4-15 The hydrolysis of ethylnitrobenzoate by hydroxyl ions

$$NO_2C_6H_4COOC_2H_5 + OH^- \rightarrow NO_2C_6H_4COO^- + C_2H_5OH$$

proceeds as follows at 15°C when the initial concentrations of both reactants are 0.05 mol L^{-1} (constant-volume batch reactor):

t/s	120	180	240	330	530	600
% hydrolyzed	32.95	41.75	48.8	58.05	69.0	70.4

Use (a) the differential method and (b) the integral method to determine the reaction order, and the value of the rate constant. Comment on the results obtained by the two methods.

4-16 The kinetics of the gas-phase reaction between nitrogen dioxide (A) and trichloroethene (B) have been investigated by Czarnowski (1992) over the range 303–362.2 K. The reaction extent, with the reaction carried out in a constant-volume BR, was determined from measurements of infrared absorption intensities, which were converted into corresponding pressures by calibration. The products of the reaction are nitrosyl chloride, NOCl (C), and glyoxyloxyl chloride, HC(O)C(O)Cl.

In a series of seven experiments at 323.1 K, the initial pressures, p_{Ao} and p_{Bo}, were varied, and the partial pressure of NOCl, p_C, was measured after a certain length of time, t. Results are as follows:

t/min	182.2	360.4	360.8	435.3	332.8	120.0	182.1
p_{Ao}/kPa	3.97	5.55	3.99	2.13	3.97	2.49	2.08
p_{Bo}/kPa	7.16	7.66	6.89	6.77	3.03	8.57	9.26
p_C/kPa	0.053	0.147	0.107	0.067	0.040	0.027	0.040

(a) Write the chemical equation representing the stoichiometry of the reaction.
(b) Can the course of the reaction be followed by measuring (total) pressure rather than by the method described above? Explain.
(c) Determine the form of the rate law and the value of the rate constant (in units of L, mol, s) at 323.1 K, with respect to NO_2.
(d) From the following values of the rate constant, with respect to NO_2 (units of kPa, min), given by Czarnowski, determine values of the Arrhenius parameters, and specify the units of each:

T/K	303.0	323.1	343.1	362.2
$10^6 \, k_p$ (units of kPa, min)	4.4	10.6	20.7	39.8

4-17 A La(Cr, Ni) O_x catalyst was tested for the cleanup of residual hydrocarbons in combustion streams by measuring the rate of methane oxidation in a differential laboratory flow reactor containing a sample of the catalyst. The following conversions were measured as a function of temperature with a fixed initial molar flow rate of methane. The inlet pressure was 1 bar and the methane mole fraction was 0.25. (Note that the conversions are small, so that the data approximately represent initial rates.) The rate law for methane oxidation is first-order with respect to methane concentration.

T/°C	250	300	350	400	450
% conversion	0.11	0.26	0.58	1.13	2.3

(a) Explain why initial methane molar concentrations are not constant for the different runs.
(b) Calculate k (s^{-1}) and k'_p (mol s^{-1} L^{-1} bar^{-1}) for each temperature, given that the void volume in the bed was 0.5 cm^3 and the methane molar flow rate into the reactor was 1 mmol min^{-1}.
(c) Show whether these data obey the Arrhenius rate expression for both k and k'_p data. What are the values of E_A and E'_{Ap}? (Indicate the units.)
(d) Explain why, if one of the Arrhenius plots of either k or k'_p is linear, the other deviates from linearity. Is this effect significant for these data? Explain.
(e) Calculate the pre-exponential factors A and A'_p. Comment on the relative magnitudes of A and A'_p as temperature approaches infinity.
(f) How would you determine if factors involving the reaction products (CO_2 and H_2O) should be included in the rate expression?

4-18 The Ontario dairy board posted the following times for keeping milk without spoilage.

T/°C	Safe storage time before spoilage
0	30 days
3	14 days
15	2 days
22	16 hours
30	3 hours

(a) Does the spoilage of milk follow the Arrhenius relation? Assume spoilage represents a given "fractional conversion" of the milk. Construct an Arrhenius plot of the data.
(b) What value of activation energy (E_A) characterizes this process? (State the units.)

4-19 The reactions of the ground-state oxygen atom O(^{3}P) with symmetric aliphatic ethers in the gas phase were investigated by Liu et al. (1990) using the flash photolysis resonance fluorescence technique. These reactions were found to be first-order with respect to each reactant. The rate constants for three ethers at several temperatures are as follows:

| | $10^{14} k$/cm^3 molecule^{-1} s^{-1} | | | | |
Ether	240 K	298 K	330 K	350 K	400 K
diethyl	17.0	38.1	55.8	66.1	98.6
di-*n*-propyl	25.8	58.2	75.3	90.0	130
di-*n*-butyl	36.0	68.9	89.7	114	153

Determine the Arrhenius parameters A and E_A for each diether and specify the units of each.

4-20 Nowak and Skrzypek (1989) have measured the rates of decomposition *separately* of (1) NH$_4$HCO$_3$ (A) (to (NH$_4$)$_2$CO$_3$), and (2) (NH$_4$)$_2$CO$_3$ (B) in aqueous solution. They used an open, isothermal BR with continuous removal of gaseous products (CO$_2$ in case (1) and NH$_3$ in (2)) so that each reaction was irreversible. They measured c_A in case (1) and c_B in case (2) at predetermined times, and obtained the following results at 323 K for (1) and 353 K for (2).

$10^{-3} t$/s	$10 c_A$/mol L^{-1}	$10 c_B$/mol L^{-1}
0	8.197	11.489
1.8	6.568	6.946
3.6	5.480	4.977
5.4	4.701	3.878
7.2	4.116	3.177
9.0	3.660	2.690
10.8	3.295	2.332
12.6	2.996	2.059
14.4	2.748	1.843
16.2	2.537	1.668
18.0	2.356	1.523

(a) Write the chemical equations for the two cases (H$_2$O is also a product in each case).
(b) Determine the best form of the rate law in each case, including the numerical value of the rate constant.

Chapter 5

Complex Systems

In previous chapters, we deal with "simple" systems in which the stoichiometry and kinetics can each be represented by a single equation. In this chapter we deal with "complex" systems, which require more than one equation, and this introduces the additional features of product distribution and reaction network. Product distribution is not uniquely determined by a single stoichiometric equation, but depends on the reactor type, as well as on the relative rates of two or more simultaneous processes, which form a reaction network. From the point of view of kinetics, we must follow the course of reaction with respect to more than one species in order to determine values of more than one rate constant. We continue to consider only systems in which reaction occurs in a single phase. This includes some catalytic reactions, which, for our purpose in this chapter, may be treated as "pseudohomogeneous." Some development is done with those famous fictitious species A, B, C, etc. to illustrate some features as simply as possible, but real systems are introduced to explore details of product distribution and reaction networks involving more than one reaction step.

We first outline various types of complexities with examples, and then describe methods of expressing product distribution. Each of the types is described separately in further detail with emphasis on determining kinetics parameters and on some main features. Finally, some aspects of reaction networks involving combinations of types of complexities and their construction from experimental data are considered.

5.1 TYPES AND EXAMPLES OF COMPLEX SYSTEMS

Reaction complexities include reversible or opposing reactions, reactions occurring in parallel, and reactions occurring in series. The description of a reacting system in terms of steps representing these complexities is called a reaction network. The steps involve only species that can be measured experimentally.

5.1.1 Reversible (Opposing) Reactions

Examples of reversible reacting systems, the reaction networks of which involve opposing reactions, are:

(1) Isomerization of butane (A)

$$n\text{-}C_4H_{10} \rightleftarrows i\text{-}C_4H_{10}$$

(2) Oxidation of SO_2

$$SO_2 + \frac{1}{2}O_2 \rightleftarrows SO_3$$

88 Chapter 5: Complex Systems

(3) Hydrolysis of methyl acetate or its reverse, esterification of acetic acid

$$CH_3COOCH_3 + H_2O \rightleftarrows CH_3COOH + CH_3OH$$

5.1.2 Reactions in Parallel

Examples of reacting systems with networks made up of parallel steps are:

(1) Dehydration and dehydrogenation of C_2H_5OH **(B)**

$$C_2H_5OH \rightarrow C_2H_4 + H_2O$$
$$C_2H_5OH \rightarrow C_2H_4O + H_2$$

(2) Nitration of nitrobenzene to dinitrobenzene

$$C_6H_5NO_2 + HNO_3 \rightarrow o\text{-}C_6H_4(NO_2)_2 + H_2O$$
$$C_6H_5NO_2 + HNO_3 \rightarrow m\text{-}C_6H_4(NO_2)_2 + H_2O$$
$$C_6H_5NO_2 + HNO_3 \rightarrow p\text{-}C_6H_4(NO_2)_2 + H_2O$$

5.1.3 Reactions in Series

An example of a reacting system with a network involving reactions in series is the decomposition of acetone (series with respect to ketene) **(C)**

$$(CH_3)_2CO \rightarrow CH_4 + CH_2CO\text{(ketene)}$$
$$CH_2CO \rightarrow \tfrac{1}{2}C_2H_4 + CO$$

5.1.4 Combinations of Complexities

(1) Series—reversible; decomposition of N_2O_5 **(D)**

$$N_2O_5 \rightarrow N_2O_4 + \tfrac{1}{2}O_2$$
$$N_2O_4 \rightleftarrows 2NO_2$$

(2) Series—parallel

- Partial oxidation of methane to formaldehyde **(E)**

$$CH_4 + O_2 \rightarrow HCHO + H_2O$$
$$HCHO + \tfrac{1}{2}O_2 \rightarrow CO + H_2O$$
$$CH_4 + 2O_2 \rightarrow CO_2 + 2H_2O$$

(This network is series with respect to HCHO and parallel with respect to CH_4 and O_2.)

- Chlorination of CH_4 **(F)**

$$CH_4 + Cl_2 \rightarrow CH_3Cl + HCl$$
$$CH_3Cl + Cl_2 \rightarrow CH_2Cl_2 + HCl$$
$$CH_2Cl_2 + Cl_2 \rightarrow CHCl_3 + HCl$$
$$CHCl_3 + Cl_2 \rightarrow CCl_4 + HCl$$

(This network is series with respect to the chlorinated species and parallel with respect to Cl_2.)

- Hepatic metabolism of lidocaine (LID, $C_{14}H_{22}N_2O$) **(G)**

This follows a series-parallel network, corresponding to either hydroxylation of the benzene ring, or de-ethylation of the tertiary amine, leading to MEGX, to hydroxylidocaine, and ultimately to hydroxyMEGX:

$$LID \xrightarrow{-C_2H_5} MEGX\ (C_{12}H_{18}N_2O)$$

$$LID \xrightarrow{+OH} hydroxylidocaine\ (C_{14}H_{22}N_2O_2)$$

$$MEGX \xrightarrow{+OH} hydroxyMEGX\ (C_{12}H_{18}N_2O_2)$$

$$hydroxylidocaine \xrightarrow{-C_2H_5} hydroxyMEGX$$

5.1.5 Compartmental or Box Representation of Reaction Network

In addition, or as an alternative, to actual chemical reaction steps, a network may be represented by compartments or boxes, with or without the reacting species indicated. This is illustrated in Figure 5.1 for networks (A) to (G) in Sections 5.1.1 to 5.1.4. This method provides a pictorial representation of the essential features of the network.

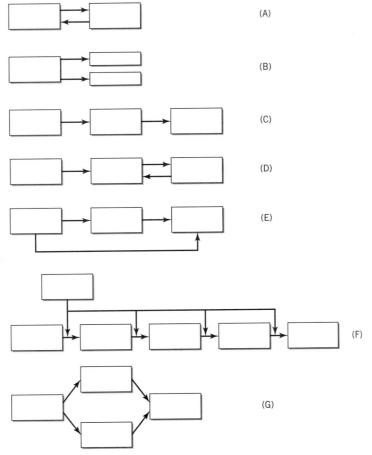

Figure 5.1 Compartmental or box representation of reaction networks (A) to (G) in Sections 5.1.1 to 5.1.4

5.2 MEASURES OF REACTION EXTENT AND SELECTIVITY

5.2.1 Reaction Stoichiometry and Its Significance

For a complex system, determination of the stoichiometry of a reacting system in the form of the maximum number (R) of linearly independent chemical equations is described in Examples 1-3 and 1-4. This can be a useful preliminary step in a kinetics study once all the reactants and products are known. It tells us the minimum number (usually) of species to be analyzed for, and enables us to obtain corresponding information about the remaining species. We can thus use it to construct a stoichiometric table corresponding to that for a simple system in Example 2-4. Since the set of equations is not unique, the individual chemical equations do not necessarily represent reactions, and the stoichiometric model does not provide a reaction network without further information obtained from kinetics.

EXAMPLE 5-1

Spencer and Pereira (1987) studied the kinetics of the gas-phase partial oxidation of CH_4 over a MoO_3-SiO_2 catalyst in a differential PFR. The products were HCHO (formaldehyde), CO, CO_2, and H_2O.

(a) Obtain a set of R linearly independent chemical equations to represent the stoichiometry of the reacting system.
(b) What is the minimum number of species whose concentrations must be measured experimentally for a kinetics analysis?

SOLUTION

(a) The system may be represented by

$$\{(CH_4, O_2, H_2O, CO, CO_2, HCHO), (C, O, H)\}$$

Using manipulations by hand or *Mathematica* as described in Example 1-3, we obtain the following set of 3 (R) equations in canonical form with CH_4, O_2, and H_2O as components, and CO, CO_2, and HCHO as noncomponents:

$$CH_4 + \frac{3}{2}O_2 = 2H_2O + CO \qquad (1)$$

$$CH_4 + 2O_2 = 2H_2O + CO_2 \qquad (2)$$

$$CH_4 + O_2 = H_2O + HCHO \qquad (3)$$

These chemical equations may be combined indefinitely to form other equivalent sets of three equations. They do not necessarily represent chemical reactions in a reaction network. The network deduced from kinetics results by Spencer and Pereira (see Example 5-8) involved (3), (1)–(3), and (2) as three reaction steps.

(b) The minimum number of species is $R = 3$, the same as the number of equations or noncomponents. Spencer and Pereira reported results in terms of CO, CO_2, and HCHO, but also analyzed for O_2 and CH_4 by gas chromatography. Measurements above the minimum number allow for independent checks on element balances, and also more data for statistical determination of rate parameters.

5.2.2 Fractional Conversion of a Reactant

Fractional conversion of a reactant, f_A for reactant A, say, is the ratio of the amount of A reacted at some point (time or position) to the amount introduced into the system, and is a measure of consumption of the reactant. It is defined in equation 2.2-3 for a batch system, and in equation 2.3-5 for a flow system. The definition is the same whether the system is simple or complex.

In complex systems, f_A is not a unique parameter for following the course of a reaction, unlike in simple systems. For both kinetics and reactor considerations (Chapter 18), this means that rate laws and design equations cannot be uniquely expressed in terms of f_A, and are usually written in terms of molar concentrations, or molar flow rates or extents of reaction. Nevertheless, f_A may still be used to characterize the overall reaction extent with respect to reactant A.

5.2.3 Yield of a Product

The yield of a product is a measure of the reaction extent at some point (time or position) in terms of a specified product and reactant. The most direct way of calculating the yield of a product in a complex system from experimental data is by means of a stoichiometric model in canonical form, with the product as a noncomponent. This is because that product appears only once in the set of equations, as illustrated for each of CO, CO_2, and HCHO in Example 5-1.

Consider reactant A and (noncomponent) product D in the following set of stoichiometric equations:

$$|\nu_A|_D A + \ldots = \nu_D D + \ldots$$

+other equations not involving D

The yield of D with respect to A, $Y_{D/A}$, is

$$Y_{D/A} = \frac{\text{moles A reacted to form D}}{\text{mole A initially}} \quad (5.2\text{-}1a)$$

$$= \frac{\text{moles A reacted to form D}}{\text{mole D formed}} \times \frac{\text{moles D formed}}{\text{mole A initially}}$$

$$= \frac{|\nu_A|_D}{\nu_D} \frac{n_D - n_{Do}}{n_{Ao}} \quad (\text{BR, constant or variable } \rho) \quad (5.2\text{-}1b)$$

$$= \frac{|\nu_A|_D}{\nu_D} \frac{F_D - F_{Do}}{F_{Ao}} \quad (\text{flow reactor, constant or variable } \rho) \quad (5.2\text{-}1c)$$

$$= \frac{|\nu_A|_D}{\nu_D} \frac{c_D - c_{Do}}{c_{Ao}} \quad (\text{BR or flow reactor, constant } \rho) \quad (5.2\text{-}1d)$$

where $|\nu_A|_D$ is the absolute value of ν_A in the equation involving D, and n_{Do}, F_{Do}, c_{Do} refer to product D initially (each may be zero).

The sum of the yields of all the noncomponents is equal to the fractional conversion of A:

$$\sum_{k=C+1}^{N} Y_{k/A} = \sum_{k=C+1}^{N} \frac{|\nu_A|_k}{\nu_k} \frac{n_k - n_{ko}}{n_{Ao}} = \frac{n_{Ao} - n_A}{n_{Ao}} = f_A \quad (5.2\text{-}2)$$

where k is a noncomponent index, C is the number of components, and N is the number of species.

For a simple system with only one noncomponent, say D,

$$Y_{D/A} = f_A \quad \text{(simple system)} \qquad (5.2\text{-}2a)$$

As defined above, $Y_{D/A}$ is normalized so that

$$0 \le Y_{D/A} \le 1 \qquad (5.2\text{-}3)$$

5.2.4 Overall and Instantaneous Fractional Yield

The *fractional* yield of a product is a measure of how selective a particular reactant is in forming a particular product, and hence is sometimes referred to as selectivity.[1] Two ways of representing selectivity are (1) the overall fractional yield (from inlet to a particular point such as the outlet); and (2) the instantaneous fractional yield (at a point). We consider each of these in turn.

For the stoichiometric scheme in Section 5.2.3, the *overall* fractional yield of D with respect to A, $\hat{S}_{D/A}$, is

$$\hat{S}_{D/A} = \frac{\text{moles A reacted to form D}}{\text{mole A reacted}} \qquad (5.2\text{-}4a)$$

$$= \frac{|\nu_A|_D}{\nu_D} \frac{n_D - n_{Do}}{n_{Ao} - n_A} \quad \text{(BR, constant or variable } \rho) \qquad (5.2\text{-}4b)$$

$$= \frac{|\nu_A|_D}{\nu_D} \frac{F_D - F_{Do}}{F_{Ao} - F_A} \quad \text{(flow reactor, constant or variable } \rho) \qquad (5.2\text{-}4c)$$

$$= \frac{|\nu_A|_D}{\nu_D} \frac{c_D - c_{Do}}{c_{Ao} - c_A} \quad \text{(BR, or flow reactor, constant } \rho) \qquad (5.2\text{-}4d)$$

From the definitions of f_A, $Y_{D/A}$, and $\hat{S}_{D/A}$, it follows that

$$Y_{D/A} = f_A \hat{S}_{D/A} \qquad (5.2\text{-}5)$$

The sum of the overall fractional yields of the noncomponents is unity:

$$\sum_{k=C+1}^{N} \hat{S}_{k/A} = \sum_{k=C+1}^{N} \frac{|\nu_A|_k}{\nu_k} \frac{n_k - n_{ko}}{n_{Ao}} = \frac{n_{Ao} - n_A}{n_{Ao} - n_A} = 1 \qquad (5.2\text{-}6)$$

As in the cases of f_A and $Y_{D/A}$, $\hat{S}_{D/A}$ is normalized in the definitions so that

$$0 \le \hat{S}_{D/A} \le 1 \qquad (5.2\text{-}7)$$

---

[1] Other definitions and notation may be used for selectivity by various authors.

The *instantaneous* fractional yield of D with respect to A is

$$\hat{s}_{D/A} = \frac{\text{rate of formation of D}}{\text{rate of disappearance of A}} = \frac{r_D}{(-r_A)} \quad (5.2\text{-}8)$$

5.2.5 Extent of Reaction

Another stoichiometric variable that may be used is the extent of reaction, ξ, defined by equation 2.3-6 for a simple system. For a complex system involving N species and represented by R chemical equations in the form

$$\sum_{i=1}^{N} \nu_{ij} A_i = 0; \; j = 1, 2, \ldots, R \quad (1.4\text{-}10)$$

where ν_{ij} is the stoichiometric coefficient of the ith species (A_i) in the jth equation, we may extend the definition to (for a flow system):

$$\nu_{ij}\xi_j = (F_i - F_{io})_j; \; i = 1, 2, \ldots N; \; j = 1, 2, \ldots, R \quad (5.2\text{-}9)$$

Since

$$\sum_{j=1}^{R} \nu_{ij}\xi_j = \sum_{j=1}^{R} (F_i - F_{io})_j = F_i - F_{io}; \; i = 1, 2, \ldots N \quad (5.2\text{-}10)$$

the flow rate of any species at any point may be calculated from measured values of ξ_j, one for each equation, at that point:

$$F_i = F_{io} + \sum_{j=1}^{R} \nu_{ij}\xi_j; \; i = 1, 2, \ldots N \quad (5.2\text{-}11)$$

or, for molar amounts in a batch system

$$n_i = n_{io} + \sum_{j=1}^{R} \nu_{ij}\xi_j; \; i = 1, 2, \ldots N \quad (5.2\text{-}12)$$

If the R equations are in canonical form with one noncomponent in each equation, it is convenient to calculate ξ_j from experimental information for the noncomponents. The utility of this is illustrated in the next section.

5.2.6 Stoichiometric Table for Complex System

A stoichiometric table for keeping track of the amounts or flow rates of all species during reaction may be constructed in various ways, but here we illustrate, by means of an example, the use of ξ_j, the extent of reaction variable. We divide the species into components and noncomponents, as determined by a stoichiometric analysis (Section 5.2.1), and assume experimental information is available for the noncomponents (at least).

Table 5.1 Stoichiometric table in terms of ξ_j for Example 5-2

Species i	Initial	Change	ξ_j	F_i
noncomponents				
CO	0	F_{CO}	$\xi_1 = F_{CO}/1$	ξ_1
CO_2	0	F_{CO_2}	$\xi_2 = F_{CO_2}/1$	ξ_2
HCHO	0	F_{HCHO}	$\xi_3 = F_{HCHO}/1$	ξ_3
components				
CH_4	$F_{CH_4,o}$			$F_{CH_4,o} - \xi_1 - \xi_2 - \xi_3$
O_2	$F_{O_2,o}$			$F_{O_2,o} - \frac{3}{2}\xi_1 - 2\xi_2 - \xi_3$
H_2O	0			$2\xi_1 + 2\xi_2 + \xi_3$
total:	$F_{CH_4,o} + F_{O_2,o}$			$F_{CH_4,o} + F_{O_2,o} + \frac{1}{2}\xi_1$

EXAMPLE 5-2

Using the chemical system and equations (1), (2), and (3) of Example 5-1, construct a stoichiometric table, based on the use of ξ_j, to show the molar flow rates of all six species. Assume experimental data are available for the flow rates (or equivalent) of CO, CO_2, and HCHO as noncomponents.

SOLUTION

The table can be displayed as Table 5.1, with both ξ_j and F_i obtained from equation 5.2-11, applied to noncomponents and components in turn.

5.3 REVERSIBLE REACTIONS

5.3.1 Net Rate and Forms of Rate Law

Consider a reversible reaction involving reactants A, B, ... and products C, D, ... written as:

$$|\nu_A|A + |\nu_B|B + \ldots \underset{r_r}{\overset{r_f}{\rightleftarrows}} \nu_C C + \nu_D D + \ldots \quad (5.3\text{-}1)$$

We assume that the experimental (net) rate of reaction, r, is the difference between the forward rate, r_f, and the reverse rate, r_r:

$$r = \frac{r_A}{\nu_A} = \ldots = \frac{r_D}{\nu_D} = r_f(c_i, T, \ldots) - r_r(c_i, T, \ldots) \quad (5.3\text{-}2)$$

If the effects of T and c_i are separable, then equation 5.3-2 may be written

$$r = k_f(T)g_f(c_i) - k_r(T)g_r(c_i) \quad (5.3\text{-}3)$$

where k_f and k_r are forward and reverse rate constants, respectively.

If, further, a power rate law of the form of equation 4.1-3 is applicable, then

$$r = k_f(T)\prod_{i=1}^{N} c_i^{\alpha_i} - k_r(T)\prod_{i=1}^{N} c_i^{\alpha_i'} \quad (5.3\text{-}4)$$

In this form, the sets of exponents α_i and α_i' are related to each other by restrictions imposed by thermodynamics, as shown in the next section.

5.3.2 Thermodynamic Restrictions on Rate and on Rate Laws

Thermodynamics imposes restrictions on both the rate r and the form of the rate law representing it. Thus, at given (T, P), for a system reacting spontaneously (but not at equilibrium),

$$\Delta G_{T,P} < 0 \quad \text{and} \quad r > 0 \tag{5.3-5}$$

At equilibrium,

$$\Delta G_{T,P} = 0 \quad \text{and} \quad r = 0 \tag{5.3-6}$$

The third possibility of $r < 0$ cannot arise, since $\Delta G_{T,P}$ cannot be positive for spontaneous change.

Equation 5.3-6 leads to a necessary relation between α_i and α_i' in equation 5.3-4. From this latter equation, at equilibrium,

$$\frac{k_f(T)}{k_r(T)} = \frac{\prod_{i=1}^{N} c_{i,eq}^{\alpha_i'}}{\prod_{i=1}^{N} c_{i,eq}^{\alpha_i}} \tag{5.3-7}$$

Also, at equilibrium, the equilibrium constant is

$$K_{c,eq}(T) = \prod_{i=1}^{N} c_{i,eq}^{\nu_i} \tag{5.3-8}$$

Since k_f/k_r and $K_{c,eq}$ are both functions of T only, they are functionally related (Denbigh, 1981, p. 444):

$$\frac{k_f(T)}{k_r(T)} = \phi(K_{c,eq}) \tag{5.3-9}$$

or

$$\frac{\prod_{i=1}^{N} c_{i,eq}^{\alpha_i'}}{\prod_{i=1}^{N} c_{i,eq}^{\alpha_i}} = \phi\left(\prod_{i=1}^{N} c_{i,eq}^{\alpha_i}\right) \tag{5.3-10}$$

It follows necessarily (Blum and Luus, 1964; Aris, 1968) that ϕ is such that

$$\frac{k_f(T)}{k_r(T)} = (K_{c,eq})^n \quad (n > 0) \tag{5.3-11}$$

where

$$n = (\alpha_i' - \alpha_i)/\nu_i; \quad i = 1, 2, \ldots \tag{5.3-12}$$

as obtained from equation 5.3-10 (rewritten to correspond to 5.3-11) by equating exponents species by species. (n is not to be confused with reaction order itself.)

If we use 5.3-11 to eliminate $k_r(T)$ in equation 5.3-4, we obtain

$$r = k_f(T)\left[\prod_{i=1}^{N} c_i^{\alpha_i} - \frac{\prod_{i=1}^{N} c_i^{\alpha_i'}}{(K_{c,eq})^n}\right] \qquad (5.3\text{-}13)$$

If the effects of T and c_i on r are separable, but the individual rate laws for r_f and r_r are *not* of the power-law form, equation 5.3-13 is replaced by the less specific form (from 5.3-3),

$$r = k_f(T)\left[g_f(c_i) - \frac{g_r(c_i)}{(K_{c,eq})^n}\right] \qquad (5.3\text{-}14)$$

The value of n must be determined experimentally, but in the absence of such information, it is usually assumed that $n = 1$.

EXAMPLE 5-3

The gas-phase synthesis of methanol (M) from CO and H_2 is a reversible reaction:

$$CO + 2H_2 \rightleftharpoons CH_3OH$$

(a) If, at low pressure with a rhodium catalyst, $r_f = k_f p_{CO}^{-0.3} p_{H_2}^{1.3}$, and $r_r = k_r p_{CO}^{a'} p_{H_2}^{b'} p_M$, what is the value of n in equation 5.3-12, and what are the values of a' and b'?

(b) Repeat (a) if $r_r = k_r p_{CO}^{a'} p_{H_2}^{b'} p_M^{0.5}$.

SOLUTION

(a) If we apply equation 5.3-12 to CH_3OH, with c, c' replacing α_i, α_i', the exponents are $c = 0$ and $c' = 1$. Then

$$n = \frac{c' - c}{\nu_M} = \frac{1 - 0}{1} = 1$$

$$a' = a + \nu_{CO} n = -0.3 - 1(1) = -1.3$$

$$b' = b + \nu_{H_2} n = 1.3 - 2(1) = -0.7$$

As a check, with $n = 1$, from equations 5.3-1 and -4

$$K_{eq} = \frac{k_f}{k_r} = \frac{p_{CO}^{-1.3} p_{H_2}^{-0.7} p_M}{p_{CO}^{-0.3} p_{H_2}^{1.3}} = \frac{p_M}{p_{CO} p_{H_2}^2}$$

(b)
$$n = \frac{0.5 - 0}{1} = 0.5$$
$$a' = -0.3 - 1(0.5) = -0.8$$
$$b' = 1.3 - 2(0.5) = 0.3$$
$$K_{eq}^{0.5} = \frac{p_M^{0.5}}{p_{CO}^{0.5} p_{H_2}}$$

5.3.3 Determination of Rate Constants

The experimental investigation of the form of the rate law, including determination of the rate constants k_f and k_r, can be done using various types of reactors and methods, as discussed in Chapters 3 and 4 for a simple system. Use of a batch reactor is illustrated here and in Example 5-4, and use of a CSTR in problem 5-2.

Consider the esterification of ethyl alcohol with formic acid to give ethyl formate (and water) in a mixed alcohol-water solvent, such that the alcohol and water are present in large excess. Assume that this is pseudo-first-order in both esterification (forward) and hydrolysis (reverse) directions:

$$C_2H_5OH(\text{large excess}) + HCOOH(A) \underset{k_r}{\overset{k_f}{\rightleftharpoons}} HCOOC_2H_5(D) + H_2O(\text{large excess})$$

For the reaction carried out isothermally in a batch reactor (density constant), the values of k_f and k_r may be determined from experimental measurement of c_A with respect to t, in the following manner.

The postulated rate law is

$$r_D = (-r_A) = k_f c_A - k_r c_D \quad (5.3\text{-}15)$$
$$= k_f c_{Ao}(1 - f_A) - k_r c_{Ao} f_A \quad (5.3\text{-}15a)$$
$$= k_f c_{Ao}[1 - (1 + 1/K_{c,eq}) f_A] \quad (5.3\text{-}16)$$

from equation 5.3-11 (with $n = 1$), which is 5.3-19 below. From the material balance for A,

$$(-r_A) = c_{Ao} df_A/dt \quad (2.2\text{-}4)$$

Combining equations 2.2-4 and 5.3-16, we obtain the governing differential equation:

$$\frac{df_A}{dt} = k_f[1 - (1 + 1/K_{c,eq}) f_A] \quad (5.3\text{-}17)$$

The equivalent equation in terms of c_A is

$$-\frac{dc_A}{dt} = k_f c_A - k_r c_D = k_f c_A - k_r(c_{Ao} - c_A) \quad (5.3\text{-}17a)$$

Integration of equation 5.3-17 with $f_A = 0$ at $t = 0$ results in

$$\ln\left(\frac{1}{1 - (1 + 1/K_{c,eq}) f_A}\right) = \left(1 + \frac{1}{K_{c,eq}}\right) k_f t \quad (5.3\text{-}18)$$

from which k_f can be determined from measured values of f_A (or c_A) at various times t, if $K_{c,eq}$ is known. Then k_r is obtained from

$$k_r = k_f/K_{c,eq} \tag{5.3-19}$$

If the reaction is allowed to reach equilibrium ($t \to \infty$), $K_{c,eq}$ can be calculated from

$$K_{c,eq} = c_{D,eq}/c_{A,eq} \tag{5.3-20}$$

As an alternative to this traditional procedure, which involves, in effect, linear regression of equation 5.3-18 to obtain k_f (or a corresponding linear graph), a nonlinear regression procedure can be combined with simultaneous numerical integration of equation 5.3-17a. Results of both these procedures are illustrated in Example 5-4. If the reaction is carried out at other temperatures, the Arrhenius equation can be applied to each rate constant to determine corresponding values of the Arrhenius parameters.

EXAMPLE 5-4

Assuming that the isomerization of A to D and its reverse reaction are both first-order:

$$A \underset{k_r}{\overset{k_f}{\rightleftarrows}} D$$

calculate the values of k_f and k_r from the following data obtained at a certain temperature in a constant-volume batch reactor:

t/h	0	1	2	3	4	∞
$100 c_A/c_{Ao}$	100	72.5	56.8	45.6	39.5	30

(a) Using the linear procedure indicated in equation 5.3-18; and
(b) Using nonlinear regression applied to equation 5.3-17 by means of the E-Z Solve software.

SOLUTION

(a) From the result at $t = \infty$,

$$K_{c,eq} = \frac{c_{D,eq}}{c_{A,eq}} = \frac{c_{Ao}f_{A,eq}}{c_{Ao}(1 - f_{A,eq})} = \frac{1 - c_{A,eq}/c_{Ao}}{c_{A,eq}/c_{Ao}} = 0.7/0.3 = 2.33$$

In the simplest use of equation 5.3-18, values of k_f may be calculated from the four measurements at $t = 1, 2, 3, 4$ h; the average of the four values gives $k_f = 0.346$ h^{-1}. Then, from equation 5.3-19, $k_r = 0.346/2.33 = 0.148$ h^{-1}.
(b) The results from nonlinear regression (see file ex5-4.msp) are: $k_f = 0.345$ h^{-1} and $k_f = 0.147$ h^{-1}. The values of $100\ c_A/c_{Ao}$ calculated from these parameters, in comparison with the measured values are:

t/h	1	2	3	4	∞
$(100 c_A/c_{Ao})_{exp}$	72.5	56.8	45.6	39.5	30
$(100 c_A/c_{Ao})_{calc}$	72.8	56.1	45.9	39.7	29.9

There is close agreement, the (absolute) mean deviation being 0.3.

5.3.4 Optimal T for Exothermic Reversible Reaction

An important characteristic of an *exothermic* reversible reaction is that the rate has an optimal value (a maximum) with respect to T at a given composition (e.g., as measured by f_A). This can be shown from equation 5.3-14 (with $n = 1$ and $K_{eq} \equiv K_{c,eq}$). Since g_f and g_r are independent of T, and $r = r_D/\nu_D$ (in equation 5.3-1),

$$\frac{1}{\nu_D}\left(\frac{\partial r_D}{\partial T}\right)_{f_A} = \left(g_f - \frac{g_r}{K_{eq}}\right)\frac{dk_f}{dT} + \frac{g_r k_f}{K_{eq}^2}\frac{dK_{eq}}{dT} \quad (5.3\text{-}21)$$

$$= \left(\frac{g_f/g_r}{g_{f,eq}/g_{r,eq}} - 1\right)\frac{g_r}{K_{eq}}\frac{dk_f}{dT} + k_r g_r \frac{d\ln K_{eq}}{dT} \quad (5.3\text{-}22)$$

since $K_{eq} = g_{r,eq}/g_{f,eq}$, and $dK_{eq}/K_{eq} = d\ln K_{eq}$. Since dk_f/dT is virtually always positive, and $(g_f/g_r)/(g_{f,eq}/g_{r,eq}) > 1$ ($g_f > g_{f,eq}$ and $g_r < g_{r,eq}$), the first term on the right in equation 5.3-22 is positive. The second term, however, may be positive (endothermic reaction) or negative (exothermic reaction), from equation 3.1-5.

Thus, for an *endothermic* reversible reaction, the rate increases with increase in temperature at constant conversion; that is,

$$(\partial r_D/\partial T)_{f_A} > 0 \quad \text{(endothermic)} \quad (5.3\text{-}23)$$

For an *exothermic* reversible reaction, since $\Delta H°$ is negative, $(\partial r_D/\partial T)_{f_A}$ is positive or negative depending on the relative magnitudes of the two terms on the right in equation 5.3-22. This suggests the possibility of a maximum in r_D, and, to explore this further, it is convenient to return to equation 5.3-3. That is, for a maximum in r_D,

$$\partial r_D/\partial T = 0, \text{ and} \quad (5.3\text{-}24)$$

$$g_f \frac{dk_f}{dT} = g_r \frac{dk_r}{dT} \quad (5.3\text{-}25)$$

Using equation 3.1-8, $k = A\exp(-E_A/RT)$ for k_f and k_r in turn, we can solve for the temperature at which this occurs:

$$\boxed{T_{opt} = \frac{E_{Ar} - E_{Af}}{R}\left[\ln\left(\frac{g_r A_r E_{Ar}}{g_f A_f E_{Af}}\right)\right]^{-1}} \quad (5.3\text{-}26)$$

EXAMPLE 5-5

(a) For the reversible exothermic first-order reaction $A \rightleftarrows D$, obtain T_{opt} in terms of f_A, and, conversely, the "locus of maximum rates" expressing f_A (at $r_{D,max}$) as a function of T. Assume constant density and no D present initially.
(b) Show that the rate (r_D) decreases monotonically as f_A increases at constant T, whether the reaction is exothermic or endothermic.

SOLUTION

(a) For this case, equation 5.3-3 (with $r = r_D$) becomes

$$r_D = k_f c_A - k_r c_D \quad (5.3\text{-}15)$$

That is,

$$g_f = c_A = c_{Ao}(1 - f_A)$$

and

$$g_r = c_D = c_{Ao} f_A$$

Hence, from equation 5.3-26,

$$T_{opt} = M_1 \left[\ln\left(\frac{f_A M_2}{1 - f_A}\right) \right]^{-1} \quad (5.3\text{-}27)$$

where $f_A = f_A(r_{D,max})$, and on solving equation (5.3-27) for f_A, we have

$$f_A(\text{at } r_{D,max}) = [1 + M_2 \exp(-M_1/T)]^{-1} \quad (5.3\text{-}28)$$

where

$$M_1 = (E_{Ar} - E_{Af})/R \quad (5.3\text{-}29)$$

and

$$M_2 = A_r E_{Ar}/A_f E_{Af} \quad (5.3\text{-}30)$$

(b) Whether the reaction is exothermic or endothermic, equation 5.3-15a can be written

$$r_D = c_{Ao}[k_f - (k_f + k_r)f_A] \quad (5.3\text{-}31)$$

from which

$$(\partial r_D/\partial f_A)_T = -c_{Ao}(k_f + k_r) < 0 \quad (5.3\text{-}32)$$

That is, r_D decreases as f_A increases at constant T.

The optimal rate behavior with respect to T has important consequences for the design and operation of reactors for carrying out reversible, exothermic reactions. Examples are the oxidation of SO_2 to SO_3 and the synthesis of NH_3.

This behavior can be shown graphically by constructing the r_D–T–f_A relation from equation 5.3-16, in which k_f, k_r, and K_{eq} depend on T. This is a surface in three-dimensional space, but Figure 5.2 shows the relation in two-dimensional contour form, both for an exothermic reaction and an endothermic reaction, with f_A as a function of T and $(-r_A)$ (as a parameter). The full line in each case represents equilibrium conversion. Two constant-rate $(-r_A)$ contours are shown in each case (note the direction of increase in $(-r_A)$ in each case). As expected, each rate contour exhibits a maximum for the exothermic case, but not for the endothermic case.

5.4 PARALLEL REACTIONS

A reaction network for a set of reactions occurring in parallel with respect to species A may be represented by

5.4 Parallel Reactions

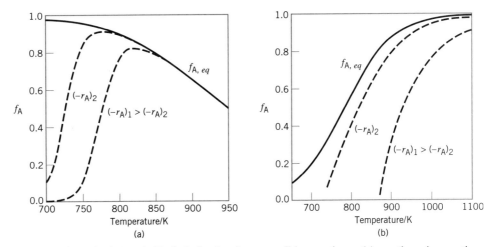

Figure 5.2 Typical $(-r_A)$–T–f_A behavior for reversible reactions: (a) exothermic reaction; (b) endothermic reaction

$$|\nu_{A1}|A + \ldots \xrightarrow{k_{A1}} \nu_D D + \ldots$$
$$|\nu_{A2}|A + \ldots \xrightarrow{k_{A2}} \nu_E E + \ldots \quad (5.4\text{-}1)$$
$$\vdots \qquad \qquad \vdots$$

The product distribution is governed by the relative rates at which these steps occur. For example, if the rate laws for the first two steps are given by

$$r_D/\nu_D = (-r_{A1})/|\nu_{A1}| = k_{A1}(T)g_{A1}(c_A, \ldots)/|\nu_{A1}| \quad (5.4\text{-}2a)$$

and

$$r_E/\nu_E = (-r_{A2})/|\nu_{A2}| = k_{A2}(T)g_{A2}(c_A, \ldots)/|\nu_{A2}| \quad (5.4\text{-}2b)$$

the relative rate at which D and E are formed is

$$\boxed{\frac{r_D}{r_E} = \frac{\nu_D \nu_{A2} k_{A1}(T) g_{A1}(c_A, \ldots)}{\nu_E \nu_{A1} k_{A2}(T) g_{A2}(c_A, \ldots)}} \quad (5.4\text{-}3)$$

The product distribution depends on the factors $(c_A, \ldots, T)$ that govern this ratio, and the design and operation of a reactor is influenced by the requirement for a favorable distribution.

From the point of view of kinetics, we illustrate here how values of the rate constants may be experimentally determined, and then used to calculate such quantities as fractional conversion and yields.

EXAMPLE 5-6

For the kinetics scheme

$$\begin{aligned} A &\to B + C; \quad r_B = k_{A1} c_A \\ A &\to D + E; \quad r_D = k_{A2} c_A \end{aligned} \quad (5.4\text{-}4)$$

(a) Describe how experiments may be carried out in a constant-volume BR to measure k_{A1} and k_{A2}, and hence confirm the rate laws indicated (the use of a CSTR is considered in problem 5-5);

(b) If $k_{A1} = 0.001$ s^{-1} and $k_{A2} = 0.002$ s^{-1}, calculate (i) f_A, (ii) the product distribution (c_A, c_B, etc.), (iii) the yields of B and D, and (iv) the overall fractional yields of B and D, for reaction carried out for 10 min in a constant-volume BR, with only A present initially at a concentration $c_{Ao} = 4$ mol L^{-1}.

(c) Using the data in (b), plot c_A, c_B and c_D versus t.

SOLUTION

(a) Since there are two independent reactions, we use two independent material balances to enable the two rate constants to be determined. We may choose A and B for this purpose.

A material balance for A results in

$$-dc_A/dt = k_{A1}c_A + k_{A2}c_A \tag{5.4-5}$$

This integrates to

$$\ln c_A = \ln c_{Ao} - (k_{A1} + k_{A2})t \tag{5.4-6}$$

In other words, if we follow reaction with respect to A, we can obtain the sum of the rate constants, but not their individual values.

If, in addition, we follow reaction with respect to B, then, from a material balance for B,

$$dc_B/dt = k_{A1}c_A \tag{5.4-7}$$

From equations 5.4-5 and -7,

$$-dc_A/dc_B = (k_{A1} + k_{A2})/k_{A1}$$

which integrates to

$$c_A = c_{Ao} - (1 + k_{A2}/k_{A1})(c_B - c_{Bo}) \tag{5.4-8}$$

From the slopes of the linear relations in equations 5.4-6 and -8, k_{A1} and k_{A2} can be determined, and the linearity would confirm the forms of the rate laws postulated.

(b) (i) From equation 5.4-6,

$$c_A = 4\exp[-(0.001 + 0.002)10(60)] = 0.661 \text{ mol L}^{-1}$$

$$f_A = (4 - 0.661)/4 = 0.835$$

(ii) c_A is given in (i).
From equation 5.4-8,

$$c_B = c_C = (4 - 0.661)/(1 + 0.002/0.001) = 1.113 \text{ mol L}^{-1}$$

From an overall material balance,

$$c_D = c_E = c_{Ao} - c_A - c_B = 2.226 \text{ mol L}^{-1}$$

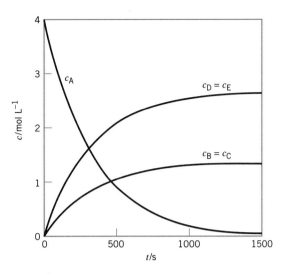

Figure 5.3 Concentration profiles for parallel reaction network in Example 5-6

(iii) $Y_B = 1.113/4 = 0.278$
$Y_D = 2.226/4 = 0.557$
(iv) $\hat{S}_{B/A} = 1.113/(4 - 0.661) = 0.333$
$\hat{S}_{D/A} = 2.226/(4 - 0.661) = 0.667$

(c) From equations 5.4-6, 5.4-8, and 5.4-7, together with $dc_D/dt = k_{A2}c_A$,

$$c_A = c_{Ao}\exp[-(k_{A1} + k_{A2})t] = 4e^{-0.003t}$$

$$c_B = (c_{Ao} - c_A)/(1 + k_{A2}/k_{A1}) = (4 - c_A)/3$$

$$c_D = (k_{A2}/k_{A1})c_B = 2c_B$$

In Figure 5.3, c_A, $c_B(= c_C)$, and $c_D(= c_E)$ are plotted for $t = 0$ to 1500 s; as $t \to \infty$, $c_A \to 0$, $c_B \to 1.33$, and $c_D \to 2.67$ mol L^{-1}.

5.5 SERIES REACTIONS

A kinetics scheme for a set of (irreversible) reactions occurring in series with respect to species A, B, and C may be represented by

$$|\nu_A|A + \ldots \xrightarrow{k_1} \nu_B B + \ldots \xrightarrow{k_2} \nu_C C + \ldots \quad (5.5\text{-}1)$$

in which the two sequential steps are characterized by rate constants k_1 and k_2. Such a scheme involves two corresponding stoichiometrically independent chemical equations, and two species such as A and B must be followed analytically to establish the complete product distribution at any instant or position.

We derive the kinetics consequences for this scheme for reaction in a constant-volume batch reactor, the results also being applicable to a PFR for a constant-density system. The results for a CSTR differ from this, and are explored in Example 18-4.

Consider the following simplified version of scheme 5.5-1, with each of the two steps being first-order:

$$A \xrightarrow{k_1} B \xrightarrow{k_2} C \quad (5.5\text{-}1a)$$

For reaction in a constant-volume BR, with only A present initially, the concentrations of A, B and C as functions of time t are governed by the following material-balance equations for A, B and C, respectively, incorporating the two independent rate laws:

$$-dc_A/dt = k_1 c_A \tag{5.5-2}$$

$$dc_B/dt = k_1 c_A - k_2 c_B \tag{5.5-3}$$

$$c_C = c_{Ao} - c_A - c_B \tag{5.5-4}$$

The first two equations can be integrated to obtain $c_A(t)$ and $c_B(t)$ in turn, and the results used in the third to obtain $c_C(t)$. Anticipating the quantitative results, we can deduce the general features of these functions from the forms of the equations above. The first involves only A, and is the same for A decomposing by a first-order process to B, since A has no direct "knowledge" of C. Thus, the $c_A(t)$ profile is an exponential decay. The concentration of B initially increases as time elapses, since, for a sufficiently short time (with $c_B \to 0$), $k_1 c_A > k_2 c_B$ (equation 5.5-3). Eventually, as c_B continues to increase and c_A to decrease, a time is reached at which $k_1 c_A = k_2 c_B$, and c_B reaches a maximum, after which it continuously decreases. The value of c_C continuously increases with increasing time, but, since, from equations 5.5-2 to -4, $d^2 c_C/dt^2 \propto dc_B/dt$, there is an inflection point in $c_C(t)$ at the time at which c_B is a maximum. These results are illustrated in Figure 5.4 for the case in which $k_1 = 2$ min^{-1} and $k_2 = 1$ min^{-1}, as developed below. For the vertical scale, the normalized concentrations c_A/c_{Ao}, c_B/c_{Ao} and c_C/c_{Ao} are used, their sum at any instant being unity.

The integration of equation 5.5-2 results in

$$c_A = c_{Ao} \exp(-k_1 t) \tag{3.4-10}$$

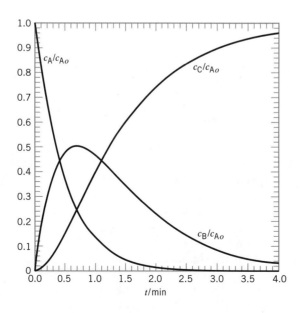

Figure 5.4 Concentration-time profiles (product distribution) for $A \xrightarrow{k_1} B \xrightarrow{k_2} C$ in a batch reactor; $k_1 = 2$ min^{-1}; $k_2 = 1$ min^{-1}.

This result may be used to eliminate c_A in equation 5.5-3, to give a differential equation from which $c_B(t)$ may be obtained:

$$dc_B/dt + k_2 c_B = k_1 c_{Ao} \exp(-k_1 t) \quad (5.5\text{-}5)$$

This is a linear, first-order differential equation, the solution of which is

$$c_B = [k_1 c_{Ao}/(k_2 - k_1)](e^{-k_1 t} - e^{-k_2 t}) \quad (5.5\text{-}6)$$

Finally, c_C may be obtained from equation 5.5-4 together with equations 3.4-10 and 5.5-6:

$$c_C = [c_{Ao}/(k_2 - k_1)]\left[k_2(1 - e^{-k_1 t}) - k_1(1 - e^{-k_2 t})\right] \quad (5.5\text{-}7)$$

The features of the behavior of c_A, c_B, and c_C deduced qualitatively above are illustrated quantitatively in Figure 5.4. Other features are explored in problem 5-10.

Values of the rate constants k_1 and k_2 can be obtained from experimental measurements of c_A and c_B at various times in a BR. The most sophisticated procedure is to use either equations 5.5-2 and -3 or equations 3.4-10 and 5.5-6 together in a nonlinear parameter-estimation treatment (as provided by the E-Z Solve software; see Figure 3.11). A simpler procedure is first to obtain k_1 from equation 3.4-10, and second to obtain k_2 from k_1 and either of the coordinates of the maximum value of c_B (t_{max} or $c_{B,max}$). These coordinates can be related to k_1 and k_2, as shown in the following example.

EXAMPLE 5-7

Obtain expressions relating t_{max} and $c_{B,max}$ in a BR to k_1 and k_2 in reaction 5.5-1a.

SOLUTION

Differentiating equation 5.5-6, we obtain

$$\frac{d(c_B/c_{Ao})}{dt} = \frac{k_1}{k_2 - k_1}(k_2 e^{-k_2 t} - k_1 e^{-k_1 t})$$

Setting $d(c_B/c_{Ao})/dt = 0$ for $t = t_{max}$, we obtain

$$t_{max} = \frac{\ln(k_2/k_1)}{k_2 - k_1} \quad (5.5\text{-}8)$$

From equation 5.5-3 with $dc_B/dt = 0$ at $c_{B,max}$,

$$c_{B,max} = \frac{k_1}{k_2} c_A(t_{max}) = \frac{k_1}{k_2} c_{Ao} e^{-k_1 t_{max}}$$

Thus, the maximum yield of B, on substitution for t_{max} from 5.5-8, is

$$(Y_{B/A})_{max} = \frac{c_{B,max}}{c_{Ao}} = \frac{k_1}{k_2} \exp\left[\frac{-k_1 \ln(k_2/k_1)}{k_2 - k_1}\right] = \frac{k_1}{k_2}\left(\frac{k_2}{k_1}\right)^{\frac{-k_1}{k_2-k_1}} = \left(\frac{k_2}{k_1}\right)^{\frac{k_2}{k_1-k_2}} \quad (5.5\text{-}9)$$

5.6 COMPLEXITIES COMBINED

5.6.1 Concept of Rate-Determining Step (*rds*)

In a kinetics scheme involving more than one step, it may be that one change occurs much faster or much slower than the others (as determined by relative magnitudes of rate constants). In such a case, the overall rate, and hence the product distribution, may be determined almost entirely by this step, called the rate-determining step (*rds*).

For reactions in parallel, it is the "fast" step that governs. Thus, if $A \xrightarrow{k_{AB}} B$ and $A \xrightarrow{k_{AC}} C$ are two competing reactions, and if $k_{AB} \gg k_{AC}$, the rate of formation of B is much higher than that of C, and very little C is produced. Chemical rates can vary by very large factors, particularly when different catalysts are involved. For example, a metal catalyst favors dehydrogenation of an alcohol to an aldehyde, but an oxide catalyst often favors dehydration.

For reactions in series, conversely, it is the "slow" step that governs. Thus, for the scheme $A \xrightarrow{k_1} B \xrightarrow{k_2} C$, if $k_1 \gg k_2$, the formation of B is relatively rapid, and the formation of C waits almost entirely on the rate at which B forms C. On the other hand, if $k_2 \gg k_1$, then B forms C as fast as B is formed, and the rate of formation of C is determined by the rate at which B is formed from A. These conclusions can be obtained quantitatively from equation 5.5-7. Thus, if $k_1 \gg k_2$,

$$dc_C/dt = [k_2 k_1 c_{Ao}/(k_2 - k_1)](e^{-k_1 t} - e^{-k_2 t}) \quad (5.6\text{-}1)$$

$$= k_2 c_{Ao} e^{-k_2 t} \quad (k_1 \gg k_2) \quad (5.6\text{-}1a)$$

so that the rate of formation of C is governed by the rate constant for the second (slow) step. If $k_2 \gg k_1$,

$$dc_C/dt = k_1 c_{Ao} e^{-k_1 t} \quad (k_2 \gg k_1) \quad (5.6\text{-}1b)$$

and the rate of formation of C is governed by the rate constant for the first step.

Since the rates of reaction steps in series may vary greatly, the concept of the slow step as the governing factor in the overall rate of reaction is very important. It is also a matter of everyday experience. If you are in a long, slowly moving lineup getting into the theater (followed by a relatively rapid passage past a ticket-collector and thence to a seat), the rate of getting seated is largely determined by the rate at which the lineup moves.

5.6.2 Determination of Reaction Network

A reaction network, as a model of a reacting system, may consist of steps involving some or all of: opposing reactions, which may or may not be considered to be at equilibrium, parallel reactions, and series reactions. Some examples are cited in Section 5.1.

The determination of a realistic reaction network from experimental kinetics data may be difficult, but it provides a useful model for proper optimization, control, and improvement of a chemical process. One method for obtaining characteristics of the

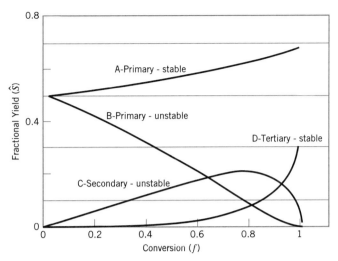

Figure 5.5 Fractional yield behavior of primary, secondary, and tertiary products

network is by analysis of the behavior of the fractional yields, $\hat{S}$, of products as functions of the conversion of a reactant. Figure 5.5 shows some of the possible types of behavior.

As indicated in Figure 5.5, products may be divided into primary, secondary, and tertiary products. Primary products are those made directly from reactants. Since they are the first products formed, they have finite fractional yields at very low conversion. Products A and B are primary products. If these products are stable (do not react further to other products), the fractional yields of these products increase with increasing conversion (product A). The fractional yields of products which react further eventually decrease (to zero if the second reaction is irreversible) as conversion increases (product B). Secondary products arise from the second reaction in a series, and, since they cannot be formed until the intermediate product is formed, have zero fractional yields at low conversion, which increase as conversion increases but eventually decrease if the product is unstable; the initial slope of the fractional yield curve is finite (product C). Finally, tertiary products (i.e., those that are three steps from reactants) have zero initial fractional yields, and zero initial slopes (product D). A possible network that fits the behavior in Figure 5.5 is shown in Figure 5.6. The increase in the fractional yield of A may be a result of it being a byproduct of the reaction that produces C (such as CO_2 formation at each step in selective oxidation reactions), or could be due to different rate laws for the formation of A and B. The verification of a proposed reaction network experimentally could involve obtaining data on the individual steps, such as studying the conversion of C to D, to see if the behavior is consistent. Since a large variety of possible networks exists, the investigator responsible for developing the reaction network for a process must obtain as much kinetics information as possible, and build a kinetics model that best fits the system under study.

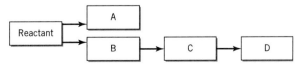

Figure 5.6 Compartmental diagram to illustrate possible reaction network for behavior in Figure 5.5

EXAMPLE 5-8

In their study of the kinetics of the partial oxidation of methane to HCHO, along with CO, CO_2, and H_2O (Example 5-1), Spencer and Pereira (1987) observed the following:

(1) $\hat{S}_{HCHO/CH_4} = 0.89$ when extrapolated to $f_{CH_4} = 0$, and decreased as f_{CH_4} increased.
(2) $\hat{S}_{CO/CH_4} = 0$ at $f_{CH_4} = 0$, and increased as f_{CH_4} increased.
(3) $\hat{S}_{CO_2/CH_4} = 0.11$ at $f_{CH_4} = 0$ and remained constant, independent of f_{CH_4}.
(4) There was no change in the observed selectivity or conversion when the initial molar ratio of CH_4 to O_2 was varied over a wide range.
(5) In separate experiments on HCHO oxidation over the same catalyst, CO was formed (but very little CO_2).

Construct a reaction network that is consistent with these observations.

SOLUTION

The five points listed above lead to the following corresponding conclusions:

(1) HCHO is a primary unstable product (like B in Figure 5.5); see also (5).
(2) CO is a secondary stable product (similar to C in Figure 5.5, but with no maximum or drop-off); see also (5).
(3) CO_2 is a primary stable product (like A in Figure 5.5, but remaining constant).
(4) The rate of any step involving O_2 is independent of c_{CO_2} (zero-order).
(5) CO is a primary product of HCHO oxidation.

A reaction network could then consist of two steps in series in which CH_4 forms HCHO, which subsequently oxidizes to CO, together with a third step in parallel in which CH_4 oxidizes to CO_2. Thus,

$$CH_4 + O_2 \xrightarrow{k_1} HCHO + H_2O$$
$$HCHO + \frac{1}{2} O_2 \xrightarrow{k_2} CO + H_2O$$
$$CH_4 + 2O_2 \xrightarrow{k_3} CO_2 + 2H_2O$$

The corresponding rate laws (tested by means of experimental measurements from a differential PFR) are:

$$(-r_{CH_4}) = (k_1 + k_3)c_{CH_4}$$
$$r_{HCHO} = k_1 c_{CH_4} - k_2 c_{HCHO}$$
$$r_{CO_2} = k_3 c_{CH_4}$$

(Values of the rate constants, together with those of corresponding activation energies, are given by the authors.)

5.7 PROBLEMS FOR CHAPTER 5

5-1 Consider a reacting system in which species B and C are formed from reactant A. How could you determine from rudimentary experimental information whether the kinetics scheme should be represented by

(i) A → B + C

or (ii) A → B,

A → C

or (iii) A → B → C

5-2 Suppose the reaction in Example 5-4 was studied in a CSTR operated at steady-state, and the results given below were obtained. Calculate the values of k_f and k_r, and hence write the rate law. Assume T to be the same, constant density, and no D in the feed.

$\bar{t}/h$	1	2	3	4
$100\, c_A/c_{Ao}$	76.5	65.9	57.9	53.3

5-3 The liquid-phase hydrolysis of methyl acetate (A) to acetic acid and methyl alcohol is a reversible reaction (with rate constants k_f and k_r, as in equation 5.3-3). Results of an experiment carried out at a particular (constant) temperature in a BR in terms of the fraction hydrolyzed (f_A) measured at various times (t), with $c_{Ao} = 0.05$ mol L^{-1} (no products present initially), are as follows (Coulson et al., 1982, p. 616):

t/s	0	1350	3060	5340	7740	∞
f_A	0	0.21	0.43	0.60	0.73	0.90

(a) Write the chemical equation representing the reaction.
(b) Obtain a rate law for this reaction, including values of the rate constants. State any assumption(s) made.

5-4 In an experiment (Williams, 1996) to evaluate a catalyst for the selective oxidation of propene (C_3H_6) to various products, 1 g of catalyst was placed in a plug-flow reactor operated at 450°C and 1 bar. The feed consisted of propene and air (21 mole % O_2, 79% N_2 (inert)). GC analysis of the inlet and outlet gas gave the following results, the outlet being on a water-free basis (H_2O is formed in the oxidation):

Substance	Inlet mole %	Outlet mole %
propene (C_3H_6)	10.0	?
oxygen (O_2)	18.9	?
nitrogen (N_2, inert)	71.1	78.3
acrolein (C_3H_4O)	0	3.17
propene oxide (C_3H_6O)	0	0.40
acetaldehyde (C_2H_4O)	0	0.59
carbon dioxide	0	7.91

(a) If the feed rate of C_3H_6 is $F_{C_3H_6,o} = 1$ mmol min^{-1}, at what rate do (i) C_3H_6, (ii) O_2, and (iii) H_2O leave the reactor?
(b) What is $f_{C_3H_6}$, the fractional conversion of C_3H_6?
(c) What is the selectivity or fractional yield of each of acrolein, propene oxide, and acetaldehyde with respect to propene?
(d) What is the rate of reaction expressed as (i) $(-r_{C_3H_6})$; (ii) $r_{C_3H_4O}$ (in mmol min^{-1} (g cat)$^{-1}$)? Assume that the reactor acts as a differential reactor (Section 3.4.1.3.1).

5-5 Repeat Example 5-6 for a CSTR with $V = 15$ L and $q = 1.5$ L min^{-1}.

5-6 Suppose the liquid-phase decomposition of A takes place according to the following kinetics scheme with rate laws as indicated:

$$A \to B + E; \; r_B = k_1 c_A$$
$$A \to D + E; \; r_D = k_2 c_A$$

Reaction is carried out isothermally in a batch reactor with only A present initially at a concentration $c_{Ao} = 4$ mol L^{-1} in an inert solvent. At $t = 1200$ s, $c_A = 1.20$ mol L^{-1} and $c_B = 0.84$ mol L^{-1}. Calculate (a) the values of k_1 and k_2 (specify the units), and (b) the values of c_D and c_E at $t = 1200$ s.

5-7 For reaction according to the kinetics scheme

$$A \to B + C; \; r_B = k_1 c_A$$
$$A \to D; \; r_D = k_2 c_A$$

data are as follows:

t/min	x_B	x_D
t	0.20	x_{D1}
$t + 20$	0.30	0.20

Assuming that only A is present at $t = 0$, and that reaction occurs at constant T in a constant-volume batch reactor, calculate x_{D1}, k_1 and k_2.

5-8 The following data are for the kinetics scheme:

$$A \to B + C; \; r_B = k_1 c_A$$
$$A \to D; \; r_D = k_2 c_A$$

t/min	c_A	c_B	c_D/mol L^{-1}
0	?	?	0
20	0.060	?	0.010
40	0.036	0.088	?

Assuming that reaction occurs in a constant-volume batch reactor at a fixed temperature, and that at time zero only A and B are present, calculate (not necessarily in the order listed): (a) k_1 and k_2; (b) c_{Ao} and c_{Bo} at time zero; (c) c_D at 40 min; (d) c_B at 20 min.

5-9 Suppose a substance B decomposes to two sets of products according to the kinetics scheme

$$B \xrightarrow{k_1} P_1 + \ldots; \; k_1 = A_1 \exp(-E_{A1}/RT)$$
$$B \xrightarrow{k_2} P_2 + \ldots; \; k_2 = A_2 \exp(-E_{A2}/RT)$$

such that the rate laws for both steps are of the same form (e.g., same order). What is the overall activation energy, E_A, for the decomposition of B, in terms of the Arrhenius parameters for the individual steps? (Giralt and Missen, 1974.)

(a) Consider E_A to be defined by $E'_A = RT^2 d\ln k/dT$, where k is the overall rate constant.
(b) Consider E_A to be defined by $k = A\exp(-E''_A/RT)$, where A is the overall pre-exponential factor.
(c) If there is any difference between E'_A and E''_A, how are they related?

5-10 For the kinetics scheme $A \xrightarrow{k_1} B \xrightarrow{k_2} C$, each step being first-order, for reaction occurring in a constant-volume batch reactor (only A present initially),
(a) At what time, t, in terms of k_1 and k_2, are c_A and c_B equal (other than $t \to \infty$), and what is the condition for this to happen?
(b) What is the value of t_{max} when $k_1 = k_2$?
(c) Show that c_B has an inflection point at $2t_{max}$.
(d) Calculate $k_1 t_{max}$ and c_{Bmax}/c_{Ao} for each of the cases (i) $K = k_2/k_1 = 10$, (ii) $K = 1$, and (iii) $K = 0.1$.
(e) From the results in (d), describe how t_{max} and c_{Bmax}/c_{Ao} change with decreasing K.

5-11 The following liquid-phase reactions take place in a CSTR operating at steady state.

$$2A \to B + C; \quad r_C = k_1 c_A^2$$
$$A + B \to 2D; \quad r_D = 2k_2 c_A c_B$$

The inlet concentration of A is 2.50 mol L^{-1}. The outlet concentrations of A and C are respectively 0.45 mol L^{-1} and 0.75 mol L^{-1}. Assuming that there is no B, C, or D in the feed, and that the space time (τ) is 1250 s, calculate:
(a) The outlet concentrations of B and D; and
(b) k_1 and k_2.

5-12 The following data are for the kinetics scheme:

$$A + B \to C + E; (-r_B) = k_1 c_A c_B; \quad k_1 = ?$$
$$A + C \to D + E; r_D = k_2 c_A c_C; \quad k_2 = 3.0 \times 10^{-3} \text{ L mol}^{-1}\text{ min}^{-1}$$

t/min	Concentration/mol L^{-1}				
	c_A	c_B	c_C	c_D	c_E
0	5.0	0.040	?	0	0
23	–	0.020	?	–	–
∞	–	0	0	0.060	?

Assuming that the reactions occur at constant T in a constant-volume batch reactor, calculate:
(a) The concentration of C at time zero and the concentration of E at time ∞;
(b) The second-order rate constant k_1; and
(c) The concentration of C at time 23 min.

5-13 Consider a liquid-phase reaction taking place in a CSTR according to the following kinetics scheme:

$$A \to B + C; r_B = k_1 c_A$$
$$A + C \to 2D; r_D = 2k_2 c_A c_C$$

The inlet concentration of A is $c_{Ao} = 3$ mol L^{-1}, and there is no B, C, or D in the feed. If, for a space time $\tau = 10$ min, the outlet concentrations of A and B are $c_A = 1.25$ and $c_B = 1.50$ mol L^{-1} at steady-state, calculate the values of (a) k_1, (b) k_2, (c) c_C, and (d) c_D (not necessarily in the order listed). Include the units of k_1 and k_2 in your answer.

5-14 For reaction according to the kinetics scheme

$$A + B \to C + D; \quad r_D = k_1 c_A c_B$$
$$A + C \to 2E; \quad r_E = 2k_2 c_A c_C$$

data are as follows:

t/s	c_A	c_B	c_C	$(-r_A)$	r_C
	mol L^{-1}			mol L^{-1} s^{-1}	
0	0.20	0.10	0	4.0×10^{-4}	?
t	0.08	?	?	6.4×10^{-5}	0

Assuming that reaction occurs at constant T in a constant-volume batch reactor, calculate k_1, c_C at t, and k_2; state the units of k_1 and k_2.

5-15 The decomposition of N_2O_5 in the gas phase to N_2O_4 and O_2 is complicated by the subsequent decomposition of N_2O_4 to NO_2 (presence indicated by brown color) in a rapidly established equilibrium. The reacting system can then be modeled by the kinetics scheme

$$N_2O_5(A) \xrightarrow{k_A} N_2O_4(B) + \frac{1}{2}O_2(C)$$
$$N_2O_4 \underset{}{\overset{K_p}{\rightleftharpoons}} 2\,NO_2(D)$$

Some data obtained in an experiment at 45°C in a constant-volume BR are as follows (Daniels and Johnston, 1921):

t/s	P/kPa	p_A	p_B	p_C	p_D
0	?	?	0	0	0
3600	83.0	?	?	?	?
∞	89.0	0	?	?	?

where the partial pressures p_A, ... are also in kPa.
(a) Confirm that the kinetics scheme corresponds to the stoichiometry.
(b) Calculate the values indicated by ?, if $K_p = 0.558$ bar.
(c) If the decomposition of N_2O_5 is first-order, calculate the value of k_A.

5-16 The following data (p in bar) were obtained for the oxidation of methane over a supported molybdena catalyst in a PFR at a particular T (Mauti, 1994). The products are CO_2, HCHO, and H_2O.

t/ms	p_{CH_4}	p_{HCHO}	p_{CO_2}
0	0.25	0	0
8	0.249	0.00075	0.00025
12	0.2485	0.00108	0.00042
15	0.248125	0.001219	0.000656
24	0.247	0.00177	0.00123
34	0.24575	0.00221	0.00204
50	0.24375	0.002313	0.003938
100	0.2375	0.00225	0.01025

Construct a suitable reaction network for this system, and estimate the values of the rate constants involved (assume a first-order rate law for each reaction).

5-17 In pulp and paper processing, anthraquinone (AQ) accelerates the delignification of wood and improves liquor selectivity. The kinetics of the liquid-phase oxidation of anthracene (AN) to AQ with NO_2 in acetic acid as solvent has been studied by Rodriguez and Tijero (1989) in a semibatch reactor (batch with respect to the liquid phase), under conditions such that the kinetics of the overall gas-liquid process is controlled by the rate of the liquid-phase reaction. This reaction proceeds through the formation of the intermediate compound anthrone (ANT):

$$C_{14}H_{10}\ (AN) \xrightarrow[k_1]{NO_2} C_{14}H_9O\ (ANT) \xrightarrow[k_2]{NO_2} C_{14}H_8O_2\ (AQ)$$

The following results (as read from a graph) were obtained for an experiment at 95°C, in which $c_{AN,o} = 0.0337$ mol L^{-1}:

t/min	c_{AN}	c_{ANT}	c_{AQ}
	mol L^{-1}		
0	0.0337	0	0
10	0.0229	0.0104	0.0008
20	0.0144	0.0157	0.0039
30	0.0092	0.0181	0.0066
40	0.0058	0.0169	0.0114
50	0.0040	0.0155	0.0144
60	0.0030	0.0130	0.0178
70	0.0015	0.0114	0.0209
80	0.0008	0.0088	0.0240
90	0.0006	0.0060	0.0270

If each step in the series network is first-order, determine values of the rate constants k_1 and k_2 in s^{-1}.

5-18 Duo et al. (1992) studied the kinetics of reaction of NO, NH_3 and (excess) O_2 in connection with a process to reduce NO_x emissions. They used an isothermal PFR, and reported measured ratios $c_{NO}/c_{NO,o}$ and $c_{NH_3}/c_{NH_3,o}$ for each of several residence times, t. For $T = 1142$ K, and inlet concentrations $c_{NO,o} = 5.15 \times 10^{-3}$, $c_{NH_3,o} = 8.45 \times 10^{-3}$, and $c_{O_2,o} = 0.405$ mol m^{-3}, they obtained results as follows (as read from graphs):

t/s:	0.039	0.051	0.060	0.076	0.102	0.151	0.227
$c_{NO}/c_{NO,o}$:	0.756	0.699	0.658	0.590	0.521	0.435	0.315
$c_{NH_3}/c_{NH_3,o}$:	0.710	0.721	0.679	0.607	0.579	0.476	0.381

(a) If the other species involved are N_2 and H_2O, determine a permissible set of chemical equations to represent the system stoichiometry.
(b) Construct a reaction network consistent with the results in (a), explaining the basis and interpretation.
(c) Calculate the value of the rate constant for each step in (b), assuming (i) constant density; (ii) constant c_{O_2}; (iii) each step is irreversible and of order indicated by the form of the step. Comment on the validity of assumptions (i) and (ii).

5-19 Vaidyanathan and Doraiswamy (1968) studied the kinetics of the gas-phase partial oxidation of benzene (C_6H_6, B) to maleic anhydride ($C_4H_2O_3$, M) with *air* in an integral PFR containing a catalyst of V_2O_5 - MoO_3 on silica gel. In a series of experiments, they varied the space time $\tau \equiv W/F$, where W is the weight of catalyst and F is the total molar flow rate of gas (τ in (g cat) h mol^{-1}), and analyzed for M and CO_2 (C) in the outlet stream. (W/F is analogous to the space time V/q_o in equation 2.3-2.) For one series at 350°C and an inlet ratio $(F_{air}/F_B)_o = 140$, they reported the following results, with partial pressure p in atm:

$\tau = W/F$	$10^2 p_B$	$10^3 p_M$	$10^2 p_C$	$10^2 p_{H_2O}$
0	1.83	0	0	0
61	1.60	1.36	0.87	0.57
99	1.49	1.87	1.30	0.84
131	1.42	2.20	1.58	1.01
173	1.34	2.71	1.82	1.18
199	1.32	2.86	1.93	1.25
230	1.30	3.10	1.97	1.30
313	1.23	3.48	2.24	1.47

In the following, state any assumptions made and comment on their validity.

(a) Since there are six species involved, determine, from a stoichiometric analysis, how many of the partial pressures (p_i) are independent for given (T, P), that is, the smallest number from which all the others may be calculated. Confirm by calculation for $W/F = 313$.

(b) For $W/F = 313$, calculate (i) f_B; (ii) $Y_{M/B}$ and $Y_{C/B}$; (iii) $\hat{S}_{M/B}$ and $\hat{S}_{C/B}$.

(c) From the data in the table, determine whether $C_4H_2O_3$(M) and CO_2 are primary or secondary products.

(d) From the data given and results above, construct a reaction network, together with corresponding rate laws, and determine values of the rate constants.

(e) The authors used a three-step reaction network to represent all their experimental data (only partial results are given above):

$$C_6H_6(B) + \frac{9}{2}O_2 \rightarrow C_4H_2O_3(M) + 2CO_2 + 2H_2O; \; r_1 = k_1 p_B$$

$$C_4H_2O_3 + 3O_2 \rightarrow 4CO_2 + H_2O; \; r_2 = k_2 p_M$$

$$C_6H_6 + \frac{15}{2}O_2 \rightarrow 6CO_2 + 3H_2O; \; r_3 = k_3 p_B$$

Values of the rate constants at 350°C reported are: $k_1 = 1.141 \times 10^{-3}$; $k_2 = 2.468 \times 10^{-3}$; $k_3 = 0.396 \times 10^{-3}$ mol h^{-1} (g cat)$^{-1}$.

(i) Obtain expressions for p_B and p_M as functions of τ.
(ii) Calculate the five quantities in (b) and compare the two sets of results.
(iii) Does this kinetics model predict a maximum in M? If so, calculate values of τ_{max} and $p_{M,max}$.
(iv) Are there features of this kinetics model that are not reflected in the (partial) data given in the table above? (Compare with results from (c) and (d).)

Chapter 6

Fundamentals of Reaction Rates

In the preceding chapters, we are primarily concerned with an empirical macroscopic description of reaction rates, as summarized by rate laws. This is without regard for any description of reactions at the molecular or microscopic level. In this chapter and the next, we focus on the fundamental basis of rate laws in terms of theories of reaction rates and reaction "mechanisms."

We first introduce the idea of a reaction mechanism in terms of elementary reaction steps, together with some examples of the latter. We then consider various aspects of molecular energy, particularly in relation to energy requirements in reaction. This is followed by the introduction of simple forms of two theories of reaction rates, the *collision theory* and the *transition state theory,* primarily as applied to gas-phase reactions. We conclude this chapter with brief considerations of reactions in condensed phases, surface phenomena, and photochemical reactions.

6.1 PRELIMINARY CONSIDERATIONS

6.1.1 Relating to Reaction-Rate Theories

As a model of real behavior, the role of a theory is twofold: (1) to account for observed phenomena in relatively simple terms (hindsight), and (2) to predict hitherto unobserved phenomena (foresight).

What do we wish to account for and predict? Consider the form of the rate law used for the model reaction A + ... → products (from equations 3.1-8 and 4.1-3):

$$(-r_A) = A \exp(-E_A/RT) \prod_{i=1}^{N} c_i^{\alpha_i} \quad \text{(6.1-1)}$$

We wish to account for (i.e., interpret) the Arrhenius parameters A and E_A, and the form of the concentration dependence as a product of the factors $c_i^{\alpha_i}$ (the order of reaction). We would also like to predict values of the various parameters, from as simple and general a basis as possible, without having to measure them for every case. The first of these two tasks is the easier one. The second is still not achieved despite more than a century of study of reaction kinetics; the difficulty lies in quantum mechanical calculations—not in any remaining scientific mystery. However, the current level of theoretical understanding has improved our ability to estimate many kinetics parameters, and has sharpened our intuition in the search for improved chemical processes.

In many cases, reaction rates cannot be adequately represented by equation 6.1-1, but are more complex functions of temperature and composition. Theories of reaction kinetics should also explain the underlying basis for this phenomenon.

6.1.2 Relating to Reaction Mechanisms and Elementary Reactions

Even a "simple" reaction usually takes place in a "complex" manner involving multiple steps making up a reaction mechanism. For example, the formation of ammonia, represented by the simple reaction $N_2 + 3H_2 \rightarrow 2NH_3$, does not take place in the manner implied by this chemical statement, that is, by the simultaneous union of one molecule of N_2 and three molecules of H_2 to form two of NH_3. Similarly, the formation of ethylene, represented by $C_2H_6 \rightarrow C_2H_4 + H_2$, does not occur by the disintegration of one molecule of C_2H_6 to form one of C_2H_4 and one of H_2 directly.

The original reaction mechanism (Rice and Herzfeld, 1934) proposed for the formation of C_2H_4 from C_2H_6 consists of the following five steps:[1]

$$C_2H_6 \rightarrow 2CH_3^\bullet$$
$$CH_3^\bullet + C_2H_6 \rightarrow CH_4 + C_2H_5^\bullet$$
$$C_2H_5^\bullet \rightarrow C_2H_4 + H^\bullet$$
$$H^\bullet + C_2H_6 \rightarrow H_2 + C_2H_5^\bullet$$
$$H^\bullet + C_2H_5^\bullet \rightarrow C_2H_6$$

where the "dot" denotes a free-radical species.

We use this example to illustrate and define several terms relating to reaction fundamentals:

Elementary reaction: a chemical reaction step that takes place in a single molecular encounter (each of the five steps above is an elementary reaction); it involves one, two, or (rarely) three molecular entities (atoms, molecules, ions, radicals, etc.). Only a small number of chemical bonds is rearranged.

Reaction mechanism: a postulated sequence of elementary reactions that is consistent with the observed stoichiometry and rate law; these are necessary but not sufficient conditions for the correctness of a mechanism, and are illustrated in Chapter 7.

Reactive intermediate: a transient species introduced into the mechanism but not appearing in the stoichiometric equation or the rate law; the free atomic and free radical species $H^\bullet$, $CH_3^\bullet$, and $C_2H_5^\bullet$ are reactive intermediates in the mechanism above. Such species must ultimately be identified experimentally to justify their inclusion.

Molecularity of a reaction: the number of reacting partners in an elementary reaction: unimolecular (one), bimolecular (two), or termolecular (three); in the mechanism above, the first and third steps are unimolecular as written, and the remainder are bimolecular. Molecularity (a mechanistic concept) is to be distinguished from order (algebraic). Molecularity must be integral, but order need not be; there is no necessary connection between molecularity and order, except for an elementary reaction: the numbers describing molecularity, order, and stoichiometry of an elementary reaction are all the same.

---

[1] In the dehydrogenation of C_2H_6 to produce C_2H_4, CH_4 is a minor coproduct; this is also reflected in the second step of the mechanism; hence, both the overall reaction and the proposed mechanism do not strictly represent a simple system.

It is the combination of individual elementary reaction steps, each with its own rate law, that determines the overall kinetics of a reaction. Elementary reactions have simple rate laws of the form

$$r = k(T) \prod_i^{N_R} c_i^{\alpha_i} \qquad (6.1\text{-}2)$$

where the temperature dependence of rate constant k is Arrhenius-like, and the reaction orders α_i are equal to the absolute values of the stoichiometric coefficients $|\nu_i|$ of the *reactants* (number N_R).

This chapter presents the underlying fundamentals of the rates of elementary chemical reaction steps. In doing so, we outline the essential concepts and results from physical chemistry necessary to provide a basic understanding of how reactions occur. These concepts are then used to generate expressions for the rates of elementary reaction steps. The following chapters use these building blocks to develop intrinsic rate laws for a variety of chemical systems. Rather complicated, nonseparable rate laws for the overall reaction can result, or simple ones as in equation 6.1-1 or -2.

6.2 DESCRIPTION OF ELEMENTARY CHEMICAL REACTIONS

An elementary step must necessarily be simple. The reactants are together with sufficient energy for a very short time, and only simple rearrangements can be accomplished. In addition, complex rearrangements tend to require more energy. Thus, almost all elementary steps break and/or make one or two bonds. In the combustion of methane, the following steps (among many others) occur as elementary reactions:

$$CH_4 + O_2 \rightarrow CH_3^\bullet + HO_2^\bullet$$
$$OH^\bullet + CO \rightarrow CO_2 + H^\bullet$$

These two steps are simple rearrangements. The overall reaction

$$CH_4 + 2O_2 \rightarrow CO_2 + 2H_2O$$

cannot occur in a single step; too much would have to transpire in a single encounter.

6.2.1 Types of Elementary Reactions

The following list of elementary reactions, divided into various categories, allows us to understand and build rate laws for a wide variety of chemical systems.

6.2.1.1 *Elementary Reactions Involving Neutral Species (Homogeneous Gas or Liquid Phase)*

This is the most common category of elementary reactions and can be illustrated by unimolecular, bimolecular, and termolecular steps.

Unimolecular Steps:

- Fragmentation/dissociation—the molecule breaks into two or more fragments:

$$C_4H_9O\text{--}OH \rightarrow C_4H_9O^\bullet + OH^\bullet$$

- Rearrangements—the internal bonding of a molecule changes:

$$HCN \rightarrow HNC$$

Bimolecular Steps:

- Bimolecular association/recombination—two species combine:

$$H_3C^\bullet + CH_3^\bullet \rightarrow C_2H_6$$

- Bimolecular exchange reactions—atoms or group of atoms transferred:

$$OH^\bullet + C_2H_6 \rightarrow H_2O + C_2H_5^\bullet$$

- Energy transfer—this is not actually a reaction; there is no change in bonding; but it is nevertheless an important process involving another molecule M:

$$R^* + M \rightarrow R + M^*$$

The asterisk denotes an excited state—a molecule with excess energy (more than enough energy to enable it to undergo a specific reaction step).

Termolecular Steps:

- Termolecular steps are rare, but may appear to arise from two rapid bimolecular steps in sequence.

6.2.1.2 Photochemical Elementary Reactions

Light energy (absorbed or emitted in a quantum or photon of energy, $h\nu$, where h is Planck's constant (6.626×10^{-34} J s), and ν is the frequency of the light, s^{-1}) can change the energy content of a molecule enough to produce chemical change.

- Absorption of light (photon):

$$Hg + h\nu \rightarrow Hg^*$$

- Photodissociation:

$$O_3 + h\nu \rightarrow O_2 + O^\bullet$$

- Photoionization (electron ejected from molecule):

$$CH_4 + h\nu \rightarrow CH_4^+ + e^-$$

- Light (photon) emission (reverse of absorption):

$$Ne^* \rightarrow Ne + h\nu$$

6.2.1.3 Elementary Reactions Involving Charged Particles (Ions, Electrons)

These reactions occur in plasmas, or other high-energy situations.

- Charge exchange:

$$X^- + A \rightarrow A^- + X$$
$$M^+ + A \rightarrow M + A^+$$

- Electron attachment:
$$e^- + X \rightarrow X^-$$

- Electron-impact ionization:
$$e^- + X \rightarrow X^+ + 2e^-$$

- Ion-molecule reactions:
$$C_3H_7^+ + C_4H_8 \rightarrow C_3H_6 + C_4H_9^+ \text{ (bimolecular exchange)}$$
$$H^+ + C_3H_6 \rightarrow C_3H_7^+ \text{ (bimolecular association)}$$

6.2.1.4 Elementary Reactions on Surfaces

Surface reactions are important in heterogeneous reactions and catalysis.

- Adsorption/desorption—molecules or fragments from gas or liquid bond to solid surface.
  - Simple adsorption—molecule remains intact.

$$\underset{\text{Ni Ni Ni}}{} + \underset{}{O{=}C} \rightarrow \underset{\text{Ni Ni Ni}}{\overset{O{=}C}{|}}$$

  - Dissociative adsorption—molecule forms two or more surface-bound species.

$$H{-}H + \underset{\text{Cu Cu Cu}}{} \rightarrow \underset{\text{Cu Cu Cu}}{\overset{H\ \ H}{|\ \ |}}$$

- Site hopping—surface-bound intermediates move between binding sites on surface.
- Surface reactions—similar to gas-phase arrangements, but occur while species bonded to a solid surface.
  - Dissociation:

(CH₃–C≡C on Pt Pt Pt → CH₂=C=C on Pt Pt Pt + H on Pt Pt Pt)

  - Combination:

(O=C– on Pd Pd Pd + H on Pd Pd Pd → O=C(–H) on Pd Pd Pd)

Rearrangements of the adsorbed species are also possible.

6.2.2 General Requirements for Elementary Chemical Reactions

The requirements for a reaction to occur are:

(1) The reaction partners must encounter one another.
(2) The encounter must be successful. This in turn requires:
  (i) the geometry of the encounter to be correct (e.g., the atoms in the proper position to form the new bonds) and,
  (ii) sufficient energy to be available to overcome any energy barriers to this transformation.

The simple theories of reaction rates involve applying basic physical chemistry knowledge to calculate or estimate the rates of successful molecular encounters. In Section 6.3 we present important results from physical chemistry for this purpose; in subsequent sections, we show how they are used to build rate theories, construct rate laws, and estimate the values of rate constants for elementary reactions.

6.3 ENERGY IN MOLECULES

Energy in molecules, as in macroscopic objects, can be divided into potential energy (the energy which results from their position at rest) and kinetic energy (energy associated with motion). Potential energy in our context deals with the energy associated with chemical bonding. The changes in bond energy often produce energy barriers to reaction as the atoms rearrange. The kinetic energy of a group of molecules governs (1) how rapidly reactants encounter one another, and (2) how much energy is available in the encounter to surmount any barriers to reaction. Research has led to a detailed understanding of how these factors influence the rates of elementary reactions, and was recognized by the award of the Nobel prize in chemistry to Lee, Herschbach, and Polanyi in 1986.

6.3.1 Potential Energy in Molecules—Requirements for Reaction

6.3.1.1 Diatomic Molecules

The potential energy of a pair of atoms (A and B) is shown schematically in Figure 6.1 as a function of the distance between them, r_{AB}. As the atoms approach one another, the associated electron orbitals form a bonding interaction which lowers the potential energy (i.e., makes the system more stable than when the two atoms are far apart).

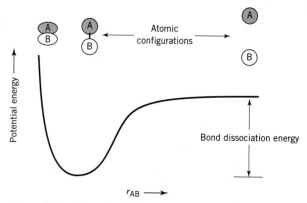

Figure 6.1 Potential energy of a two-atom system

The minimum energy on the curve corresponds to the most stable configuration where the bonding is most effective, and thus to the stable A–B diatomic molecule. In the specific case of a pair of iodine atoms, this minimum is 149 kJ mol^{-1} below that of the separated atoms. Therefore, to dissociate an isolated I_2 molecule at rest, $I_2 \rightarrow 2I^{\bullet}$, 149 kJ mol^{-1} must be supplied from outside the molecule. This elementary reaction is said to be endoergic (energy absorbing) by this amount, also known as the bond dissociation energy. This energy can be supplied by absorption of light energy, or by transfer of kinetic energy from other molecules. This energy can also be thought of as the height of an energy barrier to be scaled in order for reaction to occur. The path along the potential energy curve can be thought of as a path or trajectory leading to reaction, which is described as the "reaction coordinate".

Now consider the reverse reaction, $2I^{\bullet} \rightarrow I_2$. The reaction coordinate in this case is just the reverse of that for the dissociation reaction. The reaction is exoergic (energy releasing), and for the I_2 molecule to come to rest in its most stable configuration, an amount of energy equal to the bond energy must be given off to the rest of the system. If not, the molecule has enough energy (converted to internal kinetic energy) to dissociate again very quickly. This requirement to "offload" this excess energy (usually through collisions with other molecules) is important in the rates of these bimolecular association reactions. The input of additional energy is not required along the reaction coordinate for this reaction to occur; the two atoms only have to encounter each other; that is, there is no energy barrier to this reaction. These concepts form a useful basis for discussing more complicated systems.

6.3.1.2 Triatomic Systems: Potential Energy Surface and Transition State

Consider a system made up of the atoms A, B, and C. Whereas the configuration of a diatomic system can be represented by a single distance, the internal geometry of a triatomic system requires three independent parameters, such as the three interatomic distances r_{AB}, r_{BC}, and r_{CA}, or r_{AB}, r_{BC}, and the angle ϕ_{ABC}. These are illustrated in Figure 6.2.

The potential energy is a function of all three parameters, and is a surface (called the potential energy surface) in three-dimensional (3-D) space. If we simplify the system by constraining the atoms to remain in a straight line in the order A-B-C, the potential energy depends only on two parameters (i.e., r_{AB} and r_{BC}), and we can conveniently represent it as a 2-D "topographical map" in Figure 6.3(a), or as a 3-D perspective drawing in Figure 6.3(b). At the lower-left corner of Figure 6.3(a), all three atoms are far apart: there are no bonding interactions. As A approaches B while C remains distant (equivalent to moving up the left edge of Figure 6.3(a)), a stable AB molecule is formed (like the I_2 case). Similarly, a B–C bond is formed if B approaches C with A far away (moving right along the bottom edge of Figure 6.3(a)). When all three atoms are near each other, the molecular orbitals involve all three atoms. If additional bonding is possible, the energy is lowered when this happens, and a stable triatomic molecule can be formed. This is not the case shown in Figure 6.3(a), since in all configurations where A, B, and C are close together, the system is less stable than AB + C or A + BC. This is typical for many systems where AB (and BC) are stable molecules with saturated bonding. The two partial bonds A–B and B–C are weaker than either complete bond.

Figure 6.2 Representation of configuration of three-atom system

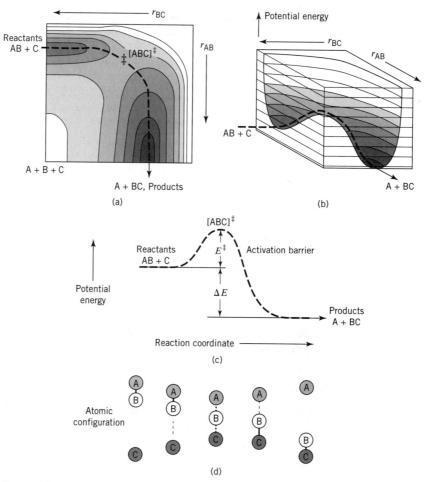

Figure 6.3 Potential energy surface for colinear reaction AB + C → A + BC; (a) 2-D topographical representation; (b) 3-D representation; (c) potential energy along reaction coordinate; (d) atomic configurations along reaction coordinate

Now consider the reaction

$$AB + C \rightarrow A + BC \qquad (6.3\text{-}1)$$

For the reaction to occur, the atoms must trace out a path on this surface from the configuration, in Figure 6.3(a), labeled "reactants" (AB + C), to the point labeled "products" (A + BC). The path which requires the minimum energy is shown by the dashed line. In this example, the energy rises as C approaches A–B and there is an energy barrier (marked "‡"). As a result, for the reaction to occur, the reactants must have at least enough additional (kinetic) energy to "get over the pass" at "‡". This critical configuration of the atoms, [ABC‡], is called the "transition state" of the system (or "activated complex"). This minimum energy path describes the most likely path for reaction, and is the reaction coordinate, although other paths are possible with additional energy. Plotting the potential energy E as a function of distance along this reaction coordinate, we obtain Figure 6.3(c) (corresponding to Figure 6.1 for the diatomic case). This figure shows the energy barrier $E^{\ddagger}$ at the transition state and that the reaction is exoergic. The height of the energy barrier, $E^{\ddagger}$, corresponds approximately to the Arrhenius activation energy, E_A, of the reaction. Figure 6.3(d) indicates atomic configurations along the reaction coordinate.

In the elementary reaction

$$O^\bullet + H_2 \rightarrow OH^\bullet + H^\bullet \tag{6.3-1a}$$

which is part of the reaction mechanism in hydrogen flames and the space shuttle main rocket engine, the transition state would resemble:

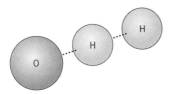

The energy barrier for this reaction is quite low, 37 kJ mol^{-1}. There are many schemes for the estimation of the barrier height, $E^\ddagger$. The simplest of these are based on empirical correlations. For details see Steinfeld et al., 1989, p. 231.

The reverse reaction (BC + A $\rightarrow$ AB + C) follows the same reaction coordinate in the opposite direction. The barrier for the reverse reaction occurs at the same place. The barrier height in the reverse direction is related to the barrier height in the forward direction by

$$E^\ddagger(\text{reverse}) = E^\ddagger(\text{forward}) - \Delta E(\text{forward}) \tag{6.3-2}$$

where ΔE (forward) is the reaction energy change in the forward direction. For example, reaction 6.3-1a is endoergic by approximately 9 kJ mol^{-1}, and so the energy barrier for the reverse reaction is $37 - 9 = 28$ kJ mol^{-1}.

6.3.1.3 Relationship Between Barrier Height and Reaction Energy

In reaction 6.3-1, the A–B bond weakens as the B–C bond is formed. If there is a barrier, these two effects do not cancel. However, if the B–C bond is much stronger than the A–B bond (very exoergic reaction), even partial B–C bond formation compensates for the weakening of the A–B bond. This explains the observation that for a series of similar reactions, the energy barrier (activation energy) is lower for the more exoergic reactions. A correlation expressing this has been given by Evans and Polanyi (1938):

$$E^\ddagger = E_o^\ddagger + q\Delta E(\text{reaction}) \tag{6.3-3}$$

where $E_o^\ddagger$ is the barrier for an energetically neutral reaction (such as $CH_3^\bullet + CD_4 \rightarrow CH_3D + CD_3^\bullet$). The correlation predicts the barriers ($E^\ddagger$) for similar exoergic/endoergic reactions to be smaller/larger by a fraction, q, of the reaction energy (ΔE (reaction)). For one set of H transfer reactions, the best value of q is 0.4. This correlation holds only until the barrier becomes zero, in the case of sufficiently exoergic reactions; or until the barrier becomes equal to the endoergicity, in the case of sufficiently endoergic reactions. Figure 6.4 shows reaction coordinate diagrams for a hypothetical series of reactions, and the "data" for these reactions are indicated in Figure 6.4, along with the Evans-Polanyi correlation (dashed line). This and other correlations allow unknown rate constant parameters to be estimated from known values.

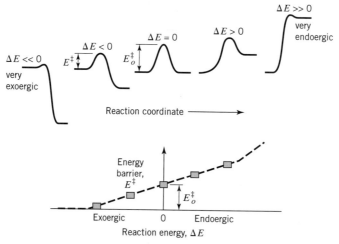

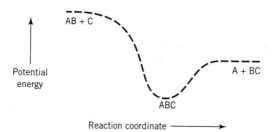

Figure 6.4 Reaction coordinate diagrams showing various types of energy-barrier behavior

Figure 6.5 Potential energy diagram for stable ABC molecule

If a stable ABC molecule exists, the reaction coordinate may appear as in Figure 6.5. In this case, there is no barrier to formation of the ABC molecule in either direction. Just like the diatomic case, energy must be removed from this molecule, because not only does it have enough internal energy to form reactants again, it has more than enough to form products. In the reverse direction, additional energy must be carried into the reaction if the system is to form AB + C. There can also be barriers to formation of triatomic molecules, particularly if the AB bond must be broken, for example, to form the molecule ACB. The reactions of ions with molecules rarely have intrinsic barriers because of the long-range attractive force (ion-induced dipole) between such species.

6.3.1.4 *Potential Energy Surface and Transition State in More Complex Systems*

For a system containing a larger number of atoms, the general picture of the potential energy surface and the transition state also applies. For example, in the second reaction step in the mechanism of ethane pyrolysis in Section 6.1.2,

$$CH_3^\bullet + C_2H_6 \rightarrow CH_4 + C_2H_5^\bullet \tag{6.3-4}$$

the transition state should resemble:

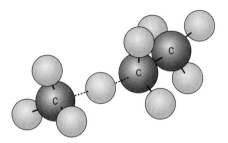

Here, the CH_3–H bond is formed as the C_2H_5–H bond is broken. For this system, the other bond lengths and angles also affect the potential energy, and the potential energy surface therefore depends on all other coordinates (3N − 6 or 30 in all). This system, however, is similar to the triatomic case above, where A = $C_2H_5^\bullet$, B = $H^\bullet$, and C = $CH_3^\bullet$. Again note that the transition state for the reverse reaction is the same.

The notion of the transition state is central to both theories discussed in this chapter. The transition state is the atomic configuration that must be reached for reaction to occur, and the bonding dictates the *energy* required for the reaction. The configuration or shape of the transition state indicates how probable it is for the reactants to "line up" properly or have the correct *orientation* to react. *The rate of a reaction is the rate at which these requirements are achieved.* A quantitative interpretation of both these issues, as treated by the two theories, is the subject of Sections 6.4 and 6.5.

In reactions which occur on solid surfaces, it is acceptable to think of the surface as a large molecule capable of forming bonds with molecules or fragments. Because of the large number of atoms involved, this is theoretically complicated. However, the binding usually occurs at specific sites on the surface, and very few surface atoms have their bonding coordination changed. Therefore, the same general concepts are useful in the discussion of surface reactions. For example, the nondissociated adsorption of CO on a metal surface (Section 6.2.1.4) can be thought of as equivalent to bimolecular association reactions, which generally have no barrier. Desorption is similar to unimolecular dissociation reactions, and the barrier equals the bond strength to the surface. Some reactions involving bond breakage, such as the dissociative adsorption of H_2 on copper surfaces, have energy barriers.

6.3.1.5 Other Electronic States

If the electrons occupy orbitals different from the most stable (ground) electronic state, the bonding between the atoms also changes. Therefore, an entirely different potential energy surface is produced for each new electronic configuration. This is illustrated in Figure 6.6 for a diatomic molecule.

The most stable (ground state) potential energy curve is shown (for AB) along with one for an electronically excited state (AB*) and also for a positive molecular ion (AB$^+$, with one electron ejected from the neutral molecule). Both light absorption and electron-transfer reactions produce a change in the electronic structure. Since electrons move so much faster than the nuclei in molecules, the change in electronic state is complete before the nuclei have a chance to move, which in turn means that the initial geometry of the final electronic state in these processes must be the same as in the initial state. This is shown by the arrow symbolizing the absorption of light to produce an electronically excited molecule. The r_{AB} distance is the same after the transition as before, although this is not the most stable configuration of the excited-state molecule. This has the practical implication that the absorption of light to promote a molecule from its stable bonding configuration to an excited state often requires more energy

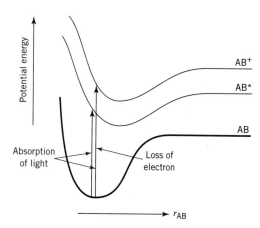

Figure 6.6 Potential energy diagrams for various electronic configurations

than is required to make the most stable configuration of the excited state. Similarly, charge-exchange reactions, in which an electron is transferred between molecules, often require more energy than the minimum required to make the products. This is one of the reasons for overpotentials in electrochemical reactions. The extra energy in the new molecule appears as internal energy of motion (vibration), or, if there is enough energy to dissociate the molecule, as translational energy.

6.3.2 Kinetic Energy in Molecules

Energy is also stored in the motion of atoms, and for a molecule, this takes the form of translational motion, where the whole molecule moves, and internal motion, where the atoms in the molecule move with respect to each other (vibration and rotation). These modes are illustrated in Figure 6.7.

All forms of kinetic energy, including relative translational motion, can be used to surmount potential energy barriers during reaction. In Figure 6.3, C can approach AB with sufficient kinetic energy to "roll up the barrier" near the transition state. Alternatively, A–B vibrational motion can scale the barrier from a different angle. The actual trajectories must obey physical laws (e.g., momentum conservation), and the role of different forms of energy in reactions has been investigated in extensive computer calculations for a variety of potential energy surfaces. In addition to its role in topping the energy barrier, translational motion governs the rate that reactants encounter each other.

6.3.2.1 Energy States

All forms of energy are subject to the rules of quantum mechanics, which allow only certain (discrete) energy levels to exist. Therefore, an isolated molecule cannot contain

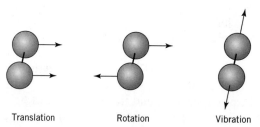

Figure 6.7 Modes of molecular motion

any arbitrary amount of vibrational energy, but must have one of a relatively small number of discrete quantities of vibrational energy. This is also true for rotational energy, although many more states are available. For translational energy, there are usually so many allowed translational energy states that a continuous distribution is assumed. Extra energy can also be stored in the electrons, by promoting an electron from an occupied orbital to an unoccupied orbital. This changes the bonding interactions and can be thought of as an entirely separate potential energy surface at higher energy. These energy states are not usually encountered in thermal reactions, but are an important part of photochemistry and high-energy processes which involve charged species.

6.3.2.2 Distribution of Molecular Energy

In a group of molecules in thermal equilibrium at temperature T, the distribution of energy among the various modes of energy and among the molecules is given by the Boltzmann distribution, which states that the probability of finding a molecule within a narrow energy range around ϵ is proportional to the number of states in that energy range times the "Boltzmann factor," $e^{-\epsilon/k_BT}$:

$$P(\epsilon) = g(\epsilon)e^{-\epsilon/k_BT} \tag{6.3-5}$$

where k_B is the Boltzmann constant:

$$k_B = R/N_{Av} = 1.381 \times 10^{-23} \text{J K}^{-1} \tag{6.3-6}$$

and $g(\epsilon)$, the number of states in the energy range ϵ to $\epsilon + d\epsilon$, is known as the "density of states" function. This function is derived from quantum mechanical arguments, although when many levels are accessible at the energy (temperature) of the system, classical (Newtonian) mechanics can also give satisfactory results. This result arises from the concept that energy is distributed randomly among all the types of motion, subject to the constraint that the total energy and the number of molecules are conserved. This relationship gives the probability that any molecule has energy above a certain quantity (like a barrier height), and allows one to derive the distribution of molecular velocities in a gas. The randomization of energy is accomplished by energy exchange in encounters with other molecules in the system. Therefore, each molecule spends some time in high-energy states, and some time with little energy. The energy distribution over time of an individual molecule is equal to the instantaneous distribution over the molecules in the system. We can use molar energy (E) in 6.3-5 to replace molecular energy (ϵ), if R is substituted for k_B.

6.3.2.3 Distribution of Molecular Translational Energy and Velocity in a Gas

In an ideal gas, molecules spend most of the time isolated from the other molecules in the system and therefore have well defined velocities. In a *liquid*, the molecules are in a constant state of collision. The derivation of the translational energy distribution from equation 6.3-5 (which requires obtaining $g(\epsilon)$) gives the distribution (expressed as dN/N, the fraction of molecules with energy between ϵ and $\epsilon + d\epsilon$):

$$dN(\epsilon)/N = 2\pi^{-1/2}(k_BT)^{-3/2}\epsilon^{1/2}e^{-\epsilon/k_BT}d\epsilon \tag{6.3-7}$$

which is Boltzmann's law of the distribution of energy (Moelwyn-Hughes, 1957, p. 37). The analogous velocity distribution in terms of molecular velocity, $u = (2\epsilon/m)^{1/2}$, where m is the mass per molecule, is:

128 Chapter 6: Fundamentals of Reaction Rates

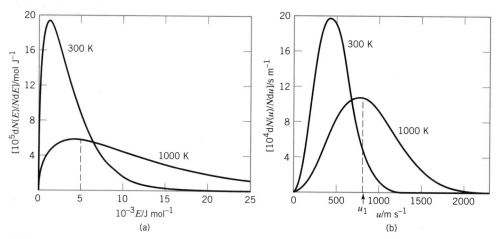

Figure 6.8 (a) Translational kinetic energy distribution for an ideal gas (equation 6.3-7); (b) velocity distribution for N_2 molecules (equation 6.3-8)

$$dN(u)/N = (2/\pi)^{1/2}(m/k_BT)^{3/2}u^2 e^{-mu^2/2k_BT}\,du \quad (6.3\text{-}8)$$
$$= g(u)\,du \quad (6.3\text{-}9)$$

which is Maxwell's law of the distribution of velocities (Moelwyn-Hughes, 1957, p. 38).

These distributions are shown in Figure 6.8. The energy distribution, Figure 6.8(a), is independent of the molecular mass and is shown for $T = 300$ K and 1000 K. The fraction of molecules with translational kinetic energy in excess of a particular value increases as T increases. The increase is more dramatic for energies much higher than the average. By comparing the scale in Figure 6.8(a) with values for even modest energy barriers (e.g., 10 kJ mol^{-1}), we see that a very small fraction of the molecules at either temperature has enough translational energy to overcome such a barrier. The average translational energy is

$$\bar{\epsilon} = (3/2)k_BT \quad (6.3\text{-}10)$$

The velocity distribution for N_2 at these two temperatures is shown in Figure 6.8(b). The average velocity is (Moelwyn-Hughes, 1957, p. 38):

$$\bar{u} = (8k_BT/\pi m)^{1/2} \quad (6.3\text{-}11)$$

6.4 SIMPLE COLLISION THEORY OF REACTION RATES

The collision theory of reaction rates in its simplest form (the "simple collision theory" or SCT) is one of two theories discussed in this chapter. Collision theories are based on the notion that only when reactants encounter each other, or *collide*, do they have the chance to react. The reaction rate is therefore based on the following expressions:

$$\text{reaction rate} \equiv \text{number of } effective \text{ collisions m}^{-3}\text{s}^{-1} \quad (6.4\text{-}1)$$

or, reaction rate $\equiv$

(number of collisions m^{-3} s^{-1}) $\times$ (probability of success (energy, orientation, etc.))

$$(6.4\text{-}2)$$

6.4.1 Simple Collision Theory (SCT) of Bimolecular Gas-Phase Reactions

6.4.1.1 Frequency of Binary Molecular Collisions

In this section, we consider the total rate of molecular collisions without considering whether they result in reaction. This treatment introduces many of the concepts used in collision-based theories; the criteria for success are included in succeeding sections.

Consider a volume containing c'_A molecules of A (mass m_A) and c'_B molecules of B (mass m_B) per unit volume. A simple estimate of the frequency of A-B collisions can be obtained by assuming that the molecules are hard spheres with a finite size, and that, like billiard balls, a collision occurs if the center of the B molecule is within the "collision diameter" d_{AB} of the center of A. This distance is the arithmetic mean of the two molecular diameters d_A and d_B:

$$d_{AB} = (d_A + d_B)/2 \qquad (6.4\text{-}3)$$

and is shown in Figure 6.9(a). The area of the circle of *radius* d_{AB}, $\sigma = \pi d_{AB}^2$, is the collision target area (known as the collision "cross-section"). If the A molecules move at average velocity $\bar{u}$ (equation 6.3-11) and the B molecules are assumed to be stationary, then each A sweeps out a volume $\sigma \bar{u}$ per unit time (Figure 6.9(b)) such that every B molecule inside is hit. The frequency of A-B collisions for each A molecule is then $\sigma \bar{u} c'_B$. By multiplying by the concentration of A, we obtain the frequency of A-B collisions per unit volume:

$$Z_{AB} = \sigma \bar{u} c'_A c'_B \qquad (6.4\text{-}4)$$

This simple calculation gives a result close to that obtained by integrating over the three-dimensional Maxwell velocity distributions for both A and B. In this case, the same expression is obtained with the characteristic velocity of approach between A and B given by

$$\bar{u} = (8k_B T/\pi \mu)^{1/2} \qquad (6.4\text{-}5)$$

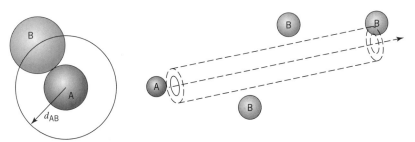

Figure 6.9 (a) Collision diameter d_{AB}; (b) simplified basis for calculating frequency of A-B collisions

130   Chapter 6: Fundamentals of Reaction Rates

where μ is the reduced molecular mass defined by:

$$\mu = m_A m_B/(m_A + m_B) \qquad (6.4\text{-}6)$$

The collision frequency of like molecules, Z_{AA}, can be obtained similarly, but the collision cross-section is $\sigma = \pi d_A^2$, the reduced mass is $\mu = m_A/2$, and we must divide by 2 to avoid counting collisions twice:

$$Z_{AA} = (1/2)\sigma \bar{u}(c'_A)^2 \qquad (6.4\text{-}7)$$

EXAMPLE 6-1

(a) Calculate the rate of collision (Z_{AB}) of molecules of N_2 (A) and O_2 (B) in air (21 mol % O_2, 78 mol % N_2) at 1 bar and 300 K, if $d_A = 3.8 \times 10^{-10}$ m and $d_B = 3.6 \times 10^{-10}$ m.

(b) Calculate the rate of collision (Z_{AA}) of molecules of N_2 (A) with each other in air.

SOLUTION

(a) From equations 6.4-4 and -5, with $\sigma = \pi d_{AB}^2$,

$$Z_{AB} = d_{AB}^2 c'_A c'_B (8\pi k_B T/\mu)^{1/2} \qquad (6.4\text{-}4a)$$

with

$$d_{AB} = (3.8 + 3.6) \times 10^{-10}/2 = 3.7 \times^{-10} \text{ m}$$

From equation 4.2-3a,

$$c'_A = N_{Av} c_A = N_{Av} p_A/RT = 6.022 \times 10^{23}(0.78)10^5/8.314(300)$$
$$= 1.88 \times 10^{25} \text{ molecules m}^{-3}$$

Similarly,

$$c'_B = 0.507 \times 10^{25} \text{ molecules m}^{-3}$$
$$\mu = m_A m_B/(m_A + m_B) = 28.0(32.0)/(28.0 + 32.0)(6.022 \times 10^{23})1000$$
$$= 2.48 \times 10^{-26} \text{ kg}$$
$$Z_{AB} = (3.7 \times 10^{-10})^2(1.88 \times 10^{25})(0.507 \times 10^{25})[8\pi(1.381 \times 10^{-23})300/2.48 \times 10^{-26}]^{1/2}$$
$$= 2.7 \times 10^{34} \text{m}^{-3}\text{s}^{-1}$$

(b) From equation 6.4-7, together with 6.4-5 and -6 (giving $\mu = m_A/2$), and with $\sigma = \pi d_A^2$,

$$Z_{AA} = 2d_A^2(c'_A)^2(\pi k_B T/m_A)^{1/2} \qquad (6.4\text{-}7a)$$

From (a),

$$c'_A = 1.88 \times 10^{25} \text{ molecules m}^{-3}$$
$$m_A = 28.0/(6.022 \times 10^{23})1000 = 4.65 \times 10^{-26} \text{ kg molecule}^{-1}$$
$$Z_{AA} = 2(3.8 \times 10^{-10})^2(1.88 \times 10^{25})^2[\pi(1.381 \times 10^{-23})300/4.65 \times 10^{-26}]^{1/2}$$
$$= 5.4 \times 10^{34} \text{m}^{-3}\text{s}^{-1}$$

Both parts (a) and (b) of Example 6-1 illustrate that rates of molecular collisions are extremely large. If "collision" were the only factor involved in chemical reaction, the rates of all reactions would be virtually instantaneous (the "rate" of N_2-O_2 collisions in air calculated in Example 6-1(a) corresponds to 4.5×10^7 mol L^{-1} s^{-1}!). Evidently, the energy and orientation factors indicated in equation 6.4-2 are important, and we now turn attention to them.

6.4.1.2 Requirements for Successful Reactive Collision

The rate of reaction in collision theories is related to the number of "successful" collisions. A successful reactive encounter depends on many things, including (1) the speed at which the molecules approach each other (relative translational energy), (2) how close they are to a head-on collision (measured by a miss distance or impact parameter, b, Figure 6.10), (3) the internal energy states of each reactant (vibrational (v), rotational (J)), (4) the timing (phase) of the vibrations and rotations as the reactants approach, and (5) orientation (or steric aspects) of the molecules (the H atom to be abstracted in reaction 6.3-4 must be pointing toward the radical center).

Detailed theories include all these effects in the reaction cross-section, which is then a function of all the various dynamic parameters:

$$\sigma_{reaction} = \sigma(\bar{u}, b, v_A, J_A, \ldots) \tag{6.4-8}$$

The SCT treats the reaction cross-section as a separable function,

$$\sigma_{reaction} = \sigma_{hard\ sphere} f(E) p \tag{6.4-9}$$
$$= \pi d_{AB}^2 f(E) p \tag{6.4-10}$$

where the energy requirements, $f(E)$, and the steric requirements, p, are multiplicative factors.

6.4.1.3 Energy Requirements

The energy barrier $E^{\ddagger}$ is the minimum energy requirement for reaction. If only this amount of energy is available, only one orientation out of all the possible collision orientations is successful. The probability of success rises rapidly if extra energy is

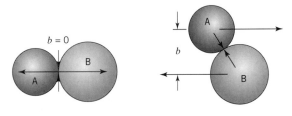

Figure 6.10 Illustration of (a) a head-on collision ($b = 0$), and (b) a glancing collision ($0 < b < d_{AB}$)

available, since other configurations around the transition state (at higher energy) can be reached, and the geometric requirements of the collision are not as precise. Therefore, the best representation of the "necessary" amount of energy is somewhat higher than the barrier height. Because the Boltzmann factor decreases rapidly with increasing energy, this difference is not great. Nevertheless, in the simplified theory, we call this "necessary" energy E^* to distinguish it from the barrier height. The simplest model for the collision theory of rates assumes that the molecules are hard spheres and that only the component of kinetic energy between the molecular centers is effective. As illustrated in Figure 6.10, in a head-on collision ($b = 0$), all of the translational energy of approach is available for internal changes, whereas in a grazing collision ($b = d_{AB}$) none is. By counting only collisions where the intermolecular component at the moment of collision exceeds the "necessary" energy E^*, we obtain a simple expression from the tedious, but straightforward, integration over the joint Maxwell velocity distributions and b (Steinfeld et al., 1989, pp. 248–250). Thus, for the reaction A + B → products, if there are no steric requirements, the rate of reaction is

$$r \equiv (-r_A) = Z_{AB} e^{-E^*/RT} \qquad (6.4\text{-}11)$$

that is, the function $f(E)$ in equation 6.4-9 (in molar units) is $\exp(-E^*/RT)$.

Similarly, for the reaction 2A → products,

$$r \equiv (-r_A)/2 = Z_{AA} e^{-E^*/RT} \qquad (6.4\text{-}12)$$

6.4.1.4 Orientation or Steric Factors

The third factor in equation 6.4-9, p, contains any criteria other than energy that the reactants must satisfy to form products. Consider a hydrogen atom and an ethyl radical colliding in the fifth step in the mechanism in Section 6.1.2. If the hydrogen atom collides with the wrong (CH_3) end of the ethyl radical, the new C–H bond in ethane cannot be formed; a fraction of the collisions is thus ineffective. Calculation of the real distribution of successful collisions is complex, but for simplicity, we use the steric factor approach, where all orientational effects are represented by p as a constant. This factor can be estimated if enough is known about the reaction coordinate: in the case above, an estimate of the fraction of directions given by the H–CH_2–CH_3 bond angle which can form a C–H bond. A reasonable, but uncertain, estimate for p in this case is 0.2. Alternatively, if the value of the rate constant is known, the value of p, and therefore some information about the reaction coordinate, can be estimated by comparing the measured value to that given by theory. In this case $p(\text{derived}) = r(\text{observed})/r(\text{theory})$. Reasonable values of p are equal to or less than 1; however, in some cases the observed rate is much greater than expected ($p \gg 1$); in such cases a chain mechanism is probably involved (Chapter 7), and the reaction is not an elementary step.

6.4.1.5 SCT Rate Expression

We obtain the SCT rate expression by incorporating the steric factor p in equation 6.4-11 or -12. Thus,

$$r_{SCT}/\text{molecules m}^{-3}\,\text{s}^{-1} = pZ e^{-E^*/RT} \qquad (6.4\text{-}13)$$

where $Z = Z_{AB}$ for A+B → products, or $Z = Z_{AA}$ for A+A → products. We develop the latter case in more detail at this point; a similar treatment for A + B → products is left to problem 6-3.

6.4 Simple Collision Theory of Reaction Rates

For the bimolecular reaction $2A \to$ products, by combining equations 6.4-12 and -13, using equation 6.4-7a to eliminate Z_{AA}, and converting completely to a molar basis, with (r_{SCT}) in mol L^{-1} s^{-1}, $c'_A = 1000 N_{Av} c_A$, where c_A is in mol L^{-1}, and $k_B/m_A = R/M_A$, where M_A is the molar mass of A, we obtain

$$r_{SCT} = 2000 p N_{Av} d_A^2 (\pi R/M_A)^{1/2} T^{1/2} e^{-E^*/RT} c_A^2 \equiv k_{SCT} c_A^2 \quad (6.4\text{-}14)$$

where

$$k_{SCT} = 2000 p N_{Av} d_A^2 (\pi R/M_A)^{1/2} T^{1/2} e^{-E^*/RT} \quad (6.4\text{-}15)$$

We may compare these results with a second-order rate law which exhibits Arrhenius temperature dependence:

$$r_{obs} = k_{obs} c_A^2 = A e^{-E_A/RT} c_A^2 \quad (6.1\text{-}1)$$

We note that the concentration dependence (c_A^2) is the same, but that the temperature dependence differs by the factor $T^{1/2}$ in r_{SCT}. Although we do not have an independent value for E^* in equations 6.4-14 and -15, we may compare E^* with E_A by equating r_{SCT} and r_{obs}; thus,

$$k_{obs} = k_{SCT}$$
$$d \ln k_{obs}/dT = d \ln k_{SCT}/dT$$

and, from the Arrhenius equation, 3.1-6,

$$E_A/RT^2 = 1/2T + E^*/RT^2$$

or

$$E_A = \frac{1}{2} RT + E^* \quad (6.4\text{-}16)$$

Similarly, the pre-exponential factor A_{SCT} can be obtained by substitution of E^* from 6.4-16 into 6.4-15:

$$A_{SCT} = 2000 p N_{Av} d_A^2 (\pi R/M_A)^{1/2} e^{1/2} T^{1/2} \quad (6.4\text{-}16a)$$

According to equations 6.4-16 and -16a, E_A and A are somewhat dependent on T. The calculated values for A_{SCT} usually agree with measured values within an order of magnitude, which, considering the approximations made regarding the cross-sections, is satisfactory support for the general concepts of the theory. SCT provides a basis for the estimation of rate constants, especially where experimental values exist for related reactions. Then, values of p and E^* can be estimated by comparison with the known system.

EXAMPLE 6-2

For the reaction $2HI \to H_2 + I_2$, the observed rate constant ($2k$ in $r_{HI} = 2k c_{HI}^2$) is 2.42×10^{-3} L mol^{-1} s^{-1} at 700 K, and the observed activation energy, E_A, is 186 kJ mol^{-1} (Moelwyn-Hughes, 1957 p. 1109). If the collision diameter, d_{HI}, is 3.5×10^{-10} m for HI ($M = 128$), calculate the value of the ("steric") p factor necessary for agreement between the observed rate constant and that calculated from the SCT.

SOLUTION

From equation 6.4-15, with E^* given by equation 6.4-17, and $M_A = (128/1000)$ kg mol^{-1},

$$k_{SCT}/p = 2.42 \times 10^{-3} \text{L mol}^{-1}\text{s}^{-1}$$

This is remarkably coincident with the value of k_{obs}, with the result that $p = 1$. Such closeness of agreement is rarely the case, and depends on, among other things, the correctness and interpretation of the values given above for the various parameters.

For the bimolecular reaction A + B → products, as in the reverse of the reaction in Example 6-2, equation 6.4-15 is replaced by

$$k_{SCT} = 1000 p N_{Av} d_{AB}^2 [8\pi R(M_A + M_B)/M_A M_B]^{1/2} T^{1/2} e^{-E^*/RT} \qquad (6.4\text{-}17)$$

The proof of this is left to problem 6-3.

6.4.1.6 Energy Transfer in Bimolecular Collisions

Collisions which place energy into, or remove energy from, internal modes in one molecule without producing any chemical change are very important in some processes. The transfer of this energy into reactant A is represented by the bimolecular process

$$M + A \rightarrow M + A^*$$

where A^* is a molecule with a critical amount of internal energy necessary for a subsequent process, and M is any collision partner. For example, the dissociation of I_2 discussed in Section 6.3 requires 149 kJ mol^{-1} to be deposited into the interatomic bond. The SCT rate of such a process can be expressed as the rate of collisions which meet the energy requirements to deposit the critical amount of energy in the reactant molecule:

$$r = Z_{AM} \exp(-E^*/RT) = k_{ET} c_A c_M$$

where E^* is approximately equal to the critical energy required. However, this simple theory underestimates the rate constant, because it ignores the contribution of internal energy distributed in the A molecules. Various theories which take this into account provide more satisfactory agreement with experiment (Steinfeld et al., 1989, pp. 352–357). The deactivation step

$$A^* + M \rightarrow A + M$$

is assumed to happen on every collision, if the critical energy is much greater than $k_B T$.

6.4.2 Collision Theory of Unimolecular Reactions

For a unimolecular reaction, such as $I_2 \rightarrow 2I^\bullet$, there are apparently no collisions necessary, but the overwhelming majority of molecules do not have the energy required for this dissociation. For those that have enough energy (> 149 kJ mol^{-1}), the reaction occurs in the time for energy to become concentrated into motion along the reaction coordinate, and for the rearrangement to occur (about the time of a molecular vibration, 10^{-13} s). The internal energy can be distributed among all the internal modes, and so the time required for the energy to become concentrated in the critical reaction coordinate is greater for complex molecules than for smaller ones. Those that do not have

6.4 Simple Collision Theory of Reaction Rates 135

enough energy must wait until sufficient energy is transferred by collision, as in Section 6.4.1.6. Therefore, as Lindemann (1922) recognized, three separate basic processes are involved in this reaction:

(1) Collisions which transfer the critical amount of energy:

$$I_2 + M \text{ (any molecule in the mixture)} \xrightarrow{k_1} I_2^* \text{(energized molecule)} + M \quad \text{(A)}$$

(2) The removal of this energy (deactivation) by subsequent collisions (reverse of (A)):

$$I_2^* + M \xrightarrow{k_{-1}} I_2 + M \quad \text{(B)}$$

(3) The dissociation reaction:

$$I_2^* \xrightarrow{k_2} 2I^\bullet \quad \text{(C)}$$

Steps (A), (B), and (C) constitute a reaction mechanism from which a rate law may be deduced for the overall reaction. Thus, if, in a generic sense, we replace I_2 by the reactant A, I_2^* by A^*, and $2I^\bullet$ by the product P, the rate of formation of A^* is

$$r_{A^*} = -k_2 c_{A^*} + k_1 c_A c_M - k_{-1} c_{A^*} c_M \quad (6.4\text{-}18)$$

and the rate of reaction to form product P, r_P, is:

$$r_P = k_2 c_{A^*} = \frac{k_2(k_1 c_A c_M - r_{A^*})}{k_2 + k_{-1} c_M} \quad (6.4\text{-}19)$$

if we use equation 6.4-18 to eliminate c_{A^*}. Equation 6.4-19 contains the unknown r_{A^*}. To eliminate this we use the *stationary-state hypothesis* (SSH): an approximation used to simplify the derivation of a rate law from a reaction mechanism by eliminating the concentration of a reactive intermediate (RI) on the assumption that its rate of formation and rate of disappearance are equal (i.e., net rate $r_{RI} = 0$).

By considering A^* as a reactive intermediate, we set $r_{A^*} = 0$ in equations 6.4-18 and -19, and the latter may be rewritten as

$$r_P = \left(\frac{k_1 k_2 c_M}{k_2 + k_{-1} c_M} \right) c_A \quad (6.4\text{-}20)$$

$$= k_{uni} c_A \quad (6.4\text{-}20a)$$

where k_{uni} is an effective first-order rate constant that depends on c_M. There are two limiting cases of equation 6.4-20, corresponding to relatively high c_M ("high pressure" for a gas-phase-reaction), $k_{-1} c_M \gg k_2$, and low c_M ("low pressure"), $k_2 \gg k_{-1} c_M$:

$$r_P = (k_1 k_2 / k_{-1}) c_A \quad \text{("high-pressure" limit)} \quad (6.4\text{-}21)$$
$$r_P = k_1 c_M c_A \quad \text{("low-pressure" limit)} \quad (6.4\text{-}22)$$

Thus, according to this (Lindemann) mechanism, a unimolecular reaction is first-order at relatively high concentration (c_M) and second-order at low concentration. There is a transition from first-order to second-order kinetics as c_M decreases. This is referred to as the "fall-off regime," since, although the order increases, k_{uni} decreases as c_M decreases (from equations 6.4-20 and -20a).

This mechanism also illustrates the concept of a *rate-determining step* (*rds*) to designate a "slow" step (relatively low value of rate constant; as opposed to a "fast" step), which then controls the overall rate for the purpose of constructing the rate law.

At low c_M, the rate-determining step is the second-order rate of activation by collision, since there is sufficient time between collisions that virtually every activated molecule reacts; only the rate constant k_1 appears in the rate law (equation 6.4-22). At high c_M, the rate-determining step is the first-order disruption of A* molecules, since both activation and deactivation are relatively rapid and at virtual equilibrium. Hence, we have the additional concept of a *rapidly established equilibrium* in which an elementary process and its reverse are assumed to be at equilibrium, enabling the introduction of an equilibrium constant to replace the ratio of two rate constants.

In equation 6.4-21, although all three rate constants appear, the ratio k_1/k_{-1} may be considered to be a virtual equilibrium constant (but it is not usually represented as such).

A test of the Lindemann mechanism is normally applied to observed apparent first-order kinetics for a reaction involving a single reactant, as in A → P. The test may be used in either a differential or an integral manner, most conveniently by using results obtained by varying the initial concentration, c_{Ao} (or partial pressure for a gas-phase reaction). In the differential test, from equations 6.4-20 and -20a, we obtain, for an initial concentration $c_{Ao} \equiv c_M$, corresponding to the initial rate r_{Po},

$$k_{uni} = \frac{k_1 k_2 c_{Ao}}{k_2 + k_{-1} c_{Ao}}$$

or

$$\frac{1}{k_{uni}} = \frac{k_{-1}}{k_1 k_2} + \frac{1}{k_1} \frac{1}{c_{Ao}} = \frac{1}{k_\infty} + \frac{1}{k_1} \frac{1}{c_{Ao}} \tag{6.4-23}$$

where k_∞ is the asymptotic value of k_{uni} as $c_{Ao} \to \infty$. Thus k_{uni}^{-1} should be a linear function of c_{Ao}^{-1}, from the intercept and slope of which k_∞ and k_1 can be determined. This is illustrated in the following example. The integral method is explored in problem 6-4.

EXAMPLE 6-3

For the gas-phase unimolecular isomerization of cyclopropane (A) to propylene (P), values of the observed first-order rate constant, k_{uni}, at various initial pressures, P_o, at 470° C in a batch reactor are as follows:

P_o/kPa	14.7	28.2	51.8	101.3
$10^5 k_{uni}$/s^{-1}	9.58	10.4	10.8	11.1

(a) Show that the results are consistent with the Lindemann mechanism.
(b) Calculate the rate constant for the energy transfer (activation) step.
(c) Calculate k_∞.
(d) Suggest a value of E_A for the deactivation step.

SOLUTION

(a) In this example, P_o is the initial pressure of cyclopropane (no other species present), and is a measure of c_{Ao}. Expressing c_{Ao} in terms of P_o by means of the ideal-gas law,

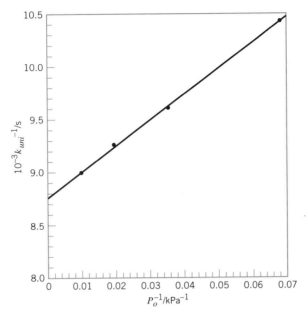

Figure 6.11 Test of Lindemann mechanism in Example 6-3

equation 4.2.3a, we rewrite equation 6.4-23 as:

$$\frac{1}{k_{uni}} = \frac{1}{k_\infty} + \frac{1}{k_1}\frac{RT}{P_o} \tag{6.4-23a}$$

The linear relation is shown in Figure 6.11.
(b) From the slope of the fitted linear form, $k_1 = 0.253$ L mol^{-1} s^{-1}.
(c) Similarly, from the intercept, $k_\infty = 11.4 \times 10^{-5}$ s^{-1}.
(d) E_A (deactivation) → 0, since A* is an activated state (energetically), and any collision should lead to deactivation.

6.4.3 Collision Theory of Bimolecular Combination Reactions; Termolecular Reactions

A treatment similar to that for unimolecular reactions is necessary for recombination reactions which result in a single product. An example is the possible termination step for the mechanism for decomposition of C_2H_6, $H^\bullet + C_2H_5^\bullet \rightarrow C_2H_6$ (Section 6.1.2). The initial formation of ethane in this reaction can be treated as a bimolecular event. However, the newly formed molecule has enough energy to redissociate, and must be stabilized by transfer of some of this energy to another molecule.

Consider the recombination reaction

$$A + B \rightarrow P$$

A three-step mechanism is as follows:

(1) Reaction to form P* (an activated or energized form of P):

$$A + B \xrightarrow{k_1} P^* \tag{A}$$

(2) Unimolecular dissociation of P* (reverse of (A)):

$$P^* \xrightarrow{k_{-1}} A + B \qquad \textbf{(B)}$$

(3) Stabilization of P* by collision with M (any other molecule):

$$P^* + M \xrightarrow{k_2} P + M \qquad \textbf{(C)}$$

Treatment of steps (A), (B), and (C) similar to that for the steps in a unimolecular reaction, including application of the SSH to P*, results in

$$r_P = \left(\frac{k_1 k_2 c_M}{k_{-1} + k_2 c_M}\right) c_A c_B \qquad (6.4\text{-}24)$$

$$\equiv k_{bi} c_A c_B \qquad (6.4\text{-}25)$$

where k_{bi} is an effective second-order rate constant that depends on c_M. Just as for a unimolecular reaction, there are two limiting cases for equation 6.4-24, corresponding to relatively high and low c_M:

$$r_P = k_1 c_A c_B \qquad (\text{"high-pressure" limit}) \qquad (6.4\text{-}26)$$

$$r_P = (k_1 k_2 / k_{-1}) c_M c_A c_B \qquad (\text{"low-pressure" limit}) \qquad (6.4\text{-}27)$$

Thus, according to this three-step mechanism, a bimolecular recombination reaction is second-order at relatively high concentration (c_M), and third-order at low concentration. There is a transition from second- to third-order kinetics as c_M decreases, resulting in a "fall-off" regime for k_{bi}.

The low-pressure third-order result can also be written as a termolecular process:

$$A + B + M \to P + M$$

which implies that all three species must collide with one another at the same time. In the scheme above, this is pictured as taking place in two sequential bimolecular events, the second of which must happen within a very short time of the first. In the end, the distinction is a semantic one which depends on how collision is defined. There are few termolecular elementary reactions of the type

$$A + B + C \to P + Q$$

and the kinetics of these can also be thought of as sequences of bimolecular events.

The "fall-off" effects in unimolecular and recombination reactions are important in modern low-pressure processes such as chemical vapor deposition (CVD) and plasma-etching of semiconductor chips, and also for reactions in the upper atmosphere.

The importance of an "energized" reaction complex in bimolecular reactions is illustrated by considering in more detail the termination step in the ethane dehydrogenation mechanism of Section 6.1.2:

$$H^\bullet + C_2H_5^\bullet \to C_2H_6$$

The formation of C_2H_6 must first involve the formation of the "energized" molecule $C_2H_6^*$:

$$H^\bullet + C_2H_5^\bullet \rightarrow C_2H_6^*$$

which is followed by collisional deactivation:

$$C_2H_6^* + M \rightarrow C_2H_6 + M$$

However, $C_2H_6^*$ may convert to other possible sets of products:

(1) Redissociation to $H^\bullet$ and $C_2H_5^\bullet$:

$$C_2H_6^* \rightarrow H^\bullet + C_2H_5^\bullet$$

(2) Dissociation into two methyl radicals:

$$C_2H_6^* \rightarrow 2CH_3^\bullet$$

(3) Formation of stable products:

$$C_2H_6^* \rightarrow H_2 + C_2H_4$$

The overall process for this last possibility

$$H^\bullet + C_2H_5^\bullet \rightarrow [C_2H_6^*] \rightarrow H_2 + C_2H_4$$

can be thought of as a bimolecular reaction with a stable molecule on the reaction coordinate ($C_2H_6^*$), as illustrated in Figure 6.5. The competition of these other processes with the formation of ethane can substantially influence the overall rate of ethane dehydrogenation. These and similar reactions have a substantial influence in reactions at low pressures and high temperatures.

6.5 TRANSITION STATE THEORY (TST)

6.5.1 General Features of the TST

While the collision theory of reactions is intuitive, and the calculation of encounter rates is relatively straightforward, the calculation of the cross-sections, especially the steric requirements, from such a dynamic model is difficult. A very different and less detailed approach was begun in the 1930s that sidesteps some of the difficulties. Variously known as *absolute rate theory, activated complex theory,* and *transition state theory* (TST), this class of model ignores the rates at which molecules encounter each other, and instead lets thermodynamic/statistical considerations predict how many combinations of reactants are in the transition-state configuration under reaction conditions.

Consider three atomic species A, B, and C, and reaction represented by

$$AB + C \rightarrow A + BC \tag{6.5-1}$$

The TST considers this reaction to take place in the manner

$$AB + C \underset{}{\overset{K_c^\ddagger}{\rightleftharpoons}} ABC^\ddagger \overset{\nu^\ddagger}{\rightarrow} A + BC \tag{6.5-2}$$

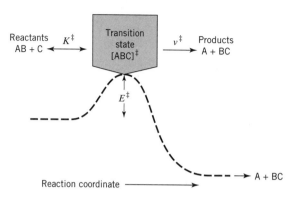

Figure 6.12 Potential energy along the reaction coordinate for reaction 6.5-2

in which ABC‡ represents the transition state described in Section 6.3. The potential energy along the reaction coordinate, showing the energy barrier, is illustrated in Figure 6.12 (cf. Figure 6.3(c)).

The two main assumptions of the TST are:

(1) The transition state is treated as an unstable molecular species in equilibrium with the reactants, as indicated by the equilibrium constant for its formation, $K_c^\ddagger$, where, for reaction 6.5-2,

$$K_c^\ddagger = c_{ABC^\ddagger}/c_{AB}c_C \qquad (6.5\text{-}3)$$

and $c_{ABC^\ddagger}$ is the concentration of these "molecules"; it is implied in this assumption that the transition state and the reactants are in thermal equilibrium (i.e., their internal energy distributions are given by the Boltzmann distribution).

(2) The frequency with which the transition state is transformed into products, $\nu^\ddagger$, can be thought of as a typical unimolecular rate constant; no barrier is associated with this step. Various points of view have been used to calculate this frequency, and all rely on the assumption that the internal motions of the transition state are governed by thermally equilibrated motions. Thus, the motion along the reaction coordinate is treated as thermal translational motion between the product fragments (or as a vibrational motion along an unstable potential). Statistical theories (such as those used to derive the Maxwell-Boltzmann distribution of velocities) lead to the expression:

$$\nu^\ddagger = k_B T/h \qquad (6.5\text{-}4)$$

where k_B is the Boltzmann constant and h is Planck's constant. In some variations of TST, an additional factor (a transmission coefficient, κ) is used to allow for the fact that not all decompositions of the transition state lead to products, but this is seldom used in the estimation of rate constants by the TST.

Thus, from equations 6.5-3 and -4, the rate of formation of products (P) in reaction 6.5-2 is written as

$$r_P = \nu^\ddagger c_{ABC^\ddagger} = (k_B T/h) K_c^\ddagger c_{AB} c_C \qquad (6.5\text{-}5)$$

If we compare equation 6.5-5 with the usual form of rate law, then the rate constant is given by

$$k = (k_B T/h) K_c^\ddagger \qquad (6.5\text{-}6)$$

In the TST, molecularity (m) is the number of reactant molecules forming one molecule of the transition state. In reaction 6.5-2, $m = 2$ (AB and C); that is, the formation is bimolecular. Other possibilities are $m = 1$ (unimolecular) and $m = 3$ (termolecular). The molecularity of formation of the transition state affects the form of $K_c^\ddagger$, and the order of the reaction equals m.

6.5.2 Thermodynamic Formulation

The reaction isotherm of classical thermodynamics applied to the formation of the transition state relates $K_c^\ddagger$ to $\Delta G^{\circ\ddagger}$, the standard Gibbs energy of formation of the activated complex:

$$\Delta G^{\circ\ddagger} = -RT \ln K_c^\ddagger \qquad (6.5\text{-}7)$$

Also

$$\Delta G^{\circ\ddagger} = \Delta H^{\circ\ddagger} - T\Delta S^{\circ\ddagger} \qquad (6.5\text{-}8)$$

where $\Delta H^{\circ\ddagger}$ and $\Delta S^{\circ\ddagger}$ are, respectively, the (standard) enthalpy of activation and (standard) entropy of activation. Combining equations 6.5-6 to -8, we obtain

$$k = (k_B T/h) e^{\Delta S^{\circ\ddagger}/R} e^{-\Delta H^{\circ\ddagger}/RT} \qquad (6.5\text{-}9)$$

for the rate constant according to the TST. As with the SCT, we may compare this expression with observed behavior

$$k_{obs} = A e^{-E_A/RT} \qquad (3.1\text{-}8)$$

to obtain interpretations of the Arrhenius parameters A and E_A in terms of the TST quantities.

We first relate E_A to $\Delta H^{\circ\ddagger}$. From equation 6.5-6,

$$\frac{d \ln k}{dT} = \frac{1}{T} + \frac{d \ln K_c^\ddagger}{dT} = \frac{1}{T} + \frac{\Delta U^{\circ\ddagger}}{RT^2} \qquad (6.5\text{-}10)$$

where $\Delta U^{\circ\ddagger}$ is the internal energy of activation, and we have used the analogue of the van't Hoff equation (3.1-5) for the temperature-dependence of $K_c^\ddagger$ (Denbigh, 1981, p.147). For the activation step as a gas-phase reaction of molecularity m involving ideal gases, from the definition $H = U + PV$,

$$\Delta H^{\circ\ddagger} = \Delta U^{\circ\ddagger} + (1 - m)RT. \qquad (6.5\text{-}11)$$

From equations 3.1-8 (i.e., from 3.1-6), and 6.5-10 and -11,

$$E_A = \Delta H^{\circ\ddagger} + mRT \qquad (6.5\text{-}12)$$

We next relate the pre-exponential factor A to $\Delta S^{\circ\ddagger}$. From equations 6.5-9 and 6.5-12,

$$k = (k_B T/h) e^{\Delta S^{\circ\ddagger}/R} e^m e^{-E_A/RT} \qquad (6.5\text{-}13)$$

Table 6.1 Expected (approximate) values of $\Delta S^{\circ\ddagger}$ for different values of molecularity (m) at 500 K

m	$A/(\text{L mol}^{-1})^{m-1} \text{ s}^{-1}$	$\Delta S^{\circ\ddagger}/\text{J mol}^{-1} \text{ K}^{-1}$
1	10^{13} to 10^{14}	0
2	10^{11} to 10^{12}	-45
3	10^{9} to 10^{10}	-90

Comparing equations 6.5-13 and 3.1-8, we obtain

$$A = (k_B T/h) e^{\Delta S^{\circ\ddagger}/R} e^m \qquad (6.5\text{-}14)$$

or

$$\Delta S^{\circ\ddagger} = R[\ln(Ah/k_B T) - m] \qquad (6.5\text{-}15)$$

$$= 8.314(-23.76 + \ln A - \ln T - m) \text{ J mol}^{-1} \text{ K}^{-1} \qquad (6.5\text{-}15\text{a})$$

on substitution of numerical values for the constants.

From equation 6.5-15a and typical experimental values of A, we may estimate expected values for $\Delta S^{\circ\ddagger}$. The results are summarized in Table 6.1.

EXAMPLE 6-4

If the Arrhenius parameters for the gas-phase unimolecular decomposition of ethyl chloride (C_2H_5Cl) to ethylene (C_2H_4) and HCl are $A = 4 \times 10^{14} \text{ s}^{-1}$ and $E_A = 254 \text{ kJ mol}^{-1}$, calculate the entropy of activation ($\Delta S^{\circ\ddagger}$ /J mol^{-1} K^{-1}), the enthalpy of activation ($\Delta H^{\circ\ddagger}$ /J mol^{-1}), and the Gibbs energy of activation ($\Delta G^{\circ\ddagger}$ /J mol^{-1}) at 500 K. Comment on the value of $\Delta S^{\circ\ddagger}$ in relation to the normally "expected" value for a unimolecular reaction.

SOLUTION

From equation 6.5-15,

$$\Delta S^{\circ\ddagger} = R\left(\ln \frac{Ah}{k_B T} - m\right)$$

$$= 8.314 \left[\ln\left(\frac{4 \times 10^{14} \times 6.626 \times 10^{-34}}{1.381 \times 10^{-23} \times 500}\right) - 1\right]$$

$$= 22 \text{ J mol}^{-1} \text{ K}^{-1}$$

From equation 6.5-12,

$$\Delta H^{\circ\ddagger} = E_A - mRT$$

$$= 254,000 - 1(8.314)500$$

$$= 250,000 \text{ J mol}^{-1}$$

$$\Delta G^{\circ\ddagger} = \Delta H^{\circ\ddagger} - T\Delta S^{\circ\ddagger} \qquad (6.5\text{-}8)$$
$$= 250{,}000 - 500(22)$$
$$= 239{,}000 \text{ J mol}^{-1}$$

(Comment: the normally expected value of $\Delta S^{\circ\ddagger}$ for a unimolecular reaction, based on $A \approx 10^{13}$ to 10^{14}, is ≈ 0 (Table 6.1); the result here is greater than this.)

A method for the estimation of thermodynamic properties of the transition state and other unstable species involves analyzing parts of the molecule and assigning separate properties to functional groups (Benson, 1976). Another approach stemming from statistical mechanics is outlined in the next section.

6.5.3 Quantitative Estimates of Rate Constants Using TST with Statistical Mechanics

Quantitative estimates of $E_o^{\ddagger}$ are obtained the same way as for the collision theory, from measurements, or from quantum mechanical calculations, or by comparison with known systems. Quantitative estimates of the A factor require the use of statistical mechanics, the subject that provides the link between thermodynamic properties, such as heat capacities and entropy, and molecular properties (bond lengths, vibrational frequencies, etc.). The transition state theory was originally formulated using statistical mechanics. The following treatment of this advanced subject indicates how such estimates of rate constants are made. For more detailed discussion, see Steinfeld et al. (1989).

Statistical mechanics yields the following expression for the equilibrium constant, $K_c^{\ddagger}$,

$$K_c^{\ddagger} = (Q^{\ddagger}/Q_r)\exp(-E_o^{\ddagger}/RT) \qquad (6.5\text{-}16)$$

The function $Q^{\ddagger}$ is the *partition function* for the transition state, and Q_r is the product of the partition functions for the reactant molecules. The partition function essentially counts the number of ways that thermal energy can be "stored" in the various modes (translation, rotation, vibration, etc.) of a system of molecules, and is directly related to the number of quantum states available at each energy. This is related to the freedom of motion in the various modes. From equations 6.5-7 and -16, we see that the entropy change is related to the ratio of the partition functions:

$$\Delta S^{\circ\ddagger} = R\ln(Q^{\ddagger}/Q_r) \qquad (6.5\text{-}17)$$

An increase in the number of ways to store energy increases the entropy of a system. Thus, an estimate of the pre-exponential factor A in TST requires an estimate of the ratio $Q^{\ddagger}/Q_r$. A common approximation in evaluating a partition function is to separate it into contributions from the various modes of energy storage, translational (tr), rotational (rot), and vibrational (vib):

$$Q = Q_{tr}Q_{rot}Q_{vib}Q(\text{electronic, symmetry}) \qquad (6.5\text{-}18)$$

This approximation is valid if the modes of motion are completely independent—an assumption that is often made. The ratio in equation 6.5-17 can therefore be written as a product of ratios:

$$(Q^{\ddagger}/Q_r) = (Q_{tr}^{\ddagger}/Q_{tr})(Q_{rot}^{\ddagger}/Q_{rot})(Q_{vib}^{\ddagger}/Q_{vib})\cdots \qquad (6.5\text{-}19)$$

Furthermore, each Q factor in equation 6.5-18 can be further factored for each individual mode, if the motions are independent; for example,

Table 6.2 Forms for translational, rotational, and vibrational contributions to the molecular partition function

Mode	Partition function	Model
Q_{tr}/V = translational (per unit volume)	$(2\pi m k_B T/h^2)^{3/2}$	particle of mass m in 3D box of volume V; increases if mass increases
Q_{rot} = rotational	$(8\pi^2 I k_B T/h^2)^{1/2}$	rigid rotating body with moment of inertia I per mode; increases if moment of inertia increases
Q_{vib} = vibrational	$(1 - \exp(-hc\nu/k_B T))^{-1}$	harmonic vibrator with frequency ν per mode; increases if frequency decreases (force constant decreases)

$$Q_{vib} = Q_{vib,\ mode\ 1} Q_{vib,\ mode\ 2} \cdots \qquad (6.5\text{-}20)$$

with a factor for each normal mode of vibration. The A factor can then be evaluated by calculating the individual ratios. For the translational, rotational, and vibrational modes of molecular energy, the results obtained from simplified models for the contributions to the molecular partition function are shown in Table 6.2.

Generally, $Q_{tr} > Q_{rot} > Q_{vib}$, reflecting the decreasing freedom of movement in the modes. Evaluating the partition functions for the reactants is relatively straightforward, since the molecular properties (and the related thermodynamic properties) can be measured. The same parameters for the transition state are not available, except in a few simple systems where the full potential energy surface has been calculated. The problem is simplified by noting that if a mode is unchanged in forming the transition state, the ratio for that mode is equal to 1. Therefore, only the modes that change need to be considered in calculating the ratio. The following two examples illustrate how estimates of rate constants are made, for unimolecular and bimolecular reactions.

EXAMPLE 6-5

For the unimolecular reaction in Example 6-4, $C_2H_5Cl \rightarrow HCl + C_2H_4$, the transition state should resemble the configuration below, with the C-Cl and C-H bonds almost broken, and HCl almost formed:

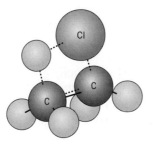

The ratio of translational partition functions $(Q_{tr}^{\ddagger}/Q_{tr})$ is 1 here, and for all unimolecular reactions, because the mass and number of molecules of the reactants is the same as for the transition state. The rotational ratio $(Q_{rot}^{\ddagger}/Q_{rot})$ is given by the ratio of the moments of inertia: $(I_1^{\ddagger} I_2^{\ddagger} I_3^{\ddagger}/I_1 I_2 I_3)^{1/2}$. The moments of inertia are probably slightly higher in the transition state because the important Cl–C bond is stretched. The increased C–C–Cl bond angle also increases the value of the smallest moment of inertia. Thus, the ratio $Q^{\ddagger}_{rot}/Q_{rot}$ is greater than 1. An exact calculation requires a quantitative estimate of the bond lengths and angles. The transition state has the same number of vibrational modes, but several of the vibrational frequencies in the transition state are expected to be somewhat lower, particularly those involving both the weakened C–Cl bond stretch and the affected C–H bond. It is also possible to form the transition state with any of the three hydrogen atoms on the CH_3 group, and so a symmetry number of 3 accrues to the transition state. The internal rotation around the C–C bond is inhibited in the transition state, which decreases the contribution of this model to $Q^{\ddagger}$, but the rest of the considerations increase it, and the net effect is that $(Q^{\ddagger}/Q_r) > 1$. From the value of the A factor in Example 6-4, $A/(kT/h) = (Q^{\ddagger}/Q_r) = 38.4$. As with many theories, the information flows two ways: (1) measured rate constants can be used to study the properties of transition states, and (2) information about transition states gained in such studies, as well as in calculations, can be used to estimate rate constants.

EXAMPLE 6-6

Consider a bimolecular reaction, A + B → products. Confining two molecules A and B to be together in the transition state in a bimolecular reaction always produces a loss of entropy. This is dominated by the ratio of the translational partition functions:

$$(Q^{\ddagger}_{tr}/V)/(Q_{trA}/V)(Q_{trB}/V) = (2\pi m_{A+B}k_BT/h^2)^{3/2}/[2\pi m_A k_BT/h^2)^{3/2}(2\pi m_B k_BT/h^2)^{3/2}]$$
$$= (2\pi \mu k_BT/h^2)^{-3/2}$$

where μ is the reduced mass, equation 6.4-6. This ratio introduces the volume units to the rate constant, and is always less than 1 for a bimolecular (and termolecular) reaction. At 500 K, and for a reduced mass of 30 g mol^{-1}, this factor is 1.7×10^{-6} L mol^{-1} s^{-1}, and corresponds to an entropy change of -110 J mol^{-1} K^{-1}. The number of internal modes (rotation and vibration) is increased by 3, which partly compensates for this loss of entropy.

If A and B are atoms, the two rotational modes in the transition state add 70 J mol^{-1} K^{-1} to the entropy of the transition state. The total $\Delta S^{\circ \ddagger}$ is therefore approximately -40 J mol^{-1} K^{-1}, a value in agreement with the typical value given in Table 6.1. For each of the two rotational modes, the moment of inertia cited in Table 6.2 is $I = \mu d^2_{AB}$; the value above is calculated using $d_{AB} = 3 \times 10^{-10}$ m.

6.5.4 Comparison of TST with SCT

Qualitatively, both the TST and the SCT are in accord with observed features of kinetics:

(1) Both theories yield laws for elementary reactions in which order, molecularity, and stoichiometry are the same (Section 6.1.2).
(2) The temperature dependence of the reaction rate constant closely (but not exactly) obeys the Arrhenius equation. Both theories, however, predict non-Arrhenius behavior. The deviation from Arrhenius behavior can usually be ignored over a small temperature range. However, non-Arrhenius behavior is common (Steinfeld et al., 1989, p. 321). As a consequence, rate constants are often fitted to the more general expression $k = BT^n \exp(-E/RT)$, where B, n, and E are empirical constants.

The activation energy in both theories arises from the energy barrier at the transition state, and is treated similarly in both. The relationship between the pre-exponential factors in the two theories is not immediately obvious, since many of the terms which arise from the intuitive dynamical picture in SCT are "hidden" in the partition functions in TST. Nevertheless, the ratio of partition functions (thermodynamics) tells how easy (probable) the achievement of the transition state is. This ratio contains many of the notions in collision theories, for example, (1) how close the reactants must approach to react (equivalent to the hard-sphere cross-section in SCT), and (2) the precision of alignment of the atoms in the transition state (equivalent to the p factor in SCT). The combination of a smaller cross-section and more demanding configuration is equivalent to a smaller entropy in the transition state. All of the dynamics in TST is contained in kT/h, which in turn is contained in the velocity of approach in bimolecular reactions in SCT. The assumption that the transition state is in thermal equilibrium with the reactants is central to a discussion of the merits of TST. On the one hand, this assumption allows a relatively simple statistical (thermodynamic) calculation to replace the detailed dynamics. This has made transition state theory the more useful of the two for the estimation of unmeasured rate constants. This considerable advantage of TST is also its main weakness, and TST must fail when the assumption of thermal equilibrium is grossly wrong. Such an example is the behavior of unimolecular reactions at low pressure, where the supply of energy is rate limiting. Both theories have been very useful in the understanding of kinetics, and in building detailed mechanisms of important chemical processes.

6.6 ELEMENTARY REACTIONS INVOLVING OTHER THAN GAS-PHASE NEUTRAL SPECIES

The two simple theories SCT and TST have been developed in the context of neutral gas-phase reactions. In this section, we consider other types of elementary reactions listed in Section 6.2.1, and include reactions in condensed phases. The rates of this diverse set of reactions, including photochemistry, can be understood with the concepts developed for gas-phase reactions.

6.6.1 Reactions in Condensed Phases

Reactions in solution proceed in a similar manner, by elementary steps, to those in the gas phase. Many of the concepts, such as reaction coordinates and energy barriers, are the same. The two theories for elementary reactions have also been extended to liquid-phase reactions. The TST naturally extends to the liquid phase, since the transition state is treated as a thermodynamic entity. Features not present in gas-phase reactions, such as solvent effects and activity coefficients of ionic species in polar media, are treated as for stable species. Molecules in a liquid are in an almost constant state of collision so that the collision-based rate theories require modification to be used quantitatively. The energy distributions in the jostling motion in a liquid are similar to those in gas-phase collisions, but any reaction trajectory is modified by interaction with neighboring molecules. Furthermore, the frequency with which reaction partners approach each other is governed by diffusion rather than by random collisions, and, once together, multiple encounters between a reactant pair occur in this molecular traffic jam. This can modify the rate constants for individual reaction steps significantly. Thus, several aspects of reaction in a condensed phase differ from those in the gas phase:

- **(1)** *Solvent interactions*: Because all species in solution are surrounded by solvent, the solvation energies can dramatically shift the energies of the reactants, products, and the transition state. The most dramatic changes in energies are for ionic species, which are generally unimportant in gas-phase chemistry, but are prominent in polar solvents. Solvation energies for other species can also be large enough to change the reaction mechanism. For example, in the alkylation of naphthol by methyl iodide, changes in solvent can shift the site of alkylation from oxygen to carbon. The TST is altered by allowing the thermodynamic properties to be modified by activity coefficients.

(2) *Encounter frequency*: Between two reactive species in solution, the encounter frequency is slower than in the gas phase at the same concentration. The motion in a liquid is governed by diffusion, and in one version, which assumes that there are no long-range forces between the reactants (too simple for ionic species), the collision rate is given by $Z_{AB} = 4\pi Dd_{AB}c'_A c'_B$, where D is the sum of the diffusion coefficients of the two species. If reaction occurs on every collision, then the rate constant is lower in solution (even with no appreciable solvent interactions) than in the gas phase. If reaction does not occur on every collision, but is quite slow, then the probability of finding the two reactants together is similar to that in the gas phase, and the rate constants are also similar. One way to think of this is that diffusion in the liquid slows the rate at which the reactants move away from each other to the same degree that it slows the rate of encounters, so that each encounter lasts longer in a liquid. This "trapping" of molecules near each other in condensed phases is sometimes referred to as the "cage effect," and is important in photochemical reactions in liquids, among others.

(3) *Energy transfer*: Because the species are continually in collision, the rate of energy transfer is never considered to be the rate-limiting step, unlike in unimolecular gas-phase reactions.

(4) *Pressure effects*: The diffusion through liquids is governed by the number of "defects" or atomic-sized holes in the liquid. A high external pressure can reduce the concentration of holes and slow diffusion. Therefore, in a liquid, a diffusion-controlled rate constant also depends on the pressure.

6.6.2 Surface Phenomena

Elementary reactions on solid surfaces are central to heterogeneous catalysis (Chapter 8) and gas-solid reactions (Chapter 9). This class of elementary reactions is the most complex and least understood of all those considered here. The simple quantitative theories of reaction rates on surfaces, which begin with the work of Langmuir in the 1920s, use the concept of "sites," which are atomic groupings on the surface involved in bonding to other atoms or molecules. These theories treat the sites as if they are stationary gas-phase species which participate in reactive collisions in a similar manner to gas-phase reactants.

6.6.2.1 Adsorption

Adsorption can be considered to involve the formation of a "bond" between the surface and a gas-phase or liquid-phase molecule. The surface "bond" can be due to physical forces, and hence weak, or can be a chemical bond, in which case adsorption is called chemisorption. Adsorption is therefore like a bimolecular combination reaction:

$$A + s \xrightarrow{k_a} A \bullet s \tag{6.6-1}$$

where "s" is an "open" surface site without a molecule bonded to it, and $A \bullet s$ is a surface-bound molecule of A. By analogy with gas-phase reactions, the collision rate of molecules of A with a site with a reaction cross-section σ on a flat surface, Z_A, can be calculated by integration of the Maxwell-Boltzmann velocity distributions over the possible angles of impingement:

$$Z_A/\text{molecules site}^{-1}\,\text{s}^{-1} = (1/4)\sigma \bar{u} c'_A \tag{6.6-2}$$

where $\bar{u}$ is the average velocity $(8k_BT/\pi m_A)^{1/2}$. If the reaction requires a direct impingement on an open surface site (one with no molecules bonded to it), then the rate of adsorption per unit area on the surface should be proportional to the number of open sites on the surface:

$$r_a/\text{mol m}^{-2}\text{s}^{-1} = Z_A N\theta_\square/N_{Av} = (2.5 \times 10^{-4}\sigma\bar{u}N)\theta_\square c_A \equiv k_a\theta_\square c_A \qquad (6.6\text{-}3)$$

where N is the number of sites m^{-2} of surface, $\theta_\square$ is the fraction of sites which are open, and c_A is the gas-phase concentration in mol L^{-1}. This "bimolecular" type of adsorption kinetics, where the cross-section does not depend on the amount of adsorbed material, is said to obey Langmuir adsorption kinetics. The factor in parentheses is the SCT expression for the adsorption rate constant k_a. Like bimolecular combination reactions, no activation energy is expected, unless bond-breaking must take place in the solid or in the adsorbing molecule.

6.6.2.2 Desorption

Desorption, the reverse of reaction 6.6-1, that is,

$$A\bullet s \xrightarrow{k_d} A + s \qquad (6.6\text{-}4)$$

is a unimolecular process, which, like gas-phase analogues, requires enough energy to break the bond to the surface. Similar to reactions in liquids, energy is transferred through the solid, making collisions unnecessary to supply energy to the adsorbed molecule. If the sites are independent, the rate is proportional to the amount of adsorbed material:

$$r_d/\text{mol m}^{-2}\text{s}^{-1} = k_d\theta_A$$

where k_d is the unimolecular desorption rate constant, which is expected to have an activation energy similar to the adsorption bond strength, and θ_A is the fraction of the sites which have A adsorbed on them, often called the "coverage" of the surface by A.

6.6.2.3 Surface Reactions

The simplest theories of reactions on surfaces also predict surface rate laws in which the rate is proportional to the amount of each adsorbed reactant raised to the power of its stoichiometric coefficient, just like elementary gas-phase reactions. For example, the rate of reaction of adsorbed carbon monoxide and hydrogen atoms on a metal surface to produce a formyl species and an open site,

$$CO\bullet s + H\bullet s \rightarrow HCO\bullet s + s \qquad (6.6\text{-}5)$$

is assumed to exhibit the following rate law:

$$r/\text{mol m}^{-2}\text{ s}^{-1} = k\theta_{CO}\theta_H \qquad (6.6\text{-}6)$$

This behavior arises, as in the gas phase, from assuming statistical encounter rates of the reactants on the surface. Because the motion of adsorbed species on surfaces is not well understood, however, quantitative prediction of this encounter rate is not generally possible.

6.6.2.4 General Observations

Simple theories provide useful rate expressions for reactions involving solid surfaces (Chapter 8). In fundamental studies, there are examples of adsorption kinetics which obey the simple Langmuir rate expressions. However, many others are more complex and do not show first-order dependence on the number of open sites. These variations can be appreciated, if we accept the notion that a solid can be thought of as a giant molecule which presents a large number of locations where bonds can be made, and that changes in the bonding at one site on this molecule can change the bonding at other locations. As a result, the site properties can depend on whether molecules are adsorbed on neighboring sites. Furthermore, molecules can "pre-adsorb" weakly even on occupied sites and "hunt" for an open site. The desorption rate constant can vary with the amount of adsorbed material, if, for instance, the surface bond strength depends on the amount of adsorbed material. For these reasons, and because of the difficulty in obtaining reliable information on the structure of surface-adsorbed reaction intermediates, quantitative theories of surface reactions are not generally available.

6.6.3 Photochemical Elementary Reactions

Light energy interacts with matter in quantum units called photons which contain energy $E = h\nu$ (Section 6.2.1.2). The frequency ν is related to the wavelength λ by

$$\lambda = c/\nu \tag{6.6-7}$$

where c is the speed of light (3×10^8 m s^{-1}). The energy of photons can be expressed in units, such as J mol^{-1}, to compare with chemical energies:

$$E/\text{J mol}^{-1} \equiv N_{Av}h\nu \equiv N_{Av}hc/\lambda = 0.1196/\lambda \tag{6.6-8}$$

where λ is in m. Low-energy photons (infrared wavelengths and longer, $\lambda > \approx 0.8$ μm, $E_{photon} < 150$ kJ mol^{-1}) are generally only capable of exciting *vibrational* levels in the molecules. In photochemistry, we are usually concerned with photons with enough energy to produce changes in *electronic* states (visible wavelengths and shorter, $\lambda < \approx 0.8$ μm, $E_{photon} > 150$ kJ mol^{-1}), and therefore to disrupt chemical bonds.

6.6.3.1 Light Absorption

Although light behaves like both waves and particles, photons can be thought of as particles which participate in elementary reactions analogous to those for neutral molecules. Furthermore, the language of collision theories is often used to describe the rates of these reactions. For example, the absorption of light can be treated in a collision theory as a "bimolecular" process in which light particles (photons) collide with the molecules, and are absorbed to produce a higher-energy "excited" state in the molecule:

$$h\nu + A \rightarrow A^* \tag{6.6-9}$$

There is a cross-section for absorption, σ, which characterizes the size of the "target" a photon has to hit to be absorbed. The rate of absorption is given a little differently, since the photons travel much faster than the A molecules (which can be treated as stationary). If the flux of photons (number traversing a given area per unit time) is I, then the rate of absorption per unit volume is

$$r/\text{events m}^{-3}\text{ s}^{-1} = (I/(\text{photons m}^{-2}\text{ s}^{-1})) \times (c'_A/\text{molecules m}^{-3}) \times (\sigma/\text{m}^2) \tag{6.6-10}$$

The attenuation of a light beam as it traverses a volume of light-absorbing material of thickness dl can be expressed as

$$r = -dI/dl = I\sigma c'_A \qquad (6.6\text{-}11)$$

The integration of equation 6.6-11 with the boundary condition that $I = I_o$ at $l = 0$ gives the Beer-Lambert law (with c_A/mol L^{-1} = c'_A/N_{Av}):

$$I = I_o \exp(-ac_A l) \qquad (6.6\text{-}12)$$

where $a(= \sigma N_{Av}/1000)$ is called the molar extinction coefficient of the medium. The cross-section is highly energy dependent and produces characteristic absorption spectra for each molecule.

6.6.3.2 Elementary Reactions of Molecules in Excited States

An electronically excited molecule can undergo several subsequent reaction steps. In addition to dissociation and rearrangements, there are processes involving light. These are:

Light emission (fluorescence): The reverse of reaction 6.6-9

$$A^* \to A + h\nu \qquad (6.6\text{-}13)$$

is called fluorescence and can be thought of as another unimolecular reaction, with a first-order rate expression:

$$r = k_e c_A^* \qquad (6.6\text{-}14)$$

The rate constant k_e corresponds to the reciprocal of the lifetime of the excited state.

Internal conversion: The excited state can do other things, such as convert some of the original electronic excitation to a mixture of vibration and a different electronic state. These are also treated as unimolecular processes with associated rate constants:

$$A^* \to A^{*\prime} \qquad (6.6\text{-}15)$$

Often, the second state formed this way is longer-lived, thus giving the excited molecule a longer time to undergo other reactions.

Stimulated emission: Another form of photon emission is called stimulated emission, where a photon of the right energy can cause an excited state to emit an additional identical photon, that is,

$$A^* + h\nu \to A + 2h\nu \qquad (6.6\text{-}16)$$

The waves of the two "product" photons are in phase; this process is the basis of laser operation.

6.6.4 Reactions in Plasmas

In specialized processes associated with the materials science industry, a reactive atmosphere is generated by reactions in which charged species are participants. A gaseous system wherein charged particles (electrons, ions) are important species is called a *plasma*, and the response of charged particles to an external field is used to increase

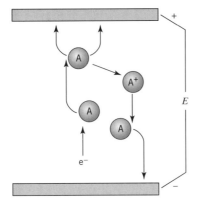

Figure 6.13 Illustration of collisional processes in a plasma

their translational energy. Consider a gas which has an electric field E (V cm^{-1}) applied across it, as illustrated in Figure 6.13.

An electron (or ion) in the gas is accelerated (gains kinetic energy) in the electric field until it collides with a gas molecule (A). In this collision, kinetic energy is transferred to the collision partner and eventually randomized to the rest of the gas. The electron is again accelerated until the next collision, and so on. The average energy attained before each collision is

$$E_{kin} = E\lambda \qquad (6.6\text{-}17)$$

where λ is the mean free path (average distance between collisions). For illustration, consider a gas at one bar (P) and an applied electric field E of 1000 V cm^{-1}: $\lambda \approx 1$ μm, and the average kinetic energy of the electrons is 0.1 eV (electron volt) or about 10 kJ mol^{-1}. This is not enough to disrupt any chemical bonds and only serves to increase the gas temperature. The average energy can be increased by increasing the field strength or the mean free path (by decreasing P). As the average energy rises, more can be accomplished in the collisions. At an average energy of a few hundred kJ mol^{-1}, bonds can be broken and electronic excitations achieved in the collisions:

$$e^- + O_2 \rightarrow 2O + e^- \qquad (6.6\text{-}18)$$
$$e^- + Ar \rightarrow Ar^* + e^- \qquad (6.6\text{-}19)$$

The reactive species produced in these reactions can then participate in chemical processes. At slightly higher energy, it is possible to ionize the neutral species in the gas in the collisions:

$$e^- + Ar \rightarrow Ar^+ + 2e^- \qquad (6.6\text{-}20)$$

Figure 6.13 schematically shows this event. The control of a plasma then relies on control of pressure and voltage/current. Although plasma chemistry takes place in the gas phase, the reactive intermediates are often used to accomplish the production or etching of solid materials, as in chemical vapor deposition (CVD).

6.7 SUMMARY

This chapter contains basic information for at least partial understanding of reaction kinetics. Some main points are summarized as follows:

(1) Almost all chemical reactions involve a sequence of elementary steps, and do not occur in a single step.

(2) The elementary steps in gas-phase reactions have rate laws in which reaction order for each species is the same as the corresponding molecularity. The rate constants for these elementary reactions can be understood quantitatively on the basis of simple theories. For our purpose, reactions involving photons and charged particles can be understood in the same way.

(3) Elementary steps on surfaces and in condensed phases are more complex because the environment for the elementary reactions can change as the composition of the reaction mixture changes, and, in the case of surface reactions, there are several types of reactive sites on solid surfaces. Therefore, the rate constants of these elementary steps are not really constant, but can vary from system to system. Despite this complexity, the approximation of a single type of reaction step is useful and often generally correct.

In the following chapter, rate laws based on reaction mechanisms are developed. Although some of these are of the simple "generic" form described in Chapters 3 and 4, others are more complex. In some cases of reactor design, only an approximate fit to the real reaction kinetics is required, but more often the precision of the correct law is desirable, and the underlying mechanistic information can be useful for the rational improvement of chemical processes.

6.8 PROBLEMS FOR CHAPTER 6

6-1 In each of the following cases, state whether the reaction written could be an elementary reaction, as defined in Section 6.1.2; explain briefly.
(a) $SO_2 + \frac{1}{2}O_2 \rightarrow SO_3$

(b) $I^\bullet + I^\bullet + M \rightarrow I_2 + M$

(c) $2C_3H_6 + 2NH_3 + 3O_2 \rightarrow 2C_3H_3N + 6H_2O$

(d) $C_2H_4 + H_2 \rightarrow C_2H_5^\bullet + H^\bullet$

6-2 Calculate the fraction of ideal-gas molecules with translational kinetic energy equal to or greater than 5000 J mol^{-1} (a) at 300 K, and (b) at 1000 K.

6-3 Show that, for the bimolecular reaction $A + B \rightarrow$ products, k_{SCT} is given by equation 6.4-17.

6-4 Some of the results obtained by Hinshelwood and Askey (1927) for the decomposition of dimethyl ether, $(CH_3)_2O$ (A), to CH_4, CO and H_2 at 777.2 K in a series of experiments in a constant-volume batch reactor are as follows:

P_o/kPa	7.7	12.1	22.8	34.8	52.5	84.8
t_{31}/s	1500	1140	824	670	590	538

Each pair of points, P_o and t_{31}, refers to one experiment. P_o is the initial pressure of ether (no other species present initially), and t_{31} is the time required for 31% of the ether to decompose.

(a) If the reaction is first-order, calculate the value of the rate constant k_{uni}/s^{-1} for each experiment.

(b) Test, using the differential method, whether the experimental data conform to the Lindemann hypothesis for a unimolecular reaction, and, if appropriate, calculate the values of the rate constants in the unimolecular mechanism as far as possible; use units of L, mol, s.

6-5 Repeat problem 6-4 using an integral method. For this purpose, substitute the rate law into the material balance for a constant-volume BR, and integrate the resulting expression to relate f_A and t. Then, with c_{Ao} as a parameter (corresponding to P_o in problem 6-4), show that, for a constant value of f_A (0.31 in problem 6-4), t_{f_A} (t_{31} above) is a linear function of $1/c_{Ao}$, from the slope and intercept of which k_1 and k_∞ can be determined. Compare the values with those obtained in problem 6-4.

6-6 (a) Is the experimental quantity E_A in the Arrhenius equation intensive or extensive? Does its numerical value depend on the way in which the stoichiometry of reaction is expressed (cf. ΔH of reaction)?
(b) The dimensions of E_A are energy mol^{-1}. To what does "mol" refer?

6-7 The isomerization of cyclopropane to propylene has Arrhenius parameters $A = 1.6 \times 10^{15}$ s^{-1} and $E_A = 270$ kJ mol^{-1}.
(a) Calculate the entropy of activation, $\Delta S^{o\ddagger}$/J mol^{-1} K^{-1}, at 500 K.
(b) Comment on the answer in (a) in comparison with the "expected" result for a unimolecular reaction.
(c) Calculate the enthalpy of activation, $\Delta H^{o\ddagger}$/kJ mol^{-1}, at 500 K.

6-8 Rowley and Steiner (1951) have obtained the result

$$k = A \exp(-E_A/RT) = 3.0 \times 10^7 \exp(-115{,}000/RT),$$

where A is in L mol^{-1} s^{-1} and E_A is in J mol^{-1}, for the rate constant for the reaction

$$C_2H_4 + C_4H_6 \rightarrow C_6H_{10} \text{ (cyclohexene)}.$$

(a) Calculate the entropy of activation for this reaction at 800 K.
(b) Comment on the answer in (a) in comparison with the "expected" result for a bimolecular reaction.
(c) Calculate the enthalpy of activation in kJ mol^{-1}.

6-9 (a) If the Arrhenius parameters for the gas-phase reaction

$$CH_2=CH-CH=CH_2 + CH_2=CH-CHO \rightarrow$$

(structure: cyclohexene ring with H and CHO substituents)

are $A = 1.5 \times 10^6$ L mol^{-1} s^{-1} and $E_A = 82.8$ kJ mol^{-1}, calculate, at 500 K,
  (i) the entropy of activation ($\Delta S^{o\ddagger}$/J mol^{-1} K^{-1}), and
  (ii) the enthalpy of activation ($\Delta H^{o\ddagger}$/kJ mol^{-1}).
(b) Comment on the value of $\Delta S^{o\ddagger}$ calculated.
(c) Corresponding to the value of $\Delta S^{o\ddagger}$ calculated in (a) for the transition state theory, would you expect the value of the steric factor p in the simple collision theory to be ≈ 1, > 1, or < 1? Explain briefly—detailed calculations or proofs are not necessary.

6-10 Show that, for the bimolecular reaction A + B $\rightarrow$ P, where A and B are hard spheres, k_{TST} is given by the same result as k_{SCT}, equation 6.4-17. A and B contain no internal modes, and the transition state is the configuration in which A and B are touching (at distance d_{AB} between centers). The partition functions for the reactants contain only translational modes (one factor in Q_r for each reactant), while the transition state has one translation mode and two rotational modes. The moment of inertia (I in Table 6.2) of the transition state (the two spheres touching) is μd_{AB}^2, where μ is reduced mass (equation 6.4-6).

Chapter 7

Homogeneous Reaction Mechanisms and Rate Laws

This chapter provides an introduction to several types of homogeneous (single-phase) reaction mechanisms and the rate laws which result from them. The concept of a reaction mechanism as a sequence of elementary processes involving both analytically detectable species (normal reactants and products) and transient reactive intermediates is introduced in Section 6.1.2. In constructing the rate laws, we use the fact that the elementary steps which make up the mechanism have individual rate laws predicted by the simple theories discussed in Chapter 6. The resulting rate law for an overall reaction often differs significantly from the type discussed in Chapters 3 and 4.

There are several benefits which arise from knowledge of the reaction mechanism. The first benefit of practical value is that the functional form of the rate law derived from the correct mechanism is more precise, enabling better reactor modeling and optimization, and more confident extrapolation to conditions outside the database. The second benefit is that a better understanding of the mechanism reveals the steps in the mechanism which limit the overall rate or selectivity in the reaction, and thus provides guidance to improve the process. Important examples where knowledge of the reaction mechanisms is critical can be found (1) in atmospheric-chemistry models, including the stratospheric ozone problem, air pollution, and nitrogen oxide formation in combustion, and (2) in an industrial process like ethane dehydrogenation, where detailed molecular models of the free-radical chemistry are required to predict the influence of feed composition and reactor parameters on product selectivity.

Constructing a reaction mechanism is a way of modeling a chemical reaction. There is no fixed set of rules to follow, but a proposed mechanism must be consistent with the overall stoichiometry and observed rate law. It is difficult to verify the mechanism of a given reaction. Testing the predicted rate laws against observations is a key step in gaining confidence in a proposed mechanism, but proof requires identifying the reaction intermediates (often in very small concentrations) under reaction conditions, or measurements of the kinetics of all the individual elementary reactions involving all the intermediates. Other techniques used to provide information about reaction mechanisms include isotope-substitution and stereochemical studies. Rate constants for many elementary chemical reactions have been measured. Despite the difficulty, an incomplete or imprecise mechanism which contains the essence of the reaction pathways is often more valuable than a purely empirical kinetics rate law.

7.1 SIMPLE HOMOGENEOUS REACTIONS

7.1.1 Types of Mechanisms

A reaction mechanism may involve one of two types of sequence, open or closed (Wilkinson, 1980, pp. 40, 176). In an open sequence, each reactive intermediate is produced in only one step and disappears in another. In a closed sequence, in addition to steps in which a reactive intermediate is initially produced and ultimately consumed, there are steps in which it is consumed and reproduced in a cyclic sequence which gives rise to a chain reaction. We give examples to illustrate these in the next sections. Catalytic reactions are a special type of closed mechanism in which the catalyst species forms reaction intermediates. The catalyst is regenerated after product formation to participate in repeated (catalytic) cycles. Catalysts can be involved in both homogeneous and heterogeneous systems (Chapter 8).

7.1.2 Open-Sequence Mechanisms: Derivation of Rate Law from Mechanism

The derivation of a rate law from a postulated mechanism is a useful application of reaction mechanisms. It shows how the kinetics of the elementary reaction steps are reflected in the kinetics of the overall reaction. The following example illustrates this for a simple, gas-phase reaction involving an open sequence. The derivations typically employ the stationary-state hypothesis (SSH) to eliminate unknown concentrations of reactive intermediates.

EXAMPLE 7-1

The decomposition of N_2O_5 to NO_2 and O_2 is a simple system (if we ignore dimerization of NO_2 to N_2O_4) and a first-order reaction:

$$2N_2O_5 \longrightarrow 4NO_2 + O_2 \quad \text{(A)}$$

$$r_{O_2} = k_{obs} c_{N_2O_5}$$

A proposed mechanism (Ogg, 1953) is as follows:

$$N_2O_5 \underset{k_{-1}}{\overset{k_1}{\rightleftharpoons}} NO_2 + NO_3 \quad (1)$$

$$NO_2 + NO_3 \overset{k_2}{\longrightarrow} NO + O_2 + NO_2 \quad (2)$$

$$NO + NO_3 \overset{k_3}{\longrightarrow} 2NO_2 \quad (3)$$

(a) Show how the mechanism can be made consistent with the observed (overall) stoichiometry.
(b) Derive the rate law for this mechanism so as to show consistency with the observed form, and to interpret k_{obs} in terms of the rate constants for the individual steps.
(c) Relate the experimental activation energy, $E_{A,obs}$, to the activation energies of the individual steps, if (i) step (2) is fast, and (ii) step (2) is the rate-determining step.

SOLUTION

(a) We note first that the reactive intermediates in the mechanism are NO_3 and NO, which do not appear either in the overall stoichiometry (reaction (A)) or in the observed rate law.

If we simply add the three steps, we do not recapture (A). To get around this, we introduce the *stoichiometric number*, s, for each step, as the number by which that step must be multiplied so that addition of the steps results in (A):

$$s_1(1) + s_2(2) + s_3(3) \equiv (A) \qquad (7.1\text{-}1)$$

where s_1, s_2, and s_3 are the stoichiometric numbers for the three steps. To determine their values systematically, we utilize the stoichiometric coefficients in the three steps for each species in turn so as to correspond to the coefficient in (A):

$$
\begin{aligned}
N_2O_5: & \quad -1s_1 + 0s_2 + 0s_3 \text{ (from the three steps)} = -2 \text{ (from (A))} \\
NO_2: & \quad 1s_1 + (-1+1)s_2 + 2s_3 = 4 \\
O_2: & \quad 0s_1 + 1s_2 + 0s_3 = 1 \\
NO_3: & \quad 1s_1 - 1s_2 - 1s_3 = 0 \\
NO: & \quad 0s_1 + 1s_2 - 1s_3 = 0
\end{aligned}
$$

This set of linear equations can be solved by inspection, or, more formally, by Gauss-Jordan reduction of the augmented coefficient matrix:

$$
\begin{pmatrix}
-1 & 0 & 0 & -2 \\
1 & 0 & 2 & 4 \\
0 & 1 & 0 & 1 \\
1 & -1 & -1 & 0 \\
0 & 1 & -1 & 0
\end{pmatrix}
\rightarrow
\begin{pmatrix}
1 & 0 & 0 & 2 \\
0 & 1 & 0 & 1 \\
0 & 0 & 1 & 1 \\
0 & 0 & 0 & 0 \\
0 & 0 & 0 & 0
\end{pmatrix}
$$

with the result $s_1 = 2$, $s_2 = 1$, and $s_3 = 1$. (Note that in this case the last two of the five equations are redundant in obtaining values of the three stoichiometric numbers.) Thus, the three steps are consistent with (A) if added as

$$2(1) + 1(2) + 1(3) \equiv (A)$$

(b) From the mechanism, step (2),

$$r_{O_2} = k_2 c_{NO_2} c_{NO_3} \qquad (B)$$

We eliminate c_{NO_3} (not allowed in the final rate law) by applying the stationary-state hypothesis to NO_3, $r_{NO_3} = 0$ (and subsequently to NO):

$$r_{NO_3} = k_1 c_{N_2O_5} - k_{-1} c_{NO_2} c_{NO_3} - k_2 c_{NO_2} c_{NO_3} - k_3 c_{NO} c_{NO_3} = 0 \qquad (C)$$

$$r_{NO} = k_2 c_{NO_2} c_{NO_3} - k_3 c_{NO} c_{NO_3} = 0 \qquad (D)$$

from (D),

$$c_{NO} = (k_2/k_3) c_{NO_2} \qquad (E)$$

from (C) and (E),

$$c_{NO_3} = \frac{k_1 c_{N_2O_5}}{(k_{-1} + 2k_2) c_{NO_2}} \qquad (F)$$

from (B) and (F),

$$r_{O_2} = \frac{k_1 k_2}{k_{-1} + 2k_2} c_{N_2O_5} \qquad (G)$$

Thus, the mechanism provides a first-order rate law with

$$k_{obs} = \frac{k_1 k_2}{k_{-1} + 2k_2} \quad \text{(H)}$$

(c) Note that, although a simple reaction order arises from this mechanism, the observed rate constant is a combination of elementary rate constants for steps (1) and (2), and can exhibit non-Arrhenius temperature dependence. The effective activation energy varies from one extreme, (i), in which step (2) is relatively fast (large k_2), to the other, (ii), in which step (2) is so slow (small k_2) as to be the rate-determining step (*rds*).

(i) In the first case, $k_2 \gg k_{-1}$, and equation (G) becomes

$$r_{O_2} = (k_1/2)c_{N_2O_5} \quad (k_2 \text{ large}) \quad \text{(J)}$$

with the result that the experimental activation energy is the same as that for forward step (1); that is, applying the Arrhenius equation, 3.1-6, to $k_{obs} = k_1/2$, we obtain

$$E_{A,obs} = E_{A1} \quad (k_2 \text{ large}) \quad \text{(K)}$$

(ii) In the other extreme, $k_2 \ll k_{-1}$, and equation (G) becomes

$$r_{O_2} = (k_1 k_2/k_{-1})c_{N_2O_5} \quad (k_2 \text{ small}) \quad \text{(L)}$$

This implies that step (1) is so rapid as to be in virtual equilibrium. Then, from equation 5.3-11 (with $n = 1$),

$$k_1/k_{-1} = K_{eq1} \quad \text{(M)}$$

where K_{eq1} is the equilibrium constant for step (1). From the Arrhenius equation, 3.1-6, applied to $k_{obs} = k_1 k_2/k_{-1} \equiv k_2 K_{eq1}$, we obtain

$$E_{A,obs} = E_{A2} + E_{A1} - E_{A,-1} \equiv E_{A2} + \Delta H_1 \quad \text{(N)}$$

where $E_{A,-1}$ and E_{A2} are the activation energies for reverse step (1) and step (2), respectively, and ΔH_1 is the enthalpy of reaction for step (1); the second part of equation (N) comes from the van't Hoff equation 3.1-5, $d \ln K_{eq1}/dT = \Delta H_1/RT^2$.

Many mechanisms involve reversible steps which are rapid (and therefore in virtual equilibrium) followed by the critical *rds*. In these cases, the equilibrium constant for each of the rapid steps appears as a multiplicative factor in the rate law. The effective activation energy is the sum of the enthalpies of the equilibrium steps and the activation energy of the *rds*.

7.1.3 Closed-Sequence Mechanisms; Chain Reactions

In some reactions involving gases, the rate of reaction estimated by the simple collision theory in terms of the usually inferred species is much *lower* than observed. Examples of these reactions are the oxidation of H_2 and of hydrocarbons, and the formation of HCl and of HBr. These are examples of chain reactions in which very reactive species (chain carriers) are initially produced, either thermally (i.e., by collision) or photochemically (by absorption of incident radiation), and regenerated by subsequent steps, so that reaction can occur in chain-fashion relatively rapidly. In extreme cases these become "explosions," but not all chain reactions are so rapid as to be termed explosions. The chain constitutes a closed sequence, which, if unbroken, or broken relatively infrequently, can result in a very rapid rate overall.

The experimental detection of a chain reaction can be done in a number of ways:

(1) The rate of a chain reaction is usually sensitive to the ratio of surface to volume in the reactor, since the surface serves to allow chain-breaking reactions (recombination of chain carriers) to occur. Thus, if powdered glass were added to a glass vessel in which a chain reaction occurred, the rate of reaction would decrease.

(2) The rate of a chain reaction is sensitive to the addition of any substance which reacts with the chain carriers, and hence acts as a chain breaker. The addition of NO sometimes markedly decreases the rate of a chain reaction.

Chain carriers are usually very reactive molecular fragments. Atomic species such as $H^\bullet$ and $Cl^\bullet$, which are electrically neutral, are in fact the simplest examples of "free radicals," which are characterized by having an unpaired electron, in addition to being electrically neutral. More complex examples are the methyl and ethyl radicals, $CH_3^\bullet$ and $C_2H_5^\bullet$, respectively.

Evidence for the existence of free-radical chains as a mechanism in chemical reactions was developed about 1930. If lead tetraethyl is passed through a heated glass tube, a metallic mirror of lead is formed on the glass. This is evidently caused by decomposition according to $Pb(C_2H_5)_4 \rightarrow Pb + 4C_2H_5^\bullet$, for if the ensuing gas passes over a previously deposited mirror, the mirror disappears by the reverse recombination: $4C_2H_5^\bullet + Pb \rightarrow Pb(C_2H_5)_4$. The connection with chemical reactions was made when it was demonstrated that the same mirror-removal action occurred in the thermal decomposition of a number of substances such as ethane and acetone, thus indicating the presence of free radicals during the decomposition. More recently, spectroscopic techniques using laser probes have made possible the in-situ detection of small concentrations of transient intermediates.

We may use the reaction mechanism for the formation of ethylene from ethane ($C_2H_6 \rightarrow C_2H_4 + H_2$), Section 6.1.2, to illustrate various types of steps in a typical chain reaction:

chain initiation:	$C_2H_6 \rightarrow 2CH_3^\bullet$	(1)
chain transfer:	$CH_3^\bullet + C_2H_6 \rightarrow CH_4 + C_2H_5^\bullet$	(2)
chain propagation:	$C_2H_5^\bullet \rightarrow C_2H_4 + H^\bullet$	(3)
	$H^\bullet + C_2H_6 \rightarrow H_2 + C_2H_5^\bullet$	(4)
chain breaking or termination:	$H^\bullet + C_2H_5^\bullet \rightarrow C_2H_6$	(5)

In the first step, $CH_3^\bullet$ radicals are formed by the rupture of the C–C bond in C_2H_6. However, $CH_3^\bullet$ is not postulated as a chain carrier, and so the second step is a chain-transfer step, from $CH_3^\bullet$ to $C_2H_5^\bullet$, one of the two chain carriers. The third and fourth steps constitute the chain cycle in which $C_2H_5^\bullet$ is first used up to produce one of the products (C_2H_4) and another chain carrier ($H^\bullet$), and then is reproduced, to continue the cycle, along with the other product (H_2). The last (fifth) step interrupts a chain by removing two chain carriers by recombination. For a rapid reaction overall, the chain propagation steps occur much more frequently than the others. An indication of this is given by the average chain length, CL:

$$CL = \frac{\text{number of (reactant) molecules reacting}}{\text{number of (reactant) molecules activated}}$$
$$= \text{rate of overall reaction/rate of initiation} \qquad (7.1\text{-}2)$$

7.1 Simple Homogeneous Reactions

Chain mechanisms may be classified as linear-chain mechanisms or branched-chain mechanisms. In a linear chain, one chain carrier is produced for each chain carrier reacted in the propagation steps, as in steps (3) and (4) above. In a branched chain, more than one carrier is produced. It is the latter that is involved in one type of explosion (a thermal explosion is the other type). We treat these types of chain mechanisms in turn in the next two sections.

7.1.3.1 Linear-Chain Mechanisms

We use the following two examples to illustrate the derivation of a rate law from a linear-chain mechanism.

EXAMPLE 7-2

(a) A proposed free-radical chain mechanism for the pyrolysis of ethyl nitrate, $C_2H_5ONO_2$ (A), to formaldehyde, CH_2O (B), and methyl nitrite, CH_3NO_2 (D), $A \rightarrow B + D$, is as follows (Houser and Lee, 1967):

$$A \xrightarrow{k_1} C_2H_5O^\bullet + NO_2 \quad (1)$$

$$C_2H_5O^\bullet \xrightarrow{k_2} CH_3^\bullet + B \quad (2)$$

$$CH_3^\bullet + A \xrightarrow{k_3} D + C_2H_5O^\bullet \quad (3)$$

$$2C_2H_5O^\bullet \xrightarrow{k_4} CH_3CHO + C_2H_5OH \quad (4)$$

Apply the stationary-state hypothesis to the free radicals $CH_3^\bullet$ and $C_2H_5O^\bullet$ to derive the rate law for this mechanism.

(b) Some of the results reported in the same investigation from experiments carried out in a CSTR at 250°C are as follows:

c_A/mol m^{-3}	0.0713	0.0759	0.0975	0.235	0.271
$(-r_A)$/mol m^{-3}s^{-1}	0.0121	0.0122	0.0134	0.0209	0.0230

Do these results support the proposed mechanism in (a)?

(c) From the result obtained in (a), relate the activation energy for the pyrolysis, E_A, to the activation energies for the four steps, E_{A1} to E_{A4}.

(d) Obtain an expression for the chain length CL.

SOLUTION

(a) The first step is the chain initiation forming the ethoxy free-radical chain carrier, $C_2H_5O^\bullet$, and NO_2, which is otherwise unaccounted for, taking no further part in the mechanism. The second and third steps are chain propagation steps in which a second chain carrier, the methyl free radical, $CH_3^\bullet$, is first produced along with the product formaldehyde (B) from $C_2H_5O^\bullet$, and then reacts with ethyl nitrate (A) to form the other product, methyl nitrite (D), and regenerate $C_2H_5O^\bullet$. The fourth step is a chain-breaking step, removing $C_2H_5O^\bullet$. In a chain reaction, addition of the chain-propagation steps typically gives the overall reaction. This may be interpreted in terms of stoichiometric numbers (see Example 7-1) by the assignment of the value 1 to the stoichiometric number for each propagation step and 0 to the other steps.

To obtain the rate law, we may use $(-r_A)$ or r_B or r_D. Choosing r_B, we obtain, from step (2),

$$r_B = k_2 c_{C_2H_5O^\bullet}$$

We eliminate $c_{C_2H_5O^\bullet}$ by applying the stationary-state hypothesis to $C_2H_5O^\bullet$, $r_{C_2H_5O^\bullet} = 0$, and also to the other chain carrier, $CH_3^\bullet$.

$$r_{C_2H_5O^\bullet} = k_1 c_A - k_2 c_{C_2H_5O^\bullet} + k_3 c_A c_{CH_3^\bullet} - 2k_4 c_{C_2H_5O^\bullet}^2 = 0$$

$$r_{CH_3^\bullet} = k_2 c_{C_2H_5O^\bullet} - k_3 c_A c_{CH_3^\bullet} = 0$$

Addition of these last two equations results in

$$c_{C_2H_5O^\bullet} = (k_1/2k_4)^{1/2} c_A^{1/2}$$

and substitution for $c_{C_2H_5O^\bullet}$ in the equation for r_B gives

$$r_B = k_2 (k_1/2k_4)^{1/2} c_A^{1/2}$$

which is the rate law predicted by the mechanism. According to this, the reaction is half-order.

(b) If we calculate the value of $k_{obs} = (-r_A)/c_A^{1/2}$ for each of the five experiments, we obtain an approximately constant value of 0.044 (mol m^{-3})$^{1/2}$ s^{-1}. Testing other reaction orders in similar fashion results in values of k_{obs} that are not constant. We conclude that the experimental results support the proposed mechanism.

(c) From (b), we also conclude that

$$k_{obs} = k_2 (k_1/2k_4)^{1/2}$$

from which

$$\frac{d \ln k_{obs}}{dT} = \frac{d \ln k_2}{dT} + \frac{1}{2}\frac{d \ln k_1}{dT} - \frac{1}{2}\frac{d \ln k_4}{dT}$$

or, from the Arrhenius equation, 3.1-6,

$$E_A = E_{A_2} + \frac{1}{2}(E_{A_1} - E_{A_4})$$

(d) From equation 7.1-2, the chain length is

$$CL = k_2 (k_1/2k_4)^{1/2} c_A^{1/2} / k_1 c_A$$
$$= k_2 (2 k_1 k_4 c_A)^{-1/2}$$

The rate law obtained from a chain-reaction mechanism is not necessarily of the power-law form obtained in Example 7-2. The following example for the reaction of H_2 and Br_2 illustrates how a more complex form (with respect to concentrations of reactants and products) can result. This reaction is of historical importance because it helped to establish the reality of the free-radical chain mechanism. Following the experimental determination of the rate law by Bodenstein and Lind (1907), the task was to construct a mechanism consistent with their results. This was solved independently by Christiansen, Herzfeld, and Polanyi in 1919–1920, as indicated in the example.

EXAMPLE 7-3

The gas-phase reaction between H_2 and Br_2 to form HBr is considered to be a chain reaction in which the chain is initiated by the thermal dissociation of Br_2 molecules. The chain is propagated first by reaction between Br$^\bullet$ and H$_2$ and second by reaction of H$^\bullet$ (released in the previous step) with Br$_2$. The chain is inhibited by reaction of HBr with H$^\bullet$ (i.e., HBr competes with Br$_2$ for H$^\bullet$). Chain termination occurs by recombination of Br$^\bullet$ atoms.

(a) Write the steps for a chain-reaction mechanism based on the above description.
(b) Derive the rate law (for r_{HBr}) for the mechanism in (a), stating any assumption made.

SOLUTION

(a) The overall reaction is

$$H_2 + Br_2 \rightarrow 2HBr$$

The reaction steps are:

$$\text{initiation:} \quad Br_2 \xrightarrow{k_1} 2Br^\bullet \tag{1}$$

$$\text{propagation:} \quad Br^\bullet + H_2 \xrightarrow{k_2} HBr + H^\bullet \tag{2}$$

$$H^\bullet + Br_2 \xrightarrow{k_3} HBr + Br^\bullet \tag{3}$$

$$\text{inhibition (reversal of (2)):} \quad H^\bullet + HBr \xrightarrow{k_{-2}} H_2 + Br^\bullet \tag{4}$$

$$\text{termination (reversal of (1)):} \quad 2\,Br^\bullet \xrightarrow{k_{-1}} Br_2 \tag{5}$$

(b) By constructing the expression for r_{HBr} from steps (2), (3), and (4), and then eliminating $c_{Br^\bullet}$ and $c_{H^\bullet}$ from this by means of the SSH ($r_{Br^\bullet} = r_{H^\bullet} = 0$), we obtain the rate law (see problem 7-5):

$$r_{HBr} = \frac{2k_3(k_2/k_{-2})(k_1/k_{-1})^{1/2} c_{H_2} c_{Br_2}^{1/2}}{(k_3/k_{-2}) + (c_{HBr}/c_{Br_2})} \tag{7.1-3}$$

This has the same form as that obtained experimentally by Bodenstein and Lind earlier.

This rate law illustrates several complexities:

(1) The effects on the rate of temperature (through the rate constants) and concentration are not separable, as they are in the power-law form of equation 6.1-1.
(2) Product inhibition of the rate is shown by the presence of c_{HBr} in the denominator.
(3) At a given temperature, although the rate is first-order with respect to H$_2$ at all conditions, the order with respect to Br$_2$ and HBr varies from low conversion ($k_3/k_{-2} \gg c_{HBr}/c_{Br_2}$), (1/2) order for Br$_2$ and zero order for HBr, to high conversion ($k_3/k_{-2} \ll c_{HBr}/c_{Br_2}$), (3/2) order for Br$_2$ and negative first-order for HBr. It was such experimental observations that led Bodenstein and Lind to deduce the form of equation 7.1-3 (with empirical constants replacing the groupings of rate constants).

7.1.3.2 Branched-Chain Mechanisms; Runaway Reactions (Explosions)

In a branched-chain mechanism, there are elementary reactions which produce more than one chain carrier for each chain carrier reacted. An example of such an elementary reaction is involved in the hydrogen-oxygen reaction:

$$O^\bullet + H_2 \rightarrow OH^\bullet + H^\bullet$$

Two radicals (OH$^\bullet$ and H$^\bullet$) are produced from the reaction of one radical (O$^\bullet$). This allows the reaction rate to increase without limit if it is not balanced by corresponding radical-destruction processes. The result is a "runaway reaction" or explosion. This can be demonstrated by consideration of the following simplified chain mechanism for the reaction A + ... → P.

$$\text{initiation:} \quad A \xrightarrow{k_1} R^\bullet$$
$$\text{chain branching:} \quad R^\bullet + A \xrightarrow{k_2} P + nR^\bullet \ (n > 1)$$

(If $n = 1$, this is a linear-chain step)

$$\text{termination:} \quad R^\bullet \xrightarrow{k_3} X$$

The rate of production of R$^\bullet$ is

$$\begin{aligned} r_{R^\bullet} &= k_1 c_A + (n-1)k_2 c_A c_{R^\bullet} - k_3 c_{R^\bullet} \\ &= k_1 c_A + [(n-1)k_2 c_A - k_3] c_{R^\bullet} \end{aligned} \quad (7.1\text{-}4)$$

A runaway reaction occurs if

$$\partial r_{R^\bullet}/\partial c_{R^\bullet} [= (n-1)k_2 c_A - k_3] > 0$$

$$\text{or } (n-1)k_2 c_A > k_3$$

which can only be the case if $n > 1$. In such a case, a rapid increase in $c_{R^\bullet}$ and in the overall rate of reaction ($r_P = k_2 c_A c_{R^\bullet}$) can take place, and an explosion results.

Note that the SSH *cannot* be applied to the chain carrier R$^\bullet$ in this branched-chain mechanism. If it *were* applied, we would obtain, setting $r_{R^\bullet} = 0$ in equation 7.1-4,

$$c_{R^\bullet}(\text{SSH}) = \frac{k_1 c_A}{k_3 - (n-1)k_2 c_A} < 0$$
$$\text{if } (n-1)k_2 c_A > k_3$$

which is a nonsensical result.

The region of unstable explosive behavior is influenced by temperature, in addition to pressure (concentration). The radical destruction processes generally have low activation energies, since they are usually recombination events, while the chain-branching reactions have high activation energies, since more species with incomplete bonding are produced. As a consequence, a system that is nonexplosive at low T becomes explosive above a certain threshold T. A species Y that interferes with a radical-chain mechanism by deactivating reactive intermediates (R$^\bullet$ + Y → Q) can be used (1) to increase the stability of a runaway system, (2) to quench a runaway system (e.g., act as a fire retardant), and (3) to slow undesirable reactions.

Another type of explosion is a thermal explosion. Instability in a reacting system can be produced if the energy of reaction is not transferred to the surroundings at a sufficient rate to prevent T from rising rapidly. A rise in T increases the reaction rate, which reinforces the rise in T. The resulting very rapid rise in reaction rate can cause an explosion. Most explosions that occur probably involve both chain-carrier and thermal instabilities.

7.1.4 Photochemical Reactions

In the mechanism of a photochemical reaction, at least one step involves photons. The most important such step is a reaction in which the absorption of light (ultraviolet or visible) provides a reactive intermediate by activating a molecule or atom. The mechanism is usually divided into primary photochemical steps and secondary processes that are initiated by the primary steps.

Consider as an example the use of mercury vapor in a photoactivated hydrocarbon process, and the following steps:

(1) Absorption of light to produce an energetically excited atom:

$$Hg + h\nu \rightarrow Hg^*$$

(2) Reaction of excited atom with a hydrocarbon molecule to produce a radical (desired):

$$Hg^* + RH \xrightarrow{k_2} HgH + R^\bullet$$

(3) Parallel (competing) reaction(s) in which excitation energy is lost (undesired):
(3a) Re-emission of energy as light (fluorescence):

$$Hg^* \xrightarrow{k_3} Hg + h\nu$$

(3b) Nonreactive energy transfer to another species (including reactant):

$$Hg^* + M \xrightarrow{k_4} Hg + M$$

(In a gas phase, the loss of energy requires collisions, whereas in a condensed phase, it can be considered a unimolecular process.)

The fraction of absorbed photons which results in the desired chemical step is called the *quantum yield*, Φ. In this case,

$$\Phi = \frac{k_2 c_{RH}}{k_2 c_{RH} + k_3 + k_4 c_M} \tag{7.1-5}$$

If all the re-emitted photons remain available to be reabsorbed (e.g., trapped by the use of mirrors),

$$\Phi = \frac{k_2 c_{RH}}{k_2 c_{RH} + k_4 c_M} \tag{7.1-6}$$

In this example, the Hg atom is the primary absorber of light. If the primary absorber is regenerated, it can participate in subsequent cycles, and is called a *photosensitizer*. In other cases, the photoactive species yields the active species directly. Thus, chlorine molecules can absorb light and dissociate into chlorine atomic radicals:

$$Cl_2 + h\nu \rightarrow 2Cl^\bullet$$

The competing process which determines Φ in this case is the recombination process:

$$2Cl^\bullet + M \rightarrow Cl_2 + M$$

164 Chapter 7: Homogeneous Reaction Mechanisms and Rate Laws

Re-emission of a photon in the reversal of the excitation step photodissociation is unimportant.

If the reactive species in the chemical activation step initiates a radical chain with a chain length CL, then the overall quantum yield based on the ultimate product is $\Phi \times CL$, and can be greater than 1. Photons are rather expensive reagents, and are only used when the product is of substantial value or when the overall quantum yield is large. Examples are the use of photoinitiators for the curing of coatings (a radical-polymerization process (Section 7.3.1)), and the transformation of complex molecules as medications.

Sources of radiation other than ultraviolet or visible light, such as high-energy ions, electrons, and much higher-energy photons, can also generate reactive species. Such processes are usually much less selective, however, since reactive fragments can be generated from all types of molecules. The individual absorption characteristics of molecules subjected to radiation in the ultraviolet and visible range lead to greater specificity.

7.2 COMPLEX REACTIONS

7.2.1 Derivation of Rate Laws

A complex reaction requires more than one chemical equation and rate law for its stoichiometric and kinetics description, respectively. It can be thought of as yielding more than one set of products. The mechanisms for their production may involve some of the same intermediate species. In these cases, their rates of formation are coupled, as reflected in the predicted rate laws.

For illustration, we consider a simplified treatment of methane oxidative coupling in which ethane (desired product) and CO_2 (undesired) are produced (Mims et al., 1995). This is an example of the effort (so far not commercially feasible) to convert CH_4 to products for use in chemical syntheses (so-called "C_1 chemistry"). In this illustration, both C_2H_6 and CO_2 are stable primary products (Section 5.6.2). Both arise from a common intermediate, $CH_3^{\bullet}$, which is produced from CH_4 by reaction with an oxidative agent, MO. Here, MO is treated as another gas-phase molecule, although in practice it is a solid. The reaction may be represented by parallel steps as in Figure 7.1(a), but a mechanism for it is better represented as in Figure 7.1(b).

A mechanism corresponding to Figure 7.1(b) is:

$$CH_4 + MO \xrightarrow{k_1} CH_3^{\bullet}(+\text{reduced MO})$$

$$2CH_3^{\bullet} \xrightarrow{k_2} C_2H_6$$

$$CH_3^{\bullet} + MO \xrightarrow{k_3} P \xrightarrow{\text{fast}} CO_2(+\text{reduced MO})$$

Application of the SSH to $CH_3^{\bullet}$ results in the *two* rate laws (see problem 7-12):

$$r_{C_2H_6} = k_2 c_{CH_3^{\bullet}}^2 = \frac{\{[(k_3 c_{MO})^2 + 8k_1 k_2 c_{MO} c_{CH_4}]^{1/2} - k_3 c_{MO}\}^2}{16 k_2} \quad (7.2\text{-}1)$$

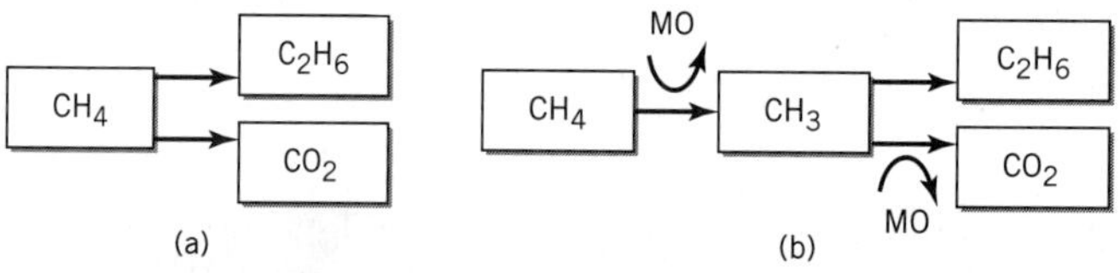

Figure 7.1 Representations of CH_4 oxidative-coupling reaction to produce C_2H_6 and CO_2

$$r_{CO_2} = k_3 c_{MO} c_{CH_3^\bullet} = \frac{(k_3 c_{MO})^2 \{[1 + 8k_1 k_2 c_{CH_4}/(k_3^2 c_{MO})]^{1/2} - 1\}}{4k_2} \quad (7.2\text{-}2)$$

Furthermore, the rate of disappearance of CH_4 is

$$(-r_{CH_4}) = 2r_{C_2H_6} + r_{CO_2} = k_1 c_{MO} c_{CH_4} \quad (7.2\text{-}3)$$

which is also the limiting rate for either product, if the competing reaction is completely suppressed.

7.2.2 Computer Modeling of Complex Reaction Kinetics

In the examples in Sections 7.1 and 7.2.1, explicit analytical expressions for rate laws are obtained from proposed mechanisms (except branched-chain mechanisms), with the aid of the SSH applied to reactive intermediates. In a particular case, a rate law obtained in this way can be used, if the Arrhenius parameters are known, to simulate or model the reaction in a specified reactor context. For example, it can be used to determine the concentration–(residence) time profiles for the various species in a BR or PFR, and hence the product distribution. It may be necessary to use a computer-implemented numerical procedure for integration of the resulting differential equations. The software package E-Z Solve can be used for this purpose.

It may not be possible to obtain an explicit rate law from a mechanism even with the aid of the SSH. This is particularly evident for complex systems with many elementary steps and reactive intermediates. In such cases, the numerical computer modeling procedure is applied to the full set of differential equations, including those for the reactive intermediates; that is, it is not necessary to use the SSH, as it is in gaining the advantage of an analytical expression in an approximate solution. Computer modeling of a reacting system in this way can provide insight into its behavior; for example, the effect of changing conditions (feed composition, T, etc.) can be studied. In modeling the effect of man-made chemicals on atmospheric chemistry, where reaction-coupling is important to the net effect, hundreds of reactions can be involved. In modeling the kinetics of ethane dehydrogenation to produce ethylene, the relatively simple mechanism given in Section 6.1.2 needs to be expanded considerably to account for the formation of a number of coproducts; even small amounts of these have significant economic consequences because of the large scale of the process. The simulation of systems such as these can be carried out with E-Z Solve or more specific-purpose software. For an example of the use of *CHEMKIN*, an important type of the latter, see Mims et al. (1994).

The inverse problem to simulation from a reaction mechanism is the determination of the reaction mechanism from observed kinetics. The process of building a mechanism is an interactive one, with successive changes followed by experimental testing of the model predictions. The purpose is to be able to explain why a reacting system behaves the way it does in order to control it better or to improve it (e.g., in reactor performance).

7.3 POLYMERIZATION REACTIONS

Because of the ubiquitous nature of polymers and plastics (synthetic rubbers, nylon, polyesters, polyethylene, etc.) in everyday life, we should consider the kinetics of their formation (the focus here is on kinetics; the significance of some features of kinetics in relation to polymer characteristics for reactor selection is treated in Chapter 18).

Polymerization, the reaction of monomer to produce polymer, may be self-polymerization (e.g., ethylene monomer to produce polyethylene), or copolymerization (e.g., styrene monomer and butadiene monomer to produce SBR type of synthetic rubber). These may both be classified broadly into chain-reaction polymerization and step-reaction (condensation) polymerization. We consider a simple model of each, by way of introduction to the subject, but the literature on polymerization and polymerization kinetics is very extensive (see, e.g., Billmeyer, 1984). Many polymerization reactions are catalytic.

7.3.1 Chain-Reaction Polymerization

Chain-reaction mechanisms differ according to the nature of the reactive intermediate in the propagation steps, such as free radicals, ions, or coordination compounds. These give rise to radical-addition polymerization, ionic-addition (cationic or anionic) polymerization, etc. In Example 7-4 below, we use a simple model for radical-addition polymerization.

As for any chain reaction, radical-addition polymerization consists of three main types of steps: initiation, propagation, and termination. Initiation may be achieved by various methods: from the monomer thermally or photochemically, or by use of a free-radical initiator, a relatively unstable compound, such as a peroxide, that decomposes thermally to give free radicals (Example 7-4 below). The rate of initiation (r_{init}) can be determined experimentally by labeling the initiator radioactively or by use of a "scavenger" to react with the radicals produced by the initiator; the rate is then the rate of consumption of the initiator. Propagation differs from previous consideration of linear chains in that there is no recycling of a chain carrier; polymers may grow by addition of monomer units in successive steps. Like initiation, termination may occur in various ways: combination of polymer radicals, disproportionation of polymer radicals, or radical transfer from polymer to monomer.

EXAMPLE 7-4

Suppose the chain-reaction mechanism for radical-addition polymerization of a monomer M (e.g., CH_2CHCl), which involves an initiator I (e.g., benzoyl peroxide), at low concentration, is as follows (Hill, 1977, p. 124):

$$\text{initiation:} \quad I \xrightarrow{k_d} 2\,R^\bullet \quad (1)$$

$$R^\bullet + M \xrightarrow{k_i} P_1^\bullet \quad (2)$$

$$\text{propagation:} \quad P_1^\bullet + M \xrightarrow{k_p} P_2^\bullet \quad (P1)$$

$$P_2^\bullet + M \xrightarrow{k_p} P_3^\bullet \quad (P2)$$

$$\vdots$$

$$P_{r-1}^\bullet + M \xrightarrow{k_p} P_r^\bullet \quad (Pr)$$

$$\vdots$$

$$\text{termination:} \quad P_k^\bullet + P_\ell^\bullet \xrightarrow{k_t} P_{k+\ell} \quad k, \ell = 1, 2, \ldots \quad (3)$$

in which it is assumed that rate constant k_p is the same for all propagation steps, and k_t is the same for all termination steps; $P_{k+\ell}$ is the polymer product; and $P_r^\bullet$, $r = 1, 2, \ldots$, is a radical, the growing polymer chain.

(a) By applying the stationary-state hypothesis (SSH) to each radical species (including $R^\bullet$), derive the rate law for the rate of disappearance of monomer, $(-r_M)$, for the mechanism above, in terms of the concentrations of I and M, and f, the efficiency of utilization of the $R^\bullet$ radicals; f is the fraction of $R^\bullet$ formed in (1) that results in initiating chains in (2).

(b) Write the special cases for $(-r_M)$ in which (i) f is constant; (ii) $f \propto c_M$; and (iii) $f \propto c_M^2$.

SOLUTION

(a)

$$r_{R\bullet} = 2fk_d c_I - k_i c_{R\bullet} c_M = 0 \tag{4}$$

$$r_{init}(= r[\text{step (2)}]) = k_i c_{R\bullet} c_M = 2fk_d c_I \text{ [from (4)]} \tag{5}$$

$$r_{P_1^\bullet} = r_{init} - k_p c_M c_{P_1^\bullet} - k_t c_{P_1^\bullet} \sum_{k=1}^{\infty} c_{P_k^\bullet} = 0 \tag{6}$$

where the last term is from the rate of termination according to step (3). Similarly,

$$r_{P_2^\bullet} = k_p c_M c_{P_1^\bullet} - k_p c_M c_{P_2^\bullet} - k_t c_{P_2^\bullet} \sum_{k=1}^{\infty} c_{P_k^\bullet} = 0 \tag{7}$$

$$\cdots$$

$$r_{P_r^\bullet} = k_p c_M c_{P_{r-1}^\bullet} - k_p c_M c_{P_r^\bullet} - k_t c_{P_r^\bullet} \sum_{k=1}^{\infty} c_{P_k^\bullet} = 0 \tag{8}$$

From the summation of (6), (7), ..., (8), with the assumption that $k_p c_M c_{P_r^\bullet}$ is relatively small (since $c_{P_r^\bullet}$ is very small),

$$r_{init} = k_t \left(\sum_{k=1}^{\infty} c_{P_k^\bullet} \right)^2 \tag{9}$$

which states that the rate of initiation is equal to the rate of termination. For the rate law, the rate of polymerization, the rate of disappearance of monomer, is

$$(-r_M) = r_{init} + k_p c_M \sum_{k=1}^{\infty} c_{P_k^\bullet}$$

$$= k_p c_M \sum_{k=1}^{\infty} c_{P_k^\bullet} \quad [\text{if } r_{init} \ll (-r_M)]$$

$$= k_p c_M (r_{init}/k_t)^{1/2} \quad [\text{from (9)}]$$

$$= k_p c_M (2fk_d c_I/k_t)^{1/2} \quad [\text{from (5)}]$$

We write this finally as

$$(-r_M) = k f^{1/2} c_I^{1/2} c_M \tag{7.3-1}$$

where $\quad k = k_p (2k_d/k_t)^{1/2} \tag{7.3-2}$

(b)

(i) $\quad (-r_M) = k' c_I^{1/2} c_M \tag{7.3-1a}$

(ii) $\quad (-r_M) = k'' c_I^{1/2} c_M^{3/2} \tag{7.3-1b}$

(iii) $\quad (-r_M) = k''' c_I^{1/2} c_M^2 \tag{7.3-1c}$

7.3.2 Step-Change Polymerization

Consider the following mechanism for step-change polymerization of monomer M (P_1) to $P_2, P_3, \ldots, P_r, \ldots$. The mechanism corresponds to a complex series-parallel scheme: series with respect to the growing polymer, and parallel with respect to M. Each step is a second-order elementary reaction, and the rate constant k (defined for each *step*)[1] is the same for all steps.

$$M + M \xrightarrow{k} P_2 \qquad (1)$$

$$M + P_2 \xrightarrow{k} P_3 \qquad (2)$$

$$\cdot \quad \cdot \quad \cdot$$

$$M + P_{r-1} \xrightarrow{k} P_r \qquad (r-1)$$

$$\cdot \quad \cdot \quad \cdot$$

where r is the number of monomer units in the polymer. This mechanism differs from a chain-mechanism polymerization in that there are no initiation or termination steps. Furthermore, the species P_2, P_3, etc. are product species and not reactive intermediates. Therefore, we cannot apply the SSH to obtain a rate law for the disappearance of monomer (as in the previous section for equation 7.3-1), independent of c_{P_2}, c_{P_3}, etc.

From the mechanism above, the rate of disappearance of monomer, $(-r_M)$, is

$$\begin{aligned}(-r_M) &= 2kc_M^2 + kc_M c_{P_2} + \ldots + kc_M c_{P_r} + \ldots \\ &= kc_M\left(2c_M + \sum_{r=2}^{\infty} c_{P_r}\right)\end{aligned} \qquad (7.3\text{-}3)$$

The rates of appearance of dimer, trimer, etc. correspondingly are

$$r_{P_2} = kc_M(c_M - c_{P_2}) \qquad (7.3\text{-}4)$$

$$r_{P_3} = kc_M(c_{P_2} - c_{P_3}) \qquad (7.3\text{-}5)$$

$$\cdot \quad \cdot \quad \cdot$$

$$r_{P_r} = kc_M(c_{P_{r-1}} - c_{P_r}), \text{ etc.} \qquad (7.3\text{-}6)$$

These rate laws are coupled through the concentrations. When combined with the material-balance equations in the context of a particular reactor, they lead to uncoupled equations for calculating the product distribution. For a constant-density system in a CSTR operated at steady-state, they lead to algebraic equations, and in a BR or a PFR at steady-state, to simultaneous nonlinear ordinary differential equations. We demonstrate here the results for the CSTR case.

For the CSTR case, illustrated in Figure 7.2, suppose the feed concentration of monomer is c_{Mo}, the feed rate is q, and the reactor volume is V. Using the material-balance equation 2.3-4, we have, for the monomer:

$$c_{Mo}q - c_M q + r_M V = 0$$

---

[1] The interpretation of k as a *step* rate constant (see equations 1.4-8 and 4.1-3) was used by Denbigh and Turner (1971, p. 123). The interpretation of k as the *species* rate constant k_M was used subsequently by Denbigh and Turner (1984, p. 125). Details of the consequences of the model, both here and in Chapter 18, differ according to which interpretation is made. In any case, we focus on the use of the model in a general sense, and not on the correctness of the interpretation of k.

7.3 Polymerization Reactions

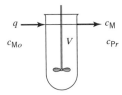

Figure 7.2 Polymerization of monomer M in a CSTR at steady-state

or

$$(-r_M) = (c_{Mo} - c_M)/(V/q) = (c_{Mo} - c_M)/\tau \qquad (7.3\text{-}7)$$

where τ is the space time.

Similarly, for the dimer, P_2,

$$0 - c_{P_2}q + r_{P_2}V = 0$$

or

$$r_{P_2} = c_{P_2}/\tau = kc_M(c_M - c_{P_2}) \qquad \text{(from 7.3-4)}$$

and

$$c_{P_2} = kc_M\tau(c_M - c_{P_2}) \qquad (7.3\text{-}8)$$

Similarly, it follows that

$$c_{P_3} = kc_M\tau(c_{P_2} - c_{P_3}) \qquad (7.3\text{-}9)$$

$$\cdot \quad \cdot \quad \cdot$$

$$c_{P_r} = kc_M\tau(c_{P_{r-1}} - c_{P_r}) \qquad (7.3\text{-}10)$$

and, thus, on summing 7.3-8 to 7.3-10, we obtain

$$\sum_{r=2}^{\infty} c_{P_r} = kc_M\tau(c_M - c_{P_2} + c_{P_2} - c_{P_3} + c_{P_3} + \ldots - c_{P_{r-1}} + c_{P_{r-1}} - c_{P_r})$$

$$= kc_M\tau(c_M - c_{P_r}) = kc_M^2\tau \qquad (7.3\text{-}11)$$

since $c_{P_r} \to 0$ as $r \to \infty$.

Substitution of 7.3-7 and -11 in 7.3-3 results in

$$c_{Mo} - c_M = kc_M\tau(2c_M + kc_M^2\tau) \qquad (7.3\text{-}11a)$$

from which a cubic equation in c_M arises:

$$c_M^3 + (2/k\tau)c_M^2 + (1/k^2\tau^2)c_M - c_{Mo}/k^2\tau^2 = 0 \qquad (7.3\text{-}12)$$

Solution of equation 7.3-12 for c_M leads to the solution for c_{P_2}, c_{P_3}, etc.:
From equation 7.3-8,

$$c_{P_2} = \frac{kc_M^2\tau}{1 + kc_M\tau} \qquad (7.3\text{-}13)$$

Similarly, from 7.3-9 and -13,

$$c_{P_3} = \frac{kc_M \tau c_{P_2}}{1 + kc_M \tau} = \frac{c_M(kc_M\tau)^2}{(1 + kc_M\tau)^2} \quad (7.3\text{-}14)$$

Proceeding in this way, from 7.3-10, we obtain in general:

$$c_{P_r} = \frac{c_M(kc_M\tau)^{r-1}}{(1 + kc_M\tau)^{r-1}} = c_M[1 + (kc_M\tau)^{-1}]^{1-r} \quad (7.3\text{-}15)$$

Thus, the product distribution (distribution of polymer species P_r) leaving the CSTR can be calculated, if c_{M_o}, k, and τ are known.

For a BR or a PFR in steady-state operation, corresponding differential equations can be established to obtain the product distribution (problem 7-15).

7.4 PROBLEMS FOR CHAPTER 7

7-1 The rate of production of urea, $(NH_2)_2CO$, from ammonium cyanate increases by a factor of 4 when the concentration of ammonium cyanate is doubled. Show whether this is accounted for by the following mechanism:

$$NH_4^+ + CNO^- \underset{}{\overset{K_{eq}}{\rightleftharpoons}} NH_3 + HNCO; \text{ fast}$$

$$NH_3 + HNCO \overset{k}{\longrightarrow} (NH_2)_2CO; \text{ slow}$$

Note that ammonium cyanate is virtually completely dissociated in solution.

7-2 What rate law (in terms of r_{O_2}) is predicted for the reaction

$$2O_3 \rightarrow 3O_2$$

from the following mechanism:

$$O_3 \underset{k_{-1}}{\overset{k_1}{\rightleftharpoons}} O_2 + O^\bullet$$

$$O^\bullet + O_3 \overset{k_2}{\longrightarrow} 2O_2$$

Clearly state any assumption(s) made.

7-3 The gas-phase reaction between nitric oxide and hydrogen, which can be represented stoichiometrically by

$$2NO + 2H_2 = N_2 + 2H_2O$$

is a third-order reaction with a rate law given by

$$(-r_{NO}) = k_{NO} c_{NO}^2 c_{H_2}$$

(a) If the species $(NO)_2$ and H_2O_2 are allowed as reactive intermediates, construct a reaction mechanism in terms of elementary processes or steps. Clearly indicate any features such as equilibrium, and "fast" and rate-determining ("slow") steps. Use only *bimolecular* steps.

(b) Derive the rate law from the mechanism constructed to show that it is consistent with the observed order of reaction.

(c) Express k_{NO} in terms of the constants in the rate law derived.

7-4 The oxidation of NO to NO_2, which is an important step in the manufacture of nitric acid by the ammonia-oxidation process, is an unusual reaction in having an observed *third-order* rate constant (k_{NO} in $(-r_{NO}) = k_{NO}c_{NO}^2 c_{O_2}$) which *decreases* with increase in temperature. Show how the order and sign of temperature dependence could be accounted for by a simple mechanism which involves the formation of $(NO)_2$ in a rapidly established equilibrium, followed by a relatively slow bimolecular reaction of $(NO)_2$ with O_2 to form NO_2.

7-5 (a) Verify the rate law obtained in Example 7-3, equation 7.1-3.
(b) For the $H_2 + Br_2$ reaction in Example 7-3, if the initiation and termination steps involve a third body (M), $Br_2 + M \rightarrow 2Br^\bullet + M$, and $2Br + M \rightarrow Br_2 + M$, respectively, what effect does this have on the rate law in equation 7.1-3? (The other steps remain as in Example 7-3.)

7-6 The rate of decomposition of ethylene oxide, $C_2H_4O(A)$, to CH_4 and CO, has been studied by Crocco et al. (1959) at 900–1200 K in a flow reactor. They found the rate constant to be given by

$$k_A = 10^{11} \exp(-21{,}000/T)$$

in s^{-1} (with T in K). They proposed a free-radical chain mechanism which involves the initial decomposition of C_2H_4O into radicals ($C_2H_3O^\bullet$ and $H^\bullet$), and propagation steps which involve the radicals $C_2H_3O^\bullet$ and $CH_3^\bullet$ (but not $H^\bullet$) in addition to the reactant and products; termination involves recombination of the chain carriers to form products that can be ignored.

(a) Write the following:
  (i) an equation for the overall stoichiometry;
  (ii) the initiation step in the mechanism;
  (iii) the propagation steps;
  (iv) the termination step.
(b) Derive the rate law from the steps of the mechanism, and state whether the form agrees with that observed. Clearly state any assumption(s) made.
(c) Estimate the activation energy (E_{A1}) for the initiation step, if the *sum* of the activation energies for the propagation steps is 126,000 J mol^{-1}, and E_A for the termination step is 0.

7-7 Suppose the mechanism for the thermal decomposition of dimethyl ether to methane and formaldehyde

$$CH_3OCH_3 \rightarrow CH_4 + HCHO \qquad (A)$$

is a chain reaction as follows:

$$CH_3OCH_3 \xrightarrow{k_1} CH_3^\bullet + OCH_3^\bullet \qquad E_1$$

$$CH_3^\bullet + CH_3OCH_3 \xrightarrow{k_2} CH_4 + CH_2OCH_3^\bullet \qquad E_2$$

$$CH_2OCH_3^\bullet \xrightarrow{k_3} CH_3^\bullet + HCHO \qquad E_3$$

$$CH_3^\bullet + CH_2OCH_3^\bullet \xrightarrow{k_4} CH_3CH_2OCH_3 \qquad E_4$$

(a) Show how the mechanism is consistent with the stoichiometry for (A).
(b) Identify any apparent deficiencies in the mechanism, and how these are allowed for by the result in (a).
(c) Derive the rate law from the mechanism, clearly justifying any assumption(s) made to simplify it.
(d) Relate the activation energy, E_A, of the reaction (A) to the activation energies of the individual steps.

7-8 A possible free-radical chain mechanism for the thermal decomposition of acetaldehyde (to CH_4 and CO) is the Rice-Herzfeld mechanism (Laidler and Liu, 1967):

$$CH_3CHO \xrightarrow{k_1} CH_3^\bullet + CHO^\bullet$$

$$CHO^\bullet \xrightarrow{k_2} CO + H^\bullet$$

$$H^\bullet + CH_3CHO \xrightarrow{k_3} CH_3CO^\bullet + H_2$$

$$CH_3^\bullet + CH_3CHO \xrightarrow{k_4} CH_4 + CH_3CO^\bullet$$

$$CH_3CO^\bullet \xrightarrow{k_5} CH_3^\bullet + CO$$

$$2CH_3^\bullet \xrightarrow{k_6} C_2H_6$$

(a) Which species are the *chain carriers*?
(b) Classify each step in the mechanism.
(c) Derive the rate law from the mechanism for $CH_3CHO \to CH_4 + CO$, and state the order of reaction predicted. Assume H_2 and C_2H_6 are *minor* species.

7-9 From the mechanism given in problem 7-8 for the decomposition of acetaldehyde, derive a rate law or set of independent rate laws, as appropriate, if H_2 and C_2H_6 are *major* products (in addition to CH_4 and CO).

7-10 From the mechanism given in Section 6.1.2 for the dehydrogenation of C_2H_6, obtain the rate law for $C_2H_6 \to C_2H_4 + H_2$ (assign rate constants $k_1, \ldots, k_5$ to the five steps in the order given, and assume CH_4 is a *minor* product).

7-11 Repeat problem 7-10 for a rate law or set of independent rate laws, as appropriate, if CH_4 is a *major* product.

7-12 (a) For the CH_4 oxidative-coupling mechanism described in Section 7.2, verify the rate laws given in equations 7.2-1 and -2, and show that 7.2-3 is consistent with these two equations.
(b) Show (i) that equation 7.2-1 reduces to 7.2-3 if CO_2 is not formed; and (ii) that 7.2-2 reduces to 7.2-3 if C_2H_6 is not formed.
(c) From the rate laws in (a), derive an expression for the *instantaneous* fractional yield (selectivity) of C_2H_6 (with respect to CH_4).
(d) Does the selectivity in (c) increase or decrease with increase in c_{MO}?

7-13 In a certain radical-addition polymerization reaction, based on the mechanism in Example 7-4, in which an initiator, I, is used, suppose measured values of the rate, $(-r_M)$, at which monomer, M, is used up at various concentrations of monomer, c_M, and initiator, c_I, are as follows (Hill, 1977, p. 125):

c_M/kmol m^{-3}	c_I/mol m^{-3}	$(-r_M)$/mol m^{-3} s^{-1}
9.04	0.235	0.193
8.63	0.206	0.170
7.19	0.255	0.165
6.13	0.228	0.129
4.96	0.313	0.122
4.75	0.192	0.0937
4.22	0.230	0.0867
4.17	0.581	0.130
3.26	0.245	0.0715
2.07	0.211	0.0415

(a) Determine the values of k and n in the rate law $(-r_M) = k c_I^{1/2} c_M^n$.

(b) What is the order of the dependence of the efficiency (f) of radical conversion to P_1 on c_M?

7-14 In the comparison of organic peroxides as free-radical polymerization initiators, one of the measures used is the temperature (T) required for the half-life ($t_{1/2}$) to be 10 h. If it is desired to have a *lower T*, would $t_{1/2}$ be greater or smaller than 10 h? Explain briefly.

7-15 Starting from equations 7.3-3 to -6 applied to a constant-volume BR, for polymerization represented by the step-change mechanism in Section 7.3.2, show that the product distribution can be calculated by sequentially solving the differential equations:

$$\frac{d^2 c_M}{dt^2} - \frac{1}{c_M}\left(\frac{dc_M}{dt}\right)^2 + 2kc_M \frac{dc_M}{dt} + k^2 c_M^3 = 0 \quad (7.3\text{-}16)$$

$$\frac{dc_{P_r}}{dt} + kc_M c_{P_r} = kc_M c_{P_{r-1}} \ ; \ r = 2, 3, \ldots \quad (7.3\text{-}17)$$

7-16 This problem is an extension of problems 7-10 and 7-11 on the dehydrogenation of ethane to produce ethylene. It can be treated as an open-ended, more realistic exercise in reaction mechanism investigation. The choice of reaction steps to include, and many aspects of elementary gas-phase reactions discussed in Chapter 6 (including energy transfer) are significant to this important industrial reaction. Solution of the problem requires access to a computer software package which can handle a moderately stiff set of simultaneous differential equations. E-Z Solve may be used for this purpose.

(a) Use the mechanism in Section 6.1.2 and the following values of the rate constants (units of mol, L, s, J, K):

(1) $C_2H_6 \rightarrow 2CH_3^\bullet$; $\quad k_1 = 5 \times 10^{14} \exp(-334000/RT)$
(2) $CH_3^\bullet + C_2H_6 \rightarrow C_2H_5^\bullet + CH_4$; $\quad k_2 = 4 \times 10^{13} \exp(-70300/RT)$
(3) $C_2H_5^\bullet \rightarrow C_2H_4 + H^\bullet$; $\quad k_3 = 5.7 \times 10^{11} \exp(-133000/RT)$
(4) $H^\bullet + C_2H_6 \rightarrow C_2H_5^\bullet + H_2$; $\quad k_4 = 7.4 \times 10^{14} \exp(-52800/RT)$
(5) $H^\bullet + C_2H_5^\bullet \rightarrow C_2H_6$; $\quad k_5 = 3.2 \times 10^{13}$

 (i) Solve for the concentration of C_2H_5 radicals using the SSH, and obtain an expression for the rate of ethylene production.
 (ii) Obtain a rate expression for methane production as well as an expression for the reaction chain length.
 (iii) Integrate these rate expressions to obtain ethane conversion and product distribution for a residence time (t) of 1 s at 700°C (1 bar, pure C_2H_6). Assume an isothermal, constant-volume batch reactor, although the industrial reaction occurs in a flow system with temperature change and pressure drop along the reactor.
 (iv) From initial rates, what is the reaction order with respect to ethane?
 (v) What is the overall activation energy?

(b) Integrate the full set of differential equations.
 (i) Compare the conversion and integral selectivities in this calculation with those in part (a).
 (ii) Compare the ethyl radical concentrations calculated in the simulation with those predicted by the SSH.
 (iii) Approximately how long does it take for the ethyl radicals to reach their pseudo-steady-state values in this calculation?
 (iv) Run two different simulations with different ethane pressures and take the initial rates (evaluated at 100 ms) to obtain a reaction order. Compare with part (a).
 (v) Run two different simulations with two different temperatures: take the initial rates (evaluated at 3% conversion) and calculate the activation energy. Compare with the answer from part (a).

(c) At temperatures near 700°C and pressures near 1 bar, the overall reaction rate is observed to be first-order in ethane pressure with a rate constant $k = 1.1 \times 10^{15} \exp(-306000/RT)$. How well does this model reproduce these results?

(d) Now improve the model and test the importance of other reactions by including them in the computer model and examining the results. Use the following cases.

(d1) Reversible reaction steps.

(i) Include the reverse of step (3) in the mechanism and rerun the simulation—does it affect the calculated rates?

(6) $\qquad C_2H_4 + H^\bullet \rightarrow C_2H_5^\bullet \, ; \qquad k_6 = 10^{13}$

(ii) How else might one estimate the significance of this reaction without running the simulation again?

(d2) Steps involving energy transfer.

How many of the reactions in this mechanism might be influenced by the rate of energy transfer? One of them is the termination step, which can be thought of as a three-step process (reactions (7) to (9) below). As described in Section 6.4.3, there are possible further complications, since two other product channels are possible (reactions (10) and (11)).

(7) $\qquad C_2H_5 + H^\bullet \rightarrow C_2H_6^* \, ; \qquad k_7 = 6 \times 10^{13}$
(8) $\qquad C_2H_6^* + M \rightarrow C_2H_6 + M \, ; \qquad k_8 = 3 \times 10^{13}$
(9) $\qquad C_2H_6^* \rightarrow C_2H_5^\bullet + H^\bullet \, ; \qquad k_9 = 2 \times 10^{13}$
(10) $\qquad C_2H_6^* \rightarrow 2CH_3^\bullet \, ; \qquad k_{10} = 3 \times 10^{12}$
(11) $\qquad C_2H_6^* \rightarrow C_2H_4 + H_2 \, ; \qquad k_{11} = 3 \times 10^{12}$

Include these reactions in the original model in place of the original reaction (5). (You can assume that M is an extra species at the initial ethane concentration for this simulation.) Use the values of the rate constants indicated, and run the model simulation. What influence does this chemistry have on the conversion and selectivity? How would you estimate the rate constants for these reactions?

(d3) The initiation step.

The initiation step also requires energy input.

(12) $\qquad C_2H_6 + M \rightarrow C_2H_6^* + M \, ; \qquad k_{12} = 2 \times 10^{13} \exp(-340,000/RT)$

The other reactions, (8) and (10), have already been included.
At 1 bar and 700°C, is this reaction limited by energy transfer (12) or by decomposition (10)?

(d4) Termination steps.

Termination steps involving two ethyl radicals are also ignored in the original mechanism. Include the following reaction:

(13) $\qquad 2\,C_2H_5^\bullet \rightarrow C_2H_4 + C_2H_6 \, ; \qquad k_{13} = 6 \times 10^{11}$

Does this make a significant difference? Could you have predicted this result from the initial model calculation?

(d5) Higher molecular-weight products.

Higher molecular-weight products also are made. While this is a complex process, estimate the importance of the following reaction to the formation of higher hydrocarbons by including it in the model and calculating the C_4H_8 product selectivity.

(14) $\quad C_2H_4 + C_2H_5^{\bullet} \rightarrow C_4H_8 + H^{\bullet}$; $\quad k_{14} = 2 \times 10^{11}$

Plot the selectivity to C_4H_8 as a function of ethane conversion. Does it behave like a secondary or primary product? Consult the paper by Dean (1990), and describe additional reactions which lead to molecular weight growth in hydrocarbon pyrolysis systems. While some higher molecular weight products are valuable, the heavier tars are detrimental to the process economics.

Much of the investigation you have been doing was described originally by Wojciechowski and Laidler (1960), and by Laidler and Wojciechowski (1961). Compare your findings with theirs.

Chapter 8

Catalysis and Catalytic Reactions

Many reactions proceed much faster in the presence of a substance that is not a product (or reactant) in the usual sense. The substance is called a *catalyst,* and the process whereby the rate is increased is *catalysis.* It is difficult to exaggerate the importance of catalysis, since most life processes and industrial processes would not practically be possible without it.

Some industrially important catalytic reactions (with their catalysts) which are the bases for such large-scale operations as the production of sulfuric acid, agricultural fertilizers, plastics, and fuels are:

$$SO_2 + \frac{1}{2}O_2 \rightleftarrows SO_3 \quad \text{(promoted } V_2O_5 \text{ catalyst)}$$

$$N_2 + 3H_2 \rightleftarrows 2NH_3 \quad \text{(promoted Fe catalyst)}$$

$$C_8H_{10} \rightleftarrows C_8H_8 + H_2 \quad (K_2CO_3, \text{ Fe oxide catalyst)}$$

$$CO + 2H_2 \rightleftarrows CH_3OH \quad \text{(Cu, Zn oxide catalyst)}$$

$$\text{ROOH (organic hydroperoxide)} + C_3H_6 \rightarrow$$
$$C_3H_6O + ROH \quad \text{(soluble Mo organometallic catalyst)}$$

$$CH_3CHCH_2 + C_6H_6 \rightarrow \text{cumene (solution or solid acid catalyst)}$$

In this chapter, we first consider the general concepts of catalysis and the intrinsic kinetics, including forms of rate laws, for several classes of catalytic reactions (Sections 8.1 to 8.4). We then treat the influence of mass and heat transport on the kinetics of catalytic reactions taking place in porous catalyst particles (Section 8.5). Finally, we provide an introduction to aspects of catalyst deactivation and regeneration (Section 8.6). The bibliography in Appendix B gives references for further reading in this large and important field.

8.1 CATALYSIS AND CATALYSTS

8.1.1 Nature and Concept

The following points set out more clearly the qualitative nature and concept of catalysis and catalysts:

(1) The primary characteristic is that a catalyst increases the rate of a reaction, relative to that of the uncatalyzed reaction.
(2) A catalyst does not appear in the stoichiometric description of the reaction, although it appears directly or indirectly in the rate law and in the mechanism. It is not a reactant or a product of the reaction in the stoichiometric sense.
(3) The amount of catalyst is unchanged by the reaction occurring, although it may undergo changes in some of its properties.
(4) The catalyst does not affect the chemical nature of the products. This must be qualified if more than one reaction (set of products) is possible, because the catalyst usually affects the selectivity of reaction.
(5) Corresponding to (4), the catalyst does not affect the thermodynamic affinity of a given reaction. That is, it affects the rate but not the tendency for reaction to occur. It does not affect the free energy change (ΔG) or equilibrium constant (K_{eq}) of a given reaction. If a catalyst did alter the position of equilibrium in a reaction, this would be contrary to the first law of thermodynamics, as pointed out by Ostwald many years ago, since we would then be able to create a perpetual-motion machine by fitting a piston and cylinder to a gas-phase reaction in which a change in moles occurred, and by periodically exposing the reacting system to the catalyst.
(6) Since a catalyst hastens the attainment of equilibrium, it must act to accelerate both forward and reverse reactions. For example, metals are good hydrogenation and dehydrogenation catalysts.
(7) Although it may be correct to say that a catalyst is not involved in the stoichiometry or thermodynamics of a reaction, it *is* involved in the mechanism of the reaction. In increasing the rate of a reaction, a catalyst acts by providing an easier path, which can generally be represented by the formation of an intermediate between catalyst and reactant, followed by the appearance of product(s) and regeneration of the catalyst. The easier path is usually associated with a lower energy barrier, that is, a lower E_A.

Catalysis is a special type of closed-sequence reaction mechanism (Chapter 7). In this sense, a catalyst is a species which is involved in steps in the reaction mechanism, but which is regenerated after product formation to participate in another catalytic cycle. The nature of the catalytic cycle is illustrated in Figure 8.1 for the catalytic reaction used commercially to make propene oxide (with Mo as the catalyst), cited above.

This proposed catalytic mechanism (Chong and Sharpless, 1977) requires four reaction steps (3 bimolecular and 1 unimolecular), which take place on a molybdenum metal center (titanium and vanadium centers are also effective), to which various nonreactive ligands (L) and reactive ligands (e.g., O–R) are bonded. Each step around the catalytic cycle is an elementary reaction and one complete cycle is called a turnover.

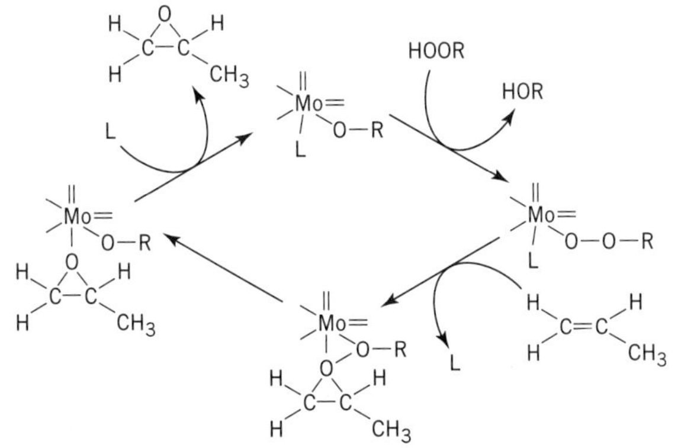

Figure 8.1 Representation of proposed catalytic cycle for reaction to produce C_3H_6O (Chong and Sharpless, 1977)

8.1.2 Types of Catalysis

We may distinguish catalysis of various types, primarily on the basis of the nature of the species responsible for the catalytic activity:

(1) *Molecular catalysis.* The term molecular catalysis is used for catalytic systems where identical molecular species are the catalytic entity, like the molybdenum complex in Figure 8.1, and also large "molecules" such as enzymes. Many molecular catalysts are used as homogeneous catalysts (see (5) below), but can also be used in multiphase (heterogeneous) systems, such as those involving attachment of molecular entities to polymers.

(2) *Surface catalysis.* As the name implies, surface catalysis takes place on the surface atoms of an extended solid. This often involves different properties for the surface atoms and hence different types of sites (unlike molecular catalysis, in which all the sites are equivalent). Because the catalyst is a solid, surface catalysis is by nature heterogeneous (see (6) below). The extended nature of the surface enables reaction mechanisms different from those with molecular catalysts.

(3) *Enzyme catalysis.* Enzymes are proteins, polymers of amino acids, which catalyze reactions in living organisms—biochemical and biological reactions. The systems involved may be colloidal—that is, between homogeneous and heterogeneous. Some enzymes are very specific in catalyzing a particular reaction (e.g., the enzyme sucrase catalyzes the inversion of sucrose). Enzyme catalysis is usually molecular catalysis. Since enzyme catalysis is involved in many biochemical reactions, we treat it separately in Chapter 10.

(4) *Autocatalysis.* In some reactions, one of the products acts as a catalyst, and the rate of reaction is experimentally observed to increase and go through a maximum as reactant is used up. This is autocatalysis. Some biochemical reactions are autocatalytic. The existence of autocatalysis may appear to contradict point (2) in Section 8.1.1. However, the catalytic activity of the product in question is a consequence of its formation and not the converse.

A further classification is based on the number of phases in the system: homogeneous (1 phase) and heterogeneous (more than 1 phase) catalysis.

(5) *Homogeneous catalysis.* The reactants and the catalyst are in the same phase. Examples include the gas-phase decomposition of many substances, including diethyl ether and acetaldehyde, catalyzed by iodine, and liquid-phase esterification reactions, catalyzed by mineral acids (an example of the general phenomenon of acid-base catalysis). The molybdenum catalyst in Figure 8.1 and other molecular catalysts are soluble in various liquids and are used in homogeneous catalysis. Gas-phase species can also serve as catalysts. Homogeneous catalysis is molecular catalysis, but the converse is not necessarily true. Homogeneous catalysis is responsible for about 20% of the output of commercial catalytic reactions in the chemical industry.

(6) *Heterogeneous catalysis.* The catalyst and the reactants are in different phases. Examples include the many gas-phase reactions catalyzed by solids (e.g., oxidation of SO_2 in presence of V_2O_5). Others involve two liquid phases (e.g., emulsion copolymerization of styrene and butadiene, with the hydrocarbons forming one phase and an aqueous solution of organic peroxides as catalysts forming the other phase). Heterogeneous, molecular catalysts are made by attaching molecular catalytic centers like the molybdenum species to solids or polymers, but heterogeneous catalysts may be surface catalysts. An important implication of heterogeneous catalysis is that the observed rate of reaction may include effects of the rates of transport processes in addition to intrinsic

reaction rates (this is developed in Section 8.5). Approximately 80% of commercial catalytic reactions involve heterogeneous catalysis. This is due to the generally greater flexibility compared with homogeneous catalysis, and to the added cost of separation of the catalyst from a homogeneous system.

8.1.3 General Aspects of Catalysis

8.1.3.1 *Catalytic Sites*

Central to catalysis is the notion of the catalytic "site." It is defined as the catalytic center involved in the reaction steps, and, in Figure 8.1, is the molybdenum atom where the reactions take place. Since all catalytic centers are the same for molecular catalysts, the elementary steps are bimolecular or unimolecular steps with the same rate laws which characterize the homogeneous reactions in Chapter 7. However, if the reaction takes place in solution, the individual rate constants may depend on the nonreactive ligands and the solution composition in addition to temperature.

For catalytic reactions which take place on surfaces, the term "catalytic site" is used to describe a location on the surface which bonds with reaction intermediates. This involves a somewhat arbitrary division of the continuous surface into smaller ensembles of atoms. This and other points about surface catalysts can be discussed by reference to the rather complex, but typical, type of metal catalyst shown in Figure 8.2. In this example, the desired catalytic sites are on the surface of a metal. In order to have as many surface metal atoms as possible in a given volume of catalyst, the metal is in the form of small crystallites (to increase the exposed surface area of metal), which are in turn supported on an inert solid (to increase the area on which the metal crystallites reside). In the electron micrograph in Figure 8.2(a), the metal crystallites show up as the small angular dark particles, and the support shows up as the larger, lighter spheres. Such a material would be pressed (with binders) into the form of a pellet for use in a reactor. Figure 8.2(b) is a closeup of several of the metal particles (showing rows of atoms). A schematic drawing of the atomic structure of one such particle is shown in Figure 8.2(c).

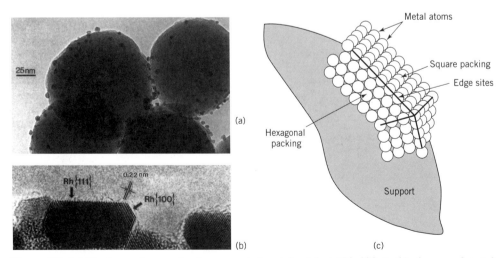

Figure 8.2 (a) Electron micrograph of a supported metal catalyst (Rh-SiO$_2$); (b) closeup of metal particles ((a) and (b) courtesy of Professor A. Datye); (c) schematic drawing of the atomic structure of a metal crystallite

Figures 8.2(b) and (c) illustrate two important aspects of surface catalysis that distinguish it from molecular catalysis:

(1) A distribution of "sites" exists on surfaces. By contrast with homogeneous and/or molecular catalysts in which all the sites are the same, the catalytic sites on solid surfaces can have a distribution of reactivities. The metal crystallites (which are the molecular catalytic entity) are of different sizes. They also have several different types of surface metal atoms available for catalytic reactions. The metal atoms are in a hexagonal packing arrangement on one face, while other faces consist of the metal atoms arranged in a square pattern. The bonding of reaction intermediates to these two surfaces is different. Further variety can be found by considering the atoms at the edges between the various faces. Finally, as discussed in Chapter 6, bonding of an intermediate to a site can be influenced by the bonding to nearby intermediates. Reaction mechanisms on surfaces are not usually known in sufficient detail to discriminate among these possibilities. Nevertheless, the simplifying assumption that a single type of site exists is often made despite the fact that the situation is more complex.

(2) Intermediates on adjacent sites can interact because of the extended nature of the surface. This option is not available to the isolated molecular catalytic entities. This allows more possibilities for reactions between intermediates.

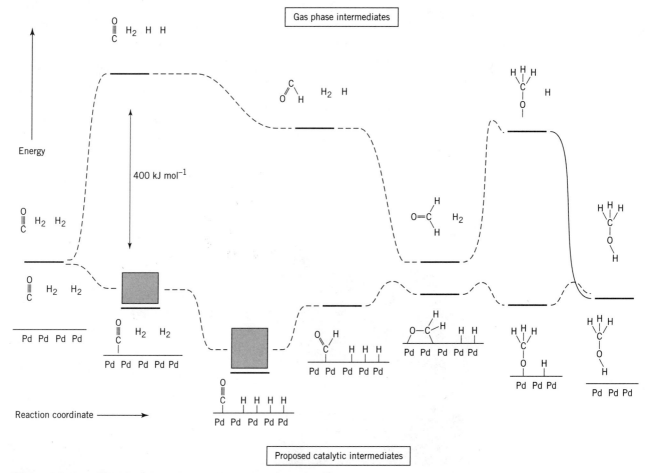

Figure 8.3 Proposed reaction mechanism for methanol synthesis on Pd and comparison with gas-phase mechanism; surface intermediates are speculative and associated energies are estimates

8.1.3.2 Catalytic Effect on Reaction Rate

Catalysts increase the reaction rate by lowering the energy requirements for the reaction. This, in turn, results from the ability of the catalyst to form bonds to reaction intermediates to offset the energy required to break reactant bonds. An example of a catalyst providing energetically easier routes to products is illustrated in the multistep reaction coordinate diagram in Figure 8.3, for the methanol-synthesis reaction, $CO + 2H_2 \rightarrow CH_3OH$. The energies of the intermediate stages and the activation energies for each step are indicated schematically.

For this reaction to proceed by itself in the gas-phase, a high-energy step such as the breakage of H–H bonds is required, and this has not been observed. Even with H_2 dissociation, the partially hydrogenated intermediates are not energetically favored. Also, even if an efficient radical-chain mechanism existed, the energetic cost to accomplish some of the steps make this reaction too slow to measure in the absence of a catalyst. The catalytic palladium metal surface also breaks the H–H bonds, but since this reaction is exoergic (Pd–H bonds are formed), it occurs at room temperature. The exact details of the catalytic reaction mechanism are unclear, but a plausible sequence is indicated in Figure 8.3. The energy scale is consistent with published values of the energies, where available. Notice how the bonding to the palladium balances the bonding changes in the organic intermediates. A good catalyst must ensure that all steps along the way are energetically possible. Very strongly bonded intermediates are to be avoided. Although their formation would be energetically favorable, they would be too stable to react further.

In general, the reaction rate is proportional to the amount of catalyst. This is true if the catalytic sites function independently. The number of turnovers per catalytic site per unit time is called the *turnover frequency*. The reactivity of a catalyst is the product of the number of sites per unit mass or volume and the turnover frequency.

8.1.3.3 Catalytic Control of Reaction Selectivity

In addition to accelerating the rates of reactions, catalysts control reaction selectivity by accelerating the rate of one (desired) reaction much more than others. Figure 8.4 shows schematically how different catalysts can have markedly different selectivities. Nickel surfaces catalyze the formation of methane from CO and H_2 but methanol is the major product on palladium surfaces. The difference in selectivity occurs because CO dissociation is relatively easy on nickel surfaces, and the resulting carbon and oxygen atoms are hydrogenated to form methane and water. On palladium, CO dissociation is difficult (indicated by a high activation energy and unfavorable energetics caused by weaker bonds to oxygen and carbon), and this pathway is not possible.

8.1.3.4 Catalyst Effect on Extent of Reaction

A catalyst increases only the rate of a reaction, not the thermodynamic affinity. Since the presence of the catalyst does not affect the Gibbs energy of reactants or products, it does not therefore affect the equilibrium constant for the reaction. It follows from this that a catalyst must accelerate the rates of both the forward and reverse reactions, since the rates of the two reactions must be equal once equilibrium is reached. From the energy diagram in Figure 8.4, if a catalyst lowers the energy requirement for the reaction in one direction, it must lower the energy requirement for the reverse reaction.

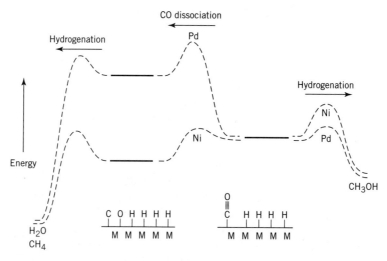

Figure 8.4 Hypothetical reaction coordinate diagrams for CO hydrogenation on Pd and Ni; the dissociation of CO is more difficult on Pd, making methanol synthesis more favorable than methane formation, which requires C–O dissociation, and is the preferred pathway on Ni

8.2 MOLECULAR CATALYSIS

8.2.1 Gas-Phase Reactions

An example of a catalytic gas-phase reaction is the decomposition of diethyl ether catalyzed by iodine (I_2):

$$(C_2H_5)_2O(A) \rightarrow C_2H_6 + CH_4 + CO$$

For the catalyzed reaction

$$(-r_A) = kc_A c_{I_2}; E_A = 142 \text{ kJ mol}^{-1}$$

and for the uncatalyzed reaction

$$(-r_A) = kc_A; E_A = 222 \text{ kJ mol}^{-1}$$

Another example of gas-phase catalysis is the destruction of ozone (O_3) in the stratosphere, catalyzed by Cl atoms. Ultraviolet light in the upper atmosphere causes the dissociation of molecular oxygen, which maintains a significant concentration of ozone:

$$O_2 + h\nu \rightarrow 2O^\bullet$$
$$O^\bullet + O_2 + M \rightarrow O_3 + M$$

Ozone in turn absorbs a different band of life-threatening ultraviolet light. The rate of ozone destruction in the pristine atmosphere is slow and is due to a reaction such as

$$O^\bullet + O_3 \rightarrow 2O_2$$

Chlorine-containing organic compounds, which are not destroyed in the troposphere, are photolyzed in the stratosphere:

$$RCl + h\nu \rightarrow Cl^\bullet + R$$

Chlorine atoms catalyze the destruction of ozone in the following two-step cycle:

$$Cl^\bullet + O_3 \rightarrow OCl + O_2$$
$$OCl + O_3 \rightarrow Cl^\bullet + 2O_2$$

with the overall result:

$$2O_3 \rightarrow 3O_2$$

In this cycle, $Cl^\bullet$ is regenerated, and each Cl atom can destroy a large number of O_3 molecules in chain-like fashion.

8.2.2 Acid-Base Catalysis

In aqueous solution, the rates of many reactions depend on the hydrogen-ion (H^+ or H_3O^+) concentration and/or on the hydroxyl-ion (OH^-) concentration. Such reactions are examples of acid-base catalysis. An important example of this type of reaction is esterification and its reverse, the hydrolysis of an ester.

If we use the Brönsted concept of an acid as a proton donor and a base as a proton acceptor, consideration of acid-base catalysis may be extended to solvents other than water (e.g., NH_3, CH_3COOH, and SO_2). An acid, on donating its proton, becomes its conjugate base, and a base, on accepting a proton, becomes its conjugate acid:

$$\text{acid} + \text{base} = \text{conjugate base} + \text{conjugate acid}$$

For proton transfer between a monoprotic acid HA and a base B,

$$HA + B = A^- + BH^+ \tag{8.2-1}$$

and for a diprotic acid,

$$H_2A + B = HA^- + BH^+$$
$$HA^- + B = A^{2-} + BH^+ \tag{8.2-2}$$

In this connection, water, an amphoteric solvent, can act as an acid (monoprotic, with, say, NH_3 as a base):

$$H_2O + NH_3 = OH^- + NH_4^+$$

or as a base (with, say, CH_3COOH as an acid):

$$CH_3COOH + H_2O = CH_3COO^- + H_3O^+$$

Acid-base catalysis can be considered in two categories: (1) specific acid-base catalysis, and (2) general acid-base catalysis. We illustrate each of these in turn in the next two sections, using aqueous systems as examples.

8.2.2.1 Specific Acid-Base Catalysis

In specific acid-base catalysis in aqueous systems, the observed rate constant, k_{obs}, depends on c_{H^+} and/or on c_{OH^-}, but not on the concentrations of other acids or bases present:

$$k_{obs} = k_o + k_{H^+}c_{H^+} + k_{OH^-}c_{OH^-} \tag{8.2-3}$$

where k_o is the rate constant at sufficiently low concentrations of both H^+ and OH^- (as, perhaps, in a neutral solution at pH = 7), k_{H^+} is the hydrogen-ion catalytic rate constant, and k_{OH^-} is the hydroxyl-ion catalytic rate constant. If only the $k_{H^+}c_{H^+}$ term is important, we have specific hydrogen-ion catalysis, and correspondingly for the $k_{OH^-}c_{OH^-}$ term. Since the ion-product constant of water, K_w, is

$$K_w = c_{H^+}c_{OH^-} \tag{8.2-4}$$

equation 8.2-3 may be written as

$$k_{obs} = k_o + k_{H^+}c_{H^+} + k_{OH^-}K_w/c_{H^+} \tag{8.2-5}$$

where the value of K_w is 1.0×10^{-14} mol² L⁻² at 25°C.

If only one term in equation 8.2-3 or 8.2-5 predominates in a particular region of pH, various cases can arise, and these may be characterized or detected most readily if equation 8.2-5 is put into logarithmic form:

$$\log_{10} k_{obs} = (\text{constant}) \pm \log_{10} c_{H^+} \tag{8.2-6}$$
$$= (\text{constant}) \mp \text{pH} \tag{8.2-6a}$$

In equation 8.2-6a, the slope of -1 with respect to pH refers to specific hydrogen-ion catalysis (type B, below) and the slope of $+1$ refers to specific hydroxyl-ion catalysis (C); if k_o predominates, the slope is 0 (A). Various possible cases are represented schematically in Figure 8.5 (after Wilkinson, 1980, p. 151). In case (a), all three types are evident: B at low pH, A at intermediate pH, and C at high pH; an example is the mutarotation of glucose. Cases (b), (c), and (d) have corresponding interpretations involving two types in each case; examples are, respectively, the hydrolysis of ethyl orthoacetate, of β-lactones, and of γ-lactones. Cases (e) and (f) involve only one type each; examples are, respectively, the depolymerization of diacetone alcohol, and the inversion of various sugars.

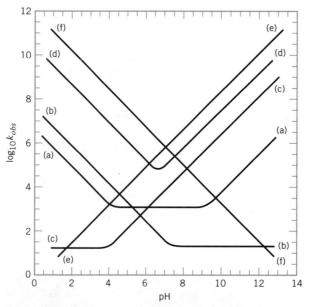

Figure 8.5 Acid-base catalysis: dependence of rate constant on pH (see text for explanation of cases (a) to (f))

A mechanism for a pseudo-first-order reaction involving the hydrolysis of substrate S catalyzed by acid HA that is consistent with the observed rate law $r_S = k_{obs} c_S$, is as follows:

$$S + HA \underset{}{\overset{K_1}{\rightleftharpoons}} SH^+ + A^- \qquad (fast)$$

$$SH^+ + H_2O \overset{k_2}{\rightarrow} products \qquad (slow)$$

This gives

$$\begin{aligned}(-r_S) &= k_2 c_{SH^+} c_{H_2O} \\ &= k_2 K_1 c_{H_2O} c_S c_{HA}/c_{A^-} \\ &= (k_2 K_1/K_a) c_{H_2O} c_{H^+} c_S \end{aligned} \qquad (8.2\text{-}7)$$

if, in addition, the acid HA is at dissociation equilibrium, characterized by K_a, the acid dissociation constant. According to this, the observed catalytic rate constant is

$$k_{obs} = (k_2 K_1/K_a) c_{H_2O} c_{H^+} + k_{H^+} c_{H^+} \qquad (8.2\text{-}8)$$

which, with c_{H_2O} virtually constant (H_2O in great excess), has the characteristics of specific hydrogen-ion catalysis (type B above), with the second term on the right of equation 8.2-5 predominating.

8.2.2.2 General Acid-Base Catalysis

In general acid-base catalysis, the observed rate constant depends on the concentrations of all acids and bases present. That is, in aqueous systems,

$$k_{obs} = k_o + k_{H^+} c_{H^+} + k_{OH^-} c_{OH^-} + \sum k_{HA} c_{HA} + \sum k_{A^-} c_{A^-} \qquad (8.2\text{-}9)$$

The systematic variation of c_{H^+}, c_{OH^-}, etc. allows the experimental determination of each rate constant. If the terms in the first summation on the right of equation 8.2-9 predominate, we have general acid catalysis; if those in the second summation do so, we have general base catalysis; otherwise, the terminology for specific acid-base catalysis applies, as in the previous section.

The mechanism in the previous section with a single acid can be used to show the features of, say, general acid catalysis, if the second step is not rate-determining but fast, and the first step is not a rapidly established equilibrium but involves a slow (rate-determining) step in the forward direction characterized by the rate constant k_{HA}. Then,

$$(-r_S) = k_{HA} c_{HA} c_S \equiv k_{obs} c_S \qquad (8.2\text{-}10)$$

which would result in

$$k_{obs} = \sum k_{HA} c_{HA} \qquad (8.2\text{-}11)$$

if more than one acid were present as catalyst, corresponding to the first summation on the right in equation 8.2-9.

Acid-base catalysis is important for reactions of hydrocarbons in the petrochemical industry. Acids, either as solids or in solution, react with hydrocarbons to form reactive carbocation intermediates:

$$H^+ + R'\text{-CH=CH-R} \rightarrow R'\text{-CH}_2\text{-CH}^+\text{-R}$$

which then participate in a variety of reactions such as alkylation, rearrangements, and cracking.

8.2.3 Other Liquid-Phase Reactions

Apart from acid-base catalysis, homogeneous catalysis occurs for other liquid-phase reactions. An example is the decomposition of H_2O_2 in aqueous solution catalyzed by iodide ion (I^-). The overall reaction is

$$2H_2O_2(A) \rightarrow 2H_2O + O_2$$

and the rate law is

$$(-r_A) = k_A c_A c_{I^-}; E_A = 59 \text{ kJ mol}^{-1}$$

A possible mechanism is

$$H_2O_2 + I^- \xrightarrow{k_1} H_2O + IO^- \quad \text{(slow)}$$
$$IO^- + H_2O_2 \xrightarrow{k_2} H_2O + O_2 + I^- \quad \text{(fast)}$$

with I^- being used in the first step to form hypoiodite ion, and being regenerated in the second step. If the first step is rate-determining, the rate law is as above with $k_A = k_1$. For the uncatalyzed reaction, $E_A = 75$ kJ mol^{-1}.

This reaction can be catalyzed in other ways: by the enzyme catalase (see enzyme catalysis in Chapter 10), in which E_A is 50 kJ mol^{-1}, and by colloidal Pt, in which E_A is even lower, at 25 kJ mol^{-1}.

Another example of homogeneous catalysis in aqueous solution is the dimerization of benzaldehyde catalyzed by cyanide ion, CN^- (Wilkinson, 1980, p.28):

$$2C_6H_5CHO(A) \rightarrow C_6H_5CH(OH)COC_6H_5$$
$$(-r_A) = k_A c_A^2 c_{CN^-}$$

Redox cycles involving metal cations are used in some industrial oxidations.

8.2.4 Organometallic Catalysis

Many homogeneous catalytic chemical processes use organometallic catalysts (Parshall and Ittel, 1992). These, like the example in Figure 8.1, consist of a central metal atom (or, rarely, a cluster) to which is bonded a variety of ligands (and during reaction, reaction intermediates). These catalysts have the advantage of being identifiable, identical molecular catalysts and the structures of the catalytic sites can be altered by use of specific ligands to change their activity or selectivity. With the addition of specific ligands, it is possible to make reactions stereoselective (i.e., only one of a possible set of enantiomers is produced). This feature has extensive application in polymerization catalysis, where the polymer properties depend on the stereochemistry, and in products related to biology and medicine, such as drug manufacture and food chemistry.

Reaction kinetics involving such catalysts can be demonstrated by the following mechanism:

$$L \bullet M \stackrel{K}{\rightleftharpoons} L + M \qquad (1)$$

$$M + A \stackrel{k_2}{\rightarrow} M \bullet A \qquad (2)$$

$$M \bullet A \stackrel{k_3}{\rightarrow} M + B \qquad (3)$$

Here, L is a mobile ligand which can leave the metal site (M) open briefly for reaction with A in the initial step of the catalytic cycle. The transformation of the M $\bullet$ A complex into products completes the cycle. The equilibrium in step (1) lies far to the left in most cases, because the ligands protect the metal centers from agglomeration. Thus, the concentration of M is very small, and the total concentration of catalyst is $c_{Mt} = c_{M \bullet A} + c_{M \bullet L}$. The rate law which arises from this mechanism is

$$(-r_A) = \frac{k_2 k_3 K c_A c_{Mt}}{k_3 c_L + k_2 K c_A} \qquad (8.2\text{-}12)$$

This rate expression has a common feature of catalysis—that of rate saturation. The (nonseparable) rate is proportional to the amount of catalyst. If reaction step (2) is slow (k_2 is small and the first term in the denominator of 8.2-12 is dominant), the rate reduces to

$$(-r_A) = (k_2 K c_{Mt}/c_L) c_A \qquad (8.2\text{-}13)$$

In this limit, the reaction is first order with respect to A, and most of the catalyst is in the form of M $\bullet$ L. Notice the inhibition by the ligand. If reaction step (3) is slow (k_3 small), the rate simplifies to

$$(-r_A) = k_3 c_{Mt} \qquad (8.2\text{-}14)$$

In this case, most of the catalyst is in the form of M $\bullet$ A and the reaction is zero order with respect to A. Thus, the kinetics move from first order at low c_A toward zero order as c_A increases. This feature of the rate "saturating" or reaching a plateau is common to many catalytic reactions, including surface catalysis (Section 8.4) and enzyme catalysis (Chapter 10).

8.3 AUTOCATALYSIS

Autocatalysis is a special type of molecular catalysis in which one of the products of reaction acts as a catalyst for the reaction. As a consequence, the concentration of this product appears in the observed rate law with a positive exponent if a catalyst in the usual sense, or with a negative exponent if an inhibitor. A characteristic of an autocatalytic reaction is that the rate increases initially as the concentration of catalytic product increases, but eventually goes through a maximum and decreases as reactant is used up. The initial behavior may be described as "abnormal" kinetics, and has important consequences for reactor selection for such reactions.

Examples of autocatalytic reactions include the decomposition of $C_2H_4I_2$ either in the gas phase or in solution in CCl_4 (Arnold and Kistiakowsky, 1933), hydrolysis of an ester, and some microbial fermentation reactions. The first of these may be used to illustrate some observed and mechanistic features.

188 Chapter 8: Catalysis and Catalytic Reactions

EXAMPLE 8-1

The rate of decomposition of gaseous ethylene iodide ($C_2H_4I_2$) into ethylene (C_2H_4) and molecular iodine is proportional to the concentration of $C_2H_4I_2$ and to the square root of the concentration of molecular iodine. Show how this can be accounted for if the reaction is catalyzed by iodine *atoms*, and if there is equilibrium between molecular iodine and iodine atoms at all times.

SOLUTION

The decomposition of $C_2H_4I_2$ is represented overall by

$$C_2H_4I_2(A) \rightarrow C_2H_4 + I_2$$

and the observed rate law is

$$(-r_A) = k_A c_A c_{I_2}^{1/2}$$

A possible mechanism to account for this involves the rapid establishment of dissociation-association equilibrium of molecules and atoms, followed by a slow bimolecular reaction between $C_2H_4I_2$ and I atoms:

$$I_2 + M \xrightleftharpoons{K_{eq}} 2I + M \quad \text{(fast)}$$

$$C_2H_4I_2 + I \xrightarrow{k_2} C_2H_4 + I_2 + I \quad \text{(slow)}$$

where M is a "third body" and the catalyst is atomic I. The rate law, based on the second step as the *rds*, is

$$(-r_A) = k_2 c_A c_I = k_A c_A c_{I_2}^{1/2}$$

as above, where $k_A = k_2 K_{eq}^{1/2}$

To illustrate quantitatively the kinetics characteristics of autocatalysis in more detail, we use the model reaction

$$A + \ldots \rightarrow B + \ldots \tag{8.3-1}$$

with the observed rate law

$$(-r_A) = k_A c_A c_B \tag{8.3-2}$$

That is, the reaction is autocatalytic with respect to product B. If the initial concentrations are c_{Ao} and c_{Bo} (which may be zero), and, since

$$c_B = c_{Bo} + c_{Ao} - c_A = M_o - c_A \tag{8.3-3}$$

where

$$M_o = c_{Ao} + c_{Bo} \tag{8.3-3a}$$

the rate law may be written in terms of c_A only:

$$(-r_A) = k_A c_A (M_o - c_A) \tag{8.3-4}$$

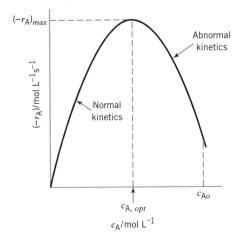

Figure 8.6 The optimal behavior of the rate of an autocatalytic reaction (T constant)

From this

$$d(-r_A)/dc_A = k_A(M_o - 2c_A) \qquad (8.3\text{-}5)$$
$$d^2(-r_A)/dc_A^2 = -2k_A < 0 \qquad (8.3\text{-}6)$$

Since the second derivative is negative, the optimal value of $(-r_A)$, obtained by setting the first derivative to zero, is a maximum. From equation 8.3-5, the maximum rate, $(-r_A)_{max}$, occurs at the optimal concentration

$$c_{A,opt} = M_o/2 \qquad (8.3\text{-}7)$$

and from this and equation 8.3-4, the maximum rate is

$$(-r_A)_{max} = k_A M_o^2/4 \qquad (8.3\text{-}8)$$

In principle, from equation 8.3-2, c_{B_o} must be > 0 for the reaction to start, but the rate of the uncatalyzed reaction (occurring in parallel with the catalyzed reaction) may be sufficient for this effectively to be the case.

The behavior of the rate (at constant T) of an autocatalytic reaction, such as represented by 8.3-1, -2, is shown schematically in Figure 8.6 with $(-r_A)$ as a function of c_A. With reaction occurring from high to low c_A, that is, from right to left in Figure 8.6, $(-r_A)$ increases from c_{A_o} to $c_{A,opt}$ ("abnormal" kinetics) to the maximum value, $(-r_A)_{max}$, after which it decreases as c_A decreases ("normal" kinetics).

EXAMPLE 8-2

Suppose reaction 8.3-1 with rate law given by equation 8.3-2 is carried out in a constant-volume batch reactor (or a constant-density PFR) at constant T.

190 Chapter 8: Catalysis and Catalytic Reactions

(a) Using the integral method of experimental investigation (Section 3.4.1.1.2), obtain a linear form of the c_A–t relation from which k_A may be determined.
(b) What is the value of t_{max}, the time at which the rate is $(-r_A)_{max}$, in terms of the parameters c_{Ao}, c_{Bo}, and k_A?
(c) How is f_A related to t? Sketch the relation to show the essential features.

SOLUTION

(a) Integration (e.g., by partial fractions) of the material-balance equation for A with the rate law included,

$$-dc_A/dt = k_A c_A (M_o - c_A) \tag{8.3-9}$$

results in

$$\ln(c_A/c_B) = \ln(c_{Ao}/c_{Bo}) - M_o k_A t \tag{8.3-10}$$

Thus, $\ln(c_A/c_B)$ is a linear function of t; from the slope, k_A may be determined. (Compare equation 8.3-10 with equation 3.4-13 for a second-order reaction with $\nu_A = \nu_B = -1$.) Note the implication of the comment following equation 8.3-8 for the application of equation 8.3-10.

(b) Rearrangement of equation 8.3-10 to solve for t and substitution of $c_{A,opt}$ from equation 8.3-7, together with 8.3-3, results in $c_A = c_B$ at t_{max}, and thus,

$$t_{max} = (1/M_o k_A) \ln(c_{Ao}/c_{Bo}) \tag{8.3-11}$$

This result is valid only for $c_{Ao} > c_{Bo}$, which is usually the case; if $c_{Ao} < c_{Bo}$, this result suggests that there is no maximum in $(-r_A)$ for reaction in a constant-volume BR. This is examined further through f_A in part (c) below. A second conclusion is that the result of equation 8.3-11 is also of practical significance only for $c_{Bo} \neq 0$. The first of these conclusions can also be shown to be valid for reaction in a CSTR, but the second is not (see problem 8-4).

(c) Since $f_A = 1 - (c_A/c_{Ao})$ for constant density, equation 8.3-10 can be rearranged to eliminate c_A and c_B so as to result in

$$f_A = \frac{1 - \exp(-M_o k_A t)}{1 + C_o \exp(-M_o k_A t)} \tag{8.3-12}$$

where

$$C_o = c_{Ao}/c_{Bo} \tag{8.3-12a}$$

Some features of the f_A–t relation can be deduced from equation 8.3-12 and the first and second derivatives of f_A. Thus, as $t \to 0$, $f_A \to 0$; df_A/dt (slope) $\to k_A c_{Bo} > 0$ (but $= 0$, if $c_{Bo} = 0$); $d^2 f_A/dt^2 \to k_A^2 c_{Bo}(c_{Ao} - c_{Bo}) > 0$, usually, with $c_{Ao} > c_{Bo}$, but < 0 otherwise. As $t \to \infty$, $f_A \to 1$; $df_A/dt \to 0$; $d^2 f_A/dt^2 \to 0(-)$. The usual shape, that is, with $c_{Ao} > c_{Bo}$, as in part (b), is sigmoidal, with an inflection at $t = t_{max}$ given by equation 8.3-11. This can be confirmed by setting $d^2 f_A/dt^2 = 0$.

The usual case, $c_{Ao} > c_{Bo}$, is illustrated in Figure 8.7 as curve A. Curve C, with no inflection point, illustrates the unusual case of $c_{Ao} < c_{Bo}$, and curve B, with $c_{Ao} = c_{Bo}$, is the boundary between these two types of behavior (it has an incipient inflection point at $t = 0$). In each case, $k_A = 0.6$ L mol^{-1} min^{-1} and $M_o = 10/6$ mol L^{-1}; $C_o = 9$, 1, and 1/9 in curves A, B, and C, respectively.

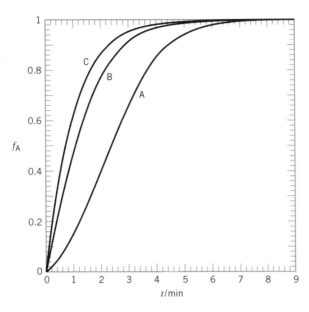

Figure 8.7 f_A as a function of t for reaction 8.3-1 in a constant-volume BR according to equation 8.3-12; curve A, $c_{Ao} > c_{Bo}$; curve B, $c_{Ao} = c_{Bo}$; curve C, $c_{Ao} < c_{Bo}$; see text for values of parameters

8.4 SURFACE CATALYSIS: INTRINSIC KINETICS

Surface catalysis is involved in a large majority of industrial catalytic reactions. The rate laws developed in this section are based on the following assumptions:

(1) The surface of the catalyst contains a fixed number of sites.
(2) All the catalytic sites are identical.
(3) The reactivities of these sites depend only on temperature. They do not depend on the nature or amounts of other materials present on the surface during the reaction.

These assumptions are the basis of the simplest rational explanation of surface catalytic kinetics and models for it. The preeminent of these, formulated by Langmuir and Hinshelwood, makes the further assumption that for an overall (gas-phase) reaction, for example, A(g) + ... → product(s), the rate-determining step is a surface reaction involving adsorbed species, such as A•s. Despite the fact that reality is known to be more complex, the resulting rate expressions find wide use in the chemical industry, because they exhibit many of the commonly observed features of surface-catalyzed reactions.

8.4.1 Surface-Reaction Steps

Central to surface catalysis are reaction steps involving one, or more than one, surface-bound (adsorbed) intermediate species. We consider three types.

(1) Unimolecular surface reaction, for example,

$$A \bullet s \rightarrow B \bullet s \tag{8.4-1}$$

where A•s is a surface-bound species involving A and site s (similarly for B•s). The rate of this reaction is given by

$$(-r_A) = k\theta_A \tag{8.4-2}$$

where θ_A is the fraction of the surface covered by adsorbed species A.

(2) Bimolecular surface reaction, for example,

$$A\bullet s + B\bullet s \rightarrow C\bullet s + s \tag{8.4-3}$$

where the rate is given by

$$(-r_A) = k\theta_A\theta_B \tag{8.4-4}$$

The rates (and rate constants) can be expressed on the basis of catalyst mass (e.g., mol kg^{-1}h^{-1}), or of catalyst surface area (e.g., μmol m^{-2} s^{-1}), or as a turnover frequency (molecules site^{-1} s^{-1}), if a method to count the sites exists.

(3) Eley-Rideal reaction, wherein a gas-phase species reacts directly with an adsorbed intermediate without having to be bound to the surface itself; thus,

$$A\bullet s + B \rightarrow C + s \tag{8.4-5}$$

Here, the rate is given by

$$(-r_A) = k\theta_A c_B \tag{8.4-6}$$

where c_B is the gas-phase concentration of B.

8.4.2 Adsorption without Reaction: Langmuir Adsorption Isotherm

We require expressions for the surface coverages, θ, for use in the equations in Section 8.4.1 to obtain catalytic rate laws in terms of the concentrations of gas-phase species. Langmuir-Hinshelwood (LH) kinetics is derived by assuming that these coverages are given by the equilibrium coverages which exist in the absence of the surface reactions. The required expressions were obtained by Langmuir in 1916 by considering the rate of adsorption and desorption of each species.

8.4.2.1 Adsorption of Undissociated Single Species

The reversible adsorption of a single species A, which remains intact (undissociated) on adsorption, can be represented by

$$A + s \underset{k_{dA}}{\overset{k_{aA}}{\rightleftharpoons}} A\bullet s \tag{8.4-7}$$

The rate of adsorption of A, r_{aA}, is proportional to the rate at which molecules of A strike the surface, which in turn is proportional to their concentration in the bulk gas, and to the fraction of unoccupied sites, $1 - \theta_A$:

$$r_{aA} = k_{aA} c_A (1 - \theta_A) \tag{8.4-8}$$

where k_{aA} is an adsorption rate constant which depends on temperature. (If the units of r_{aA} are mol m^{-2} s^{-1} and of c_A are mol m^{-3}, the units of k_{aA} are m s^{-1}.) A molecule which strikes a site already occupied may reflect without adsorption or may displace the occupying molecule; in either case, there is no net effect.

The rate of desorption of A, r_{dA}, is proportional to the fraction of surface covered, θ_A:

$$r_{dA} = k_{dA} \theta_A \tag{8.4-9}$$

where k_{dA} is a desorption rate constant which also depends on temperature. (The units of k_{dA} are mol m^{-2} s^{-1}.)

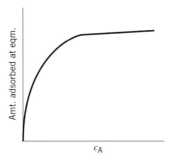

Figure 8.8 Langmuir adsorption isotherm

At adsorption equilibrium, with $r_{aA} = r_{dA}$,

$$k_{aA} c_A (1 - \theta_A) = k_{dA} \theta_A \quad (8.4\text{-}10)$$

or the fraction covered, which is proportional to the amount of gas adsorbed, is

$$\theta_A = k_{aA} c_A / (k_{dA} + k_{aA} c_A) = (k_{aA}/k_{dA}) c_A / [1 + (k_{aA}/k_{dA}) c_A]$$
$$= K_A c_A / (1 + K_A c_A) \quad (8.4\text{-}11)$$

where $K_A = k_{aA}/k_{dA}$, the ratio of the two rate constants, in m³ mol⁻¹. The equation (resulting from equation 8.4-11) expressing the (equilibrium) amount of A adsorbed on the surface as a function of c_A at constant T is called the Langmuir adsorption isotherm.

The shape of the Langmuir isotherm is shown schematically in Figure 8.8. The amount of A adsorbed increases as the (gas-phase) concentration c_A increases (at a given T), but approaches a limiting ("saturation") value at sufficiently high c_A.

8.4.2.2 Adsorption of Dissociated Single Species

If the adsorbing molecule dissociates into two or more fragments, each requiring a site, the fraction covered (coverage) differs from that given by equation 8.4-11. For example, consider the adsorption of a dissociating diatomic molecule, B_2:

$$B_2 + 2s \rightarrow 2B \bullet s \quad (8.4\text{-}12)$$

Here, the rate of adsorption is assumed to be given by

$$r_{aB_2} = k_{aB_2} c_{B_2} (1 - \theta_B)^2 \quad (8.4\text{-}13)$$

The quadratic term in open sites reflects the statistical likelihood of there being two adjacent open sites. The rate of desorption is given by

$$r_{dB_2} = k_{dB_2} \theta_B^2 \quad (8.4\text{-}14)$$

and the coverage obtained by equating the adsorption and desorption rates is

$$\theta_B = (K_{B_2} c_{B_2})^{1/2} / [1 + (K_{B_2} c_{B_2})^{1/2}] \quad (8.4\text{-}15)$$

where $K_{B_2} = k_{aB_2}/k_{dB_2}$. Similarly, if n sites are required for n fragments, the exponent 1/2 becomes $1/n$.

By measuring the amount of adsorption of reactive molecules under conditions where they do not react further and where desorption is very slow (low temperature), we can "count" the number of catalytic sites. The type of adsorption considered here is chemical in nature—chemical bonds are formed with the catalytically active surface—and is known as chemisorption. A weaker type of adsorption due to physical forces (like the forces which hold liquids together) can also occur. This latter type of adsorption (which can occur anywhere on the surface—not just on catalytic sites) is used to measure the surface area of porous materials.

8.4.2.3 Adsorption of Competing Species

For a surface reaction between two adsorbed gaseous species, A and B, we need to consider the simultaneous adsorption of the two species, competing for the available sites. For species A, the rate of adsorption is

$$r_{aA} = k_{aA} c_A (1 - \theta_A - \theta_B) \tag{8.4-16}$$

For this expression, it is assumed that a molecule of A from the bulk gas striking a site occupied by a B molecule is reflected, and does not displace the adsorbed B molecule. The rate of desorption, as for a single species, is

$$r_{dA} = k_{dA} \theta_A \tag{8.4-17}$$

At adsorption equilibrium, $r_{aA} = r_{dA}$, and

$$k_{aA} c_A (1 - \theta_A - \theta_B) = k_{dA} \theta_A \tag{8.4-18}$$

or, if $K_A = k_{aA}/k_{dA}$,

$$K_A c_A (1 - \theta_A - \theta_B) = \theta_A \tag{8.4-19}$$

Similarly,

$$K_B c_B (1 - \theta_A - \theta_B) = \theta_B \tag{8.4-20}$$

where $K_B = k_{aB}/k_{dB}$, the ratio of adsorption and desorption rate constants for B.

From equations 8.4-19 and 20, we obtain expressions for θ_A and θ_B:

$$\theta_A = K_A c_A / (1 + K_A c_A + K_B c_B) \tag{8.4-21}$$
$$\theta_B = K_B c_B / (1 + K_A c_A + K_B c_B) \tag{8.4-22}$$

According to these equations, each adsorbed species inhibits the adsorption of the other, as indicated by the term $K_B c_B$ in the denominator of the equation for θ_A and conversely for θ_B.

In a more general form of equation 8.4-21 or -22, a $K_i c_i$ term appears in the denominator for each adsorbing species i in competition for the adsorption sites. Furthermore, if the species dissociates into n_i fragments, the appropriate term is $(K_i c_i)^{1/n_i}$ as in equation 8.4-15 for $n = 2$. Therefore, in the most general case, the expression for Langmuir adsorption of species i from a multispecies gas mixture is:

$$\theta_i = \frac{(K_i c_i)^{1/n_i}}{1 + \sum_j (K_j c_j)^{1/n_j}} \quad ; i, j = 1, 2, \ldots, N \tag{8.4-23}$$

8.4.3 Langmuir-Hinshelwood (LH) Kinetics

By combining surface-reaction rate laws with the Langmuir expressions for surface coverages, we can obtain Langmuir-Hinshelwood (LH) rate laws for surface-catalyzed reactions. Although we focus on the intrinsic kinetics of the surface-catalyzed reaction, the LH model should be set in the context of a broader kinetics scheme to appreciate the significance of this.

A kinetics scheme for an overall reaction expressed as

$$A(g) \rightarrow B(g)$$

where A is a gas-phase reactant and B a gas-phase product, is as follows:

$$A(g) \xrightarrow{k_{Ag}} A \text{ (surface vicinity)}; \quad \text{mass transfer (fast)} \quad (1)$$

$$A(\text{surface vicinity}) + s \underset{k_{dA}}{\overset{k_{aA}}{\rightleftarrows}} A \bullet s; \quad \text{adsorption–desorption (fast)} \quad (2)$$

$$A \bullet s \xrightarrow{k} B(\text{surface vicinity}) + s; \quad \text{surface reaction (slow, } rds) \quad (3)$$

$$B(\text{surface vicinity}) \xrightarrow{k_{Bg}} B(g); \quad \text{mass transfer (fast)} \quad (4)$$

Here $A(g)$ and $B(g)$ denote reactant and product in the bulk gas at concentrations c_A and c_B, respectively; k_{Ag} and k_{Bg} are mass-transfer coefficients, s is an adsorption site, and $A \bullet s$ is a surface-reaction intermediate. In this scheme, it is assumed that B is not adsorbed. In focusing on step (3) as the rate-determining step, we assume k_{Ag} and k_{Bg} are relatively large, and step (2) represents adsorption-desorption equilibrium.

8.4.3.1 Unimolecular Surface Reaction (Type I)

For the overall reaction $A \rightarrow B$, if the *rds* is the unimolecular surface reaction given by equation 8.4-1, the rate of reaction is obtained by using equation 8.4-21 for θ_A in 8.4-2 to result in:

$$(-r_A) = \frac{kK_A c_A}{1 + K_A c_A + K_B c_B} \quad (8.4\text{-}24)^1$$

---

[1] The equations of the LH model can be expressed in terms of partial pressure p_i (replacing c_i). For example, equation 8.4-23 for fractional coverage of species i may be written as (with $n_i = n_j = 1$)

$$\theta_i = \frac{K_{ip} p_i}{1 + \sum_j K_{jp} p_j}; i, j = 1, 2, \ldots, N \quad (8.4\text{-}23a)$$

where K_{ip} is the ratio of adsorption and desorption rate constants in terms of (gas-phase) partial pressure, $K_{ip} = k_{api}/k_{di}$. Similarly the rate law in equation 8.4-24 may be written as

$$(-r_A) = \frac{kK_{Ap} p_A}{1 + K_{Ap} p_A + K_{Bp} p_B} \quad (8.4\text{-}24a)$$

Some of the problems at the end of the chapter are posed in terms of partial pressure.

Appropriate differences in units for the various quantities must be taken into account. If $(-r_A)$ is in mol m^{-2} s^{-1} and p_i is in kPa, the units of k_{api} are mol m^{-2} s^{-1} kPa^{-1}, and of K_{ip} are kPa^{-1}; the units of k_{di} are the same as before.

Two common features of catalytic rate laws are evident in this expression.

(1) *Saturation kinetics:* The rate is first order with respect to A at low concentrations of A (such that $K_A c_A \ll 1 + K_B c_B$), but becomes zero order at higher concentrations when $K_A c_A \gg 1 + K_B c_B$. In the high-concentration limit, all the catalytic sites are saturated with A($\theta_A = 1$), and the rate is given by the number of catalytic sites times the rate constant, k.

(2) *Product inhibition:* If the term $K_B c_B$ is significant compared to $1 + K_A c_A$, the rate is inhibited by the presence of product. In the extreme case of $K_B c_B \gg 1 + K_A c_A$, equation 8.4-24 becomes

$$(-r_A) = k' c_A c_B^{-1} \tag{8.4-25}$$

where $k' = k K_A K_B^{-1}$. Note that the inhibition of the rate by B has nothing to do with the reversibility of the reaction (which is assumed to be irreversible).

8.4.3.2 Bimolecular Surface Reaction (Type II)

For the overall reaction A + B → C, if the *rds* is the bimolecular surface reaction given by equation 8.4-3, the rate of reaction is obtained by using equation 8.4-23 (applied to A and B, with A, B, and C adsorbable) in equation 8.4-4 (for θ_A and θ_B) to result in:

$$(-r_A) = \frac{k K_A K_B c_A c_B}{(1 + K_A c_A + K_B c_B + K_C c_C)^2} \tag{8.4-26}$$

This rate law contains another widely observed feature in surface catalysis:

(3) *Inhibition by one of the reactants:* Similar to Type I kinetics, the rate is first order in c_A when $K_A c_A \ll 1 + K_B c_B + K_C c_C$, but instead of reaching a plateau in the other limit ($K_A c_A \gg 1 + K_B c_B + K_C c_C$), the rate becomes inhibited by A. The limiting rate law in this case is

$$(-r_A) = k' c_B c_A^{-1} \tag{8.4-27}$$

where $k' = k K_B / K_A$. A maximum in the rate is achieved at intermediate values of c_A, and the ultimate maximum rate occurs when $k\theta_A = \theta_B = 1/2$. Many CO hydrogenation reactions, such as the methanol synthesis reaction, exhibit rate laws with negative effective orders in CO and positive effective orders in H_2. This reflects the fact that CO is adsorbed more strongly than H_2 on the metal surface involved ($K_{CO} > K_{H_2}$).

Also apparent from equation 8.4-26 is that *product* inhibition can have a more serious effect in Type II kinetics because of the potential negative second-order term.

EXAMPLE 8-3

For the surface-catalyzed reaction A(g) + B(g) → products (C), what is the form of the rate law if

(a) Both reactants are weakly adsorbed, and products are not adsorbed, and
(b) Reactant A is weakly adsorbed, B is moderately adsorbed, and products are not adsorbed?

SOLUTION

(a) From equation 8.4-26, with $K_A c_A \ll 1 \gg K_B c_B$,

$$(-r_A) = k' c_A c_B \qquad (8.4\text{-}28)$$

which is a second-order reaction, with $k' = k K_A K_B$.

(b) From equation 8.4-26, with $K_A c_A \ll 1$,

$$(-r_A) = k c_A c_B/(1 + K_B c_B)^2 \qquad (8.4\text{-}29)$$

and the reaction is first-order with respect to A, but not with respect to B. As c_B increases, B occupies more of the surface, and its presence inhibits the rate of reaction.

8.4.4 Beyond Langmuir-Hinshelwood Kinetics

The two rate laws given by equations 8.4-24 and -26 (Types I and II) are used extensively to correlate experimental data on surface-catalyzed reactions. Nevertheless, there are many surface-catalyzed reaction mechanisms which have features not covered by LH kinetics.

Multiple surface steps: The basic LH mechanisms involve a single surface reaction, while many surface-catalyzed reactions, like the methanol synthesis mechanism in Figure 8.3, involve a series of surface steps. The surface sites are shared by the intermediates and the adsorbed reactants and products; thus, the coverages are altered from those predicted by adsorption of gas-phase species alone. The steady-state coverages are obtained from analyses identical to those used for gas-phase mechanisms involving reactive intermediates (Chapter 7). Although it is possible to obtain analytical rate laws from some such mechanisms, it often becomes impossible for complex mechanisms. In any case, the rate laws are modified from those of the standard LH expressions. For example, the following mechanistic sequence, involving the intermediate species I

$$A + B + 2s \rightleftarrows A \bullet s + B \bullet s \rightleftarrows I \bullet s \rightarrow P \bullet s \rightleftarrows P + s$$

exhibits zero-order kinetics, if the irreversible unimolecular step $I \bullet s \rightarrow P \bullet s$ is rate-determining. In this case, the surface is filled with I ($\theta_I = 1$), and the competition among A, B, and C for the remaining sites becomes unimportant. In a similar manner, if a series of initial steps which are in equilibrium is followed by a slow step, extra factors appear in the rate law.

Irreversible adsorption: The LH mechanisms assume that the adsorption of all gas-phase species is in equilibrium. Some mechanisms, however, occur by irreversible steps. In these cases, the intermediates are again treated in the same manner as reactive intermediates in homogeneous mechanisms. An example is the Mars-van Krevelen (1954) mechanism for oxidation, illustrated by the following two steps:

$$O_2 + 2s \rightarrow 2O \bullet s$$
$$O \bullet s + CO \rightarrow CO_2 + s$$

Eley-Rideal mechanisms: If the mechanism involves a direct reaction between a gas-phase species and an adsorbed intermediate (Eley-Rideal step, reaction 8.4-5), the competition between the reactants for surface sites does not occur. From equations 8.4-6 and -21, since one reactant does not have to adsorb on a site in order to react, the rate is given by

$$(-r_A) = \frac{kK_A c_A c_B}{1 + K_A c_A + K_C c_C} \tag{8.4-30}$$

Even though the reaction is bimolecular, reactant inhibition does not occur for this type of reaction.

Variable site characteristics: Sites which have variable properties have been observed. These have been treated in several ways, including (1) *distribution of site types*, which can be thought of as equivalent to having a distribution of catalysts operating independently; and (2) *site properties which change with the presence of other adsorbates*, although they are all the same at a given condition. In the latter case, for example, the rate constants for adsorption or surface reactions can depend on the amounts of other adsorbed intermediates: $k_a = f(\theta_A, \theta_B, \ldots)$. An example is the well-studied dependence of the heat of adsorption of CO on various metals, which decreases as the coverage of the surface by CO increases.

8.5 HETEROGENEOUS CATALYSIS: KINETICS IN POROUS CATALYST PARTICLES

8.5.1 General Considerations

For a solid-catalyzed gas-phase reaction, the catalyst is commonly in the form of particles or pellets of various possible shapes and sizes, and formed in various ways. Such particles are usually porous, and the interior surface accessible to the reacting species is usually much greater than the gross exterior surface.

The porous nature of the catalyst particle gives rise to the possible development of significant gradients of both concentration and temperature across the particle, because of the resistance to diffusion of material and heat transfer, respectively. The situation is illustrated schematically in Figure 8.9 for a spherical or cylindrical (viewed end-on) particle of radius R. The gradients on the left represent those of c_A, say, for A(g) + $\ldots \rightarrow$ product(s), and those on the right are for temperature T; the gradients in each case, however, are symmetrical with respect to the centerline axis of the particle.

First, consider the gradient of c_A. Since A is consumed by reaction inside the particle, there is a spontaneous tendency for A to move from the bulk gas (c_{A_g}) to the interior of the particle, first by mass transfer to the exterior surface (c_{A_s}) across a supposed film, and then by some mode of diffusion (Section 8.5.3) through the pore structure of the particle. If the surface reaction is "irreversible," all A that enters the particle is reacted within the particle and none leaves the particle as A; instead, there is a counterdiffusion of product (for simplicity, we normally assume equimolar counterdiffusion). The concentration, c_A, at any point is the gas-phase concentration at that point, and not the surface concentration.

Next, consider the gradients of temperature. If the reaction is exothermic, the center of the particle tends to be hotter, and conversely for an endothermic reaction. Two sets of gradients are thus indicated in Figure 8.9. Heat transfer through the particle is primarily by conduction, and between exterior particle surface (T_s) and bulk gas (T_g) by combined convection-conduction across a thermal boundary layer, shown for convenience in Figure 8.9 to coincide with the gas film for mass transfer. (The quantities T_o, ΔT_p, ΔT_f, and ΔT_{ov} are used in Section 8.5.5.)

The kinetics of surface reactions described in Section 8.4 for the LH model refer to reaction at a point in the particle at particular values of c_A (or p_A) and T. To obtain a rate law for the particle as a whole, we must take into account the variation of c_A

8.5 Heterogeneous Catalysis: Kinetics in Porous Catalyst Particles

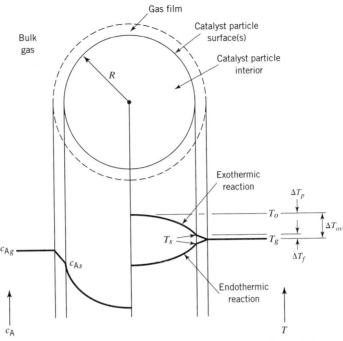

Figure 8.9 Concentration (c_A) and temperature (T) gradients (schematic) in a porous catalyst particle (spherical or end-on cylindrical)

and T in some manner, which means taking into account the diffusional and thermal effects within the particle, and between particle and bulk gas. This is the subject of the remainder of this section.

8.5.2 Particle Density and Voidage (Porosity)

Particle density, ρ_p, is defined by

$$\rho_p = m_p/v_p \tag{8.5-1}$$

where m_p and v_p are the mass and volume of the particle, respectively. Particle voidage, ϵ_p, is the ratio of the volume of void space (pores) in the particle, v_V, to the volume of the particle, v_p:

$$\epsilon_p = v_V/v_p \tag{8.5-2}$$

Because of the voidage, the particle density is less than the intrinsic density of the solid catalyst material, $\rho_S = m_p/v_S$, where v_S is the volume of solid in the particle, but is related to it by

$$\rho_p = \rho_S(1 - \epsilon_p) \tag{8.5-3}$$

since $v_p = v_V + v_S$.

8.5.3 Modes of Diffusion; Effective Diffusivity

Diffusion is the spontaneous migration of a species in space, relative to other species, as a result of a variation in its chemical potential, in the direction of decreasing potential.

The variation of chemical potential may arise as a result of variation of concentration or temperature or by other means, but we consider only the effect of concentration here.

From a molecular point of view inside a catalyst particle, diffusion may be considered to occur by three different modes: molecular, Knudsen, and surface. Molecular diffusion is the result of molecular encounters (collisions) in the void space (pores) of the particle. Knudsen diffusion is the result of molecular collisions with the walls of the pores. Molecular diffusion tends to dominate in relatively large pores at high P, and Knudsen diffusion tends to dominate in small pores at low P. Surface diffusion results from the migration of adsorbed species along the surface of the pore because of a gradient in surface concentration.

Since we don't usually know enough about pore structure and other matters to assess the relative importance of these modes, we fall back on the phenomenological description of the rate of diffusion in terms of Fick's (first) law. According to this, for steady-state diffusion in one dimension (coordinate x) of species A, the molar flux, N_A, in, say, mol m^{-2} (cross-sectional area of diffusion medium) s^{-1}, through a particle is

$$N_A = -D_e dc_A/dx \qquad (8.5\text{-}4)$$

where D_e is the *effective diffusivity* for A.

The effective diffusivity D_e is a characteristic of the particle that must be measured for greatest accuracy. However, in the absence of experimental data, D_e may be estimated in terms of molecular diffusivity, D_{AB} (for diffusion of A in the binary system A + B), Knudsen diffusivity, D_K, particle voidage, ϵ_p, and a measure of the pore structure called the particle tortuosity, τ_p.

An estimate for D_{AB} is (Reid et al., 1987, p. 582):

$$D_{AB} = \frac{0.00188 T^{3/2}[(M_A + M_B)/M_A M_B]^{1/2}}{P \Omega_D d_{AB}^2} \qquad (8.5\text{-}4a)$$

where D_{AB} is in cm^2 s^{-1}, T is in K, M_A and M_B are the molar masses of A and B, respectively, in g mol^{-1}, P is pressure in kPa, d_{AB} is the collision diameter, $(d_A + d_B)/2$, in nm, and Ω_D is the so-called collision integral.

The Knudsen diffusivity may be estimated (Satterfield, 1991, p. 502) from

$$D_K = 9700 r_e (T/M)^{1/2} \qquad (8.5\text{-}4b)$$

where D_K is in cm^2 s^{-1}, r_e is the average pore radius in cm, and M is molar mass. Equation 8.5-4b applies rigorously to straight, cylindrical pores, and is an approximation for other geometries.

The overall diffusivity, D^*, is obtained from D_{AB} and D_K by means of the conventional expression for resistances in series:

$$\frac{1}{D^*} = \frac{1}{D_{AB}} + \frac{1}{D_K} \qquad (8.5\text{-}4c)$$

The effective diffusivity is obtained from D^*, but must also take into account the two features that (1) only a portion of the catalyst particle is permeable, and (2) the diffusion path through the particle is random and tortuous. These are allowed for by the particle voidage or porosity, ϵ_p, and the tortuosity, τ_p, respectively. The former must also be measured, and is usually provided by the manufacturer for a commercial catalyst. For a straight cylinder, $\tau_p = 1$, but for most catalysts, the value lies between 3 and 7; typical values are given by Satterfield.

The final expression for estimating D_e is

$$D_e = D^* \epsilon_p / \tau_p \tag{8.5-4d}$$

Equation 8.5-4d reveals the "true" units of D_e, m³ (void space) m⁻¹ (particle) s⁻¹, as opposed to the "apparent" units in equation 8.5-4, m² s⁻¹.

8.5.4 Particle Effectiveness Factor η

8.5.4.1 Definition of η

Since c_A and T may vary from point to point within a catalyst particle (see Figure 8.9), the rate of reaction also varies. This may be translated to say that the effectiveness of the catalyst varies within the particle, and this must be taken into account in the rate law.

For this purpose, we introduce the particle effectiveness factor η, the ratio of the observed rate of reaction for the particle as a whole to the intrinsic rate at the surface conditions, c_{As} and T_s. In terms of a reactant A,

$$\eta = r_A \, (observed) / r_A(c_{As}, T_s) \tag{8.5-5}$$

We consider the effects of c_A and T separately, deferring the latter to Section 8.5.5. In focusing on the *particle* effectiveness factor, we also ignore the effect of any difference in concentration between bulk gas and exterior surface (c_{Ag} and c_{As}); in Section 8.5.6, we introduce the *overall* effectiveness factor to take this into account.

We then wish to discover how η depends on reaction and particle characteristics in order to use equation 8.5-5 as a rate law in operational terms. To do this, we first consider the relatively simple particle shape of a rectangular parallelepiped (flat plate) and simple kinetics.

8.5.4.2 η for Flat-Plate Geometry

EXAMPLE 8-4

For a flat-plate porous particle of diffusion-path length L (and infinite extent in other directions), and with only one face permeable to diffusing reactant gas A, obtain an expression for η, the particle effectiveness factor defined by equation 8.5-5, based on the following assumptions:

(1) The reaction $A(g) \rightarrow product(s)$ occurs within the particle.
(2) The surface reaction is first order.
(3) The reaction is irreversible.
(4) The particle is isothermal.
(5) The gas is of constant density.
(6) The overall process is in steady-state.
(7) The diffusion of A in the particle is characterized by the effective diffusivity D_e, which is constant.
(8) There is equimolar counterdiffusion (reactants and products).

202 Chapter 8: Catalysis and Catalytic Reactions

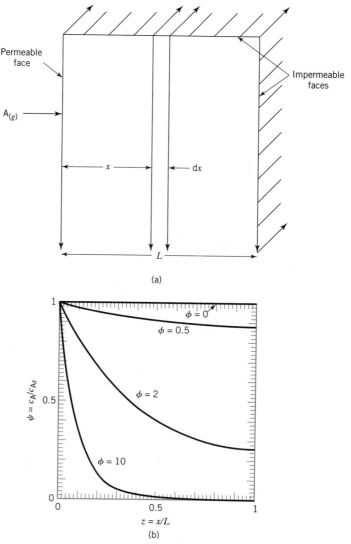

Figure 8.10 (a) Representation of flat-plate geometry; (b) concentration profile $\psi(\phi, z)$ (dimensionless) for various values of Thiele modulus ϕ

The particle shape is illustrated in Figure 8.10(a), with reactant A entering the particle through the permeable face on the left.

SOLUTION

To obtain an expression for η, we first derive the continuity equation governing steady-state diffusion of A through the pores of the particle. This is based on a material balance for A across the control volume consisting of the thin strip of width dx shown in Figure 8.10(a). We then solve the resulting differential equation to obtain the concentration profile for A through the particle (shown in Figure 8.10(b)), and, finally, use this result to obtain an expression for η in terms of particle, reaction, and diffusion characteristics.

In words, the diffusion or material-balance equation for A is:

$$\begin{pmatrix} \text{rate of} \\ \text{input by} \\ \text{diffusion} \end{pmatrix} = \begin{pmatrix} \text{rate of} \\ \text{output by} \\ \text{diffusion} \end{pmatrix} + \begin{pmatrix} \text{rate of} \\ \text{disappearance} \\ \text{in control volume} \end{pmatrix}$$

8.5 Heterogeneous Catalysis: Kinetics in Porous Catalyst Particles

That is, on applying equation 8.5-4 to both faces of the strip, we have

$$-D_e A_c \frac{dc_A}{dx} = -D_e A_c \left[\frac{dc_A}{dx} + \frac{d}{dx}\left(\frac{dc_A}{dx}\right)dx\right] + (-r_A)A_c dx \qquad (8.5\text{-}6)$$

for any surface kinetics, where A_c is the cross-sectional area perpendicular to the direction of diffusion of A (A_c is constant here and cancels). The rate law for $(-r_A)$ is not specified, but the units of $(-r_A)$ are mol A m^{-3} (particle) s^{-1}. If we introduce first-order kinetics $((-r_A) = k_A c_A)$, equation 8.5-6 becomes

$$d^2 c_A/dx^2 - k_A c_A/D_e = 0 \qquad (8.5\text{-}7)$$

To obtain a nondimensional form of this equation, we define dimensionless concentration, ψ, and length, z, respectively, as

$$\psi = c_A/c_{As} \qquad (8.5\text{-}8)$$
$$z = x/L \qquad (8.5\text{-}9)$$

Equation 8.5-7 in nondimensional form is then

$$d^2\psi/dz^2 - (k_A L^2/D_e)\psi = 0 \qquad (8.5\text{-}10)$$

The coefficient of ψ in equation 8.5-10 is used to define a dimensionless group called the Thiele modulus,[2] ϕ:

$$\phi = L(k_A/D_e)^{1/2} \qquad (n = 1) \qquad (8.5\text{-}11)$$

so that equation 8.5-10 becomes

$$d^2\psi/dz^2 - \phi^2\psi = 0 \qquad (8.5\text{-}12)$$

The importance of ϕ is that its magnitude is a measure of the ratio of intrinsic reaction rate (through k_A) to diffusion rate (through D_e). Thus, for a given value of k_A, a large value of ϕ corresponds to a relatively low value of D_e, and hence to relatively high diffusional resistance (referred to as "strong pore-diffusion" resistance). Conversely, a small value of ϕ corresponds to "negligible pore-diffusion" resistance.

The solution of equation 8.5-12 provides the concentration profile for ψ as a function of z, $\psi(z)$. On integrating the equation twice, we obtain

$$\psi = C_1 e^{\phi z} + C_2 e^{-\phi z} \qquad (8.5\text{-}12a)$$

where C_1 and C_2 are integration constants to be obtained from the boundary conditions:

$$\text{at } z = 0, \quad \psi = 1 \qquad (8.5\text{-}12b)$$
$$\text{at } z = 1, \quad d\psi/dz = 0 \qquad (8.5\text{-}12c)$$

---

[2]Equation 8.5-11 applies to a first-order surface reaction for a particle of flat-plate geometry with one face permeable. In the next two sections, the effects of shape and reaction order on ϕ are described. A general form independent of kinetics and of shape is given in Section 8.5.4.5. The units of k_A are such that ϕ is dimensionless. For catalytic reactions, the rate constant may be expressed per unit mass of catalyst $(k_A)_m$. To convert to k_A for use in equation 8.5-11 or other equations for ϕ, $(k_A)_m$ is multiplied by ρ_p, the particle density.

The second boundary condition is not known definitely, but is consistent with reactant A not penetrating the impermeable face at $z = 1$. From equations 8.5-12a to c,

$$C_1 = e^{-\phi}/(e^\phi + e^{-\phi}) \qquad (8.5\text{-}12\text{d})$$

$$C_2 = e^{\phi}/(e^\phi + e^{-\phi}) \qquad (8.5\text{-}12\text{e})$$

Then equation 8.5-12a becomes, on substitution for C_1 and C_2:

$$\psi = \frac{e^{-\phi(1-z)} + e^{\phi(1-z)}}{e^\phi + e^{-\phi}} = \frac{\cosh[\phi(1-z)]}{\cosh \phi} \qquad (8.5\text{-}13)$$

where $\cosh \phi = (e^\phi + e^{-\phi})/2$.

Figure 8.10(b) shows a plot of $\psi = c_A/c_{As}$ as a function of z, the fractional distance into the particle, with the Thiele modulus ϕ as parameter. For $\phi = 0$, characteristic of a very porous particle, the concentration of A remains the same throughout the particle. For $\phi = 0.5$, characteristic of a relatively porous particle with almost negligible pore-diffusion resistance, c_A decreases slightly as $z \to 1$. At the other extreme, for $\phi = 10$, characteristic of relatively strong pore-diffusion resistance, c_A drops rapidly as z increases, indicating that reaction takes place mostly in the outer part (on the side of the permeable face) of the particle, and the inner part is relatively ineffective.

The effectiveness factor η, defined in equation 8.5-5, is a measure of the effectiveness of the interior surface of the particle, since it compares the observed rate through the particle as a whole with the intrinsic rate at the exterior surface conditions; the latter would occur if there were no diffusional resistance, so that all parts of the interior surface were equally effective (at $c_A = c_{As}$). To obtain η, since all A entering the particle reacts (irreversible reaction), the observed rate is given by the rate of diffusion across the permeable face at $z = 0$:

$$\eta = \frac{\text{rate with diff. resist.}}{\text{rate with no diff. resist.}} = \frac{(-r_A) \text{ observed}}{(-r_A) \text{ intrinsic}}$$

$$= \frac{\text{rate of diffusion of A at } z = 0}{\text{total rate of reaction at } c_{As}} = \frac{(N_A \text{ at } z = 0)A_c}{(-R_A)_{int}}$$

$$= \frac{-D_e A_c (dc_A/dx)_{x=0}}{LA_c k_A c_{As}} = \frac{-D_e c_{As}(d\psi/dz)_{z=0}}{L^2 k_A c_{As}}$$

$$= -\frac{1}{\phi^2}\left(\frac{d\psi}{dz}\right)_{z=0} = -\frac{1}{\phi^2}\left\{\frac{-\phi \sinh[\phi(1-z)]}{\cosh \phi}\right\}_{z=0}$$

That is,

$$\eta = (\tanh \phi)/\phi \qquad \text{(flat plate)} \qquad (8.5\text{-}14)$$

where $\tanh \phi = \sinh \phi / \cosh \phi = (e^\phi - e^{-\phi})/(e^\phi + e^{-\phi})$.

Note that $\eta \to 1$ as $\phi \to 0$ and $\eta \to 1/\phi$ as $\phi \to$ large. (Obtaining the former result requires an application of L'Hôpital's rule, but the latter follows directly from $\tanh \phi \to 1$ as $\phi \to$ large.) These limiting results are shown in Figure 8.11, which is a plot of η as a function of ϕ according to equation 8.5-14, with both coordinates on logarithmic scales. The two limiting results and the transition region between may arbitrarily be considered as three regions punctuated by the points marked by G and H:

8.5 Heterogeneous Catalysis: Kinetics in Porous Catalyst Particles

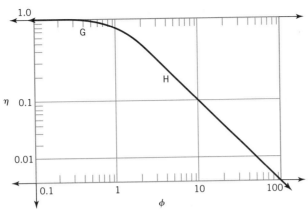

Figure 8.11 Effectiveness factor (η) as a function of Thiele modulus (ϕ) for an isothermal particle; three regions indicated:

$$\leftarrow G: \phi < 0.5; \eta \rightarrow 1$$
$$G\text{--}H: 0.5 < \phi < 5$$
$$H \rightarrow : \phi > 5; \eta \rightarrow 1/\phi$$

(1) Negligible pore-diffusion resistance (up to point G):

$$\phi < 0.5; \quad \eta \rightarrow 1 \qquad (8.5\text{-}14a)$$

(2) Significant pore-diffusion resistance (G–H):

$$0.5 < \phi < 5; \quad \eta = (\tanh \phi)/\phi \qquad (8.5\text{-}14b)$$

(3) Strong pore-diffusion resistance (beyond point H):

$$\phi > 5; \quad \eta \rightarrow 1/\phi \qquad (8.5\text{-}14c)$$

Because of the logarithmic scales used, the coordinates in Figure 8.11 extend indefinitely in all directions except that, for normal kinetics, $0 < \eta \leq 1$ for an *isothermal* particle (can η be greater than 1 for a *nonisothermal* particle?).

Substitution of the result given by equation 8.5-14 into the definition of η given by equation 8.5-5 yields the modified first-order rate law for an isothermal particle of this geometry:

$$(-r_A)_{obs} = \eta k_A c_{As} = \frac{\tanh \phi}{\phi} k_A c_{As} \qquad (8.5\text{-}15)$$

where ϕ is given by equation 8.5-11. Equation 8.5-15 is in terms of η and c_{As}. The form in terms of the *observable* concentration of A(c_{Ag}) requires consideration of the (additional) resistance to mass transfer exterior to the particle, and is developed in Section 8.5.6 dealing with the *overall* effectiveness factor η_o.

8.5.4.3 Effect of Particle Geometry (Shape) on η

The procedure described in Example 8-4 may be used to obtain analytical solutions for concentration profiles and η for other shapes of particles, such as spherical and cylindrical shapes indicated in Figure 8.9. Spherical shape is explored in problem 8-13. The solution for a cylinder is more cumbersome, requiring a series solution in terms of certain Bessel functions, details of which we omit here. The results for the dimensionless concentration gradient ψ and for η are summarized in Table 8.1 in terms of a Thiele modulus appropriate to each shape, as dictated by the form of the diffusion equation in each case. Table 8.1 includes the case of a flat plate with *two* faces permeable.

The results for spherical and cylindrical shapes are approximately in accordance with those shown in Figure 8.11, and in the limit of $\phi \to$ large, become the same, if the Thiele modulus is normalized in terms of a common effective diffusion-path-length parameter, L_e, defined by

$$L_e = \frac{\text{volume of particle}}{\text{exterior permeable surface area}} \tag{8.5-16}$$

Then the Thiele modulus normalized for shape is, for first-order kinetics:

$$\phi' = L_e(k_A/D_e)^{1/2} \qquad (n = 1)^3 \tag{8.5-17}$$

The consequences of this normalization are summarized for the various shapes in Table 8.2. In Table 8.2, subscripts FP1 and FP2 refer to a flat plate with 1 and 2 faces permeable, respectively, and subscripts s and c refer to sphere and cylinder, respectively, all as given in Table 8.1. The main consequence is that, if ϕ' replaces ϕ in Figure 8.11, η for all shapes lies approximately on the one line shown. The results become exactly the same for large values of ϕ' ($\eta \to 1/\phi'$, independent of shape). In the transition region between points G and H, the results differ slightly (about 17% at the most).

Table 8.1 Effectiveness factor (η) for various particle shapes (assumptions in Example 8-4)

Shape	ϕ	ψ	η
flat plate[a]	$L(k_A/D_e)^{1/2}$	$\cosh[\phi(1-z)]/\cosh\phi$	$(\tanh\phi)/\phi$
flat plate[b]	$L(k_A/D_e)^{1/2}$	$\cosh[\phi(\frac{1}{2}-z)]/\cosh(\phi/2)$	$\tanh(\phi/2)/(\phi/2)$
sphere[c]	$R(k_A/D_e)^{1/2}$	$\dfrac{R}{r}\dfrac{\sinh(\phi r/R)}{\sinh\phi}$	$\dfrac{3}{\phi}\left(\dfrac{1}{\tanh\phi} - \dfrac{1}{\phi}\right)$
cylinder[c]	$R(k_A/D_e)^{1/2}$	(in terms of Bessel functions (BF))	$\dfrac{2}{\phi}$ (ratio of BF)

[a] One face permeable as in Example 8-4; see Figure 8.10(a).
[b] Two faces permeable.
[c] R is particle radius; r is radial coordinate ($r = 0$ at center of particle).

Table 8.2 Thiele modulus (ϕ') normalized with respect to shape and asymptotic value of η

Shape	L_e	ϕ'	Asymptotic value of η	
			$\phi \to \infty$	$\phi' \to \infty$
flat plate (1)	L	ϕ_{FP1}	$1/\phi_{FP1}$	$1/\phi'$
flat plate (2)	$L/2$	$\phi_{FP2}/2$	$2/\phi_{FP2}$	$1/\phi'$
sphere	$R/3$	$\phi_s/3$	$3/\phi_s$	$1/\phi'$
cylinder	$R/2$	$\phi_c/2$	$2/\phi_c$	$1/\phi'$

---

[3] See footnote 2

8.5 Heterogeneous Catalysis: Kinetics in Porous Catalyst Particles

Table 8.3 Thiele modulus (ϕ'') normalized with respect to order of reaction (n) and asymptotic value of η

			Asymptotic value of η	
n	ϕ'	ϕ''	$\phi' \to \infty$	$\phi'' \to \infty$
0	$L_e(k_A/c_{As}D_e)^{1/2}$	$\phi'/2^{1/2}$	$2^{1/2}/\phi'$	$1/\phi''$
1	$L_e(k_A/D_e)^{1/2}$	ϕ'	$1/\phi'$	$1/\phi''$
2	$L_e(k_A c_{As}/D_e)^{1/2}$	$\phi'/(2/3)^{1/2}$	$(2/3)^{1/2}/\phi'$	$1/\phi''$

8.5.4.4 Effect of Order of Reaction on η

The development of an analytical expression for η in Example 8-4 is for a first-order reaction and a particular particle shape (flat plate). Other orders of reaction can be postulated and investigated. For a zero-order reaction, analytical results can be obtained in a relatively straightforward way for both η and ψ (problems 8-14 for a flat plate and 8-15 for a sphere). Corresponding results can be obtained, although not so easily, for an nth-order reaction in general; an exact result can be obtained for ψ and an approximate one for η. Here, we summarize the results without detailed justification.

For an nth-order reaction, the diffusion equation corresponding to equation 8.5-12 is

$$d^2\psi/dz^2 - \phi^2\psi^n = 0 \tag{8.5-18}$$

where the Thiele modulus, ϕ, is

$$\phi = L(k_A c_{As}^{n-1}/D_e)^{1/2} \tag{8.5-19}$$

The asymptotic solution ($\phi \to$ large) for η is $[2/(n+1)]^{1/2}/\phi$, of which the result given by 8.5-14c is a special case for a first-order reaction. The general result can thus be used to normalize the Thiele modulus for order so that the results for strong pore-diffusion resistance all fall on the same limiting straight line of slope -1 in Figure 8.11. The normalized Thiele modulus for this purpose is

$$\phi'' = \left(\frac{n+1}{2}\right)^{1/2} \phi' \tag{8.5-20}$$

$$= \left(\frac{n+1}{2}\right)^{1/2} \frac{L_e}{L} \phi \tag{8.5-20a}$$

$$= L_e \left(\frac{n+1}{2} \frac{k_A c_{As}^{n-1}}{D_e}\right)^{1/2} \tag{8.5-20b}$$

As a result,

$$\eta \to 1/\phi'' \text{ as } \phi'' \to \text{large} \tag{8.5-21}$$

regardless of order n. The results for orders 0, 1, and 2 are summarized in Table 8.3.

8.5.4.5 General Form of Thiele Modulus

The conclusions about asymptotic values of η summarized in Tables 8.2 and 8.3, and the behavior of η in relation to Figure 8.11, require a generalization of the definition of the Thiele modulus. The result for ϕ'' in equation 8.5-20 is generalized with respect to particle geometry through L_e, but is restricted to power-law kinetics. However, since surface reactions may follow other kinetics, such as Langmuir-Hinshelwood kinetics, there is a need to define a general Thiele modulus (ϕ_G) applicable to all forms of kinetics as well as shape.

The form of ϕ_G developed by Petersen (1965) in terms of reactant A, and for constant D_e, is:

$$\phi_G = \frac{L_e(-r_A)_{int|c_{As}}}{[2D_e \int_0^{c_{As}}(-r_A)_{int}dc_A]^{1/2}} \quad (8.5\text{-}22)$$

where $(-r_A)_{int}$ is the intrinsic rate given by the rate law, and $(-r_A)_{int|c_{As}}$ is the rate evaluated at the concentration at the exterior surface of the particle, c_{As}. All forms of Thiele modulus given previously may be obtained from this general expression.

8.5.4.6 Identifying the Presence of Diffusion Resistance

The presence (or absence) of pore-diffusion resistance in catalyst particles can be readily determined by evaluation of the Thiele modulus and subsequently the effectiveness factor, if the intrinsic kinetics of the surface reaction are known. When the intrinsic rate law is not known completely, so that the Thiele modulus cannot be calculated, there are two methods available. One method is based upon measurement of the rate for differing particle sizes and does not require any knowledge of the kinetics. The other method requires only a single measurement of rate for a particle size of interest, but requires knowledge of the order of reaction. We describe these in turn.

8.5.4.6.1 *Effect of particle size.* If the rate of reaction, $(-r_A)_{obs}$, is measured for two or more particle sizes (values of L_e), while other conditions are kept constant, two extremes of behavior may be observed.

(1) The rate is independent of particle size. This is an indication of negligible pore-diffusion resistance, as might be expected for either very porous particles or sufficiently small particles such that the diffusional path-length is very small. In either case, $\eta \to 1$, and $(-r_A)_{obs} = (-r_A)_{int}$ for the surface reaction.
(2) The rate is inversely proportional to particle size. This is an indication of strong pore-diffusion resistance, in which $\eta \to 1/\phi''$ as $\phi'' \to$ large. Since $\phi'' \propto L_e$ for fixed other conditions (surface kinetics, D_e, and c_{As}), if we compare measured rates for two particle sizes (denoted by subscripts 1 and 2), for strong pore-diffusion resistance,

$$\frac{(-r_A)_{obs,1}}{(-r_A)_{obs,2}} = \frac{\eta_1}{\eta_2} = \frac{\phi_2''}{\phi_1''} = \frac{L_{e2}}{L_{e1}} \quad (8.5\text{-}23)$$

8.5.4.6.2 *Weisz-Prater criterion.* The relative significance of pore-diffusion resistance can be assessed by a criterion, known as the Weisz-Prater (1954) criterion, which requires only a single measurement of the rate, together with knowledge of D_e, L_e, c_{As} and the order of the surface reaction (but not of the rate constant).

For an nth-order surface reaction of species A, the rate and Thiele modulus, respectively are

$$(-r_A)_{obs} = \eta k_A c_{As}^n \quad (8.5\text{-}24)$$

8.5 Heterogeneous Catalysis: Kinetics in Porous Catalyst Particles 209

$$\phi'' = L_e \left(\frac{n+1}{2} \frac{k_A c_{As}^{n-1}}{D_e} \right)^{1/2} \tag{8.5-20b}$$

Eliminating k_A from these two equations, and grouping measurable quantities together on the left side, we have

$$\frac{(n+1)}{2} \frac{(-r_A)_{obs} L_e^2}{D_e c_{As}} = \eta (\phi'')^2 = \Phi \tag{8.5-25}$$

where Φ is referred to as the observable modulus and is evaluated by the dimensionless group on the left.

For negligible pore-diffusion resistance, $\eta \to 1$ and $\phi'' \to$ small, say < 0.5. Thus,

$$\Phi < 0.25, \text{ say (negligible diffusion resistance)} \tag{8.5-26}$$

For strong pore-diffusion resistance, $\eta \to 1/\phi''$, and $\phi'' \to$ large, say > 5. Thus,

$$\Phi > 5, \text{ say (strong diffusion resistance)} \tag{8.5-27}$$

8.5.4.7 Strong Pore-Diffusion Resistance: Some Consequences

Here, we consider the consequences of being in the region of strong pore-diffusion resistance ($\eta \to 1/\phi''$ as $\phi'' \to$ large) for the apparent order of reaction and the apparent activation energy; ϕ'' is given by equation 8.5-20b.

Consider an nth-order surface reaction, represented by A(g) $\to$ product(s), occurring in a catalyst particle, with negligible external resistance to mass transfer so that $c_{As} \approx c_{Ag}$. Then the observed rate of reaction is

$$(-r_A)_{obs} = \eta k_A c_{Ag}^n = (1/\phi'') k_A c_{Ag}^n = \frac{1}{L_e} \left(\frac{2}{n+1} \frac{D_e}{k_A c_{Ag}^{n-1}} \right)^{1/2} k_A c_{Ag}^n = k_{obs} c_{Ag}^{(n+1)/2} \tag{8.5-28}$$

where

$$k_{obs} = \frac{1}{L_e} \left(\frac{2}{n+1} \right)^{1/2} (k_A D_e)^{1/2} \tag{8.5-29}$$

According to equation 8.5-28, the nth-order surface reaction becomes a reaction for which the observed order is $(n+1)/2$. Thus, a zero-order surface reaction becomes one of order 1/2, a first-order reaction remains first-order, and second-order becomes order 3/2. This is the result if D_e is independent of concentration, as would be the case if Knudsen diffusion predominated. If molecular diffusion predominates, for pure A, $D_e \propto c_{Ag}^{-1}$, and the observed order becomes $n/2$, with corresponding results for particular orders of surface reaction (e.g., a first-order surface reaction is observed to have order 1/2).

Consider next the apparent E_A. From equation 8.5-29,

$$\frac{d \ln k_{obs}}{dT} = \frac{1}{2} \left(\frac{d \ln k_A}{dT} + \frac{d \ln D_e}{dT} \right) \tag{8.5-30}$$

If k_{obs}, k_A, and D_e all follow Arrhenius-type behavior,

$$E_{A,obs} = \frac{1}{2}[E_A(\text{surface reaction}) + E_A(\text{diffusion})] \approx \frac{1}{2}E_A(\text{surface reaction}) \quad (8.5\text{-}31)$$

since the activation energy for diffusion ($\approx RT$) is usually small compared to the (true) activation energy for a reaction (say 50 to 200 kJ mol^{-1}). The result is that, if reaction takes places in the catalyst particle in the presence of strong pore-diffusion resistance, the observed E_A is about 1/2 the true E_A for the surface reaction. This effect may be observed on an Arrhenius plot (ln k_{obs} versus $1/T$) as a change in slope, if conditions are such that there is a change from reaction-rate control (negligible pore-diffusion resistance) at relatively low temperatures to strong pore-diffusion resistance at higher temperatures.

8.5.5 Dependence of η on Temperature

The definition of the particle effectiveness factor η involves the intrinsic rate of reaction, $(-r_A)_{int}$, for reaction A → products, at the exterior surface conditions of gas-phase concentration (c_{As}) and temperature (T_s). Thus, from equation 8.5-5,

$$(-r_A)_{obs} = \eta(-r_A)_{int, c_{As}, T_s} \quad (8.5\text{-}32)$$

So far, we have assumed that the particle is isothermal and have focused only on the diffusional characteristics and concentration gradient within the particle, and their effect on η. We now consider the additional possibility of a temperature gradient arising from the thermal characteristics of the particle and the reaction, and its effect on η.

The existence of a temperature gradient is illustrated schematically in Figure 8.9 for a spherical or cylindrical (end-on) particle, and for both an exothermic and an endothermic reaction. The overall drop in temperature ΔT_{ov} from the center of the particle to bulk gas may be divided into two parts:

$$\Delta T_{ov} = \Delta T_p + \Delta T_f \quad (8.5\text{-}33)$$

where ΔT_p is the drop across the particle itself, and ΔT_f is that across the gas film or the thermal boundary layer. It is the gradient across the particle, corresponding to ΔT_p, that influences the particle effectiveness factor, η. The gradient across the film influences the overall effectiveness factor, η_o (Section 8.5.6).

Two limiting cases arise from equation 8.5-33:

(1) Rate of intraparticle heat conduction is rate controlling:

$$\Delta T_f \to 0; T_s \to T_g \quad (8.5\text{-}33a)$$

The result is a nonisothermal particle with an exterior surface at T_g.

(2) Rate of heat transfer across gas film is rate controlling:

$$\Delta T_p \to 0; T(\text{throughout}) \to T_s \quad (8.5\text{-}33b)$$

The result is an isothermal particle, but hotter (exothermic case) or colder (endothermic case) than the bulk gas at T_g.

8.5 Heterogeneous Catalysis: Kinetics in Porous Catalyst Particles

For a catalyst particle to be isothermal while reaction is taking place within it, the enthalpy generated or consumed by reaction must be balanced by enthalpy (heat) transport (mostly by conduction) through the particle. This is more likely to occur if the enthalpy of reaction is small and the effective thermal conductivity (k_e, analogous to D_e) of the catalyst material is large. However, should this balance not occur, a temperature gradient exists. For an *exothermic* reaction, T increases with increasing distance into the particle, so that the average rate of reaction within the particle is greater than that at T_s. This is the opposite of the usual effect of concentration: the average rate is less than that at c_{As}. The result is that $\eta_{exo} > \eta_{isoth}$. Since the effect of increasing T on rate is an exponential increase, and that of decreasing c_A is usually a power-law decrease, the former may be much more significant than the latter, and η_{exo} may be > 1 (even in the presence of a diffusional resistance). For an isothermal particle, $\eta_{isoth} < 1$ because of the concentration effect alone. For an *endothermic* reaction, the effect of temperature is to reinforce the concentration effect, and $\eta_{endo} < \eta_{isoth} < 1$.

The dependence of η on T has been treated quantitatively by Weisz and Hicks (1962). We outline the approach and give some of the results for use here, but omit much of the detailed development.

For a first-order reaction, A $\rightarrow$ products, and a spherical particle, the material-balance equation corresponding to equation 8.5-7, and obtained by using a thin-shell control volume of inside radius r, is

$$\frac{d^2 c_A}{dr^2} + \frac{2}{r}\frac{dc_A}{dr} - \frac{k_A}{D_e}c_A = 0 \tag{8.5-34}$$

(the derivation is the subject of problem 8-13). The analogous energy-balance equation is

$$\frac{d^2 T}{dr^2} + \frac{2}{r}\frac{dT}{dr} + \frac{(-\Delta H_{RA})k_A}{k_e}c_A = 0 \tag{8.5-35}$$

Boundary conditions for these equations are:

At particle surface: $r = R$; $T = T_s$; $c_A = c_{As}$ (8.5-36)

At particle center: $r = 0$; $dT/dr = 0$; $dc_A/dr = 0$ (8.5-37)

Equations 8.5-34 and -35 are nonlinearly coupled through T, since k_A depends exponentially on T. The equations cannot therefore be treated independently, and there is no exact analytical solution for $c_A(r)$ and $T(r)$. A numerical or approximate analytical solution results in η expressed in terms of three dimensionless parameters:

$$\eta(T) = \eta(\phi, \gamma, \beta) \tag{8.5-38}$$

where ϕ ($= R(k_A/D_e)^{1/2}$, Table 8.1) is the Thiele modulus, and γ and β are defined as follows:

$$\gamma = E_A/RT_s \tag{8.5-39}$$

$$\beta = \frac{\Delta T_{p,max}}{T_s} = \frac{D_e(-\Delta H_{RA})c_{As}}{k_e T_s} \tag{8.5-40}$$

where $\Delta T_{p,max}$ is the value of ΔT_p when $c_A(r = 0) = 0$. For an exothermic reaction, $\beta > 0$; for an endothermic reaction, $\beta < 0$; for an isothermal particle, $\beta = 0$, since $\Delta T_p = 0$.

The result for $\Delta T_{p,max}$ contained in equation 8.5-40 can be obtained from the following energy balance for a control surface or a core of radius r:

rate of thermal conduction across control surface
$$= \text{rate of enthalpy consumption/generation within core}$$
$$\equiv \text{rate of diffusion of A across control surface} \times (-\Delta H_{RA}) \quad (8.5\text{-}41)$$

That is, from Fourier's and Fick's laws,

$$k_e \frac{dT}{dr} = D_e(-\Delta H_{RA})\frac{dc_A}{dr} \quad (8.5\text{-}42)$$

Integration of equation 8.5-42 from the center of the particle ($r = 0$, $T = T_o$, $c_A = c_{Ao}$) to the surface ($r = R$, $T = T_s$, $c_A = c_{As}$), with k_e, D_e, and $(-\Delta H_{RA})$ constant, results in

$$\Delta T_p = T_s - T_o = \frac{D_e(-\Delta H_{RA})}{k_e}(c_{As} - c_{Ao}) \quad (8.5\text{-}43)$$

or, with $c_{Ao} \to 0$, and $\Delta T_p \to \Delta T_{p,max}$,

$$\Delta T_{p,max} = \frac{D_e(-\Delta H_{RA})c_{As}}{k_e} \quad (8.5\text{-}44)$$

as used in equation 8.5-40.

Some of the results of Weisz and Hicks (1962) are shown in Figure 8.12 for $\gamma = 20$, with η as a function of ϕ and β (as a parameter). Figure 8.12 confirms the conclusions reached qualitatively above. Thus, η_{exo} ($\beta > 0$) $> \eta_{isoth}$ ($\beta = 0$), and $\eta_{exo} > 1$ for relatively high values of β and a sufficiently low value of ϕ; $\eta_{endo} < \eta_{isoth} < 1$. At high values of β and low values of ϕ, there is the unusual phenomenon of three solutions for η for a given value of β and of ϕ; of these, the high and low values represent stable steady-state solutions, and the intermediate value represents an unstable solution. The region in which this occurs is rarely encountered. Some values of the parameters are given by Hlaváček and Kubiček (1970).

8.5.6 Overall Effectiveness Factor η_o

The particle effectiveness factor η defined by equation 8.5-5 takes into account concentration and temperature gradients within the particle, but neglects any gradients from bulk fluid to the exterior surface of the particle. The overall effectiveness factor η_o takes both into account, and is defined by reference to bulk gas conditions (c_{Ag}, T_g) rather than conditions at the exterior of the particle (c_{As}, T_s):

$$\eta_o = r_A(\text{observed})/r_A(c_{Ag}, T_g) \quad (8.5\text{-}45)$$

Here, as in Section 8.5.4, we treat the isothermal case for η_o, and relate η_o to η. η_o may then be interpreted as the ratio of the (observed) rate of reaction with pore diffusion and external mass transfer resistance to the rate with neither of these present.

We first relate η_o to η, k_A, and k_{Ag}, the last two characterizing surface reaction and mass transfer, respectively; mass transfer occurs across the gas film indicated in Figure 8.9. Consider a first-order surface reaction. If $(-r_A)$ is the observed rate of reaction,

8.5 Heterogeneous Catalysis: Kinetics in Porous Catalyst Particles

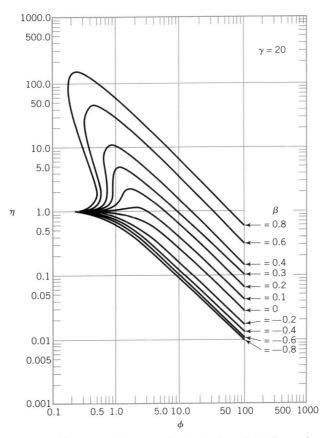

Figure 8.12 $\eta(\phi, \beta)$ for $\gamma = 20$; spherical particle, first-order reaction (reprinted from *Chemical Engineering Science*, **17**, Weisz, P.B., and Hicks, J.S., The behaviour of porous catalyst particles in view of internal mass and heat diffusion effects, pp. 265–275, 1962, with permission from Elsevier Science.)

from the definition of η_o,

$$(-r_A) = \eta_o k_A c_{Ag} \tag{8.5-46}$$

and, from the definition of η

$$(-r_A) = \eta k_A c_{As} \tag{8.5-47}$$

Furthermore, at steady-state, $(-r_A)$ is also the rate of mass transfer of A across the exterior film, such mass transfer being in series with the combined intraparticle processes of diffusion and reaction; hence, from the definition of k_{Ag},

$$(-r_A) = k_{Ag}(c_{Ag} - c_{As}) \tag{8.5-48}$$

On eliminating $(-r_A)$ and c_{As} from the three equations 8.5-46 to -48, for example, by first obtaining an expression for c_{As} from 8.5-47 and -48, and then substituting for c_{As} back in equation 8.5-47 and comparing the resulting equation with 8.5-46, we obtain

$$\eta_o = \frac{1}{(k_A/k_{Ag}) + (1/\eta)} \tag{8.5-49}$$

214 Chapter 8: Catalysis and Catalytic Reactions

and the rate law, from equations 8.5-46 and -49, may be written as

$$(-r_A) = \frac{c_{Ag}}{(1/k_{Ag}) + (1/\eta k_A)} \tag{8.5-50}$$

Special forms of equation 8.5-50 arise depending on the relative importance of mass transfer, pore diffusion, and surface reaction; in such cases, one or two of the three may be the rate-controlling step or steps. These cases are explored in problem 8-18. The result given there for problem 8-18(a) is derived in the following example.

EXAMPLE 8-5

If the surface reaction is rate controlling, what is the form of the rate law from equation 8.5-50, and what does this mean for k_{Ag}, c_{As}, η, and η_o?

SOLUTION

If the surface reaction is the rate-controlling step, any effects of external mass transfer and pore-diffusion are negligible in comparison. The interpretation of this, in terms of the various parameters, is that $k_{Ag} \gg k_A$, $c_{As} \to c_{Ag}$, and η and η_o both approach the value of 1. Thus, the rate law, from equation 8.5-50, is just that for a homogeneous gas-phase reaction:

$$(-r_A) = k_A c_{Ag} \tag{8.5-51}$$

The concentration profile for reactant A in this case is a horizontal line at $c_A = c_{Ag}$; this can be visualized from Figure 8.9.

8.6 CATALYST DEACTIVATION AND REGENERATION

Despite advances in catalyst design, all catalysts are subject to a reduction in activity with time (deactivation). The rate at which the catalyst is deactivated may be very fast, such as for hydrocarbon-cracking catalysts, or may be very slow, such as for promoted iron catalysts used for ammonia synthesis, which may remain on-stream for several years without appreciable loss of activity. Nonetheless, the design engineer must account for the inevitable loss of catalyst activity, allowing for either regeneration of the catalyst or its periodic replacement. Since these remedial steps are costly, both in terms of capital cost and loss of production during shutdown, it is preferable to minimize catalyst deactivation if possible. In this section, we explore the processes which cause deactivation, and how deactivation can affect the performance of a catalyst. We also discuss methods for preventing deactivation, and for regeneration of deactivated catalysts.

8.6.1 Fouling

Fouling occurs when materials present in the reactor (reactants, or products, or intermediates) are deposited upon the surface of the catalyst, blocking active sites. The most common form of fouling is by carbonaceous species, a process known as "coking." Coke may be deposited in several forms, including laminar graphite, high-molecular-weight polycyclic aromatics (tars), polymer aggregates, and metal carbides. The form of the coke depends upon the catalyst, the temperature, and the partial pressure of the carbonaceous compound. Very little coke forms on silica or carbon supports, but acidic supports or catalysts are especially prone to coking.

To minimize coking, the reactor may be operated at short residence times, or hydrogen may be added to the process stream to convert gas-phase carbon into methane. It is also advantageous to minimize the temperature upstream of the catalyst bed, since gas-phase carbon is less readily formed at low temperatures.

8.6.2 Poisoning

Poisoning is caused by chemisorption of compounds in the process stream; these compounds block or modify active sites on the catalyst. The poison may cause changes in the surface morphology of the catalyst, either by surface reconstruction or surface relaxation, or may modify the bond between the metal catalyst and the support. The toxicity of a poison (P) depends upon the enthalpy of adsorption for the poison, and the free energy for the adsorption process, which controls the equilibrium constant for chemisorption of the poison (K_P). The fraction of sites blocked by a reversibly adsorbed poison (θ_P) can be calculated using a Langmuir isotherm (equation 8.4-23a):

$$\theta_P = \frac{K_P p_P}{1 + K_A p_A + K_P p_P} \qquad (8.6-1)$$

where K_A and K_P are the adsorption constants for the reactant (A) and the poison, respectively, and p_A and p_P are the partial pressures of the reactant and poison. The catalyst activity remaining is proportional to the fraction of unblocked sites, $1 - \theta_P$.

The compound responsible for poisoning is usually an impurity in the feed stream; however, occasionally, the products of the desired reaction may act as poisons. There are three main types of poisons:

(1) Molecules with reactive heteroatoms (e.g., sulfur);
(2) Molecules with multiple bonds between atoms (e.g., unsaturated hydrocarbons);
(3) Metallic compounds or metal ions (e.g., Hg, Pd, Bi, Sn, Cu, Fe).

The strength of the bond between the poison and the catalyst (or support) may be relatively weak, or exceptionally strong. In the latter case, poisoning leads to an irreversible loss of activity. However, if the chemisorption bond is very weak, the observed loss of activity can be reversed by eliminating the impurity (poison) from the feed stream. Poisons may be eliminated by physical separation, or in the case of a type (1) or type (2) poison, the poison may be converted to a nontoxic compound by chemical treatment (oxidation for type (1), and hydrogenation for type (2)). If a product is responsible for poisoning, it may be helpful to operate the reactor at low conversion, and/or selectively remove product at intermediate stages of a multistage reactor.

8.6.3 Sintering

Sintering is caused by growth or agglomeration of small crystals which make up the catalyst or its support. The structural rearrangement observed during sintering leads to a decrease in surface area of the catalyst, and, consequently, an irreversible reduction in catalyst sites. Sintering generally occurs if the local temperature of the catalyst exceeds approximately one-third to one-half of its melting temperature (T_m). The upper limit (i.e., $(1/2)T_m$) applies under "dry" conditions, whereas the lower temperature limit (i.e., $(1/3)T_m$) applies if steam is present, since steam facilitates reorganization of

Table 8.4 Sintering temperatures for common metals

Metal	Sintering temperature/°C;[$(1/3)T_m$]
Cu	360
Fe	500
Ni	500
Pt	570
Pd	500

many metals, aluminas, and silicas. Table 8.4 lists some common metal catalysts and the temperature at which the onset of sintering is expected to occur.

To prevent sintering, catalysts may be doped with stabilizers which may have a high melting point and/or prevent agglomeration of small crystals. For example, chromia, alumina, and magnesia, which have high melting points, are often added as stabilizers of finely divided metal catalysts. Furthermore, there is evidence that sintering of platinum can be prevented by adding trace quantities of chlorinated compounds to the gas stream. In this case, chlorine increases the activation energy for the sintering process, and, thus, reduces the sintering rate.

8.6.4 How Deactivation Affects Performance

Catalyst deactivation may affect the performance of a reactor in several ways. A reduction in the number of catalyst sites can reduce catalytic activity and decrease fractional conversion. However, some reactions depend solely on the presence of metal, while others depend strongly on the configuration of the metal. Thus, the extent to which performance is affected depends upon the chemical reaction to be catalyzed, and the way in which the catalyst has been deactivated. For example, deposition/chemisorption of sulfur, nitrogen, or carbon on the catalyst generally affects hydrogenation reactions more than exchange reactions. Consequently, if parallel reactions are to be catalyzed, deactivation may cause a shift in selectivity to favor nonhydrogenated products. Similarly, heavy metals (e.g., Ni, Fe) present in the feed stream of catalytic crackers can deposit on the catalyst, and subsequently catalyze dehydrogenation reactions. In this case, the yield of gasoline is reduced, and more light hydrocarbons and hydrogen produced.

Another way in which catalyst deactivation may affect performance is by blocking catalyst pores. This is particularly prevalent during fouling, when large aggregates of materials may be deposited upon the catalyst surface. The resulting increase in diffusional resistance may dramatically increase the Thiele modulus, and reduce the effectiveness factor for the reaction. In extreme cases, the pressure drop through a catalyst bed may also increase dramatically.

8.6.5 Methods for Catalyst Regeneration

In some cases, it is possible to restore partially or completely the activity of a catalyst through chemical treatment. The regeneration process may be slow, either because of thermodynamic limitations or diffusional limitations arising from blockage of catalyst pores. Although the rate of desorption generally increases at high temperatures, prolonged exposure of the catalyst to a high-temperature gas stream can lead to sintering, and irreversible loss of activity. If the bound or deposited species cannot be gasified at temperatures lower than the sintering temperature (see Table 8.4), then the poisoning or fouling is considered to be irreversible.

8.6 Catalyst Deactivation and Regeneration

For catalysts poisoned by sulfur, the metal-sulfur bond is usually broken in the presence of steam, as shown for nickel:

$$Ni\text{–}S + H_2O \rightarrow NiO + H_2S$$
$$H_2S + 2H_2O \rightleftharpoons SO_2 + 3H_2$$

The equilibrium for the second reaction favors H_2S until extremely high temperatures are reached ($> 700°C$). Thus, sintering of the catalyst could be a problem. Furthermore, SO_2 can act as a poison for some catalysts. If sintering or SO_2 poisoning precludes steam treatment, it is usually possible to remove deposited sulfur by passing a sulfur-free gas stream over the catalyst at moderate temperatures for an extended period of time.

Regeneration of coked catalysts may be accomplished by gasification with oxygen, steam, hydrogen, or carbon dioxide:

$$C + O_2 \rightarrow CO_2$$
$$C + H_2O \rightarrow CO + H_2$$
$$C + 2H_2 \rightarrow CH_4$$
$$C + CO_2 \rightarrow 2CO$$

The first reaction is strongly exothermic, and may lead to high local temperatures within the catalyst. Thus, temperature must be carefully controlled to avoid sintering.

EXAMPLE 8-6

A coked porous catalyst is to be regenerated by passage of a stream of CO_2 over it at 1000 K for reaction according to $C(s) + CO_2(A) \rightarrow 2CO(B)$. From the data given below (Austin and Walker, 1963), calculate the following characteristics of the regeneration process at the conditions given: (a) the Thiele modulus, (b) the effectiveness factor, and (c) the (actual) rate of regeneration, $(-r_A)_{obs}$.

Data: For the catalyst, $D_e = 0.10$ cm^2 s^{-1}, $L_e = 0.7$ cm; c_{As} (exterior surface concentration) $= 0.012$ mol L^{-1}. The reaction follows LH kinetics, with the intrinsic rate given by

$$(-r_A)_{int} = kc_A/(1 + K_A c_A + K_B c_B)$$

where $k = 3.8 \times 10^{-4}$ s^{-1}, $K_A = 340$ L mol^{-1}, $K_B = 4.2 \times 10^6$ L mol^{-1}, and c_i is in mol L^{-1}.

SOLUTION

This example illustrates calculation of the rate of a surface reaction from an intrinsic-rate law of the LH type in conjunction with determination of the effectiveness factor (η) from the generalized Thiele modulus (ϕ_G) and Figure 8.11 as an approximate representation of the η–ϕ_G relation. We first determine ϕ_G, then η, and finally $(-r_A)_{obs}$.

(a) From equation 8.5-22,

$$\phi_G = \frac{L_e(-r_A)_{int|c_{As}}}{[2D_e \int_0^{c_{As}} (-r_A)_{int} dc_A]^{1/2}} \qquad (8.5\text{-}22)$$

where $(-r_A)_{int|c_{As}}$ is the intrinsic rate evaluated at c_{As}. Since, from the stoichiometry, $c_B = 2(c_{As} - c_A)$, we can eliminate c_B from the LH expression, and express the

218 Chapter 8: Catalysis and Catalytic Reactions

integral in 8.5-22 as:

$$\int_0^{c_{As}} (-r_A)_{int} dc_A = \int_0^{c_{As}} \left[\frac{kc_A}{1 + 2K_B c_{As} + (K_A - 2K_B)c_A} \right] dc_A$$

The integral may be evaluated either numerically by means of E-Z Solve, or analytically by means of the substitution $x = 1 + 2K_B c_{As} + (K_A - 2K_B)c_A$. The latter results in

$$\int_0^{c_{As}} (-r_A)_{int} dc_A = \frac{k}{(K_A - 2K_B)^2} \left[(K_A - 2K_B)c_{As} + (1 + 2K_B c_{As}) \ln \left(\frac{1 + 2K_B c_{As}}{1 + K_A c_{As}} \right) \right]$$

$$= \frac{3.8 \times 10^{-4}}{[340 - 2(4.2 \times 10^6)]^2} \{[340 - 2(4.2 \times 10^6)]0.012 +$$

$$[1 + 2(4.2 \times 10^6)0.012] \ln \left[\frac{1 + 2(4.2 \times 10^6)0.012}{1 + 340(0.012)} \right] \} = 4.8 \times 10^{-12} \text{mol}^2 \text{ L}^{-2} \text{ s}^{-1}$$

From the LH rate expression and the stoichiometry, since $c_A = c_{As}$,

$$(-r_A)_{int|c_{As}} = \frac{kc_{As}}{1 + K_A c_{As} + 2K_B(c_{As} - c_{As})} = \frac{kc_{As}}{1 + K_A c_{As}}$$

$$= \frac{3.8 \times 10^{-4}(0.012)}{1 + 340(0.012)} = 9.0 \times 10^{-7} \text{mol L}^{-1} \text{ s}^{-1}$$

Substitution of numerical values in 8.5-22 gives

$$\phi_G = \frac{0.7(9.0 \times 10^{-7})}{[2(0.10)4.8 \times 10^{-12}]^{1/2}} = 0.64$$

(b) From Figure 8.11,

$$\eta = 0.85 \text{ to } 0.90$$

which implies a slight but significant effect of diffusional resistance on the process.
(c) $(-r_A)_{obs} = \eta(-r_A)_{int} = 8 \times 10^{-7} \text{mol L}^{-1} \text{ s}^{-1}$.

8.7 PROBLEMS FOR CHAPTER 8

8-1 The hydrolysis of ethyl acetate catalyzed by hydrogen ion,

$$CH_3COOC_2H_5(A) + H_2O \rightarrow CH_3COOH + C_2H_5OH$$

in dilute aqueous solution, is first-order with respect to ethyl acetate at a given pH. The apparent first-order rate constant, $k(= k_A c_{H^+}^\alpha)$, however, depends on pH as indicated by:

pH	3	2	1
$10^4 k / \text{s}^{-1}$	1.1	11	110

What is the order of reaction with respect to hydrogen ion H^+, and what is the value of the rate constant k_A, which takes both c_A and c_{H^+} into account? Specify the units of k_A.

8-2 (a) The Goldschmidt mechanism (Smith, 1939) for the esterification of methyl alcohol (M) with acetic acid (A) catalyzed by a strong acid (e.g., HCl), involves the following steps:

$$M + H^+ \rightarrow CH_3OH_2^+(C); \text{rapid} \quad (1)$$

$$A + C \xrightarrow{k} CH_3COOCH_3(E) + H_3O^+; \text{slow} \quad (2)$$

$$M + H_3O^+ \xrightleftharpoons{K} C + H_2O(W); \text{rapid} \quad (3)$$

Show that the rate law for this mechanism, with M present in great excess, is

$$r_E = 2kLc_A c_{HCl}/(L + c_W), \quad (4)$$

where

$$L = c_M K = c_C c_W/c_{H_3O^+}. \quad (5)$$

Assume all H^+ is present in C and in H_3O^+.

(b) Show that the integrated form of equation (4) for a constant-volume batch reactor operating isothermally with a fixed catalyst concentration is

$$k = [(L + c_{Ao}) \ln (c_{Ao}/c_A) - (c_{Ao} - c_A)]/c_{HCl} Lt.$$

This is the form used by Smith (1939) to calculate k and L.

(c) Smith found that L depends on temperature and obtained the following values (what are the units of L?):

t/°C:	0	20	30	40	50
L:	0.11	0.20	0.25	0.32	0.42

Does L follow an Arrhenius relationship?

8-3 Brönsted and Guggenheim (1927) in a study of the mutarotation of glucose report data on the effect of the concentration of hydrogen ion and of a series of weak acids and their conjugate bases. The reaction is first-order with respect to glucose, and the rate constant (k_{obs}) is given by equation 8.2-9 (assume $k_{OH^-} = 10^4$ L mol^{-1} min^{-1}). Some of their data for three separate sets of experiments at 18°C are as follows:

(1)
$10^3 c_{HClO_4}$/mol L^{-1}	1	9.9	20	40
$10^3 k_{obs}$/min^{-1}	5.42	6.67	8.00	11.26

(2) $c_{HCO_2Na} = 0.125$ mol L^{-1} (constant)

$10^3 c_{HCO_2H}$/mol L^{-1}	5	124	250
$10^3 k_{obs}$/min^{-1}	7.48	7.86	8.50

(3) $c_{HCO_2H} = 0.005$ mol L^{-1} (constant)

$10^3 c_{HCO_2Na}$/mol L^{-1}	40	60	100	125
$10^3 k_{obs}$/min^{-1}	6.0	6.23	6.92	7.48

Calculate: (a) k_o and k_{H^+}; (b) k_{HA}; (c) k_{A^-}.

Note that HClO$_4$ is a strong acid and that HCO$_2$H (formic acid) is a weak acid ($K_a = 2.1 \times 10^{-4}$). At 18°C, $K_w = 1.5 \times 10^{-14}$.

8-4 Repeat part (b) of Example 8-2 for a CSTR, and comment on the result.

8-5 Propose a rate law based on the Langmuir-Hinshelwood model for each of the following heterogeneously catalyzed reactions:

(a) In methanol synthesis over a Cu-ZnO-Cr$_2$O$_3$ catalyst, the rate-controlling process appears to be a termolecular reaction in the adsorbed phase:

$$CO \bullet s + 2H \bullet s \rightarrow CH_3OH \bullet s + 2\ s$$

Consider two cases: (i) the product is strongly adsorbed and inhibits the reaction; and (ii) it is very weakly adsorbed.

(b) The decomposition of acetaldehyde on Pt at temperatures between 960 and 1200°C, and at pressures between 3.33 and 40.0 kPa, appears to be a bimolecular reaction with no inhibition by reaction products.

(c) A study of the kinetics of ethanol dehydrogenation over Cu in the presence of water vapor, acetone, or benzene showed that any one of these three inhibited the reaction.

(d) In the reaction of nitrous oxide (N_2O) with hydrogen over Pt (507–580°C, p_{H_2} = 7 to 53 kPa, p_{N_2O} = 40 to 53 kPa), it has been observed that N_2O is weakly adsorbed and H_2 is very strongly adsorbed.

8-6 For the surface-catalyzed gas-phase reaction $A(g) + B(g) \to$ products, what is the form of the rate law, according to the LH model, if A is strongly adsorbed and B is weakly adsorbed? Assume there is no adsorption of product(s). Interpret the results beyond what is already specified.

8-7 For the reaction in problem 8-6, suppose there is one product P which can be adsorbed. Derive the form of the rate law according to the LH model, if

(a) A, B, and P are all moderately adsorbed;

(b) A and B are weakly adsorbed and P is strongly adsorbed. (Interpret the result further.)

8-8 Consider the reaction mechanism for methanol synthesis proposed in Figure 8.3:

$$CO\bullet s + H\bullet s \to HCO\bullet s + s \quad (1)$$

$$HCO\bullet s + H\bullet s \to H_2CO\bullet s + s \quad (2)$$

$$H_2CO\bullet s + H\bullet s \to H_3CO\bullet s + s \quad (3)$$

$$H_3CO\bullet s + H\bullet s \to H_3COH\bullet s + s \quad (4)$$

Assume that the coverages of H, CO, and methanol are given by the Langmuir adsorption isotherm in which CO, H_2, and methanol adsorption compete for the same sites, and the intermediates $H_xCO\bullet s$ are present in negligible quantities.

(a) Assume that step (1) is rate limiting, and write the general rate expression.

(b) Assume that (3) is rate limiting (steps (1) and (2) are in equilibrium), and write the general rate expression.

(c) Experimental data are represented by

$$r_{CH_3OH} = k p_{CO}^{-0.5} p_{H_2}^{1.3}$$

To obtain this rate law, which of the surface steps above is rate limiting?

(d) How would the rate law change if the $H_xCO\bullet s$ intermediates were allowed to cover a substantial fraction of sites? (This can be attempted analytically, or you may resort to simulations.)

8-9 (a) Rate laws for the decomposition of PH_3 (A) on the surface of Mo (as catalyst) in the temperature range 843–918 K are as follows:

pressure, p_A/kPa	rate law
$\to 0$	$(-r_A) = k p_A$
8×10^{-3}	$(-r_A) = k p_A/(a + b p_A)$
2.6×10^{-2}	$(-r_A)$ = constant

Interpret these results in terms of a Langmuir-Hinshelwood mechanism.

(b) In the decomposition of N_2O on Pt, if N_2O is weakly adsorbed and O_2 is moderately adsorbed, what form of rate law would be expected based on a Langmuir-Hinshelwood mechanism? Explain briefly.

8-10 (a) For the decomposition of NH$_3$ (A) on Pt (as catalyst), what is the form of the rate law, according to the Langmuir-Hinshelwood model, if NH$_3$ (reactant) is weakly adsorbed and H$_2$ (product) strongly adsorbed on Pt? Explain briefly. Assume N$_2$ does not affect the rate.

(b) Do the following experimental results, obtained by Hinshelwood and Burk (1925) in a constant-volume batch reactor at 1411 K, support the form used in (a)?

t/s	0	10	60	120	240	360	720
P/kPa	26.7	30.4	34.1	36.3	38.5	40.0	42.7

P is total pressure, and only NH$_3$ is present initially. Justify your answer quantitatively, for example, by using the experimental data in conjunction with the form given in (a). Use partial pressure as a measure of concentration.

8-11 (a) For a zero-order catalytic reaction, if the catalyst particle effectiveness factor is η, what is the overall effectiveness factor, η_o (in terms of η)? Justify your answer.

(b) For a solid-catalyst, gas-phase reaction A(g) → product(s), if the gas phase is pure A and the (normalized) Thiele modulus is 10, what is the value of the overall effectiveness factor? Explain briefly.

8-12 Swabb and Gates (1972) have studied pore-diffusion/reaction phenomena in crystallites of H(hydrogen)-mordenite catalyst. The crystallites were approximate parallelepipeds, the long dimension of which was assumed to be the pore length. Their analysis was based on straight, parallel pores in an isothermal crystallite (2 faces permeable). They measured (initial) rates of dehydration of methanol (A) to dimethyl ether in a differential reactor at 101 kPa using catalyst fractions of different sizes. Results (for two sizes) are given in the table below, together with quantities to be calculated, indicated by (?).

Catalyst/reaction in general:	Value
n, order of reaction (assumed)	1
$T/°C$	205
c_{As}/mol cm^{-3}	2.55×10^{-5}
ρ_p, catalyst (particle) density/g cm^{-3}	1.7
ϵ_p, catalyst (particle) void fraction	0.28
k_A, intrinsic rate constant/s^{-1}	(?)
D_e, effective diffusivity of A/cm^2 s^{-1}	(?)

Catalyst fraction:	1	2
L, mean pore length/cm	5.9×10^{-4}	11.3×10^{-4}
$(-r_A)$/moles A g cat^{-1} s^{-1}	7.33×10^{-4}	6.17×10^{-4}
ϕ, Thiele modulus	(?)	(?)
η, effectiveness factor	(?)	(?)

8-13 Derive an expression for the catalyst effectiveness factor (η) for a spherical catalyst particle of radius R. The effective diffusivity is D_e and is constant; the reaction (A → product(s)) is first-order $[(-r_A) = k_A c_A]$ and irreversible. Assume constant density, steady-state and equimolar counterdiffusion. Clearly state the boundary conditions and the form of the Thiele modulus (ϕ). If the diffusion or continuity equation is solved in terms of r (variable radius from center) and c_A, the substitution $y = rc_A$ is helpful.

8-14 Consider a gas-solid (catalyst) reaction, A(g) → products, in which the reaction is zero-order, and the solid particles have "slab" or "flat-plate" geometry with one face permeable to A.

(a) Derive the continuity or diffusion (differential) equation in nondimensional form for A, together with the expression for the Thiele modulus, ϕ.

(b) Solve the equation in (a) to give the nondimensional concentration profile $\psi(\phi, z)$, on the assumption that $\psi > 0$ for all values of z.
(c) Derive the result for the catalyst effectiveness factor η from (b).
(d) At what value of ϕ does the concentration of A drop to zero at the impermeable face?
(e) What does it mean for both ψ and η if ϕ is greater than the value, $\phi(d)$, obtained in part (d)? To illustrate this, sketch (on the same plot for comparison) three concentration profiles (ψ versus z) for (i) $\phi < \phi(d)$; (ii) $\phi = \phi(d)$; and (iii) $\phi > \phi(d)$. Completion of part (e) leads to a value of η in terms of ϕ for the case, (iii), of $\phi > \phi(d)$. (The result from part (c) applies for cases (i) and (ii).)

8-15 Consider a gas-solid (catalyst) reaction, $A(g) \rightarrow$ products, in which the reaction is zero-order, and the solid particles are spherical with radius R.
(a) Derive the diffusion equation for A, together with the expression for the Thiele modulus, ϕ.
(b) Solve the equation in (a) to give the nondimensional concentration profile $\psi(\phi, r)$, on the assumption that $\psi > 0$ throughout the particle, where $\psi = c_A/c_{As}$. (Hint: Use the substitution $y = dc_A/dr$.)
(c) Derive the result for the catalyst effectiveness factor η from (b).
(d) At what value of ϕ does the concentration of A drop to zero at the center of the particle ($r = 0$)?
(e) In terms of ϕ, under what condition does ψ become zero at r^*, where $0 < r^* < R$? Relate (i) ϕ and r^*, and (ii) η and r^* for this situation.

8-16 (a) For a solid-catalyzed reaction (e.g., A $\rightarrow$ products), calculate the value of the catalyst effectiveness factor (η) for the following case: $E_A = 83$ kJ mol^{-1}; A is a gas at 500 K, 2.4 bar (partial pressure); the Thiele modulus (ϕ) = 10; $k_e = 1.2 \times 10^{-3}$ J s^{-1} cm^{-1} K^{-1}; $D_e = 0.03$ cm^2 s^{-1}; $\Delta H_{RA} = +135$ kJ mol^{-1}. Use the Weisz-Hicks solution (Figure 8.12) for a first-order reaction with a spherical particle. Assume gas-film resistance is negligible for both heat and mass transfer.
(b) Repeat (a), if $\Delta H_{RA} = -135$ kJ mol^{-1}.
(c) Compare the results in (a) and (b) with the result for the case of an isothermal particle.

8-17 In the use of the observable modulus, Φ, defined by equation 8.5-25, in the Weisz-Prater criterion, c_{As} must be assessed. If c_{As} is replaced by c_{Ag}, the directly measurable gas-phase concentration, what assumption is involved?

8-18 For a first-order, gas-solid (catalyst) reaction, $A(g) \rightarrow$ product(s), the (isothermal) overall effectiveness factor (η_o) is related to the catalyst effectiveness factor (η) by

$$\frac{1}{\eta_o} = \frac{k_A}{k_{Ag}} + \frac{1}{\eta} \quad \text{(from 8.5-49)}$$

where k_A is the reaction rate constant, and k_{Ag} is the gas-film mass transfer coefficient. From this and other considerations, complete the table below for the following cases, with a brief justification for each entry, and a sketch of the concentration profile for each case:
(a) The surface reaction is rate controlling.
(b) Gas-film mass transfer is rate controlling.
(c) The combination of surface reaction and intraparticle diffusion is rate controlling.
(d) The combination of surface reaction and gas-film mass transfer is rate controlling.

Case	k_{Ag} vs. k_A	$c_{As} \rightarrow$?	$\eta \rightarrow$?	$\eta_o \rightarrow$?	$(-r_A) \rightarrow$?
(a)	$k_{Ag} \gg k_A$	c_{Ag}	1	1	$k_A c_{Ag}$
(b)			(indet.)		
(c)			η		
(d)	(comparable)	(ignore)			

8-19 Experimental values for the rate constant k in the Eklund equation (1956) for the oxidation of SO_2 over V_2O_5 catalyst are as follows for two different sizes/shapes of particle (A and B, described below):

$t/°C$	416	420	429	437	455	458	474	488	504	525	544
$10^6 k*$ (for A⁺)	-	6.7	-	17.7	47	-	99	-	-	-	-
$10^6 k*$ (for B†)	1.43	-	2.23	5.34	-	11.1	-	17.7	28.0	38.6	37.7

*units of k: moles SO_2 reacted (g cat)$^{-1}$ s^{-1} atm^{-1}
⁺A particles are spherical with diameter of 0.67 mm
†B particles are cylindrical, 8 mm in diameter and 25 mm in length

(a) From these data, what activation energy is indicated for the *surface* reaction?
(b) Do the data for the cylindrical particles suggest significant pore-diffusion resistance? If they do, what is the apparent activation energy for this range? See also Jensen-Holm and Lyne (1994).
(c) One particular plant used cylindrical pellets 5 mm in diameter and 5 to 10 mm in length. What value of the rate constant should be used for these pellets at (i) 525°C, and (ii) 429°C?
N.B. For a cylindrical pellet, L (i.e., L_e) in the Thiele modulus is $R/2$, where R is the radius.

8-20 Suppose experiments were conducted to characterize the performance of a catalyst for a certain reaction (A → products) that is first-order. The following data refer to experiments with several sizes of spherical catalyst particles of diameter d_p, with $c_A = 0.025$ mol L^{-1}:

d_p/mm	0.1	0.5	1	5	10	20	25
$10^4(-r_A)_{obs}$/mol L^{-1} s^{-1}	5.8	5.9	5.3	2.4	1.3	0.74	0.59

Determine the following:
(a) the intrinsic reaction rate, $(-r_A)_{int}$, and k_A;
(b) the effectiveness factor η for the 1, 5, 20, and 25-mm particles;
(c) the Thiele modulus (ϕ') for the 5, 20, and 25-mm particles;
(d) the effective diffusivity D_e.
State any other assumptions you make.

8-21 (a) For an nth-order, solid-catalyzed, gas-phase reaction, A → products, obtain an expression for the (catalyst) particle effectiveness factor (η) in terms of the overall effectiveness factor (η_o) and other relevant quantities.
(b) From the result in (a), obtain explicit expressions for η_o in terms of η and the other quantities, for reaction orders $n = 0, 1$ (see equation 8.5-49), and 2.

8-22 Consider the second-order reaction A → products involving a catalyst with relatively porous particles ($\eta \to 1$). If the ratio k_{Ag}/k_A is 20 mol m^{-3}, by what factor does the presence of external (film) mass-transfer resistance decrease the rate of reaction at 600 K and $p_A = 0.2$ MPa?

8-23 Activated carbon has been studied as a means for removal of organic molecules from wastewater by adsorption. Using the following data for benzene (A) adsorption on activated carbon (Leyva-Ramos and Geankopolis, 1994, as read from several points from a graph), determine the adsorption coefficients m_{max} and b, assuming that the data follow a Langmuir isotherm with the form $m_{Aa} = m_{max}bc_A/(1 + bc_A)$. Comment on your results.

$c_{C_6H_6}$/mg cm^{-3}	0.055	0.10	0.14	0.26	0.32	0.60
$m_{C_6H_6a}$/mg (gC)$^{-1}$	12.2	13.4	14.4	15.4	16.9	17.7

where $m_{C_6H_6a}$ is the amount of benzene adsorbed in mg g^{-1} (carbon).

Chapter 9

Multiphase Reacting Systems

In this chapter, we consider multiphase (noncatalytic) systems in which substances in different phases react. This is a vast field, since the systems may involve two or three (or more) phases: gas, liquid, and solid. We restrict our attention here to the case of two-phase systems to illustrate how the various types of possible rate processes (reaction, diffusion, and mass and heat transfer) are taken into account in a reaction model, although for the most part we treat isothermal situations.

The types of systems we deal with are primarily gas-solid (Section 9.1) and gas-liquid (Section 9.2). In these cases, we assume first- or second-order kinetics for the intrinsic reaction rate. This enables analytical expressions to be developed in some situations for the overall rate with transport processes taken into account. Such reaction models are incorporated in reactor models in Chapters 22 and 24.

In Section 9.3, we focus more on the intrinsic rates for reactions involving solids, since there are some modern processes in which mass transport rates play a relatively small role. Examples in materials engineering are chemical vapor deposition (CVD) and etching operations. We describe some mechanisms associated with such heterogeneous reactions and the intrinsic rate laws that arise.

9.1 GAS-SOLID (REACTANT) SYSTEMS

9.1.1 Examples of Systems

Two types of gas-solid reacting systems may be considered. In one type, the solid is reacted to another solid or other solids, and in the other, the solid disappears in forming gaseous product(s).

Examples of the first type are:

$$2ZnS(s) + 3O_2(g) \rightarrow 2ZnO(s) + 2SO_2(g) \quad \textbf{(A)}$$

$$Fe_3O_4(s) + 4H_2(g) \rightarrow 3Fe(s) + 4H_2O(g) \quad \textbf{(B)}$$

$$CaC_2(s) + N_2(g) \rightarrow CaCN_2(s) + C(s) \quad \textbf{(C)}$$

$$2CaO(s) + 2SO_2(g) + O_2(g) \rightarrow 2CaSO_4(s) \quad \textbf{(D)}$$

Although these examples do not all fit the category of the following model reaction, in the reaction models to be developed, we write a standard form as

$$A(g) + bB(s) \rightarrow \text{products}[(s), (g)] \quad (9.1\text{-}1)$$

in which, for ease of notation, the stoichiometric coefficient b replaces ν_B used elsewhere; $b \equiv |\nu_B| > 0$.

Examples of the other type in which the products are all gaseous, and the solid shrinks and may eventually disappear are:

$$C(s) + O_2(g) \rightarrow CO_2(g) \tag{E}$$

$$C(s) + H_2O(g) \rightarrow CO(g) + H_2(g) \tag{F}$$

We write a standard form of this type as

$$A(g) + bB(s) \rightarrow \text{products}(g) \tag{9.1-2}$$

The first type of reaction is treated in Section 9.1.2, and the second in Section 9.1.3.

9.1.2 Constant-Size Particle

9.1.2.1 General Considerations for Kinetics Model

To develop a kinetics model (i.e., a rate law) for the reaction represented in 9.1-1, we focus on a single particle, initially all substance B, reacting with (an unlimited amount of) gaseous species A. This is the local macroscopic level of size, level 2, discussed in Section 1.3 and depicted in Figure 1.1. In Chapter 22, the kinetics model forms part of a reactor model, which must also take into account the movement or flow of a *collection* of particles (in addition to flow of the gas), and any particle-size distribution. We assume that the particle size remains constant during reaction. This means that the integrity of the particle is maintained (it doesn't break apart), and requires that the densities of solid reactant B and solid product (surrounding B) be nearly equal. The size of particle is thus a parameter but not a variable. Among other things, this assumption of constant size simplifies consideration of rate of reaction, which may be normalized with respect to a constant unit of external surface area or unit volume of particle.

The single particle acts as a batch reactor in which conditions change with respect to time t. This unsteady-state behavior for a reacting particle differs from the steady-state behavior of a catalyst particle in heterogeneous catalysis (Chapter 8). The treatment of it leads to the development of an integrated rate law in which, say, the fraction of B converted, f_B, is a function of t, or the inverse.

A kinetics or reaction model must take into account the various individual processes involved in the overall process. We picture the reaction itself taking place on solid B surface somewhere within the particle, but to arrive at the surface, reactant A must make its way from the bulk-gas phase to the interior of the particle. This suggests the possibility of gas-phase resistances similar to those in a catalyst particle (Figure 8.9): external mass-transfer resistance in the vicinity of the exterior surface of the particle, and interior diffusion resistance through pores of both product formed and unreacted reactant. The situation is illustrated in Figure 9.1 for an isothermal spherical particle of radius R at a particular instant of time, in terms of the general case and two extreme cases. These extreme cases form the bases for relatively simple models, with corresponding concentration profiles for A and B.

In Figure 9.1, a gas film for external mass transfer of A is shown in all three cases. A further significance of a constant-size particle is that any effect of external mass transfer is the same in all cases, regardless of the situation within the particle.

In Figure 9.1(b), the general case is shown in which the reactant and product solids are both relatively porous, and the concentration profiles for A and B with respect to radial position (r) change continuously, so that c_B, shown on the left of the central axis,

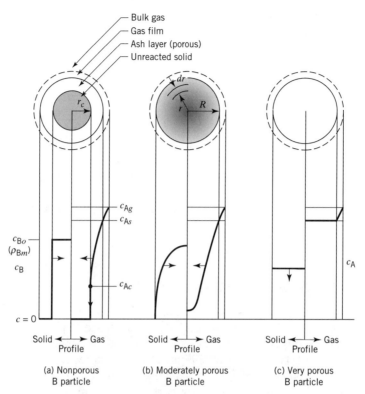

Figure 9.1 Constant-size particle (B) in reaction $A(g) + bB(s) \rightarrow$ products: instantaneous concentration profiles for isothermal spherical particle illustrating general case (b) and two extreme cases (a) and (c); solid product porous; arrows indicate direction of movement of profile with respect to time

increases, and c_A, on the right, decreases from the exterior surface to the center of the particle. The "concentration" of B is the (local) number of moles of B (unreacted) per unit volume of particle,

$$c_B = n_B/v_p \qquad (9.1\text{-}3)$$
$$= \rho_{Bm}(\text{pure B}) \qquad (9.1\text{-}3\text{a})$$

where ρ_{Bm} is the molar density (e.g., mol m^{-3}) of a particle of pure B with the same porosity; it corresponds to the (specific) particle density ρ_p in equation 8.5-3. (The concentration c_A is the usual gas-phase concentration for a single-phase fluid.) This situation is explored in a general model in Section 9.1.2.2. Solutions to obtain results for the general model are beyond our scope, but we can treat simplified models.

In Figure 9.1(a), the extreme case of a nonporous solid B is shown. In this case, reactant A initially reacts with the exterior surface of B, and as product solid (assumed to be porous) is formed, A must diffuse through a progressively increasing thickness of porous product to reach a progressively receding surface of B. There is a sharp boundary between the porous outer layer of product and the nonporous unreacted or shrinking core of reactant B. The concentration profiles reflect this: the value of c_B is either zero (completely reacted outer layer) or ρ_{Bm} (unreacted core of pure B); c_A decreases continuously because of increasing diffusional resistance through the outer layer, but is zero within the unreacted core. This case is the basis for a simplified model called the shrinking-core model (SCM), developed in Section 9.1.2.3, for which explicit solutions (integrated forms of rate laws) can be obtained for various particle shapes.

In Figure 9.1(c), the opposite extreme case of a very porous solid B is shown. In this case, there is no internal diffusional resistance, all parts of the interior of B are equally accessible to A, and reaction occurs uniformly (but not instantaneously) throughout the particle. The concentration profiles are "flat" with respect to radial position, but c_B decreases with respect to time, as indicated by the arrow. This model may be called a uniform-reaction model (URM). Its use is equivalent to that of a "homogeneous" model, in which the rate is a function of the intrinsic reactivity of B (Section 9.3), and we do not pursue it further here.

9.1.2.2. A General Model

9.1.2.2.1. Isothermal spherical particle. Consider the isothermal spherical particle of radius R in Figure 9.1(b), with reaction occurring (at the bulk-gas temperature) according to 9.1-1. A material balance for reactant A(g) around the thin shell (control volume) of (inner) radius r and thickness dr, taking both reaction and diffusion into account, yields the continuity equation for A:

$$\begin{pmatrix} \text{rate of input} \\ \text{of A} \\ \text{by diffusion} \\ \text{at } r+dr \end{pmatrix} - \begin{pmatrix} \text{rate of output} \\ \text{of A} \\ \text{by diffusion} \\ \text{at } r \end{pmatrix} - \begin{pmatrix} \text{rate of output} \\ \text{of A} \\ \text{by disappearance} \\ \text{within the shell} \end{pmatrix} = \begin{pmatrix} \text{rate of accumulation} \\ \text{of A} \\ \text{within the} \\ \text{control volume} \end{pmatrix}$$

that is,

$$4\pi r^2 D_e \frac{\partial c_A}{\partial r} + \frac{\partial}{\partial r}\left(4\pi r^2 D_e \frac{\partial c_A}{\partial r}\right)dr - 4\pi r^2 D_e \frac{\partial c_A}{\partial r} - (-r_A)4\pi r^2 dr = \frac{\partial n_A}{\partial t} = \frac{\partial}{\partial t}(c_A 4\pi r^2 dr) \quad (9.1\text{-}4)$$

where Fick's law, equation 8.5-4, has been used for diffusion, with D_e as the effective diffusivity for A through the pore structure of solid, and $(-r_A)$ is the rate of disappearance of A; with $(-r_A)$ normalized with respect to volume of particle, each term has units of mol (A) s^{-1}. If the pore structure is uniform throughout the particle, D_e is constant; otherwise it depends on radial position r. With D_e constant, we simplify equation 9.1-4 to

$$D_e\left(\frac{\partial^2 c_A}{\partial r^2} + \frac{2}{r}\frac{\partial c_A}{\partial r}\right) - (-r_A) = \frac{\partial c_A}{\partial t} \quad (9.1\text{-}5)$$

The continuity equation for B, written for the whole particle, is

$$(-R_B) = -\frac{\partial n_B}{\partial t} = -v_p \frac{\partial c_B}{\partial t} \quad (9.1\text{-}6)$$

or

$$(-r_B) = \frac{R_B}{v_p} = -\frac{\partial c_B}{\partial t} \quad (9.1\text{-}7)$$

From the stoichiometry of reaction 9.1-1, $(-r_A)$ and $(-r_B)$ are related by

$$(-r_B) = b(-r_A) \tag{9.1-8}$$

or

$$(-R_B) = b(-R_A) \tag{9.1-8a}$$

where $(-R_A)$ is the extensive rate of reaction of A for the whole particle corresponding to $(-R_B)$.

Equations 9.1-5 and -7 are two coupled partial differential equations with initial and boundary conditions as follows:

$$\text{at } t = 0, \quad c_B \equiv c_{Bo} = \rho_{Bm} \tag{9.1-9}$$

$$c_A = c_{Ag} \tag{9.1-10}$$

$$\text{at } r = R, \quad D_e \left(\frac{\partial c_A}{\partial r}\right)_{r=R} = k_{Ag}(c_{Ag} - c_{As}) \tag{9.1-11}$$

which takes the external-film mass transfer into account; k_{Ag} is a mass transfer coefficient (equation 9.2-3); the boundary condition states that the rate of diffusion of A across the exterior surface of the particle is equal to the rate of transport of A from bulk gas to the solid surface by mass transfer;

$$\text{at } r = 0, \quad (\partial c_A/\partial r)_{r=0} = 0 \tag{9.1-12}$$

corresponding to no mass transfer through the center of the particle, from consideration of symmetry.

In general, there is no analytical solution for the partial differential equations above, and numerical methods must be used. However, we can obtain analytical solutions for the simplified case represented by the shrinking-core model, Figure 9.1(a), as shown in Section 9.1.2.3.

9.1.2.2.2. Nonisothermal spherical particle. The energy equation describing the profile for T through the particle, equivalent to the continuity equation 9.1-5 describing the profile for c_A, may be derived in a similar manner from an energy (enthalpy) balance around the thin shell in Figure 9.1(b). The result is

$$k_e \left(\frac{\partial^2 T}{\partial r^2} + 2r\frac{\partial T}{\partial r}\right) - \Delta H_{RA}(-r_A) = \rho_{Bm} C_{PB} \frac{\partial T}{\partial t} \tag{9.1-13}$$

where k_e is an effective thermal conductivity for heat transfer through the particle (in the Fourier equation), analogous to D_e for diffusion, ΔH_{RA} is the enthalpy of reaction with respect to A, and C_{PB} is the molar heat capacity for solid B (each term has units of J m^{-3} s^{-1}, say). The initial and boundary conditions for the solution of equation 9.1-13 correspond to those for the continuity equations:

$$\text{at } t = 0, \quad T = T_g \tag{9.1-14}$$

$$\text{at } r = R, \quad k_e(\partial T/\partial r)_{r=R} = h(T_s - T_g) \tag{9.1-15}$$

Equation 9.1-15 equates the rate of heat transfer by conduction at the surface to the rate of heat transfer by conduction/convection across a thermal boundary layer exterior to the particle (corresponding to the gas film for mass transfer), expressed in terms of a film coefficient, h, and the difference in temperature between bulk gas at T_g and particle surface at T_s;

$$\text{at } r = 0, \quad (\partial T / \partial r)_{r=0} = 0 \qquad (9.1\text{-}16)$$

Equation 9.1-16 implies no heat transfer through the center of the particle, from consideration of symmetry.

Taken together with the continuity equations, the energy equation complicates the solution further, since c_A and T are nonlinearly coupled through $(-r_A)$.

9.1.2.3 Shrinking-Core Model (SCM)

9.1.2.3.1. Isothermal spherical particle.
The shrinking core model (SCM) for an isothermal spherical particle is illustrated in Figure 9.1(a) for a particular instant of time. It is also shown in Figure 9.2 at two different times to illustrate the effects of increasing time of reaction on the core size and on the concentration profiles.

Figure 9.2(a) or (b) shows the essence of the SCM, as discussed in outline in Section 9.1.2.1, for a partially reacted particle. There is a sharp boundary (the reaction surface) between the nonporous unreacted core of solid B and the porous outer shell of solid product (sometimes referred to as the "ash layer," even though the "ash" is desired product). Outside the particle, there is a gas film reflecting the resistance to mass transfer of A from the bulk gas to the exterior surface of the particle. As time increases, the reaction surface moves progressively toward the center of the particle; that is, the unreacted core of B shrinks (hence the name). The SCM is an idealized model, since the boundary between reacted and unreacted zones would tend to be blurred, which could be revealed by slicing the particle and examining the cross-section. If this

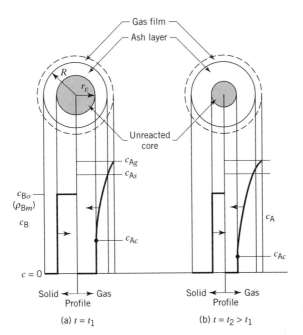

Figure 9.2 The shrinking-core model (SCM) for an isothermal spherical particle showing effects of increasing reaction time t

"blurring" is significant, a more general model (Section 9.1.2.2) may have to be used. The SCM differs from a general model in one important aspect, as a consequence of the sharp boundary. According to the SCM, the three processes involving mass transfer of A, diffusion of A, and reaction of A with B at the surface of the core are in series. In the general case, mass transfer is in series with the other two, but these last two are not in series with each other—they occur together throughout the particle in some manner. This, together with a further assumption about the relative rates of diffusion and movement of the reaction surface, allows considerable simplification of the solution of equation 9.1-5. This is achieved in analytical form for a spherical particle in Example 9-1. Results for other shapes can also be obtained, and are explored in problems at the end of this chapter. The results are summarized in Section 9.1.2.3.2.

In Figure 9.2, c_{Ag} is the (gas-phase) concentration of A in the bulk gas surrounding the particle, c_{As} that at the exterior surface of the particle, and c_{Ac} that at the surface of the unreacted core of B in the interior of the particle; R is the (constant) radius of the particle and r_c is the (variable) radius of the unreacted core. The concentrations c_{Ag} and c_{As} are constant, but c_{Ac} decreases as t increases, as does r_c, with corresponding consequences for the positions of the profiles for c_B, on the left in Figure 9.2(a) and (b), and c_A, on the right.

EXAMPLE 9-1

For a spherical particle of species B of radius R undergoing reaction with gaseous species A according to 9.1-1, derive a relationship to determine the time t required to reach a fraction of B converted, f_B, according to the SCM. Assume the reaction is a first-order surface reaction.

SOLUTION

To obtain the desired result, $t = t(f_B)$, we could proceed in either of two ways. In one, since the three rate processes involved are in series, we could treat each separately and add the results to obtain a total time. In the other, we could solve the simplified form of equation 9.1-5 for all three processes together to give one result, which would also demonstrate the additivity of the individual three results. In this example, we use the second approach (the first, which is simpler, is used for various shapes in the next example and in problems at the end of the chapter).

The basis for the analysis using the SCM is illustrated in Figure 9.3. The gas film, outer product (ash) layer, and unreacted core of B are three distinct regions. We derive the continuity equation for A by means of a material balance across a thin spherical shell in the ash layer at radial position r and with a thickness dr. The procedure is the same as that leading up to equation 9.1-5, *except* that there is no reaction term involving $(-r_A)$, since no reaction occurs in the ash layer. The result corresponding to equation 9.1-5 is

$$D_e \left(\frac{\partial^2 c_A}{\partial r^2} + \frac{2}{r} \frac{\partial c_A}{\partial r} \right) = \frac{\partial c_A}{\partial t} \qquad (9.1\text{-}17)$$

Equation 9.1-17 is the continuity equation for unsteady-state diffusion of A through the ash layer; it is unsteady-state because $c_A = c_A(r, t)$. To simplify its treatment further, we assume that the (changing) concentration gradient for A through the ash layer is established rapidly relative to movement of the reaction surface (of the core). This means that for an instantaneous "snapshot," as depicted in Figure 9.3, we may treat the diffusion as steady-state diffusion for a fixed value of r_c; i.e., $c_A = c_A(r)$. The partial differential equation,

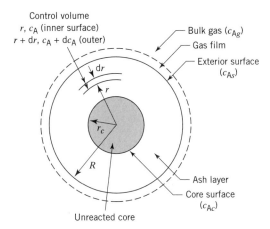

Figure 9.3 Spherical particle for Example 9-1

9.1-17, then becomes an ordinary differential equation, with $\partial c_A/\partial t = 0$:

$$\frac{d^2 c_A}{dr^2} + \frac{2}{r}\frac{dc_A}{dr} = 0 \qquad (9.1\text{-}18)$$

The assumption made is called the quasi-steady-state approximation (QSSA). It is valid here mainly because of the great difference in densities between the reacting species (gaseous A and solid B). For liquid-solid systems, this simplification cannot be made.

The solution of equation 9.1-17 is then obtained from a two-step procedure:

Step (1): Solve equation 9.1-18 in which the variables are c_A and r (t and r_c are fixed). This results in an expression for the *flux* of A, N_A, as a function of r_c; N_A, in turn, is related to the rate of reaction at r_c.

Step (2): Use the result of step (1), together with equations 9.1-7 and -8 to obtain $t = t(r_c)$, which can be translated to the desired result, $t = t(f_B)$. In this step the variables are t and r_c.

In step (1), the solution of equation 9.1-18 requires two boundary conditions, each of which can be expressed in two ways; one of these ways introduces the other two rate processes, equating the rate of diffusion of A to the rate of transport of A at the particle surface (equation 9.1-11), and also the rate of diffusion at the core surface to the rate of reaction on the surface (9.1-20), respectively. Thus,

$$\text{at } r = R, \ N_A = D_e\left(\frac{\partial c_A}{\partial r}\right)_{r=R} = k_{Ag}(c_{Ag} - c_{As}) \qquad (9.1\text{-}11)$$

$$\text{or} \quad c_A = c_{As} \qquad (9.1\text{-}19)$$

$$\text{at } r = r_c, \ N_A = D_e\left(\frac{\partial c_A}{\partial r}\right)_{r=r_c} = k_{As} c_{Ac} \qquad (9.1\text{-}20)$$

$$\text{or} \quad c_A = c_{Ac} \qquad (9.1\text{-}21)$$

where k_{As} is the rate constant for the first-order surface reaction, with the rate of reaction given by

$$(-R_A) = 4\pi r_c^2 k_{As} c_{Ac} \qquad (9.1\text{-}22)$$

$$= 4\pi r_c^2 (-r_{As}) \qquad (9.1\text{-}22a)$$

where the total rate (for the particle) $(-R_A)$ is in mol s^{-1}, k_{As} is in m s^{-1}, and the specific rate $(-r_{As})$, normalized with respect to the area of the core, is in mol m^{-2} s^{-1}. The solution of equation 9.1-18, to obtain an expression for c_{Ac} for use in equation 9.1-22, is straightforward (but tedious), if we use the substitution $y = dc_A/dr$. We integrate twice, first to obtain dc_A/dr and second to obtain c_A, using the boundary conditions to evaluate the integration constants, and eliminate c_{As} to obtain c_{Ac}. The first integration, together with equation 9.1-11, results in

$$\frac{dc_A}{dr} = \frac{k_{Ag}R^2}{D_e}(c_{Ag} - c_{As})\frac{1}{r^2} \tag{A}$$

Integration of (A), together with equation 9.1-21, gives

$$c_A = c_{Ac} + \frac{k_{Ag}R^2}{D_e}(c_{Ag} - c_{As})\left(\frac{1}{r_c} - \frac{1}{r}\right) \tag{B}$$

Applying equation (A) at the surface of the core ($r = r_c$), together with equation 9.1-20, we obtain one expression for c_{As} in terms of c_{Ac}:

$$c_{As} = c_{Ag} - (k_{As}r_c^2/k_{Ag}R^2)c_{Ac} \tag{C}$$

Similarly, from equation (B), together with equation 9.1-19, at the particle surface, we obtain another expression:

$$c_{As} = c_{Ac} + \frac{k_{Ag}R^2}{D_e}(c_{Ag} - c_{As})\left(\frac{1}{r_c} - \frac{1}{R}\right) \tag{D}$$

On elimination of c_{As} from equations (C) and (D), and substitution of the resulting expression for c_{Ac} in equation 9.1-22, we obtain, with rearrangement,

$$(-R_A) = \frac{4\pi c_{Ag}}{\dfrac{1}{k_{Ag}R^2} + \dfrac{R - r_c}{D_e R r_c} + \dfrac{1}{k_{As}r_c^2}} \tag{9.1-23}$$

This is the end of step (1), resulting in an expression for $(-R_A)$ in terms of (fixed) r_c.

In step (2) of the solution of equation 9.1-17, we allow the core surface, fixed in step (1), to move, and integrate the continuity equation for B, using the first part of equation 9.1-6. For this purpose, we substitute both equations 9.1-23 and 9.1-6, the latter written in the form

$$(-R_B) = -\frac{dn_B}{dt} = -\frac{d}{dt}\left(\rho_{Bm}\frac{4}{3}\pi r_c^3\right) = -4\pi\rho_{Bm}r_c^2\frac{dr_c}{dt} \tag{9.1-24}$$

into the stoichiometric relation 9.1-8a, resulting in

$$dt = -\frac{\rho_{Bm}}{bc_{Ag}}\left[\frac{r_c^2}{k_{Ag}R^2} + \frac{1}{D_e}\left(r_c - \frac{r_c^2}{R}\right) + \frac{1}{k_{As}}\right]dr_c \tag{9.1-25}$$

Integration of equation 9.1-25, from $t = 0$, $r_c = R$ to t, r_c, results in

$$t = \frac{\rho_{Bm}R}{bc_{Ag}}\left\{\frac{1}{3k_{Ag}}\left[1 - \left(\frac{r_c}{R}\right)^3\right] + \frac{R}{6D_e}\left[1 - 3\left(\frac{r_c}{R}\right)^2 + 2\left(\frac{r_c}{R}\right)^3\right] + \frac{1}{k_{As}}\left(1 - \frac{r_c}{R}\right)\right\} \tag{9.1-26}$$

We can eliminate r_c from this equation in favor of f_B, from a relation based on the shrinking volume of a sphere:

$$f_B = \frac{n_{Bo} - n_B}{n_{Bo}} = 1 - \left(\frac{r_c}{R}\right)^3 \qquad (9.1\text{-}27)$$

to obtain

$$t = \frac{\rho_{Bm} R}{bc_{Ag}} \left\{ \frac{f_B}{3k_{Ag}} + \frac{R}{6D_e}\left[1 - 3(1-f_B)^{2/3} + 2(1-f_B)\right] + \frac{1}{k_{As}}\left[1 - (1-f_B)^{1/3}\right] \right\} \qquad (9.1\text{-}28)$$

If we denote the time required for complete conversion of the particle ($f_B = 1$) by t_1, then, from equation 9.1-28,

$$t_1 = \frac{\rho_{Bm} R}{bc_{Ag}} \left(\frac{1}{3k_{Ag}} + \frac{R}{6D_e} + \frac{1}{k_{As}} \right) \qquad (9.1\text{-}29)$$

t_1 is a kinetics parameter, characteristic of the reaction, embodying the three parameters characteristic of the individual rate processes, k_{Ag}, D_e, and k_{As}, and particle size, R.

Equations 9.1-28 and -29 both give rise to special cases in which either one term (i.e., one rate process) dominates or two terms dominate. For example, if D_e is small compared with either k_{Ag} or k_{As}, this means that ash-layer diffusion is the rate-determining or controlling step. The value of t or t_1 is then determined entirely by the second term in each equation. Furthermore, since each term in each equation refers only to one rate process, we may write, for the overall case, the additive relation:

$$t = \Big[t(\text{film-mass-transfer control}) + t(\text{ash-layer-diffusion control}) + t(\text{surface-reaction-rate control}) \Big] \qquad (9.1\text{-}30)$$

and similarly for t_1.

EXAMPLE 9-2

For the situation in Example 9-1, derive the result for $t(f_B)$ for reaction-rate control,[1] that is, for the surface reaction as the rate-determining step (*rds*), and confirm that it is the

---

[1] As noted by Froment and Bischoff (1990, p. 209), the case of surface-reaction-rate control is not consistent with the existence of a sharp core boundary in the SCM, since this case implies that diffusional transport could be slow with respect to the reaction rate.

same as the corresponding part of equation 9.1-28. (This can be repeated for each of the two other cases of single-process control, gas-film control and ash-layer control (the latter requires use of the QSSA introduced in Example 9-1); see problem 9-1 for these and also the comment about other cases involving two of the three processes as "resistances.")

SOLUTION

Referring to the concentration profiles for A in Figure 9.2, we realize that if there is no resistance to the transport of A in either the gas film or the ash layer, c_A remains constant from the bulk gas to the surface of the unreacted core. That is,

$$c_{Ag} = c_{As} = c_{Ac}$$

and, as a result, from equation 9.1-22,

$$(-R_A) = k_{As}(4\pi r_c^2)c_{Ag}$$

Combining this with equation 9.1-24 for $(-R_B)$ and equation 9.1-8a for the stoichiometry, we obtain

$$-4\pi \rho_{Bm} r_c^2 (dr_c/dt) = bk_{As}(4\pi r_c^2)c_{Ag}$$

or

$$dt = -(\rho_{Bm}/bk_{As}c_{Ag})dr_c$$

Integration from $r_c = R$ at $t = 0$ to r_c at t results in

$$t = t_1(1 - r_c/R)$$
$$= t_1[1 - (1 - f_B)^{1/3}] \tag{9.1-31}$$

from equation 9.1-27, where the kinetics parameter t_1, the time required for complete reaction, is given by

$$t_1 = \rho_{Bm}R/bk_{As}c_{Ag} \tag{9.1-31a}$$

These results are, of course, the same as those obtained from equations 9.1-28 and -29 for the special case of reaction-rate control.

9.1.2.3.2. Summary of $t(f_B)$ for various shapes. The methods used in Examples 9-1 and 9-2 may be applied to other shapes of isothermal particles (see problems 9-1 to 9-3). The results for spherical, cylindrical, and flat-plate geometries are summarized in Table 9.1. The flat plate has one face permeable (to A) as in Figure 8.10(a), and the variable l, corresponding to r_c, is the length of the unreacted zone (away from the permeable face), the total path length for diffusion of A and reaction being L. For the cylinder, the symbols r_c and R have the same significance as for the sphere; the ends of the cylinder are assumed to be impermeable, and hence the length of the particle is not involved in the result (alternatively, we may assume the length to be $\gg r_c$).

In Table 9.1, in the third column, the relation between f_B and the particle size parameters (second column), corresponding to, and including, equation 9.1-27, is given for each shape. Similarly, in the fourth column, the relation between t and f_B, corresponding to, and including, equation 9.1-28 is given.

Table 9.1 SCM: Summary of $t(f_B)$ for various shapes of particle[1]

Particle shape	Size parameters	f_B (size parameters)	$t(f_B)$
flat plate (one face permeable)	length l (of zone unreacted) L (of particle)	$1 - \dfrac{l}{L}$	$\dfrac{\rho_{Bm} L f_B}{b c_{Ag}} \left(\dfrac{1}{k_{Ag}} + \dfrac{L f_B}{2 D_e} + \dfrac{1}{k_{As}} \right)$
cylinder (ends sealed)	radius r_c (of core unreacted) R (of particle)	$1 - \left(\dfrac{r_c}{R}\right)^2$	$\dfrac{\rho_{Bm} R}{b c_{Ag}} \left\{ \dfrac{f_B}{2 k_{Ag}} + \dfrac{R}{4 D_e} \left[f_B + (1 - f_B)\ln(1 - f_B) \right] + \dfrac{1}{k_{As}} \left[1 - (1 - f_B)^{1/2} \right] \right\}$
sphere	same as for cylinder	$1 - \left(\dfrac{r_c}{R}\right)^3$	$\dfrac{\rho_{Bm} R}{b c_{Ag}} \left\{ \dfrac{f_B}{3 k_{Ag}} + \dfrac{R}{6 D_e} \left[1 - 3(1 - f_B)^{2/3} + 2(1 - f_B) \right] + \dfrac{1}{k_{As}} \left[1 - (1 - f_B)^{1/3} \right] \right\}$

(9.1-28)

[1] Reaction: $A(g) + bB(s) \rightarrow$ products $[(s),(g)]$; first order with respect to A at core surface.
Particle(B): constant-size (L, R constant); isothermal.
For t_1 (time for complete reaction of particle), set $f_B = 1$; $[(1 - f_B)\ln(1 - f_B) \rightarrow 0]$.
Symbols: see text and Nomenclature.

For each shape and each rate-process-control special case in Table 9.1, the result for t_1 may be obtained by setting $f_B = 1$. For the cylinder, in the term for ash-layer diffusion, it may be shown that $(1 - f_B)\ln(1 - f_B) \to 0$ as $f_B \to 1$ (by use of L'Hôpital's rule).

9.1.2.3.3. Rate-process parameters; estimation of k_{Ag} for spherical particle.

The three rate-process parameters in the expressions for $t(f_B)$ (k_{Ag}, D_e, and k_{As}), may each require experimental measurement for a particular situation. However, we consider one correlation for estimating k_{Ag} for spherical particles given by Ranz and Marshall (1952).

For a free-falling spherical particle of radius R, moving with velocity u relative to a fluid of density ρ and viscosity μ, and in which the molecular diffusion coefficient (for species A) is D_A, the Ranz-Marshall correlation relates the Sherwood number (Sh), which incorporates k_{Ag}, to the Schmidt number (Sc) and the Reynolds number (Re):

$$\text{Sh} = 2 + 0.6\,\text{Sc}^{1/3}\text{Re}^{1/2} \tag{9.1-32}$$

That is,

$$2Rk_{Ag}/D_A = 2 + 0.6(\mu/\rho D_A)^{1/3}(2Ru\rho/\mu)^{1/2} \tag{9.1-33}$$

This correlation may be used to estimate k_{Ag} given sufficient information about the other quantities.

For a given fluid and relative velocity, we may write equation 9.1-33 so as to focus on the dependence of k_{Ag} on R as a parameter:

$$k_{Ag} = \frac{K_1}{R} + \frac{K_2}{R^{1/2}} \tag{9.1-34}$$

where K_1 and K_2 are constants. There are two limiting cases of 9.1-34, in which the first or the second term dominates (referred to as "small" and "large" particle cases, respectively). One consequence of this, and of the correlation in general, stems from reexamining the reaction time $t(f_B)$, as follows.

In Table 9.1, or equations 9.1-28 and -29 for a sphere, k_{Ag} appears to be a constant, independent of R. This is valid for a particular value of R. However, if R changes from one particle size to another as a parameter, we can compare the effect on $t(f_B)$ of such a change.

Suppose, for simplicity, that gas-film mass transfer is rate controlling. From Table 9.1, in this case, for a sphere,

$$t = \frac{\rho_{Bm} R f_B}{3bc_{Ag} k_{Ag}} \tag{9.1-35}$$

$$= \frac{\rho_{Bm} R^2 f_B}{3bc_{Ag} K_1} \quad \text{("small" particle)} \tag{9.1-35a}$$

$$= \frac{\rho_{Bm} R^{3/2} f_B}{3bc_{Ag} K_2} \quad \text{("large" particle)} \tag{9.1-35b}$$

from equation 9.1-34. Thus, depending on the particle-region of change, the dependence of t on R from the gas-film contribution may be R^2 or $R^{3/2}$.

The significance of this result and of other factors in identifying the existence of a rate-controlling process is explored in problem 9-4.

9.1.3 Shrinking Particle

9.1.3.1 General Considerations

When a solid particle of species B reacts with a gaseous species A to form only gaseous products, the solid can disappear by developing internal porosity, while maintaining its macroscopic shape. An example is the reaction of carbon with water vapor to produce activated carbon; the intrinsic rate depends upon the development of sites for the reaction (see Section 9.3). Alternatively, the solid can disappear only from the surface so that the particle progressively shrinks as it reacts and eventually disappears on complete reaction ($f_B = 1$). An example is the combustion of carbon in air or oxygen (reaction (E) in Section 9.1.1). In this section, we consider this case, and use reaction 9.1-2 to represent the stoichiometry of a general reaction of this type.

An important difference between a shrinking particle reacting to form only gaseous product(s) and a constant-size particle reacting so that a product layer surrounds a shrinking core is that, in the former case, there is no product or "ash" layer, and hence no ash-layer diffusion resistance for A. Thus, only two rate processes, gas-film mass transfer of A, and reaction of A and B, need to be taken into account.

9.1.3.2 A Simple Shrinking-Particle Model

We can develop a simple shrinking-particle kinetics model by taking the two rate-processes involved as steps in series, in a treatment that is simpler than that used for the SCM, although some of the assumptions are the same:

(1) The reacting particle is isothermal.
(2) The particle is nonporous, so that reaction occurs only on the exterior surface.
(3) The surface reaction between gas A and solid B is first-order.

In the following example, the treatment is illustrated for a spherical particle.

EXAMPLE 9-3

For a reaction represented by $A(g) + bB(s) \rightarrow \text{product}(g)$, derive the relation between time (t) of reaction and fraction of B converted (f_B), if the particle is spherical with an initial radius R_o, and the Ranz-Marshall correlation for $k_{Ag}(R)$ is valid, where R is the radius at t. Other assumptions are given above.

SOLUTION

The rate of reaction of A, $(-R_A)$, can be expressed independently in terms of the rate of transport of A by mass transfer *and* the rate of the surface reaction:

$$(-R_A) = k_{Ag}(R) 4\pi R^2 (c_{Ag} - c_{As}) \qquad (9.1\text{-}36)$$

$$(-R_A) = k_{As} 4\pi R^2 c_{As} \qquad (9.1\text{-}37)$$

The rate of reaction of B is (cf. equation 9.1-24)

$$(-R_B) = -4\pi \rho_{Bm} R^2 (dR/dt) \qquad (9.1\text{-}38)$$

and $(-R_A)$ and $(-R_B)$ are related by

$$(-R_B) = b(-R_A) \tag{9.1-8a}$$

We eliminate $(-R_A)$, $(-R_B)$ and c_{As} (concentration at the surface) from these four equations to obtain a differential equation involving dR/dt, which, on integration, provides the desired relation. The equation resulting for dR/dt is

$$\frac{dR}{dt} = \frac{bc_{Ag}/\rho_{Bm}}{\dfrac{1}{k_{Ag}(R)} + \dfrac{1}{k_{As}}} \tag{9.1-39}$$

which leads to

$$t = -\frac{\rho_{Bm}}{bc_{Ag}} \int_{R_o}^{R} \left[\frac{1}{k_{Ag}(R)} + \frac{1}{k_{As}} \right] dR$$

$$= \frac{\rho_{Bm}}{bc_{Ag}} \int_{R}^{R_o} \left[\left(\frac{K_1}{R} + \frac{K_2}{R^{1/2}} \right)^{-1} + \frac{1}{k_{As}} \right] dR \tag{9.1-40}$$

from equation 9.1-34. For simplicity, we consider results for the two limiting cases of equation 9.1-34 (ignoring $K_2/R^{1/2}$ or K_1/R), as described in Section 9.1.2.3.3.

For *small* particles, integration of 9.1-40 without the term $K_2/R^{1/2}$ results in

$$t = \frac{\rho_{Bm} R_o}{bc_{Ag}} \left\{ \frac{R_o}{2K_1} \left[1 - \left(\frac{R}{R_o} \right)^2 \right] + \frac{1}{k_{As}} \left[1 - \frac{R}{R_o} \right] \right\} \tag{9.1-41}$$

or, since, for a spherical particle

$$f_B = 1 - (R/R_o)^3 \tag{9.1-42}$$

$$t = \frac{\rho_{Bm} R_o}{bc_{Ag}} \left\{ \frac{R_o}{2K_1} \left[1 - (1 - f_B)^{2/3} \right] + \frac{1}{k_{As}} \left[1 - (1 - f_B)^{1/3} \right] \right\} \tag{9.1-43}$$

The time t_1 for complete reaction ($f_B = 1$) is

$$t_1 = \frac{\rho_{Bm} R_o}{bc_{Ag}} \left(\frac{R_o}{2K_1} + \frac{1}{k_{As}} \right) \tag{9.1-44}$$

For *large* particles, the corresponding results (with the term K_1/R in equation 9.1-40 dropped) are:

$$t = \frac{\rho_{Bm} R_o}{bc_{Ag}} \left\{ \frac{2R_o^{1/2}}{3K_2} \left[1 - \left(\frac{R}{R_o} \right)^{3/2} \right] + \frac{1}{k_{As}} \left[1 - \frac{R}{R_o} \right] \right\} \tag{9.1-45}$$

$$t = \frac{\rho_{Bm} R_o}{bc_{Ag}} \left\{ \frac{2R_o^{1/2}}{3K_2} \left[1 - (1 - f_B)^{1/2} \right] + \frac{1}{k_{As}} \left[1 - (1 - f_B)^{1/3} \right] \right\} \tag{9.1-46}$$

$$t_1 = \frac{\rho_{Bm} R_o}{bc_{Ag}}\left(\frac{2R_o^{1/2}}{3K_2} + \frac{1}{k_{As}}\right) \quad (9.1\text{-}47)$$

Corresponding equations for the two special cases of gas-film mass-transfer control and surface-reaction-rate control may be obtained from these results (they may also be derived individually). The results for the latter case are of the same form as those for reaction-rate control in the SCM (see Table 9.1, for a sphere) with R_o replacing (constant) R (and (variable) R replacing r_c in the development). The footnote in Example 9-2 does not apply here (explain why).

9.2 GAS-LIQUID SYSTEMS

9.2.1 Examples of Systems

Gas-liquid reacting systems may be considered from one of two points of view, depending on the purpose of the reaction: (1) as a separation process or (2) as a reaction process.

In case (1), the reaction is used for the removal of an undesirable substance from a gas stream. In this sense, the process is commonly referred to as "gas absorption with reaction." Examples are removal of H_2S or CO_2 from a gas stream by contact with an ethanolamine (e.g., monoethanolamine (MEA) or diethanolamine (DEA)) in aqueous solution, represented by:

$$H_2S(g) + HOCH_2CH_2NH_2(MEA, \ell) \rightarrow HS^- + HOCH_2CH_2NH_3^+ \quad (A)$$

$$CO_2(g) + 2(HOCH_2CH_2)_2NH(DEA, \ell) \rightarrow (HOCH_2CH_2)_2NCOO^- + (HOCH_2CH_2)_2NH_2^+ \quad (B)$$

In case (2), the reaction is used to yield a desirable product. Examples are found in the manufacture of nitric acid, phenol, and nylon 66, represented, respectively, by:

$$3NO_2(g) + H_2O(\ell) \rightarrow 2HNO_3(\ell) + NO(g) \quad (C)$$

$$\left.\begin{array}{l} C_7H_8(\ell) + (3/2)O_2(g) \rightarrow C_6H_5COOH + H_2O \\ C_6H_5COOH + (1/2)O_2 \rightarrow C_6H_5OH + CO_2 \end{array}\right\} \quad (D)$$

$$C_6H_{12}(\ell) + O_2(g) \rightarrow \text{adipic acid} \quad (E)$$

The types of reactors and reactor models used for such reactions are considered in Chapter 24. In this chapter, we are concerned with the kinetics of these reactions, and hence with reaction models, which may have to include gas-liquid mass transfer as well as chemical reaction.

Similar to the case of gas-solid reactions, we represent the stoichiometry of a gas-liquid reaction in a model or generic sense by

$$A(g) + bB(\ell) \rightarrow \text{products} \quad (9.2\text{-}1)$$

$B(\ell)$ may refer to pure liquid B or, more commonly, to B dissolved in a liquid solvent. Furthermore, we assume throughout that $B(\ell)$ is nonvolatile; that is, B occurs only in the liquid-phase, whereas A may be present in both phases. This assumption implies that chemical reaction occurs only in the liquid phase.

In the treatment to follow, we first review the two-film model for gas-liquid mass transfer, without reaction, in Section 9.2.2, before considering the implications for kinetics in Section 9.2.3.

9.2.2 Two-Film Mass-Transfer Model for Gas-Liquid Systems

Consider the transport of gaseous species A from a bulk gas to a bulk liquid, in which it has a measurable solubility, because of a difference of chemical potential of A in the two phases (higher in the gas phase). The difference may be manifested by a difference in concentration of A in the two phases. At any point in the system in which gas and liquid phases are in contact, there is an interface between the phases. The two-film model (Whitman, 1923; Lewis and Whitman, 1924) postulates the existence of a stagnant gas film on one side of the interface and a stagnant liquid film on the other, as depicted in Figure 9.4. The concentration of A in the gas phase is represented by the partial pressure, p_A, and that in the liquid phase by c_A. Subscript i denotes conditions at the interface and δ_g and δ_ℓ are the thicknesses of the gas and liquid films, respectively. The interface is real, but the two films are imaginary, and are represented by the dashed lines in Figure 9.4; hence, δ_g and δ_ℓ are unknown.

In the two-film model, the following assumptions are made:

(1) The two-film model is a steady-state model; that is, the concentration profiles indicated in Figure 9.4 are established instantaneously and remain unchanged.
(2) The steady-state transport of A through the stagnant gas film is by molecular diffusion, characterized by the molecular diffusivity D_{Ag}. The rate of transport, normalized to refer to unit area of interface, is given by Fick's law, equation 8.5-4, in the integrated form

$$N_A = D_{Ag}(p_A - p_{Ai})/RT\delta_g \quad (9.2\text{-}2)$$
$$= k_{Ag}(p_A - p_{Ai}) \quad (9.2\text{-}3)$$

where N_A is the molar flux of A, mol m^{-2} s^{-1}, and k_{Ag} is the gas-film mass transfer coefficient defined by

$$k_{Ag} = D_{Ag}/RT\delta_g \quad (9.2\text{-}4)$$

and introduced to cover the fact that δ_g is unknown.
(3) Similarly, the transport of A through the liquid film is by molecular diffusion, characterized by $D_{A\ell}$, and the flux (the same as that in equations 9.2-2 and -3 at steady-state) is

$$N_A = D_{A\ell}(c_{Ai} - c_A)/\delta_\ell \quad (9.2\text{-}5)$$
$$= k_{A\ell}(c_{Ai} - c_A) \quad (9.2\text{-}6)$$

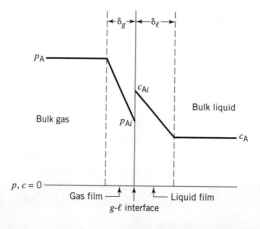

Figure 9.4 Two-film model (profiles) for mass transfer of A from gas phase to liquid phase (no reaction)

where the liquid-film mass transfer coefficient is defined by

$$k_{A\ell} = D_{A\ell}/\delta_\ell \tag{9.2-7}$$

(4) There is equilibrium at the interface, which is another way of assuming that there is no resistance to mass transfer at the interface. The equilibrium relation may be expressed by means of Henry's law:

$$p_{Ai} = H_A c_{Ai} \tag{9.2-8}$$

where H_A is the Henry's law constant for species A.

The rate of mass transfer of A may also be characterized in terms of overall mass transfer coefficients K_{Ag} and $K_{A\ell}$ defined by

$$N_A = K_{Ag}(p_A - p_A^*) \tag{9.2-9}$$

$$= K_{A\ell}(c_A^* - c_A) \tag{9.2-10}$$

where p_A^* is the (fictitious) partial pressure of A in equilibrium with a liquid phase of concentration c_A,

$$p_A^* = H_A c_A \tag{9.2-11}$$

and, correspondingly, c_A^* is the liquid-phase concentration of A in equilibrium with a gas-phase partial pressure of p_A,

$$p_A = H_A c_A^* \tag{9.2-12}$$

K_{Ag} and $K_{A\ell}$ may each be related to k_{Ag} and $k_{A\ell}$. From equations 9.2-3, -6, and -9,

$$\frac{1}{K_{Ag}} = \frac{1}{k_{Ag}} + \frac{H_A}{k_{A\ell}} \tag{9.2-13}$$

and from equations 9.2-3, -6, and -10,

$$\frac{1}{K_{A\ell}} = \frac{1}{H_A k_{Ag}} + \frac{1}{k_{A\ell}} \tag{9.2-14}$$

Each of these last two equations represents the additive contribution of gas- and liquid-film resistances (on the right) to the overall resistance (on the left). (Each mass transfer coefficient is a "conductance" and its reciprocal is a "resistance.")

Special cases arise from each of equations 9.2-13 and -14, depending on the relative magnitudes of k_{Ag} and $k_{A\ell}$. For example, from equation 9.2-13, if k_{Ag} is relatively large so that $1/k_{Ag} \ll H_A/k_{A\ell}$, then K_{Ag} (and hence N_A) is determined entirely by $k_{A\ell}$, and we have the situation of "liquid-film control." An important example of this is the situation in which the gas phase is pure A, in which case there is no gas film for A to diffuse through, and $k_{Ag} \to \infty$. Conversely, we may have "gas-film control." Similar conclusions may be reached from consideration of equation 9.2-14 for $K_{A\ell}$. In either case, we obtain the following results:

$$N_A = k_{A\ell}(p_A/H_A - c_A); \text{ liquid-film control} \tag{9.2-15}$$

$$N_A = k_{Ag}(p_A - H_A c_A); \text{ gas-film control} \tag{9.2-16}$$

9.2.3 Kinetics Regimes for Two-Film Model

9.2.3.1 Classification in Terms of Location of Chemical Reaction

The rate expressions developed in this section for gas-liquid systems, represented by reaction 9.2-1, are all based on the two-film model. Since liquid-phase reactant B is assumed to be nonvolatile, for reaction to occur, the gas-phase reactant A must enter the liquid phase by mass transfer (see Figure 9.4). Reaction between A and B then takes place at some "location" within the liquid phase. At a given point, as represented in Figure 9.4, there are two possible locations: the liquid film and the bulk liquid. If the rate of mass transfer of A is relatively fast compared with the rate of reaction, then A reaches the bulk liquid before reacting with B. Conversely, for a relatively fast rate of reaction ("instantaneous" in the extreme), A reacts with B in the liquid film before it reaches the bulk liquid. Since the intermediate situation is also possible, we may initially classify the kinetics into three regimes:

(1) Reaction in bulk liquid only;
(2) Reaction in liquid film only;
(3) Reaction in both liquid film and bulk liquid.

We treat each of these three cases in turn to obtain, as far as possible, analytical or approximate analytical rate expressions, taking both mass transfer and reaction into account. Each of these cases gives rise to important subcases, some of which are developed further, and some of which are left to problems at the end of the chapter. In treating the cases in the order above, we are proceeding from special, relatively simple, situations to more general ones, the reverse of the approach taken in Section 9.1 for gas-solid systems.

It is necessary to distinguish among three rate quantities. We use the symbol N_A to represent the flux of A, in mol m^{-2} s^{-1}, through gas and/or liquid film; if reaction takes place in the liquid film, N_A includes the effect of reaction (loss of A). We use the symbol $(-r_A)$, in mol m^{-2} s^{-1}, to represent the intensive rate of reaction per unit interfacial area. Dimensionally, $(-r_A)$ corresponds to N_A, but $(-r_A)$ and N_A are equal only in the two special cases (1) and (2) above. In case (3), they are not equal, because reaction occurs in the bulk liquid (in which there is no flux) as well as in the liquid film. In this case, furthermore, we need to distinguish between the flux of A into the liquid film at the gas-liquid interface, $N_A(z = 0)$, and the flux from the liquid film to the bulk liquid, $N_A(z = 1)$, where z is the relative distance into the film from the interface; these two fluxes differ because of the loss of A by reaction in the liquid film. The third rate quantity is $(-r_A)_{int}$ in mol m^{-3} s^{-1}, the intrinsic rate of reaction per unit volume of liquid in the bulk liquid. $(-r_A)$ and $(-r_A)_{int}$ are related as shown in equation 9.2-17 below.

9.2.3.2 Reaction in Bulk Liquid Only; Relatively Slow Reaction

If chemical reaction occurs only in the bulk liquid, but resistance to mass transfer of A through gas and liquid films is not negligible, the concentration profiles could be as shown in Figure 9.5. This is essentially the same as Figure 9.4, except that a horizontal line is added for c_B.

We assume that the reaction is intrinsically second-order (first-order with respect to each reactant), so that

$$N_A = (-r_A) = (-r_A)_{int}/a_i = (k_A/a_i)c_A c_B = k'_A c_A c_B \qquad (9.2\text{-}17)$$

where, for consistency with the units of $(-r_A)$ used above, the interfacial area a_i (e.g., m^2 (interfacial area) m^{-3} (liquid)) is introduced to relate $(-r_A)$ to $(-r_A)_{int}$ (in mol m^{-3} (liquid) s^{-1}).

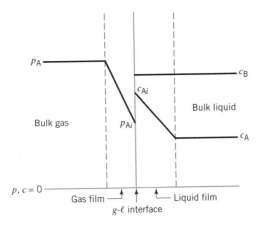

Figure 9.5 Two-film model (profiles) for relatively slow reaction A(g) + bB(ℓ) → products (nonvolatile B)

Since the system is usually specified in terms of p_A and c_B, rather than c_A and c_B, we transform equation 9.2-17 into a more useful form by elimination of c_A in favor of p_A; c_A then becomes a dependent variable.

Since the three rate processes

Mass transfer of A through gas film
Mass transfer of A through liquid film
Reaction of A and B in bulk liquid

are in series, the steady-state rate of transport or reaction, N_A or $(-r_A)$, is given independently by equations 9.2-3, -6, and -17; a fourth relation is the equilibrium connection between p_{Ai} and c_{Ai} given by Henry's law, equation 9.2-8. These four equations may be solved simultaneously to eliminate c_{Ai}, p_{Ai}, and c_A (in favor of p_A to represent the concentration of A) to obtain the following result for $(-r_A)$:

$$(-r_A) = \frac{p_A}{\dfrac{1}{k_{Ag}} + \dfrac{H_A}{k_{A\ell}} + \dfrac{H_A}{k'_A c_B}} \qquad (9.2\text{-}18)$$

The summation in the denominator represents the additivity of "resistances" for the three series processes. From 9.2-17 and -18, we obtain c_A in terms of p_A and c_B:

$$c_A = \frac{p_A}{k'_A c_B \left(\dfrac{1}{k_{Ag}} + \dfrac{H_A}{k_{A\ell}}\right) + H_A} \qquad (9.2\text{-}18a)$$

Three special cases of equation 9.2-18 arise, depending on the relative magnitudes of the two mass-transfer terms in comparison with each other and with the reaction term (which is always present for reaction in bulk liquid only). In the extreme, if all mass-transfer resistance is negligible, the situation is the same as that for a homogeneous liquid-phase reaction, $(-r_A)_{int} = k_A c_A c_B$.

If the liquid-phase reaction is pseudo-first-order with respect to A (c_B constant and $\gg c_A$),

$$(-r_A) = (-r_A)_{int}/a_i = (k'_A c_B) c_A = k''_A c_A \qquad (9.2\text{-}17a)$$

where $k''_A = k'_A c_B = k_A c_B/a_i$. Equations 9.2-18 and -18a apply with $k'_A c_B$ replaced by k''_A.

9.2.3.3 Reaction in Liquid Film Only; Relatively Fast Reaction

9.2.3.3.1. Instantaneous reaction. If the rate of reaction between A and B is so high as to result in instantaneous reaction, then A and B cannot *coexist* anywhere in the liquid phase. Reaction occurs at some point in the liquid film, the location of which is determined by the relative concentrations and diffusivities of A and B. This is shown as a reaction "plane" in Figure 9.6. At this point, c_A and c_B both become zero. The entire process is mass-transfer controlled, with A diffusing to the reaction plane from the bulk gas, first through the gas film and then through the portion of the liquid film of thickness δ, and B diffusing from the bulk liquid through the remaining portion of the liquid film of thickness $\delta_\ell - \delta$.

The three diffusion steps can be treated as series processes, with the fluxes or rates given by, respectively,

$$N_A = k_{Ag}(p_A - p_{Ai}) \tag{9.2-3}$$

$$N_A = \frac{D_{A\ell}}{\delta}(c_{Ai} - 0) = \frac{\delta_\ell}{\delta} k_{A\ell} c_{Ai} \tag{9.2-19}$$

$$N_B = \frac{D_{B\ell}}{\delta_\ell - \delta}(c_B - 0) = \frac{\delta_\ell}{\delta_\ell - \delta} k_{B\ell} c_B \tag{9.2-20}$$

The first part of equation 9.2-19 corresponds to the integrated form of Fick's law in equation 9.2-5, and the second part incorporates equation 9.2-7; equation 9.2-20 applies in similar fashion to species B. The rates N_A and N_B are related through stoichiometry by

$$N_B = bN_A \tag{9.2-20a}$$

and the concentrations p_{Ai} and c_{Ai} are related by Henry's law:

$$p_{Ai} = H_A c_{Ai} \tag{9.2-8}$$

Finally, the liquid-phase diffusivities and mass-transfer coefficients are related, as a consequence of equation 9.2-7, by

$$k_{A\ell}/k_{B\ell} = D_{A\ell}/D_{B\ell} \tag{9.2-21}$$

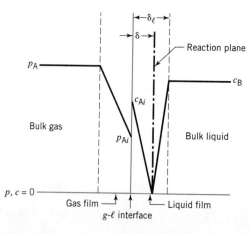

Figure 9.6 Two-film model (profiles) for instantaneous reaction $A(g) + bB(\ell) \rightarrow$ products (nonvolatile B)

9.2 Gas-Liquid Systems

These six governing equations may be solved for N_A with elimination of p_{Ai}, c_{Ai}, N_B, $k_{B\ell}$ and δ_ℓ/δ to result in the rate law, in terms of $(-r_A) \equiv N_A$,

$$N_A \equiv (-r_A) = \frac{p_A + \dfrac{D_{B\ell}H_A}{D_{A\ell}b}c_B}{\dfrac{1}{k_{Ag}} + \dfrac{H_A}{k_{A\ell}}} \qquad (9.2\text{-}22)$$

or

$$(-r_A) = K_{Ag}\left(p_A + \frac{D_{B\ell}H_A}{D_{A\ell}b}c_B\right) \qquad (9.2\text{-}22\text{a})$$

$$= K_{A\ell}\left(\frac{p_A}{H_A} + \frac{D_{B\ell}}{D_{A\ell}b}c_B\right) \qquad (9.2\text{-}22\text{b})$$

The last two forms come from the definitions of K_{Ag} and $K_{A\ell}$ in equations 9.2-13 and -14, respectively.

Two extreme cases of equation 9.2-22 or -22a or -22b arise, corresponding to gas-film control and liquid-film control, similar to those for mass transfer without chemical reaction (Section 9.2.2). The former has implications for the location of the reaction plane (at distance δ from the interface in Figure 9.6) and the corresponding value of c_B. These points are developed further in the following two examples.

EXAMPLE 9-4

What is the form of equation 9.2-22 for gas-film control, and what are the implications for the location of the reaction plane (value of δ) and c_B (and hence for $(-r_A)$ in equation 9.2-22)?

SOLUTION

In Figure 9.6, the position of the reaction plane at distance δ from the gas-liquid interface is shown for a particular value of c_B. If c_B changes (as a parameter), the position of the reaction plane changes, δ decreasing as c_B increases. This may be realized intuitively from Figure 9.6, or can be shown from equations 9.2-20, -20a, and -22a. Elimination of N_A and N_B from these three equations provides a relation between δ and c_B:

$$\delta = \delta_\ell - \frac{D_{A\ell}D_{B\ell}c_B}{K_{Ag}(D_{A\ell}bp_A + D_{B\ell}H_A c_B)} \qquad \text{(A)}$$

from which

$$\frac{\partial \delta}{\partial c_B} = -\frac{D_{A\ell}^2 D_{B\ell}bp_A}{K_{Ag}(D_{A\ell}bp_A + D_{B\ell}H_A c_B)^2} < 0 \qquad \text{(B)}$$

That is, δ decreases as c_B increases. δ can only decrease to zero, since the reaction plane cannot occur in the gas film (species B is nonvolatile). At this condition, the reaction plane coincides with the gas-liquid interface, and p_{Ai}, c_{Ai}, and c_B (at the interface) are all zero. This corresponds to gas-film control, since species A does not penetrate the liquid film.

The value of c_B in the bulk liquid is the largest value of c_B that can influence the rate of reaction, and may be designated $c_{B,max}$. From equation (A), with $\delta = 0$, and $\delta_\ell = D_{A\ell}/k_{A\ell}$, equation 9.2-7,

$$c_{B,max} = \frac{K_{Ag}D_{A\ell}bp_A}{D_{B\ell}(k_{A\ell} - K_{Ag}H_A)} \quad (9.2\text{-}23)$$
$$= k_{Ag}bp_A/k_{B\ell}$$

if we use the condition for gas-film control, from equation 9.2-13, $1/k_{Ag} \gg H_A/k_{A\ell}$ or $k_{A\ell} \gg H_A k_{Ag}$, so that $K_{Ag} = k_{Ag}$, together with equation 9.2-21 to eliminate $D_{A\ell}$, $D_{B\ell}$, and $k_{A\ell}$ in favor of $k_{B\ell}$.

To obtain $(-r_A)$ for gas-film control, we may substitute $c_{B,max}$ from equation 9.2-23 into equation 9.2-22, and again use $k_{A\ell} \gg H_A k_{Ag}$, together with equation 9.2-21:

$$(-r_A) = k_{Ag}p_A; \quad c_B \geq c_{B,max} \quad (9.2\text{-}24)$$

Thus, when $c_B \geq c_{B,max}$, the rate $(-r_A)$ depends only on p_A.

This result also has implications for $(-r_A)$ given by equation 9.2-22. This equation applies as it stands only if $c_B \leq c_{B,max}$. For higher values of c_B, equation 9.2-24 governs.

EXAMPLE 9-5

What is the form of equation 9.2-22 for liquid-film control?

SOLUTION

For liquid-film control, there is no gas-phase resistance to mass transfer of A. Thus, in equation 9.2-14, $1/k_{A\ell} \gg 1/H_A k_{Ag}$, and $K_{A\ell} = k_{A\ell}$, so that equation 9.2-22b may be written

$$(-r_A) = \left(1 + \frac{D_{B\ell}H_A}{D_{A\ell}b}\frac{c_B}{p_A}\right)\frac{k_{A\ell}}{H_A}p_A \quad (9.2\text{-}25)$$

9.2.3.3.2. Fast reaction. If chemical reaction is sufficiently fast, even though not instantaneous, it is possible for it to occur entirely within the liquid-film, but not at a point or plane. This case is considered as a special case of reaction in both bulk liquid and liquid film in Section 9.2.3.4. In this situation, $N_A \equiv (-r_A)$.

9.2.3.3.3. Enhancement factor E. For reaction occurring *only* in the liquid film, whether instantaneous or fast, the rate law may be put in an alternative form by means of a factor that measures the enhancement of the rate relative to the rate of physical absorption of A in the liquid without reaction. Reaction occurring only in the liquid film is characterized by $c_A \to 0$ somewhere in the liquid film, and the enhancement factor E is defined by

$$E = \frac{\text{rate of reaction or flux of A}}{\text{maximum rate of mass transfer of A through the liquid film}}$$
$$= (-r_A)/k_{A\ell}(c_{Ai} - c_A) = (-r_A)/k_{A\ell}c_{Ai}$$
$$(9.2\text{-}26)$$

since $c_A = 0$ for maximum rate of mass transfer. Thus, in terms of E, the rate law is

$$(-r_A) = k_{A\ell} E c_{Ai} \quad (9.2\text{-}26a)$$

The interfacial concentration c_{Ai} can be eliminated by means of equations 9.2-3 and -8 to give

$$(-r_A) = \frac{p_A}{\dfrac{1}{k_{Ag}} + \dfrac{H_A}{k_{A\ell} E}} \quad (9.2\text{-}27)$$

For an instantaneous reaction, equation 9.2-27 is an alternative to equation 9.2-22. An expression for E can be obtained from these two equations, and special cases can be examined in terms of E (see problems 9-11 and 9-12).

9.2.3.4 Reaction in Liquid Film and Bulk Liquid

The cases above, reaction in bulk liquid only and instantaneous reaction in the liquid film, have been treated by considering rate processes in series. We can't use this approach if diffusion and reaction of A and B are both spread over the liquid film. Instead, we consider solution of the continuity equations for A and B, through the liquid film.

Figure 9.7 shows concentration profiles schematically for A and B according to the two-film model. Initially, we ignore the presence of the gas film and consider material balances for A and B across a thin strip of width dx in the liquid film at a distance x from the gas-liquid interface. (Since the gas-film mass transfer is in series with combined diffusion and reaction in the liquid film, its effect can be added as a resistance in series.)

For A at steady-state, the rate of diffusion into the thin strip at x is equal to the rate of diffusion out at $(x + dx)$ plus the rate of disappearance within the strip. That is, for unit cross-sectional area,

$$-D_{A\ell}\frac{dc_A}{dx} = -D_{A\ell}\left[\frac{dc_A}{dx} + \frac{d}{dx}\left(\frac{dc_A}{dx}\right)dx\right] + (-r_A)_{int} dx$$

which becomes

$$D_{A\ell}\frac{d^2 c_A}{dx^2} - (-r_A)_{int} = 0 \quad (9.2\text{-}28)$$

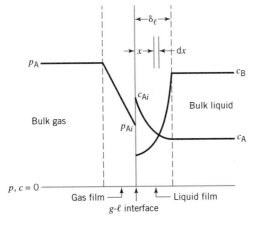

Figure 9.7 Two-film model (profiles) for reaction $A(g) + bB(\ell) \to$ products in bulk liquid and liquid film (nonvolatile B)

For B, similarly, we have

$$D_{B\ell}\frac{d^2c_B}{dx} - (-r_B)_{int} = 0 \qquad (9.2\text{-}29)$$

The intrinsic rate of reaction is a function of c_A, c_B, and T

$$(-r_A)_{int} = r_A(c_A, c_B, T) \qquad (9.2\text{-}30)$$

and $(-r_A)_{int}$ and $(-r_B)_{int}$ are related by

$$(-r_B)_{int} = b(-r_A)_{int} \qquad (9.1\text{-}8)$$

The boundary conditions for the two simultaneous second-order ordinary differential equations, 9.2-28 and -29, may be chosen as:

$$\text{at } x = 0, \quad c_A = c_{Ai} \qquad (9.2\text{-}31)$$

$$\frac{dc_B}{dx} = 0 \qquad (9.2\text{-}32)$$

(consistent with no transport of B through the interface, since B is nonvolatile)

$$\text{at } x = \delta_\ell, \quad c_A = c_A(\text{in bulk liquid}) \equiv c_{Ab} \qquad (9.2\text{-}33)$$

$$c_B = c_B(\text{in bulk liquid}) \equiv c_{Bb} \qquad (9.2\text{-}34)$$

Equations 9.2-28 and -29, in general, are coupled through equation 9.2-30, and analytical solutions may not exist (numerical solution may be required). The equations can be uncoupled only if the reaction is first-order or pseudo-first-order with respect to A, and exact analytical solutions are possible for reaction occurring in bulk liquid and liquid film together and in the liquid film only. For second-order kinetics with reaction occurring only in the liquid film, an approximate analytical solution is available. We develop these three cases in the rest of this section.

9.2.3.4.1. First-order or pseudo-first-order isothermal, irreversible reaction with respect to A; reaction in liquid film and bulk liquid. For a reaction that is first-order or pseudo-first-order with respect to A, we have

$$(-r_A)_{int} = k_A c_A \qquad (9.2\text{-}35)$$

or

$$(-r_A)_{int} = k_A''' c_A \qquad (9.2\text{-}35a)$$

where $k_A''' = k_A c_B$, which is of the same form as 9.2-35.

Use of either one of these for $(-r_A)_{int}$ has the effect of uncoupling equation 9.2-28 from -29, and we need only solve the former in terms of A. Thus, equation 9.2-28 becomes:

$$D_{A\ell}\frac{d^2c_A}{dx^2} - k_A c_A = 0 \qquad (9.2\text{-}36)$$

with boundary conditions given by equations 9.2-31 and -33. We solve equation 9.2-28 in dimensionless form by letting

$$\lambda = c_A/c_{Ai} \qquad (9.2\text{-}37)$$

and

$$z = x/\delta_\ell \qquad (9.2\text{-}38)$$

Then equation 9.2-36 becomes

$$\frac{d^2\lambda}{dz^2} - \text{Ha}^2\lambda = 0 \qquad (9.2\text{-}39)$$

where the Hatta number, Ha, is a dimensionless group defined by

$$\boxed{\text{Ha} = \delta_\ell(k_A/D_{A\ell})^{1/2} = (D_{A\ell}k_A)^{1/2}/k_{A\ell} \quad (n=1) \qquad (9.2\text{-}40)}$$

if we use equation 9.2-7 to eliminate δ_ℓ. The solution of equation 9.2-39 gives the concentration profile for A through the liquid film in terms of λ and z. The boundary conditions 9.2-31 and -33, in terms of λ and z, become

$$\text{at } z = 0, \quad \lambda = 1 \qquad (9.2\text{-}31a)$$

$$\text{at } z = 1, \quad \lambda = c_{Ab}/c_{Ai} = \lambda_1 \qquad (9.2\text{-}33a)$$

The solution of equation 9.2-39 may be written as

$$\lambda = K_1 \exp(\text{Ha } z) + K_2 \exp(-\text{Ha } z) \qquad (9.2\text{-}41)$$

and the integration constants evaluated from the boundary conditions are

$$K_1 = \frac{\lambda_1 - \exp(-\text{Ha})}{2\sinh(\text{Ha})}$$

$$K_2 = \frac{\exp(\text{Ha}) - \lambda_1}{2\sinh(\text{Ha})}$$

Elimination of K_1 and K_2 from equation 9.2-41 results in

$$\lambda = \frac{\lambda_1 \sinh(\text{Ha } z) + \sinh[\text{Ha}(1-z)]}{\sinh(\text{Ha})} \qquad (9.2\text{-}42)$$

To obtain the steady-state rate of transfer or flux of A into the liquid film, $N_A(z=0)$, we note that this flux is equal to the rate of diffusion at the gas-liquid interface:

$$N_A(z=0) = -D_{A\ell}\left(\frac{dc_A}{dx}\right)_{x=0} = -k_{A\ell}c_{Ai}\left(\frac{d\lambda}{dz}\right)_{z=0} \qquad (9.2\text{-}43)$$

250 Chapter 9: Multiphase Reacting Systems

$d\lambda/dz$ is determined by differentiating equation 9.2-42, and evaluating the derivative at $z = 0$. Thus,

$$N_A(z = 0) = \frac{\text{Ha}}{\tanh(\text{Ha})}\left[1 - \frac{c_{Ab}}{c_{Ai}\cosh(\text{Ha})}\right]k_{A\ell}c_{Ai} \quad (9.2\text{-}44)$$

on replacement of λ_1 by c_{Ab}/c_{Ai}, where c_{Ab} is the bulk liquid concentration (in Figure 9.7). Equation 9.2-44 was first obtained by Hatta (1932). It is based on liquid-film resistance only. To take the gas-film resistance into account and eliminate c_{Ai}, we use equation 9.2-44 together with equations 9.2-3 and -8

$$N_A \equiv N_A(z = 0) = k_{Ag}(p_A - p_{Ai}) \quad (9.2\text{-}3)$$

$$c_{Ai} = p_{Ai}/H_A \quad (9.2\text{-}8)$$

to obtain

$$N_A(z = 0) = \frac{p_A - \dfrac{H_A c_{Ab}}{\cosh(\text{Ha})}}{\dfrac{1}{k_{Ag}} + \dfrac{H_A \tanh(\text{Ha})}{k_{A\ell}\text{Ha}}} \quad (9.2\text{-}45)$$

Special cases of equation 9.2-45 result from the two extremes of Ha → large and Ha → 0 (see problem 9-13).

Similarly, to obtain $N_A(z = 1)$, the flux or rate of transfer of A *from the liquid film to the bulk liquid*, we evaluate the rate of diffusion at $z = 1$ from equation 9.2-42:

$$N_A(z = 1) = -k_{A\ell}c_{Ai}\left(\frac{d\lambda}{dz}\right)_{z=1} = \frac{\text{Ha}}{\sinh(\text{Ha})}\left[1 - \frac{c_{Ab}}{c_{Ai}}\cosh(\text{Ha})\right]k_{A\ell}c_{Ai}$$

$$= \frac{k_{A\ell}\text{Ha}}{\tanh(\text{Ha})}\left[\frac{c_{Ai}}{\cosh(\text{Ha})} - c_{Ab}\right] \quad (9.2\text{-}44\text{a})$$

To eliminate c_{Ai} and take gas-film resistance into account, we again use 9.2-3 and 9.2-8. Thus, eliminating c_{Ai}, p_{Ai}, and $N_A(z = 0)$ from 9.2-3, -8, -44a, and -45, we obtain the following rather cumbersome expression:

$$N_A(z = 1) = \frac{k_{A\ell}\text{Ha}}{\tanh(\text{Ha})}\left\{\frac{\dfrac{k_{Ag}\tanh(\text{Ha})}{k_{A\ell}\text{Ha}}p_A + \dfrac{c_{Ab}}{\cosh(\text{Ha})}}{\cosh(\text{Ha})[1 + \dfrac{k_{Ag}H_A\tanh(\text{Ha})}{k_{A\ell}\text{Ha}}]} - c_{Ab}\right\} \quad (9.2\text{-}45\text{a})$$

Equations 9.2-45 and -45a are used in the continuity equations for reactor models in Chapter 24.

9.2.3.4.2. Fast first-order or pseudo-first-order reaction in liquid film only. If the chemical reaction itself is sufficiently fast, without being instantaneous, c_A drops to zero in the liquid film, and reaction takes place only in the liquid film, since A does not reach the bulk liquid. The difference between this case and that in Section 9.2.4.3.1 above is in the boundary condition at $z = 1$; equation 9.2-33a is replaced by

$$\text{at } z = 1, \quad c_{Ab} = 0; \quad \lambda_1 = 0 \quad (9.2\text{-}33\text{b})$$

In this case, the solution of equation 9.2-39 with boundary conditions 9.2-31a and 9.2-33b is

$$\lambda = \frac{\sinh[\text{Ha}(1-z)]}{\sinh(\text{Ha})} \qquad (9.2\text{-}46)$$

From equation 9.2-46, using equation 9.2-43, we obtain an expression for $N_A\,(z=0)$, which also represents $(-r_A)$, since all reaction takes place in the liquid film:

$$N_A(z=0) \equiv (-r_A) = \frac{\text{Ha}}{\tanh(\text{Ha})} k_{A\ell} c_{Ai} \qquad (9.2\text{-}47)$$

and, on elimination of c_{Ai} by means of equations 9.2-3 and -8 to take the gas-film resistance into account, as in Section 9.2.3.4.1,

$$(-r_A) = \frac{p_A}{\dfrac{1}{k_{Ag}} + \dfrac{H_A \tanh(\text{Ha})}{k_{A\ell}\text{Ha}}} \qquad (9.2\text{-}48)$$

Comparison of equation 9.2-48 with 9.2-27 shows that, for this case, the enhancement factor is

$$E = \text{Ha}/\tanh(\text{Ha}) \qquad (9.2\text{-}49)$$

9.2.3.4.3. Fast second-order reaction in liquid film only. For a reaction that is second-order, first-order with respect to each of A and B, we have

$$(-r_A)_{int} = k_A c_A c_B \qquad (9.2\text{-}50)$$

In this case, an exact analytical solution of the continuity equations for A and B does not exist. An approximate solution has been developed by Van Krevelen and Hoftijzer (1948) in terms of E. Results of a numerical solution could be fitted approximately by the implicit relation

$$E = \psi/\tanh\psi \qquad (9.2\text{-}51)$$

where

$$\psi = \text{Ha}\left(\frac{E_i - E}{E_i - 1}\right)^{1/2} \qquad (9.2\text{-}52)$$

E_i is the enhancement factor for an instantaneous reaction (see problem 9-11(c)):

$$E_i = 1 + \frac{D_{B\ell} c_B}{D_{A\ell} b c_{Ai}} \qquad (9.2\text{-}53)$$

and, for a second-order reaction,

$$\text{Ha} = (k_A D_{A\ell} c_B)^{1/2}/k_{A\ell} \quad (n = 2) \quad (9.2\text{-}54)$$

For given values of Ha and E_i, E may be calculated by solving equations 9.2-51 and -52 numerically using the E-Z Solve software (or by trial). The results of such calculations are shown graphically in Figure 9.8, as a plot of E versus Ha with $E_i - 1$ as a parameter. Figure 9.8 can be used to obtain approximate values of E.

As shown in Figure 9.8, the solution for E for a fast second-order reaction given by equation 9.2-51 is bounded by E for two other cases: fast first-order or pseudo-first-order reaction, given by equation 9.2-49 and instantaneous reaction, given by equation 9.2-53. Thus, from equation 9.2-51, as $E_i \to \infty$ for constant Ha, $E \to$ Ha/tanh (Ha), and as Ha $\to \infty$ for constant E_i, $E \to E_i$ (see problem 9-17). In Figure 9.8, equation 9.2-49 is represented by the diagonal line that becomes curved for Ha < 3, equation 9.2-51 by the family of curved lines between this "diagonal" line and the dashed line, and equation 9.2-53 by the family of horizontal lines to the right of the dashed line. The dashed line is the locus of points at which E becomes horizontal, and divides the region for fast second-order reaction from that for instantaneous reaction.

9.2.3.5 Interpretation of Hatta Number (Ha); Criterion for Kinetics Regime

The Hatta number Ha, as a dimensionless group, is a measure of the maximum rate of reaction in the liquid film to the maximum rate of transport of A through the liquid

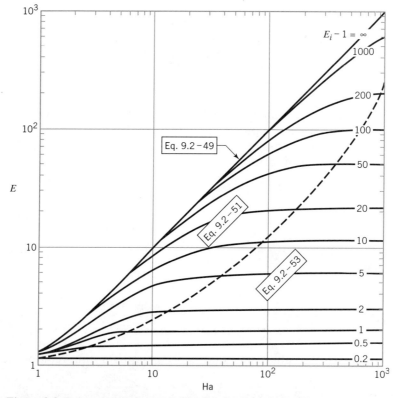

Figure 9.8 Enhancement factor, E (Ha, E_i), for fast gas-liquid reaction (in liquid film); reaction: $A(g) + bB(\ell) \to$ products (B nonvolatile)

film, and as such is analogous to the Thiele modulus ϕ in catalytic reactions (Chapter 8).

This interpretation of Ha may be deduced from its definition. For example, for a first-order reaction, from equation 9.2-40,

$$\text{Ha}^2 = \delta_\ell^2 \frac{k_A}{D_{A\ell}} = \delta_\ell \frac{k_A a_i c_{Ai}}{k_{A\ell} a_i c_{Ai}} = \frac{v_f k_A c_{Ai}}{a_i k_{A\ell} c_{Ai}} \quad (9.2\text{-}55)$$

$$= \frac{\text{maximum rate of reaction or flux in liquid film}}{\text{maximum rate of transport through liquid film}}$$

In equation 9.2-55, a_i is the interfacial area, for example, in m² m⁻³ (liquid), and v_f is the volume of the liquid film. The rates are maximum rates because c_{Ai} is the highest value of c_A in the film (for numerator), and the form of the denominator is consistent with c_A (bulk liquid) = 0, giving the largest possible driving force for mass transfer.

Thus, if Ha (strictly, Ha²) $\gg$ 1, reaction occurs in the liquid film only, and if Ha or Ha² $\ll$ 1, reaction occurs in bulk liquid. Numerically, these values for Ha may be set at about 3 and 0.1, respectively.

EXAMPLE 9-6

Suppose pure CO_2 (A) at 1 bar is absorbed into an aqueous solution of NaOH (B) at 20° C. Based on the data given below and the two-film model, how should the rate of absorption be characterized (instantaneous, fast pseudo-first-order, fast second-order), if c_B = (a) 0.1 and (b) 6 mol L⁻¹?

Data (Danckwerts, 1970, p. 118, in part): k_A = 10,000 L mol⁻¹ s⁻¹; $D_{A\ell}$ = 1.8 × 10⁻⁵ cm² s⁻¹; $k_{A\ell}$ = 0.04 cm s⁻¹; $D_{B\ell}$ = 3.1 × 10⁻⁵ cm² s⁻¹; H_A = 36 bar L mol⁻¹.

SOLUTION

The chemical reaction (in the liquid phase) is

$$CO_2(g) + 2NaOH(\ell) \rightarrow Na_2CO_3 + H_2O$$

Hence, b = 2.

Since a finite value of the rate constant k_A is specified, the reaction is not instantaneous. The units given for k_A indicate that the reaction is second-order, presumably of the form in equation 9.2-50. To obtain further insight, we calculate values of Ha and E_i, from equations 9.2-54 and 9.2-53, respectively. For E_i, since the gas phase is pure CO_2, there is no gas-phase resistance, and $c_{Ai} = p_A/H_A$, from equation 9.2-8, with $p_{Ai} = p_A$ (= 1 bar). The results for the two cases are given in the following table (together with values of E, as discussed below)

case	c_B	Ha	E_i	E	
	mol L⁻¹	(9.2-54)	(9.2-53)	(9.2-51)	(9.2-49)
(a)	0.1	3.4	4.1	2.5	3.4
(b)	6.0	26	187	24.3	26

254 Chapter 9: Multiphase Reacting Systems

The values of Ha are sufficiently large (Ha > 3) to indicate that reaction occurs entirely within the liquid film. This suggests that we need consider only the two possibilities: fast second-order reaction and fast pseudo-first-order reaction. From the values of Ha and E_i in the table, Figure 9.8 can be used to assess the kinetics model and obtain approximate values of E. Alternatively, we calculate values of E from equations 9.2-51 and -49 for second-order and pseudo-first-order, respectively, using E-Z Solve (file ex9-6.msp). The results are given in the table above. The conclusions are:

(a) In Figure 9.8, the value of E is in the region governed by equation 9.2-51 for second-order kinetics. This is shown more precisely by the two calculated values; 2.5 is only 74% of the value (3.4) for pseudo-first-order kinetics.

(b) The relatively high value of c_B suggests that the reaction may be pseudo-first-order (with respect to A), and this is confirmed by Figure 9.8 and the calculations. In the former, E is virtually on the diagonal line representing equation 9.2-49, and from the latter, the

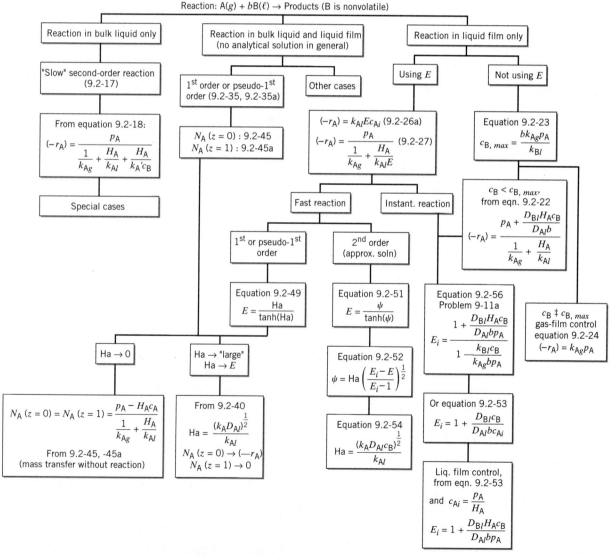

Figure 9.9 Summary of rate or flux expressions for gas-liquid reactions (two-film model)

(approximate) value of E (24.3) calculated from equation 9.2-51 is 94% of the (exact) value (26) calculated from equation 9.2-49 for pseudo-first-order. The simpler form of the latter could be used with relatively little error.

9.2.3.6 *Summary of Rate Expressions*

Application of the two-film model to a particular type of gas-liquid reaction, represented by equation 9.2-1, results in a proliferation of rate expressions for the various possible cases for which analytical solutions of the continuity equations are obtained here. These are summarized in Figure 9.9, as a branching chart, for convenient reference and to show the structure of their relationship to each other. The chart is divided at the top into three main branches according to the supposed location of chemical reaction, and progresses in individual cases from "slow" reaction on the left to instantaneous reaction on the right. The equation number from the text is given along with the rate expression or other relevant quantity such as E or Ha. The first expression in the chart for E for an instantaneous reaction is not given in the text, but is developed in problem 9-11, along with E given by equation 9.2-53.

Note that the enhancement factor E is relevant only for reaction occurring in the liquid film. For an instantaneous reaction, the expressions may or may not involve E, except that for liquid-film control, it is convenient, and for gas-film control, its use is not practicable (see problem 9-12(a)). The Hatta number Ha, on the other hand, is not relevant for the extremes of slow reaction (occurring in bulk liquid only) and instantaneous reaction. The two quantities are both involved in rate expressions for "fast" reactions (occurring in the liquid film only).

9.3 INTRINSIC KINETICS OF HETEROGENEOUS REACTIONS INVOLVING SOLIDS

The mechanisms, and hence theoretically derived rate laws, for noncatalytic heterogeneous reactions involving solids are even less well understood than those for surface-catalyzed reactions. This arises because the solid surface changes as the reaction proceeds, unlike catalytic surfaces which usually reach a steady-state behavior. The examples discussed here are illustrative.

Gasification reactions of solids: The reactions of solids with gas-phase reactants to form gaseous products are generally described in the same manner as are surface-catalyzed reactions. The reaction of carbon with water vapor is an example:

$$C(s) + H_2O \rightarrow H_2 + CO \tag{9.3-1}$$

This reaction is important in such processes as the decoking of catalysts, the manufacture of activated carbon for adsorption, and the gasification of carbonaceous materials for production of hydrogen or fuel gas.

A two-step mechanism and resulting rate law can be developed as follows. Reactive carbon sites, C^* (total number N_{C^*}), are assumed to exist on the surface of the solid. These can be oxidized reversibly by water vapor:

$$C^* + H_2O \underset{k_{-1}}{\overset{k_1}{\rightleftharpoons}} C^*O + H_2 \tag{1}$$

where C*O is an oxidized carbon site. The oxidized site can then "decompose" to produce CO(g):

$$C^*O \xrightarrow{k_2} CO(g) \tag{2}$$

In addition to CO(g) formation, step (2) exposes a variable number, n, of previously inactive carbon atoms, C, thus producing C* to continue the reaction. The average value of n is close to unity, so that N_{C^*} varies slowly as reaction proceeds.

If elementary rate laws are assumed for each step, and if N_{C^*} is essentially constant over a short time, a (pseudo-) steady-state rate law can be developed:

$$r = \frac{N_{C^*}}{N_C} \frac{k_1 k_2 c_{H_2O}}{k_1 c_{H_2O} + k_{-1} c_{H_2} + k_2} \tag{9.3-2}$$

(Note the similarity to Langmuir-Hinshelwood kinetics.) The rate is expressed on the basis of the instantaneous number of solid carbon atoms, N_C. The rate r (measured at one gas composition) typically goes through a maximum as the carbon is converted. This is the result of a maximum in the intrinsic activity (related to the fraction of reactive carbon atoms, N_{C^*}/N_C) because of both a change in N_{C^*} and a decrease in N_C.

Since both N_{C^*} and N_C change as the reaction proceeds, r can be expressed as a function of fractional conversion of carbon, f_C, or of time, t:

$$r(t) = \frac{k_{eff}(t) c_{H_2O}}{k_1 c_{H_2O} + k_{-1} c_{H_2} + k_2} \tag{9.3-3}$$

where $k_{eff}(t)$ is a time-dependent rate constant given by

$$k_{eff}(t) = \frac{N_{C^*}(t)}{N_C(t)} k_1 k_2 \tag{9.3-4}$$

Other models allow for a distribution of site reactivities. Similar considerations apply for reactions of solids in liquid solution.

Solid deposition from gas- or liquid-phase reactants: Solid-deposition reactions are important in the formation of coatings and films from reactive vapors (called chemical vapor deposition or CVD) and of pure powders of various solids. Examples are:

$$Si_2H_6 \rightarrow Si(s) + 3H_2 \tag{9.3-5}$$

$$3TiCl_4 + 4NH_3 \rightarrow Ti_3N_4(s) + 12HCl \tag{9.3-6}$$

These are the reverse of gasification reactions and proceed through initial adsorption of intermediates on an existing solid surface. The kinetics of the deposition and decomposition of the surface intermediates are often treated in Langmuir-Hinshelwood-type models, where a dynamic balance of deposition sites is maintained, just as in surface-catalyzed reactions. Further rearrangements or motion on the surface are often usually necessary to form the desired solid (such as in silicon films where sufficient crystallinity is required for semiconductor properties), and the kinetics of these arrangements can be rate controlling. It is thought that decomposition of SiH_x surface-intermediate species to form hydrogen is the rate-limiting step during growth of silicon from disilane in reaction 9.3-5. Problem 9-18 deals with the kinetics of a CVD reaction.

Solid-solid reactions: Most reactions of one solid with another require the existence of mobile intermediates, because mixtures of solid particles have very few points of contact. At the very least, mobility in one of the solids is required to bring the reactants together. In the reduction of a metal oxide, MO, by carbon, carbon oxides are thought to mediate this "solid-state" reaction through the two-step mechanism:

$$MO(s) + CO \rightarrow M(s \text{ or } \ell) + CO_2 \quad (1)$$
$$CO_2 + C(s) \rightarrow 2CO \quad (2)$$

This provides a means whereby the gas pressure can influence the kinetics, but these reactions are sufficiently complex that rarely are mechanistic rate laws used in practice.

9.4 PROBLEMS FOR CHAPTER 9

9-1 For an isothermal spherical particle (at the surrounding bulk-gas temperature) of species B reacting with gaseous species A as in Example 9-1, derive the time (t)–conversion (f_B) relation from the SCM for each of the following three cases individually, and show that additivity of the three results for t agrees with the overall result reached in Example 9-1:
   (a) gas-film mass transfer is the *rds*;
   (b) surface reaction (first-order) is the *rds*; (already done in Example 9-2)
   (c) ash-layer diffusion of A is the *rds*.
   (Any combination of two of the three may be similarly considered to obtain a corresponding total time in each case.)

9-2 Repeat Example 9-1 and problem 9-1 for an isothermal particle of "flat-plate" geometry represented in Figure 8.10, assuming only one face permeable.

9-3 Repeat Example 9-1 and problem 9-1 for an isothermal cylindrical particle of radius R and length L; assume that only the circumferential area is permeable ("ends sealed").

9-4 In the use of the shrinking-core model for a gas-solid reaction, what information could be obtained about the possible existence of a rate-controlling step (gas-film mass transfer or ash-layer diffusion or surface reaction) from each of the following:
   (a) t for given f_B decreases considerably with increase in temperature (series of experiments);
   (b) linear relationship between t and f_B;
   (c) change of relative fluid-particle velocity;
   (d) effect on t of change of particle size (series of experiments).

9-5 For a certain fluid-particle reaction, represented by $A(g) + bB(s) \rightarrow$ products, it is proposed to change some of the operating parameters as follows: the particle size R_1 is to be tripled to R_2, and the temperature is to be increased from $T_1 = 800$ K to $T_2 = 900$ K. What would the partial pressure (p_{Ag2}) be, if the original partial pressure (p_{Ag1}) was 2 bar, in order that the fractional conversion (f_B) be unchanged for a given reaction time? The particles are spherical, and reaction rate is controlling for the shrinking-core model. For the reaction, $E_A/R = 12{,}000$ K.

9-6 For a certain fluid-particle reaction, represented by $A(g) + bB(s) \rightarrow$ products, suppose the time required for complete reaction of cylindrical particles of radius R_1, at $T_1 = 800$ K and partial pressure p_{Ag1}, is t_1. If it is proposed to use particles of double the size at one-third the partial pressure (R_2, p_{Ag2}), what should the temperature (T_2) be to maintain the same value of t_1? Assume reaction control with the shrinking-core model, and for the reaction, $E_A/R = 10{,}000$ K.

9-7 An experimental reactor for a gas-solid reaction, $A(g) + bB(s) \rightarrow$ products, is used in which solid particles are carried continuously at a steady flow-rate on a horizontal grate 5 m long and moving at a constant speed. Pure gas reactant A is in continuous cross-flow upward through the solid particles, which form a relatively thin layer on the moving grate.
   (a) For a solid consisting of cylindrical particles 1 mm in radius, calculate the value of each appropriate kinetics parameter t_1 (that is, for each appropriate term in the expression given in Table 9.1), specifying the units, if, at a certain T and P, f_B was 0.8 when the

grate moved at 0.3 m min^{-1}, and 0.5 when the grate moved at 0.7 m min^{-1}. Clearly state any assumptions made.

    (b) What is the value of each t_1 for 2-mm particles at the same T and P?

    (c) For the particles in (b), what is the speed of the grate, if $f_B = 0.92$?

9-8 Consider the reduction of relatively small spherical pellets of iron ore (assume $\rho_{Bm} = 20$ mol L^{-1}) by hydrogen at 900 K and 2 bar partial pressure, as represented by the shrinking-core model, and

$$4H_2 + Fe_3O_4(s) = 4H_2O + 3Fe(s)$$

    (a) Show whether gas-film resistance is likely to be significant in comparison with ash-layer resistance at relatively high conversion ($f_B \to 1$). For diffusion of H_2, assume $D = 1$ cm^2 s^{-1} at 300 K and $D \propto T^{3/2}$; assume also that $D_e = 0.03$ cm^2 s^{-1} for diffusion through the ash layer. For relatively small, free-falling particles, $k_{Ag} = D/R$.

    (b) Repeat (a) for $f_B \to 0$.

9-9 **(a)** According to the shrinking-particle model (SPM) for a gas-solid reaction [$A(g) + bB(s) \to$ gaseous products], does k_{Ag} for gas-film mass transfer increase or decrease with time of reaction? Justify briefly but quantitatively.

    (b) **(i)** For given c_{Ag}, fluid properties and gas-solid relative velocity (u), what does the result in (a) imply for the change of c_{As} (exterior-surface, gas-phase concentration of A) with increasing time? Justify.

        (ii) What is the value of c_{As} as $t \to t_1$, the time for complete reaction? Justify.

        (iii) To illustrate your answers to (i) and (ii), draw sketches of a particle together with concentration profiles of A at 3 times: $t = 0$, $0 < t < t_1$, and $t \to t_1$.

Assume that the particle is spherical and isothermal, that both gas-film mass transfer resistance and reaction "resistance" are significant, and that the Ranz-Marshall correlation for k_{Ag} is applicable. Do not make an assumption about particle "size," but assume the reaction is first-order.

9-10 For a gas-liquid reaction, represented by 9.2-1, which occurs only in the bulk liquid, the rate law resulting from the two-film model, and given by equation 9.2-18, has three special cases. Write the special form of equation 9.2-18 for each of these three cases, (a), (b), and (c), and describe what situation each refers to.

9-11 For $A(g) + bB(\ell) \to$ products (B nonvolatile) as an instantaneous reaction:

    (a) Obtain an expression for the enhancement factor E (i.e., E_i) in terms of observable quantities [exclusive of $(-r_A)$]; see Figure 9.9, equation numbered 9.2-56.

    (b) Show $E_i = \delta_\ell/\delta$.

    (c) Show, in addition to the result in (a), that

$$E_i = 1 + \frac{D_{B\ell}c_B}{D_{A\ell}bc_{Ai}} \tag{9.2-53}$$

    (d) Show that the same result for E_i for liquid-film control is obtained from the expressions in (a) and (c).

9-12 **(a)** For the instantaneous reaction $A(g) + bB(\ell) \to$ products, in which B is nonvolatile, it has been shown that, according to the two-film model, the significance of the reaction plane moving to the gas-liquid interface (i.e., $\delta \to 0$, where δ is the distance from the interface to the plane) is that the gas-film resistance controls the rate. What is the significance of this for E_i, the enhancement factor?

    (b) What is the significance (i) in general, and (ii) for E_i, if the reaction plane moves to the (imaginary) inside film boundary in the liquid phase (i.e., $\delta \to \delta_\ell$, the film thickness)?

9-13 From equations 9.2-45 and 9.2-45a, show the significance of each of the two limiting cases (a) Ha $\to$ large and (b) Ha $\to 0$.

9-14 For a gas-liquid reaction $A(g) + bB(\ell) \rightarrow$ products (B nonvolatile), sketch concentration profiles for A and B, according to the two-film model as in Figures 9.5 to 9.7, for each of the following cases:
(a) instantaneous reaction for (i) gas-film control and (ii) liquid-film control;
(b) "slow" reaction for (i) gas-film + reaction control; (ii) liquid-film + reaction control; (iii) reaction control; can there be gas-film and/or liquid-film control in this case?
(c) "fast" reaction in liquid film only.

9-15 For the (irreversible) gas-liquid reaction $A(g) + bB(\ell) \rightarrow$ products, based on the information below,
(a) sketch the concentration profiles for both A and B according to the two-film model, and
(b) derive the form of the rate law.
The reaction is zero-order with respect to A, and second-order with respect to B, which is nonvolatile $[(-r_A)_{int} = kc_B^2]$. Assume reaction takes place only in the liquid film, that gas-film mass-transfer resistance for A is negligible, and that c_B is uniform throughout the liquid phase.
(c) Obtain an expression for the enhancement factor, E, for this case.
(d) Obtain an expression for the Hatta number, Ha, and relate E and Ha for this case.

9-16 Compare the (steady-state) removal of species A from a gas stream by the absorption of A in a liquid (i) containing nonvolatile species B such that the reaction $A(g) + bB(\ell) \rightarrow$ products takes place, and (ii) containing no species with which it (A) can chemically react, in the following ways:
(a) Assuming the reaction in case (i) is instantaneous, sketch (separately) concentration profiles for (i) and (ii) in accordance with the two-film model.
(b) Write the rate law for case (i) in terms of the enhancement factor (E), and the corresponding form for case (ii).
(c) Explain briefly, in words, in terms of the film theory, why the addition of B to the liquid increases the rate of absorption of A in case (i) relative to that in case (ii).
(d) What is the limit of the behavior underlying the explanation in (c), how may it be achieved, and what is the rate law in this situation?

9-17 An approximate implicit solution (van Krevelen and Hoftijzer, 1948) for the enhancement factor (E) for a second-order, "fast" gas-liquid reaction, $A(g) + bB(\ell) \rightarrow$ products (B nonvolatile), occurring in the liquid film, according to the two-film theory, is given by equations 9.2-51 and -52.
(a) Show that, under a certain condition, this solution may be put in a form explicit in E (in terms of Ha and E_i); that is, $E = E(\text{Ha}, E_i)$. State what the condition is, and derive the explicit form from equations 9.2-51 and -52.
(b) Confirm the result obtained in (a) for Ha = 100 and E_i = 100.
(c) Show whether (i) equation 9.2-49 and (ii) equation 9.2-53 can each be obtained as a limiting case from the result derived in (a) for "fast" pseudo-first-order and instantaneous reaction, respectively. (cf. de Santiago and Farena, 1970).

9-18 Boron, because of its high hardness and low density, is important in providing protective coatings for cutting tools and fibers for use in composite materials. The kinetics of deposition of boron by CVD from H_2 and BBr_3 was studied by Haupfear and Schmidt (1994). The overall reaction is $2BBr_3 + 3H_2 \rightarrow 2B(s) + 6HBr$.
(a) Assuming that the reaction occurs on open deposition sites on the growing boron film, write as many mechanistic steps as possible for this reaction.
(b) Write the LH rate law that results from a mechanism in which the rds is the reaction of adsorbed H atoms and adsorbed BBr_3. Assume that the coverages of these species are given by Langmuir competitive isotherms.
(c) For the proposed rate law in (b), obtain values of the adsorption constants and the surface rate constant from the following data obtained with a deposition temperature of 1100 K (regression or graphical techniques may be used). r_t is the rate of growth of the thickness

of the film. Comment on how well the proposed rate law fits the data. For example, is the order at high p_{H_2} too high (or too low)?

p_{BBr_3} = 3.33 Pa		p_{BBr_3} = 26.7 Pa		p_{H_2} = 2.67 Pa	
p_{H_2}/Pa	$r_t/\mu m\ min^{-1}$	p_{H_2}/Pa	$r_t/\mu m\ min^{-1}$	p_{BBr_3}/Pa	$r_t/\mu m\ min^{-1}$
0.42	0.0302	1.00	0.1023	2.12	0.0413
1.35	0.0481	3.72	0.2203	3.72	0.0673
2.86	0.0585	7.91	0.2716	7.91	0.1450
7.62	0.0721	16.80	0.4128	13.92	0.1787
14.73	0.0889	43.08	0.4427	24.49	0.1450
26.90	0.0509	62.79	0.5852	35.69	0.1023
55.03	0.0359	91.50	0.5458		
110.46	0.0192	133.36	0.7736		

(d) Propose an improved rate law based on your observations in (c).

9-19 (a) Confirm the instantaneous reaction rate expression in equation 9.3-2 by suitably applying the SSH to the two-step mechanism leading to the equation.

(b) Determine the simplified rate law resulting from the assumption that step (2) (C*O decomposition) is the *rds*.

(c) Suppose the rate law $r = k_{eff} c_{H_2O}/c_{H_2}$ was obtained experimentally for reaction 9.3-1. Under what conditions could this rate law be obtained from equation 9.3-2? (How does this k_{eff} differ from that in 9.3-4?)

9-20 (a) Determine the rate of reaction for a particular gas-liquid reaction A(g) + 2B(ℓ) → products, which is pseudo-first-order with respect to A. The following data are available:

$H_A = 300$ MPa L mol^{-1} $k_A = 450$ m^4 mol^{-1} min^{-1}
$D_{A\ell} = 4.2 \times 10^{-6}$ cm^2 s^{-1} $k_{A\ell} = 0.08$ m s^{-1}
$D_{B\ell} = 1.7 \times 10^{-5}$ cm^2 s^{-1} $k_{Ag} = 0.68$ mol kPa^{-1} m^{-2} s^{-1}

The partial pressure of A is 250 kPa, and the concentration of B is 0.25 mol L^{-1}.

(b) If an enhancement factor can be used for this system, determine its value.

(c) If a Hatta number is appropriate for this system, determine its value.

9-21 A kinetics study was performed to examine the rate-controlling steps in a gas-solid reaction governed by the shrinking-core model:

$$A(g) + 2B(s) \rightarrow products(s, g)$$

Experiments were conducted with four different sizes of nonporous spherical particles with the temperature and partial pressure of A kept constant at 700 K and 200 kPa, respectively. The time to complete reaction, t_1, was measured in each case, and the following data were obtained, with $\rho_{Bm} = 2.0 \times 10^5$ mol m^{-3}:

particle radius/m:	0.005	0.010	0.015	0.020
$10^{-5} t_1$/s:	2.31	4.92	7.83	11.0

In a separate set of experiments, it was determined that the gas-film mass transfer coefficient (k_{Ag}) was 9.2×10^{-5} s^{-1}.

(a) Determine the rate constant for the intrinsic surface reaction (k_{As}/m s^{-1}) and the effective ash-layer diffusion coefficient for A (D_e/m^2 s^{-1}).

(b) Is one process clearly rate limiting? Justify your answer.

(c) Determine the time required for 80% conversion of a 0.010 m particle, given the same flow conditions, T, and p_A used in the kinetics experiments.

Chapter 10

Biochemical Reactions: Enzyme Kinetics

The subject of biochemical reactions is very broad, covering both cellular and enzymatic processes. While there are some similarities between enzyme kinetics and the kinetics of cell growth, cell-growth kinetics tend to be much more complex, and are subject to regulation by a wide variety of external agents. The enzymatic production of a species via enzymes in cells is inherently a complex, coupled process, affected by the activity of the enzyme, the quantity of the enzyme, and the quantity and viability of the available cells. In this chapter, we focus solely on the kinetics of enzyme reactions, without considering the source of the enzyme or other cellular processes. For our purpose, we consider the enzyme to be readily available in a relatively pure form, "off the shelf," as many enzymes are.

Reactions with soluble enzymes are generally conducted in batch reactors (Chapter 12) to avoid loss of the catalyst (enzyme), which is usually expensive. If steps are taken to prevent the loss of enzyme, or facilitate its reuse (by entrapment or immobilization onto a support), flow reactors may be used (e.g., CSTR, Chapter 14). More comprehensive treatments of biochemical reactions, from the point of view of both kinetics and reactors, may be found in books by Bailey and Ollis (1986) and by Atkinson and Mavituna (1983).

10.1 ENZYME CATALYSIS

10.1.1 Nature and Examples of Enzyme Catalysis

Enzymes are proteins that catalyze many reactions, particularly biochemical reactions, including many necessary for the maintenance of life. The catalytic action is usually very specific, and may be affected by the presence of other substances both as inhibitors and as coenzymes.[1]

---

[1] A coenzyme is an organic compound that activates the primary enzyme to a catalytically active form. A coenzyme may act as a cofactor (see footnote 2), but the converse is not necessarily true. For example, the coenzyme nicotinamide adenine dinucleotide, in either its oxidized or reduced forms (NAD^+ or NADH), often participates as a cofactor in enzyme reactions.

Enzymes are commonly grouped according to the type of reaction catalyzed. Six classes of enzymes have been identified:

(1) oxidoreductases, which catalyze various types of oxidation-reduction reactions;
(2) transferases, which catalyze the transfer of functional groups, such as aldehydic or acyl groups;
(3) hydrolases, which catalyze hydrolysis reactions;
(4) isomerases, which catalyze isomerization;
(5) ligases, which, with ATP (adenosine triphosphate) as a cofactor,[2] lead to the formation of bonds between carbon and other atoms, including carbon, oxygen, nitrogen, and sulfur; and
(6) lyases, which catalyze the addition of chemical groups onto double bonds.

Examples of some common enzyme-catalyzed reactions are as follows:

(1) hydrolysis of urea with the enzyme (E) urease

$$(NH_2)_2CO + H_2O \xrightarrow{E} CO_2 + 2\,NH_3 \qquad \text{(A)}$$

(the specific nature of the catalytic action is indicated by the fact that urease has no effect on the rate of hydrolysis of substituted ureas, e.g., methyl urea);

(2) hydrolysis of sucrose with invertase to form glucose and fructose

$$C_{12}H_{22}O_{11} + H_2O \xrightarrow{E} 2C_6H_{12}O_6 \qquad \text{(B)}$$

(3) hydrolysis of starch with amylase to form glucose (this is a key step in the conversion of corn to fuel-grade ethanol)

$$(C_6H_{10}O_5)_n + nH_2O \xrightarrow{E} nC_6H_{12}O_6 \qquad \text{(C)}$$

(this may also be carried out by acid-catalyzed hydrolysis with HCl, but at a higher T and probably with a lower yield);

(4) decomposition of hydrogen peroxide in aqueous solution with catalase (which contains iron), an example of oxidation-reduction catalyzed by an enzyme;

$$2H_2O_2 \xrightarrow{E} 2H_2O + O_2 \qquad \text{(D)}$$

(5) treatment of myocardial infarction with aldolase;
(6) treatment of Parkinson's disease with L-DOPA, produced via tyrosinase (Pialis et al., 1996):

$$\text{L-tyrosine} + O_2 \xrightarrow{E} \text{dihydroxyphenylalanine (L-DOPA)} + H_2O$$

As a model reaction, we represent an enzyme-catalyzed reaction by

$$S + E \rightarrow P + E \qquad (10.1\text{-}1)$$

where S is a substrate (reactant), and P is a product.

---

[2] A cofactor is a nonprotein compound that combines with an inactive enzyme to generate a complex that is catalytically active. Metal ions are common cofactors for enzymatic processes. A cofactor may be consumed in the reaction, but may be regenerated by a second reaction unrelated to the enzymatic process.

Since enzymes are proteins, which are polymers of amino acids, they have relatively large molar masses ($> 15{,}000$ g mol^{-1}), and are in the colloidal range of size. Hence, enzyme catalysis may be considered to come between homogeneous catalysis, involving molecular dispersions (gas or liquid), and heterogeneous catalysis, involving particles (solid particles or liquid droplets). For the purpose of developing a kinetics model or rate law, the approach involving formation of an intermediate complex may be used, as in molecular catalysis, or that involving sites on the catalyst, as in the Langmuir-Hinshelwood model in surface catalysis. We use the former in Section 10.2, but see problem 10-2 for the latter.

10.1.2 Experimental Aspects

Factors that affect enzyme activity, that is, the rates of enzyme-catalyzed reactions, include:

(1) the nature of the enzyme (E), including the presence of inhibitors and coenzymes;
(2) the nature of the substrate;
(3) the concentrations of enzyme and substrate;
(4) temperature;
(5) pH; and
(6) external factors such as irradiation (photo, sonic or ionizing) and shear stress.

Some of the main experimental observations with respect to concentration effects, point (3) above, are:

(1) The rate of reaction, $(-r_S)$ or r_P, is proportional to the total (initial) enzyme concentration, c_{E_o}.
(2) At low (initial) concentration of substrate, c_S, the initial rate, $(-r_{S_o})$, is first-order with respect to substrate.
(3) At high c_S, $(-r_{S_o})$ is independent of (initial) c_S.

The effect of temperature generally follows the Arrhenius relation, equations 3.1-6 to -8, but the applicable range is relatively small because of low- and high-temperature effects. Below, say, 0°C, enzyme-catalyzed biochemical reactions take place slowly, as indicated by the use of refrigerators and freezers for food preservation. At sufficiently high temperatures (usually 45 to 50°C), the enzyme, as a protein with a helical or random coil structure, becomes "denatured" by the unwinding of the coil, resulting in loss of its catalytic activity. It is perhaps not surprising that the catalytic activity of enzymes, in general, is usually at a maximum with respect to T at about "body" temperature, 37°C. However, some recently developed thermophilic enzymes may remain active up to 100°C.

The effect of extreme pH values can be similar to that of T in denaturing the enzyme. This is related to the nature of enzymatic proteins as polyvalent acids and bases, with acid and basic groups (hydrophilic) concentrated on the outside of the protein.

Mechanical forces such as shear and surface tension affect enzyme activity by disturbing the shape of the enzyme molecule. Since the shape of the active site of the enzyme is specifically "engineered" to correspond to the shape of the substrate, even small changes in structure may drastically affect enzyme activity. Consequently, fluid flow rates, stirrer speeds, and foaming must be carefully controlled in order to ensure that an enzyme's productivity is maintained.

For these reasons, in the experimental study of the kinetics of enzyme-catalyzed reactions, T, shear and pH are carefully controlled, the last by use of buffered solutions. In the development, examples, and problems to follow, we assume that both T and pH

264 Chapter 10: Biochemical Reactions: Enzyme Kinetics

are constant, and that shear effects are negligible, and focus on the effects of concentration (of substrate and enzyme) in a rate law. These are usually studied by means of a batch reactor, using either the initial-rate method or the integrated-form-of-rate-law method (Section 3.4.1.1). The initial-rate method is often used because the enzyme concentration is known best at $t = 0$, and the initial enzyme activity and concentration are reproducible.

10.2 MODELS OF ENZYME KINETICS

10.2.1 Michaelis-Menten Model

The kinetics of the general enzyme-catalyzed reaction (equation 10.1-1) may be simple or complex, depending upon the enzyme and substrate concentrations, the presence/absence of inhibitors and/or cofactors, and upon temperature, shear, ionic strength, and pH. The simplest form of the rate law for enzyme reactions was proposed by Henri (1902), and a mechanism was proposed by Michaelis and Menten (1913), which was later extended by Briggs and Haldane (1925). The mechanism is usually referred to as the Michaelis-Menten mechanism or model. It is a two-step mechanism, the first step being a rapid, reversible formation of an enzyme-substrate complex, ES, followed by a slow, rate-determining decomposition step to form the product and reproduce the enzyme:

$$S + E \underset{k_{-1}}{\overset{k_1}{\rightleftharpoons}} ES \quad \text{(fast)} \tag{1}$$

$$ES \overset{k_r}{\rightarrow} P + E \quad \text{(slow)} \tag{2}$$

Henri and Michaelis and Menten assumed that the first step is a fast reaction virtually at equilibrium, such that the concentration of the complex, ES, may be represented by:

$$c_{ES} = \frac{k_1 c_S c_E}{k_{-1}} \tag{10.2-1}$$

c_{ES} and c_E are related by a material balance on the total amount of enzyme:

$$c_E + c_{ES} = c_{Eo} \tag{10.2-2}$$

Thus, combining 10.2-1 and 10.2-2, we obtain the following expression for c_{ES}:

$$c_{ES} = \frac{k_1 c_S c_{Eo}}{k_1 c_S + k_{-1}} = \frac{c_S c_{Eo}}{\frac{k_{-1}}{k_1} + c_S} \tag{10.2-3}$$

The ratio k_{-1}/k_1 is effectively the dissociation equilibrium constant[3] for ES in step (1), and is usually designated by K_m (Michaelis constant, with units of concentration). The rate of formation of the product, P, is determined from step (2):[4]

$$v \equiv r_P = k_r c_{ES} \tag{10.2-4}$$

---

[3] In the biochemical literature, equilibrium constants are expressed as *dissociation* constants (for complexes), rather than as association constants.

[4] In the biochemical literature, the rate of reaction (e.g., r_P) is expressed as a reaction "velocity" (v). In this chapter, we continue to use r to represent reaction rate.

Combining equation 10.2-4 with 10.2-3 and the definition of K_m, we obtain

$$r_P = \frac{k_r c_{E_o} c_S}{K_m + c_S} \qquad (10.2\text{-}5)$$

The form of the equation 10.2-5 is consistent with the experimental observations (Section 10.1.2) about the dependence of rate in general on c_{E_o}, and about the dependence of the initial rate on c_S. Thus, r_P [or $(-r_S)$] $\propto c_{E_o}$, and the initial rate, given by

$$r_{P_o} = (-r_{S_o}) = \frac{k_r c_{E_o} c_{S_o}}{K_m + c_{S_o}} \qquad (10.2\text{-}6)$$

is proportional to c_{S_o} at sufficiently low values of c_{S_o} ($c_{S_o} \ll K_m$), and reaches a limiting (maximum) value independent of c_{S_o} at sufficiently high values of c_{S_o} ($c_{S_o} \gg K_m$). The form of equation 10.2-6 is shown schematically in Figure 10.1, together with the following interpretations of the parameters involved, for given c_{E_o} at fixed T and pH: The maximum initial rate[5] is obtained for $c_{S_o} \gg K_m$:

$$r_{P_o,max} = k_r c_{E_o} = V_{max} \qquad (10.2\text{-}7)$$

and, if $K_m = c_{S_o}$,

$$r_{P_o} = \frac{1}{2} k_r c_{E_o} = \frac{1}{2} V_{max}; \quad c_{S_o} = K_m \qquad (10.2\text{-}8)$$

That is, the Michaelis constant K_m is equal to the value of c_{S_o} that makes the initial rate equal to one-half the maximum rate.

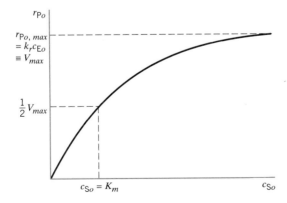

Figure 10.1 Initial-rate plot for S + E → P + E showing interpretation of $r_{P_o,max}$ in terms of rate parameters; constant c_{E_o}, T, pH

---

[5]In the biochemical literature, the maximum rate $r_{P_o,max}$ is designated by V_{max}, and we use this henceforth in this chapter.

Chapter 10: Biochemical Reactions: Enzyme Kinetics

With V_{max}, the maximum reaction velocity, replacing $k_r c_{Eo}$, equation 10.2-5 is rewritten as:

$$r_P = \frac{V_{max} c_S}{K_m + c_S} \tag{10.2-9}$$

Equation 10.2-9 is known as the Michaelis-Menten equation. It represents the kinetics of many simple enzyme-catalyzed reactions which involve a single substrate (or if other substrates are in large excess). The interpretation of K_m as an equilibrium constant is not universally valid, since the assumption that step (1) is a fast equilibrium process often does not hold. An extension of the treatment involving a modified interpretation of K_m is given next.

10.2.2 Briggs-Haldane Model

Briggs and Haldane (1925) proposed an alternative mathematical description of enzyme kinetics which has proved to be more general. The Briggs-Haldane model is based upon the assumption that, after a short initial startup period, the concentration of the enzyme-substrate complex is in a pseudo-steady state. Derivation of the model is based upon material balances written for each of the four species S, E, ES, and P.

For a constant-volume batch reactor operated at constant T and pH, an exact solution can be obtained numerically (but not analytically) from the two-step mechanism in Section 10.2.1 for the concentrations of the four species S, E, ES, and P as functions of time t, without the assumptions of "fast" and "slow" steps. An approximate analytical solution, in the form of a rate law, can be obtained, applicable to this and other reactor types, by use of the stationary-state hypothesis (SSH). We consider these in turn.

From material balances on S as "free" substrate, ES, E, and S as total substrate, we obtain the following four independent equations, for any time t:

$$r_S = dc_S/dt = k_{-1} c_{ES} - k_1 c_S c_E \tag{10.2-10}$$

$$r_{ES} = dc_{ES}/dt = k_1 c_S c_E - k_{-1} c_{ES} - k_r c_{ES} \tag{10.2-11}$$

$$c_{Eo} = c_E + c_{ES} \tag{10.2-2}$$

$$c_{So} = c_S + c_{ES} + c_P \tag{10.2-12}$$

Elimination of c_E from the first two equations using 10.2-2 results in two simultaneous first-order differential equations to solve for c_S and c_{ES} as functions of t:

$$dc_S/dt = k_{-1} c_{ES} - k_1 c_S (c_{Eo} - c_{ES}) \tag{10.2-13}$$

$$dc_{ES}/dt = k_1 c_S (c_{Eo} - c_{ES}) - (k_{-1} + k_r) c_{ES} \tag{10.2-14}$$

with the two initial conditions:

$$\text{at } t = 0, \quad c_S = c_{So} \quad \text{and} \quad c_{ES} = 0 \tag{10.2-15}$$

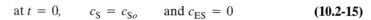

These can be solved numerically (e.g., with the E-Z Solve software), and the results used to obtain $c_E(t)$ and $c_P(t)$ from equations 10.2-2 and 10.2-12, respectively.

For an approximate solution resulting in an analytical form of rate law, we use the SSH applied to the intermediate complex ES. (If $c_{Eo} \ll c_{So}$, the approximation is close to the exact result.) Using the SSH, we set $dc_{ES}/dt = 0$, and then, from equation 10.2-14,

$$c_{ES} = \frac{k_1 c_{E_o} c_S}{k_{-1} + k_r + k_1 c_S} \qquad (10.2\text{-}16)$$

Substituting for c_{ES} in the rate of reaction from the rate-determining step, 10.2-4, we again obtain

$$r_P = k_r c_{ES} = \frac{k_r c_{E_o} c_S}{K_m + c_S} \qquad (10.2\text{-}5)$$

where

$$K_m = (k_{-1} + k_r)/k_1 \qquad (10.2\text{-}17)$$

This is a more general definition of the Michaelis constant than that given in Section 10.2.1. If $k_{-1} \gg k_r$, it simplifies to the form developed in 10.2.1.

Again, V_{max} may be substituted for $k_r c_{E_o}$, producing the Michaelis-Menten form of the rate law, that is,

$$r_P = \frac{V_{max} c_S}{K_m + c_S} \qquad (10.2\text{-}9)$$

10.3 ESTIMATION OF K_m AND V_{max}

The nonlinear form of the Michaelis-Menten equation, 10.2-9, does not permit simple estimation of the kinetic parameters (K_m and V_{max}). Three approaches may be adopted:

(1) use of initial-rate data with a linearized form of the rate law;
(2) use of concentration-time data with a linearized form of the integrated rate law; and
(3) use of concentration-time data with the integrated rate law in nonlinear form.

These approaches are described in the next three sections.

10.3.1 Linearized Form of the Michaelis-Menten Equation

The Michaelis-Menten equation, 10.2-9, in initial-rate form, is

$$r_{P_o} = \frac{V_{max} c_{S_o}}{K_m + c_{S_o}} \qquad (10.3\text{-}1)$$

Inverting equation 10.3-1, we obtain

$$\frac{1}{r_{P_o}} = \frac{1}{V_{max}} + \frac{K_m}{V_{max}} \frac{1}{c_{S_o}} \qquad (10.3\text{-}2)$$

This is a linear expression for $1/r_{P_o}$ as a function of $1/c_{S_o}$, and was first proposed by Lineweaver and Burk (1934). A plot of $1/r_{P_o}$ against $1/c_{S_o}$, known as a Lineweaver-Burk plot, produces a straight line with intercept $1/V_{max}$ and slope K_m/V_{max}.

Although, in principle, values of both V_{max} and K_m can be obtained from a Lineweaver-Burk plot, the value of K_m obtained from the slope is often a poor one. The most accurately known values of r_{P_o} are those near V_{max} (at high values of c_{S_o}); these are grouped near the origin on the Lineweaver-Burk plot (at $1/c_{S_o} = 0$), and lead to a very good value of V_{max} itself. The least accurately known values are at low values of c_{S_o}; these are far removed from the origin, and greatly influence the value of the slope, resulting in a possibly poor estimate of K_m. Thus, use of linear regression with unweighted data may lead to a good estimate of V_{max} and a poor estimate of K_m. A better estimate of the latter may result from using equation 10.2-8, from the value of c_{S_o} corresponding to $V_{max}/2$. Nevertheless, the experimental determination of Michaelis rate parameters, V_{max} and K_m, is shown in the following example for the initial-rate method with the Lineweaver-Burk expression. Other linear forms of equation 10.3-1 are possible, and are explored in problem 10-5.

EXAMPLE 10-1

The hydrolysis of sucrose (S) catalyzed by the enzyme invertase has been studied by measuring the initial rate, r_{P_o}, at a series of initial concentrations of sucrose (c_{S_o}). At a particular temperature and enzyme concentration, the following results were obtained (Chase et al., 1962):

c_{S_o}/mol L^{-1}	0.0292	0.0584	0.0876	0.117	0.146	0.175	0.234
r_{P_o}/mol L^{-1} s^{-1}	0.182	0.265	0.311	0.330	0.349	0.372	0.371

Determine the values of V_{max} and the Michaelis constant K_m in the Michaelis-Menten rate law.

SOLUTION

Linear regression of the data given, according to equation 10.3-2, results in $V_{max} = 0.46$ mol L^{-1} s^{-1}, and $K_m = 0.043$ mol L^{-1}. Figure 10.2 shows the given experimental data plotted according to equation 10.3-2. The straight line is that from the linear regression; the intercept at $1/c_{S_o} = 0$ is $1/V_{max} = 2.17$ mol^{-1} L s, and the slope is $K_m/V_{max} = 0.093$ s.

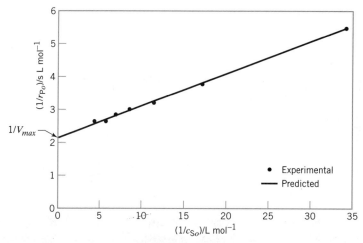

Figure 10.2 Lineweaver-Burk plot for Example 10-1

10.3.2 Linearized Form of the Integrated Michaelis-Menten Equation

For a constant-volume BR, integration of the Michaelis-Menten equation leads to a form that can also be linearized. Thus, from equation 10.2-9,

$$r_P = (-r_S) = -\frac{dc_S}{dt} = \frac{V_{max} c_S}{K_m + c_S} \tag{10.3-3}$$

or

$$\frac{K_m + c_S}{c_S} dc_S = -V_{max}\, dt \tag{10.3-4}$$

or

$$\frac{K_m}{c_S} dc_S + dc_S = -V_{max}\, dt \tag{10.3-5}$$

which, with the boundary condition $c_S = c_{S_o}$ at $t = 0$, integrates to

$$K_m \ln(c_S/c_{S_o}) + (c_S - c_{S_o}) = -V_{max} t \tag{10.3-6}$$

Equation 10.3-6 may be written as

$$\frac{\ln(c_S/c_{S_o})}{c_{S_o} - c_S} = \frac{1}{K_m} - \frac{V_{max}}{K_m} \frac{t}{c_{S_o} - c_S} \tag{10.3-7}$$

According to equation 10.3-7, $\ln(c_S/c_{S_o})/(c_{S_o} - c_S)$ is a linear function of $t/(c_{S_o} - c_S)$. K_m and V_{max} can be determined from equation 10.3-7 with values of c_S measured as a function of t for a given c_{S_o}.

10.3.3 Nonlinear Treatment

A major limitation of the linearized forms of the Michaelis-Menten equation is that none provides *accurate* estimates of *both* K_m and V_{max}. Furthermore, it is impossible to obtain meaningful error estimates for the parameters, since linear regression is not strictly appropriate. With the advent of more sophisticated computer tools, there is an increasing trend toward using the integrated rate equation and nonlinear regression analysis to estimate K_m and V_{max}. While this type of analysis is more complex than the linear approaches, it has several benefits. First, accurate nonbiased estimates of K_m and V_{max} can be obtained. Second, nonlinear regression may allow the errors (or confidence intervals) of the parameter estimates to be determined.

To determine K_m and V_{max}, experimental data for c_S versus t are compared with values of c_S predicted by numerical integration of equation 10.3-3; estimates of K_m and V_{max} are subsequently adjusted until the sum of the squared residuals is minimized. The E-Z Solve software may be used for this purpose. This method also applies to other complex rate expressions, such as Langmuir-Hinshelwood rate laws (Chapter 8).

10.4 INHIBITION AND ACTIVATION IN ENZYME REACTIONS

The simple Michaelis-Menten model does not deal with all aspects of enzyme-catalyzed reactions. The model must be modified to treat the phenomena of inhibition and

270 Chapter 10: Biochemical Reactions: Enzyme Kinetics

activation, which decrease and increase the observable enzyme activity, respectively. These effects may arise from features inherent in the enzyme-substrate system ("internal" or substrate effects), or from other substances which may act on the enzyme as poisons (inhibitors) or as coenzymes or cofactors (activators). In the following two sections, we consider examples of these substrate and external effects in turn, by introducing simple extensions of the model, and interpreting the resulting rate laws to account for inhibition and activation.

10.4.1 Substrate Effects

Substrates may affect enzyme kinetics either by activation or by inhibition. Substrate activation may be observed if the enzyme has two (or more) binding sites, and substrate binding at one site enhances the affinity of the substrate for the other site(s). The result is a highly active ternary complex, consisting of the enzyme and two substrate molecules, which subsequently dissociates to generate the product. Substrate inhibition may occur in a similar way, except that the ternary complex is *nonreactive*. We consider first, by means of an example, inhibition by a single substrate, and second, inhibition by multiple substrates.

10.4.1.1 *Single-Substrate Inhibition*

EXAMPLE 10-2

The following mechanism relates to an enzyme E with two binding sites for the substrate S. Two complexes are formed: a reactive binary complex ES, and a nonreactive ternary complex ESS.

$$E + S \underset{k_{-1}}{\overset{k_1}{\rightleftarrows}} ES \qquad (1)$$

$$ES + S \underset{k_{-2}}{\overset{k_2}{\rightleftarrows}} ESS; \quad K_2 = k_{-2}/k_2 \qquad (2)$$

$$ES \overset{k_r}{\rightarrow} E + P \qquad (3)$$

(a) Derive the rate law for this mechanism.
(b) Show that inhibition occurs in the rate of formation of P, relative to that given by the Michaelis-Menten equation for the two-step mechanism for a single binding site in which only ES is formed.
(c) What is the maximum rate of reaction (call it the apparent V_{max}, or $V_{max,app}$), and how does it compare with the parameter $V_{max} = k_r c_{E_o}$? At what value of c_S does it occur?
(d) Sketch r_P versus c_S for comparison with Figure 10.1.

SOLUTION

(a) We apply the SSH to the complexes ES and ESS; in the latter case, this is equivalent to assumption of equilibrium with the dissociation constant for ESS given by $K_2 = k_{-2}/k_2$:

$$r_{ES} = k_1 c_E c_S - k_{-1} c_{ES} - k_2 c_{ES} c_S + k_{-2} c_{ESS} - k_r c_{ES} = 0 \qquad (1)$$

$$r_{ESS} = k_2 c_{ES} c_S - k_{-2} c_{ESS} = 0 \qquad (2)$$

$$c_{E_o} = c_E + c_{ES} + c_{ESS} \qquad (3)$$

$$r_P = k_r c_{ES} \qquad (4)$$

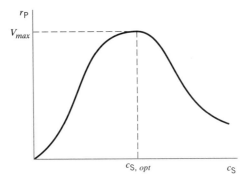

Figure 10.3 Substrate inhibition from mechanism in Example 10-2

Solving for c_E and c_{ESS} in terms of c_{ES} and c_S from (1) and (2), respectively, and substituting the results in (3) and rearranging to obtain c_{ES} in terms of c_{Eo} and c_S, we have

$$c_{ES} = \frac{c_{Eo}}{\frac{K_m}{c_S} + 1 + \frac{c_S}{K_2}} \tag{5}$$

where K_m is the Michaelis constant (equation 10.2-17). Substituting the result for c_{ES} from (5) into (4), we obtain the rate law:

$$r_P = \frac{k_r c_{Eo} c_S}{K_m + c_S + c_S^2/K_2} = \frac{V_{max} c_S}{K_m + c_S + c_S^2/K_2} \tag{10.4-1}$$

(b) Inhibition arises if ESS is nonreactive, so that there is a reduction in the quantity of the enzyme available to generate P. This is shown by the form of equation 10.4-1, which is similar to the Michaelis-Menten equation, except for the third term c_S^2/K_2 in the denominator. The inclusion of this term means that the rate is reduced. The extent of inhibition depends on the relative magnitudes of K_m, c_S, and K_2. Inhibition is significant at relatively large c_S and relatively small K_2.

(c) From equation 10.4-1, setting $dr_P/dc_S = 0$, we obtain the maximum rate at

$$c_S = (K_m K_2)^{1/2} \tag{10.4-2}$$

with a value

$$V_{max,app} = \frac{k_r c_{Eo}}{1 + 2(K_m/K_2)^{1/2}} = \frac{V_{max}}{1 + 2(K_m/K_2)^{1/2}} \tag{10.4-3}$$

The maximum rate for the inhibited reaction is lower than V_{max}, which represents the maximum rate for the uninhibited reaction.

(d) Since $V_{max,app}$ occurs at a finite value of c_S, r_P exhibits a maximum (at the value of c_S given in part (c)), and, for this type of inhibition, schematically behaves as shown in Figure 10.3.

10.4.1.2 Multiple-Substrate Inhibition

In some cases, the enzyme may act upon more than one substrate present in the system. This typically occurs with hydrolytic enzymes, which may act upon structurally similar compounds with identical types of bonds. One example is lysozyme, which hydrolyzes starch, a complex mucopolysaccharide. To illustrate the effect of multiple substrates upon enzyme kinetics, we consider the reaction of two substrates (S_1 and S_2) competing for the same active site on a single enzyme (E). The substrates compete for E in forming the complexes ES_1 and ES_2 en route to the final products P_1 and P_2 in the following mechanism:

$$S_1 + E \underset{k_{-1}}{\overset{k_1}{\rightleftarrows}} ES_1$$

$$S_2 + E \underset{k_{-2}}{\overset{k_2}{\rightleftarrows}} ES_2$$

$$ES_1 \overset{k_{r1}}{\to} E + P_1$$

$$ES_2 \overset{k_{r2}}{\to} E + P_2$$

The rates of production of P_1 and P_2 may be determined by invoking the SSH for both ES_1 and ES_2 in a procedure similar to that in Example 10-2. The resulting rate laws are:

$$r_{P_1} = \frac{V_{max1} c_{S1}}{K_{m1} + c_{S1} + (K_{m1}/K_{m2}) c_{S2}} \qquad (10.4\text{-}4)$$

$$r_{P_2} = \frac{V_{max2} c_{S2}}{K_{m2} + c_{S2} + (K_{m2}/K_{m1}) c_{S1}} \qquad (10.4\text{-}5)$$

where $V_{max1} = k_{r1} c_{Eo}$ and $V_{max2} = k_{r2} c_{Eo}$; K_{m1} and K_{m2} are the Michaelis constants $(k_{-1} + k_{r1})/k_1$ and $(k_{-2} + k_{r2})/k_2$, respectively.

The individual reaction rates in 10.4-4 and 10.4-5 are slower in the presence of both substrates than in the presence of a single substrate. For example, if c_{S2} is zero, equation 10.4-4 simplifies to:

$$r_{P1} = \frac{V_{max1} c_{S1}}{K_{m1} + c_{S1}} \qquad (10.4\text{-}6)$$

which is identical in form to equation 10.2-9, the Michaelis-Menten equation. It is also worth noting that the presence of other substrates can influence the results of studies aimed at determining the kinetics parameters for a given enzyme process. If a detailed compositional analysis is not performed, and S_2 is present, the predicted Michaelis constant for S_1 is inaccurate; the effective (i.e., predicted) Michaelis constant is:

$$K_{m,eff} = K_{m1}[1 + (c_{S2}/K_{m2})] \qquad (10.4\text{-}7)$$

In the absence of S_2, the true Michaelis constant, K_{m1}, is obtained. Consequently, the values of $K_{m,eff}$ and V_{max} determined in a kinetics study depend upon the composition (c_{S1} and c_{S2}) and total substrate concentration within the system.

10.4.2 External Inhibitors and Activators

Although substrates may enhance or inhibit their own conversion, as noted in Section 10.4.1, other species may also affect enzyme activity. *Inhibitors* are compounds that decrease observable enzyme activity, and *activators* increase activity. The combination of an inhibitor or activator with an enzyme may be irreversible, reversible, or partially reversible. Irreversible inhibitors (*poisons*), such as lead or cyanide, completely and irreversibly inactivate an enzyme. Reversible inhibitors reduce enzyme activity, but allow enzyme activity to be restored when the inhibitor is removed. Partial reversibility occurs when some, but not all, of the enzyme's activity is restored on removal of the inhibitor. If the modification of activity is irreversible, the process is known as *inactivation*. Thus, the term "inhibition" is normally reserved for fully reversible or partially reversible processes.

Inhibitors are usually classified according to their effect upon V_{max} and K_m. Competitive inhibitors, such as substrate analogs, compete with the substrate for the same binding site on the enzyme, but do not interfere with the decomposition of the enzyme-substrate complex. Therefore, the primary effect of a competitive inhibitor is to increase the apparent value of K_m. The effect of a competitive inhibitor can be reduced by increasing the substrate concentration relative to the concentration of the inhibitor.

Noncompetitive inhibitors, conversely, do not affect substrate binding, but produce a ternary complex (enzyme-substrate-inhibitor) which either decomposes slowly, or fails to decompose (i.e., is inactive). Consequently, the primary effect of a noncompetitive inhibitor is to reduce the apparent value of V_{max}.

The inhibition process in general may be represented by the following six-step scheme (a similar scheme may be used for activation—see problem 10-12), in which I is the inhibitor, EI is a binary enzyme-inhibitor complex, and EIS is a ternary enzyme-inhibitor-substrate complex.

$$E + S \underset{k_{-1}}{\overset{k_1}{\rightleftarrows}} ES \tag{1}$$

$$E + I \underset{k_{-2}}{\overset{k_2}{\rightleftarrows}} EI \tag{2}$$

$$ES + I \underset{k_{-3}}{\overset{k_3}{\rightleftarrows}} EIS \tag{3}$$

$$EI + S \underset{k_{-4}}{\overset{k_4}{\rightleftarrows}} EIS \tag{4}$$

$$ES \overset{k_{r_1}}{\rightarrow} E + P \tag{5}$$

$$EIS \overset{k_{r_2}}{\rightarrow} EI + P \tag{6}$$

In steps (1) and (2), S and I compete for (sites on) E to form the binary complexes ES and EI. In steps (3) and (4), the ternary complex EIS is formed from the binary complexes. In steps (5) and (6), ES and EIS form the product P; if EIS is inactive, step (6) is ignored. Various special cases of competitive, noncompetitive, and mixed (competitive and noncompetitive) inhibition may be deduced from this general scheme, according to the steps allowed, and corresponding rate laws obtained.

EXAMPLE 10-3

Competitive inhibition involves (only) the substrate (S) and the inhibitor (I) competing for one type of site on the enzyme (E), in fast, reversible steps, followed by the slow decomposition of the complex ES to form product (P); the complex EI is assumed to be inactive. The fact that there is only one type of binding site on the enzyme implies that a ternary complex EIS cannot be formed.

(a) Derive the rate law for competitive inhibition.
(b) Show what effect, if any, competitive inhibition has on V_{max} and K_m, relative to the uninhibited case described in Section 10.2.

274 Chapter 10: Biochemical Reactions: Enzyme Kinetics

SOLUTION

(a) Only steps (1), (2), and (5) of the general scheme above are involved in competitive inhibition. We apply the SSH to the complexes ES and EI to obtain the rate law:

$$r_{ES} = k_1 c_E c_S - k_{-1} c_{ES} - k_{r1} c_{ES} = 0 \tag{1}$$

$$r_{EI} = k_2 c_E c_I - k_{-2} c_{EI} = 0 \tag{2}$$

$$c_{Eo} = c_E + c_{ES} + c_{EI} \tag{3}$$

(In the corresponding material balance for I, it is usually assumed that $c_I \gg c_{EI}$, because c_E itself is usually very low; c_I is retained in the final expression for the rate law.)

$$r_P = k_{r1} c_{ES} \tag{4}$$

From (1),

$$c_E = \frac{(k_{-1} + k_{r1}) c_{ES}}{k_1 c_S} = \frac{K_m c_{ES}}{c_S} \tag{5}$$

where K_m is the Michaelis constant, equation 10.2-17.
From (2),

$$c_{EI} = \frac{k_2 c_E c_I}{k_{-2}} = \frac{K_m c_{ES} c_I}{K_2 c_S} \tag{6}$$

(using (5) to eliminate c_E), where $K_2 = k_{-2}/k_2$, the dissociation constant for EI.
From (3), (5), and (6), on elimination of c_E and c_{EI},

$$c_{ES} = \frac{c_{Eo}}{\frac{K_m}{c_S} + 1 + \frac{K_m c_I}{K_2 c_S}} \tag{7}$$

Substituting (7) in (4), we obtain the rate law:

$$r_P = \frac{k_{r1} c_{Eo}}{\frac{K_m}{c_S} + 1 + \frac{K_m c_I}{K_2 c_S}} = \frac{V_{max} c_S}{c_S + K_m(1 + c_I/K_2)} \tag{10.4-8}$$

The effect of inhibition is to decrease r_P relative to r_P given by the Michaelis-Menten equation 10.2-9 ($c_I = 0$). The extent of inhibition is a function of c_I.
(b) To show the effect of inhibition on the Michaelis parameters V_{max} and K_m, we compare equation 10.4-8 with the (uninhibited form of the) Michaelis-Menten equation, 10.2-9. V_{max} is the same, but if we write 10.4-8 in the form of 10.2-9 as

$$r_P = \frac{V_{max} c_S}{c_S + K_{m,app}} \tag{10.4-9}$$

the apparent value $K_{m,app}$ is given by

$$K_{m,app} = K_m(1 + c_I/K_2) \tag{10.4-10}$$

and $K_{m,app} > K_m$.

10.4 Inhibition and Activation in Enzyme Reactions

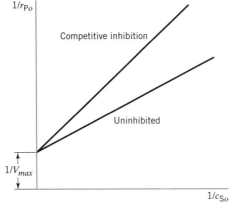

Figure 10.4 Lineweaver-Burk plot illustrating comparison of competitive inhibition with no inhibition of enzyme activity

The same conclusions can be reached by means of the linearized Lineweaver-Burk form of the rate law. From 10.4-8, for the initial rate,

$$\frac{1}{r_{P_o}} = \frac{1}{V_{max}} + \frac{K_m}{V_{max}}\left(1 + \frac{c_I}{K_2}\right)\frac{1}{c_{S_o}} \quad (10.4\text{-}11)$$

In comparison with equation 10.3-2, the intercept, $1/V_{max}$, remains the same, but the slope is increased by the factor $(1 + c_I/K_2)$. This is illustrated schematically in Figure 10.4.

A case of noncompetitive inhibition is represented by steps (1), (2), (3), and (5) of the general scheme above:

$$\text{E} + \text{S} \underset{k_{-1}}{\overset{k_1}{\rightleftarrows}} \text{ES} \quad (1)$$

$$\text{E} + \text{I} \underset{k_{-2}}{\overset{k_2}{\rightleftarrows}} \text{EI (inactive)} \quad (2)$$

$$\text{ES} + \text{I} \underset{k_{-3}}{\overset{k_3}{\rightleftarrows}} \text{EIS (inactive)} \quad (3)$$

$$\text{ES} \overset{k_r}{\rightarrow} \text{E} + \text{P} \quad (5)$$

By applying the SSH to ES, EI, and EIS, we may obtain the rate law in the form

$$r_P = \frac{V_{max} c_S}{c_S(1 + c_I/K_3) + K_m(1 + c_I/K_2)} \quad (10.4\text{-}12)$$

where $K_2 = k_{-2}/k_2$ (the dissociation constant for EI) and $K_3 = k_{-3}/k_3$ (the dissociation constant for EIS to ES and I). Equation 10.4-12 again shows that r_P (inhibited) $< r_P$ (uninhibited, $c_I = 0$), and that the extent of inhibition depends on c_I.

If we further assume that $K_2 = K_3$, that is, that the affinity for the inhibitor is the same for both E and ES, equation 10.4-12 simplifies to

$$r_P = \frac{V_{max,app} c_S}{c_S + K_m} \quad (10.4\text{-}13)$$

where

$$V_{max,app} = V_{max}/(1 + c_I/K_2) < V_{max} \quad (10.4\text{-}14)$$

Equations 10.4-13 and -14 illustrate that, relative to the uninhibited case, V_{max} changes (decreases), but K_m remains the same.

If we do not make the assumption leading to $K_2 = K_3$, then the four-step mechanism above also represents mixed (competitive and noncompetitive) inhibition, and both K_m and V_{max} change. In this case, equation 10.4-12 may be written as

$$r_P = \frac{V_{max,app} c_S}{c_S + K_{m,app}} \quad (10.4\text{-}15)$$

where

$$V_{max,app} = V_{max}/(1 + c_I/K_3) \quad (10.4\text{-}16)$$
$$K_{m,app} = K_m(1 + c_I/K_2)/(1 + c_I/K_3) \quad (10.4\text{-}17)$$

Treatment of the full six-step kinetic scheme above with the SSH leads to very cumbersome expressions for c_E, c_{EI}, etc., such that it would be better to use a numerical solution. These can be simplified greatly to lead to a rate law in relatively simple form, if we assume (1) the first four steps are at equilibrium, and (2) $k_{r1} = k_{r2}$:

$$r_P = \frac{V_{max} c_S}{c_S + K_m(1 + c_I/K_2)/(1 + c_I/K_3)} \quad (10.4\text{-}18)$$

This represents competitive inhibition in the sense that V_{max} is unchanged (relative to the uninhibited reaction), but K_m is changed.

10.5 PROBLEMS FOR CHAPTER 10

10-1 If the activation energy for the decomposition of H_2O_2 in aqueous solution catalyzed by the enzyme catalase is 50 kJ mol^{-1}, and that for the uncatalyzed reaction is 75 kJ mol^{-1}, calculate the ratio of the rate of the catalyzed reaction to that of the uncatalyzed reaction at 300 K. What assumptions have you made in your calculation?

10-2 The Michaelis-Menten equation, 10.2-9, is developed in Section 10.2.1 from the point of view of homogeneous catalysis and the formation of an intermediate complex. Use the Langmuir-Hinshelwood model of surface catalysis (Chapter 8), applied to the substrate in liquid solution and the enzyme as a "colloidal particle" with active sites, to obtain the same form of rate law.

10-3 Ouellet et al. (1952) have reported a kinetics analysis of the enzymatic diphosphorylation of adenosine triphosphate (ATP). Because of the suggestion that myosin might be the transducer which, in muscles, converts the free energy of ATP into external mechanical work, the system chosen for study was the hydrolysis of ATP (S) in the presence of myosin to give ADP (adenosine diphosphate) and phosphate. Their initial-rate data obtained at 25°C are as follows

(c_{Eo} for each experiment was 0.039 g L^{-1}):

c_{So}/mmol L^{-1}	0.0097	0.016	0.021	0.034	0.065	0.14	0.32
$(-r_{So})$/μmol L^{-1} s^{-1}	0.067	0.093	0.12	0.15	0.17	0.19	0.20

Calculate the maximum rate parameter, V_{max}, and the Michaelis constant, K_m, from these data.

10-4 Phosphofructokinase is an enzyme that catalyzes one of the steps in the degradation of carbohydrates. Initial rates of the reaction that converts fructose-6-phosphate (S) to fructose-1,6-di-phosphate (P) as a function of c_{So} (the initial fructose-6-diphosphate concentration) are as follows (the concentration of enzyme added, c_{Eo}, is the same in each case):

c_{So}/ μmol L^{-1}	50	60	80	100	150	200	250	300	400
$(-r_{So})$/ μmol L^{-1} min^{-1}	61	65	72	77	84	88	90	93	95

Calculate the maximum rate parameter, V_{max}, and the Michaelis constant, K_m, from these data (Bromberg, 1984, p. 923).

10-5 Repeat Example 10-1 using a linearized form of equation 10.3-1 that is alternative to the linearized form 10.3-2 (the Lineweaver-Burk form). Comment on any advantage(s) of one form over the other. (There is more than one possible alternative form.)

10-6 Suppose, at a particular temperature, results for the hydrolysis of sucrose, S, catalyzed by the enzyme invertase (c_{Eo} = 1 × 10^{-5} mol L^{-1}) in a batch reactor are given by:

c_S/mmol L^{-1}	1	0.84	0.68	0.53	0.38	0.27	0.16	0.09	0.04	0.018	0.006	0.0025
t/h	0	1	2	3	4	5	6	7	8	9	10	11

Show that the results conform to the Michaelis-Menten rate law, and determine the values of the kinetics parameters V_{max}, K_m, and k_r.

10-7 Benzyl alcohol can be produced from benzaldehyde (S) by a dehydrogenation reaction catalyzed by yeast alcohol dehydrogenase (YADH). Nikolova et al. (1995) obtained initial-rate data for this reaction using immobilized YADH immersed in iso-octane with 1% v/v water. The following data were obtained:

c_{So}/mmol L^{-1}	0.50	1.0	2.0	3.0	5.0	7.5
$(-r_{So})$/μmol mg(YADH)$^{-1}$ h^{-1}	0.58	1.2	2.7	3.3	4.6	5.2

Determine the kinetics parameters K_m and V_{max}, assuming that the standard Michaelis-Menten model applies to this system, (a) by nonlinear regression, and (b) by linear regression of the Lineweaver-Burk form.

10-8 Confirm (by derivation) the results given by
(a) equation 10.4-12;
(b) equations 10.4-13, -14;
(c) equations 10.4-15, -16, -17;
(d) equation 10.4-18.

10-9 Compare the Lineweaver-Burk linear forms resulting from the four cases in problem 10-8, together with those given by equations 10.3-2 and 10.4-11. Note which cases have the same and different intercepts and slopes. Sketch plots to show the relative magnitudes of both intercepts and slopes.

10-10 In the production of L-DOPA from L-tyrosine (S) using tyrosinase (E) (Pialis, 1996; Pialis and Saville, 1998), the following data were obtained:

t/h	0	1	2	3	4	5	6	7
c_S/mmol L^{-1}	2.50	1.77	1.42	1.14	1.04	0.96	0.81	0.77

The reaction, S + O$_2$ → L-DOPA, was conducted under a constant oxygen partial pressure, such that the reaction was pseudo-zero order with respect to oxygen. Two possible kinetics models were considered: (1) standard Michaelis-Menten kinetics, equation 10.2-9, and (2) competitive production inhibition, in which the product L-DOPA (P) acts as inhibitor (I), and the rate law is given by equation 10.4-8. Determine V_{max} for each model given $K_m = 3.9$ mmol L^{-1}, and comment on the quality of the model predictions. Assume that $c_I = c_{S_o} - c_S$. In model (2), assume that the inhibition constant K_2 is equal to 0.35 mmol L^{-1}.

10-11 The decomposition of 1-kestose is a key step in the production of fructo-oligosaccharides, which are found in many health foods because of their noncaloric and noncariogenic nature. Duan et al. (1994) studied the decomposition of 1-kestose (S) using β-fructosfuranosidase (E), both in the presence and absence of the competitive inhibitor glucose (G).

(a) The following initial rate data were obtained in the absence of glucose:

c_{S_o}/g L^{-1}	66.7	100	150	200	250	325	500
$(-r_{S_o})$/g L^{-1} h^{-1}	4.4	6.9	10.0	10.6	13.3	16.6	18.1

Determine the maximum reaction velocity (V_{max}) and the Michaelis constant (K_m) from these data, using (i) Lineweaver-Burk analysis and (ii) nonlinear regression. Comment on any differences between the parameter values obtained. The authors cite values of 30.7 g L^{-1} h^{-1} and 349.5 g L^{-1} for V_{max} and K_m, respectively.

(b) The following initial rate data were obtained in the presence of 100 g L^{-1} glucose:

c_{S_o}/ g L^{-1}	75.0	100	150	225	325	500
$(-r_{S_o})$/ g L^{-1} h^{-1}	1.5	2.3	3.0	4.0	6.2	8.6

Using the values of K_m and V_{max} estimated in part (a), estimate the value of the inhibition constant, K_2, in equation 10.4-8.

10-12 As a model for enzyme activation (as opposed to inhibition), a six-step kinetics scheme corresponding to that in Section 10.4.2 may be used, with activator A replacing inhibitor I. A special case of this may involve different sites for the substrate S and A, and in which S only binds to the EA complex. The simplified model is then

$$A + E \underset{k_{-1}}{\overset{k_1}{\rightleftharpoons}} AE$$

$$AE + S \underset{k_{-2}}{\overset{k_2}{\rightleftharpoons}} EAS$$

$$EAS \overset{k_r}{\rightarrow} AE + P$$

Derive the rate law for this model by applying the SSH to AE and AES. Show how the rate law indicates activation (i.e., enhanced rate relative to the unactivated reaction).

Chapter 11

Preliminary Considerations in Chemical Reaction Engineering

In this chapter, we return to the main theme of this book, chemical reaction engineering (CRE). We amplify some of the general considerations introduced in Chapter 1, before the detailed consideration of quantitative design methods in Chapter 12 and subsequent chapters.

We begin by considering general aspects of reactor selection, performance, and design, primarily from the point of view of process design, but with passing reference to mechanical design. We then present a number of equipment/flow diagrams, some generically schematic and some relating to specific industrial processes, to illustrate many of these aspects (see also Figure 1.4). In this way, we attempt to develop a general appreciation for reactors and CRE before the detailed consideration of their design and performance.

11.1 PROCESS DESIGN AND MECHANICAL DESIGN

Process design has to do with specifying matters relating to the process itself, such as operating conditions, and size, configuration, and mode of operation of the reactor. Mechanical design has to do with specifying matters relating to the equipment itself in the sense of structural and mixing requirements, among others. Both are necessary for complete design, but our scope in this book is confined to process design, which is primarily the domain of the chemical engineer.

11.1.1 Process Design

11.1.1.1 Nature of Process Design in CRE

The problem of process design in CRE typically stems from the requirement to produce a specified product at a particular rate (e.g., 1000 tonnes day^{-1} of NH_3). The substance(s) from which the product is made may be specified or may have to be chosen from more than one option. Process design then involves making decisions, as quantitatively as possible, about the type of reactor and its mode of operation (e.g., batch or continuous), its size (e.g., volume or amount of catalyst), and processing conditions (e.g., T, P, product distribution, if relevant). The criteria constraining these decisions relate to technical feasibility (e.g., can the required fractional conversion be achieved in the size specified?), and to socio-economic feasibility (cost, safety, and environmental considerations). The factors to be taken into account are listed in more detail in the next section.

The design problem usually fits into the spectrum ranging from (1) the rational design of a completely new reactor for a new process, to (2) the analysis of performance of an existing reactor for an existing process. A common situation, between these extremes, even for a new plant, is the modification of an existing type of reactor, the design of which has evolved over time.

11.1.1.2 *Matters for Consideration*

Process design involves making decisions about a series of matters, on as rational and quantitative a basis as possible, given the information available. The following is an illustrative list of such matters but not an exhaustive one; the items are not all mutually exclusive.

(1) type of processing
    batch
    continuous
    semibatch or semicontinuous
        batch with respect to (specified phase or phases)
        continuous with respect to (specified phase or phases)

(2) type and nature of reacting system
    reactant(s) and product(s)
    simple
    complex (desirable, undesirable products)
    stoichiometry
    phases, number of phases
    catalytic (choice of catalyst) or noncatalytic
    endothermic or exothermic
    possibility of equilibrium limitation

(3) type and size of reactor
    batch
    continuous
      stirred tank
      tubular, multitubular
      tower/column
        spray
        packed, plate
        bubble
      bed
        fixed
        fluidized
        spouted
        trickle
      furnace

(4) mode of operation
    configurational
      single-stage or multistage (number of)
      parallel (e.g., multitubular)
      axial or radial flow (through fixed bed)
      arrangement of heat transfer surface (if any)

flow pattern (backmix flow, tubular flow, etc.)
contacting pattern (cocurrent, countercurrent, crosscurrent flow of phases; method of addition of reactants—all at once? in stages? which phase dispersed? continuous?)
thermal
  adiabatic
  isothermal
  nonisothermal, nonadiabatic
use of recycle

(5) process conditions
  $T(T$ profile)
  $P(P$ profile)
  feed (composition, rate)
  product (composition, rate)

(6) optimality
  of process conditions
  of size
  of product distribution
  of conversion
  of cost (local, global context)

(7) control and stability of operation
  instrumentation
  control variables
  sensitivity analysis
  catalyst life, deactivation, poisons

(8) socioeconomic
  cost
  environmental
  safety

(9) materials of construction; corrosion

(10) startup and shutdown procedures

Most of these matters are subjects for exploration in the following chapters, but some are outside our scope. Some may be specified *ab initio*, and others may provide constraints in various ways.

11.1.1.3 *Data Required*

Data required include those specific to the situation in hand (design specifications) and those of a more general nature:

(1) specifications
  reactants
  products
  throughput or capacity

(2) general data
  rate data/parameters relating to
    reaction (rate law(s))
    heat transfer
    diffusion
    mass transfer
    pressure drop

282   Chapter 11: Preliminary Considerations in Chemical Reaction Engineering

> equilibrium data
>   thermodynamic
>   equation of state
> thermochemical data
>   enthalpy of reaction
>   heat capacities
> other physical property data
>   density
>   viscosity
> cost data

The lack of appropriate information, particularly rate data over the entire range of operating conditions, is often a major impediment to a complete rational design. If sufficiently important, new experimental data may have to be developed for the situation at hand.

11.1.1.4 *Tools Available*

The rational design of a chemical reactor is perhaps the most difficult equipment-design task of a chemical engineer. All the tools in the workshop (or, to use a more belligerent metaphor, all the weapons in the arsenal) of the chemical engineer may have to be brought to bear on the problem. These include those relating to all of the following:

(1) rate processes and rate laws
  reaction kinetics
    development of a reaction model/kinetics scheme
  diffusion and mass transfer
    diffusion equation
    mass transfer model (e.g., two-film model)
  heat transfer
    as part of thermal characteristics, including T profile
  fluid mechanics
    flow patterns
    mixing
    pressure drop
(2) conservation and balance equations
  mass balances, including stoichiometry, continuity equation
  energy balance, including energetics of reaction, thermochemistry
(3) equilibrium
  reaction equilibrium, single or multiphase systems; equilibrium conversion for comparison
  phase equilibrium, multiphase systems
(4) mathematics
  development of a reactor model from above considerations
  analytical or numerical methods for solution of equations, simulation
  statistical analysis of rate data
(5) computers and computer software
  use of a PC, workstation, etc., coupled with software packages to solve sets of algebraic and/or differential equations, and to perform statistical analyses necessary for implementation of a reactor model for design or for assessment of reactor performance
  software: spreadsheet packages, simulation software, numerical equation solvers, computer algebra system (e.g., E-Z Solve and Excel)
(6) process economics

11.1.2 Mechanical Design

The distinction between process design (as outlined in Section 11.1.1) and "mechanical" design (most other aspects!) is somewhat arbitrary. However, in the latter we include the following, which, although important, are outside our scope in this book (see Perry, et al., 1984; Peters and Timmerhaus, 1991):

impeller or agitator design (as in a stirred tank)
power requirement (for above)
reactor-as-pressure-vessel design
    wall thickness
    over-pressure relief
    fabrication
support-structure design
maintenance features

11.2 EXAMPLES OF REACTORS FOR ILLUSTRATION OF PROCESS DESIGN CONSIDERATIONS

Reactors exist in a variety of types with differing modes of operation for various processes and reacting systems. Figures 11.1 to 11.8 illustrate a number of these, some in a schematic, generic sense, and some for specific processes.

11.2.1 Batch Reactors

A schematic representation of a batch reactor in Figure 11.1(a) shows some of the essential features. A cylindrical vessel is provided with nozzles for adding and removing reactor contents. To ensure adequate mixing, the vessel has a stirrer (turbine) equipped with an external drive, and several vertical baffles to break up vortices around the impeller tips. Temperature control may be achieved with internal heating/cooling coils, as shown, or with an external heat exchanger or jacket. The vessel may be designed as a pressure vessel, in which case a pressure-relief device (e.g., a rupture disc) is required in case over-pressure develops, or as an "atmospheric" vessel, in which case a vent is used. Batch reactors are discussed in Chapter 12.

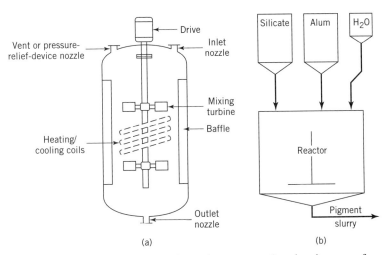

Figure 11.1 Batch reactors: (a) schematic representation showing some features; (b) pilot plant reactor for production of sodium aluminosilicate (Courtesy of National Silicates Ltd.)

Figure 11.1(b) illustrates a pilot plant batch reactor used in the early 1990s for the production of sodium aluminosilicate, from alum and sodium silicate, for use in the pulp and paper industry. The ratio of SiO_2 to Al_2O_3 in the product is controlled by adjustment of the feed amounts from the hoppers above the reactor. Efficient mixing is required for the reacting system (a non-Newtonian slurry) to produce the desired amorphous form. For this, baffles on the walls and mixing paddles (not shown) on the central shaft are used. After an appropriate reaction time, the pigment slurry (intermediate product) is removed for further processing.

11.2.2 Stirred-Tank Flow Reactors

A stirred-tank flow reactor may be single-stage or multistage. As an ideal backmix flow reactor, it is referred to as a CSTR or multistage CSTR; this is treated in Chapter 14. Nonideal flow effects are discussed in Chapter 20.

A three-stage stirred-tank flow reactor is illustrated in Figure 11.2. Mixing and heat transfer features may be similar to those of a batch reactor, but there may be more use of external heat exchangers as preheaters/coolers, interstage heaters/coolers, and afterheaters/coolers. In an extreme case of heat-transfer requirement, the reactor may resemble a shell-and-tube heat exchanger, as in a "Stratco" contactor for HF-alkylation of hydrocarbons (Perry et al., 1984, p. 21–61). A multistage reactor may be contained within a single vessel, as in a Kellogg cascade alkylator (Perry et al., 1984, p. 21–60).

The major difference between a stirred-tank batch reactor and a stirred-tank flow reactor is that, in the latter, provision must be made for continuous flow of material into and out of the reactor by gravity flow or forced-circulation flow with a pump, together with appropriate block and relief valves.

11.2.3 Tubular Flow Reactors

The term "tubular flow reactor" is used generically to refer to a reactor in which the flow of fluid is essentially in one direction (e.g., axial flow in a cylindrical vessel) without any attempt to promote backmixing by stirring. Idealized forms are a plug-flow reactor and a laminar-flow reactor, as discussed in Chapter 2. The configuration may vary widely from a very high to a very low length-to-diameter (L/D) ratio, as shown schematically in Figure 11.3. Reactors shown in Figures 11.4 to 11.6 below are examples of tubular reactors, which are essentially plug-flow reactors. Design aspects of tubular reactors, mainly as PF reactors, are introduced in Chapter 15 and continued in some subsequent chapters. Effects of nonideal flow are considered in Chapter 20.

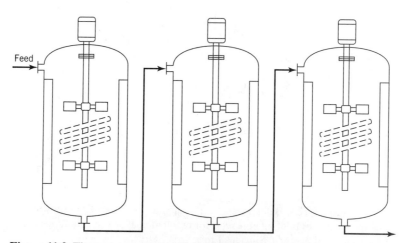

Figure 11.2 Three-stage stirred-tank flow reactor

11.2 Examples of Reactors for Illustration of Process Design Considerations

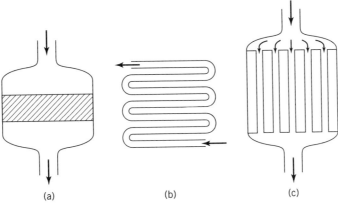

Figure 11.3 Tubular flow reactor—some possible configurations: (a) low L/D; (b) high L/D; (c) tubular arrangement

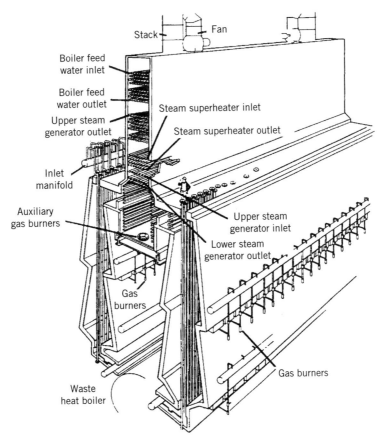

Figure 11.4 Schematic arrangement of a top-fired furnace for steam-reforming of natural gas (Twigg, 1996, p. 262; reproduced with permission from Catalyst Handbook, ed. M.V. Twigg, Manson Publishing Company, London, 1996)

286 Chapter 11: Preliminary Considerations in Chemical Reaction Engineering

11.2.3.1 Furnace Reactors

Some endothermic reactions require reactors that can provide high rates of heat transfer at relatively high temperatures (900 to 1100 K). Examples are the dehydrogenation of C_2H_6 to produce C_2H_4 (noncatalytic, low P), and the steam-reforming of natural gas, $CH_4 + H_2O \rightarrow CO + 3H_2$, to produce H_2 for ammonia and methanol syntheses (catalytic, $P \approx 30$ bar). The reaction typically takes place as the reacting system (gas) flows through long coils of tubes contained in a combustion chamber (furnace). Heat transfer occurs by radiation and convection in corresponding sections of the reactor. A fuel gas is burned in the combustion chamber. Figure 11.4 shows a schematic arrangement of a "top-fired" furnace for steam reforming, in which the gas burners are located at the top of the combustion chamber.

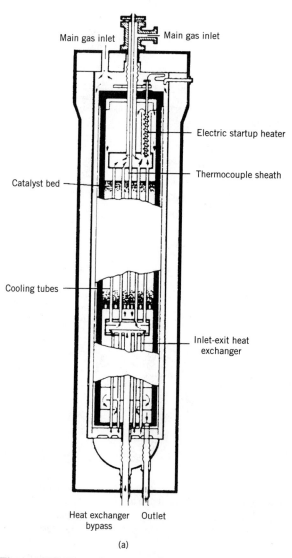

Figure 11.5 Examples of ammonia synthesis converters: (a) tube-cooled, axial-flow converter (Twigg, 1996, p. 438; reproduced with permission from Catalyst Handbook, ed. M.V. Twigg, Manson Publishing Company, London, 1996.)

11.2 Examples of Reactors for Illustration of Process Design Considerations

11.2.3.2 *Fixed-Bed Catalytic Reactors: Ammonia Synthesis*

The synthesis of ammonia, $N_2 + 3H_2 = 2NH_3$, like the oxidation of SO_2 (Section 1.5.4 and Figure 1.4), is an exothermic, reversible, catalytic reaction carried out in a multistage tubular flow reactor in which each stage consists of a (fixed) bed of catalyst particles. Unlike SO_2 oxidation, it is a high-pressure reaction (150–350 bar, at an average temperature of about 450°C). The usual catalyst is metallic Fe.

The reactors used can be classified in two main ways (Twigg, 1996, p. 424):

(1) by type of flow of the gas through the beds: axial, radial, or cross-flow; and
(2) by method used to control T and recover the energy of reaction: indirect cooling with heat exchanger surface or quench cooling.

Figure 11.5 illustrates four of the many variations of these converters. Figures 11.5(a) and (b) both show axial-flow converters, but the first uses heat exchange and the

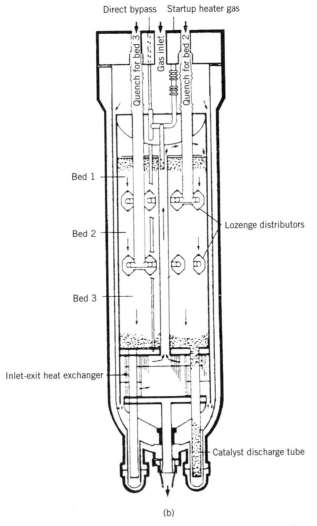

Figure 11.5 (cont'd) (b) ICI quench converter with axial flow; quench gas is injected and mixed by means of lozenges (Twigg, 1996, p. 429; reproduced with permission from Catalyst Handbook, ed. M.V. Twigg, Manson Publishing Company, London, 1996.)

288 Chapter 11: Preliminary Considerations in Chemical Reaction Engineering

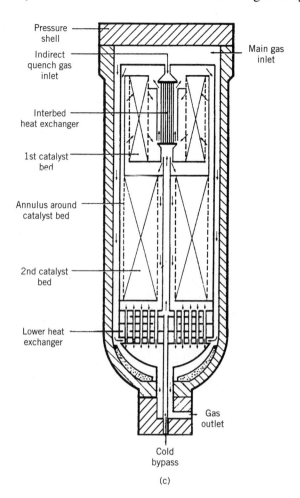

(c)

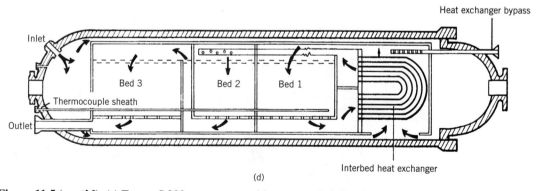

(d)

Figure 11.5 (cont'd) (c) Topsøe S 200 converter combines two radial-flow beds with interstage cooling by heat exchange; (d) Kellogg horizontal converter with cross-flow and interstage heat exchange (Twigg, 1996, pp. 437, 432; reproduced with permission from Catalyst Handbook, ed. M.V. Twigg, Manson Publishing Company, London, 1996)

11.2 Examples of Reactors for Illustration of Process Design Considerations

second uses quench cooling. In the latter, part of the cool reactor feed gas is added to the main gas stream at intervals along the length of a bed, usually between stages. This is achieved by means of the "lozenge" distributors shown. Figure 11.5(c) shows a converter employing radial flow of gas through each of two beds (together with cooling by heat exchangers); the flow is radially inward from the outer perimeter. The advantage of radial flow relative to axial flow is that the resulting pressure drop (and hence, power consumption) is lower, because the flow area is greater and the path length smaller. Figure 11.5(d) illustrates "cross-flow" of gas through three beds in a horizontal converter with cooling by heat exchange.

Some of the many design aspects of fixed-bed catalytic reactors are indicated in Figure 11.5. These and other aspects are taken up in Chapter 21.

11.2.3.3 Fixed-Bed Catalytic Reactors: Methanol Synthesis

The synthesis of methanol, $CO + 2H_2 = CH_3OH$, like ammonia synthesis, is an exothermic, reversible, catalytic reaction. Unlike a catalyst for ammonia synthesis, a catalyst for

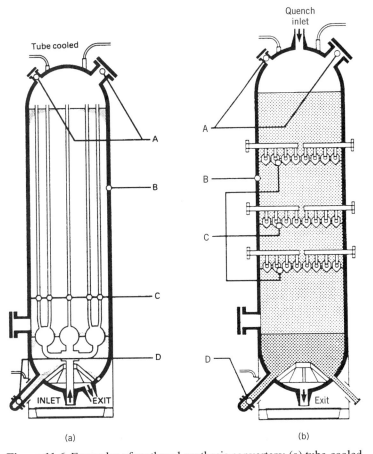

Figure 11.6 Examples of methanol synthesis converters: (a) tube-cooled, low-pressure reactor; A: nozzles for charging and inspecting catalyst; B: outer wall of reactor as a pressure vessel; C: thin-walled cooling tubes; D: port for catalyst discharge by gravity; (b) quench-cooled, low-pressure reactor; A,B,D, as in (a); C: ICI lozenge quench distributors (Twigg, 1996, pp. 450, 449; reproduced with permission from Catalyst Handbook, ed. M.V. Twigg, Manson Publishing Company, London, 1996.)

methanol synthesis must be selective (as well as active), to avoid formation of species such as CH_4 and C_2H_5OH. The process was formerly carried out at high pressure (100–300 bar) with a ZnO/Cr_2O_3 catalyst, but a catalyst of $Cu/ZnO/Al_2O_3$ enables the reaction to be carried out at 50–100 bar, which is referred to as low-pressure synthesis.

Figure 11.6 gives two examples of converters for low-pressure synthesis of methanol with axial flow. Figure 11.6(a) shows a converter using heat exchanger surface (tube cooling) within a single bed of catalyst. Figure 11.6(b) shows a converter using quench cooling with quench gas added at three intervals in the bed through lozenge distributors.

11.2.4 Fluidized-Bed Reactors

In a fluidized-bed reactor, fluid is introduced at many points and at a sufficiently high velocity that the upward flow through a bed of particles causes the particles to lift and fall back in a recirculating pattern. The result is an expanded "fluidized-bed" of particles + fluid holdup, behaving somewhat like a vigorously boiling liquid. The fluid completes its passage through the bed of particles by disengaging at the upper "surface."

The fluidized-bed reactor was originally developed for catalytic "cracking" in petroleum processing to enable continuous operation in a situation in which rapid fouling of catalyst particles occurs. It may also be used for noncatalytic reactions in which the particulate material is a reactant. Kunii and Levenspiel (1991, Chapter 2) illustrate many types of fluidized-bed reactors for various applications.

Figure 11.7 is a schematic diagram of a fluidized-bed "roaster" for oxidizing zinc "concentrate" (ZnS) to ZnO as part of the process for making Zn metal:

$$2ZnS + 3O_2 \rightarrow 2ZnO + 2SO_2$$

The fluidized bed occupies the lower part of the vessel, and is supported by a grate containing many openings through which air, entering at the bottom, flows to bring about fluidization. The greater part of the vessel is "freeboard," in which lower gas

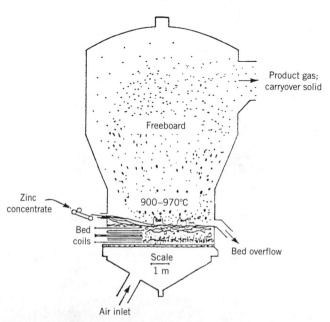

Figure 11.7 Fluidized-bed reactor for roasting zinc concentrate (after Themelis and Freeman, 1983)

11.2 Examples of Reactors for Illustration of Process Design Considerations

velocity allows for partial disengagement of entrained particles from the overhead gas stream. The solid reactant (zinc concentrate) enters at the left, and solid product leaves at the right. Product gas, with some carryover solid, leaves at the top right (to proceed to a cyclone for further removal of solid particles from the gas). The "coils" in the bed are for a heat exchanger to control T.

Design aspects of fluidized-bed reactors are considered in Chapter 23.

11.2.5 Other Types of Reactors

There are many other reactors of various types not included among those discussed above. These include tower reactors (Chapter 24), which may be modeled as PF or modified PF reactors. We describe one further example in this section.

Figure 11.8 shows the Kvaerner Chemetics electrolytic process for the production of sodium chlorate (in a concentrated solution of "strong liquor") and hydrogen from sodium chloride solution ("brine"). The overall reaction is $NaCl + 3H_2O \rightarrow NaClO_3 + H_2(g)$. The part of the system shown consists primarily of brine electrolyzers, a degasifier, a chlorate reactor, and an electrolyte cooler.

The reactor is designed to provide sufficient residence time (for recirculating liquid) for the reaction producing chlorate (started in the electrolyzers) to be completed. This involves further reaction of intermediates formed by the complex reactions in the electrolyzer, such as hypochlorite and hypochlorous acid, to produce chlorate. The reactor receives weak chlorate liquor from a crystallizer (not shown), fresh brine feed (also not

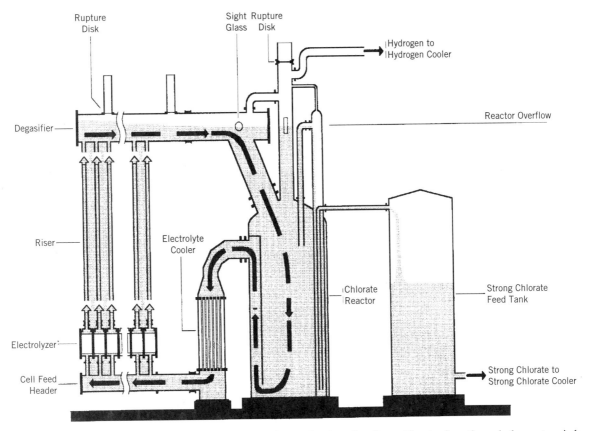

Figure 11.8 Kvaerner Chemetics electrolytic system for production of sodium chlorate; flow through the system is by natural convection (Courtesy of Kvaerner Chemetics Inc.)

indicated), and very strong chlorate liquor from the degasifier. The reactor produces strong chlorate liquor (overflow to tank on the right), and feeds liquor to the electrolyzers through an electrolyte cooler. No pumps are required to circulate liquor through the reactor and electrolyzers. A high natural circulation rate (shown by the dashed line and arrows) is established by the difference in density between the higher density liquid in the reactor and the lower density liquid + gas (H_2) in the electrolyzer and riser. The chlorate reactor contains an internal baffle to prevent short-circuiting of the circulating liquor. The resulting liquid flow pattern in the reactor is nonideal, between BMF and PF.

11.3 PROBLEMS FOR CHAPTER 11

11-1 For each of the examples shown in Figures 11.1 to 11.8, relate the features, as far as possible, to the points listed in Section 11.1.1.2.

The following reactor problems illustrate some process calculations that involve material and/or energy balances, and/or equilibrium considerations. They do not require rate considerations (or data) involved with reactor sizing and product distribution, which are taken up in later chapters for these same processes.

11-2 For a 1000-tonne day^{-1} sulfuric acid plant (100% H_2SO_4 basis), calculate the total molar flow rate (mol s^{-1}) of gas entering the SO_2 converter (for oxidation to SO_3), for steady-state operation, if the fractional conversion (f_{SO_2}) in the converter is 0.98, and the feed to the converter is 9.5 mol % SO_2. (1 tonne = 1000 kg.)

11-3 Calculate the (total) volumetric flow rate (m^3 s^{-1}) of gas leaving the reactor of a 1000-tonne day^{-1} ammonia plant, if the gas originates from H_2 and N_2 in the stoichiometric ratio, and 20% conversion to ammonia occurs. T = 450°C, P = 300 bar, and the compressibility factor z = 1.09. (1 tonne = 1000 kg.)

Table 11.1 Data for problems 11-4 and 11-5 (JANAF, 1986)[a]

Species/property/units	T/K				
	298.15	700	800	900	1000
ΔG_f°/kJ mol^{-1}					
SO_2	−300.125	−299.444	−298.370	−296.051	−288.725
SO_3	−371.016	−322.365	−321.912	−310.528	−293.639
ΔH_f°/kJ mol^{-1}					
SO_2	−296.842	−306.291	−307.667	−362.026	−361.940
SO_3	−395.765	−405.014	−406.068	−460.062	−459.581
C_P°/J mol^{-1} K^{-1}					
SO_2	39.878	50.961	52.434	53.580	54.484
SO_3	50.661	70.390	72.761	74.570	75.968
O_2	29.376	32.981	33.733	34.355	34.870
N_2	29.124	30.754	31.433	32.090	32.697

	T/K					
(additional C_P° data)	300	400	500	600	1100	1200
SO_2	39.945	43.493	46.576	49.049	55.204	55.794
SO_3	50.802	57.672	63.100	67.255	77.067	77.937
O_2	29.385	30.106	31.091	32.090	35.300	35.667
N_2	29.125	29.249	29.580	30.110	33.241	33.723

[a] Standard-state pressure, P° = 0.1 MPa.

11-4 The first chemical step in making sulfuric acid from elemental sulfur is burning the sulfur (completely) to form SO_2. Calculate the temperature of the gas leaving the burner based on the following: the burner operates adiabatically, sulfur enters as liquid at 140°C, and excess air (79% mol N_2, 21% O_2) enters at 25°C; ΔH ($S(\ell)$, 140°C $\rightarrow$ $S(s, \text{rh})$, 25°C) = -5.07 kJmol^{-1}; the outlet gas contains 9.5 mol % SO_2; other data required are given in Table 11.1.

11-5 The gas from the sulfur burner (problem 11-4), after reduction of T in a "waste-heat boiler," enters a catalytic converter for oxidation of SO_2 to SO_3 in several stages (see Figure 1.4). If the gas enters the converter at 700 K, calculate f_{SO_2} and the temperature (T/K) of the gas at the end of the first stage, assuming the reaction is adiabatic and equilibrium is attained. Assume $P = 1$ bar. See Table 11.1 for data.

11-6 The gas leaving an ammonia oxidation unit in a continuous process is cooled rapidly to 20°C and contains 9 mol % NO, 1% NO_2, 8% O_2, and 82% N_2 (all the water formed by reaction is assumed to be condensed). It is desirable to allow oxidation of NO to NO_2 in a continuous reactor to achieve a molar ratio of NO_2 to NO of 5 before absorption of the NO_2 to make HNO_3. Determine the outlet temperature of the reactor, if it operates adiabatically (at essentially 6.9 bar). The following data (JANAF, 1986) are to be used; state any assumptions made. (Based on Denbigh and Turner, 1984, pp. 57–64.)

Species	$\Delta H^\circ_{f,298}$/ J mol^{-1}	$\bar{C}_P$ (approx.)/ J mol^{-1} K^{-1}
O_2	0	29.4
NO	+90,291	29.9
NO_2	+33,095	39.0
N_2	0	29.1

Chapter 12

Batch Reactors (BR)

The general characteristics of a batch reactor (BR) are introduced in Chapter 2, in connection with its use in measuring rate of reaction. The essential picture (Figure 2.1) in a BR is that of a well-stirred, closed system that may undergo heat transfer, and be of constant or variable density. The operation is inherently unsteady-state, but at any given instant, the system is uniform in all its properties.

In this chapter, we first consider uses of batch reactors, and their advantages and disadvantages compared with continuous-flow reactors. After considering what the essential features of process design are, we then develop design or performance equations for both isothermal and nonisothermal operation. The latter requires the energy balance, in addition to the material balance. We continue with an example of optimal performance of a batch reactor, and conclude with a discussion of semibatch and semicontinuous operation. We restrict attention to simple systems, deferring treatment of complex systems to Chapter 18.

12.1 USES OF BATCH REACTORS

Batch reactors are used both in the laboratory for obtaining design and operating data and in industrial processes for production of chemicals.

The use of batch reactors in the laboratory is described in Section 2.2.2 for the interpretation of rate of reaction, in Section 3.4.1.1 for experimental methodology, and in Chapter 4 and subsequent chapters for numerical treatment of kinetics experimental data for various types of reacting systems.

The use of batch reactors in commercial processes is usually most suitable for small-volume production, particularly for situations in which switching from one process or product to another is required, as in the manufacture of pharmaceuticals. Typically, in such processes, the value of the products is relatively large compared with the cost of production. However, batch reactors may also be used for large-volume production (England, 1982). Examples are the production of "vinyl" (polyvinyl chloride or PVC) involving suspension polymerization, and of emulsion-polymerized latex. In some cases, it may be desirable to deviate from a strictly batch process. For example, it may be necessary to replenish periodically a limiting reactant which has been consumed, or, if products of different purity are required, portions of the batch may be removed at different times. In this type of operation, referred to as a semibatch process (Section 12.4), the system is no longer closed, since some mass is added to, or removed from, the system during the process. If addition of reactants and removal of products occur continuously, the process is classified as continuous, and specific flow rates at the inlet and outlet of the reactor may be identified.

12.2 BATCH VERSUS CONTINUOUS OPERATION

Some advantages (A) and disadvantages (D) of batch operation compared with continuous operation are given in Table 12.1.

Batch and continuous processes may also be compared by examining their governing mass-balance relations. As an elaboration of equation 1.5-1, a general mass balance may be written with respect to a control volume as:

$$\begin{pmatrix} \text{rate} \\ \text{of} \\ \text{input} \\ \text{by} \\ \text{flow} \end{pmatrix} - \begin{pmatrix} \text{rate} \\ \text{of} \\ \text{output} \\ \text{by} \\ \text{flow} \end{pmatrix} + \begin{pmatrix} \text{rate of} \\ \text{formation} \\ \text{by reaction} \end{pmatrix} = \begin{pmatrix} \text{rate of} \\ \text{accumulation} \end{pmatrix} \quad (12.2\text{-}1)$$

In a batch reactor, the first two terms in equation 12.2-1 are absent. In a semibatch reactor, one of these two terms is usually absent. In a semicontinuous reactor for a multiphase system, both flow terms may be absent for one phase and present for another. In a continuous reactor, the two terms are required to account for the continuous inflow to and outflow from the reactor, whether the system is single-phase or multiphase.

Table 12.1 Comparison of batch and continuous operation

	Batch operation	Continuous operation
1.	Usually better for small-volume production (A)	Better for indefinitely long production runs of one product or set of products (A)
2.	More flexible for multiproduct (multiprocess) operation (A)	
3.	Capital cost usually relatively low (A)	Capital cost usually relatively high (D)
4.	Easy to shut down and clean for fouling service (A)	
5.	Inherent down-time between batches (D)	No down-time except for scheduled and emergency maintenance (A); but loss of production in lengthy stoppages can be costly (D)
6.	Operating cost may be relatively high (D)	Operating cost relatively low (A)
7.	Unsteady-state operation means process control and obtaining uniformity of product more difficult (D) (but see England, 1982)	Steady-state operation means process control and obtaining uniformity of product less difficult (A)

12.3 DESIGN EQUATIONS FOR A BATCH REACTOR

12.3.1 General Considerations

The process design of a batch reactor may involve determining the time (t) required to achieve a specified fractional conversion (f_A, for limiting reactant A, say) in a single batch, or the volume (V) of reacting system required to achieve a specified rate of production on a continual basis. The phrase "continual basis" means an ongoing operation, that is, operation "around the clock" with successive batches. This includes allowance for the down-time (t_d) during operation for loading, unloading, and cleaning. The operation may involve constant or varying density (ρ), and constant or varying temperature (T). The former requires an equation of state to determine V, and latter requires the energy balance to determine T. We consider various cases in the following sections.

For a single-phase system, V always refers to the volume of the reacting system, but is not necessarily the volume of the reactor. For example, for a liquid-phase reaction in a BR, allowance is made for additional "head-space" above the liquid, so that the actual reactor volume is larger than the system volume. In any case, we use V conventionally to refer to "reactor volume," with this proviso.

12.3.1.1 Time of Reaction

Consider the reaction

$$A + \ldots \rightarrow \nu_C C + \ldots \tag{12.3-1}$$

Interpretation of the mass balance, equation 12.2-1, leads to equation 2.2-4, which may be rewritten, to focus on t, as

$$t = n_{Ao} \int_{f_{A1}}^{f_{A2}} \frac{df_A}{(-r_A)V} \tag{12.3-2}$$

where t is the time required for reaction in an uninterrupted batch from fractional conversion f_{A1} to f_{A2}, and n_{Ao} is the initial number of moles of A. Equation 12.3-2 is general in the sense that it allows for variable density and temperature. For specified n_{Ao}, f_{A1}, and f_{A2}, the unknown quantities are t, V, $(-r_A)$, and T (on which $(-r_A)$ and V may depend). To determine all four quantities, equation 12.3-2 may have to be solved simultaneously with the rate law,

$$(-r_A) = r_A(f_A, T) \tag{12.3-3}$$

the energy balance, which gives

$$T = T(f_A, V) \tag{12.3-4}$$

and an equation of state (incorporating the stoichiometry),

$$V = V(n_A, T, P) \tag{2.2-9}$$

in which case, we assume P is also specified.

In equation 12.3-2, the quantity t/n_{Ao} is interpreted graphically as the area under the curve of a plot of $1/(-r_A)V$ against f_A, as shown schematically in Figure 12.1.

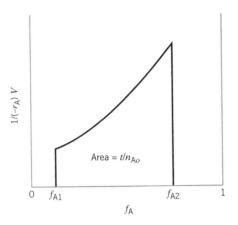

Figure 12.1 Interpretation of t/n_{Ao} in equation 12.3-2

12.3.1.2 Rate of Production; Volume of Reactor

Suppose reaction 12.3-1 is carried out in a batch reactor of volume V on a continual basis. To determine the rate of production, we must take into account the time of reaction (t in equation 12.3-2) and the down-time (t_d) between batches. The total time per batch, or cycle time, is

$$t_c = t + t_d \qquad (12.3\text{-}5)$$

The rate of production (formation) of C on a continual basis is then

$$Pr(C) = \frac{\text{moles of C formed}}{\text{batch}} \times \frac{\text{batch}}{\text{time}}$$

$$= \frac{n_{C2} - n_{C1}}{t_c} = \frac{\Delta n_C}{t_c} = \frac{-\nu_C \Delta n_A}{t + t_d}$$

That is, in terms of fractional conversion of A

$$Pr(C) = \frac{\nu_C n_{Ao}(f_{A2} - f_{A1})}{t + t_d} \qquad (12.3\text{-}6)$$

The volume of reactor (V) is related to n_{Ao} through the equation of state, 2.2-9. In many cases, $f_{A1} = 0$, and f_{A2} is simply designated f_A. In equation 12.3-6, t is obtained from the material balance (12.3-2). The other quantities, n_{Ao}, f_A, and t_d, must be specified or considered as parameters.

12.3.1.3 Energy Balance; Temperature Change

The energy balance in general, from equation 1.5-1, is

$$\begin{pmatrix} \text{rate of} \\ \text{input of} \\ \text{energy} \end{pmatrix} - \begin{pmatrix} \text{rate of} \\ \text{output of} \\ \text{energy} \end{pmatrix} = \begin{pmatrix} \text{rate of} \\ \text{accumulation} \\ \text{of energy} \end{pmatrix} \qquad (12.3\text{-}7)$$

Each term refers to a control volume, which for a BR is the volume of the reacting system. The input of energy may be by heat transfer from a heating coil or jacket, and/or by generation by reaction. Similarly, the output of energy may be by heat transfer to a coil or jacket, and/or by loss by reaction. The accumulation is the net result of the inputs and outputs, and may result in an increase or decrease in T of the reacting system.

The rate of heat transfer, $\dot{Q}$, whether it is an input or output, may be expressed as:

$$\dot{Q} = UA_c(T_c - T)_m \quad (12.3\text{-}8)$$

where
- U = the overall heat transfer coefficient, J m^{-2} s^{-1} K^{-1} or W m^{-2} K^{-1}, obtained experimentally or from an appropriate correlation;
- A_c = the area of the heating or cooling coil, m^2;
- T_c = the temperature of the coil, K (not necessarily constant with respect to position or time);
- $(T_c - T)_m$ = the appropriate mean temperature difference ΔT_m for heat transfer (question: what is the interpretation of the form of ΔT_m for a BR?).

If $\dot{Q} > 0$ ($T_c > T$), the heat transfer is an input (to the reacting system), and the converse applies if $\dot{Q} < 0$.

There may also be a significant input or output because of the energetics of the reaction. An exothermic reaction implies a generation (an input) of energy. An endothermic reaction implies a loss (an output) of energy. The rate of generation or loss is the product of the energy of reaction and the (extensive) rate of reaction. The energy of reaction is represented by the enthalpy of reaction (ΔH_R) for a constant-pressure process, and by the internal energy (ΔU_R) for a constant-volume (or constant-density) process. For a reaction represented by equation 12.3-1, with both the energy and rate of reaction in terms of reactant A, then

$$\text{rate of generation or loss of energy by reaction} = (-\Delta H_{RA})(-r_A)V \text{ or } (-\Delta U_{RA})(-r_A)V \quad (12.3\text{-}9)$$

Whether equation 12.3-9 represents an input or output depends on the sign of ΔH_{RA} (or ΔU_{RA}). If $\Delta H_{RA} > 0$ (endothermic reaction), it is an output (negative; recall that $(-r_A)$ is positive); if $\Delta H_{RA} < 0$ (exothermic reaction), it is an input.

Since equation 12.3-7 represents an enthalpy (or internal energy) balance, the rate of accumulation of energy is the rate of change of enthalpy, H (or internal energy, U) of the reacting system:

$$\text{rate of accumulation} = dH/dt$$
$$= n_t C_P dT/dt \quad (12.3\text{-}10)$$
$$= m_t c_P dT/dt \quad (12.3\text{-}11)$$

where n_t is the total number of moles, including moles of inert substances:

$$n_t = \sum_{i=1}^{N} n_i \quad (12.3\text{-}12)$$

C_P is the (total) molar heat capacity of the system at constant pressure, usually approximated, as though for an ideal solution, by[1]

$$C_P \doteq \sum_{i=1}^{N} x_i C_{Pi} \quad (12.3\text{-}13)$$

---

[1] For a nonideal solution, C_{Pi} is replaced by the partial molar heat capacity, $\bar{C}_{Pi}$, but such information may not be available.

where x_i is the mole fraction of species i, and C_{Pi} is its molar heat capacity as a pure species; m_t is the total (specific) mass of the system:

$$m_t = \sum_{i=1}^{N} m_i \qquad (12.3\text{-}14)$$

and c_P is the specific heat capacity of the system, approximated by

$$c_P \doteq \sum_{i=1}^{N} w_i c_{Pi} \qquad (12.3\text{-}15)$$

where w_i is the weight (mass) fraction of species i, and c_{Pi} is its specific heat capacity as a pure species.

Equations similar to 12.3-10 to -15 may be written in terms of internal energy, U, with C_V, the heat capacity at constant volume, replacing C_P. For liquid-phase reactions, the difference between the two treatments is small. Since most single-phase reactions carried out in a BR involve liquids, we continue to write the energy balance in terms of H, but, if required, it can be written in terms of U. In the latter case, it is usually necessary to calculate ΔU from ΔH and C_V from C_P, since ΔH and C_P are the quantities listed in a database. Furthermore, regardless of which treatment is used, it may be necessary to take into account the dependence of ΔH (or ΔU) and C_P (or C_V) on T.[2]

From equations 12.3-8, -9, and -10, the energy balance for a BR, equation 12.3-7, becomes

$$UA_c(T_c - T)_m + (-\Delta H_{RA})(-r_A)V = n_t C_P \frac{dT}{dt} \qquad (12.3\text{-}16)$$

Equation 12.3-16 is valid whether heat is transferred to or from the system, and whether the reaction is exothermic or endothermic. Note that each term on the left side of equation 12.3-16 may be an input or an output. Furthermore, C_P is the molar heat capacity of the *system*, and is given by equation 12.3-13; as such, it may depend on both T and composition (through f_A). The right side of equation 12.3-16 may also be expressed on a specific-mass basis (12.3-11). This does not affect the consistency of the units of the terms in the energy balance, which are usually $J\ s^{-1}$.

---

[2]
$$\Delta H = \Delta U + \Delta(PV) \qquad (12.3\text{-}17)$$
$$C_P - C_V = \alpha^2 VT/\kappa_T \qquad (12.3\text{-}18)$$

where α is the coefficient of cubical expansion, and κ_T is the isothermal compressibility;

$$d\Delta H/dT = \Delta C_P \qquad (12.3\text{-}19)$$

where ΔC_P is the heat capacity change corresponding to that of ΔH. The dependence of C_P on T is usually given by an empirical expression such as

$$C_P = a + bT + cT^2 \qquad (12.3\text{-}20)$$

in which the coefficients a, b, and c depend on the species, and must be given.

12.3.2 Isothermal Operation

12.3.2.1 Constant-Density System

For a liquid-phase reaction, or gas-phase reaction at constant temperature and pressure with no change in the total number of moles, the density of the system may be considered to remain constant. In this circumstance, the system volume (V) also remains constant, and the equations for reaction time (12.3-2) and production rate (12.3-6) may then be expressed in terms of concentration, with $c_A = n_A/V$:

$$t = c_{Ao} \int_{f_{A1}}^{f_{A2}} \frac{df_A}{(-r_A)} \quad \text{(constant density)} \qquad (12.3\text{-}21)$$

$$Pr(C) = \frac{\nu_C c_{Ao} V(f_{A2} - f_{A1})}{t + t_d} \quad \text{(constant density)} \qquad (12.3\text{-}22)$$

Equation 12.3-21 may be interpreted graphically in a manner similar to that of equation 12.3-2 in Figure 12.1 (see problem 12-16).

The following two examples illustrate the use of equations 12.3-21 and -22.

EXAMPLE 12-1

Determine the time required for 80% conversion of 7.5 mol A in a 15-L constant-volume batch reactor operating isothermally at 300 K. The reaction is first-order with respect to A, with $k_A = 0.05$ min^{-1} at 300 K.

SOLUTION

In equation 12.3-21,

$$c_{Ao} = n_{Ao}/V = 7.5/15 = 0.5$$
$$f_{A1} = 0; \quad f_{A2} = 0.80$$
$$(-r_A) = k_A c_A = k_A c_{Ao}(1 - f_A)$$
$$t = c_{Ao} \int_0^{f_A} \frac{df_A}{k_A c_{Ao}(1 - f_A)} = \frac{1}{k_A} \int_0^{f_A} \frac{df_A}{1 - f_A}$$
$$= \frac{-\ln(1 - f_A)}{k_A} = \frac{-\ln(0.2)}{0.05} = 32.2 \text{ min}$$

Note that for a first-order reaction, t is independent of c_{Ao}.

EXAMPLE 12-2

A liquid-phase reaction between cyclopentadiene (A) and benzoquinone (B) is conducted in an isothermal batch reactor, producing an adduct (C). The reaction is first-order with respect to each reactant, with $k_A = 9.92 \times 10^{-3}$ L mol^{-1} s^{-1} at 25°C. Determine the reactor volume required to produce 175 mol C h^{-1}, if $f_A = 0.90$, $c_{Ao} = c_{Bo} = 0.15$ mol L^{-1}, and the down-time t_d between batches is 30 min. The reaction is A + B → C.

12.3 Design Equations for a Batch Reactor 301

SOLUTION

This example illustrates the use of the design equations to determine the volume of a batch reactor (V) for a specified rate of production $Pr(C)$, and fractional conversion (f_A) in each batch. The time for reaction (t) in each batch in equation 12.3-22 is initially unknown, and must first be determined from equation 12.3-21.

In equation 12.3-21, from the stoichiometry, since $c_{Ao} = c_{Bo}$,

$$(-r_A) = k_A c_A c_B = k_A c_A^2 = k_A c_{Ao}^2 (1 - f_A)^2$$

Then, with $f_{A1} = 0$, and $f_{A2} = f_A$,

$$t = c_{Ao} \int_0^{f_A} \frac{df_A}{k_A c_{Ao}^2 (1 - f_A)^2} = \frac{1}{k_A c_{Ao}} \frac{f_A}{1 - f_A}$$

$$= \frac{1}{9.92 \times 10^{-3} \times 0.15} \frac{0.9}{0.1} = 6050 \text{ s} = 1.68 \text{ h}$$

From equation 12.3-22,

$$V = \frac{(t + t_d) Pr(C)}{\nu_C c_{Ao} f_A} = \frac{(1.68 + 0.5) 175}{1 (0.15) 0.9}$$

$$= 2830 \text{ L or } 2.83 \text{ m}^3$$

This example may also be solved using the E-Z Solve software (see file ex12-2.msp).

12.3.2.2 Variable-Density System

If the system is not of constant density, we must use the more general form of the equation for reaction time (12.3-2) to determine t for a specified conversion, together with a rate law, equation 12.3-3, and an equation of state, equation 2.2-9. Variable density implies that the volume of the reactor or reacting system is not constant. This may be visualized as a vessel equipped with a piston; V changes with the position of the piston. Systems of variable density usually involve a gas phase. The density may vary if any one of T, P or n_t (total number of moles) changes (so as to alter the position of the piston).

EXAMPLE 12-3

A gas-phase reaction A → B + C is to be conducted in a 10-L (initially) isothermal batch reactor at 25°C at constant P. The reaction is second-order with respect to A, with $k_A = 0.023$ L mol^{-1} s^{-1}. Determine the time required for 75% conversion of 5 mol A.

SOLUTION

The time required is given by equation 12.3-2:

$$t = n_{Ao} \int_0^{f_A} \frac{df_A}{(-r_A) V} \qquad (12.3\text{-}2)$$

In this expression, both $(-r_A)$ and V vary and must be related to f_A, through the rate law and an equation of state incorporating the stoichiometry, respectively; for the equation of state, we assume ideal-gas behavior. Thus, for $(-r_A)$,

$$(-r_A) = k_A c_A^2 = k_A(n_A/V)^2 = \frac{k_A n_{Ao}^2 (1 - f_A)^2}{V^2} \quad \textbf{(A)}$$

Since this is a gas-phase reaction, and the total number of moles changes, the volume changes as the reaction progresses. We use a stoichiometric table to determine the effect of f_A on V.

Species	Initial moles	Change Δn	Final moles
A	n_{Ao}	$-n_{Ao} f_A$	$n_{Ao}(1 - f_A)$
B	0	$n_{Ao} f_A$	$n_{Ao} f_A$
C	0	$n_{Ao} f_A$	$n_{Ao} f_A$
total:	n_{Ao}	$n_{Ao} f_A$	$n_{Ao}(1 + f_A)$

If the gas phase is ideal, $V = n_t RT/P$, and in this case, R, T, and P are constant. Therefore,

$$\frac{V}{V_o} = \frac{n_{Ao}(1 + f_A)}{n_{Ao}}$$

or

$$V = V_o(1 + f_A), \quad \textbf{(B)}$$

where V_o is the initial volume of the system.

Substituting for $(-r_A)$ from (A) and for V from (B) in equation 12.3-2, and simplifying, we obtain:

$$t = \frac{V_o}{k_A n_{Ao}} \int_0^{f_A} \frac{(1 + f_A) df_A}{(1 - f_A)^2}$$

To integrate, we let $\alpha = 1 - f_A$; then, $f_A = 1 - \alpha$, and $df_A = -d\alpha$. The integral is:

$$\int_1^{0.25} \frac{\alpha - 2}{\alpha^2} d\alpha = \ln(0.25) + 6 = 4.61$$

Therefore, $t = 10 \text{ L} \times 4.61/(0.023 \text{ L mol}^{-1} \text{ s}^{-1} \times 5.0 \text{ mol}) = 400 \text{ s}$

12.3.2.3 Control of Heat Transfer to Maintain Isothermal Conditions

In certain circumstances, it may be desirable to maintain nearly isothermal conditions, even if the reaction is significantly exothermic or endothermic. In the absence of any attempt to control T, it may become too high for product stability or too low for reaction rate. If control of T is required, a cooling or heating coil or jacket can be added to the reactor to balance the energy generated or consumed by the reaction. The coil temperature (T_c) is adjusted to control the rate of heat transfer ($\dot{Q}$) to achieve (nearly) isothermal conditions. T_c usually varies as the reaction proceeds, because the rate of reaction, $(-r_A)$, and hence ΔH, is a function of time. The relationship between T_c or $\dot{Q}$ and f_A can be determined by combining the material and energy balances.

12.3 Design Equations for a Batch Reactor

In the energy balance, for isothermal operation of the reaction A + ... → product(s), $dT/dt = 0$, so that equation 12.3-16 becomes, for the required rate of heat transfer:

$$\dot{Q} = UA_c(T_c - T)_m = -(-\Delta H_{RA})(-r_A)V \quad \text{(isothermal operation)} \quad \textbf{(12.3-23)}$$

From the material balance in terms of f_A, equation 2.2-4 becomes

$$(-r_A) = (n_{Ao}/V)(df_A/dt) \quad \textbf{(2.2-4)}$$

Eliminating $(-r_A)$ from equation 12.3-23 with equation 2.2-4, we have

$$\dot{Q} = UA_c(T_c - T)_m = -(-\Delta H_{RA})n_{Ao}(df_A/dt) \quad \textbf{(12.3-24)}$$

For simplicity, if we assume that T_c is constant throughout the coil at a given instant, then T_c depends on t, from equation 12.3-24, through

$$T_c = T - \frac{(-\Delta H_{RA})n_{Ao}}{UA_c} \frac{df_A}{dt} \quad \textbf{(12.3-25)}$$

EXAMPLE 12-4

Determine $\dot{Q}$ and T_c (as functions of time, t) required to maintain isothermal conditions in the reactor in Example 12-1, if $\Delta H_{RA} = -47{,}500$ J mol^{-1}, and $UA_c = 25.0$ W K^{-1}. Does $\dot{Q}$ represent a rate of heat removal or heat addition?

SOLUTION

To use equations 12.3-24 and -25, we first obtain df_A/dt for a first-order reaction, and use this to eliminate f_A in terms of t.

From the material balance, equation 2.2-4,

$$\frac{df_A}{dt} = \frac{V}{n_{Ao}}(-r_A) = \frac{V}{n_{Ao}} k_A \frac{n_{Ao}}{V}(1 - f_A) = k_A(1 - f_A) \quad \textbf{(A)}$$

From (A), on integration,

$$f_A = 1 - e^{-k_A t} \quad \textbf{(B)}$$

Substituting (A) and (B) into equation 12.3-24, we obtain (with t in min)

$$\dot{Q} = -(-\Delta H_{RA})n_{Ao}k_A e^{-k_A t} = -(47{,}500)7.5(0.05/60)e^{-0.05t} = -297e^{-0.05t} \text{ J s}^{-1} \text{ or W}$$

Since $\dot{Q} < 0$, it represents heat removal from the system, which is undergoing an exothermic reaction.

Similarly, from equation 12.3-25, we obtain for the coolant temperature T_c in the coil

$$T_c = 300 - \frac{47{,}500(7.5)}{25.0} \frac{0.05}{60} e^{-0.05t} = 300 - 11.9 e^{-0.05t}$$

with T_c in K and t in min.

304 Chapter 12: Batch Reactors (BR)

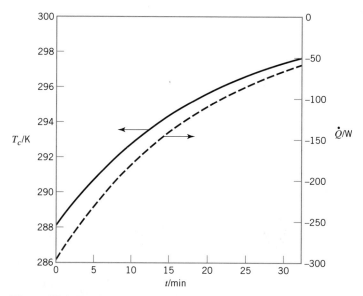

Figure 12.2 Profiles for coil temperature (T_c) and heat transfer rate ($\dot{Q}$) for an isothermal batch reactor (Example 12-4)

Figure 12.2 shows the changes in $\dot{Q}$ and T_c with respect to time t during the course of reaction to achieve $f_A = 0.8$, a conversion that requires 32.2 min (from Example 12-1). The decrease in magnitude of $\dot{Q}$ as the reaction progresses is expected, since $(-r_A)$ decreases with time. As a consequence, T_c must gradually increase. As $t \to \infty$, $(-r_A) \to 0$, $T_c \to 300$, and $\dot{Q} \to 0$.

12.3.3 Nonisothermal Operation

It may not be realistic to use isothermal operation of a BR as a basis for design, particularly for a reaction that is strongly exothermic or endothermic. Although T may change considerably if left unattended, and may need to be controlled so that it does not go too high or too low, it need not be strictly constant. Furthermore, in some cases, there may be advantages, from the point of view of kinetics, if T is allowed to increase in a controlled manner.

In order to assess the design of both the reactor and the heat exchanger required to control T, it is necessary to use the material balance and the energy balance, together with information on rate of reaction and rate of heat transfer, since there is an interaction between T and f_A. In this section, we consider two cases of nonisothermal operation: adiabatic ($\dot{Q} = 0$) and nonadiabatic ($\dot{Q} \neq 0$).

12.3.3.1 *Adiabatic Operation*

In adiabatic operation, there is no attempt to cool or heat the contents of the reactor (that is, there is no heat exchanger). As a result, T rises in an exothermic reaction and falls in an endothermic reaction. This case may be used as a limiting case for nonisothermal behavior, to determine if T changes sufficiently to require the additional expense of a heat exchanger and T controller.

For an adiabatic system with $\dot{Q} = 0$, the energy balance (12.3-16) becomes

$$(-\Delta H_{RA})(-r_A)V = n_t C_P \frac{dT}{dt} \qquad (12.3\text{-}26)$$

12.3 Design Equations for a Batch Reactor 305

Substituting for $(-r_A)V$ from the material balance in terms of f_A, equation 2.2-4, we obtain

$$(-\Delta H_{RA})n_{Ao}\frac{df_A}{dt} = n_t C_P \frac{dT}{dt} \quad (12.3\text{-}27)$$

Since the relationship between df_A/dt and dT/dt is implicit with respect to t, we may write equation 12.3-27 as

$$(-\Delta H_{RA})n_{Ao}df_A = n_t C_P dT \quad (12.3\text{-}28)$$

Equation 12.3-28 may be integrated to give T as a function of f_A:

$$T = T_o + n_{Ao}\int_{f_{Ao}}^{f_A}\frac{(-\Delta H_{RA})}{n_t C_P}df_A \quad (12.3\text{-}29)$$

where T_o and f_{Ao} refer to initial conditions. The simplest result from equation 12.3-29 is for constant $(-\Delta H_{RA})$, C_P, and n_t; then

$$T = T_o + \frac{(-\Delta H_{RA})n_{Ao}}{n_t C_P}(f_A - f_{Ao}) \quad (12.3\text{-}30)$$

The time t required to achieve fractional conversion f_A is obtained by integration of the material balance, equation 2.2-4:

$$t = n_{Ao}\int_{f_{Ao}}^{f_A}\frac{df_A}{(-r_A)V} \quad (12.3\text{-}31)$$

where $(-r_A)$ is obtained from a rate law and V from an equation of state. This also requires simultaneous solution of equation 12.3-29, since the integral in equation 12.3-31 depends on T as well as on f_A.

An algorithm to calculate t from equation 12.3-31 to achieve a specified f_A, and also to obtain $f_A(T)$ is as follows:

(1) choose value of f_A; $f_{Ao} \leq f_A \leq f_A$ (specified);
(2) calculate T at f_A from equation 12.3-29;
(3) calculate $(-r_A)$ from rate law given;
(4) calculate V from equation of state given, if required;
(5) repeat steps (1) to (4) for several values of f_A;
(6) calculate t from equation 12.3-31 by numerical or graphical integration.

Alternatively, the E-Z Solve software may be used to integrate simultaneously the material- and energy-balance expressions and solve the equation of state.

EXAMPLE 12-5

A gas-phase decomposition, A $\to$ R + S, is to be conducted in a batch reactor, with initial conditions of $T_o = 300$ K, $V_o = 0.5$ m^3, and a (constant) total pressure of 500 kPa. The

306 Chapter 12: Batch Reactors (BR)

values of C_P for A, R, and S are, respectively, 185.6, 104.7, and 80.9 J mol^{-1} K^{-1}. The enthalpy of reaction is -6280 J (mol A)$^{-1}$, and the reaction is first-order with respect to A, with $k_A = 10^{14} e^{-10,000/T}$ h^{-1}. Determine the profiles of f_A and T versus t, if the process is adiabatic, and T and t for $f_A = 0.99$.

SOLUTION

We first obtain the appropriate forms of the material-balance and energy-balance equations (12.3-31 and 12-3-29, respectively) for use in the above algorithm.

From the given rate law, we have

$$(-r_A) = k_A c_A = k_A n_A/V = k_A n_{Ao}(1 - f_A)/V$$

so that equation 12.3-31 becomes

$$t = \int_0^{f_A} \frac{df_A}{k_A(1 - f_A)} \quad \textbf{(A)}$$

where

$$k_A = 10^{14} e^{-10,000/T} \quad \textbf{(B)}$$

For equation 12.3-29, we may not be able to use the simplified result in equation 12.3-30, since $n_t C_P$ may not be constant but depend on f_A; to investigate this, we form

$$n_t C_P = n_t \sum x_i C_{Pi} = n_t \sum (n_i/n_t) C_{Pi}$$
$$= \sum n_i C_{Pi} = n_A C_{PA} + n_R C_{PR} + n_S C_{PS}$$
$$= n_{Ao}(1 - f_A) C_{PA} + n_{Ao} f_A C_{PR} + n_{Ao} f_A C_{PS}$$
$$= n_{Ao} [C_{PA} + (C_{PR} + C_{PS} - C_{PA}) f_A]$$
$$= 185.6 n_{Ao}$$

Since $n_t C_P$ is independent of f_A, we *can* use equation 12.3-30:

$$T = 300 + \frac{6280 n_{Ao}}{185.6 n_{Ao}} f_A = 300 + 33.8 f_A \quad \textbf{(C)}$$

Equations (A), (B) and (C) are used in the algorithm to obtain the information required. Step (3) is used to calculate k_A from equation (B), and step (4) is not required. Results are summarized in Table 12.2, for the arbitrary step-size in f_A indicated; $G = 1/[k_A(1 - f_A)]$, and G^* represents the average of two consecutive values of G. The last column lists the time required to achieve the corresponding conversion in the second column. These times were obtained as approximations for the value of the integral in equation (A) by means of the trapezoidal rule:

$$t_j = t_{j-1} + 0.5(G_j + G_{j-1})(f_{Aj} - f_{A,j-1})$$

The estimated time required to achieve a fractional conversion of 0.99 is 1.80 h, and the temperature at this time is 333.5 K, if the reactor operates adiabatically. The $f_A(t)$ profile is given by the values listed in the second and last columns; the $T(t)$ profile is given by the third and last columns.

Table 12.2 Results for Example 12-5 using trapezoidal rule

j	f_A	T/K	k_A/h^{-1}	G	G^*	t/h
1	0.00	300.0	0.334	3.00		0.00
2	0.10	303.4	0.484	2.29	2.65	0.26
3	0.20	306.8	0.696	1.79	2.04	0.47
4	0.30	310.2	0.994	1.44	1.62	0.63
5	0.40	313.5	1.408	1.18	1.31	0.76
6	0.50	316.9	1.979	1.01	1.10	0.87
7	0.60	320.3	2.762	0.91	0.96	0.97
8	0.70	323.7	3.828	0.87	0.89	1.06
9	0.80	327.1	5.269	0.95	0.91	1.15
10	0.90	330.5	7.206	1.39	1.17	1.26
11	0.99	333.5	9.500	10.53	5.96	1.80

Alternate solution using E-Z Solve software: (see file ex12-5.msp). The results for t using the trapezoidal rule approximation (Table 12.2) may differ significantly from those using a more accurate form of numerical integration. For values of f_A up to 0.90, the values of t differ by less than 1%. However, between $f_A = 0.90$ and 0.99, the results differ considerably, primarily because of the large step size (0.09) chosen for the trapezoidal approximation, compared with the much smaller step size used in the simulation software. The simulation software predicts $t = 1.52$ h for $f_A = 0.99$, rather than 1.80 h as in the table. The results for T (at given f_A) are unaffected, since f_A and T are related algebraically by Equation (C). The output from the simulation software can be in the form of a table or graph.

12.3.3.2 Nonadiabatic Operation

If the batch reactor operation is both nonadiabatic and nonisothermal, the complete energy balance of equation 12.3-16 must be used together with the material balance of equation 2.2-4. These constitute a set of two simultaneous, nonlinear, first-order ordinary differential equations with T and f_A as dependent variables and t as independent variable. The two boundary conditions are $T = T_o$ and $f_A = f_{Ao}$ (usually 0) at $t = 0$. These two equations usually must be solved by a numerical procedure. (See problem 12-9, which may be solved using the E-Z Solve software.)

12.3.4 Optimal Performance for Maximum Production Rate

The performance of a batch reactor may be optimized in various ways. Here, we consider the case of choosing the cycle time, t_c, equation 12.3-5, to maximize the rate of production of a product. For simplicity, we assume constant density and temperature.

The greater the reaction time t (equation 12.3-21), the greater the production per batch, but the smaller the number of batches per unit time. Since the rate of production, Pr, is the product of these two, a compromise must be made between large and small values of t to maximize Pr.

This may also be seen from limiting values obtained from equation 12.3-22:

$$Pr(C) = \frac{\nu_C c_{Ao} V(f_{A2} - f_{A1})}{t + t_d} \quad (12.3\text{-}22)$$

$\lim_{t \to 0} Pr = 0$, since $f_{A2} - f_{A1} \to 0$ and t_d is a finite constant

$\lim_{t \to \infty} Pr = 0$, since $f_{A2} - f_{A1} \to$ a constant

Between these two extremes, Pr must go through a maximum with respect to t.

308 Chapter 12: Batch Reactors (BR)

To obtain the value of t for maximum Pr, consider equation 12.3-22 with $f_{A1} = 0$, $f_{A2} = f_A(t)$. Then the equation may be written as

$$Pr = K f_A(t)/(t + t_d) \qquad (12.3\text{-}32)$$

where $K (= \nu_C c_{Ao} V)$ is a constant, and Pr is $Pr(C)$. For maximum Pr,

$$\frac{dPr}{dt} = K \left[\frac{(t + t_d)(df_A(t)/dt) - f_A(t)}{(t + t_d)^2} \right] = 0$$

or

$$(t + t_d)\frac{df_A(t)}{dt} - f_A(t) = 0 \qquad (12.3\text{-}33)$$

Equation 12.3-33 is solved for t, and the result is used in equation 12.3-32 or its equivalent to obtain the maximum value of Pr.

EXAMPLE 12-6

Consider a liquid-phase, first-order reaction A → C, occurring in a reactor of volume V, with a specified down-time, t_d. The reactor initially contains 5 moles of pure A. Determine the reaction time which maximizes $Pr(C)$, given $k_A = 0.021$ min^{-1}, and $t_d = 30$ min; and calculate the maximum value of $Pr(C)$.

SOLUTION

Since this is a constant-density system, equation 12.3-33 applies. To use this, we require $f_A(t)$. From the rate law, and the material balance, equation 2.2-10,

$$(-r_A) = k_A c_A = k_A c_{Ao}(1 - f_A) = -\frac{dc_A}{dt} = c_{Ao}\frac{df_A}{dt}$$

From this,

$$\frac{df_A}{dt} = k_A(1 - f_A)$$

This integrates to

$$f_A = 1 - e^{-k_A t} \qquad \text{(A)}$$

from which, in terms of t,

$$\frac{df_A}{dt} = k_A e^{-k_A t} \qquad \text{(B)}$$

Substituting the results from (A) and (B) in equation 12.3-33, we have

$$(t + t_d) k_A e^{-k_A t} - (1 - e^{-k_A t}) = 0 \qquad \text{(C)}$$

Equation (C) can be solved by trial, or by using an equation solver. For $t_d = 30$ min and $k_A = 0.021$ min^{-1}, we obtain $t = 45$ min using the E-Z Solve software (see file ex12-6.msp). Then, from equation (A), $f_A = 0.612$ for maximum $Pr(C)$, and from equation 12.3-22, with $n_{Ao} = 5$ mol, $Pr(C) = 2.45$ mol h^{-1}.

12.4 SEMIBATCH AND SEMICONTINUOUS REACTORS

A semibatch reactor is a variation of a batch reactor in which one reactant may be added intermittently or continuously to another contained as a batch in a vessel, or a product may be removed intermittently or continuously from the vessel as reaction proceeds. The reaction may be single-phase or multiphase. As in a batch reactor, the operation is inherently unsteady-state and usually characterized by a cycle of operation, although in a more complex manner.

A semicontinuous reactor is a reactor for a multiphase reaction in which one phase flows continuously through a vessel containing a batch of another phase. The operation is thus unsteady-state with respect to the batch phase, and may be steady-state or unsteady-state with respect to the flowing phase, as in a fixed-bed catalytic reactor (Chapter 21) or a fixed-bed gas-solid reactor (Chapter 22), respectively.

In this section, we consider various modes of operation of these types of reactors, their advantages and disadvantages, and some design aspects. Since there are many variations of these reactors, it is difficult to generalize their design or analysis, and consequently we use an example for illustration.

12.4.1 Modes of Operation: Semibatch and Semicontinuous Reactors

12.4.1.1 Semibatch Reactors

Figure 12.3 illustrates some modes of operation of semibatch reactors. In Figure 12.3(a), depicting a homogeneous liquid-phase reaction of the type A + B → products, reactant A is initially charged to the vessel, and reactant B is added at a prescribed rate, as reaction proceeds. In Figure 12.3(b), depicting a "liquid-phase" reaction in which a gaseous product is formed, gas is removed as reaction proceeds; an example is the removal of $H_2O(g)$ in an esterification reaction. In Figure 12.3(c), a combination of these two modes of operation is shown.

12.4.1.2 Semicontinuous Reactors

Figure 12.4 illustrates some modes of operation of semicontinuous reactors. In Figure 12.4(a), depicting a gas-liquid reaction of the type $A(g) + B(\ell) \rightarrow$ products, reactant A is dispersed (bubbled) continuously through a batch of reactant B; an important example is an aerobic fermentation in which air (or O_2) is supplied continuously to a liquid substrate (e.g., a batch of culture, as in penicillin production). In Figure 12.4(b),

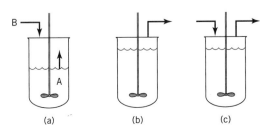

Figure 12.3 Some modes of operation of semibatch reactors: (a) addition of a reactant; (b) removal of a product; (c) simultaneous addition and removal

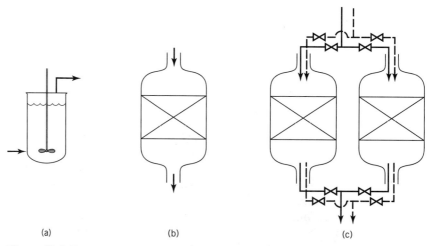

Figure 12.4 Some modes of operation of semicontinuous reactors: (a) gas-liquid reaction; (b) gas-solid (catalyst or reactant) reaction; (c) cyclic operation (reaction(—) and regeneration(- -)) for deactivating catalyst

depicting a fixed-bed catalytic reactor (Chapter 21) for a gas-solid (catalyst) reaction, the bed of catalyst is a batch phase, and the reacting gas phase flows continuously through the bed. A fluidized-bed reactor (Chapter 23) may be operated in a similar manner. A reactor for chemical vapor deposition (CVD) is also operated in this manner, although gas flows continuously over a solid substrate in the form of a surface, rather than through a bed of particles. Semicontinuous operation is appropriate for a catalyst that does not deactivate over time or deactivates slowly. If the catalyst deactivates rapidly, the reactor configuration shown in Figure 12.4(c) may be used. This consists of two identical fixed-bed reactors operating in parallel, with each alternating, in a prescribed cyclic manner, between reaction and regeneration. This is a rather complex mode of operation, requiring intricate piping and valving, and timing devices for overall continual operation. Wherever feasible, it should be replaced by inherently continuous operation for both reaction and regeneration, as in a fluidized-bed reactor.

12.4.2 Advantages and Disadvantages (Semibatch Reactor)

We focus mainly on the advantages and disadvantages of semibatch reactors. A semicontinuous reactor may be treated in many cases as either a batch reactor or a continuous reactor, depending on the overall kinetics and/or the phase of interest.

Advantages:

(1) In comparison with a batch reactor, the gradual or intermittent addition of a reactant (say, B in A + B → products, Figure 12.3(a)) in semibatch operation can result in improved control of T, particularly for a reaction with a large ΔH.

(2) Similarly, the concentration of a reactant can be kept relatively low (B in point (1)) or high (A in point (1)), if either is advantageous for suppressing undesired side reactions.

(3) The withdrawal of a product (continuously, as in Figure 12.3(b), or intermittently) can result in higher conversion of a reactant, particularly if the reaction is equilibrium-limited.

12.4 Semibatch and Semicontinuous Reactors

Disadvantages:

(1) As in the case of a batch reactor, the production rate may be limited because of the cyclic nature of the operation.
(2) Also, as in the case of a batch reactor, the operating cost may be relatively high.
(3) The design or performance analysis is complicated because of the unsteady-state operation.
(4) Semicontinuous operation shown in Figure 12.4(c) requires intricate piping and valving.

12.4.3 Design Aspects

The design or performance analysis of a semibatch or a semicontinuous reactor is complicated by the unsteady-state nature of its operation. Although, strictly speaking, this is characteristic of every such reactor, in some semicontinuous reactors, steady-state behavior is a valid approximation for the flowing phase. Examples are the gas phase in Figure 12.4(a), if the gas flow rate is sufficiently great that the composition of the gas remains nearly constant; and the gas phase in Figure 12.4(b), if the catalyst deactivates slowly or not at all. For a multiphase gas-liquid system, as in Figure 12.4(a), the overall rate, and hence the governing rate law, may be dictated by one of two extremes (Chapter 9): the rate of mass transfer of a reactant between phases, or the intrinsic rate of reaction in the liquid phase. In the latter case, the reactor may be treated as a batch reactor for the liquid phase. Similar considerations may apply to the solid substrate in a CVD reactor. In each case, the relevance of various terms in the material-balance equation needs to be taken into account, as discussed for equation 12.2-1.

The following example illustrates a combination of semibatch and semicontinuous operation for an irreversible reaction, with one reactant added intermittently and the other flowing (bubbling) continuously, that is, a combination of Figures 12.3(a) and 12.4(a). Chen (1983, pp. 168–211, 456–460) gives several examples of other situations, including reversible, series-reversible, and series-parallel reactions, and nonisothermal and autothermal operation.

EXAMPLE 12-7

Gluconic acid (P) may be produced by the oxidation of glucose (G) in a batch reactor. The reactor is provided with a continuous supply of oxygen and an intermittent supply of glucose at fixed intervals. The reaction is $C_6H_{12}O_6(G) + \frac{1}{2}O_2 \rightarrow C_6H_{12}O_7(P)$, and is catalyzed by the enzyme glucose oxidase. The overall rate is governed by the liquid-phase reaction rate. The rate law is given by equation 10.2-9:

$$r_P = \frac{V_{max} c_G}{K_m + c_G} \quad \text{(A)}$$

Since oxygen is a substrate, V_{max} and K_m are functions of the oxygen concentration c_{O_2}, which is fixed by p_{O_2} in the continuously flowing gas phase. At low values of c_G, the reaction is pseudo-first-order with respect to glucose, and (A) simplifies to

$$r_P = k_1 c_G \quad \text{(B)}$$

where, at 37°C, $k_1 = 0.016 \text{ h}^{-1} \simeq V_{max}/K_m$.

Suppose a 100-L reactor, operated isothermally, initially contains 0.050 moles of glucose, and 0.045 moles of fresh glucose is added to the reactor every 144 h for 30 d.

(a) Calculate values of c_G and c_P at the beginning and end of each interval between glucose additions. Assume that addition of (solid) glucose does not alter the volume of the liquid (reacting system).

(b) Plot c_G and c_P for the entire time period of 30 d.

SOLUTION

(a) The intermittent addition of glucose makes the operation semibatch (Figure 12.3(a)). Furthermore, the continuous supply of O_2 also makes the overall operation semicontinuous (Figure 12.4(a)), but it is the semibatch nature that governs, since the flow of O_2 mainly serves to provide a particular c_{O_2} (assumed constant).

The number of 144-h intervals in 30 d is $24(30)/144 = 5$. During each interval, the reaction follows pseudo-first-order kinetics. Thus, from equation (B) and the material balance, equation 2.2-10, applied to glucose,

$$-dc_G/dt = k_1 c_G \quad \text{(C)}$$

At the conclusion of each interval of 144 h, the glucose is replenished by addition of fresh glucose. For each interval, $i = 1, 2, \ldots, 5$, from integration of equation (C),

$$c_{Gi} = c_{Goi} \exp[-k_1(t - t_o)_i] \quad \text{(D)}$$
$$= c_{Goi} \exp[-0.016(144)] = 0.0998 c_{Goi} \quad (i = 1, 2, \ldots, 5) \quad \text{(D')}$$

where c_{Gi} is the concentration of glucose at the end of the interval of 144 h, and c_{Goi} is the initial concentration of glucose for the interval, given by

$$c_{Goi} = 0.05/100 = 5 \times 10^{-4} \text{mol L}^{-1} \quad (i = 1) \quad \text{(E)}$$

and

$$c_{Goi} = c_{G,i-1} + 0.045/100 = c_{G,i-1} + 4.5 \times 10^{-4} \quad (i = 2, 3, 4, 5) \quad \text{(E')}$$

From the reaction stoichiometry, the concentration of gluconic acid at the end of the ith interval is

$$c_{Pi} = c_{Poi} + (c_{Go} - c_G)_i$$
$$= c_{P,i-1} + (c_{Go} - c_G)_i \quad (i = 1, 2, \ldots, 5) \quad \text{(F)}$$

where c_{Poi} is the concentration at the beginning of the ith interval, which is the same as that at the end of the previous interval, $c_{P,i-1}$.

Equations (D') to (F) may be used recursively to calculate the quantities required. The results are summarized in Table 12.3.

Table 12.3 Results for Example 12-7

Interval	Initial conditions			Final conditions		
$i =$	$t_o/$ h	$10^4 c_{Go}/$ mol L^{-1}	$10^4 c_{Po}/$ mol L^{-1}	$t/$ h	$10^5 c_G/$ mol L^{-1}	$10^4 c_P/$ mol L^{-1}
1	0	5	0	144	4.99	4.50
2	144	4.999	4.50	288	4.99	9.00
3	288	4.999	9.00	432	4.99	13.50
4	432	4.999	13.50	576	4.99	18.00
5	576	4.999	18.00	720	4.99	22.50

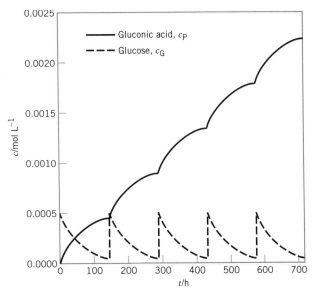

Figure 12.5 Example 12-7: c_G and c_P as functions of t

(b) In Figure 12.5, c_G and c_P are plotted against t for the whole time interval 0 to 720 h (30 d). For c_G the result is a "saw-tooth" behavior, with c_G decreasing exponentially during each interval (Equation (D)), and periodically being restored virtually to the initial value every 144 h. c_P increases continuously as t increases, but in a series of bowed curves, with cusps at points of addition of glucose; this is in accordance with equation (F), which, combined with (D), can be written for each interval as

$$c_{P_i} = c_{P,i-1} + c_{Goi}[1 - e^{-k_1(t-t_o)_i}] \quad (i = 1, 2, \ldots, 5) \quad \text{(G)}$$

The E-Z Solve software may also be used to solve Example 12-7 (see file ex12-7.msp). In this case, user-defined functions account for the addition of fresh glucose, so that a single differential equation may be solved to describe the concentration-time profiles over the entire 30-day period. This example file, with the user-defined functions, may be used as the basis for solution of a problem involving the nonlinear kinetics in equation (A), in place of the linear kinetics in (B) (see problem 12-17).

12.5 PROBLEMS FOR CHAPTER 12

12-1 At elevated temperatures, NO_2 (A) decomposes into NO and O_2. The reaction is second-order with respect to NO_2, and at 400°C the rate constant is $k_A = 5.36 \times 10^{-4}$ kPa^{-1} s^{-1}. If a sample of NO_2 is placed in a constant-volume batch reactor maintained at 673 K, and the initial pressure is 50 kPa, calculate the time required (in seconds) for the total pressure P to change by 10 kPa.

12-2 The decomposition of phosphine [$4PH_3(g) \rightarrow P_4(g) + 6H_2(g)$] is first-order with respect to phosphine with a rate constant $k = 0.0155$ s^{-1} at 953 K. If the decomposition occurs in a constant-volume batch reactor at 953 K, calculate, for 40% conversion of PH_3,
(a) the time required, s;
(b) the mole fraction of H_2 in the reaction mixture.

12-3 (a) For the gas-phase reaction

$$C_2H_4 + C_4H_6 \rightarrow C_6H_{10} \quad \text{or} \quad A + B \rightarrow C$$

carried out isothermally in a constant-volume batch reactor, what should the temperature (T/K) be to achieve 57.6% conversion of reactants, initially present in an equimolar ratio, in 4 min? The initial total pressure is 0.8 bar (only A and B present initially). The rate law (Example 4-8) is

$$(-r_A) = 3.0 \times 10^7 \exp(-115,000/RT) c_A c_B,$$

with the Arrhenius parameters, A and E_A, in L mol^{-1} s^{-1} and J mol^{-1}, respectively.

(b) Using the same temperature as in part (a), find the time required for 57.6% conversion of ethene, if the reaction is carried out isothermally and isobarically (i.e., in a variable-volume batch reactor).

12-4 A liquid-phase reaction A + B → products is conducted in an isothermal batch reactor. The reaction is first-order with respect to each reactant, with $k_A = 0.025$ L mol^{-1} s^{-1}, $c_{Ao} = 0.50$ mol L^{-1}, and $c_{Bo} = 1.0$ mol L^{-1}. Determine the time required for 75% conversion of A.

12-5 Ethyl acetate is to be produced from ethanol and acetic acid in a 10-m^3 batch reactor at 100°C:

$$C_2H_5OH(A) + CH_3COOH(B) \underset{k_r}{\overset{k_f}{\rightleftharpoons}} H_2O(C) + CH_3COOC_2H_5(D)$$

The reaction rate in the liquid phase is given by:

$$(-r_A) = k_f c_A c_B - k_r c_C c_D$$

The reaction has an equilibrium constant of 2.93, and the rate constant for the forward reaction (k_f) is 7.93×10^{-6} L mol^{-1} s^{-1} (Froment and Bischoff, 1990, p. 330). The feed contains 30 weight percent (wt%) acetic acid, 49 wt% ethanol, and the balance is water. The density of the fluid is essentially constant (1.0 g cm^{-3}).

(a) Determine the time required to achieve 25% conversion of the acetic acid.

(b) What is the maximum (i.e., equilibrium) conversion of acid that can be obtained?

12-6 A gas-phase reaction, A → 2B, is conducted at 300 K (constant) in a variable-volume reactor equipped with a piston to maintain a constant pressure of 150 kPa. Initially, 8 mol of A are present in the reactor. The reaction is second-order with respect to A, with the following rate expression:

$$(-r_A) = k c_A^2; \quad k = 0.075 \text{ L mol}^{-1} \text{ min}^{-1}$$

Determine the time of reaction, t, if f_A, the fractional conversion of A, is 0.80. State any assumption(s) made.

12-7 At 473 K, NOCl(g) decomposes into NO(g) and Cl$_2$(g). The reaction is second-order with respect to NOCl, with a rate constant of 0.078 L mol^{-1} s^{-1} (Boikess and Edelson, 1981, p. 614).

(a) If, initially, 0.25 mol NOCl is placed in a reactor, maintained at a constant pressure of 2 bar, determine the time required for 30% conversion of NOCl. The initial reactor volume is 7.5 L.

(b) What is the partial pressure of Cl$_2$(g) at the conclusion of the process?

12-8 A liquid-phase reaction A + B → C is conducted in a 50-L batch reactor. The reaction is first-order with respect to each reactant.

(a) Determine the time required for 90% conversion of A, if (i) the reaction occurs adiabatically; (ii) the reaction occurs isothermally at T_o.

(b) Determine $\dot{Q}$ and T_c (as functions of time), if a cooling coil is placed in the tank to maintain the isothermal conditions required in (a) part (ii).

(c) For (a) part (i), sketch the conversion-versus-time and temperature-versus-time profiles.

Data: c_{Ao} and c_{Bo} are 0.50 mol L^{-1} and 0.75 mol L^{-1}, respectively. The initial temperature (T_o) is 400 K, and the heat capacity of the reactor contents is 3.8 J g^{-1} K^{-1}. The fluid density is 0.75 g cm^{-3}, and the heat transfer parameter (UA_c) for part (b) is 100 W K^{-1}. The reaction is exothermic (-145 kJ (mol A)$^{-1}$), and $k_A = 1.4 \times 10^7 e^{-7700/T}$ L mol^{-1} min^{-1}.

12-9 Consider the hydrolysis of acetic anhydride carried out in dilute aqueous solution in a batch reactor:

$$(CH_3CO)_2O\ (A) + H_2O = 2CH_3COOH.$$

The initial concentration is $c_{Ao} = 0.03$ mol L^{-1}, and the initial temperature is $T_o = 15°C$. Calculate the time (t/min) required for 80% conversion of the anhydride:
(a) if the reaction is carried out isothermally at T_o;
(b) if the reaction is carried out adiabatically;
(c) if the reaction is carried out nonisothermally and nonadiabatically with $UA_c = 200$ W K^{-1}, $T_c = 300$ K, and $V = 100$ L.

Data: Arrhenius parameters are $A = 2.14 \times 10^7$ min^{-1} and $E_A = 46.5$ kJ mol^{-1}. The enthalpy of reaction is -209 kJ (mol A)$^{-1}$, the specific heat of reactor contents is 3.8 kJ kg^{-1} K^{-1}, and the density is 1.07 kg L^{-1}.

12-10 The Diels-Alder liquid-phase reaction between 1,4-benzoquinone (A, $C_6H_4O_2$) and cyclopentadiene (B, C_5H_6) to form the adduct $C_{11}H_{10}O_2$ is second-order with a rate constant $k_A = 9.92 \times 10^{-6}$ m^3 mol^{-1} s^{-1} at 25°C (Wassermann, 1936). Calculate the size (m^3) of a batch reactor required to produce adduct at the rate of 125 mol h^{-1}, if $c_{Ao} = c_{Bo} = 100$ mol m^{-3}, the reactants are 90% converted at the end of each batch (cycle), the reactor operates isothermally at 25°C, and the reactor down-time (for discharging, cleaning, charging) between batches is 30% of the total batch (cycle) time.

12-11 A second-order, liquid-phase reaction (A → products) is to take place in a batch reactor at constant temperature. The rate constant is $k_A = 0.05$ L mol^{-1} min^{-1}, and the initial concentration (c_{Ao}) is 2 mol L^{-1}. If the down-time (t_d) between batches is 20 min, what should the reaction time (t) be for each batch so that the rate of production, on a continual basis, is maximized?

12-12 (a) The liquid-phase reaction A → 2B + C takes place isothermally in a batch reactor of volume $V = 7500$ L. The batch-cycle time ($t_c = t + t_d$) is considered to be fixed by a shift period of 8 h. If the down-time (t_d) between batches is 45 min, calculate the rate of production of B (mol h^{-1}) on a continual basis, with 1 batch per shift, if the feed concentration (c_{Ao}) is 3.0 mol L^{-1}. The rate of reaction is given by $(-r_A) = k_A c_A^{1.5}$, where $k_A = 0.0025$ L$^{0.5}$ mol$^{-0.5}$ min^{-1}.

(b) You are asked to examine the possibility of increasing the rate of production of B by uncoupling t_c from the length of the shift period. If all else remains unchanged, what would you recommend? Your answer must include a quantitative assessment leading to a value of the production rate for a particular value of t_c (or t), and to a statement of any possible disadvantage(s) involved. Assume that the reactant A is relatively valuable.

12-13 For the liquid-phase reaction A → R, $k = 0.020$ min^{-1}. We wish to produce 4752 mol of R during each 10-hour production day, and 99% of A entering the batch reactor is to be converted. It takes 0.26 h to fill the reactor and heat the contents to the reaction temperature. It takes 0.90 h to empty the reactor and prepare it for the next batch. If the reactor initially contains pure A at a concentration of 8 mol L^{-1}, determine the volume of the reactor.

12-14 A liquid-phase reaction, A → products, was studied in a constant-volume isothermal batch reactor. The reaction rate expression is $(-r_A) = k_A c_A$, and $k_A = 0.030$ min^{-1}. The reaction time, t, may be varied, but the down-time, t_d, is fixed at 30 min for each cycle. If the reactor operates 24 hours per day, what is the ratio of reaction time to down-time that maximizes production for a given reactor volume and initial concentration of A? What is the fractional conversion of A at the optimum?

12-15 The aroma and flavor of wines may be enhanced by converting terpenylglycosides in the skin of grapes into volatile terpenols. This conversion can be accomplished using glycosidases (enzymes) which have β-glucosidase, α-arabinosidase, and α-rhamnosidase activities. Determine the time required to convert 80% of 4-nitrophenyl-β-glucopyranoside (NPG)

using β-glucosidase in a 10-L batch reactor. The reaction follows Michaelis-Menten kinetics (Chapter 10), with a maximum reaction velocity (V_{max}) of 36 mmol L^{-1} min^{-1}, and a Michaelis constant (K_m) of 0.60 mmol L^{-1} (Caldini et al., 1994). The initial concentration of NPG in the reactor is 4.5 mmol L^{-1}.

12-16 Sketch the graphical interpretation of equation 12.3-21 corresponding to that of equation 12.3-2 in Figure 12.1. What does the area under the curve represent?

12-17 The process in Example 12-7 was conducted by initially adding 0.35 mol glucose to the 100-L reactor, and replenishing the reaction with 0.28 mol fresh glucose every 144 h for 30 days. Determine the gluconic acid and glucose concentration-time profiles in the reactor over this period.

Data: $V_{max} = 5.8 \times 10^{-5}$ mol L^{-1} h^{-1}; $K_m = 3.6 \times 10^{-3}$ mol L^{-1}.

12-18 During the process described in problem 12-17, it was discovered that the enzyme used to produce gluconic acid was subject to deactivation, with a half-life of 12 days. It appears that the deactivation process is first-order, such that V_{max} decreases exponentially with time.

(a) What is the effect of inactivation on the production rate of gluconic acid over the 30-day process?

(b) How much of the original enzyme activity remains at the conclusion of the process?

Chapter 13

Ideal Flow

Ideal flow is introduced in Chapter 2 in connection with the investigation of kinetics in certain types of ideal reactor models, and in Chapter 11 in connection with chemical reactors as a contrast to nonideal flow. As its name implies, ideal flow is a model of flow which, in one of its various forms, may be closely approached, but is not actually achieved. In Chapter 2, three forms are described: backmix flow (BMF), plug flow (PF), and laminar flow (LF).

In this chapter, we focus on the characteristics of the ideal-flow models themselves, without regard to the type of process equipment in which they occur, whether a chemical reactor, a heat exchanger, a packed tower, or some other type. In the following five chapters, we consider the design and performance of reactors in which ideal flow occurs. In addition, in this chapter, we introduce the segregated-flow model for a reactor as one application of the flow characteristics developed.

In general, each form of ideal flow can be characterized exactly mathematically, as can the consequences of its occurrence in a chemical reactor (some of these are explored in Chapter 2). This is in contrast to nonideal flow, a feature which presents one of the major difficulties in assessing the design and performance of actual reactors, particularly in scale-up from small experimental reactors. This assessment, however, may be helped by statistical approaches, such as provided by residence-time distributions. It is these that we introduce in this chapter, since they can be derived exactly for ideal flow. For nonideal flow, residence-time distributions must be investigated experimentally, as described in Chapter 19.

The chapter begins with a reiteration and extension of terms used, and the types of ideal flow considered. It continues with the characterization of flow in general by age-distribution functions, of which residence-time distributions are one type, and with derivations of these distribution functions for the three types of ideal flow introduced in Chapter 2. It concludes with the development of the segregated-flow model for use in subsequent chapters.

13.1 TERMINOLOGY

Some of the terms used are introduced in Chapters 1 and 2, but are reviewed and extended here, and others added, as follows:

Element of fluid (Section 1.3): an amount of fluid small with respect to vessel size, but large with respect to molecular size, such that it can be characterized by values of (macroscopic) properties such as T, P, ρ, and c_i.

Age (of an element of fluid): length of time, from entry, that an element of fluid has been in a vessel at a particular instant.

Residence time (of an element of fluid)(Section 2.1): length of time that an element of fluid spends in a vessel (i.e., from entry to exit); a residence time is an age, but the converse is not necessarily true.

Residence-time distribution (RTD): relative times taken by different elements of fluid to flow through a vessel; a spread in residence times leads to a statistical treatment, in the form of a distribution; whether or not there is a spread in residence times has important implications for reactor performance.

13.2 TYPES OF IDEAL FLOW; CLOSED AND OPEN VESSELS

13.2.1 Backmix Flow (BMF)

Backmix flow (BMF) is the flow model for a CSTR, and is described in Section 2.3.1. BMF implies perfect mixing and, hence, uniform fluid properties throughout the vessel. It also implies a continuous *distribution* of residence times. The stepwise or discontinuous change in properties across the point of entry, and the continuity of property behavior across the exit are illustrated in Figure 2.3.

13.2.2 Plug Flow (PF)

Plug flow (PF, sometimes called piston flow) is the flow model for a PFR, and is a form of tubular flow (no mixing in direction of flow). As is described in Section 2.4.1, PF implies that there is no axial mixing in the vessel, but complete radial mixing (in a cylindrical vessel). The lack of axial mixing implies that all fluid elements have the same residence time (i.e., *no* distribution of residence times). Complete radial mixing implies that fluid properties, including velocity, are uniform across any plane perpendicular to the flow direction. The continuous change in properties in the axial direction is illustrated in Figure 2.4.

13.2.3 Laminar Flow (LF)

Laminar flow (LF) is also a form of tubular flow, and is the flow model for an LFR. It is described in Section 2.5. LF occurs at low Reynolds numbers, and is characterized by a lack of mixing in both axial *and* radial directions. As a consequence, fluid properties vary in both directions. There is a distribution of residence times, since the fluid velocity varies as a parabolic function of radial position.

13.2.4 Closed and Open Vessels

The type of flow occurring at the inlet and outlet of a vessel leads to the following definitions:

"*Closed*" *vessel*: fluid enters and leaves the vessel by PF, regardless of the type of flow *inside* the vessel;

"*Open*" *vessel*: the fluid may have any degree of backmixing at the inlet and outlet; that is, the flow at these points may be anywhere between the extremes of PF (no backmixing) and BMF (complete backmixing).

These definitions are illustrated in Figure 13.1. The flow conditions may be of one type at the inlet and the other at the outlet, leading to "closed-open" and "open-closed" vessels.

13.3 CHARACTERIZATION OF FLOW BY AGE-DISTRIBUTION FUNCTIONS

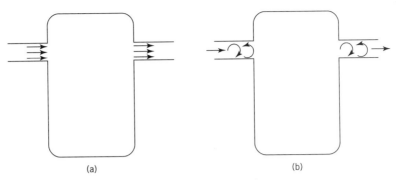

Figure 13.1 (a) "Closed" vessel; (b) "open" vessel

The characterization of flow by statistical age-distribution functions applies whether the flow is ideal or nonideal. Thus, the discussion in this section applies both in Section 13.4 below for ideal flow, and in Chapter 19 for nonideal flow.

Several age-distribution functions may be used (Danckwerts, 1953), but they are all interrelated. Some are residence-time distributions and some are not. In the discussion to follow in this section and in Section 13.4, we assume steady-flow of a Newtonian, single-phase fluid of constant density through a vessel without chemical reaction. Ultimately, we are interested in the effect of a spread of residence times on the performance of a chemical reactor, but we concentrate on the characterization of flow here.

13.3.1 Exit-Age Distribution Function E

The exit-age distribution function E is a measure of the distribution of the ages of fluid elements leaving a vessel, and hence is an RTD function. As a function of time, t, it is defined as:

> $E(t)dt$ is the fraction of the exit stream of age between t and $t + dt$

$E(t)$ itself is a distribution frequency and has dimensions of time^{-1}. For arbitrary flow, it may have the appearance shown in Figure 13.2.
A consequence of the definition is:

$$\int_0^\infty E(t)dt = 1 \qquad (13.3\text{-}1)$$

That is, the total area under the $E(t)$ curve in Figure 13.2 is unity, since all fluid elements in the exit stream are in the vessel between times of 0 and ∞. Furthermore,

$$\int_0^{t_1} E(t)dt = \text{fraction of exit stream of age} \leq t_1 \text{ (shaded area in Figure 13.2)} \qquad (13.3\text{-}2)$$

$$\int_{t_1}^\infty E(t)dt = \text{fraction of exit stream of age} \geq t_1 \text{ (unshaded area in Figure 13.2)} \qquad (13.3\text{-}3)$$

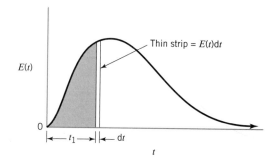

Figure 13.2 Exit-age distribution function $E(t)$ for arbitrary (nonideal) flow showing significance of area under the $E(t)$ curve

$$\int_{t_1}^{t_2} E(t)dt = \text{fraction of exit stream for which } t_1 \leq \text{age} \leq t_2 \quad (13.3\text{-}4)$$

The exit-age distribution function may also be expressed in terms of dimensionless time θ defined by

$$\theta = t/\bar{t} \quad (13.3\text{-}5)$$

where $\bar{t}$ is the mean residence time, given by equation 2.3-1 for steady-state flow. Then $E(\theta)$ is the exit-age distribution function in terms of θ, and is itself dimensionless. Its definition is similar to that of $E(t)$:

$E(\theta)d\theta$ is the fraction of the exit stream of age between θ and $\theta + d\theta$

Consequences of this definition are analogous to those from equations 13.3-1 to -4.

EXAMPLE 13-1

(a) What is the relation between $E(t)$ and $E(\theta)$?
(b) For a system with a mean-residence time ($\bar{t}$) of 20 min, calculate $E(\theta)$ at $t = 7$ min, if $E(t) = 0.25$ min^{-1} at $t = 7$ min.

SOLUTION

(a) Since the definitions of $E(\theta)d\theta$ and $E(t)dt$ refer to the same fraction in the exit stream,

$$E(\theta)d\theta = E(t)dt \quad (13.3\text{-}6)$$

Furthermore, for the steady flow of a constant-density fluid, $\bar{t}$ is constant, and, from equation (13.3-5), $d\theta = dt/\bar{t}$. Thus, on combining equations 13.3-5 and -6, we obtain

$$E(\theta) = \bar{t}E(t) \quad (13.3\text{-}7)$$

(b) $\theta = t/\bar{t} = 7/20 = 0.35$
$E(\theta) = E(0.35) = \bar{t}E(t) = 20(0.25) = 5.0$

13.3 Characterization of Flow by Age-Distribution Functions 321

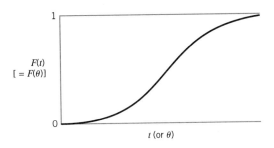

Figure 13.3 Typical form of F for arbitrary (nonideal) flow

13.3.2 Cumulative Residence-Time Distribution Function F

The cumulative residence-time distribution function $F(t)$ is defined as the fraction of exit stream that is of age 0 to t (i.e., of age $\leq t$); it is also the probability that a fluid element that entered at $t = 0$ has left at or by time t. Since it is defined as a fraction, it is dimensionless. Furthermore, since $F(0) = 0$, that is, no fluid (of age 0) leaves the vessel before time 0; and $F(\infty) = 1$, that is, all fluid leaving the vessel is of age 0 to ∞, or all fluid entering at time 0 has left by time ∞, then

$$0 \leq F(t) \leq 1 \qquad (13.3\text{-}8)$$

In terms of θ, $F(\theta)$ has a similar definition, and since $F(t)$ and $F(\theta)$ refer to the same fraction,

$$F(t) = F(\theta) \qquad (13.3\text{-}9)$$

Figure 13.3 illustrates a typical form of $F(t)$ or $F(\theta)$ for arbitrary (nonideal) flow.

EXAMPLE 13-2

What is the relation between $F(t)$ and $E(t)$?

SOLUTION

From one point of view, the definition of $F(t)$ corresponds to the probability that a fluid element has left the vessel between time 0 and t. This is equivalent to the area under the $E(t)$ curve given by equation 13.3-2. That is,

$$F(t) = \int_0^t E(t)dt \qquad (13.3\text{-}10)$$

Thus, $F(t)$ is the shaded area generated from left to right by $E(t)$ in Figure 13.2.
  From another point of view,

$F(t)$ = fraction of exit stream of age t or less
$F(t) + dF(t)$ = fraction of age $t + dt$ or less

$F(t) + dF(t) - F(t) = dF(t)$ = fraction of age between t and $t + dt$ which by definition is $E(t)dt$; that is,

$$E(t) = dF(t)/dt \qquad (13.3\text{-}11)$$

Expressions analogous to equations 13.3-10 and -11 relate $F(\theta)$ and $E(\theta)$.

Both E and F are RTD functions. Although not apparent here, each corresponds to a different way of experimentally investigating RTD for arbitrary flow (Chapter 19; see also problem 13-1), and hence each is important.

13.3.3 Washout Residence-Time Distribution Function W

The washout residence-time distribution function $W(t)$ is defined as the fraction of the exit stream of age $\geq t$ (and similarly for $W(\theta)$). It is also the probability that an element of fluid that entered a vessel at $t = 0$ has *not* left at time t. By comparison, $F(t)$ (or $F(\theta)$) is the probability that a fluid element *has* left by time t (or θ) (Section 13.3.2.)

The washout function also provides the basis for a method of experimentally measuring RTD (Chapter 19; see also problem 13-1). Consequences of the definition of W and its relation to F and E are explored in problem 13-5.

13.3.4 Internal-Age Distribution Function I

The internal-age distribution function $I(t)$ is a measure of the distribution of ages of elements of fluid *within* a vessel, and not in the exit stream. However, it is defined similarly to $E(t)$ by:

$I(t)dt$ is the fraction of fluid within a vessel of age between t and $t + dt$

Like $E(t)$, $I(t)$ is a distribution frequency, and has dimensions of time^{-1}. Unlike $E(t)$, it is not an RTD function. The dimensionless function in terms of θ, $I(\theta)$, has a corresponding definition (to $I(t)$). The relation of $I(t)$ to $I(\theta)$ is similar to that given in equation 13.3-7, which relates $E(t)$ to $E(\theta)$. The proof of this and the relation of $I(t)$ to $E(t)$ and $F(t)$ are left to problem 13-3.

13.3.5 Holdback H

The holdback H is the fraction of fluid within a vessel of age greater than $\bar{t}$, the mean residence time. As a fraction, it is dimensionless. It can be obtained from age-distribution functions (see problem 13-4).

13.3.6 Summary of Relationships Among Age-Distribution Functions

A summary of the relationships among the age-distribution functions is given in Table 13.1. This includes the results relating $E(\theta)$ and $F(\theta)$ from Section 13.3.2, together with those for $W(\theta)$, $I(\theta)$, and H (these last provide answers to problems 13-3, -4 and -5(a), (b)). Each row in Table 13.1 relates the function shown in the first column to the others. The means of converting to results in terms of $E(t)$, $F(t)$, etc. is shown in the first footnote to the table.

13.3 Characterization of Flow by Age-Distribution Functions

Table 13.1 Summary of relationships among age-distribution functions[a,b]

Function ↓ in terms of function →	$E(\theta)$	$F(\theta)$	$W(\theta)$	$I(\theta)$
$E(\theta)$	—	$dF(\theta)/d\theta$	$-dW(\theta)/d\theta$	$-dI(\theta)/d\theta$
$F(\theta)$	$\int_0^\theta E(\theta)d\theta$	—	$1 - W(\theta)$	$1 - I(\theta)$
$W(\theta)$	$1 - \int_0^\theta E(\theta)d\theta$	$1 - F(\theta)$	—	$I(\theta)$
$I(\theta)$	$1 - \int_0^\theta E(\theta)d\theta$	$1 - F(\theta)$	$W(\theta)$	—
H	c	$\int_0^1 F(\theta)d\theta$	c	c

[a] In terms of dimensionless time $\theta = t/\bar{t}$; to convert to $E(t)$, $F(t)$, etc., use $dt = \bar{t}d\theta$; $E(t) = (1/\bar{t})E(\theta)$; $F(t) = F(\theta)$; $W(t) = W(\theta)$; $I(t) = (1/\bar{t})I(\theta)$.
[b] E, F, and W are RTD functions; I and H are not RTD functions.
[c] Follows from entry for $F(\theta)$ in this row and entries in second row.

13.3.7 Moments of Distribution Functions

Distribution functions are shown graphically in Figure 6.8 for molecular velocity and kinetic energy, and schematically in Figures 13.2 and 13.3 for E and F, respectively. Such distributions can be characterized by their *moments* (see, e.g., Kirkpatrick, 1974, pp. 39–41). (For comparison, consider moments of bodies in physics: the first moment is the center of gravity, and the second moment is the moment of inertia.) A moment of a distribution may be taken about the origin or about the mean of the distribution. We consider each of these types of moments in turn, designating moments about the origin by μ, and those about the mean by M. In this chapter, we deal only with continuous distributions. Experimental data may yield discrete distributions, and the analogous treatment and results for these are discussed in Chapter 19.

13.3.7.1 Moments About the Origin

For a continuous distribution function $f(t)$, the k th moment about the origin is *defined* by:

$$\mu_k = \int_0^\infty t^k f(t) dt \, ; \, t \geq 0 \tag{13.3-12}$$

The function $f(t)$ could be $E(t)$.

The zeroth moment ($k = 0$) is simply the area under the distribution curve:

$$\mu_0 = \int_0^\infty f(t) dt = 1 \tag{13.3-13}$$

The first moment ($k = 1$) is the mean of the distribution, $\bar{t}$, a measure of the location of the distribution, or the expected (average) value $\langle t \rangle$ of the distribution $f(t)$:

$$\mu_1 = \int_0^\infty t f(t) dt = \bar{t} \tag{13.3-14}$$

Higher moments are usually taken about the mean to characterize the shape of the distribution about this centroid.

13.3.7.2 Moments about the Mean

The kth moment about the mean is *defined* by:

$$M_k = \int_0^\infty (t - \bar{t})^k f(t) dt \,;\, t \geq 0 \qquad (13.3\text{-}15)$$

The second moment ($k = 2$) about the mean is called the variance, σ_t^2, and is a measure of the spread of the distribution:

$$M_2 = \int_0^\infty (t - \bar{t})^2 f(t) dt = \sigma_t^2 \qquad (13.3\text{-}16)$$

An alternative form that may be more convenient to use is obtained by expanding the square in equation 13.3-16 and using the results of equations 13.3-13 and -14:

$$\sigma_t^2 = \left[\int_0^\infty t^2 f(t) dt \right] - \bar{t}^2 \qquad (13.3\text{-}16a)$$

The square-root of the variance is the standard deviation σ.

The third moment M_3 is a measure of the skewness or symmetry of the distribution, and the fourth moment M_4 a measure of "peakedness," but we use only μ_0, μ_1, and M_2 here.

As indicated above, age distributions may also be expressed in terms of dimensionless time, $\theta = t/\bar{t}$. The definitions of moments in terms of θ are analogous to those in equations 13.3-12 to -16(a), with θ replacing t, leading to expressions for the mean $\bar{\theta}$ and the variance σ_θ^2, both of which are dimensionless.

EXAMPLE 13-3

For $E(\theta)$, how are $\bar{\theta}$ and σ_θ^2 related to $\bar{t}$ and σ_t^2, respectively, for $E(t)$?

SOLUTION

By analogy with equation 13.3-14,

$$\bar{\theta} = \int_0^\infty \theta E(\theta) d\theta = \int_0^\infty (t/\bar{t}) \bar{t} E(t) dt/\bar{t} = (1/\bar{t}) \int_0^\infty t E(t) dt = \bar{t}/\bar{t} = 1 \quad (13.3\text{-}17)$$

By analogy with equation 13.3-16a,

$$\sigma_\theta^2 = \left[\int_0^\infty \theta^2 E(\theta) d\theta \right] - 1$$

$$= \left[\int_0^\infty (t^2/\bar{t}^2) \bar{t} E(t) dt/\bar{t} \right] - (\bar{t}^2/\bar{t}^2)$$

$$= (1/\bar{t}^2) \left\{ \left[\int_0^\infty t^2 E(t) dt \right] - \bar{t}^2 \right\}$$

That is,

$$\sigma_\theta^2 = \sigma_t^2/\bar{t}^2 \qquad (13.3\text{-}18)$$

13.4 AGE-DISTRIBUTION FUNCTIONS FOR IDEAL FLOW

In this section, we derive expressions for the age-distribution functions E and F for three types of ideal flow: BMF, PF, and LF, in that order. Expressions for the quantities W, I, and H are left to problems at the end of this chapter. The results are collected in Table 13.2 (Section 13.4.4).

13.4.1 Backmix Flow (BMF)

13.4.1.1 E for BMF

Consider the steady flow of fluid at a volumetric rate q through a stirred tank as a "closed" vessel, containing a volume V of fluid, as illustrated in Figure 13.4. We assume the flow is ideal in the form of BMF at constant density, and that no chemical reaction occurs. We wish to derive an expression for $E(t)$ describing the residence-time distribution (RTD) for this situation.

Consider a small volume of fluid q'_o entering the vessel virtually instantaneously over the time interval dt at a particular time ($t = 0$). Thus $q'_o = qdt$, such that $q'_o \ll V$ and $dt \ll \bar{t}$. We note that only the small amount q'_o enters at $t = 0$. This means that at any subsequent time t, in the exit stream, only fluid that originates from q'_o is of age t to $t + dt$; all other elements of fluid leaving the vessel in this interval are either "older" or "younger" than this. In an actual experiment to measure $E(t)$, q'_o could be a small pulse of "tracer" material, distinguishable in some manner from the main fluid. In any case, for convenience, we refer to q'_o as "tracer," and to obtain $E(t)$, we keep track of tracer by a material balance as it leaves the vessel. Note that the process is unsteady-state with respect to q'_o (which enters only once), even though the flow at rate q (which is maintained) is in steady state.

If we let q' be the amount of tracer left in the vessel at time t, then the concentration (volume fraction) of tracer in the vessel is q'/V. A material balance for the tracer around the vessel at time t is

$$\text{rate of input} = \text{rate of output} + \text{rate of accumulation}$$

That is,

$$0 = q(q'/V) + dq'/dt$$

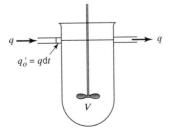

Figure 13.4 Basis for deriving $E(t)$ for BMF

or

$$dq'/q' = -(q/V)dt = -dt/\bar{t}$$

Integration with the boundary condition that $q' = q'_o$ at $t = 0$ results in

$$q' = q'_o e^{-t/\bar{t}} \tag{13.4-1}$$

This shows how the amount of tracer material in the vessel, q', that originated from q'_o, varies with time t.

To obtain $E(t)$, consider fluid leaving the vessel (exit stream) between t and $t + dt$. The total volume of fluid leaving in the interval dt is $q\,dt$ and the amount of tracer is $(q'/V)q\,dt$, since q'/V is the concentration of tracer in the vessel. Thus, the fraction that is tracer is

$$\frac{\text{tracer volume}}{\text{fluid volume}} = \frac{(q'/V)q\,dt}{q\,dt} = \frac{q'}{V} = \frac{q'_o e^{-t/\bar{t}}}{V} = \frac{q}{V}e^{-t/\bar{t}}dt = \frac{1}{\bar{t}}e^{-t/\bar{t}}dt$$

This is also the fraction that is of age between t and $t + dt$, since only tracer entered between time 0 and dt. But this fraction is also $E(t)dt$, from the definition of $E(t)$. Thus,

$$E(t) = (1/\bar{t})e^{-t/\bar{t}} \quad \text{(BMF)} \tag{13.4-2}$$

Equation 13.4-2 is the desired expression for $E(t)$ for BMF. In terms of dimensionless time θ, using $\theta = t/\bar{t}$ and $E(\theta) = \bar{t}E(t)$, we rewrite equation 13.4-2 as

$$E(\theta) = e^{-\theta} \quad \text{(BMF)} \tag{13.4-3}$$

From equation 13.4-3, $E(0) = 1$, and $E(\infty) = 0$. The exit-age RTD distribution function for BMF is shown as $E(\theta)$ in Figure 13.5. Note that the mean of the distribution is at $\bar{\theta} = 1$. The use of dimensionless time θ has advantages over the use of t, since, for BMF, $\bar{\theta} = 1$, and $E(\theta) = 1$ at $\theta = 0$.

13.4.1.2 F for BMF

The cumulative RTD function F may be obtained by combining equations 13.3-9, 13.3-10 and 13.4-2 to give:

$$F(\theta) = F(t) = \int_0^t E(t)dt = \int_0^t (1/\bar{t})e^{-t/\bar{t}}dt$$

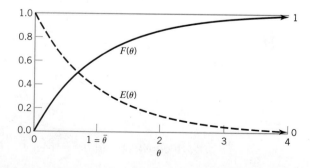

Figure 13.5 RTD functions $E(\theta)$ and $F(\theta)$ for BMF

From this,

$$F(\theta) = 1 - e^{-\theta} \quad \text{(BMF)} \quad (13.4\text{-}4)$$

Thus, for BMF,

$$F(\theta) + E(\theta) = 1 \quad \text{(BMF)} \quad (13.4\text{-}5)$$

The cumulative RTD function for BMF is shown as $F(\theta)$ in Figure 13.5.

EXAMPLE 13-4

(a) For a CSTR of volume 6 m^3 operating with a steady-state (liquid) feed rate of 0.4 m^3 min^{-1}, what fraction of the exit stream
  (i) is of age less than 10 min?
  (ii) has been in the tank longer than 30 min?
(b) For a CSTR operating as in (a), suppose a small pulse of a tracer material is added to the feed at a particular time ($t = 0$). How long (min) does it take for 80% of the tracer material to leave the tank?

SOLUTION

(a) $\bar{t} = V/q = 6/0.4 = 15$ min
  (i) $\theta = t/\bar{t} = 10/15 = 0.667$
      The required fraction is given by $F(\theta)$. From equation 13.4-4, for BMF (in a CSTR),

$$F(\theta) = 1 - e^{-0.667} = 0.487$$

  (ii) $\theta = 30/15 = 2$
      The required fraction is $1 - F(\theta)$, corresponding to the fluid of age *greater* than 30 min.

$$1 - F(\theta) = 1 - F(2) = 1 - (1 - e^{-2}) = 0.1353$$

(b) 20% of the tracer material remains in the tank. From the material balance resulting in equation 13.4-1,

$$0.20 = q'/q'_o = e^{-\theta} (= E(\theta) \text{ for BMF})$$
$$\theta = -\ln 0.2 = 1.609$$
$$t = \bar{t}\theta = 15(1.609) = 24.1 \text{ min}$$

13.4.2 Plug Flow (PF)

13.4.2.1 E for PF

Consider the steady flow of a constant-density fluid in PF at a volumetric rate q through a cylindrical vessel of constant cross-section and of volume V. We wish to obtain an expression for $E(t)$ for this situation.

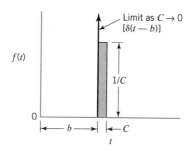

Figure 13.6 Motivation for Dirac delta function, $\delta(t-b)$

Consider an element of fluid (as "tracer") entering the vessel at $t = 0$. Visualizing what happens to the element of fluid is relatively simple, but describing it quantitatively as $E(t)$ requires an unusual mathematical expression. The element of fluid moves through the vessel without mixing with fluid ahead of or behind it, and leaves the vessel all at once at a time equal to the mean residence time ($\bar{t} = V/q$ for constant density). Thus, $E(t) = 0$ for $0 < t < \bar{t}$, but what is $E(t)$ at $t = \bar{t}$?

To answer this question, away from the context of PF, consider a characteristic function $f(t)$ that, at $t = b$, is suddenly increased from 0 to $1/C$, where C is a relatively small, but nonzero, interval of time, and is then suddenly reduced to 0 at $t = b + C$, as illustrated in Figure 13.6. The shaded area of $C(1/C)$ represents a unit amount of a pulse disturbance of a constant value $(1/C)$ for a short period of time (C). As $C \to 0$ for unit pulse, the height of the pulse increases, and its width decreases. The limit of this behavior is indicated by the vertical line with an arrow (meaning "goes to infinity") and defines a mathematical expression for an instantaneous ($C \to 0$) unit pulse, called the Dirac delta function (or unit impulse function):

$$\delta(t - b) = \lim_{C \to 0}[f(t)] = \infty; t = b$$
$$= 0; t \neq b \quad (13.4\text{-}6)$$

such that

$$\int_{-\infty}^{\infty} \delta(t - b)\,dt = 1 \quad (13.4\text{-}7)$$

For $E(t)$, $b = \bar{t}$, and hence $E(t)$ for PF is represented mathematically by

$$E(t) = \delta(t - \bar{t}) = \infty; t = \bar{t}$$
$$= 0; \; t \neq \bar{t} \quad (PF) \quad (13.4\text{-}8)$$

Graphically, $E(t)$ for PF is represented by a vertical line with an arrow at $t = \bar{t}$, as in Figure 13.6. From the point of view of $E(t)$, the unit area property of equation 13.4-7 is expressed more appropriately as

$$\int_{t_1}^{t_2} \delta(t - \bar{t})\,dt = 1, \text{ if } t_1 \leq \bar{t} \leq t_2$$
$$= 0, \text{ otherwise} \quad (13.4\text{-}9)$$

Since $0 \leq \bar{t} < \infty$, the area under the vertical line representation for PF is unity, as required by equation 13.3-1.

Another important property of the delta function is, for any function $g(t)$:

$$\int_{t_1}^{t_2} g(t)\delta(t-b)dt = g(b), \text{ if } t_1 \leq b \leq t_2$$

$$= 0, \text{ otherwise} \qquad (13.4\text{-}10)$$

Wylie (1960, pp. 341–2) presents a proof of equation 13.4-10, based on approximating $\delta(t-\bar{t})$ by $1/C$ (Figure 13.6), and using the law of the mean for integrals, but it is omitted here.

In terms of dimensionless time, θ, for PF

$$E(\theta) = \delta(\theta - 1) = \infty; \ \theta = \bar{\theta} = 1$$
$$= 0; \ \theta \neq 1 \qquad \text{(PF)} \qquad (13.4\text{-}11)$$

EXAMPLE 13-5

For $E(\theta)$ in PF, what are the mean value $\bar{\theta}$ and the variance σ_θ^2?

SOLUTION

We use the property of the delta function contained in equation 13.4-10, and the definitions of $\bar{\theta}$ and σ_θ^2 in the θ analogues of equations 13.3-14 and 13.3-16a, respectively, to obtain

$$\bar{\theta} = \int_0^\infty \theta \delta(\theta - 1) d\theta = 1$$

$$\sigma_\theta^2 = \left[\int_0^\infty \theta^2 \delta(\theta - 1) d\theta\right] - \bar{\theta}^2 = 1^2 - 1^2 = 0$$

The second result confirms our intuitive realization that there is no spread in residence time for PF in a vessel.

13.4.2.2 F for PF

For PF, the F function requires another type of special mathematical representation. For this, however, consider a sudden change in a property of the fluid flowing that is maintained (and not pulsed) (e.g., a sudden change from pure water to a salt solution). If the change occurs at the inlet at $t = 0$, it is not observed at the outlet until $t = \bar{t}$. For the exit stream, $F(t) = 0$ from $t = 0$ to $t = \bar{t}$, since the fraction of the exit stream of age less than t is 0 for $t < \bar{t}$; in other words, the exit stream is pure water. For $t > \bar{t}$, $F(t) = 1$, since all the exit stream (composed of the salt solution) is of age less than t. This behavior is represented by the unit step function $S(t - b)$ (sometimes called the Heaviside unit function), and is illustrated in Figure 13.7, in which the arbitrary constant $b = \bar{t}$. With this change, the unit step function is

330 Chapter 13: Ideal Flow

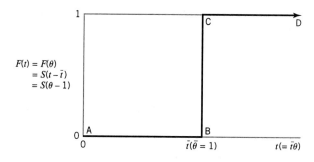

Figure 13.7 $F(t)$ for PF represented by the unit step function $S(t - \bar{t})$ (ABCD)

defined by

$$F(t) = S(t - \bar{t}) = 0; \quad t < \bar{t}$$
$$= 1; \quad t > \bar{t} \quad \text{(PF)} \quad (13.4\text{-}12)$$

Alternatively, in terms of θ,

$$F(t) = F(\theta) = S(\theta - 1) = 0; \quad \theta < 1$$
$$= 1; \quad \theta > 1 \quad \text{(PF)} \quad (13.4\text{-}13)$$

We note that the δ and S functions are related by

$$\delta(t - \bar{t}) = dS(t - \bar{t})/dt \quad (13.4\text{-}14)$$

consistent with $E(t) = dF(t)/dt$, equation 13.3-11.

13.4.3 Laminar Flow (LF)

13.4.3.1 E for LF

We consider steady-state, one-dimensional laminar flow (q) through a cylindrical vessel of constant cross-section, with no axial or radial diffusion, and no entry-length effect, as illustrated in the central portion of Figure 2.5. The length of the vessel is L and its radius is R. The parabolic velocity profile $u(r)$ is given by equation 2.5-1, and the mean velocity $\bar{u}$ by equation 2.5-2:

$$u(r) = u_o[1 - (r/R)^2] \quad (2.5\text{-}1)$$
$$\bar{u} = u_o/2 \quad (2.5\text{-}2)$$

We wish to obtain $E(t)$ for this situation.

Consider an element of fluid ("tracer") entering the vessel at $t = 0$ with the parabolic velocity profile fully established. The portion at the center travels fastest, and has a residence time $t_o = L/u_o = L/2\bar{u} = \bar{t}/2$, since $\bar{u} = u_o/2$. That is, no portion of the "tracer" entering at $t = 0$ leaves until $t = \bar{t}/2$. As a result, we conclude that

$$E(t) = 0; \quad t < \bar{t}/2 \quad \text{(LF)} \quad (13.4\text{-}15)$$

13.4 Age-Distribution Functions for Ideal Flow

The situation for $t > \bar{t}/2$ is more complex. Each thin annulus of "tracer" of radius r to $r + dr$ travels at velocity u to $u + du$ and has a residence time that is of age t to $t + dt$, where $t = L/u$; its volumetric flow is $dq = (2\pi r dr)u$, compared with the total flow rate of $q = \pi R^2 \bar{u}$. Thus, the fraction of fluid leaving between t and $t + dt$ is

$$E(t)dt = \frac{dq}{q} = \frac{(2\pi r dr)u}{\pi R^2 \bar{u}} = \frac{(2r dr)u}{R^2 \bar{u}} \quad (13.4\text{-}16)$$

We wish to have the right side of equation 13.4-16 in terms of t rather than r and u. To achieve this, we use equations 2.5-3 to -5. From equations 2.5-3 and -4

$$u/\bar{u} = \bar{t}/t \quad (13.4\text{-}17)$$

From equation 2.5-5, $\bar{t} = 2t(r)[1 - (r/R)^2]$, by differentiation,

$$2r dr/R^2 = \bar{t} dt/2t^2 \quad (13.4\text{-}18)$$

where $t \equiv t(r)$. Substituting equations 13.4-17 and -18 into equation 13.4-16, we obtain

$$E(t) = \bar{t}^2/2t^3; \quad t > \bar{t}/2 \quad (\text{LF}) \quad (13.4\text{-}19)$$

Thus, for LF, $E(t)$ is divided into two regions on either side of $\bar{t}/2$.
In terms of dimensionless time $\theta = t/\bar{t}$, with $E(\theta) = \bar{t}E(t)$,

$$E(\theta) = 0; \quad \theta < 1/2 \quad (13.4\text{-}20)$$
$$E(\theta) = 1/2\theta^3; \quad \theta > 1/2 \quad (\text{LF}) \quad (13.4\text{-}21)$$

The behavior of $E(\theta)$ for LF is shown in Figure 13.8.

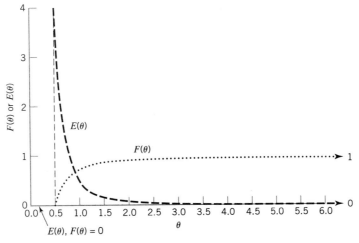

Figure 13.8 $E(\theta)$ and $F(\theta)$ for LF

Table 13.2 Summary of age-distribution functions for ideal flow[a]

Function	Type of ideal flow		
	BMF	PF	LF
$E(\theta)$	$e^{-\theta}$	$\delta(\theta - 1) = \infty; \theta = 1$ $= 0; \theta \neq 1$	$0; \theta < 1/2$ $1/2\theta^3; \theta > 1/2$
$F(\theta)$	$1 - e^{-\theta}$	$S(\theta - 1) = 0; \theta < 1$ $= 1; \theta > 1$	$0; \theta < 1/2$ $1 - (1/4\theta^2); \theta > 1/2$
$W(\theta)$	$e^{-\theta}$	$1 - S(\theta - 1) = 1; \theta < 1$ $= 0; \theta > 1$	$1; \theta < 1/2$ $1/4\theta^2; \theta > 1/2$
$I(\theta)$	$e^{-\theta}$	$1; \theta < 1$ $0; \theta > 1$	$1; \theta < 1/2$ $1/4\theta^2; \theta > 1/2$
H	e^{-1}	0	1/4

[a]In terms of dimensionless time $\theta = t/\bar{t}$; to convert to $E(t), F(t)$, etc., use $E(t) = (1/\bar{t})E(\theta); F(t) = F(\theta); W(t) = W(\theta); I(t) = (1/\bar{t})I(\theta)$.

13.4.3.2 F for LF

Using $F(\theta) = \int_0^\theta E(\theta)d(\theta)$ (Table 13.1), we can determine $F(\theta)$ for LF from equations 13.4-20 and -21:

$$F(\theta) = 0; \qquad \theta < 1/2 \qquad (13.4\text{-}22)$$
$$F(\theta) = 1 - 1/4\theta^2; \qquad \theta > 1/2 \text{ (LF)} \qquad (13.4\text{-}23)$$

The behavior of $F(\theta)$ is also shown in Figure 13.8.

13.4.4 Summary of Results for Ideal Flow

The expressions obtained above for $E(\theta)$ and $F(\theta)$ for BMF, PF, and LF are summarized in Table 13.2, together with corresponding results for $W(\theta)$, $I(\theta)$, and H (these last provide answers to problems 13-5(c), 13-6(a), and 13-7). The entries in Table 13.2 may be converted to $E(t), F(t)$, etc., as shown in the footnote to the table.

13.5 SEGREGATED FLOW

Segregated flow is an idealization (for a single-phase fluid) that can be used as a reference for both ideal and nonideal flow. As the name suggests, segregated flow implies that *no* mixing occurs between portions of fluid of different ages as they flow in an arbitrary pattern through a vessel. Since the fluid in a chemical reactor changes composition with age, inhomogeneities may develop in the fluid if mixing between the portions is incomplete.

We use the term "degree of segregation" to refer to the level of mixing of fluid of different ages. There are two extremes: complete segregation and complete dispersion (nonsegregation). In the former, or segregated flow, extreme, mixing occurs only at a macroscopic level, as though the fluid were contained in separate packets as closed systems; the size of the individual packets is not important. In segregated flow, mixing does *not* occur at the molecular or microscopic level. At the other extreme, nonsegregation, mixing occurs at the microscopic level. Between these two extremes, we may consider partial segregation, but our focus here is on segregated flow as an ideal flow model.

The *concept* of segregated flow in a single-phase fluid may be compared with the *reality* of segregated flow in a two-phase system made up of solid particles and a fluid. Flow of the solid particles is segregated flow, since the material in any one particle cannot mix with material in any other particle; instead, mixing of particles may occur, and this could be characterized by an RTD. The segregated-flow assumption (SFA) applied to a single-phase fluid is inconsistent with the concept of a phase with properties that are either uniform or vary in a continuous manner, since closed packets imply discontinuities in properties. Nevertheless, we use the segregated flow model in subsequent chapters to assess reactor performance.

Consider a reaction represented by A + ... → products, where A is the limiting reactant. Each segregated packet behaves independently as a batch system, and c_A or f_A in the packet, as it leaves the reactor, depends on t, the residence time of the packet in the reactor. In the completely mixed stream leaving the reactor, the value is $\bar{c}_A$ or $\bar{f}_A$, obtained by averaging over all the packets weighted in accordance with the RTD:

$$\bar{c}_A = \sum_{\substack{all \\ packets}} \binom{c_A \text{ in packet of}}{\text{age } t \text{ to } t + \delta t}\binom{\text{fraction of packets of}}{\text{age } t \text{ to } t + \delta t} \quad (13.5\text{-}1)$$

Thus, for the exit stream (constant-density situation), in the limit of a large number of packets, with $\delta t \to dt$, we have for both c_A and f_A (omitting the bar notation):

$$1 - f_A = \frac{c_A}{c_{Ao}} = \int_0^\infty \left[\frac{c_A(t)}{c_{Ao}}\right]_{BR} E(t)dt \quad \text{(SFM)} \quad (13.5\text{-}2)$$

$$\bar{c}_A = \int_0^\infty c_A(t) E(t) dt$$

where BR indicates the expression for $c_A(t)/c_{Ao}$ is that for a batch reactor.

Equation 13.5-2 is the segregated-flow model (SFM) with a continuous RTD, $E(t)$. To what extent does it give valid results for the performance of a reactor? To answer this question, we apply it first to ideal-reactor models (Chapters 14 to 16), for which we have derived the exact form of $E(t)$, and for which exact performance results can be compared with those obtained independently by material balances. The utility of the SFM lies eventually in its potential use in situations involving nonideal flow, where results cannot be predicted *a priori*, in conjunction with an experimentally measured RTD (Chapters 19 and 20); in this case, confirmation must be done by comparison with experimental results.

In any case, use of the SFM requires two types of information: reaction kinetics, to obtain $[c_A(t)/c_{Ao}]_{BR}$, and RTD, to obtain $E(t)$.

13.6 PROBLEMS FOR CHAPTER 13

13-1 Describe an experiment that could be used to measure each of the following:
(a) $F(t)$; (b) $W(t)$; (c) $E(t)$.

13-2 For a vessel with BMF, show that $I(t) = E(t)$. Is this a general conclusion for any type of flow?

13-3 (a) Show that $I(\theta) = \bar{t}I(t)$.
(b) Show that $I(\theta) = 1 - F(\theta)$.
(c) How is $I(t)$ related to $E(t)$? Justify.

13-4 (a) How is H related to $I(t)$? Justify.
(b) Using the results from (a) and 13-3 (b), relate H to $F(\theta)$.

13-5 The washout RTD function $W(t)$ or $W(\theta)$ is defined in Section 13.3.3.
(a) Relate $W(\theta)$ to $F(\theta)$.
(b) Relate $W(\theta)$ to $E(\theta)$.
(c) What is the functional form of $W(\theta)$ for (i) BMF; (ii) PF; (iii) LF? Justify.
(d) Sketch $W(\theta)$ for BMF, PF, and LF on the same plot to show essential features.

13-6 (a) What is the functional form of $I(\theta)$ for (i) BMF; (ii) PF; (iii) LF? Justify.
(b) Sketch $I(\theta)$ for BMF, PF, and LF on the same plot to show essential features.

13-7 What is H for (i) BMF; (ii) PF; (iii) LF? Justify.

13-8 For the case of laminar flow (LF) through a vessel, derive the values of the first ($\bar{\theta}$) and the second (σ_θ^2) moments of $E(\theta)$. The results in equations 13.4-20 and -21 for $E(\theta)$ may be used as a starting point.

13-9 For $E(t)$ and BMF, derive the result for σ_t^2 in terms of $\bar{t}$, and also obtain the value of σ_θ^2.

13-10 Consider a CSTR of volume 12 m³ operating at steady-state with a liquid feed rate of 0.3 m³ min⁻¹. Suppose a relatively small amount of contaminant (0.06 m³) accidentally entered virtually all at once in the feed to the reactor. How many hours of production would be lost if the reactor continued to operate until the concentration of contaminant dropped to an allowable level of, say, 10 ppmv? (The lost production is due to the formation of "off-specification" product.)

13-11 For a CSTR operating at steady-state, and with a stream of constant density, what fraction of the exit stream
(a) has been in the tank for a time greater than $\bar{t}$?
(b) is of age less than $2\bar{t}$?
(c) has been in the tank between $\bar{t}$ and $3\bar{t}$?

13-12 A pulse of tracer was injected into a tubular reactor to determine if the flow was laminar or plug flow. Effluent tracer concentrations were measured over time, and RTD profiles were developed.
(a) What features of the $F(\theta)$ profile distinguish laminar flow from plug flow?
(b) Assuming laminar flow and a mean residence time of 16 min, how long does it take for half of the injected tracer to pass through the reactor?

13-13 For two equal-sized, well-mixed tanks arranged in series, $E(t) = 4t/\bar{t}^2 e^{-2t/\bar{t}}$.
(a) Develop the expressions for $F(t)$ and $W(t)$ for this system.
(b) What is the value of $F(t)$ at $t = 5$ min, when the mean residence time ($\bar{t}$) is 10 min?

Problems 13-1, 13-3, 13-4, and 13-5(a), (b) and their results are general, and are not restricted to ideal flow. For confirmation of results for problems 13-2 to 13-7, apart from the sketches, see Tables 13.1 and 13.2.

Chapter 14

Continuous Stirred-Tank Reactors (CSTR)

In this chapter, we develop the basis for design and performance analysis for a CSTR (continuous stirred-tank reactor). The general features of a CSTR are outlined in Section 2.3.1, and are illustrated schematically in Figure 2.3 for both a single-stage CSTR and a two-stage CSTR. The essential features, as applied to complete dispersion at the microscopic level, i.e., nonsegregated flow, are recapitulated as follows:

(1) The flow pattern is BMF, and the CSTR is a "closed" vessel (Section 13.2.4).
(2) Although flow through the CSTR is continuous, the volumetric flow rates at the inlet and exit may differ because of a change in density.
(3) BMF involves perfect mixing within the reactor volume, which, in turn, implies that all system properties are uniform throughout the reactor.
(4) Perfect mixing also implies that, at any instant, each element of fluid within the reactor has an equal probability of leaving the vessel.
(5) As a result of (4), there is a continuous distribution of residence times, as derived in Section 13.4.1.
(6) As a result of (4), the outlet stream has the same properties as the fluid inside the vessel.
(7) As a result of (6), there is a step-change across the inlet in any property that changes from inlet to outlet.
(8) Although there is a distribution of residence times, the complete mixing of fluid at the microscopic and macroscopic levels leads to an averaging of properties across all fluid elements. Thus, the exit stream has a concentration (average) equivalent to that obtained as if the fluid existed as a single, large fluid element with a residence time of $\bar{t} = V/q$ (equation 2.3-1).

Most of these points are shown in Figure 2.3; point (5) is shown in Figure 13.5.

In the development to follow, we first consider uses of a CSTR, and then some advantages and disadvantages. After presenting design and performance equations in general, we then apply them to cases of constant-density and variable-density operation. The treatment is mainly for steady-state operation, but also includes unsteady-state operation, and the possibility of multiple stationary-states, the relationship to the segregated-flow model (Chapter 13), and an introduction to the use of a multistage CSTR (continued in Chapter 17). We restrict attention in this chapter to simple, single-phase systems.

336 Chapter 14: Continuous Stirred-Tank Reactors (CSTR)

14.1 USES OF A CSTR

As in the case of a batch reactor for commercial operation, a CSTR is normally used for a liquid-phase reaction. In the laboratory, it may also be used for a gas-phase reaction for experimental measurements, particularly for a solid-catalyzed reaction, as in Figure 1.2. The operation is normally one of steady-state, except for startup, shutdown, and operational disturbances or upsets, in which cases unsteady-state operation has to be taken into account.

A CSTR may consist of a single stage (one vessel) or multiple stages (two or more vessels) in series, as illustrated in Figure 2.3. This raises the question of what advantage can be gained by multistage operation to offset the cost of additional vessels and interconnecting piping.

The use of a single-stage CSTR for HF alkylation of hydrocarbons in a special forced-circulation shell-and-tube arrangement (for heat transfer) is illustrated by Perry et al. (1984, p. 21-6). The emulsion copolymerization of styrene and butadiene to form the synthetic rubber SBR is carried out in a multistage CSTR.

14.2 ADVANTAGES AND DISADVANTAGES OF A CSTR

Some advantages and disadvantages of a CSTR can be realized at this stage from the model outlined above, in anticipation of the quantitative development to follow:

Advantages

(1) Relatively cheap to construct;
(2) Ease of control of temperature in each stage, since each operates in a stationary state; heat transfer surface for this can be easily provided;
(3) Can be readily adapted for automatic control in general, allowing fast response to changes in operating conditions (e.g., feed rate and concentration);
(4) Relatively easy to clean and maintain;
(5) With efficient stirring and viscosity that is not too high, the model behavior can be closely approached in practice to obtain predictable performance.

Disadvantages

The most obvious disadvantage in principle stems from the fact that the outlet stream is the same as the contents of the vessel. This implies that *all* reaction takes place at the *lowest* concentration (of reactant A, say, c_A) between inlet and outlet. For normal kinetics, in which rate of reaction $(-r_A)$ decreases as c_A decreases, this means that a greater volume of reactor is needed to obtain a desired conversion. (For abnormal kinetics, the opposite would be true, but this is unusual—what is an example of one such situation?)

14.3 DESIGN EQUATIONS FOR A SINGLE-STAGE CSTR

14.3.1 General Considerations; Material and Energy Balances

The process design of a CSTR typically involves determining the volume of a vessel required to achieve a specified rate of production. Parameters to investigate include the number of stages to use for optimal operation, and the fractional conversion and temperature within each stage. We begin, however, by considering the material and energy balances for a *single* stage in this section. The treatment is extended to a multistage CSTR in Section 14.4.

14.3.1.1 *Material Balance; Volume of Reactor; Rate of Production*

For continuous operation of a CSTR as a "closed" vessel, the general material balance equation for reactant A (in the reaction $A + \ldots \rightarrow \nu_C C + \ldots$), with a control volume defined as the volume of fluid in the reactor, is written as

$$\begin{pmatrix} \text{rate of} \\ \text{input of} \\ \text{A by flow} \end{pmatrix} - \begin{pmatrix} \text{rate of} \\ \text{output of} \\ \text{A by flow} \end{pmatrix} - \begin{pmatrix} \text{rate of} \\ \text{disappearance} \\ \text{of A by} \\ \text{reaction} \end{pmatrix} = \begin{pmatrix} \text{rate of} \\ \text{accumulation} \\ \text{of A within} \\ \text{the control} \\ \text{volume} \end{pmatrix} \quad (14.3\text{-}1)$$

Operationally, we may write this word statement in various ways, as done in Section 2.3.2. Thus, in terms of molar flow rates, with only unreacted A in the feed ($f_{Ao} = 0$),

$$F_{Ao} - F_A - (-r_A)V = dn_A/dt \quad (2.3\text{-}3)$$

or, in terms of volumetric flow rates,

$$c_{Ao}q_o - c_A q - (-r_A)V = dn_A/dt \quad (14.3\text{-}2)$$

and, in terms of fractional conversion of A,

$$F_{Ao}f_A - (-r_A)V = dn_A/dt \quad (14.3\text{-}3)$$

The interpretation of the various quantities with respect to location about the reactor is shown in Figure 14.1.

The material-balance equation, in whatever form, is usually used to solve for V in steady-state operation, or to determine the changes of outlet properties with respect to time in unsteady-state operation for a particular V. Thus, for steady-state operation, with $dn_A/dt = 0$, from equations 14.3-2 and -3,

$$V = (c_{Ao}q_o - c_A q)/(-r_A) \quad (14.3\text{-}4)$$
$$= F_{Ao}f_A/(-r_A) \quad (14.3\text{-}5)$$

The mean residence time is given by

$$\bar{t} = V/q \quad (2.3\text{-}1)$$

and the space time by

$$\tau = V/q_o \quad (2.3\text{-}2)$$

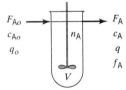

Figure 14.1 Quantities in material-balance equation for CSTR

Equation 14.3-5 may be interpreted graphically from a plot of reciprocal rate, $1/(-r_A)$, against f_A, as shown by the curve EB in Figure 14.2. Point B is the "operating point" of the reactor; that is, it represents the stationary-state condition in the reactor (and in the outlet stream). Area ABCD, the product of $1/(-r_A)$ and f_A at the operating conditions, represents the ratio V/F_{Ao} for the CSTR (from equation 14.3-5). Figure 14.2 applies to normal kinetics, and is valid, as are all the equations in this section, whether density remains constant or varies. Note that, although every point on the reciprocal-rate curve has significance in general, only the point B has significance for the particular CSTR represented in Figure 14.2.

The rate of production in terms of the product C is the molar rate of flow of C from the reactor. That is,

$$Pr(C) = F_C = \nu_C F_{Ao} f_A = c_C q \qquad (14.3\text{-}6)$$

14.3.1.2 Energy Balance

For a continuous-flow reactor, such as a CSTR, the energy balance is an enthalpy (H) balance, if we neglect any differences in kinetic and potential energy of the flowing stream, and any shaft work between inlet and outlet. However, in comparison with a BR, the balance must include the input and output of H by the flowing stream, in addition to any heat transfer to or from the control volume, and generation or loss of enthalpy by reaction within the control volume. Then the energy (enthalpy) equation in words is

$$\begin{pmatrix} \text{rate of} \\ \text{input of} \\ \text{enthalpy by} \\ \text{flow, heat} \\ \text{transfer or} \\ \text{reaction} \end{pmatrix} - \begin{pmatrix} \text{rate of} \\ \text{output} \\ \text{of enthalpy by} \\ \text{flow, heat} \\ \text{transfer or} \\ \text{reaction} \end{pmatrix} = \begin{pmatrix} \text{rate of} \\ \text{accumulation} \\ \text{of enthalpy} \\ \text{within control} \\ \text{volume} \end{pmatrix} \qquad (14.3\text{-}7)$$

To translate this into operational form, we use the total specific mass flow rate, $\dot{m}$, rather than molar flow rate, to show an alternative treatment to that used for a BR in Chapter 12. An advantage to this treatment is that in *steady-state* operation, $\dot{m}$ is constant, whereas the molar flow rate need not be. Thus, using T_{ref} as a reference temperature

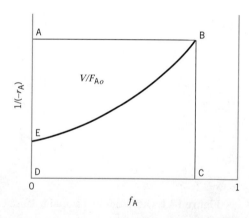

Figure 14.2 Graphical interpretation of equation 14.3-5

to evaluate the enthalpy for the inlet and outlet streams, we have

$$\int_{T_{ref}}^{T_o} \dot{m}_o c_{Po} dT - \int_{T_{ref}}^{T} \dot{m} c_P dT + UA_c(T_c - T)_m + (-\Delta H_{RA})(-r_A)V = dH/dt$$
$$= d(m_t c_P T)/dt \quad (14.3\text{-}8)$$

In equation 14.3-8, subscript "o" represents an inlet condition, c_P is the specific heat of the (total) system as indicated, and m_t is the total mass contained in the control volume at time t; the interpretation of the various quantities is shown in Figure 14.3. The first term on the left side is the input of enthalpy by flow, the second term is the output of enthalpy by flow, and the third and fourth terms represent heat transfer and enthalpy generation or consumption by reaction, respectively.

Equation 14.3-8 may be simplified in various ways. For steady-state operation, $\dot{m} = \dot{m}_o$, and the right side vanishes; furthermore, if c_P and T_c are constant, and if we choose $T_{ref} = T_o$, so that the enthalpy input by flow is zero, then

$$\dot{m} c_P(T_o - T) + UA_c(T_c - T) + (-\Delta H_{RA})(-r_A)V = 0 \quad (14.3\text{-}9)$$

or, on substitution of $F_{Ao} f_A$ for $(-r_A)V$ from equation 14.3-5,

$$\dot{m} c_p(T_o - T) + UA_c(T_c - T) + (-\Delta H_{RA})F_{Ao} f_A = 0 \quad (14.3\text{-}10)$$

Equation 14.3-10 relates f_A and T for steady-state operation; solving for f_A, we have:

$$f_A = -\frac{\dot{m} c_P T_o + UA_c T_c}{(-\Delta H_{RA})F_{Ao}} + \left[\frac{\dot{m} c_p + UA_c}{(-\Delta H_{RA})F_{Ao}}\right] T \quad (14.3\text{-}11)$$

The use of the energy balance in the design of a CSTR depends on what is specified initially. If T is specified, the energy balance is uncoupled from the material balance, and is used primarily to design the heat exchanger required to control T; V is given by the performance equation 14.3-5. If T is not specified, as in adiabatic operation, the material and energy balances are coupled (through T), and must be solved simultaneously. This gives rise to possible multiple stationary states, and is discussed in Section 14.3.4.2, below.

14.3.2 Constant-Density System

For a constant-density system, several simplifications result. First, regardless of the type of reactor, the fractional conversion of limiting reactant, say f_A, can be expressed in terms of molar concentration, c_A:

$$f_A = (c_{Ao} - c_A)/c_{Ao} \quad \text{(constant density)} \quad (14.3\text{-}12)$$

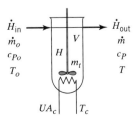

Figure 14.3 Quantities in energy-balance equation (14.3-8) for CSTR

Second, for a flow reactor, such as a CSTR, the mean residence time and space time are equal, since $q = q_o$:

$$\bar{t} = \tau \quad \text{(constant density)} \tag{14.3-13}$$

Third, for a CSTR, the accumulation term in the material-balance equation 14.3-2 or -3 becomes

$$dn_A/dt = V dc_A/dt \quad \text{(constant density)} \tag{14.3-14}$$

Finally, for a CSTR, the material-balance equation 14.3-4 can be simplified to:

$$V = (c_{Ao} - c_A)q/(-r_A) \quad \text{(constant density)} \tag{14.3-15}$$

(The material-balance for V in terms of f_A (14.3-5) remains unchanged.)

14.3.2.1 Steady-State Operation at Specified T

For steady-state operation, the accumulation term in the material-balance equation (14.3-2 or -3) vanishes:

$$dn_A/dt = 0 \quad \text{(steady-state)} \tag{14.3-16}$$

or, for constant density,

$$dc_A/dt = 0 \quad \text{(steady-state, constant density)} \tag{14.3-17}$$

If T is specified, V can be calculated from the material-balance (equation 14.3-5 or -15) without the need for the energy balance.

EXAMPLE 14-1

For the liquid-phase reaction A + B → products at 20°C, suppose 40% conversion of A is desired in steady-state operation. The reaction is pseudo-first-order with respect to A, with $k_A = 0.0257$ h^{-1} at 20°C. The total volumetric flow rate is 1.8 m^3 h^{-1}, and the inlet molar flow rates of A and B are F_{Ao} and F_{Bo} mol h^{-1}, respectively. Determine the vessel volume required, if, for safety, it can only be filled to 75% capacity.

SOLUTION

Since this is a liquid-phase reaction, we assume density is constant. The quantity V in the material-balance equation (14.3-5 or -15) is the volume of the system (liquid) in the reactor. The reactor (vessel) volume is greater than this because of the 75%-capacity requirement.

From the specified rate law,

$$(-r_A) = k_A c_A = k_A c_{Ao}(1 - f_A)$$

Thus, substituting for $(-r_A)$ in the material-balance equation 14.3-5, we obtain

$$V = F_{Ao} f_A / k_A c_{Ao}(1 - f_A)$$
$$= q_o f_A / k_A (1 - f_A)$$
$$= 1.8(0.4)/0.0257(0.6) = 46.7 \text{ m}^3$$

and the actual volume of the vessel is 46.7/0.75 = 62.3 m³. The same result is obtained from equation 14.3-15. Since the reaction is first-order, the result is independent of c_{Ao} (or F_{Ao}), and since it is pseudo-first-order with respect to A, the result is also independent of F_{Bo}. It *does* depend on the total flow rate q, the conversion f_A, and the rate constant k_A.

EXAMPLE 14-2

A liquid-phase reaction A → B is to be conducted in a CSTR at steady-state at 163°C. The temperature of the feed is 20°C, and 90% conversion of A is required. Determine the volume of a CSTR to produce 130 kg B h⁻¹, and calculate the heat load ($\dot{Q}$) for the process. Does this represent addition or removal of heat from the system?

Data:

$$c_P = 2.0 \text{ J g}^{-1} \text{ K}^{-1} \qquad M_A = M_B = 200 \text{ g mol}^{-1}$$
$$\rho = 0.95 \text{ g cm}^{-3} \qquad \Delta H_{RA} = -87 \text{ kJ mol}^{-1}$$
$$k_A = 0.80 \text{ h}^{-1} \text{ at } 163°C$$

SOLUTION

Since the reaction occurs at a specified temperature, V may be determined from the material-balance equation (14.3-5), together with the given rate law:

$$V = F_{Ao}f_A/(-r_A) = F_{Ao}f_A/k_A c_A$$
$$= F_{Ao}f_A/k_A c_{Ao}(1 - f_A) = f_A q/k_A(1 - f_A)$$

$$q = \frac{130 \text{ kg B}}{h} \times \frac{1}{0.90} \frac{\text{kg(total)}}{\text{kg B}} \times \frac{1}{0.95} \frac{\text{cm}^3}{\text{g}} \times \frac{1000}{1000} \frac{\text{g}}{\text{kg}} \frac{\text{L}}{\text{cm}^3} = 152 \text{ L h}^{-1}$$

Thus,

$$V = 0.90(152)/0.80(0.10) = 1710 \text{ L}$$

The heat load, $\dot{Q}$, is obtained from the energy-balance equation (14.3-10) for steady-state operation with constant values for the various parameters:

$$\dot{Q} = UA_c(T_c - T) = -\dot{m}c_P(T_o - T) - (-\Delta H_{RA})F_{Ao}f_A \qquad (14.3\text{-}10)$$

$$= -\frac{130}{0.90}(2.0)(20 - 163) - (87)\frac{130}{0.90}\left(\frac{1000}{200}\right)0.90$$

$$= 41,311 - 56,550 = -15,239 \text{ kJ h}^{-1} = -4.2 \text{ kW}$$

Since $\dot{Q}$ is negative, heat is removed from the system.

14.3.2.2 Unsteady-State Operation

For unsteady-state operation of a CSTR, the full form of the material balance, equation 14.3-2 or its equivalent, must be used.

EXAMPLE 14-3

Consider the startup of a CSTR for the liquid-phase reaction A → products. The reactor is initially filled with feed when steady flow of feed (q) is begun. Determine the time (t)

required to achieve 99% of the steady-state value of f_A. Data: $V = 8000$ L; $q = 2$ L s^{-1}; $c_{Ao} = 1.5$ mol L^{-1}; $k_A = 1.5 \times 10^{-4}$ s^{-1}.

SOLUTION

Although there is steady flow of feed, the reactor is in unsteady-state operation during the time t, since the outlet concentration c_A (and hence f_A) is continuously changing (c_A decreasing from c_{Ao}, and f_A increasing from 0).

We first determine the *steady-state* value of f_A. We use the general material-balance equation 14.3-2, with $q = q_o$:

$$c_{Ao} q_o - c_A q_o - (-r_A)V = dn_A/dt \qquad (14.3\text{-}2)$$

Equation 14.3-2 can be recast in terms of $\tau = V/q_o$ and f_A using $c_A = c_{Ao}(1 - f_A)$; on combination with the rate law, the result is

$$c_{Ao} - c_{Ao}(1 - f_A) - k_A c_{Ao}(1 - f_A)\tau = \tau dc_A/dt \qquad (A)$$

For steady-state operation, with $dc_A/dt = 0$, this becomes

$$f_A - k_A(1 - f_A)\tau = 0$$

from which

$$f_A = \frac{k_A \tau}{1 + k_A \tau} = \frac{1.5(10^{-4})(8000/2)}{1 + 1.5(10^{-4})(8000/2)} = 0.375$$

The time t required to achieve 99% of this value ($f_A = 0.371$) is obtained by integrating equation (A) rewritten as:

$$c_{Ao} f_A - k_A \tau c_{Ao}(1 - f_A) = -\tau c_{Ao} df_A/dt$$

or

$$k_A \tau - (1 + k_A \tau) f_A = \tau df_A/dt$$

or

$$dt = \frac{\tau df_A}{k_A \tau - (1 + k_A \tau) f_A} \qquad (B)$$

With the boundary condition $f_A = 0$ at $t = 0$, equation (B) integrates to

$$t = \frac{\tau}{1 + k_A \tau} \ln\left[\frac{k_A \tau}{k_A \tau - (1 + k_A \tau) f_A}\right]$$

This yields $t = 11{,}500$ s $= 3.2$ h for $f_A = 0.371$. (This is not necessarily the most efficient way to start up a CSTR. How may the time required be reduced?)

This problem may also be solved by numerical integration using the E-Z Solve software (file ex14-3.msp). This simulation is well-suited to the investigation of the effect of initial conditions on the time for a specified approach to steady-state. To optimize

14.3.2.3 CSTR Performance in Relation to SFM

At this stage, we raise the question as to how the performance of a CSTR depends on the degree of segregation or mixing at the molecular level of the (single-phase) fluid system. The extremes (Section 13.5) are nonsegregation (corresponding to a microfluid and micromixing at the molecular level), and complete segregation (corresponding to a macrofluid with no micromixing at the molecular level, but allowing for macromixing of "packets of fluid" at a macroscopic level). In this section, we restrict attention to steady-state behavior for a constant-density reaction, A $\rightarrow$ products, with a rate law given by $(-r_A) = k_A c_A^n$. Performance may be measured by $f_A = 1 - c_A/c_{Ao}$ for given $\bar{t}$ (or τ) and rate law.

For nonsegregation, the performance is obtained from the material-balance equation (14.3-5). Thus,

$$(-r_A)V = F_{Ao}f_A = c_{Ao}qf_A$$
$$k_A c_A^n \tau = c_{Ao}f_A \qquad (14.3\text{-}18)$$
$$k_A c_{Ao}^n \tau (1 - f_A)^n = c_{Ao}f_A$$
$$M_{An}(1 - f_A)^n = f_A \qquad (14.3\text{-}19)$$

where M_{An} is the dimensionless reaction number, introduced in Section 4.3.5, defined by

$$M_{An} = k_A c_{Ao}^{n-1} \bar{t} = k_A c_{Ao}^{n-1} \tau \qquad (4.3\text{-}4)$$

For complete segregation, the performance is obtained from the segregated flow model (SFM) (Section 13.5):

$$1 - f_A = \frac{c_A}{c_{Ao}} = \int_0^\infty \left[\frac{c_A(t)}{c_{Ao}}\right]_{BR} E(t)dt \qquad (13.5\text{-}2)$$

For a first-order reaction,

$$[c_A(t)/c_{Ao}]_{BR} = e^{-k_A t} \qquad (n = 1) \qquad (3.4\text{-}10)$$

For all other orders, the solution of the material balance equation for a BR is

$$c_A^{1-n} - c_{Ao}^{1-n} = (n - 1)k_A t \qquad (n \neq 1) \qquad (3.4\text{-}9)$$

which can be rearranged to give (equivalent to equation 4.3-5):

$$\left[\frac{c_A(t)}{c_{Ao}}\right]_{BR} = \left[\frac{1}{1 + (n - 1)k_A c_{Ao}^{n-1} t}\right]^{\frac{1}{n-1}} \qquad (n \neq 1) \qquad (14.3\text{-}20)$$

For a CSTR, $E(t)$ is given by

$$E(t) = (1/\bar{t})e^{-t/\bar{t}} \qquad (13.4\text{-}2)$$

Equation 13.4-2 is combined with equation 3.4-10 or 14.3-20 in the SFM (13.5-2) to obtain f_A, usually by numerical integration.

EXAMPLE 14-4

For a first-order reaction in a CSTR, compare the predicted performance for completely segregated flow with that for nonsegregated flow.

SOLUTION

For this comparison, we fix the volume V and the flow rate q; that is, we set $\bar{t} = \tau$.
For nonsegregated flow, from equations 14.3-18 and -19, with $n = 1$,

$$f_A = \frac{M_{A1}}{1 + M_{A1}} = \frac{k_A \tau}{1 + k_A \tau} \qquad (14.3\text{-}21)$$

For segregated flow, from equation 13.5-2, with equations 3.4-10 and 13.4-2,

$$1 - f_A = \int_0^\infty e^{-k_A t}(1/\bar{t})e^{-t/\bar{t}}dt$$

$$= \frac{1}{1 + k_A \bar{t}}$$

That is,

$$f_A = \frac{k_A \bar{t}}{1 + k_A \bar{t}}$$

which is the same result as in equation 14.3-21 above, since $\bar{t} = \tau$.

From Example 14-4, we conclude that the performance for a first-order reaction in a CSTR is independent of the degree of segregation. This conclusion does *not* apply to any other order (value of n). The results for $n = 0, 1,$ and 2 are summarized in Table 14.1, which includes equations 14.3-22 to -25.

Introduction of numerical values for M_{An} in the results given in Table 14.1 (and results for other orders) leads to the following conclusions for a CSTR:

$$\left. \begin{array}{l} f_{A,\text{nonseg}} > f_{A,\text{seg}} \,;\, n < 1 \\ f_{A,\text{nonseg}} = f_{A,\text{seg}} \,;\, n = 1 \\ f_{A,\text{nonseg}} < f_{A,\text{seg}} \,;\, n > 1 \end{array} \right\} \qquad (14.3\text{-}26)$$

In other words, an increase in segregation improves the performance of a CSTR for $n > 1$, and lowers it for $n < 1$ (Levenspiel, 1972, p. 335).

14.3.3 Variable-Density System

For a CSTR, variable density means that the inlet and outlet streams differ in density. A significant difference in density occurs only for gas-phase reactions in which there is a change in at least one of: total moles, T, or P, although the last is usually very small.

14.3 Design Equations for a Single-Stage CSTR

Table 14.1 Comparison of nonsegregated and segregated flow in a CSTR[a]

Order n	f_A Nonsegregated[b]	f_A Segregated[c]
0	M_{A0}; $M_{A0} < 1$ 1; $M_{A0} \geq 1$ (14.3-22)	$M_{A0}(1 - e^{-1/M_{A0}})$ (all values of M_{A0}) (14.3-23)
1	$\dfrac{M_{A1}}{1 + M_{A1}}$ (14.3-21)	$\dfrac{M_{A1}}{1 + M_{A1}}$ (14.3-21)
2	$1 - \dfrac{(1 + 4M_{A2})^{1/2} - 1}{2M_{A2}}$ (14.3-24)	$\dfrac{e^{1/M_{A2}}}{M_{A2}} E_1\left(\dfrac{1}{M_{A2}}\right)$[d] (14.3-25)

[a] Constant-density; A → products; $(-r_A) = k_A c_A^n$.
[b] From equation 14.3-19, with M_{An} defined by equation 4.3-4.
[c] From equation 13.5-2, with 3.4-10 or 14.3-20 and 13.4-2.
[d] E_1 is an exponential integral defined by $E_1(x) = \int_x^\infty y^{-1} e^{-y} \, dy$, where y is a dummy variable; the integral must be evaluated numerically (e.g., using E-Z Solve); tabulated values also exist.

EXAMPLE 14-5

A gas-phase reaction between ethylene (A) and hydrogen to produce ethane is carried out in a CSTR. The feed, containing 40 mol% ethylene, 40 mol% hydrogen, and 20% inert species (I), enters the reactor at a total rate of 1.5 mol min^{-1}, with $q_o = 2.5$ L min^{-1}. The reaction is first-order with respect to both hydrogen and ethylene, with $k_A = 0.25$ L mol^{-1} min^{-1}. Determine the reactor volume required to produce a product that contains 60 mol% ethane. Assume T and P are unchanged.

SOLUTION

The reaction is:

$$C_2H_4(A) + H_2(B) \rightarrow C_2H_6(C)$$

From the material balance (equation 14.3-5) applied to C$_2$H$_4$(A), together with the rate law, and the fact that $c_A = c_B$, since $F_{Ao} = F_{Bo}$ and $\nu_A = \nu_B$ in the chemical equation,

$$V = \frac{F_{Ao}f_A}{(-r_A)} = \frac{F_{Ao}f_A}{k_A c_A c_B} = \frac{F_{Ao}f_A}{k_A c_A^2} = \frac{F_{Ao}f_A}{k_A (F_A/q)^2}$$

$$= \frac{q^2 F_{Ao} f_A}{k_A F_{Ao}^2 (1 - f_A)^2} = \frac{q^2 f_A}{k_A F_{Ao}(1 - f_A)^2} \quad \textbf{(A)}$$

In equation (A), q and f_A are unknown, but may be determined from the given composition of the outlet stream, with the aid of a stoichiometric table. For a feed consisting of F_{Ao} moles of A, this is constructed as follows:

346 Chapter 14: Continuous Stirred-Tank Reactors (CSTR)

Species	Initial moles	Change	Final moles
A	F_{Ao}	$-f_A F_{Ao}$	$F_{Ao}(1 - f_A)$
B	$F_{Bo} = F_{Ao}$	$-f_A F_{Ao}$	$F_{Ao}(1 - f_A)$
C	0	$f_A F_{Ao}$	$f_A F_{Ao}$
I	$F_{Io} = F_{Ao}/2$	0	$F_{Ao}/2$
total	$2.5 F_{Ao}$		$F_{Ao}(2.5 - f_A)$

Since the outlet stream is to be 60 mol % C_2H_6(C),

$$f_A F_{Ao} / F_{Ao}(2.5 - f_A) = 0.60$$

from which $f_A = 0.9375$.

If we assume ideal-gas behavior, with T and P unchanged,

$$\frac{q}{q_o} = \frac{F_{Ao}(2.5 - f_A)}{2.5 F_{Ao}} = \frac{2.5 - f_A}{2.5}$$

from which $q = 1.563$ L min^{-1}. Substituting numerical values for f_A, q, and k_A, together with $F_{Ao} = 0.4(1.5) = 0.6$ mol min^{-1}, in (A), we obtain

$$V = \frac{(1.563)^2 0.9375}{0.25(0.6)(1 - 0.9375)^2} = 3900 \text{ L}$$

EXAMPLE 14-6

At elevated temperatures, acetaldehyde (CH$_3$CHO, A) undergoes gas-phase decomposition into methane and carbon monoxide. The reaction is second-order with respect to acetaldehyde, with $k_A = 22.2$ L mol^{-1} min^{-1} at a certain T. Determine the fractional conversion of acetaldehyde that can be achieved in a 1500-L CSTR, given that the feed rate of acetaldehyde is 8.8 kg min^{-1}, and the inlet volumetric flow rate is 2.5 m^3 min^{-1}. Assume T and P are unchanged.

SOLUTION

The reaction is

$$\text{CH}_3\text{CHO(A)} \rightarrow \text{CH}_4\text{(B)} + \text{CO(C)}$$

From the design equation, 14.3-5, and the given rate law,

$$f_A = \frac{(-r_A)V}{F_{Ao}} = \frac{k_A c_A^2 V}{F_{Ao}} = \frac{k_A (F_A/q)^2 V}{F_{Ao}}$$

$$= \frac{k_A F_{Ao}^2 (1 - f_A)^2 V}{q^2 F_{Ao}} = \frac{k_A F_{Ao}(1 - f_A)^2 V}{q^2} \quad \textbf{(A)}$$

In (A), k_A and V are known, and $F_{Ao} = \dot{m}_{Ao}/M_A = 8.8(1000)/44 = 200$ mol min^{-1}. However, since there is a change in the number of moles on reaction, $q \neq$ (given) q_o, but may be related to q_o with the aid of a stoichiometric table:

14.3 Design Equations for a Single-Stage CSTR 347

Species	Initial moles	Change	Final moles
A	F_{Ao}	$-f_A F_{Ao}$	$F_{Ao}(1 - f_A)$
B	0	$f_A F_{Ao}$	$f_A F_{Ao}$
C	0	$f_A F_{Ao}$	$f_A F_{Ao}$
total	F_{Ao}		$F_{Ao}(1 + f_A)$

If we assume ideal-gas behavior, with T and P unchanged, then, from the table,

$$\frac{q}{q_o} = \frac{F_{Ao}(1 + f_A)}{F_{Ao}} = 1 + f_A$$

Substituting this result for q in (A) and rearranging to solve for f_A, we obtain the cubic polynomial equation:

$$f_A^3 + (2 - K)f_A^2 + (1 + 2K)f_A - K = 0 \quad \textbf{(B)}$$

where $K = k_A F_{Ao} V / q_o^2 = 22.2(200)1500/(2500)^2 = 1.066$ (dimensionless). With this value inserted, (B) becomes

$$f_A^3 + 0.934 f_A^2 + 3.132 f_A - 1.066 = 0 \quad \textbf{(C)}$$

From the Descartes rule of signs,[1] since there is one change in the sign of the coefficients in (C), there is only one positive real root. (The same rule applied to $-f_A$ in (C) indicates that there may be two negative real roots for f_A, but these are not allowable values.) Solution of (C) by trial or by means of the E-Z Solve software (file ex14-6.msp) gives

$$f_A = 0.304$$

The E-Z Solve numerical solution requires combination of the material balance (14.3-5), the rate law, and the relation of q to q_o (accounting for the density change in the system).

14.3.4 Existence of Multiple Stationary States

An unusual feature of a CSTR is the possibility of multiple stationary states for a reaction with certain nonlinear kinetics (rate law) in operation at a specified T, or for an exothermic reaction which produces a difference in temperature between the inlet and outlet of the reactor, including adiabatic operation. We treat these in turn in the next two sections.

14.3.4.1 *Operation at Specified T; Rate-Law Determined*

A reaction which follows power-law kinetics generally leads to a single, unique steady state, provided that there are no temperature effects upon the system. However, for certain reactions, such as gas-phase reactions involving competition for surface active sites on a catalyst, or for some enzyme reactions, the design equations may indicate several potential steady-state operating conditions. A reaction for which the rate law includes concentrations in both the numerator and denominator may lead to multiple steady states. The following example (Lynch, 1986) illustrates the multiple steady states

---

[1] For the polynomial equation $f(x) = 0$ with real coefficients, the number of positive real roots of x is either equal to the number of changes in sign of the coefficients *or* less than that number by a positive even integer; the number of negative real roots is similarly given, if x is replaced by $-x$.

EXAMPLE 14-7

In the catalyzed gas-phase decomposition A → B + C, suppose A also acts as an inhibitor of its own decomposition. The resulting rate law (a type of Langmuir-Hinshelwood kinetics, Chapter 8) is:

$$(-r_A) = \frac{k_1 c_A}{(1 + k_2 c_A)^2}$$

where $k_1 = 0.253$ min^{-1}, and $k_2 = 0.429$ m^3 mol^{-1} at a certain T. Determine all possible steady-state concentrations (of A) for a 60.64 m^3 CSTR operating with $c_{Ao} = 68.7$ mol m^{-3} and $q_o = 116.5$ L min^{-1}. Assume T and P are constant.

SOLUTION

Since this is a gas-phase reaction, and the total number of moles changes, the density varies, and q must be linked to q_o. Using a stoichiometric table, we have:

Species	Initial moles	Change	Final moles
A	F_{Ao}	$F_A - F_{Ao}$	F_A
B	0	$F_{Ao} - F_A$	$F_{Ao} - F_A$
C	0	$F_{Ao} - F_A$	$F_{Ao} - F_A$
total	F_{Ao}	$F_{Ao} - F_A$	$2F_{Ao} - F_A$

If we assume ideal-gas behavior, with T and P constant,

$$q = q_o(2F_{Ao} - F_A)/F_{Ao} \quad \textbf{(A)}$$

Also, since $q_o = F_{Ao}/c_{Ao}$, and $F_A = c_A q$, then, from (A),

$$q = \frac{F_{Ao}(2F_{Ao} - c_A q)}{c_{Ao} F_{Ao}}$$

which, on rearrangement, becomes

$$q = 2F_{Ao}/(c_A + c_{Ao}) = F_A/c_A \quad \textbf{(B)}$$

Now, consider the given rate law, on the one hand, and the design equation (14.3-5), on the other:

$$(-r_A) = \frac{k_1 c_A}{(1 + k_2 c_A)^2} \quad \textbf{(C)}$$

$$(-r_A) = F_{Ao} f_A/V = (F_{Ao} - F_A)/V$$

$$= \frac{F_{Ao}}{V}\left(1 - \frac{2c_A}{c_A + c_{Ao}}\right) \quad \textbf{(D)}$$

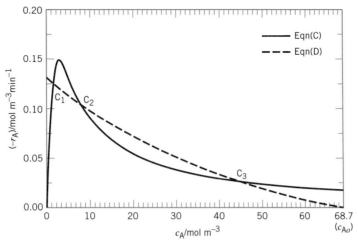

Figure 14.4 Multiple steady states in a CSTR (Example 14-7)

on elimination of F_A by means of (B). Since $(-r_A)$ in (C) and (D) must be the same, we equate the two right sides, and, after multiplication and rearrangement, obtain the following cubic equation in (the unknown) c_A:

$$k_2^2 c_A^3 + \left(\frac{k_1 V}{F_{Ao}} - k_2^2 c_{Ao} + 2k_2\right) c_A^2 + \left(\frac{k_1 c_{Ao} V}{F_{Ao}} - 2k_2 c_{Ao} + 1\right) c_A - c_{Ao} = 0$$

With numerical values inserted for k_1, k_2, V, c_{Ao}, and F_{Ao}, this becomes

$$c_A^3 - 53.6228 c_A^2 + 400.705 c_A - 373.288 = 0$$

This equation may have three real, positive roots (as indicated by the Descartes rule of signs; see Example 14-6). Three roots can be obtained by trial or by a suitable root-finding technique, for example, as provided by the E-Z Solve software (see note below):

$$c_A = 1.086,\ 7.657,\ \text{and}\ 44.88\ \text{mol m}^{-3}$$

The three roots are shown as the points of intersection C_1, C_2, and C_3 in the plot of equations (C) and (D) in Figure 14.4, $(-r_A)$ versus c_A. Thus, three stationary states are possible for steady-state operation of the CSTR at the conditions given. The stationary-state at C_1 is the one normally desired, since it represents a relatively high conversion, corresponding to a low value of c_A.

It is not possible to predict *a priori* which of the possible stationary-states is actually attained, based on steady-state operating considerations. It may be done by integrating the material-balance equation in unsteady-state form, equation 14.3-2 or equivalent, with the given rate law incorporated. For this, the initial concentration of A in the reactor $c_A(t)$ at $t = 0$ must be known; this is not necessarily the same as c_{Ao}.

Example 14-7 can also be solved using the E-Z Solve software (file ex14-7.msp). In this simulation, the problem is solved using design equation 2.3-3, which includes the transient (accumulation) term in a CSTR. Thus, it is possible to explore the effect of c_{Ao} on transient behavior, and on the ultimate steady-state solution. To examine the stability of each steady-state, solution of the differential equation may be attempted using each of the three steady-state conditions determined above. Normally, if the unsteady-state design equation is used, only stable steady-states can be identified, and unstable steady-states are missed. The best way to identify the possibility of unstable steady-states is by graphical analysis (e.g., Figure 14.4).

We can, however, consider the stability of each of the three operating points in Example 14-7 with respect to the inevitable small random fluctuations in operating conditions, including c_A, in steady-state operation. Before doing this, we note some features of the rate law as revealed in Figure 14.4. There is a maximum value of $(-r_A)$ at $c_A = 1.166$ mol m^{-3}. For $c_A < 1.166$, the rate law represents normal kinetics: $(-r_A)$ increases as c_A increases; for $c_A > 1.166$, we have abnormal kinetics: $(-r_A)$ decreases as c_A increases. We also note that $(-r_A)$ in equation (C), the rate law, represents the (positive) rate of *disappearance* of A by reaction within the CSTR, and that $(-r_A)$ in equation (D), the material balance, represents the (positive) net rate of *appearance* of A by flow into and out of the reactor. As noted above, in steady-state operation, these two rates balance.

For a stability analysis, consider first point C_1. Suppose there is a small random upward fluctuation in c_A. This is accompanied by an increase in the rate of disappearance of A by reaction, and a decrease in the appearance of A by flow, both of which tend to decrease c_A to offset the fluctuation and restore the stationary-state at C_1. Conversely, a downward fluctuation in c_A is accompanied by a decrease in the rate of disappearance by reaction, and an increase in the rate of appearance by flow, both of which tend to offset the fluctuation. Thus, the stationary-state at C_1 is stable with respect to small random fluctuations. Now, consider point C_2. An upward fluctuation in c_A is accompanied by decreases in both rates, but the rate of disappearance by reaction decreases faster, with the net result that the upward fluctuation is reinforced by a resulting increase in c_A, so that the fluctuation grows and is not damped-out, unlike at C_1. This effect continues until the point C_3 is reached, which is the next point of balance. Thus, for *any* upward fluctuation in c_A, C_2 is unstable relative to C_3. Similarly, for a downward fluctuation in c_A at C_2, the rate of disappearance by reaction increases faster, with the net result that the downward fluctuation is reinforced and grows until C_1 is reached at the next point of balance. Thus, for *any* downward fluctuation in c_A, C_2 is unstable relative to C_1. Finally, consider point C_3. For an upward fluctuation in c_A, the rate of disappearance by reaction decreases slower than the rate of appearance by flow, with the net result that the fluctuation is damped-out. A similar conclusion results for a downward fluctuation. In summary, the stationary-states at C_1 and C_3 are stable, and the state at C_2 is unstable, and cannot be achieved in steady-state operation. It can, however, be stabilized by a control system involving a feedback loop. For a simple system, as considered here, there is no reason to attempt this, as the stationary-state at C_1 (high conversion) would normally be the desired state. The initial conditions in startup, such as c_A at $t = 0$, should be set to achieve this state (see comment following Example 14-7).

14.3.4.2 *Operation at Specified $\dot{Q}$; Autothermal Operation*

If feed at a specified rate and T_o enters a CSTR, the steady-state values of the operating temperature T and the fractional conversion f_A (for A $\rightarrow$ products) are not known *a priori*. In such a case, the material and energy balances must be solved simultaneously for T and f_A. This can give rise to multiple stationary states for an exothermic reaction, but not for an endothermic reaction.

For an exothermic reaction, since a resulting stationary-state may be at either a relatively low f_A ("quench" region) or a relatively high f_A ("sustained" region), for the same operating conditions, it is important to choose conditions, including $\dot{Q}$, to avoid the quench region. A related type of behavior is autothermal operation, in which the sustained region is achieved without any energy input (heat transfer) *to* the system (operation may be adiabatic or there may be heat transfer *from* the system); an autothermal reaction is thermally self-sustaining (e.g., a bunsen-burner flame, once the gas is ignited).

EXAMPLE 14-8

A first-order liquid-phase reaction, A → products, is conducted in a 2000-L CSTR. The feed contains pure A, at a rate of 300 L min^{-1}, with an inlet concentration of 4.0 mol L^{-1}. The following additional data are available:

$$c_P = 3.5 \text{ J g}^{-1} \text{ K}^{-1}; \quad \rho = 1.15 \text{ g cm}^{-3}; \quad \Delta H_{RA} = -50 \text{ kJ mol}^{-1}$$

(We assume that these values are constants.)

$$k_A = Ae^{-E_A/RT} = 2.4 \times 10^{15} e^{-12,000/T} \text{ min}^{-1}, \text{ with } T \text{ in K}$$

Determine f_A and T at steady-state based upon adiabatic operation, if: (a) $T_o = 290$ K; (b) $T_o = 298$ K; (c) $T_o = 305$ K.

SOLUTION

We first obtain f_A explicitly in terms of T from both the material and energy balances, and then examine the solutions graphically. From equation 14.3-5, the material balance, and the given rate law:

$$f_A = \frac{(-r_A)V}{F_{Ao}} = \frac{k_A c_A V}{c_{Ao} q_o} = \frac{Ae^{-E_A/RT} c_{Ao}(1-f_A)\tau}{c_{Ao}}$$

From this,

$$f_A = \frac{1}{\frac{e^{+E_A/RT}}{A\tau} + 1} \tag{A}$$

From the energy balance (equation 14.3-10) for adiabatic operation $[\dot{Q} = UA_c(T_c - T) = 0]$,

$$f_A = \frac{\dot{m} c_P}{(-\Delta H_{RA}) F_{Ao}} (T - T_o) = \frac{\rho c_P}{(-\Delta H_{RA}) c_{Ao}} (T - T_o) \tag{B}$$

since $\dot{m} = \rho q_o$ and $F_{Ao} = c_{Ao} q_o$.

Equation (A) represents a sigmoidal curve, and equation (B) represents a straight line with a positive slope for an exothermic reaction (and a negative slope for an endothermic reaction).

With numerical values inserted, equations (A) and (B) become, respectively,

$$f_A = (6.25 \times 10^{-17} e^{12,000/T} + 1)^{-1} \tag{A'}$$
$$f_A = 0.0201(T - T_o) \tag{B'}$$

The solution of equations (A′) and (B′) gives the stationary-state values of f_A and T.

The solutions for the three cases (a), (b), and (c) are shown graphically in Figure 14.5. The sigmoid-shaped curve is constructed from (A′), and the three straight lines from (B′), each with slope 0.0201 K^{-1}, for the three values of T_o. Note that for adiabatic operation, $T = T_o$ at $f_A = 0$.

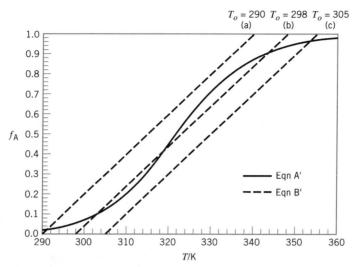

Figure 14.5 Graphical solution for Example 14-8

(a) $T_o = 290$ K. There is only one stationary-state at the intersection of the straight line and the sigmoidal curve:

$$f_A = 0.02;\ T = 291 \text{ K (quench region)}$$

(b) $T_o = 298$ K. Three possible steady-states exist:
  (i) $f_A = 0.08;\ T = 302$ K (quench region)
  (ii) $f_A = 0.42;\ T = 319$ K
  (iii) $f_A = 0.91;\ T = 343$ K (sustained region)

(c) $T_o = 305$ K. There is one steady-state:

$$f_A = 0.96;\ T = 353 \text{ K (sustained region)}$$

The values and the nature of the solution are sensitive to the value of T_o in a narrow range. For adiabatic operation of the reactor, T_o can be adjusted by means of a heat exchanger upstream of the reactor.

Both the steady-state and transient behavior of this system can be examined using the E-Z Solve software. The steady-state operating conditions can be determined for each T_o (file ex14-8.msp)). For $T_o = 298$ K, where multiple steady-states exist, different initial estimates of T and f_A lead to different solutions, corresponding to the different steady-state conditions. Transient behavior (file ex14-8.msp) can be used to predict the steady-state operating condition arising from a particular set of starting conditions ($f_A(0), T(0)$) in the reactor.

The existence of possible multiple stationary-states, as illustrated in Example 14-8, raises further questions (for adiabatic operation):

(1) What is the range of T_o values for which multiple stationary-states exist?
(2) Are all multiple states stable?
(3) Which of the possible multiple states actually occurs in steady-state operation?

The answer to question (1) is illustrated graphically in Figure 14.6 (not specifically for Example 14-8). Again, the sigmoidal curve is constructed from the material balance, equation 14.3-5 (equation (A) in Example 14-8 or its equivalent for other rate laws). The two straight lines corresponding to feed temperatures T_o' and T_o'' are constructed

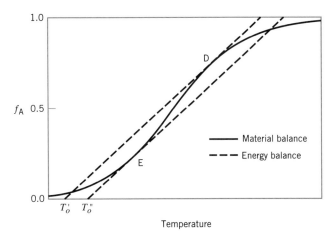

Figure 14.6 Illustration of range of feed temperatures (T'_o to T''_o) for multiple stationary-states in CSTR for adiabatic operation (autothermal behavior occurs for $T_o \geq T''_o$)

from the energy balance, equation 14.3-10 (equation (B) in Example 14-8), with slope $\rho c_P/(-\Delta H_{RA})c_{Ao}$. These two lines are drawn tangent to the sigmoidal curve at points D and E. Multiple steady-states exist only for $T'_o < T_o < T''_o$. For $T_o < T'_o$, there is one stationary-state in the quench region. For $T_o > T''_o$, there is one stationary-state in the sustained region; this is the autothermal region with respect to feed temperature, and T''_o is commonly referred to as the ignition temperature.

The answer to question (2) raised above is more easily seen if we translate Figure 14.5 into an enthalpy-temperature diagram, and then consider the stationary-states as those resulting from balancing the rate of enthalpy generation by reaction with the rate of enthalpy removal by flow (we are still considering adiabatic operation for an exothermic reaction).

$$\text{rate of enthalpy generation} = H_{gen} = (-\Delta H_{RA})(-r_A)V = (-\Delta H_{RA})F_{Ao}f_A$$
$$= \frac{(-\Delta H_{RA})F_{Ao}}{\frac{e^{E_A/RT}}{A\tau} + 1} \qquad (14.3\text{-}27)$$

from equation (A) of Example 14-8 for a first-order reaction.

$$\text{rate of enthalpy removal} = H_{rem} = \dot{m}c_P(T - T_o) \qquad (14.3\text{-}28)$$

At steady-state operation,

$$H_{gen} = H_{rem}$$

The steady-state is represented graphically by the intersection of the expressions given by equations 14.3-27 and -28. This is illustrated in Figure 14.7. The enthalpy-generation curve given by equation 14.3-27 is sigmoidal, since it corresponds to the sigmoidal curve for f_A in Figure 14.5 multiplied by the constant $(-\Delta H_{RA})F_{Ao}$. Similarly, equation 14.3-28, for enthalpy removal, generates a straight line corresponding to the line (b) in Figure 14.5. The three possible stationary-states are shown by the intersections at C_1, C_2 and C_3. We consider the stability of each of these with respect to the inevitable small fluctuations in process conditions at steady-state, of which T is one example. We ask whether a fluctuation (+ or −) tends to grow to render the (presumed) stationary-state unstable, or tends to be damped-out to confirm the stability of the stationary-state. Consider first point C_1 at T_1. If T fluctuates above T_1, H_{gen} (full line) increases, but

354 Chapter 14: Continuous Stirred-Tank Reactors (CSTR)

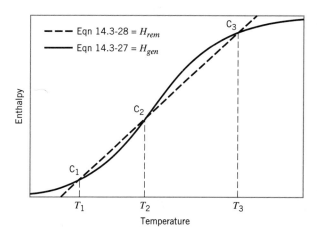

Figure 14.7 Representation of multiple stationary-states on an enthalpy-temperature diagram corresponding to (b) in Figure 14.5

H_{rem} (dashed line) increases faster, so that T is restored to T_1; if T fluctuates below T_1, H_{gen} decreases, but H_{rem} decreases faster, so that, again, T is restored to T_1. Hence C_1 represents a stable stationary-state. The same reasoning applies to point C_3 at T_3, which also represents a stable stationary-state. For point C_2 at T_2, the story is different. If T fluctuates above T_2, H_{gen} (full line) is greater than H_{rem} (dashed line), and T continues to climb until point C_3 is reached, at which point H_{gen} and H_{rem} are again equal. Thus, an upward fluctuation in T makes C_2 unstable relative to C_3. Similarly, if T fluctuates below T_2, C_2 is unstable relative to C_1. Since fluctuations in T are random with respect to direction, it is immaterial whether C_2 disappears in favor of C_1 or C_3. The important conclusion is that C_2 is not a stable stationary-state. This leaves C_1 and C_3 as the only possible stationary-states.

Question (3) raised above relates to which of the two possible stable stationary-states actually occurs. For any feed temperature between T_o' and T_o'' in Figure 14.6, it is not possible to predict which of the two would prevail for given initial conditions. The stationary-state in the sustained region (C_3 in Figure 14.7) is certainly preferred to that in the quench region (C_1), if a high value of f_A is desired. To avoid the ambiguity of multiple stationary-states, the feed temperature T_o should be chosen to avoid them altogether, and thus should be higher than the ignition temperature (T_o'' in Figure 14.6).

For nonadiabatic operation and an exothermic reaction, the overall situation is similar, but is complicated by the inclusion of the heat transfer term for $\dot{Q}$ in the energy balance in equation 14.3-10.

For an endothermic reaction, whether the operation is adiabatic or nonadiabatic, there is no possibility of multiple stationary-states because of the negative slope of the f_A versus T relation in the energy balance (see equation 14.3-11). This is illustrated schematically in Figure 14.8.

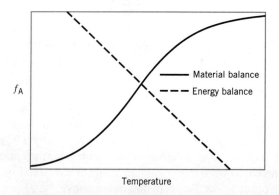

Figure 14.8 Illustration of solution of material and energy balances for an endothermic reaction in a CSTR (no multiple stationary-states possible)

14.4 MULTISTAGE CSTR

A multistage CSTR consists of two or more stirred tanks in series. Are there advantages in using a multistage CSTR consisting of two or more relatively small tanks rather than a single tank of relatively large size to achieve the same result? Recall that a major disadvantage of a CSTR is that it operates at the lowest concentration between inlet and outlet. For a single CSTR, this means that operation is at the lowest possible concentration in the system, and for normal kinetics, the required volume of reactor is the largest possible. If two tanks (operating at the same T) are arranged in series, the second operates at the same concentration as the single tank above, but the first operates at a higher concentration, so that the total volume of the two is smaller than the volume of the single vessel. If more than two tanks are used, the volume advantage becomes even greater. Furthermore, there is an opportunity to adjust temperature and/or concentration (or conversion) between stages to achieve an optimal situation with respect to total volume for a given number of stages (N). If N is chosen as a parameter and allowed to vary, there is a practical limit to the number of stages chosen (usually about five), since the cost of the interconnecting piping and valving increases. Also, although the sizes of the N stages may not be equal for an optimal V, there is an advantage in the cost of fabrication in making them all the same size. Thus, a minimum value of V is not necessarily the best answer to the design of a CSTR, but it is one to consider.

In this section, we introduce some of the basic considerations for design of a multistage CSTR, treating constant-density systems in both isothermal (all stages at same T) and nonisothermal operation. The arrangement and notation (for reaction A + ... → products, where A is the limiting reactant) are shown in Figure 14.9. We restrict attention to steady-state operation. There may be external heat exchangers between stages or internal heat exchangers within the vessels (not shown in Figure 14.9).

As for a single-stage CSTR, the volume of each stage is obtained from the material balance around that stage, together with the rate law for the system. Thus, for the ith stage, the material-balance equation (14.3-5) becomes

$$V_i = F_{Ao}(f_{Ai} - f_{A,i-1})/(-r_A)_i \qquad (14.4\text{-}1)$$

Note that, because of the definition of f_A, F_{Ao} is the original feed rate to the CSTR and *not* that to the ith stage.

A graphical interpretation to illustrate the use of equation 14.4-1, and also to demonstrate the reduction in V as N increases, is shown in Figure 14.10 for $N = 3$. This may be compared with Figure 14.2 for $N = 1$. In Figure 14.10, EB is the reciprocal-rate curve obtained from kinetics data (for normal kinetics). The shaded rectangular area FGMD represents V_1/F_{Ao} for the first stage operating at rate $(-r_A)_1$ to achieve conversion f_{A1}. Similarly, area HKLM represents V_2/F_{Ao}, and area JBCL represents V_3/F_{Ao}. The total volume required to achieve conversion f_{A3} in three stages is $V(3) = V_1 + V_2 + V_3$

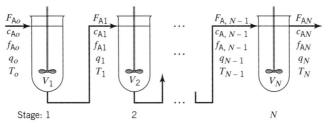

Figure 14.9 Arrangement and notation of N-stage CSTR (for limiting reactant A in A + ... → products)

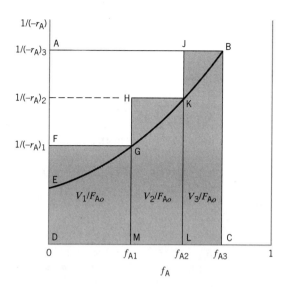

Figure 14.10 Graphical interpretation of equation 14.4-1; total volume $V = V_1 + V_2 + V_3$ (shaded areas)

(shaded areas). This may be compared with the volume $V(1)$ required to achieve the same conversion; the large rectangular area ABCD represents $V(1)/F_{Ao}$ (as in Figure 14.2). Comparison of the areas shows that $V(3) < V(1)$, the difference being proportional to the irregular area AJKHGF.

As N increases, for a given f_A, the difference $V(N) - V(1)$ increases until it approaches the area bounded by AB, AE and the reciprocal-rate curve EB as $N \to \infty$. From this limiting behavior, the curve EB can be interpreted as the locus of operating points for a multistage CSTR consisting of an infinite number of stages in series, which is equivalent to a PFR (Chapter 15).

From the discussion relating to Figure 14.10, we then conclude that

$$V(1) > V(N) \qquad \text{for given } F_{Ao}, f_A, \text{ (normal) kinetics} \qquad (14.4\text{-}2)$$

Conversely, Figure 14.10 can be used to show that

$$f_A(1) < f_A(N) \qquad \text{for given } F_{Ao}, V \text{ (total), (normal) kinetics} \qquad (14.4\text{-}3)$$

(the proof is left to problem 14-28).

The solution of equation 14.4-1 to obtain V (given f_A) or f_A (given V) can be carried out either analytically or graphically. The latter method can be used conveniently to obtain f_A for a specified set of vessels, or when $(-r_A)$ is not known in analytical form.

The basis for a graphical solution for $N = 2$ is illustrated in Figure 14.11, which shows $(-r_A)$ as a function of f_A, together with lines that represent the material balance for each stage.

For stage 1, from equation 14.4-1,

$$(-r_A)_1 = (F_{Ao}/V_1) f_{A1}$$

This is the straight line AB, relating $(-r_A)$ to f_A for a given V_1, with slope equal to F_{Ao}/V_1. The solution for the outlet conversion f_{A1} is given at E, the intersection of AB with the rate curve. Similarly, for stage 2,

$$(-r_A)_2 = (F_{Ao}/V_2)(f_{A2} - f_{A1})$$

14.4 Multistage CSTR

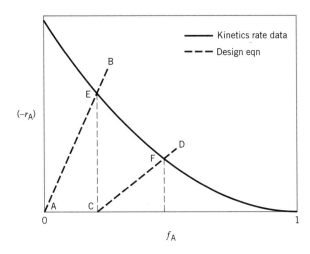

Figure 14.11 Basis for graphical solution for multistage CSTR (for A + ... → products)

This relation is represented by the straight line CD, which intersects the rate curve at F to give the value of f_{A2}. E and F are thus the operating points for the two stages.

Optimal operation of a multistage CSTR can be considered, for example, from the point of view of minimizing the total volume V for a given throughput (F_{Ao}) and fractional conversion (f_A). This involves establishing an objective function for V from the material balance together with a rate law and energy balance as required. The situation can become quite complicated, since V depends on interstage conversions and temperatures, as well as on heat transfer parameters for internal and/or interstage heat exchangers. We consider a relatively simple case as an example in the next section.

14.4.1 Constant-Density System; Isothermal Operation

The following example illustrates both an analytical and a graphical solution to determine the outlet conversion from a three-stage CSTR.

EXAMPLE 14-9

A three-stage CSTR is used for the reaction A → products. The reaction occurs in aqueous solution, and is second-order with respect to A, with $k_A = 0.040 \text{ L mol}^{-1} \text{ min}^{-1}$. The inlet concentration of A and the inlet volumetric flow rate are 1.5 mol L^{-1} and 2.5 L min^{-1}, respectively. Determine the fractional conversion (f_A) obtained at the outlet, if $V_1 = 10$ L, $V_2 = 20$ L, and $V_3 = 50$ L, (a) analytically, and (b) graphically.

SOLUTION

(a) Each stage is considered in sequence by means of equation 14.4-1 and the given rate law.

For stage 1, from the rate law,

$$(-r_A)_1 = k_A c_{A1}^2 = k_A c_{Ao}^2 (1 - f_{A1})^2$$

since density is constant.

From equation 14.4-1,

$$(-r_A)_1 = (F_{Ao}/V_1)(f_{A1} - 0)$$

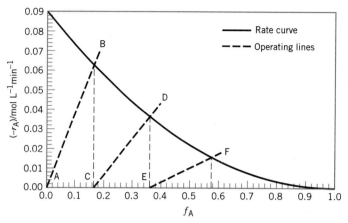

Figure 14.12 Graphical solution of Example 14-9

Equating the two right sides and rearranging, we obtain the quadratic equation

$$f_{A1}^2 - \left(\frac{F_{Ao}}{k_A c_{Ao}^2 V_1} + 2\right) f_{A1} + 1 = 0$$

or, with numerical values inserted,

$$f_{A1}^2 - 6.167 f_{A1} + 1 = 0$$

from which $f_{A1} = 0.167$.

Similarly, for stages 2 and 3, we obtain $f_{A2} = 0.362$, and $f_{A3} = 0.577$, which is the outlet fractional conversion from the three-stage CSTR.

The system of three equations based on equation 14.4-1 for f_{A1}, f_{A2}, and f_{A3} may also be solved simultaneously using the E-Z Solve software (file ex14-9.msp). The same values are obtained.

(b) The graphical solution is shown in Figure 14.12. The curve for $(-r_A)$ from the rate law is first drawn. Then the operating line AB is constructed with slope $F_{Ao}/V_1 = c_{Ao}q_o/V_1 = 0.375$ mol L^{-1} min^{-1} to intersect the rate curve at $f_{A1} = 0.167$; similarly, the lines CD and EF, with corresponding slopes 0.1875 and 0.075, respectively, are constructed to intersect the rate curve at the values $f_{A2} = 0.36$ and $f_{A3} = 0.58$, respectively. These are the same values as obtained in part (a).

14.4.2 Optimal Operation

The following example illustrates a simple case of optimal operation of a multistage CSTR to minimize the total volume. We continue to assume a constant-density system with isothermal operation.

EXAMPLE 14-10

Consider the liquid-phase reaction A + ... → products taking place in a two-stage CSTR. If the reaction is first-order, and both stages are at the same T, how are the sizes of the two stages related to minimize the total volume V for a given feed rate (F_{Ao}) and outlet conversion (f_{A2})?

SOLUTION

From the material balance, equation 14.4-1, the total volume is

$$V = V_1 + V_2 = F_{Ao}\left[\frac{f_{A1} - 0}{(-r_A)_1} + \frac{f_{A2} - f_{A1}}{(-r_A)_2}\right] \quad \text{(A)}$$

From the rate law,

$$(-r_A)_1 = k_A c_{Ao}(1 - f_{A1}) \quad \text{(B)}$$
$$(-r_A)_2 = k_A c_{Ao}(1 - f_{A2}) \quad \text{(C)}$$

Substituting (B) and (C) in (A), we obtain

$$V = \frac{F_{Ao}}{k_A c_{Ao}}\left(\frac{f_{A1}}{1 - f_{A1}} + \frac{f_{A2} - f_{A1}}{1 - f_{A2}}\right) \quad \text{(D)}$$

for minimum V,

$$\left(\frac{\partial V}{\partial f_{A1}}\right)_{f_{A2,T}} = 0 \quad \text{(E)}$$

From (E) and (D), we obtain

$$\frac{\partial V}{\partial f_{A1}} = \frac{F_{Ao}}{k_A c_{Ao}}\left[\frac{1}{(1 - f_{A1})^2} - \frac{1}{1 - f_{A2}}\right] = 0$$

from which

$$f_{A2} = f_{A1}(2 - f_{A1})$$

If we substitute this result into the material balance for stage 2 (contained in the last term in (D)), we have

$$V_2 = \frac{F_{Ao}}{k_A c_{Ao}}\left(\frac{f_{A2} - f_{A1}}{1 - f_{A2}}\right) = \frac{F_{Ao}}{k_A c_{Ao}}\left(\frac{f_{A1}}{1 - f_{A1}}\right) = V_1$$

That is, *for a first-order reaction, the two stages must be of equal size to minimize V.* The proof can be extended to an N-stage CSTR. For other orders of reaction, this result is approximately correct. The conclusion is that tanks in series should all be the same size, which accords with ease of fabrication.

Although, for other orders of reaction, equal-sized vessels do not correspond to the minimum volume, the difference in total volume is sufficiently small that there is usually no economic benefit to constructing different-sized vessels once fabrication costs are considered.

EXAMPLE 14-11

A reactor system is to be designed for 85% conversion of A (f_A) in a second-order liquid-phase reaction, A → products; $k_A = 0.075$ L mol^{-1} min^{-1}, $q_o = 25$ L min^{-1}, and $c_{Ao} =$

0.040 mol L^{-1}. The design options are:

(a) two equal-sized stirred tanks in series;
(b) two stirred tanks in series to provide a minimum total volume.

The cost of a vessel is \$290 m^{-3}, but a 10% discount applies if both vessels are the same size and geometry. Which option leads to the lower capital cost?

SOLUTION

Case (a). From the material-balance equation 14.4-1 applied to each of the two vessels 1 and 2,

$$V_1 = F_{Ao} f_{A1}/k_A c_{A1}^2 = F_{Ao} f_{A1}/k_A c_{Ao}^2 (1 - f_{A1})^2 \quad \text{(A)}$$
$$V_2 = F_{Ao}(f_{A2} - f_{A1})/k_A c_{Ao}^2 (1 - f_{A2})^2 \quad \text{(B)}$$

Equating V_1 and V_2 from (A) and (B), and simplifying, we obtain

$$(f_{A2} - f_{A1})/(1 - f_{A2})^2 = f_{A1}/(1 - f_{A1})^2$$

This is a cubic equation for f_{A1} in terms of f_{A2}:

$$f_{A1}^3 - (2 + f_{A2})f_{A1}^2 + (2 + f_{A2}^2)f_{A1} - f_{A2} = 0$$

or, on substitution of $f_{A2} = 0.85$,

$$f_{A1}^3 - 2.85 f_{A1}^2 + 2.7225 f_{A1} - 0.85 = 0$$

This equation has one positive real root, $f_{A1} = 0.69$, which can be obtained by trial or by means of the E-Z Solve software (file ex14-11.msp).

This corresponds to $V_1 = V_2 = 5.95 \times 10^4$ L (from equation (A) or (B)) and a total capital cost of $0.9(290)(5.95 \times 10^4)2/1000 = \$31,000$ (with the 10% discount taken into account).

Case (b). The total volume is obtained from equations (A) and (B):

$$V = V_1 + V_2 = \frac{F_{Ao} f_{A1}}{k_A c_{Ao}^2 (1 - f_{A1})^2} + \frac{F_{Ao}(f_{A2} - f_{A1})}{k_A c_{Ao}^2 (1 - f_{A2})^2}$$

For minimum V,

$$\left(\frac{\partial V}{\partial f_{A1}}\right)_{f_{A2}} = \frac{F_{Ao}}{k_A c_{Ao}^2} \left[\frac{1 + f_{A1}}{(1 - f_{A1})^3} - \frac{1}{(1 - f_{A2})^2}\right] = 0$$

This also results in a cubic equation for f_{A1}, which, with the value $f_{A2} = 0.85$ inserted, becomes

$$f_{A1}^3 - 3 f_{A1}^2 + 3.0225 f_{A1} - 0.9775 = 0$$

Solution by trial or by means of the E-Z Solve software (file ex14-11.msp) yields one positive real root: $f_{A1} = 0.665$. This leads to $V_1 = 4.95 \times 10^4$ L, $V_2 = 6.84 \times 10^4$ L, and a capital cost of \$34,200.

14.5 PROBLEMS FOR CHAPTER 14

14-1 Suppose the liquid-phase hydration of ethylene oxide (A) to ethylene glycol, $C_2H_4O(A) + H_2O \rightarrow C_2H_6O_2$, takes place in a CSTR of volume (V) 10,000 L; the rate constant is $k_A = 2.464 \times 10^{-3}$ min^{-1} (Brönsted et al., 1929; see Example 4-3).
(a) Calculate the steady-state fractional conversion of A (f_A), if the feed rate (q) is 0.30 L s^{-1} and the feed concentration (c_{Ao}) is 0.120 mol L^{-1}.
(b) If the feed rate suddenly drops to 70% of its original value, and is maintained at this new value, (i) what is the value of f_A after 60 min, and (ii) how close (%) is this value to the new steady-state value?
(c) What is the ratio of the *steady-state production rate* in (b) to that in (a)?

14-2 Reactant A is fed (at $t = 0$) at a constant rate of 5.0 L s^{-1} to an empty 7000-L CSTR until the CSTR is full. Then the outlet valve is opened. If the rate law for the reaction A $\rightarrow$ products is $(-r_A) = k_A c_A$, where $k_A = 8.0 \times 10^{-4}$ s^{-1}, and if the inlet and outlet rates remain constant at 5.0 L s^{-1}, calculate c_A at $t = 15$ min and at $t = 40$ min. Assume that the temperature and density of the reacting system are constant, and that $c_{Ao} = 2.0$ mol L^{-1}.

14-3 An aqueous solution of methyl acetate (A) enters a CSTR at a rate of 0.5 L s^{-1} and a concentration (c_{Ao}) of 0.2 mol L^{-1}. The tank is initially filled with 2000 L of water so that material flows out at a rate of 0.5 L s^{-1}. A negligibly small stream of HCl (catalyst) is added to the entering solution of acetate so that the concentration of acid in the tank is maintained at 0.1 mol L^{-1}, in which case the hydrolysis of acetate occurs at a rate characterized by $k_A = 1.1 \times 10^{-4}$ s^{-1}. What is the concentration of acetate in the outlet stream at the end of 30 min, and what is the eventual steady-state concentration?

14-4 For the reaction between hydrocyanic acid (HCN) and acetaldehyde (CH$_3$CHO) in aqueous solution,

$$HCN(A) + CH_3CHO(B) \rightarrow CH_3CH(OH)CN$$

the rate law at 25°C and a certain pH is $(-r_A) = k_A c_A c_B$, where $k_A = 0.210$ L mol^{-1} min^{-1} (see problem 4-6). If the reaction is carried out at steady-state at 25°C in a CSTR, how large a reactor (V/L) is required for 75% conversion of HCN, if the feed concentration is 0.04 mol L^{-1} for each reactant, and the feed rate is 2 L min^{-1}?

14-5 The liquid-phase reaction A $\rightarrow$ B + C takes place in a single-stage CSTR. The rate law for the reaction is $(-r_A) = k_A c_A c_B$.
(a) Calculate the feed rate, q_o, that maximizes the rate of production of C.
(b) For the feed rate in (a), calculate the outlet values of c_A, c_B, c_C, and f_A.
Data: $k_A = 0.005$ L mol^{-1} s^{-1}; $c_{Ao} = 0.2$ mol L^{-1}; $c_{Bo} = 0.01$ mol L^{-1}; $V = 10,000$ L.

14-6 A pure gaseous reactant A is fed at a steady rate (q_o) of 30 L h^{-1} and a concentration (c_{Ao}) of 0.1 mol L^{-1} into an experimental CSTR of volume (V) 0.1 L, where it undergoes dimerization (2A $\rightarrow$ A$_2$). If the steady-state outlet concentration (c_A) is 0.0857 mol L^{-1}, and if there is no change in T or P, calculate:
(a) the fractional conversion of A, f_A;
(b) the outlet flow rate, q;
(c) the rate of reaction, $(-r_A)$/mol L^{-1} h^{-1};
(d) the space time (τ/h) based on the feed rate;
(e) the mean residence time $\bar{t}$/h; and
(f) briefly explain any difference between τ and $\bar{t}$.

14-7 Calculate the mean residence time ($\bar{t}$) and the space time (τ) for a reaction in a CSTR for the following cases (explain any differences between τ and $\bar{t}$):
  (a) homogeneous liquid-phase reaction: $V = 75$ L, $q_o = 2.5$ L min^{-1};
  (b) homogeneous gas-phase reaction, A → B + C; $V = 75$ L, $q_o = 150$ L min^{-1} at 300 K (T_o). The temperature at the outlet of the reactor is 375 K, the pressure is constant at 100 kPa, and f_A is 0.50.

14-8 A gas-phase hydrogenation reaction, $C_2H_4(A) + H_2 \rightarrow C_2H_6$, is conducted under isothermal and isobaric conditions in a CSTR. The feed, consisting of equimolar amounts of each reactant and an inert, is fed to the reactor at a rate of 1.5 mol H_2 min^{-1}, with $c_{Ao} = 0.40$ mol L^{-1}. Determine the volumetric flow rate at the inlet of the reactor and the mean residence time, if the desired fractional conversion f_A is 0.85, and the reaction is first-order with respect to each reactant ($k_A = 0.5$ L mol^{-1} min^{-1}).

14-9 A first-order, liquid-phase endothermic reaction, A → B + C, takes place in a CSTR operating at steady-state. If the feed temperature (T_o) is 310 K, at what temperature (T_c) must a heating coil in the tank be maintained in order to achieve a conversion (f_A) of 0.75?
Given: (Arrhenius parameters) $A = 3.5 \times 10^{13}$ s^{-1}, $E_A = 100{,}000$ J mol^{-1}; $q = 8.3$ L s^{-1}; $V = 15{,}000$ L; $c_{Ao} = 0.80$ mol L^{-1}; $\Delta H_{RA} = 51{,}000$ J mol^{-1}; $\rho = 950$ g L^{-1}; $c_P = 3.5$ J g^{-1} K^{-1}; $UA = 10{,}000$ J s^{-1} K^{-1}.

14-10 A first-order, liquid-phase, endothermic reaction, A → products, takes place in a CSTR. The feed concentration (c_{Ao}) is 1.25 mol L^{-1}, the feed rate (q) is 20 L s^{-1}, and the volume of the reactor (V) is 20,000 L. The enthalpy of reaction (ΔH_{RA}) is $+50{,}000$ J mol^{-1}, the specific heat of the reacting system is 3 J g^{-1} K^{-1}, and its density is 900 g L^{-1}.
  (a) If the feed temperature (T_o) is 300 K, at what temperature (T_c) must a heating coil in the tank be maintained to keep the temperature (T) of the reaction in the tank the same as T_o? What conversion is obtained? At 300 K, the rate constant $k_A = 4 \times 10^{-3}$ s^{-1}. For the coil, assume $UA = 10{,}000$ J s^{-1} K^{-1}.
  (b) If there were no heating coil in the tank, what should the feed temperature (T_o) be to achieve the same result?

14-11 A liquid-phase, exothermic first-order reaction (A → products) is to take place in a 4000-L CSTR. The Arrhenius parameters for the reacting systems are $A = 2 \times 10^{13}$ s^{-1} and $E_A = 100$ kJ mol^{-1}; the thermal parameters are $\Delta H_{RA} = -50$ kJ mol^{-1} and $c_P = 4$ J g^{-1} K^{-1}; the density (ρ) is 1000 g L^{-1}.
  (a) For a feed concentration (c_{Ao}) of 4 mol L^{-1}, a feed rate (q) of 5 L s^{-1}, and adiabatic operation, determine f_A and T at steady-state, if the feed temperature (T_o) is (i) 290 K; (ii) 297 K; and (iii) 305 K.
  (b) What is the minimum value of T_o for autothermal behavior, and what are the corresponding values of f_A and T at steady-state?
  (c) For $T_o = 305$ K as in (a)(iii), explain, without doing any calculations, what would happen eventually if the feed rate (q) were increased.
  (d) If the result in (c) is adverse, what change could be made to offset it at the higher throughput?
  (e) Suppose the feed temperature (T_o) is 297 K, and it is desired to achieve a steady-state conversion (f_A) of 0.932 without any alternative possibility of steady-state operation in the "quench region." If a fluid stream is available at 360 K (T_c, assume constant) for use in a heat exchanger within the tank, what value of UA (J K^{-1} s^{-1}) would be required for the heat exchanger? Show that the "quench" region is avoided.

14-12 For the gas-phase, second-order reaction $C_2H_4 + C_4H_6 \rightarrow C_6H_{10}$ (or A + B → C) carried out adiabatically in a 2-liter experimental CSTR at steady-state, what should the temperature (T/K) be to achieve 40% conversion, if the (total) pressure (P) is 1.2 bar (assume constant), the feed rate (q_o) is 20 cm^3 s^{-1}, and the reactants are equimolar in the feed. The Arrhenius parameters are $E_A = 115{,}000$ J mol^{-1} and $A = 3.0 \times 10^7$ L mol^{-1} s^{-1} (Rowley and Steiner, 1951; see Example 4-8). Thermochemical data are as follows (from Stull et al., 1969):

T/K	C_P/J mol^{-1} K^{-1}			ΔH_f°/kJ mol^{-1}		
	C_2H_4	C_4H_6	C_6H_{10}	C_2H_4	C_4H_6	C_6H_{10}
600	71.5	133.2	206.9	44.35	100.88	−24.81
700	78.5	144.6	229.8	42.47	98.87	−28.28
800	84.5	154.1	248.9	40.88	97.28	−30.63
900	89.8	162.4	265.0	39.54	96.02	−32.01

14-13 Determine the volume of a CSTR required to obtain 65% conversion of A in A → P, given that $q_o = 5$ L s^{-1}, the density of the system is constant, $c_{Ao} = 4$ mol L^{-1}, and $k_A = 2.5 \times 10^{-3}$ s^{-1}.

14-14 For the reaction in problem 14-13, determine the exit fractional conversion for a two-stage CSTR, if the volume of each CSTR is equal to one-half the reactor volume calculated in problem 14-13.

14-15 For the reaction in problem 14-13, determine the minimum reactor volume required for a two-stage CSTR used to achieve 65% conversion of A. What is the volume of each tank?

14-16 A liquid-phase reaction, A + 2B → C, is to be conducted in two equal-sized CSTRs. The reaction is first order with respect to both A and B, with $k_A = 6.0 \times 10^{-2}$ L mol^{-1} s^{-1}. The volumetric feed rate is 25 L min^{-1}, $c_{Ao} = 0.1$ mol L^{-1}, and $c_{Bo} = 0.20$ mol L^{-1}; 75% conversion of A is desired.

(a) The CSTRs are arranged in series. What is the volume of each?
(b) The CSTRs are arranged in parallel. What is the volume of each, if the volumetric flow rate through each reactor is equal?

14-17 Using the data and three reactors described in Example 14-9, determine the reactor configuration (i.e., order) that maximizes the fractional conversion of A.

14-18 A second-order reaction of A → B, with $k_A = 2.4$ L mol^{-1} h^{-1}, is to be conducted in up to two CSTRs arranged in series. Determine the minimum reactor volume required to achieve 66.7% conversion of A, given that the feed rate and volumetric flow rate at the inlet are 3.75 mol min^{-1} and 2.5 L min^{-1}, respectively. The feed consists of pure A.

14-19 The liquid-phase reaction A + B → C is to be carried out in two tanks (CSTRs) in series. The rate law for the reaction is $(-r_A) = k_A c_A c_B$, where $k_A = 9.5 \times 10^{-2}$ L mol^{-1} s^{-1}. If $c_{Ao} = c_{Bo} = 0.08$ mol L^{-1} and the feed rate is 48 L min^{-1}, what is the minimum total volume required to achieve 90% conversion? What is the volume of tank 1 and of tank 2?

14-20 (a) Two CSTRs of unequal (but fixed) size are available to form a reactor system for a particular first-order, liquid-phase reaction at a given temperature. How should they be arranged to achieve the most efficient operation for a given throughput? Justify your conclusion quantitatively.
(b) Repeat (a), if the reaction is second-order.

14-21 A liquid-phase reaction takes place in two CSTRs operating (at steady-state) in parallel at the same temperature. One reactor is twice the size of the other. The (total) feed is split appropriately between the two reactors to achieve the highest fractional conversion of reactant, which is 0.70. The smaller reactor is then taken out of service. If the (total) feed rate remains the same, what is the resulting conversion from the larger reactor? Assume the reaction is first-order, and the larger reactor is operating at steady-state.

14-22 The liquid-phase reaction A → B + C takes place in a series of three CSTRs with an overall conversion (f_A) of 0.90. One of the reactors develops a leak and is removed from the series. The throughput (q) is then reduced in order to keep the overall conversion at 0.90. If the original throughput was 250 L min^{-1}, what is the new throughput? Assume: (i) steady-state operation; (ii) the reactors are of equal volume and operate at the same temperature; (iii) the reaction is first-order in A.

14-23 An aqueous solution of ethyl acetate (A) with a concentration of 0.3 mol L^{-1} and flowing at 0.5 L s^{-1} mixes with an aqueous solution of NaOH (B) with a concentration of 0.45 mol L^{-1} and flowing at 1.0 L s^{-1}. The combined stream enters a reactor system for saponification to occur at 16°C, at which temperature the rate constant (k_A) is 0.0574 L mol^{-1} s^{-1}. Conversion (f_A) is to be 80% at steady-state.
  (a) If the reactor system is a single CSTR, what volume (V) is required?
  (b) If the reactor system is made up of two equal-sized CSTRs in *series*, what is the volume of each?
  (c) If the reactor system is made up of two equal-sized CSTRs in *parallel*, what is the volume of each?

14-24 A single CSTR of volume V is to be replaced in an emergency by three tanks of volumes $V/2$, $3V/8$, and $V/8$. If the same performance with a given reaction and throughput is to be achieved, how should the tanks be arranged, and how should the total volumetric throughput (q) be distributed? Sketch a flow diagram to show these.

14-25 For a first-order liquid-phase reaction, A → products, conducted at steady-state in a series of N CSTRs, develop a mathematical expression which may be used to relate directly c_{AN}, the concentration at the outlet of the Nth reactor, with c_{Ao}, the inlet concentration to the first reactor. Can a similar expression be developed for a second-order reaction?

14-26 What effect does recycling have on the operation of a single-stage CSTR? Consider the material balance for reactant A in a liquid-phase reaction with a recycle stream (from outlet to inlet) of Rq_o, where q_o is the volumetric feed rate and R is the recycle ratio. Compare this material balance with that for no recycle ($R = 0$).

14-27 (a) Derive the expression in Table 14.1 for f_A for segregated flow in a CSTR for a zero-order reaction.
  (b) Repeat (a) for a half-order reaction ($n = 1/2$), and compare with the result for nonsegregated flow to confirm the conclusion in relation 14.3-26 (in this case, the results are not given in Table 14.1).

14-28 Prove relation 14.4-3 for $N = 2$. (An analytical/graphical proof may be based on Figure 14.10.)

Chapter 15

Plug Flow Reactors (PFR)

In this chapter, we develop the basis for design and performance analysis for a plug flow reactor (PFR). Like a CSTR, a PFR is usually operated continuously at steady-state, apart from startup and shutdown periods. Unlike a CSTR, which is used primarily for liquid-phase reactions, a PFR may be used for either gas-phase or liquid-phase reactions.

The general features of a PFR are outlined in Section 2.4.1, and are illustrated schematically in Figure 2.4. Residence-time distribution (RTD) functions are described in Section 13.4.2. The essential features are recapitulated as follows:

(1) The flow pattern is PF, and the PFR is a "closed" vessel (Section 13.2.4).
(2) The volumetric flow rate may vary continuously in the direction of flow because of a change in density.
(3) Each element of fluid is a closed system (cf. CSTR); that is, there is no axial mixing, although there is complete radial mixing (in a cylindrical vessel).
(4) As a consequence of (3), fluid properties may change continuously in the axial direction (see Figure 2.4), but are constant radially (at a given axial position).
(5) Each element of fluid has the same residence time t as any other (cf. CSTR); the RTD functions E and F are shown in Figures 13.6 and 13.7, respectively; the former is represented by the "vertical spike" of the Dirac delta function, and the latter by the step function.

In the development to follow, we first consider some uses of a PFR. After presenting design and performance equations in general, we then apply them to cases of constant-density and variable-density operations. These include both isothermal and nonisothermal operation, as well as the possibility of significantly nonisobaric operation. The case of recycle operation is then developed. The chapter concludes with considerations related to configurational aspects of series and parallel arrangement of PFRs. The treatment is mainly for steady-state operation, and also includes the relationship to the segregated-flow model. See Figure 2.4 for property profile (e.g., c_A for A → products in a PFR (at steady-state)). We restrict attention in this chapter to simple, single-phase systems.

15.1 USES OF A PFR

The PFR model is frequently used for a reactor in which the reacting system (gas or liquid) flows at relatively high velocity (high Re, to approach PF) through an otherwise empty vessel or one that may be packed with solid particles. There is no device, such as a stirrer, to promote backmixing. The reactor may be used in large-scale operation

366 Chapter 15: Plug Flow Reactors (PFR)

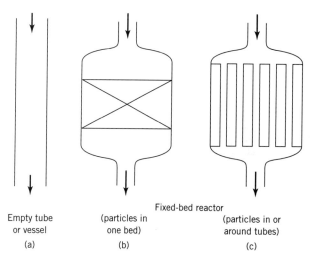

Empty tube
or vessel
(a)

Fixed-bed reactor
(particles in
one bed)
(b)

(particles in or
around tubes)
(c)

Figure 15.1 Three examples (schematic) of PFR

for commercial production, or in laboratory- or pilot-scale operation to obtain design data.

Three possible situations are illustrated schematically in Figure 15.1. In Figure 15.1(a), the reactor is represented by an empty tube or vessel, which may be vertical, as shown, or horizontal. Many single-phase reactions, such as the dehydrogenation of ethane for production of ethylene, take place in such reactors. In Figure 15.1(b), the reactor incorporates a fixed-bed of solid particles of catalyst; that is, the particles do not move. The reacting fluid moves through the void space around the particles. It may be thought at first that such an apparently tortuous flow path would be inconsistent with the occurrence of PF, but, provided certain conditions are met, such need not be the case; in any event, the departure from PF can be investigated experimentally (Chapter 19). The arrangement in Figure 15.1(b) is for adiabatic operation, as used, for example, in the dehydrogenation of ethylbenzene for the production of styrene monomer, or in the oxidation of sulfur dioxide to sulfur trioxide in the manufacture of sulfuric acid (there may be two or more such beds in series; see Figure 1.4). Figure 15.1(c) also represents a fixed-bed catalytic reactor, but one in which the catalyst particles are packed either inside or outside of tubes, so that a heat-transfer fluid on the other side may be used for control of T. This arrangement is used in reactors for ammonia synthesis and methanol synthesis (see Figures 11.5 and 11.6). This arrangement may also be used for control of T in a situation without a catalyst, as a variation of that in Figure 15.1(a). In this chapter, we consider only cases relating to Figure 15.1(a) or 15.1(c), in which no additional phase of solid particles is present. Cases corresponding to Figure 15.1(b) are taken up in Chapter 21.

15.2 DESIGN EQUATIONS FOR A PFR

15.2.1 General Considerations; Material, Energy, and Momentum Balances

The process design of a PFR typically involves determining the size of a vessel required to achieve a specified rate of production. The size is initially determined as a volume, which must then be expressed in terms of, for example, the length and diameter of a cylindrical vessel, or length and number of tubes of a given size. Additional matters to consider are effects of temperature resulting from the energetics of the reaction,

15.2 Design Equations for a PFR

and effects of pressure resulting from frictional pressure drop. If temperature effects are important, whether they involve adiabatic operation with no heat transfer surface, or nonadiabatic operation incorporating appropriate heat transfer surface, the material and energy balances must be considered together. If, in addition, pressure effects are significant, the momentum balance must be used together with the other two balances. This could be the case for a gas-phase reaction with fluid flowing at high velocity through a vessel of considerable length, as in the dehydrogenation of ethane to ethylene. We consider next the three types of balance separately in turn, and apply them in subsequent sections.

15.2.1.1 Material Balance

The material balance for a PFR is developed in a manner similar to that for a CSTR, except that the control volume is a differential volume (Figure 2.4), since properties change continuously in the axial direction. The material balance for a PFR developed in Section 2.4.2 is from the point of view of interpreting rate of reaction. Here, we turn the situation around to examine it from the point of view of the volume of reactor, V. Thus equation 2.4-4, for steady-state operation involving reaction represented by $A + \ldots \rightarrow \nu_C C + \ldots$, may be written as a differential equation for reactant A as follows:

$$\frac{dV}{df_A} - \frac{F_{Ao}}{(-r_A)} = 0 \quad (15.2\text{-}1)$$

To obtain V, this may be integrated in the form

$$V = F_{Ao} \int df_A/(-r_A) \quad (15.2\text{-}2)$$

Equation 15.2-2 may also be written in terms of space time, $\tau = V/q_o$, as

$$\tau = c_{Ao} \int df_A/(-r_A) \quad (15.2\text{-}3)$$

since $c_{Ao} = F_{Ao}/q_o$.

A graphical interpretation of equation 15.2-2 or -3 is illustrated in Figure 15.2 (for normal kinetics). The area under the curve of reciprocal rate, $1/(-r_A)$, versus f_A is equal to V/F_{Ao} or τ/c_{Ao}. Equations 15.2-1 to -3 apply whether density varies or is constant.

Equation 15.2-1 may be written in various other forms, depending on the gradient to be determined, as illustrated in Example 15-1.

EXAMPLE 15-1

Rewrite equation 15.2-1 as a differential equation representing the gradient of f_A with respect to position (x) in a PFR, df_A/dx. (Integration of this form gives the conversion profile.)

368 Chapter 15: Plug Flow Reactors (PFR)

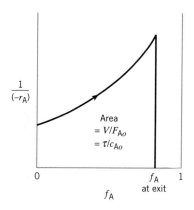

Figure 15.2 Graphical interpretation of equation 15.2-2 or -3

SOLUTION

We assume that the reactor is a cylinder of radius R. The volume of reactor from the inlet to position x is $V(x) = \pi R^2 x$, or $dV = \pi R^2 dx$. Substitution of this for dV in equation 15.2-1 and rearrangement lead to

$$\frac{df_A}{dx} - \frac{\pi R^2(-r_A)}{F_{Ao}} = 0 \qquad (15.2\text{-}4)$$

15.2.1.2 Energy Balance

The energetics of a reaction may lead to a temperature gradient along the length of a PFR, which may be modified by simultaneous heat transfer, usually through the wall of the tube or vessel. Efficient heat transfer may be required if appropriate reaction conditions are to be maintained, or if "runaway" reactions are to be avoided. To investigate these aspects, the energy balance is required.

The energy balance for a PFR, as an enthalpy balance, may be developed in a manner similar to that for a CSTR in Section 14.3.1.2, except that the control volume is a differential volume. This is illustrated in Figure 15.3, together with the symbols used.

In developing the enthalpy balance for a PFR, we consider only steady-state operation, so that the rate of accumulation vanishes. The rates of input and output of enthalpy by (1) flow, (2) heat transfer, and (3) reaction may be developed based on the differential control volume dV in Figure 15.3:

(1) rate of input of enthalpy by flow $-$ rate of output of enthalpy by flow

$$= \dot{H} - (\dot{H} + d\dot{H}) = -d\dot{H} = -\dot{m}c_P dT$$

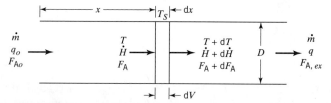

Figure 15.3 Control volume and symbols used for energy balance for PFR

15.2 Design Equations for a PFR

where $\dot{m}$ is the steady-state mass rate of flow, c_P is the specific heat of the flowing stream, and dT is the change in temperature, which may be positive or negative;

(2) rate of heat transfer to (or from) control volume

$$= \delta\dot{Q} = U(T_S - T)dA_p$$

where U is the (overall) heat transfer coefficient, T_S is the temperature of the surroundings outside the tube at the point in question, and dA_p is the differential circumferential heat transfer area;

(3) rate of input/generation (or output/absorption) of enthalpy by reaction

$$= (-\Delta H_{RA})(-r_A)dV$$

Taken together, applied to a fluid flowing inside the tube, these terms lead to the energy balance for steady-state operation in the form:

$$-\dot{m}c_P dT + U(T_S - T)dA_p + (-\Delta H_{RA})(-r_A)dV = 0 \quad (15.2\text{-}5)$$

Equation 15.2-5 may be more conveniently transformed to relate T and f_A. Since

$$dA_p = \pi D dx \quad (15.2\text{-}6)$$

and

$$dV = (\pi D^2/4)dx \quad (15.2\text{-}7)$$

where D is the diameter of the tube or vessel,

$$dA_p = (4/D)dV \quad (15.2\text{-}8)$$

If we use equations 15.2-1 and -8 to eliminate dV and dA_p, from equation 15.2-5, we obtain, on rearrangement,

$$\dot{m}c_P dT = \left[(-\Delta H_{RA}) + \frac{4U(T_S - T)}{D(-r_A)}\right] F_{Ao} df_A \quad (15.2\text{-}9)$$

Alternatively, equation 15.2-5 may be transformed to lead to the temperature profile in terms of x, the distance along the reactor from the inlet. This involves use of equations 15.2-6 and -7 to eliminate dA_p and dV. Then,

$$\dot{m}c_P dT = [(-\Delta H_{RA})(-r_A)(D/4) + U(T_S - T)]\pi D dx \quad (15.2\text{-}10)$$

We emphasize that all the forms of the energy balance developed above refer to the arrangement in which reacting fluid flows inside a single tube or vessel, as in Figure 15.1(a) or Figure 15.3, or inside a set of tubes, as in Figure 15.1(c). If the reacting fluid flows outside the tubes in the arrangement in Figure 15.1(c), the geometry is different and leads to a different form of equation.

For adiabatic operation ($\delta\dot{Q} = 0$), equations 15.2-9 and -10 can be simplified by deletion of the term involving the heat transfer coefficient U.

15.2.1.3 *Momentum Balance; Nonisobaric Operation*

In many cases, the length and diameter of a cylindrical reactor can be selected to obtain near isobaric conditions. However, in some cases, the frictional pressure drop is large enough to influence reactor performance. As a rule of thumb, for a compressible fluid, if the difference between the inlet and outlet pressures is greater than 10 to 15%, the change in pressure is likely to affect conversion, and should be considered when designing a reactor, or assessing its performance. In this situation, the pressure gradient along the reactor must be determined simultaneously with the f_A and T gradients.

To determine the pressure gradient along the reactor, the Fanning or Darcy equation for flow in cylindrical tubes may be used (Knudsen and Katz, 1958, p. 80):

$$-\frac{dP}{dx} = \frac{2\rho u^2 f}{D} = \frac{32\rho q^2 f}{\pi^2 D^5} \tag{15.2-11}$$

where P is the pressure, x is the axial position in the reactor, ρ is the fluid density, u is the linear velocity, f is the Fanning friction factor, D is the reactor diameter, and q is the volumetric flow rate; ρ, u, and q may vary with position.

The Fanning friction factor may be determined either from a chart for both rough and smooth tubes or from a variety of correlations (Knudsen and Katz, 1958, pp. 173, 176). The following correlation applies for turbulent flow in smooth tubes and for Reynolds numbers between 3,000 and 3,000,000:

$$f = 0.00140 + 0.125(\text{Re})^{-0.32} \tag{15.2-12}$$

Note that Re, and hence f, may change along the length of the reactor, if the fluid velocity and density change.

To evaluate the effect of pressure drop on performance, differential equations for the pressure drop (15.2-11), material balance (15.2-4), and energy balance (15.2-10) must be integrated simultaneously to solve for P, f_A, and T as functions of axial position, x.

15.2.2 Constant-Density System

15.2.2.1 *Isothermal Operation*

For a constant-density system, since

$$f_A = (c_{Ao} - c_A)/c_{Ao} \tag{14.3-12}$$

then

$$df_A = -(1/c_{Ao})dc_A \tag{15.2-13}$$

The residence time t and the space time τ are equal. This follows from Example 2-3 and equation 2.3-2:

$$t = \int dV/q = V/q_o = \tau \quad \text{(constant density)} \tag{15.2-14}$$

and

$$dt = dV/q_o = d\tau \tag{15.2-15}$$

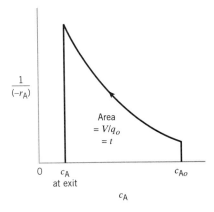

Figure 15.4 Graphical interpretation of equation 15.2-17 (constant-density system)

With these results, the general equations of Section 15.2.1 can be transformed into equations analogous to those for a constant-density BR. The analogy follows if we consider an element of fluid (of arbitrary size) flowing through a PFR as a closed system, that is, as a batch of fluid. Elapsed time (t) in a BR is equivalent to residence time (t) or space time (τ) in a PFR for a constant-density system. For example, substituting into equation 15.2-1 [$dV/df_A - F_{Ao}/(-r_A) = 0$] for dV from equation 15.2-15 and for df_A from 15.2-13, we obtain, since $F_{Ao} = c_{Ao}q_o$,

$$\frac{dc_A}{dt} + (-r_A) = 0 \qquad (15.2\text{-}16)$$

which is the same as equation 2.2-10 (space time τ may be used in place of t). Furthermore, we may similarly write equation 15.2-2 as

$$V/q_o = -\int dc_A/(-r_A) = \tau \qquad (15.2\text{-}17)$$

A graphical interpretation of this result is given in Figure 15.4.

EXAMPLE 15-2

A liquid-phase double-replacement reaction between bromine cyanide (A) and methylamine takes place in a PFR at 10°C and 101 kPa. The reaction is first-order with respect to each reactant, with $k_A = 2.22$ L mol^{-1} s^{-1}. If the residence or space time is 4 s, and the inlet concentration of each reactant is 0.10 mol L^{-1}, determine the concentration of bromine cyanide at the outlet of the reactor.

SOLUTION

The reaction is:

$$\text{BrCN(A)} + \text{CH}_3\text{NH}_2\text{(B)} \rightarrow \text{NH}_2\text{CN(C)} + \text{CH}_3\text{Br(D)}$$

Since this is a liquid-phase reaction, we assume density is constant. Also, since the inlet concentrations of A and B are equal, and their stoichiometric coefficients are also equal, at all points, $c_A = c_B$. Therefore, the rate law may be written as

$$(-r_A) = k_A c_A^2 \tag{A}$$

From equations 15.2-16 and (A),

$$-dc_A/dt = k_A c_A^2$$

which integrates to

$$k_A t = \frac{1}{c_A} - \frac{1}{c_{Ao}}$$

On insertion of the numerical values given for k_A, t, and c_{Ao}, we obtain

$$c_A = 0.053 \text{ mol L}^{-1}$$

We note the use of τ as a scaling factor for reactor size or capacity. In Example 15-2, neither V nor q_o is specified. For a given τ, if either V or q_o is specified, then the other is known. If either V or q_o is changed, the other changes accordingly, for the specified τ and performance (c_A or f_A). This applies also to a CSTR, and to either constant- or variable-density situations. The residence time t may similarly be used for constant-density, but not variable-density cases.

EXAMPLE 15-3

A gas-phase reaction between methane (A) and sulfur (B) is conducted at 600°C and 101 kPa in a PFR, to produce carbon disulfide and hydrogen sulfide. The reaction is first-order with respect to each reactant, with $k_B = 12$ m^3 mol^{-1} h^{-1} (based upon the disappearance of sulfur). The inlet molar flow rates of methane and sulfur are 23.8 and 47.6 mol h^{-1}, respectively. Determine the volume (V) required to achieve 18% conversion of methane, and the resulting residence or space time.

SOLUTION

The reaction is $CH_4 + 2S_2 \rightarrow CS_2 + 2H_2S$. Although this is a gas-phase reaction, since there is no change in T, P, or total molar flow rate, density is constant. Furthermore, since the reactants are introduced in the stoichiometric ratio, neither is limiting, and we may work in terms of B (sulphur), since k_B is given, with $f_B(= f_A) = 0.18$. It also follows that $c_A = c_B/2$ at all points. The rate law may then be written as

$$(-r_B) = k_B c_A c_B = k_B c_B^2/2 \tag{A}$$

From the material-balance equation 15.2-17 and (A),

$$V = -q_o \int_{c_{Bo}}^{c_B} dc_B/(-r_B) = -q_o \int_{c_{Bo}}^{c_B} \frac{dc_B}{k_B c_B^2/2} = \frac{2q_o}{k_B}\left(\frac{1}{c_B} - \frac{1}{c_{Bo}}\right) \tag{B}$$

Since $F_{Bo} = c_{Bo} q_o$, and, for constant-density, $c_B = c_{Bo}(1 - f_B)$, equation (B) may be written as

$$V = \frac{2q_o^2}{k_B F_{Bo}} \left(\frac{f_B}{1 - f_B} \right) \quad \text{(C)}$$

To obtain q_o in equation (C), we assume ideal-gas behavior; thus,

$$q_o = (F_{Ao} + F_{Bo})RT/P = 71.4(8.314)873/101{,}000 = 5.13 \text{ m}^3 \text{ h}^{-1}$$

From equation (C),

$$V = \frac{2(5.13)^2 0.18}{12(47.6)0.82} = 0.020 \text{ m}^3$$

From equation 15.2-14, we solve for τ:

$$t = \tau = V/q_o = 0.020/5.13 = 0.00390 \text{ h} = 14.0 \text{ s}$$

The E-Z Solve software can be used to integrate numerically the differential equation resulting from the combination of the material-balance equation (15.2-17) and the rate law [equation (A)]—see file ex15-3.msp. The same results are obtained for V and τ.

15.2.2.2 Nonisothermal Operation

To characterize the performance of a PFR subject to an axial gradient in temperature, the material and energy balances must be solved simultaneously. This may require numerical integration using a software package such as E-Z Solve. Example 15-4 illustrates the development of equations and the resulting profile for f_A with respect to position (x) for a constant-density reaction.

EXAMPLE 15-4

A liquid-phase reaction $A + B \rightarrow 2C$ is conducted in a nonisothermal multitubular PFR. The reactor tubes (7 m long, 2 cm in diameter) are surrounded by a coolant which maintains a constant wall temperature. The reaction is pseudo-first-order with respect to A, with $k_A = 4.03 \times 10^5 \, e^{-5624/T}$ s^{-1}. The mass flow rate is constant at 0.06 kg s^{-1}, the density is constant at 1.025 g cm^{-3}, and the temperature at the inlet of the reactor (T_o) is 350 K.

(a) Develop expressions for df_A/dx and dT/dx.
(b) Plot $f_A(x)$ profiles for the following wall temperatures (T_S): 350 K, 365 K, 400 K, and 425 K.

Data: $c_{Ao} = 0.50$ mol L^{-1}; $c_P = 4.2$ J g^{-1} K^{-1}; $\Delta H_{RA} = -210$ kJ mol^{-1}; $U = 1.59$ kW m^{-2} K^{-1}.

SOLUTION

(a) The rate law is

$$(-r_A) = k_A c_A = k_A c_{Ao}(1 - f_A) \quad \text{(A)}$$

where k_A is given in Arrhenius form above. Substitution of equation (A) in the material-balance equation 15.2-4, $(-r_A) = (F_{Ao}/\pi r^2) df_A/dx$, results in (with $R = D/2$ and $F_{Ao}/c_{Ao} = q_o$):

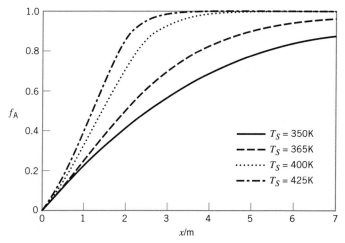

Figure 15.5 Effect of wall temperature (T_S) on conversion in a non-isothermal PFR (Example 15-4)

$$\frac{df_A}{dx} = \frac{\pi D^2 k_A (1 - f_A)}{4q_o} \quad \textbf{(B)}$$

Equation (B) gives the required expression for df_A/dx.

To develop an expression for dT/dx, we use the energy balance of equation 15.2-9, divided by dx, with $(-r_A)$ eliminated by equation (A), and df_A/dx by equation (B), to result in

$$\frac{dT}{dx} = \frac{\pi D^2 k_A c_{Ao}(1 - f_A)(-\Delta H_{RA})}{4\dot{m}c_P} + \frac{\pi U D}{\dot{m}c_P}(T_S - T) \quad \textbf{(C)}$$

where T_S is the wall temperature. Note that dT/dx is implicitly related to df_A/dx.

(b) The E-Z Solve software may be used to integrate numerically the differential equations (B) and (C) from $x = 0$ to $x = 7$ m, using the four wall temperatures specified (see file ex15-4.msp). The resulting $f_A(x)$ profiles are illustrated in Figure 15.5. Note that as the wall temperature (T_S) is increased from 350 K to 425 K, the length of reactor required to achieve a particular fractional conversion is reduced. For example, virtually 100% conversion is achieved within 4 m when T_S is 425 K, but requires nearly 5 m when T_S is 400 K.

15.2.2.3 PFR Performance in Relation to SFM

For the segregated-flow model (SFM), for the reaction A + ... → products,

$$1 - f_A = \frac{c_A}{c_{Ao}} = \int_0^\infty \left[\frac{c_A(t)}{c_{Ao}}\right]_{BR} E(t)dt \quad \textbf{(13.5-2)}$$

For a PFR,

$$E(t) = \delta(t - \bar{t}) = \infty; \; t = \bar{t}$$
$$= 0; \; t \neq \bar{t} \quad \textbf{(13.4-8)}$$

From the discussion in Section 15.2.2.1 comparing the performance of a BR and a PFR for a constant-density system, it follows that

15.2 Design Equations for a PFR

$$\left[\frac{c_A(t)}{c_{Ao}}\right]_{BR} = \left[\frac{c_A(t)}{c_{Ao}}\right]_{PF} \quad (15.2\text{-}18)$$

Furthermore, from equation 13.4-10, a property of the delta function may be written as

$$\int_{t_1}^{t_2} g(t)\delta(t-\bar{t})dt = g(\bar{t}), \text{ if } t_1 \leq \bar{t} \leq t_2 \quad (15.2\text{-}19)$$

If we substitute equations 13.4-8, and 15.2-18 and -19, with $g(t) = [c_A(t)/c_{Ao}]_{PFR}$, into equation 13.5-2, we obtain

$$(1 - f_A)_{SF} = \left(\frac{c_A}{c_{Ao}}\right)_{SF} = \int_0^\infty \left[\frac{c_A(t)}{c_{Ao}}\right]_{PF} \delta(t-\bar{t})dt$$

$$= \left[\frac{c_A(\bar{t})}{c_{Ao}}\right]_{PF} = (1 - f_A)_{PF} \quad (15.2\text{-}20)$$

since $0 \leq \bar{t} \leq \infty$. The implication of equation 15.2-20 is that the SFM applied to a PFR gives the same result for the outlet concentration or fractional conversion as does integration of the material balance, equation 15.2-16, regardless of the kinetics of the reaction. This result is not surprising when it is realized that plug flow, as occurs in a PFR, is, in fact, segregated flow (in the axial direction): an element of fluid is a closed system, not mixing with any other element of fluid. Equation 15.2-20 is illustrated more specifically for particular kinetics in the following example.

EXAMPLE 15-5

Suppose the liquid-phase reaction A → products is second-order, with $(-r_A) = k_A c_A^2$, and takes place in a PFR. Show that the SFM gives the same result for $1 - f_A = c_A/c_{Ao}$ as does the integration of equation 15.2-16, the material balance.

SOLUTION

Integration of equation 15.2-16 with the rate law given results in (for a PFR)

$$1 - f_A = \frac{c_A}{c_{Ao}} = \frac{1}{1 + ktc_{Ao}} \quad (A)$$

where t is the residence time (the same result is obtained for a BR, for a constant-density system).

Using the SFM, equation 13.5-2, with $[c_A(t)/c_{Ao}]_{BR}$ given by (A), and $E(t)$ by equation 13.4-8, we obtain

$$1 - f_A = \frac{c_A}{c_{Ao}} = \int_0^\infty \left(\frac{1}{1 + ktc_{Ao}}\right)\delta(t-\bar{t})dt = \frac{1}{1 + k\bar{t}c_{Ao}}$$

which is the same result as in (A) from the material balance, since $\bar{t}$ is the same as the residence time.

The same conclusion would be reached for any form of rate law introduced into equation 15.2-16 on the one hand, and into equation 13.5-2 on the other.

15.2.3 Variable-Density System

When the density of the reacting system is not constant through a PFR, the general forms of performance equations of Section 15.2.1 must be used. The effects of continuously varying density are usually significant only for a gas-phase reaction. Change in density may result from any one, or a combination, of: change in total moles (of gas flowing), change in T, and change in P. We illustrate these effects by examples in the following sections.

15.2.3.1 Isothermal, Isobaric Operation

EXAMPLE 15-6

Consider the gas-phase decomposition of ethane (A) to ethylene at 750°C and 101 kPa (assume both constant) in a PFR. If the reaction is first-order with $k_A = 0.534$ s^{-1} (Froment and Bischoff, 1990, p. 351), and τ is 1 s, calculate f_A. For comparison, repeat the calculation on the assumption that density is constant. (In both cases, assume the reaction is irreversible.)

SOLUTION

The reaction is $C_2H_6(A) \rightarrow C_2H_4(B) + H_2(C)$. Since the rate law is

$$(-r_A) = k_A c_A = k_A F_A/q = k_A F_{Ao}(1 - f_A)/q$$

equation 15.2-3 for the material balance is

$$\tau = c_{Ao} \int \frac{df_A}{(-r_A)} = \frac{1}{k_A q_o} \int_0^{f_A} \frac{q \, df_A}{1 - f_A} \tag{A}$$

As shown in Example 14-6, a stoichiometric table is used to relate q and q_o. The resulting expression is

$$q = (1 + f_A) q_o \tag{B}$$

With this result, equation (A) becomes

$$\tau = \frac{1}{k_A} \int_0^{f_A} \frac{(1 + f_A) df_A}{1 - f_A}$$

The integral in this expression may be evaluated analytically with the substitution $z = 1 - f_A$. The result is

$$-f_A - 2\ln(1 - f_A) = k_A \tau = 0.534 \tag{C}$$

Solution of equation (C) leads to

$$f_A = 0.361$$

If the change in density is ignored, integration of equation 15.2-17, with $(-r_A) = k_A c_A = k_A c_{Ao}(1 - f_A)$, leads to

15.2 Design Equations for a PFR

$$\ln(1 - f_A) = -k_A \tau \quad \textbf{(D)}$$

from which

$$f_A = 0.414$$

The assumption of constant density leads to a result 15% greater than that from equation (C). Desirable as this may be, it is an overestimate of the performance of the PFR based on the other assumptions made.

 This example can also be solved by numerical integration of equation (A) using the E-Z Solve software (file ex15-6.msp). For variable density, equation (B) is used to substitute for q. For constant density, $q = q_o$.

15.2.3.2 Nonisothermal, Isobaric Operation

EXAMPLE 15-7

A gas-phase reaction between butadiene (A) and ethene (B) is conducted in a PFR, producing cyclohexene (C). The feed contains equimolar amounts of each reactant at 525°C (T_o) and a total pressure of 101 kPa. The enthalpy of reaction is -115 kJ (mol A)$^{-1}$, and the reaction is first-order with respect to each reactant, with $k_A = 32,000 \, e^{-13,850/T}$ m^3 mol^{-1} s^{-1}. Assuming the process is adiabatic and isobaric, determine the space time required for 25% conversion of butadiene.

Data: $C_{P_A} = 150$ J mol^{-1} K^{-1}; $C_{P_B} = 80$ J mol^{-1} K^{-1}; $C_{P_C} = 250$ J mol^{-1} K^{-1}

SOLUTION

The reaction is $C_4H_6(A) + C_2H_4(B) \rightarrow C_6H_{10}(C)$. Since the molar ratio of A to B in the feed is 1:1, and the ratio of the stoichiometric coefficients is also 1:1, $c_A = c_B$ throughout the reaction. Combining the material-balance equation (15.2-2) with the rate law, we obtain

$$V = F_{Ao} \int \frac{df_A}{(-r_A)} = F_{Ao} \int \frac{df_A}{k_A c_A c_B} = F_{Ao} \int \frac{df_A}{k_A c_A^2}$$

$$= F_{Ao} \int \frac{df_A}{k_A (F_A/q)^2} = \frac{1}{F_{Ao}} \int_0^{f_A} \frac{q^2 df_A}{k_A (1 - f_A)^2} \quad \textbf{(A)}$$

Since k_A depends on T, it remains inside the integral, and we must relate T to f_A. Since the density (and hence q) changes during the reaction (because of changes in temperature and total moles), we relate q to f_A and T with the aid of a stoichiometric table and the ideal-gas equation of state.

Species	Initial moles	Change	Final moles
A	F_{Ao}	$-f_A F_{Ao}$	$F_{Ao}(1 - f_A)$
B	F_{Ao}	$-f_A F_{Ao}$	$F_{Ao}(1 - f_A)$
C	0	$f_A F_{Ao}$	$F_{Ao} f_A$
total	$2 F_{Ao}$		$F_t = F_{Ao}(2 - f_A)$

Chapter 15: Plug Flow Reactors (PFR)

Since at any point in the reactor, $q = F_t RT/P$, and the process is isobaric, q is related to the inlet flow rate q_o by

$$\frac{q}{q_o} = \frac{F_{Ao}(2 - f_A)T}{2F_{Ao}T_o}$$

That is,

$$q = q_o(1 - 0.5f_A)T/T_o \tag{B}$$

Substitution of equation (B) into (A) to eliminate q results in

$$\tau = \frac{V}{q_o} = \frac{1}{c_{Ao}T_o^2} \int_0^{f_A} \frac{(1 - 0.5f_A)^2 T^2}{k_A(1 - f_A)^2} df_A \tag{C}$$

To relate f_A and T, we require the energy balance (15.2-9). From this equation, for adiabatic operation (the heat transfer term on right side vanishes),

$$\dot{m}c_p dT \equiv F_t C_P dT = (-\Delta H_{RA}) F_{Ao} df_A \tag{D}$$

where C_P is the molar heat capacity of the flowing steam, and

$$\begin{aligned} F_t C_P &= F_A C_{PA} + F_B C_{PB} + F_C C_{PC} \\ &= F_{Ao}(1 - f_A)C_{PA} + F_{Ao}(1 - f_A)C_{PB} + F_{Ao} f_A C_{PC} \\ &= F_{Ao}[(C_{PA} + C_{PB}) + (C_{PC} - C_{PA} - C_{PB})f_A] \end{aligned} \tag{E}$$

Substituting equation (E) in (D), and integrating on the assumption that $(-\Delta H_{RA})$ is constant, we obtain

$$T = T_o + (-\Delta H_{RA}) \int_0^{f_A} \frac{df_A}{(C_{PA} + C_{PB}) + (C_{PC} - C_{PA} - C_{PB})f_A} \tag{F}$$

With numerical values inserted, this becomes

$$\begin{aligned} T &= 798 + 115{,}000 \int_0^{f_A} \frac{df_A}{230 + 20f_A} \\ &= 798 + \frac{115{,}000}{20}[\ln(230 + 20f_A) - \ln 230] \\ &= 798 + 5750[\ln(230 + 20f_A) - 5.438] \end{aligned} \tag{G}$$

The value of τ in equation (C) may now be determined by numerically (or graphically) evaluating the right side, with T given by equation (G), and k_A obtained from the given Arrhenius equation at this T. For use in equation (C),

$$c_{Ao} = F_{Ao}/q_o = p_{Ao}/RT_o = 0.5(101{,}000)/8.314(798) = 7.6 \text{ mol m}^{-3}$$

For numerical evaluation, we use the simple trapezoidal rule, and a stepwise procedure similar to that in Example 12-5, which can be readily implemented by a spreadsheet program. For convenience, in equation (C), we let

$$G = \frac{1}{c_{Ao}T_o^2}\left\{\frac{(1 - 0.5f_A)^2[T(f_A)]^2}{k_A(T)(1 - f_A)^2}\right\} \tag{H}$$

15.2 Design Equations for a PFR

Table 15.1 Results for Example 15-7

f_A	T/ K	k_A/ $m^3\ mol^{-1}\ s^{-1}$	G/ s	G^*/ s	$G^*\Delta f_A$/ s
0.000	798.0	0.928	141.78		
0.025	810.5	1.213	114.82	128.30	3.21
0.050	822.9	1.571	93.85	104.33	2.61
0.075	835.4	2.018	77.38	85.61	2.14
0.100	847.8	2.572	64.34	70.86	1.77
0.125	860.2	3.253	53.95	59.14	1.48
0.150	872.5	4.086	45.59	49.77	1.24
0.175	884.8	5.097	38.83	42.21	1.06
0.200	897.1	6.317	33.32	36.07	0.90
0.225	909.4	7.781	28.80	31.06	0.78
0.250	921.7	9.526	25.08	26.94	0.67
				$\tau = \Sigma G^*\Delta f_A =$	15.86 s

so that

$$\tau = \int G df_A \simeq \sum G^*\Delta f_A \quad \text{(I)}$$

where G^* is the average of two consecutive values of G, and Δf_A is the increment in f_A over that interval, chosen in the calculation to be 0.025. The following algorithm is used:

(1) Choose a value of $f_A = 0$.
(2) Calculate T from equation (G).
(3) Calculate k_A from T and the Arrhenius equation given.
(4) Calculate G using equation (H).
(5) Repeat steps (1) to (4) for $f_A = 0.025, 0.050, \ldots, 0.25$ (that is, in steps of 0.025).
(6) Calculate τ from equation (I).

The results are given in Table 15.1. Thus, the space time is 15.9 s for 25% conversion of butadiene.

The integrals in equations (C) and (F) can also be evaluated numerically using the E-Z Solve software (file ex15-7.msp). The calculated value of τ is 15.8 s, slightly less than that obtained using the trapezoidal approximation.

If simplifying conditions such as adiabatic behavior and temperature-independence of $(-\Delta H_{RA})$ and C_P are not valid, the material and energy balances may have to be integrated numerically as a system of two coupled ordinary differential equations.

15.2.3.3 Nonisobaric Operation

EXAMPLE 15-8

The dehydrogenation of ethane (A) to ethene (B) is conducted in a 0.5-m³ PFR. The reaction is first-order with respect to A, with a rate constant of 15.2 min⁻¹ at 725°C. The feed contains pure ethane at 725°C, 400 kPa, and a flow rate of 1.0 kmol min⁻¹. Compare the conversion predicted if isothermal, isobaric conditions are assumed with that if the pressure drop is accounted for with isothermal flow. The diameter of the reactor tube is 0.076 m, and the viscosity of the gas is 2.5×10^{-5} Pa s.

SOLUTION

The reaction is A → B + H_2. The reactor length corresponding to the reactor volume and diameter specified is 110 m. The $f_A(x)$ and $P(x)$ profiles can be determined by simultaneously integrating the material-balance equation (15.2-4) and the pressure-drop equation (15.2-11):

$$\frac{df_A}{dx} = \frac{\pi D^2 (-r_A)}{4 F_{Ao}} \quad (15.2\text{-}4)$$

where $(-r_A) = k_A c_A = k_A F_A/q = k_A F_{Ao}(1 - f_A)/q$, and

$$\frac{dP}{dx} = \frac{-32 f \rho q^2}{\pi^2 D^5} \quad (15.2\text{-}11)$$

Assuming ideal-gas behavior, we can determine the volumetric flow rate (q) from

$$q = F_t RT/P$$

where $F_t = F_{Ao}(1 - f_A)$ (obtained from a stoichiometric table).

The linear velocity (u) can be determined from

$$u = 4q/\pi D^2$$

The friction factor, f, can be estimated using equation 15.2-12:

$$f = 0.00140 + 0.125 \text{Re}^{-0.32} \quad (15.2\text{-}12)$$

The density may be evaluated from the ideal-gas equation:

$$\rho = (P/RT) M_{av}$$

where M_{av} is the average molar mass of the gas mixture.

The E-Z Solve software can be used to integrate equations 15.2-4 and 15.2-11 numerically, while simultaneously updating q, u, ρ, Re, and f at each step (file ex15-8.msp). The predicted conversion for isothermal, nonisobaric conditions is 0.247; the calculated pressure drop is 114 kPa. If the pressure drop is ignored (i.e., $P = 400$ kPa throughout the reactor), the resulting conversion is 0.274. Thus, for this case, it is important that the pressure drop be accounted for.

15.3 RECYCLE OPERATION OF A PFR

In a chemical process, the use of recycle, that is, the return of a portion of an outlet stream to an inlet to join with fresh feed, may have the following purposes: (1) to conserve feedstock when it is not completely converted to desired products, and/or (2) to improve the performance of a piece of equipment such as a reactor. It is the latter purpose that we consider here for a PFR (the former purpose usually involves a separation process downstream from a reactor). For a CSTR, solution of problem 14-26 shows that recycling alone has no effect on its performance, and hence is not used. However, it provides a clue as to the anticipated effect for a PFR. The recycle serves to backmix the product stream with the feed stream. The effect of backmixing is to make the performance of a PFR become closer to that of a CSTR. The degree of backmixing, and

15.3 Recycle Operation of a PFR

hence the extent of the effect of recycling, is controlled by the recycle ratio, R, defined in general by

$$R = \frac{\text{molar flow rate of A in recycle stream}}{\text{molar flow rate of A in product stream}}$$
$$= \frac{F_{AR}}{F_{A1}} = \frac{c_{A1} q_R}{c_{A1} q_1} = \frac{q_R}{q_1} \quad (15.3\text{-}1)$$
$$= \frac{\text{volumetric flow rate of recycle stream}}{\text{volumetric flow rate of product stream}}$$

where subscript R refers to recycle and subscript 1 to the vessel outlet. Equation 15.3-1 is applicable to both constant-density and variable-density systems.

R may vary from 0 (no recycle) to a very large value (virtually complete recycle). Thus, as shown quantitatively below, we expect that a recycle PFR may vary in performance between that of a PFR with no recycle and that of a CSTR (complete recycle), depending on the value of R. The question may be raised as to situations in which a recycle PFR may be used to advantage. These include cases with abnormal kinetics, for reducing volume; with complex kinetics, for control of selectivity; and with strongly exothermic reactions, for reducing the temperature gradient along the reactor and possibly avoiding "hot-spots" within it.

15.3.1 Constant-Density System

For a constant-density system, since $q_1 = q_o$, equation 15.3-1 may be written as:

$$R = q_R/q_o \quad \text{(constant density)} \quad (15.3\text{-}2)$$

A recycle PFR, operating at steady-state for the reaction A + ... → products, is shown schematically in Figure 15.6, together with associated streams and terminology. At the split point S, the exit stream is divided into the recycle stream (flow rate Rq_o) and the product stream (flow rate q_o), both at the exit concentration c_{A1}. At the mixing point M, the recycle stream joins the fresh feed stream (flow rate q_o, concentration c_{Ao}) to form the stream actually entering the reactor (flow rate $(1+R)q_o$, concentration c'_{Ao}). The inlet concentration c'_{Ao} may be related to c_{Ao}, c_{A1}, and R by a material balance for A around M:

$$q_o c_{Ao} + R q_o c_{A1} = (1+R) q_o c'_{Ao}$$

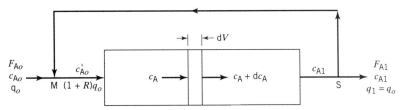

Figure 15.6 Flow diagram and terminology for recycle PFR (constant density)

from which

$$c'_{Ao} = \frac{c_{Ao} + Rc_{A1}}{1 + R} \quad (15.3\text{-}3)$$

The volume of the recycle PFR may be obtained by a material balance for A around the differential control volume dV. Equating molar flow input and output, we obtain

$$(1 + R)q_o c_A = (1 + R)q_o(c_A + dc_A) + (-r_A)dV$$

From this,

$$V/q_o = \tau = -(1 + R)\int_{c'_{Ao}}^{c_{A1}} \frac{dc_A}{(-r_A)} \quad (15.3\text{-}4)$$

which is the analog of equation 15.2-17 for a PFR without recycle.

A graphical interpretation of equation 15.3-4 is shown in Figure 15.7, which is a plot of reciprocal rate for normal kinetics (curve GE′) against concentration. In Figure 15.7, the shaded area under the curve GE′ from c'_{Ao} to c_{A1}, BFGA, represents the value of the *integral* in equation 15.3-4 (the integral has a negative value). The space time, τ, in 15.3-4 is equal to $-(1 + R)$ times the value of this integral, which is based on c'_{Ao} and c_{A1}. We can determine τ in terms of c_{Ao} and c_{A1}, as shown in the following proof:

$$(1 + R)c'_{Ao} = c_{Ao} + Rc_{A1} \quad \textbf{(from 15.3-3)}$$
$$= c_{Ao} + Rc_{A1} + Rc_{Ao} - Rc_{Ao}$$
$$= (1 + R)c_{Ao} + R(c_{A1} - c_{Ao})$$

Therefore,

$$\frac{c_{Ao} - c'_{Ao}}{c_{Ao} - c_{A1}} = \frac{R}{1 + R} \quad (15.3\text{-}5)$$

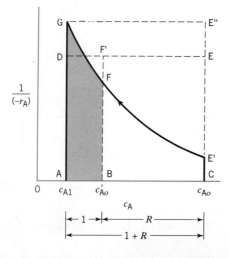

Figure 15.7 Graphical interpretation of equation 15.3-4 for recycle PFR (constant density)

The construction in Figure 15.7 uses this relation graphically. If the distance $c_{Ao} - c_{A1}$ is represented by $1 + R$, then, from equation 15.3-5, the distance $c_{Ao} - c'_{Ao}$ is represented by R, and $c'_{Ao} - c_{A1}$ by 1 (unity). $c_{Ao} - c_{A1}$ is then $(1 + R)$ times $c'_{Ao} - c_{A1}$. We can therefore relate $c_{Ao} - c_{A1}$ and $c'_{Ao} - c_{A1}$ using rectangles of equal height, as shown in Figure 15.7. We choose point D such that the rectangle DF'BA has the same area as the shaded area. Consequently, based on equation 15.3-5, rectangular area DECA is $(1 + R)$ times area DF'BA, since $c_{Ao} - c_{A1}$ is $(1 + R)$ times $c'_{Ao} - c_{A1}$.

The two limiting cases of $R \to 0$ and $R \to \infty$ can be realized qualitatively from Figure 15.7. As $R \to 0$, the point $F \to E'$, and the area under the curve represents V/q_o for a PFR without recycle, as in equation 15.2-17 and Figure 15.4. As $R \to \infty$, the point $F \to G$, and the rectangular area CE"GA represents V/q_o for the recycle reactor, the same as that for a CSTR operating at steady-state for the same conditions, as in equation 14.3-15 (with $q = q_o$) and a graph corresponding to that in Figure 14.2.

An analytical proof of the first of these limiting cases follows directly from equations 15.3-3 and -4. As $R \to 0$, $c'_{Ao} \to c_{Ao}$, and $V/q_o \to$ that for a PFR without recycle. An analytical proof of the second limiting case does *not* follow directly from these two equations. As $R \to \infty$, $c'_{Ao} \to c_{A1}$ (from equation 15.3-3), and $V/q_o \to -(\infty)(0)$ (from equation 15.3-4), which is an "indeterminant" form. The latter can be evaluated with the aid of L'Hôpital's Rule, but the proof is left to problem 15-18.

The treatment of a recycle PFR thus far, based on "normal" kinetics, does not reveal any advantage relative to a PFR itself. In fact, there is a disadvantage for the size of reactor, which is larger for any recycling, because of the backmixing effect: the rectangle CEDA in Figure 15.7, for any possible position of the line ED, is greater than the area CE'GA under the reciprocal-rate curve.

For "abnormal" kinetics, however, the situation is reversed, and allows the possibility of a smaller reactor. An example of "abnormal" kinetics, at least in part, is given by autocatalysis (Section 8.3 and Figure 8.6). Figure 15.8 is a reciprocal-rate plot corresponding to the rate plot in Figure 8.6: the maximum in Figure 8.6 becomes a minimum in Figure 15.8 (at H). The portion of the curve GE' from H to E' represents "abnormal" kinetics, and that from G to H "normal" kinetics. The remainder of the graphical construction in Figure 15.8 corresponds to that in Figure 15.7 to show the effect of recycle and interpretation of equation 15.3-4 in such a case. Since the rate law for the autocatalytic reaction $A + \ldots \to B + \ldots$ is, say, $(-r_A) = k_A c_A c_B$, the finite location of the point E' implies a nonzero value for c_{Bo} (otherwise $E' \to \infty$, and the area under the curve GE' is infinite).

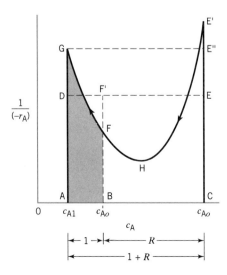

Figure 15.8 Graphical interpretation of equation 15.3-4 for recycle PFR for autocatalytic reaction (constant density)

As in Figure 15.7, the area CEDA in Figure 15.8 represents τ for a PFR with recycle ratio R. It varies in a complex manner from that for a PFR without recycle (represented by the area under the curve GE') to that for a CSTR (rectangular area CE''GA). Because of the shape of the curve GE', the question arises as to whether there is an optimal value of R for which τ is a minimum, and, if so, how this minimum compares with τ for a PFR without recycle. This is explored in Example 15-9 and illustrated numerically in Example 15-10.

EXAMPLE 15-9

(a) For the liquid-phase autocatalytic reaction A + ... → B + ... taking place isothermally at steady-state in a recycle PFR, derive an expression for the optimal value of the recycle ratio, R_{opt}, that minimizes the volume or space time of the reactor. The rate law is $(-r_A) = k_A c_A c_B$.

(b) Express the minimum volume or space time of the reactor in terms of R_{opt}.

SOLUTION

(a) With $c_{Ao} \gg c_{Bo}$, $c_B = c_{Ao} - c_A$, and the rate law becomes

$$(-r_A) = k_A c_A (c_{Ao} - c_A) \quad \text{(A)}$$

From the material-balance equation 15.3-4, on substitution for $(-r_A)$ from (A),

$$\frac{V}{q_o} = \tau = -(1 + R) \int_{c'_{Ao}}^{c_{A1}} \frac{dc_A}{k_A c_A (c_{Ao} - c_A)}$$

$$= \frac{1 + R}{k_A c_{Ao}} \ln \left[\frac{c'_{Ao}(c_{Ao} - c_{A1})}{c_{A1}(c_{Ao} - c'_{Ao})} \right]$$

$$= \frac{1 + R}{k_A c_{Ao}} \ln \left(\frac{c_{Ao} + R c_{A1}}{R c_{A1}} \right) \quad \text{(B)}$$

on substitution for c'_{Ao} from equation 15.3-3. Equation (B) gives $V(R)/q_o$ or $\tau(R)$ as a function of R for fixed c_{Ao} and c_{A1}, and the given rate law. To obtain R_{opt} for minimum V/q_o or τ, we set $d\tau(R)/dR = 0$. This leads to a transcendental, implicit equation for R_{opt}:

$$\ln \left(\frac{c_{Ao} + R_{opt} c_{A1}}{R_{opt} c_{A1}} \right) = \frac{c_{Ao}(1 + R_{opt})}{R_{opt}(c_{Ao} + R_{opt} c_{A1})} \quad (15.3\text{-}6)$$

(b) To obtain an expression for the minimum volume or space time, we combine equations (B) and 15.3-6 to obtain

$$\frac{V_{min}}{q_o} = \tau_{min} = \frac{(1 + R_{opt})^2}{k_A R_{opt}(c_{Ao} + R_{opt} c_{A1})} \quad (15.3\text{-}7)$$

EXAMPLE 15-10

A liquid-phase autocatalytic irreversible reaction, A → B + ..., is to be carried out isothermally in a continuous flow reactor. The fractional conversion of A (f_A) is to be 90%, the feed rate (q_o) is 0.5 L s^{-1}, and the feed concentration (c_{Ao}) is 1.5 mol L^{-1}. The rate law is $(-r_A) = k_A c_A c_B$, with $k_A = 0.002$ L mol^{-1} s^{-1}. There is no B in the feed.

(a) What is the volume (V/L) required for a CSTR?
(b) What is the volume required for a PFR?
(c) What is the minimum volume required for a PFR with recycle?

SOLUTION

(a) From the material-balance equation (14.3-5), for a CSTR,

$$V = \frac{F_{Ao} f_A}{(-r_A)} = \frac{q_o c_{Ao}(c_{Ao} - c_A)}{c_{Ao} k_A c_A (c_{Ao} - c_A)} = \frac{q_o}{k_A c_A}$$

$$= \frac{0.5}{0.002(1.5)0.1} = 1670 \text{ L}$$

(Here, $c_A \equiv c_{A1} = c_{Ao}(1 - f_A)$, and $c_B = c_{Ao} - c_A$.)

This steady-state analysis to obtain V is slightly misleading. According to the rate law, for any reaction to occur, there must be some B present "initially." This can be accomplished by a *one-time* addition of B to the reactor, sufficient to make c_B in the reactor initially equal to the steady-state value of $c_B = c_{Ao} - c_A = 1.35$ mol L^{-1}. Subsequently, backmixing within the reactor maintains the steady-state value of c_B.

(b) From the material-balance equation for a PFR (15.2-17) and the rate law,

$$V = -q_o \int_{c_{Ao}}^{c_{A1}} \frac{dc_A}{k_A c_A (c_{Ao} - c_A)} = \frac{q_o}{k_A c_{Ao}} \ln \left[\frac{c_{Ao}(c_{Ao} - c_{A1})}{c_{A1}(c_{Ao} - c_{Ao})} \right] \to \infty$$

(The integral can be evaluated by means of the substitution $c_A = 1/x$.) That is, with no B in the feed, the reaction doesn't get started in a PFR. The volume becomes finite if $c_{Bo} \neq 0$, but this value must be *maintained* in the feed in steady-state operation. For example, if $c_{Bo} = 0.01$ mol L^{-1}, $V = 1195$ L; if $c_{Bo} = 0.05$, $V = 909$ L; and if $c_{Bo} = 0.11$, $V = 760$ L, the value obtained in part (c), below.

(c) For the minimum volume of a recycle PFR, we first determine the value of R_{opt} from equation 15.3-6. With numerical values inserted for c_{Ao} and c_{A1}, this becomes

$$\ln \left(\frac{10}{R_{opt}} + 1 \right) = \frac{1 + R_{opt}}{R_{opt}(1 + 0.1 R_{opt})}$$

from which, $R_{opt} = 0.43$. This leads to the value of $V_{min} = 760$ L, from equation 15.3-7.

As in the case of the CSTR in part (a), there must be a *one-time* addition of B so that reaction can proceed. In this case, it can be accomplished by addition of B to the feed (at point M, say, in Figure 15.6), so that c'_{Ao} is initially equal to the steady-state value of 1.094 mol L^{-1} (from equation 15.3-3).

386 Chapter 15: Plug Flow Reactors (PFR)

For an autocatalytic reaction, Example 15-10 shows that a recycle PFR operating with an optimal value of R requires the smallest volume for the three reactor possibilities posed. (In the case of a PFR without recycle, the size disadvantage can be offset at the expense of *maintaining* a sufficient value of c_{B_o} (in the feed), but this introduces an alternative disadvantage.) A fourth possibility exists for an even smaller volume. This can be realized from Figure 15.8 (although not shown explicitly), if the favorable characteristics of both "normal" and "abnormal" kinetics are used to advantage. Since this involves a combination of reactor types, we defer consideration to Chapter 17.

15.3.2 Variable-Density System

In this section, we develop the design equation for a recycle reactor in which the density of the reacting system varies. This is a generalization of the development of equation 15.3-4 for a constant-density system, which is thus included as a special case. In the general case, it is convenient to use molar flow rates and fractional conversion (for the reaction A + ... → products taking place in a recycle PFR). The flow diagram and terminology corresponding to Figure 15.6 are shown schematically in Figure 15.9.

From a material balance for A around the mixing point M, the molar flow rate of A entering the reactor is

$$F'_{Ao} = F_{Ao} + RF_{A1} \quad (15.3\text{-}8)$$

Although the fractional conversion in the fresh feed (f_{Ao}) is indicated to be zero, the fractional conversion in the stream actually entering the reactor (f'_{Ao}) is not zero, since this stream is a combination of fresh feed and the recycled stream (conversion f_{A1}). At the exit from the system at S, or at the exit from the reactor,

$$f_{A1} = \frac{F_{Ao} - F_{A1}}{F_{Ao}} = \frac{(1+R)F_{Ao} - (1+R)F_{A1}}{(1+R)F_{Ao}}$$
$$= \frac{(1+R)F_{Ao} - F'_{A1}}{(1+R)F_{Ao}}$$

where $F'_{A1} = (1+R)F_{A1}$.

Correspondingly, at the inlet of the reactor

$$f'_{Ao} = \frac{(1+R)F_{Ao} - F'_{Ao}}{(1+R)F_{Ao}} = \frac{(1+R)F_{Ao} - (F_{Ao} + RF_{A1})}{(1+R)F_{Ao}}$$
$$= \frac{R}{1+R}\frac{F_{Ao} - F_{A1}}{F_{Ao}} = \frac{R}{1+R}f_{A1} \quad (15.3\text{-}9)$$

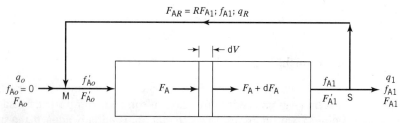

Figure 15.9 Flow diagram and terminology for recycle PFR (constant or variable density)

and at any point in the reactor,

$$f_A = \frac{(1+R)F_{Ao} - F_A}{(1+R)F_{Ao}} \tag{15.3-10}$$

where F_A is the molar flow rate of A at that point.

The volume of the recycle PFR may be obtained by a material balance for A around the differential control volume dV. Equating molar flow input and output, for steady-state operation, we have

$$F_A = F_A + dF_A + (-r_A)dV$$

$$dV = -\frac{dF_A}{(-r_A)} = (1+R)F_{Ao}\frac{df_A}{(-r_A)}$$

from equation 15.3-10. Therefore,

$$V = (1+R)F_{Ao}\int_{f'_{Ao}}^{f_{A1}} \frac{df_A}{(-r_A)} \tag{15.3-11}$$

where f'_{Ao} is given by equation (15.3-9).

The limiting cases for equation 15.3-11 are as discussed in Section 15.3.1 for constant density. That is, as $R \to 0$, V is that for a PFR without recycle; as $R \to \infty$, V is that for a CSTR (see problem 15-18 relating to an analytical proof of the latter). A graphical construction similar to that in Figure 15.8, but in which $1/(-r_A)$ is plotted against f_A, can be used to interpret equation 15.3-11 in a manner analogous to that of equation 15.3-4 for constant-density.

15.4 COMBINATIONS OF PFRs: CONFIGURATIONAL EFFECTS

A PFR may take various configurational forms, as indicated in Figure 15.1. The design or performance equation, 15.2-2 or its equivalent, provides a volume, V, which must then be interpreted geometrically for the purpose of fabrication. Thus, for cylindrical shape, the most common, it must be converted to a diameter (D) and length (L):

$$V = \frac{\pi D^2}{4}L \equiv \frac{\pi D^3}{4}\left(\frac{L}{D}\right) \tag{15.4-1}$$

The L/D ratio may vary from very large ($L/D \gg 1$), as in an ethane "cracker," which may consist of a very long length (i.e., many lengths connected in series) of 4-in pipe, to very small ($L/D \ll 1$), as in a sulfur dioxide converter in a large-scale sulfuric acid plant, which may consist of three or four shallow beds of catalyst in series in a large-diameter vessel (Figure 1.4). These situations correspond to those shown schematically in Figures 15.1(a) and (b), respectively. The parallel-tube arrangement shown in Figure 15.1(c), whether the tubes are packed or not, is also usually one with a relatively large L/D ratio; it facilitates heat transfer, for control of temperature, and is essentially a shell-and-tube heat exchanger in which reaction takes place either inside or outside the tubes.

The volume V calculated by means of equation 15.2-2 may be arranged into any L/D ratio and any series or parallel arrangement, *provided*:

(1) The resulting flow remains PF (to be consistent with the model on which the calculation of V is based).
(2) In a series arrangement, all changes in properties are taken into account, such as adjustment of T (without reaction) by a heat exchanger between stages (lengths). Such adjustment is more common for solid-catalyzed reactions than for homogeneous gas-phase reactions.
(3) In a parallel arrangement, streams which join at any point, most commonly at the outlet, have the same properties (e.g., composition and temperature).

For point (1), the value of Re ($= DG/\mu$, where G is the mass rate of flow per unit area, $\dot{m}/A_c$, and μ is viscosity) may sometimes be used as a guide. For cylindrical geometry,

$$\text{Re} = (4\dot{m}/\pi\mu)D^{-1} \tag{15.4-2}$$

$$= (2\dot{m}/\pi^{1/2}\mu V^{1/2})L^{1/2} \tag{15.4-3}$$

$$= (\dot{m}D^2/\mu V)(L/D) \tag{15.4-4}$$

Equation 15.4-2 states that, for given $\dot{m}$ and μ, Re $\propto D^{-1}$; 15.4-3 states that, for given $\dot{m}$, μ, and V, Re $\propto L^{1/2}$; and 15.4-4 states that, for given $\dot{m}$, μ, V, and D, Re $\propto L/D$. This last may be used to determine the minimum ratio, $(L/D)_{min}$, based on Re$_{min}$ for PF in an empty vessel, as illustrated in the next example. Equation 15.4-3 is more useful for a packed vessel (Chapter 21).

EXAMPLE 15-11

Devise a procedure for determining L and D for PF in an empty cylindrical vessel with a specified value of V.

SOLUTION

A stringent requirement for PF, nearly in accordance with fluid mechanics, is that it be fully developed turbulent flow. For this, there is a minimum value of Re that depends on D and on ϵ, surface roughness:

$$\text{Re}_{min} = g(\text{Re}, \epsilon/D) \tag{15.4-5}$$

A graphical correlation for this, based on the friction factor, is given by Perry et al. (1984, p. 5–24), together with values of ϵ. This can be used in conjunction with equation 15.4-2 to determine D, and hence L, by means of the following algorithm:

[1] Choose a value of D.
[2] Calculate Re from equation 15.4-2 ($\dot{m}$ and μ assumed given).
[3] Compare Re with Re$_{min}$ from 15.4-5.
[4] Repeat steps [1] to [3] until Re $\geq$ Re$_{min}$; D is then known.
[5] Calculate value of L from 15.4-1 and known V.

This procedure based on the friction factor chart is too stringent for large-diameter vessels, which behave nearly like smooth pipe, requiring inordinately large values of Re.

For illustration of a series arrangement, point (2) above, the division of a (total) volume (V) into three parts ($V_1 + V_2 + V_3 = V$) in three different situations is shown in

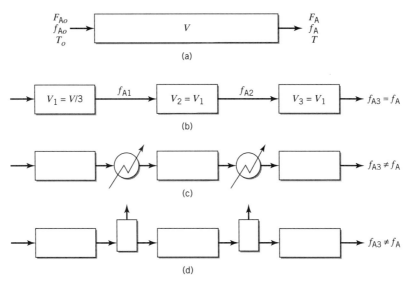

Figure 15.10 Comparison of PFR and various series arrangements (see text)

Figure 15.10. Figure 15.10(a) represents the total volume as one PFR, with reaction taking place leading to a particular fractional conversion at the outlet. Figure 15.10(b) represents the same total volume divided into three parts, not necessarily equal, in series. Since the mere division of V into three parts does not alter the operating conditions from point to point, the two configurations are equivalent, and lead to the same final conversion. That is, from equation 15.2-2,

$$\frac{V}{F_{Ao}} = \int_{f_{Ao}}^{f_A} \frac{df_A}{(-r_A)} = \int_{f_{Ao}}^{f_{A1}} \frac{df_A}{(-r_A)} + \int_{f_{A1}}^{f_{A2}} \frac{df_A}{(-r_A)} + \int_{f_{A2}}^{f_{A3}} \frac{df_A}{(-r_A)} = \frac{V_1 + V_2 + V_3}{F_{Ao}}$$

as stipulated, if $f_{A3} = f_A$. Figure 15.10(c) represents the same division, but with heat exchangers added between stages for adjustment of T. Since the T profile is not the same as in (a), the performance is not the same ($f_{A3} \neq f_A$). Similarly, Figure 15.10(d) represents the same division, but with separation units added between stages to remove a rate-inhibiting product. Since the concentration profiles are not the same as in (a), the performance is not the same ($f_{A3} \neq f_A$).

For a parallel arrangement, point (3) above, as illustrated in Figure 15.1(c), similar arguments can be used to show that if the space time and operating conditions (T, P, etc.) are the same for each tube, then each exhibits the same performance, and the total volume is the same as if no division took place (although the latter may be impracticable from a heat transfer point of view). In this case, it is also important to show that the division of the throughput (F_{Ao}) so that τ is the same in each tube leads to the most efficient operation (highest conversion for given total volume and throughput). This is considered further in Chapter 17, in connection with the similar case of CSTRs in parallel.

15.5 PROBLEMS FOR CHAPTER 15

15-1 A liquid-phase substitution reaction between aniline (A) and 2-chloroquinoxaline (B), A + B → products, is conducted in an isothermal, isobaric PFR. The reaction is first-order with respect to each reactant, with $k_A = 4.0 \times 10^{-5}$ L mol^{-1} s^{-1} at 25°C (Patel, 1992). Determine the reactor volume required for 80% conversion of aniline, if the initial concentration of each reactant is 0.075 mol L^{-1}, and the feed rate is 1.75 L min^{-1}.

15-2 A first-order liquid-phase reaction (A → P) is conducted isothermally in a 2.0-m³ PFR. The volumetric flow rate is 0.05 m³ min^{-1}, and $k_A = 0.02$ min^{-1}.
  (a) Determine the steady-state fractional conversion obtained using this reactor.
  (b) Determine the volume of a CSTR which gives the same conversion.
  (c) Determine the time required to reach steady-state in the PFR, if the reactor initially contains no A. Hint: To do this, you can either solve the partial differential equation with respect to t and V, or assume that a PFR is equivalent to 30 CSTRs arranged in series, and solve the resulting system of ordinary differential equations using E-Z Solve.

15-3 (a) The rate of the liquid-phase reaction A → B + ... is given by $(-r_A) = k_A c_A c_B$. Data obtained from the steady-state operation of a small-scale (50-liter) CSTR are to be used to determine the size of a commercial-scale recycle PFR operating isothermally at the same T as in the CSTR. The feed composition is the same for both reactors and $c_{Ao} \gg c_{Bo}$. The throughput (q_o) for the recycle PFR is 100 times that for the CSTR. If the fractional conversion (f_A) obtained in the CSTR is 0.75, what size (m³) of recycle PFR is required to obtain the same conversion with a recycle ratio (R) of 2?
  (b) What is the optimal value of R to minimize the volume in (a), and what is the minimum volume?

15-4 (a) Verify the values of V calculated in Example 15-10(b) for $c_{Bo} = 0.01, 0.05,$ and 0.11 mol L^{-1}.
  (b) For the reactor size calculated in part (c) of Example 15-10, what final conversion (f_A) is obtained, if the steady-state feed rate (q_o) is increased to 0.9 L s^{-1}? No other changes are made.

15-5 Ethyl acetate (A) reacts with OH$^-$ (B) to produce ethanol and acetate ion. The reaction is first-order with respect to each reactant, with $k_A = 0.10$ L mol^{-1} s^{-1} at 298 K (Streitwieser and Heathcock, 1981, p. 59). Determine the reactor volume for $f_A = 0.90$, given that q_o is 75 L min^{-1}, c_{Ao} is 0.050 mol L^{-1}, and c_{Bo} is 0.10 mol L^{-1}. Assume isothermal, isobaric conditions.

15-6 The decomposition of dimethyl ether (CH$_3$)$_2$O (E), to CH$_4$, H$_2$ and CO is a first-order irreversible reaction, and the rate constant at 504°C is $k_E = 4.30 \times 10^{-4}$ s^{-1}. What volume would be required for a PFR to achieve 60% decomposition of the ether, if it enters at 0.1 mol s^{-1} at 504°C and 1 bar, and the reactor operates at constant temperature and pressure?

15-7 A steady stream of SO$_2$Cl$_2$ is passed through a tube 2 cm in diameter and 1 meter long. Decomposition into SO$_2$ and Cl$_2$ occurs, and the temperature is maintained at 593 K, at which value the rate constant for the decomposition is 2.20×10^{-5} s^{-1}. Calculate the partial pressure of Cl$_2$ in the exit gas, if the total pressure is constant at 1 bar, and the feed rate of SO$_2$Cl$_2$ is 1 L min^{-1} measured at 0°C and 1 bar.

15-8 (a) Estimate the volume of a plug-flow reactor (in m³) required for production of ethylene from ethane (A) based on the following data and assumptions:
  (1) The feed is pure ethane at 1 kg s^{-1}, 1000 K and 2 bar.
  (2) The reaction is first-order and irreversible at low conversion, with rate constant $k_A = 0.254$ s^{-1} at 1000 K.
  (3) The reactor is operated isothermally and isobarically.
  (4) Conversion of ethane is 20% at the outlet.
  (b) If 4″ schedule-40 pipe is used (0.102 m I.D.), calculate the length required (m).
  (c) Comment on the most important assumptions which represent oversimplifications.

15-9 The oxidation of NO to NO$_2$ is an important step in the manufacture of nitric acid. The reaction is third-order, $(-r_{NO}) = k_{NO} c_{NO}^2 c_{O_2}$, and k_{NO} is 1.4×10^4 m^6 kmol^{-2} s^{-1} at 20°C.
  (a) Calculate the volume of a plug-flow reactor (which might actually be the lower part of the absorber) for 90% conversion of NO in a feed stream containing 11 mol % NO,

8% O_2, and 81% N_2. The reactor is to operate at 20°C and 6 bar, and the feed rate is 2000 m^3 h^{-1} at these conditions. (cf. Denbigh and Turner, 1984, pp. 57–61.)

(b) If the reaction were adiabatic instead of isothermal, would the volume required be greater or less than in (a)? Explain briefly without further calculations.

15-10 A gas-phase reaction A → R + T is carried out in an isothermal PFR. Pure A is fed to the reactor at a rate of 1 L s^{-1}, and the reactor pressure is constant at 150 kPa. The rate law is given by $(-r_A) = k_A c_A^3$, where $k_A = 1.25$ L^2 mol^{-2} s^{-1}. Determine the reactor volume required for 50% conversion of A, given that $c_{Ao} = 0.050$ mol L^{-1}.

15-11 Consider the following gas-phase reaction that occurs at 350 K and a constant pressure of 200 kPa (Lynch, 1986): A → B + C, for which the rate law is $(-r_A) = 0.253 c_A/(1 + 0.429 c_A)^2$, where $(-r_A)$ has units of mol m^{-3} s^{-1}; c_A has units of mol m^{-3}. Pure A is fed to a reactor at a rate of 8 mol s^{-1}. The desired fractional conversion, f_A, is 0.99. A recycle PFR is proposed for the reaction. When the recycle ratio, R, is zero, the recycle reactor is equivalent to a PFR. As R approaches infinity, the system is equivalent to a CSTR. However, it is generally stated that the recycle reactor behavior is close to that of a CSTR once R reaches approximately 10 to 20. Furthermore, it is often stated that, for an equivalent fractional conversion, the volume of a PFR is less than that of a CSTR. Check the generality of these statements by

(a) determining the volume of a PFR needed for 99% conversion;
(b) determining the volume of a CSTR needed for 99% conversion;
(c) determining the volume of a recycle PFR required for 99% conversion, if $R = 0, 1, 10, 20$, and 500.
(d) Comment on your findings.

Note: The equations used may be integrated numerically (using E-Z Solve) or analytically. For an analytical solution, use the substitution $Z = (1 - f_A)/(1 + f_A)$.

15-12 An isothermal PFR is used for the gas-phase reaction A → 2B + C. The feed, flowing at 2.0 L s^{-1}, contains 50 mol% A and 50 mol% inert species. The rate is first-order with respect to A; the rate constant is 2.0 s^{-1}.

(a) Determine the reactor volume required for 80% conversion of A.
(b) What is the volumetric flow rate at the exit of the reactor, if $f_A = 0.8$?
(c) What volume of CSTR would be needed to achieve the same conversion? Assume ideal-gas behavior and negligible pressure drop.

15-13 An exothermic first-order liquid-phase reaction A → R is conducted in a PFR. Determine the volume required for 90% conversion of A, if the process is adiabatic.

Data:

$C_{PA} = 143.75$ J mol^{-1} K^{-1} $q_o = 360$ L h^{-1}
$C_{PR} = 264.1$ J mol^{-1} K^{-1} $\Delta H_{RA} = -19000$ J mol^{-1}
$c_{Ao} = 2.5$ mol L^{-1} $\rho = 0.85$ g cm^{-3}
$k_A = 2.1 \times 10^7 \exp(-6500/T)$ min^{-1} $T_o = 325$ K

15-14 Calculate the space time (τ/sec) required for a PFR for the steady-state production of vinyl chloride from chlorine (A) and ethylene (B) according to $Cl_2 + C_2H_4 \rightarrow C_2H_3Cl + HCl$. Base the calculation on the following: molar feed ratio, $F_{Bo}/F_{Ao} = 20$; $T_o = 593$ K; adiabatic operation; $\Delta H_R = -96,000$ J (mol $Cl_2)^{-1}$ (assume constant); C_P (flowing stream) = 70 J mol^{-1} K^{-1} (assume constant); $P = 1$ bar (assume constant); f_A (at outlet) = 0.6. Assume the reaction is pseudo-first-order with respect to Cl_2 (C_2H_4 in great excess), with the following values of the rate constant (Subbotin et al., 1966):

T/K	593	613	633	653
k_A/s^{-1}	2.23	6.73	14.3	32.6

15-15 A first-order gas-phase reaction, A → C, is conducted in a nonisothermal PFR, operating isobarically. Equimolar amounts of A and an inert species are fed to the reactor at a total rate of 8 L min^{-1}, with $c_{Ao} = 0.25$ mol L^{-1}. Determine the residence time required for 50%

conversion of A, if operation is (a) adiabatic; and (b) nonadiabatic, using the following data:

$$k_A = 1.3 \times 10^{11} \, e^{-(10550/T)} \, \text{s}^{-1}, \text{ with } T \text{ in Kelvins}$$
$$T_o = 100°C \qquad \Delta H_{RA} = -15.5 \text{ kJ mol}^{-1}$$
$$U = 100 \text{ W m}^{-2} \text{ K}^{-1} \qquad d_t = 2 \text{ cm}$$
$$C_P = 37.0 \text{ J mol}^{-1} \text{ K}^{-1} \qquad T_S = 80°C$$

15-16 A gas-phase reaction A + B → 2C is conducted at steady-state in a PFR. The reaction is first-order with respect to both A and B. The reactor, constructed of stainless steel with a surface roughness (ϵ) of 0.0018 inches, has a diameter of 1.5 cm, and a length of 10 m. The feed contains 60 mol% A and 40 mol% B. The following data also apply:

$$c_{PA} = 0.85 \text{ J g}^{-1} \text{ K}^{-1} \qquad \mu = 0.022 \text{ cp (assume constant)}$$
$$c_{PB} = 0.95 \text{ J g}^{-1} \text{ K}^{-1} \qquad M_A = 28 \text{ g mol}^{-1}$$
$$c_{PC} = 1.0 \text{ J g}^{-1} \text{ K}^{-1} \qquad M_B = 32 \text{ g mol}^{-1}$$
$$q_o = 5.5 \text{ L s}^{-1} \qquad M_C = 30 \text{ g mol}^{-1}$$
$$T_o = 410 \text{ K} \qquad k_A = 10^6 \exp(-4900/T) \text{ L mol}^{-1} \text{ s}^{-1}$$
$$P_o = 0.800 \text{ MPa} \qquad \Delta H_{RB} = -11.0 \text{ kJ mol}^{-1}$$

A process fluid is available to cool the reactor. If it is used, the wall temperature of the reactor is constant at 410 K, and the overall heat transfer coefficient (U) is 125 W m^{-2} K^{-1}.
(a) Determine f_B if the reactor is assumed to operate isothermally and isobarically.
(b) Repeat part (a), but allow for pressure drop. (Assume the Reynolds number Re is constant and is based upon q_o.)
(c) Repeat part (b), but allow for nonisothermal behavior, assuming the process is adiabatic.
(d) Repeat part (c), but use the given heat transfer information to include $\dot{Q}$.
(To check the assumption in (b), how much does Re change from inlet to outlet of the reactor?)

15-17 For a certain homogeneous, gas-phase reaction taking place in a PFR, suppose it has been determined that a reactor of volume 0.5 m^3 is required for a particular conversion. Suggest suitable dimensions (length and diameter in m) for the reactor, and hence a reactor configuration, if the flow rate is 0.16 kg s^{-1} of material with a viscosity of 2×10^{-5} kg m^{-1} s^{-1}, and if Re must be at least 100,000. Assume steady-state, isothermal, isobaric operation.

15-18 The volume of a recycle PFR (V) for steady-state, isothermal operation involving a constant-density system is given by equation 15.3-4, and is a function of the recycle ratio R for given operating conditions (c_{Ao}, c_{A1}, q_o). Show that, as $R \to \infty$, V becomes equal to the volume of a CSTR operating at the same conditions, as given by equation 14.3-15 (with $c_A = c_{A1}$, $q = q_o$, and $(-r_A)$ the rate at c_{A1}). It is convenient to let $m = 1/R$, and transform the indeterminant form to 0/0. This can be evaluated by means of L'Hôpital's Rule. This involves, in this case, differentiating under the integral sign. Here, if the integral in equation 15.3-4 is represented by $u(m) = \int_{a(m)}^{b} v(c_A) dc_A$, then $du(m)/dm = -v[a(m)]da(m)/dm$. Note that it is unnecessary to introduce a particular form of rate law for $(-r_A)$, and hence the result is independent of the kinetics of the reaction.

Chapter 16

Laminar Flow Reactors (LFR)

In a laminar flow reactor (LFR), we assume that one-dimensional laminar flow (LF) prevails; there is no mixing in the (axial) direction of flow (a characteristic of tubular flow) and also no mixing in the radial direction in a cylindrical vessel. We assume LF exists between the inlet and outlet of such a vessel, which is otherwise a "closed" vessel (Section 13.2.4). These and other features of LF are described in Section 2.5, and illustrated in Figure 2.5. The residence-time distribution functions $E(\theta)$ and $F(\theta)$ for LF are derived in Section 13.4.3, and the results are summarized in Table 13.2.

Laminar flow in a cylindrical tube is characterized by a parabolic velocity profile or distribution:

$$u(r) = u_o[1 - (r/R)^2] \tag{2.5-1}$$

where $u(r)$ is the velocity of an element of fluid (a cylindrical shell of radius r and thickness dr in a tube of radius R—see Figure 2.5), and u_o is the velocity at the center of the tube ($r = 0$), the maximum velocity. The mean velocity, $\bar{u}$, of all elements of fluid is given by the ratio of the total volumetric flow, q, to the cross-sectional area of flow, A_c:

$$\bar{u} = \frac{q}{A_c} = \frac{\int_0^R 2\pi r u(r) dr}{\pi R^2} = \frac{u_o}{2} \tag{2.5-2}$$

on introduction of equation 2.5-1.

To simplify the treatment for an LFR in this chapter, we consider only isothermal, steady-state operation for cylindrical geometry, and for a simple system (A → products) at constant density. After considering uses of an LFR, we develop the material-balance (or continuity) equation for any kinetics, and then apply it to particular cases of power-law kinetics. Finally, we examine the results in relation to the segregated-flow model (SFM) developed in Chapter 13.

16.1 USES OF AN LFR

In the case of an LFR, it is important to distinguish between its use as a model and its occurrence in any actual case. As a model, the LFR can be treated exactly as far as the consequences for performance are concerned, but there are not many cases in which the model serves as a close approximation. In contrast, the CSTR and PFR models serve as useful and close approximations in many actual situations.

394 Chapter 16: Laminar Flow Reactors (LFR)

Since laminar flow itself occurs at low values of $Re(= Du\rho/\mu)$, the most likely situations are those characterized by low velocity (u) or high viscosity (μ), such as those involving the slow flow of polymers in extrusion reactors, or of blood in certain organs in animals. Even if not a close approximation in some cases, the predictable performance of an LFR may serve as a limiting model for actual performance.

16.2 DESIGN EQUATIONS FOR AN LFR

16.2.1 General Considerations and Material Balance

The process design of an LFR, like that of other types of reactors, may involve, for a specified throughput (q_o or F_{Ao}), calculation of the fractional conversion (f_A) achieved at the outlet of a vessel of given size, or the size of vessel required for a specified f_A.

In comparison with a PFR, the situation for an LFR is complicated by the fact that the radial parabolic velocity distribution at any axial position implies that the fluid characteristics, such as f_A (or c_A) and $(-r_A)$, depend on radial position (r) as well as on axial position (x). The material balance must take this into account. Laminar flow itself may be considered as flow of fluid through a series of concentric shells, each of radius r and thickness dr, and moving with velocity $u(r)$, as indicated for one such shell in Figure 16.1. Since there is no axial or radial mixing of fluid elements, the material in each cylindrical shell moves as through an independent PFR. Thus, we may develop the material balance in a similar manner to that in Chapter 15, always remembering, however, that the result expresses the situation at a particular radial position. Thus, consider the control volume as the "doughnut"-shaped volume $dV(r) = (2\pi r dr)dx$ shown shaded in Figure 16.1. Let the (constant) volumetric flow rate through the control volume be dq. Then

$$\text{rate of input of A by flow} = c_A dq = c_A 2\pi r dr u(r)$$
$$\text{rate of output of A by flow} = (c_A + dc_A)dq = (c_A + dc_A)2\pi r dr u(r)$$
$$\text{rate of disappearance of A by reaction} = (-r_A)dV(r) = (-r_A)2\pi r dr dx$$

The material balance at steady-state is, on cancellation of $2\pi r dr$,

$$c_A u(r) - (c_A + dc_A)u(r) - (-r_A)dx = 0$$

which may be written as

$$u(r)\frac{\partial c_A}{\partial x} + (-r_A) = 0 \quad \text{(at } r\text{)} \quad (16.2\text{-}1)$$

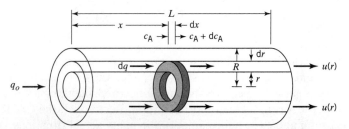

Figure 16.1 Schematic diagram of a laminar flow reactor

or, since $dx = u(r)dt$,

$$\frac{\partial c_A}{\partial t} + (-r_A) = 0 \quad \text{(at } r\text{)} \quad (16.2\text{-}2)$$

These equations are similar to the material-balance equations for a PFR, except that (for the LFR) they apply at a radial position r and not across the entire cross-section. The partial derivative is required because of this.

Each of equations 16.2-1 and -2 may be interpreted to define $(-r_A)$ at a point with respect to both radial and axial position. The average rate, $(-\bar{r}_A)$, over the entire circular cross-section at a particular axial position is determined by integration as:

$$(-\bar{r}_A) = \frac{1}{R}\int_0^R [-r_A(r)]dr = -\frac{1}{R}\int_0^R u(r)\frac{\partial c_A}{\partial x}dr \quad (16.2\text{-}3)$$

16.2.2 Fractional Conversion and Concentration (Profiles)

Since the material-balance equations, 16.2-1 and -2, derived above refer to a particular radial position, we must integrate radially to obtain an average concentration c_A at any axial position x, including at the outlet, where $x = L$. The latter is the observed outlet concentration that corresponds to the outlet fractional conversion. We develop the expression in terms of c_A.

At any position x, the rate of flow of A in terms of total and average quantities is

$$F_A = q_o c_A = \pi R^2 \bar{u} c_A$$

On integrating over the entire cross-section, we obtain for the rate of flow of A:

$$F_A = \int_0^R u(r)c_A(r)2\pi r dr$$

From these two expressions, at any axial point x,

$$c_A = \frac{2}{\bar{u}R^2}\int_0^R u(r)c_A(r)r dr \quad (16.2\text{-}4)$$

To obtain the concentration profile $c_A(x)$ as well as the outlet concentration $c_A(L)$, we express equation 16.2-4 in terms of t rather than r. To achieve this, we use equations 2.5-1 and -2 to eliminate $u(r)$ and $\bar{u}$, respectively, together with

$$t(x, r) = x/u(r) \quad (16.2\text{-}5)$$

and

$$\bar{t}(x) = x/\bar{u} \quad (16.2\text{-}6)$$

where $\bar{t}(x)$ is the mean residence time to the point at x. At the outlet, these equations become

$$t(r) = L/u(r) \quad (16.2\text{-}7)$$

and

$$\bar{t} = L/\bar{u} \qquad (16.2\text{-}8)$$

where $\bar{t}$ is the mean residence time in the vessel. We proceed by developing the concentration profile $c_A(x)$, which can be readily interpreted to obtain c_A and f_A at the outlet. From equations 16.2-5 and -6,

$$u(r)/\bar{u} = \bar{t}(x)/t(x, r) \qquad (16.2\text{-}9)$$

From this equation and equations 2.5-1 and -2, on elimination of u_o, $u(r)$ and $\bar{u}$, we obtain

$$\frac{\bar{t}(x)}{t(x,r)} = 2\left(1 - \frac{r^2}{R^2}\right)$$

On differentiation at fixed x, and letting t stand for $t(x, r)$, we have

$$-\frac{\bar{t}(x)}{t^2}dt = -\frac{4r}{R^2}dr$$

or

$$2rdr/R^2 = \bar{t}(x)dt/2t^2 \qquad (16.2\text{-}10)$$

Substitution of equations 16.2-9 and -10 in equation 16.2-4, and change of limits as follows:

$$\text{when} \quad r = 0, \quad t = \bar{t}(x)/2$$
$$\text{when} \quad r = R, \quad t = \infty$$

results in

$$c_A(x) = \frac{[\bar{t}(x)]^2}{2}\int_{\bar{t}(x)/2}^{\infty} c_A(t)\frac{dt}{t^3} \qquad (16.2\text{-}11)$$

The interpretation of $c_A(t)$ comes from the realization that each cylindrical shell passes through the vessel as an independent batch. Thus, $c_A(t)$ is obtained by integration of the material balance for a batch reactor (BR). Accordingly, we may rewrite equation 16.2-11, in terms of either $c_A(x)$ or $f_A(x)$, as

$$1 - f_A(x) = \frac{c_A(x)}{c_{Ao}} = \frac{[\bar{t}(x)]^2}{2}\int_{\bar{t}(x)/2}^{\infty}\left[\frac{c_A(t)}{c_{Ao}}\right]_{BR}\frac{dt}{t^3} \qquad (16.2\text{-}12)$$

At the outlet of the reactor, this becomes

$$1 - f_A = \frac{c_A}{c_{Ao}} = \frac{\bar{t}^2}{2}\int_{\bar{t}/2}^{\infty}\left[\frac{c_A(t)}{c_{Ao}}\right]_{BR}\frac{dt}{t^3} \qquad (16.2\text{-}13)$$

Using the E-Z Solve software, we may integrate equation 16.2-12 to develop the axial profile for c_A or f_A; equation 16.2-13 gives the value of the final (exit) point on this profile.

16.2.3 Size of Reactor

The size of an LFR for a specified f_A or c_A at the outlet is defined by the length L for a specified radius R (the latter may be used as a parameter, but values chosen must be consistent with the assumption of laminar flow). If the throughput q_o and R are given, one way to establish L is to use equation 16.2-12 to develop the $f_A(x)$ or $c_A(x)$ profile. For this purpose,

$$\bar{t}(x) = x/\bar{u} \tag{16.2-6}$$

and

$$\bar{u} = q_o/\pi R^2$$

so that

$$\bar{t}(x) = \pi R^2 x/q_o \tag{16.2-14}$$

L is the value of x for which $f_A(x)$ or $c_A(x)$ is the specified outlet value.

16.2.4 Results for Specific Rate Laws

In this section, we use two examples to illustrate the application of equations 16.2-12 and -13 to determine c_A or f_A for specified power-law forms of the rate law, $(-r_A) = k_A c_A^n$.

EXAMPLE 16-1

Derive an equation for calculating the outlet c_A or f_A for a zero-order reaction, if the inlet concentration is c_{Ao}, and the mean residence time is $\bar{t}$.

SOLUTION

For use in equation 16.2-13, we must first obtain $[c_A(t)/c_{Ao}]_{BR}$. From the integration of the material balance for constant density in a BR, equation 3.4-9, with $n = 0$,

$$\left[\frac{c_A(t)}{c_{Ao}}\right]_{BR} = 1 - \frac{k_A t}{c_{Ao}} \tag{16.2-15}$$

We note that there is a restriction on the value of t; since $c_A(t) \geq 0$,

$$t \leq c_{Ao}/k_A \quad (BR, n = 0) \tag{16.2-16}$$

With equations 16.2-15 and -16 incorporated, equation 16.2-13 becomes

$$1 - f_A = \frac{c_A}{c_{Ao}} = \frac{\bar{t}^2}{2} \int_{\bar{t}/2}^{\infty} \left(1 - \frac{k_A t}{c_{Ao}}\right) \frac{dt}{t^3}$$

$$= \frac{\bar{t}^2}{2} \left[\int_{\bar{t}/2}^{c_{Ao}/k_A} \left(1 - \frac{k_A t}{c_{Ao}}\right) \frac{dt}{t^3} + \int_{c_{Ao}/k_A}^{\infty} \left(1 - \frac{k_A t}{c_{Ao}}\right) \frac{dt}{t^3} \right]$$

$$= \frac{\bar{t}^2}{2} \int_{\bar{t}/2}^{c_{Ao}/k_A} \left(1 - \frac{k_A t}{c_{Ao}}\right) \frac{dt}{t^3} + 0$$

398 Chapter 16: Laminar Flow Reactors (LFR)

(The second integral vanishes because of equations 16.2-15 and -16.) Thus,

$$1 - f_A = 1 - \frac{k\bar{t}}{c_{Ao}} + \frac{1}{4}\left(\frac{k\bar{t}}{c_{Ao}}\right)^2 = 1 - M_{A0} + M_{A0}^2/4 \qquad (16.2\text{-}17)$$

where $M_{A0} = k\bar{t}/c_{Ao}$, the dimensionless reaction number for a zero-order reaction, from equation 4.3-4.

Equation 16.2-17 applies only for $f_A \leq 1$. This is the case for $0 \leq M_{A0} \leq 2$. For $M_{A0} \geq 2$, $f_A = 1$. Thus, from equation 16.2-17:

$$\boxed{\begin{array}{ll} \text{LFR, } n = 0 \\ f_A = M_{A0}(1 - M_{A0}/4) \text{ ; } & 0 \leq M_{A0} \leq 2 \\ f_A = 1 & \text{; } M_{A0} \geq 2 \end{array}} \qquad (16.2.18)$$

EXAMPLE 16-2

Repeat Example 16-1 for a first-order reaction.

SOLUTION

For a first-order reaction,

$$\left[\frac{c_A(t)}{c_{Ao}}\right]_{BR} = e^{-k_A t} \qquad (3.4\text{-}10)$$

(There is no restriction on the value of t in this case.)
Substitution of equation 3.4-10 in equation 16.2-13 results in

$$1 - f_A = \frac{c_A}{c_{Ao}} = \frac{\bar{t}^2}{2}\int_{\bar{t}/2}^{\infty} \frac{e^{-k_A t}}{t^3} dt \quad (\text{LFR, } n = 1) \qquad (16.2\text{-}19)$$

This integral is related to the exponential integral (see Table 14.1). It cannot be solved in closed analytical form, but it can be evaluated numerically using the E-Z Solve software; the upper limit may be set equal to $10\bar{t}$.

EXAMPLE 16-3

A first-order, liquid-phase reaction, A $\to$ products, is to be conducted in a tubular reactor to achieve 75% conversion of A (f_A). The feed rate (q_o) is 7.5×10^{-3} m^3 h^{-1} of pure A, and the rate constant (k_A) is 0.15 h^{-1}. For a reactor diameter of 0.05 m, determine the length of reactor (L/m) required, if the flow is (a) LF, and (b) PF.

SOLUTION

(a) Equation 16.2-19 may be used to generate the profile of f_A versus x (distance from inlet):

$$f_A(x) = 1 - \frac{[\bar{t}(x)]^2}{2}\int_{\bar{t}(x)/2}^{\infty} \frac{e^{-k_A t}}{t^3} dt \qquad (16.2\text{-}19)$$

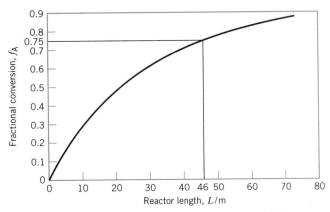

Figure 16.2 Determining reactor size for Example 16-3

with $\bar{t}(x)$ related to x by equation 16.2-14:

$$\bar{t}(x) = \pi R^2 x/q_o \quad (16.2\text{-}14)$$

For the purpose of numerical integration using E-Z Solve (file ex16.3-msp), the upper limit in the integral is assumed to be $10\,\bar{t}(x)$.

The profile of f_A versus x is shown in Figure 16.2. For $f_A = 0.75$, the reactor length is $L = 46$ m.

(b) For PF, we model the reactor as a PFR. We first use the design equation 15.2-1, together with the rate law, to calculate the residence time $\bar{t}$, and then use equation 16.2-14 to calculate the reactor length (L). Thus,

$$F_{Ao}df_A/dV = (-r_A) = k_A c_A = k_A F_A/q = k_A F_{Ao}(1 - f_A)/q_o$$

From this,

$$\int_0^{0.75} df_A/(1 - f_A) = k_A \int dV/q_o = k_A V/q_o = k_A \bar{t}$$

The solution for $\bar{t}$ is, on integration,

$$\bar{t} = (1/k_A)[-\ln(1 - f_A)]_0^{0.75} = 1.386/0.15 = 9.24\ \text{h}$$

From equation 16.2-14, with $x \equiv L$ at the outlet,

$$L = \bar{t} q_o/\pi R^2 = 9.24(7.5 \times 10^{-3})/(\pi/4)(0.05)^2 = 35.3\ \text{m}$$

Comparing the results for (a) and (b), we note that a smaller reactor is required if the flow is PF rather than LF.

16.2.5 Summary of Results for LFR

The examples above illustrate that the determination of f_A (or c_A) for an LFR can be done either analytically or numerically (using the E-Z Solve software); most situations involve the latter. Another case that can be done analytically is for a second-order reaction ($n = 2$); see problem 16-1. The results for the three cases $n = 0, 1,$ and 2 are summarized in Table 16.1.

Table 16.1 Fractional conversion (f_A) for LFR[a]

Order n	M_{An} (equation 4.3-4)	$f_A = 1 - \dfrac{c_A}{c_{Ao}}$	
0	$k_A \bar{t}/c_{Ao}$	$M_{A0}\left(1 - \dfrac{M_{A0}}{4}\right); \ 0 \leq M_{A0} \leq 2$ $1; \ M_{A0} \geq 2$	(16.2-18)
1	$k_A \bar{t}$	$1 - \dfrac{\bar{t}^2}{2}\displaystyle\int_{\bar{t}/2}^{\infty} \dfrac{e^{-k_A t}}{t^3}\,dt$	(16.2-19)[b]
2	$k_A c_{Ao} \bar{t}$	$M_{A2}\left[1 - \dfrac{M_{A2}}{2}\ln\left(1 + \dfrac{2}{M_{A2}}\right)\right]$	(16.2-20)[c]

[a] For reaction A → products; constant-density, isothermal, steady-state operation.
[b] Second term is related to exponential integral; see Table 14.1.
[c] From problem 16-1.

16.2.6 LFR Performance in Relation to SFM

For the SFM,

$$1 - f_A = \frac{c_A}{c_{Ao}} = \int_0^\infty \left[\frac{c_A(t)}{c_{Ao}}\right]_{BR} E(t)\,dt \qquad (13.5\text{-}2)$$

For an LFR,

$$E(t) = 0 \,; \ t < \bar{t}/2 \qquad (13.4\text{-}15)$$
$$E(t) = \bar{t}^2/2t^3 \,; \ t > \bar{t}/2 \qquad (13.4\text{-}19)$$

From 13.5-2 with 13.4-15 and 13.4-19 introduced, we recapture equation 16.2-13:

$$1 - f_A = \frac{c_A}{c_{Ao}} = \frac{\bar{t}^2}{2}\int_{\bar{t}/2}^\infty \left[\frac{c_A(t)}{c_{Ao}}\right]_{BR} \frac{dt}{t^3} \qquad (16.2\text{-}13)$$

Thus, the material balance for an LFR and the SFM applied to an LFR give the same result. This is because LF, like PF, is, indeed, segregated flow. The SFM is exact for an LFR for any kinetics, just as it is for a PFR.

16.3 PROBLEMS FOR CHAPTER 16

16-1 For a second-order reaction (A → products) occurring in an LFR, show that the fractional conversion of A is given by:

$$f_A = M_{A2}\left[1 - \frac{M_{A2}}{2}\ln\left(1 + \frac{2}{M_{A2}}\right)\right] \qquad (16.2\text{-}20)$$

where M_{A2} is the dimensionless reaction number $k_A c_{Ao} \bar{t}$ (equation 4.3-4), c_{Ao} is the feed concentration of A, k_A is the second-order rate constant, and $\bar{t}$ is the mean residence time.

16-2 Calculate f_A for a second-order reaction (as in problem 16-1) in an LFR, if $k_A c_{Ao} \bar{t} = M_{A2} = 4$, and compare with the result for a PFR and for a CSTR.

16-3 Determine the fractional conversion f_A of A for a zero-order reaction (A → products) in a laminar flow reactor, where $c_{Ao} = 0.25$ mol L^{-1}, $k_A = 0.0015$ mol L^{-1} s^{-1}, and $\bar{t} = 150$ s. Compare the result with the fractional conversion for a PFR and for a CSTR.

16-4 Using equation 16.2-18, develop a graph which shows the fractional conversion f_A as a function of the dimensionless reaction number M_{A0} for a zero-order reaction, where $M_{A0} = k\bar{t}/c_{Ao}$, equation 4.3-4. What are the "real" limits on M_{A0} (i.e., values for which reasonable values of f_A are obtained)? Explain.

16-5 Develop the $E(t)$ profile for a 10-m laminar-flow reactor which has a maximum flow velocity of 0.40 m min^{-1}. Consider $t = 0.5$ to 80 min. Compare the resulting profile with that for a reactor system consisting of a CSTR followed by a PFR in series, where the CSTR has the same mean residence time as the LFR, and the PFR has a residence time of 25 min. Include in the comparison a plot of the two profiles on the same graph.

Chapter 17

Comparisons and Combinations of Ideal Reactors

In this chapter, we first compare the performance of ideal reactors as single vessels, and then examine the consequences of arranging flow reactors in various parallel and series configurations, involving vessels of the same type and of different types. Performance includes determining the size of reactor for a specified throughput or rate of production or conversion, and the converse situation for a given size. Using a multiple-vessel configuration raises the possibility of optimal arrangements of size and distribution of feed, and of taking advantage of any special features of the kinetics, for example, as noted for recycling and autocatalysis (Section 15.3).

We focus attention in this chapter on simple, isothermal reacting systems, and on the four types BR, CSTR, PFR, and LFR for single-vessel comparisons, and on CSTR and PFR models for multiple-vessel configurations in flow systems. We use residence-time-distribution (RTD) analysis in some of the multiple-vessel situations, to illustrate some aspects of both performance and mixing.

17.1 SINGLE-VESSEL COMPARISONS

17.1.1 BR and CSTR

A performance comparison between a BR and a CSTR may be made in terms of the size of vessel required in each case to achieve the same rate of production for the same fractional conversion, with the BR operating isothermally at the same temperature as that in the CSTR. Since both batch reactors and CSTRs are most commonly used for constant-density systems, we restrict attention to this case, and to a reaction represented by

$$A + \ldots \rightarrow \nu_C C + \ldots \tag{12.3-1}$$

in which A is the limiting reactant.

For a BR, the rate of production of C, $Pr(C)$, is given by equation 12.3-22, which may be rewritten as, with $f_{A2} = f_A$ and $f_{A1} = 0$,

$$Pr(C) = \nu_C c_{Ao} f_A V_{BR}/(1 + a)t \tag{17.1-1}$$

where f_A is the final conversion, V_{BR} is the volume of the BR, and a is the ratio of "down-time" to operating or reaction time (i.e., $a = t_d/t$). From equation 12.3-21,

$$t = c_{Ao} \int_0^{f_A} df_A/(-r_A) \tag{17.1-2}$$

For a CSTR, in steady-state operation at a flow rate q, from equation 14.3-6, for a constant-density system,

$$Pr(C) = qc_C = q\nu_C c_{Ao} f_A = \nu_C(-r_A)V_{ST} \tag{17.1-3}$$

on substitution from equation 14.3-5. Here c_C is the concentration of product C in the exit stream, and V_{ST} is the volume of the CSTR.

The ratio of the vessel sizes for the same $Pr(C)$ (and f_A) may be determined by equating $Pr(C)$ in equations 17.1-1 and -3, and substituting for t from 17.1-2:

$$\boxed{\frac{V_{BR}}{V_{ST}} = \frac{(1+a)\int_0^{f_A} df_A/(-r_A)}{f_A/(-r_A)}} \quad \text{(constant } \rho, T) \tag{17.1-4}$$

The ratio thus depends on a, f_A and the form of $(-r_A)$, but is valid for any kinetics. If size of vessel is a major consideration, the advantage is to the CSTR for high values of a, but to the BR for high conversion, since $V_{ST} \to \infty$ faster as $f_A \to 1$. It can be shown (problem 17-7) that, for given f_A and a, the ratio decreases as order of reaction increases, and also that, at low conversion, it approaches $1 + a$, independent of order.

The following example illustrates a similar comparison through time quantities: space time τ for a CSTR, and t for a BR.

EXAMPLE 17-1

Compare τ for a CSTR and t for a BR, and interpret the comparison graphically. The comparison is for the same f_A (in reaction 12.3-1), constant-density, isothermal BR (at same T as in CSTR), and any form of rate law for $(-r_A)$.

SOLUTION

For a BR, t is given by equation 17.1-2, rewritten as

$$t_{BR}/c_{Ao} = \int_0^{f_A} df_A/(-r_A) \tag{17.1-2}$$

For a CSTR, from the definition of τ ($= V_{ST}/q_o$), and equation 14.3-15 for constant density (with $q = q_o$),

$$\tau_{ST}/c_{Ao} = f_A/(-r_A) \tag{17.1-5}$$

Dividing equation 17.1-2 by 17.1-5, we obtain

$$\frac{t_{BR}}{\tau_{ST}} = \frac{\int_0^{f_A} df_A/(-r_A)}{f_A/(-r_A)} \tag{17.1-6}$$

which is the same result as in equation 17.1-4 for the volume ratio with $a = 0$.

404 Chapter 17: Comparisons and Combinations of Ideal Reactors

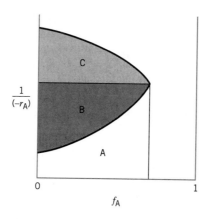

Figure 17.1 Comparison of t for a BR and τ for a CSTR; area A + B $\equiv \tau/c_{Ao}$; area A $\equiv t/c_{Ao}$ (normal kinetics); area A + B + C $\equiv t/c_{Ao}$ (abnormal kinetics)

Figure 17.1, a plot of reciprocal rate, $1/(-r_A)$, versus f_A illustrates the results expressed separately by equations 17.1-2 and 17.1-5. τ/c_{Ao} for a CSTR is equivalent to the area A + B. For normal kinetics, in which the rate decreases with increasing f_A, area A represents the integral in equation 17.1-2, that is, t/c_{Ao}; in such a case, $\tau > t$. Conversely, for abnormal kinetics, as might be experienced for an autocatalytic reaction (Chapter 8), $t > \tau$, since area A + B + C represents t/c_{Ao}.

17.1.2 BR and PFR

The performances of a BR and of a PFR may be compared in various ways, and are similar in many respects, as discussed in Section 15.2.2.1, since an element of fluid, of arbitrary size, acts as a closed system (i.e., a batch) in moving through a PFR. The residence time in a PFR, the same for all elements of fluid, corresponds to the reaction time in a BR, which is also the same for all elements of fluid. Depending on conditions, these quantities, and other performance characteristics, may be the same or different.

For constant-density and isothermal operation, the performance characteristics are the same, although allowance must be made for the down-time t_d of a BR operating on a continual basis. Thus, if we compare vessel sizes as in Section 17.1.1, for the same rate of production at the same conversion for reaction 12.3-1, a similar analysis leads to

$$V_{BR}/V_{PF} = 1 + a \quad \text{(constant } \rho, T\text{)} \tag{17.1-7}$$

where V_{PF} is the volume of a PFR. For the PFR, equation 17.1-7 is based on equation 15.2-2 and on 14.3-6, which gives $Pr(C)$ for a PFR as well as a CSTR.

Similarly, for constant ρ and T, we can compare the reaction time t_{BR} in a BR and the residence time t_{PF} (or space time τ_{PF}) in a PFR required for the same conversion. From equations 17.1-2 for t_{BR}, and 2.4-9 for t_{PF}, $t_{PF} = \int dV/q = V/q_o$ (for constant density), together with 15.2-2, $V = F_{Ao} \int_0^{f_A} df_A/(-r_A)$, we obtain

$$\frac{t_{BR}}{t_{PF}} = \frac{c_{Ao} \int_0^{f_A} df_A/(-r_A)}{\int dV/q}$$

$$= \frac{c_{Ao} \int_0^{f_A} df_A/(-r_A)}{(F_{Ao}/q_o) \int_0^{f_A} df_A/(-r_A)} = 1 \quad \text{(constant } \rho, T\text{)} \tag{17.1-8}$$

17.1.3 CSTR and PFR

In the following example, we compare the sizes of a CSTR and a PFR required to achieve the same conversion. This is for "normal" kinetics.

EXAMPLE 17-2

Calculate the ratio of the volumes of a CSTR and a PFR (V_{ST}/V_{PF}) required to achieve, for a given feed rate in each reactor, a fractional conversion (f_A) of (i) 0.5 and (ii) 0.99 for the reactant A, if the liquid-phase reaction A → products is (a) first-order, and (b) second-order with respect to A. What conclusions can be drawn? Assume the PFR operates isothermally at the same T as that in the CSTR.

SOLUTION

This example can be solved using the E-Z Solve software (file ex17-2.msp) or as follows. The volumes for a CSTR and a PFR are determined from the material-balance equations:

$$V_{ST} = F_{Ao} f_A / (-r_A) \quad (14.3\text{-}5)$$

$$V_{PF} = F_{Ao} \int_0^{f_A} df_A / (-r_A) \quad (15.2\text{-}2)$$

Thus,

$$\boxed{\frac{V_{ST}}{V_{PF}} = \frac{f_A / (-r_A)}{\int_0^{f_A} df_A / (-r_A)}} \quad (17.1\text{-}9)$$

In accordance with the discussion in Section 17.1.1, this is the same result as for V_{ST}/V_{BR} given by equation 17.1-4 with $a = 0$ (i.e., with no down-time).

(a) For a first-order reaction, $(-r_A) = k_A c_A$, occurring at the same T in each reactor, equation 17.1-9 becomes

$$\frac{V_{ST}}{V_{PF}} = \frac{f_A / k_A c_{Ao}(1 - f_A)}{\int_0^{f_A} [df_A / k_A c_{Ao}(1 - f_A)]}$$

$$= \frac{f_A}{(1 - f_A) \ln[1/(1 - f_A)]} \quad (n = 1) \quad (17.1\text{-}10)$$

(b) Similarly, for a second-order reaction, equation 17.1-9 is

$$\frac{V_{ST}}{V_{PF}} = \frac{f_A / k_A c_{Ao}(1 - f_A)^2}{\int_0^{f_A} [df_A / k_A c_{Ao}^2 (1 - f_A)^2]}$$

Table 17.1 Values of V_{ST}/V_{PF} for Example 17-2

f_A	n	
	1	2
0	1	1
0.5	1.4	2
0.99	21.5	100
1.0	∞	∞

which, on integration and simplification, becomes

$$\frac{V_{ST}}{V_{PF}} = \frac{1}{1-f_A} \quad (n = 2) \quad \text{(17.1-11)}$$

From these equations, values of V_{ST}/V_{PF} for f_A = (i) 0.5 and (ii) 0.99 are given in Table 17.1, together with the limiting values for $f_A = 0, 1$.

The conclusions illustrated in Table 17.1 are: (1) for a given order, n, the ratio increases as f_A increases, and (2) for a given f_A, the ratio increases as order increases. In any case, for normal kinetics, $V_{ST} > V_{PF}$, since the CSTR operates entirely at the lowest value of c_A, the exit value. (Levenspiel, 1972, p. 332, gives a more detailed graphical comparison for five values of n. This can also be obtained from the E-Z Solve software.)

Figure 17.1, for both normal and abnormal kinetics, can also be interpreted as a comparison between a CSTR and a PFR, with the PFR replacing the BR. Again, area A + B is a measure of τ_{ST}/c_{Ao} (or V_{ST}/F_{Ao}). Area A, for normal kinetics, is a measure of τ_{PF}/c_{Ao} or t_{PF}/c_{Ao} (or V_{PF}/F_{Ao}), where t_{PF} is the residence time of fluid in the PFR, corresponding to reaction time t in a BR. Area A + B + C corresponds to t_{PF}/c_{Ao} for abnormal kinetics.

17.1.4 PFR, LFR, and CSTR

In Table 17.2, f_A (for the reaction A → products) is compared for each of the three flow reactor models PFR, LFR, and CSTR. The reaction is assumed to take place at constant density and temperature. Four values of reaction order are given in the first column: $n = 0, 1/2, 1,$ and 2 ("normal" kinetics). For each value of n, there are six values of the dimensionless reaction number: $M_{An} = 0, 0.5, 1, 2, 4,$ and ∞, where $M_{An} = k_A c_{Ao}^{n-1} \bar{t}$, equation 4.3-4. The fractional conversion f_A is a function only of M_{An}, and values are given for three models in the last three columns. The values for a PFR are also valid for a BR for the conditions stated, with reaction time $t = \bar{t}$ and no down-time ($a = 0$), as described in Section 17.1.2.

Table 17.2 confirms that the best performance, for normal kinetics, is always obtained in a PFR, and, in most cases, the worst is in a CSTR. An LFR gives intermediate results, except for sufficiently small values of M_{An} at the lower orders, in which cases f_A for an LFR is the lowest. Note that the results are the same for three pairs of situations as follows: (1) $n = 0$: PFR and CSTR; (2) $n = 1/2$ for PFR and $n = 0$ for LFR; (3) $n = 2$ for PFR and $n = 1$ for CSTR. A more detailed comparison of the performance of an LFR and a PFR is shown graphically by Aris (1965, p. 296).

The equations used to calculate f_A in Table 17.2 are mostly contained in Chapters 14, 15, and 16. They are collected together in the three boxes below, and include those not given explicitly previously.

Table 17.2 Comparison of fractional conversion (f_A) for PFR, LFR and CSTR

n (order)	M_{An}^a	f_A^b PFR^c	LFR	CSTR
0	0	0	0	0
	0.5	0.5	0.438	0.5
	1	1	0.750	1
	2	1	1	1
	4	1	1	1
	∞	1	1	1
1/2	0	0	0	0
	0.5	0.438	0.388	0.390
	1	0.750	0.639	0.618
	2	1	0.903	0.828
	4	1	1	0.944
	∞	1	1	1
1	0	0	0	0
	0.5	0.394	0.35	0.333
	1	0.632	0.56	0.500
	2	0.865	0.78	0.667
	4	0.982	0.94	0.800
	∞	1	1	1
2	0	0	0	0
	0.5	0.333	0.299	0.268
	1	0.500	0.451	0.382
	2	0.667	0.614	0.500
	4	0.800	0.756	0.610
	∞	1	1	1

$^a M_{An} = k_A c_{Ao}^{n-1} \bar{t}$ (equation 4.3-4).

^{b}For reaction A $\rightarrow$ products; f_A is a function of M_{An}; see text for equations relating them for constant density and temperature.

^{c}Values are also valid for BR with reaction time $t = \bar{t}$ and no down-time.

PFR: $\underline{n = 0}$

$$f_A = M_{A0}; \quad M_{A0} \leq 1$$
$$= 1; \quad M_{A0} \geq 1 \qquad (17.1\text{-}12)$$

$\underline{n = 1/2}$

$$f_A = M_{A1/2}(1 - M_{A1/2}/4); \quad M_{A1/2} \leq 2$$
$$= 1; \quad M_{A1/2} \geq 2 \qquad (17.1\text{-}13)$$

$\underline{n = 1}$

$$f_A = 1 - \exp(-M_{A1}) \qquad (17.1\text{-}14)$$

$\underline{n = 2}$

$$f_A = M_{A2}/(1 + M_{A2}) \qquad (17.1\text{-}15)$$

These results for a PFR can be obtained from equations 3.4-9 ($n \neq 1$) and 3.4-10 ($n = 1$), together with 14.3-12; or, from integration of equation 15.2-16 with the appropriate rate law, together with equations 14.3-12 and 15.2-14.

LFR: $\underline{n = 0}$

$$f_A = M_{A0}(1 - M_{A0}/4); \qquad 0 \leq M_{A0} \leq 2$$

$$f_A = 1; \qquad M_{A0} \geq 2 \qquad (16.2\text{-}18)$$

$\underline{n = 1/2}$

$$f_A = M_{A1/2}\{1 - (M_{A1/2}/8)[1.5 + \ln(4/M_{A1/2})]\} \qquad ; M_{A1/2} \leq 4$$

$$= 1 \qquad ; M_{A1/2} \geq 4 \quad (17.1\text{-}16)$$

$\underline{n = 1}$

$$f_A = 1 - (\bar{t}^2/2)\int_{\bar{t}/2}^{\infty}(e^{-k_A t}/t^3)dt \qquad (16.2\text{-}19)$$

$\underline{n = 2}$

$$f_A = M_{A2}[1 - (M_{A2}/2)\ln(1 + 2/M_{A2})] \qquad (16.2\text{-}20)$$

Equation 17.1-16 can be obtained from equation 16.2-13, together with the result for c_A/c_{A_o} for a BR (or PFR) obtained by integration of, for example, equation 15.2-16. Note that the latter places restrictions on the values for $M_{A1/2}$, as noted in equation 17.1-13, and this has an implication for the allowable upper limit in equation 16.2-13.

CSTR: $\underline{n = 0}$ (nonsegregated flow)

$$f_A = M_{A0}; \qquad M_{A0} < 1$$

$$f_A = 1; \qquad M_{A0} \geq 1 \qquad (14.3\text{-}22)$$

$\underline{n = 1/2}$ (nonsegregated)

$$f_A = (M_{A1/2}^2/2)[(4/M_{A1/2}^2 + 1)^{1/2} - 1] \qquad (17.1\text{-}17)$$

$\underline{n = 1}$ (segregated or nonsegregated)

$$f_A = M_{A1}/(1 + M_{A1}) \qquad (14.3\text{-}21)$$

$\underline{n = 2}$ (nonsegregated)

$$f_A = 1 - \frac{(1 + 4M_{A2})^{1/2} - 1}{2M_{A2}} \qquad (14.3\text{-}24)$$

Equation 17.1-17 can be obtained from equation 14.3-19 with $n = 1/2$.

17.2 MULTIPLE-VESSEL CONFIGURATIONS

Some multiple-vessel configurations and consequences for design and performance are discussed previously in Section 14.4 (CSTRs in series) and in Section 15.4 (PFRs in series and in parallel). Here, we consider some additional configurations, and the residence-time distribution (RTD) for multiple-vessel configurations.

17.2.1 CSTRs in Parallel

In Section 15.4 for PFRs operated in parallel, it is noted (point (3)) that when exit streams are combined, they should be at the same state; for example, they should have the same composition. The same applies to CSTRs operated in parallel. In the following example, it is shown that this leads to the most efficient operation.

EXAMPLE 17-3

Consider two CSTRs of unequal sizes, V_1 and V_2, operating in parallel, as shown in Figure 17.2. The liquid-phase reaction, A $\rightarrow$ products, is first-order and both tanks operate at the same T. How should the total feed rate, q_o, be split (at S) so as to maximize the rate of production in the combined exit streams (at M)?

SOLUTION

For a given throughput (q_o), the maximum rate of production is obtained if c_A at M is minimized. We let the ratio of the feed to the two tanks be $r = q_{o2}/q_{o1}$ so that $q_{o2} = rq_{o1}$, where $q_{o1} + q_{o2} = q_o$. A material balance around M serves to relate c_A to c_{A1} and c_{A2}, each of which can be related to the variable r by means of the material balance for a CSTR and the rate law. We then minimize c_A with respect to r, all other quantities being fixed.

Material balance for A around point M:

$$q_o c_A = q_{o1} c_{A1} + q_{o2} c_{A2}$$

$$(1 + r) q_{o1} c_A = q_{o1} c_{A1} + r q_{o1} c_{A2}$$

$$c_A = \frac{c_{A1}}{1 + r} + \frac{r c_{A2}}{1 + r} \tag{A}$$

Material balance for A around V_1, from equation 14.3-15:

$$V_1 = (c_{Ao} - c_{A1}) q_{o1}/(-r_A) = (c_{Ao} - c_{A1})[q_o/(1 + r)]/k_A c_{A1}$$

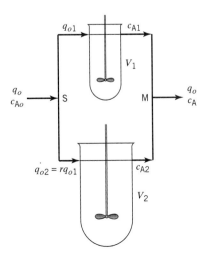

Figure 17.2 Flow diagram for CSTRs in parallel in Example 17-3

This results in:

$$\frac{c_{A1}}{c_{Ao}} = \frac{1}{1 + \frac{k_A V_1}{q_o}(1+r)} \quad \text{(B)}$$

Similarly, from a material balance for A around V_2:

$$\frac{c_{A2}}{c_{Ao}} = \frac{1}{1 + \frac{k_A V_2}{q_o} \frac{1+r}{r}} \quad \text{(C)}$$

Substituting (B) and (C) into (A), we obtain

$$\frac{c_A}{c_{Ao}} = \frac{1}{1+r}\left[\frac{1}{(1+a_1)+a_1 r} + \frac{r^2}{a_2 + (1+a_2)r}\right] \quad \text{(D)}$$

where

$$a_1 = k_A V_1/q_o; \quad a_2 = k_A V_2/q_o \quad \text{(E)}$$

Setting $d(c_A/c_{Ao})/dr = 0$ for minimum c_A, we solve for r in terms of a_1 and a_2 to obtain:

$$r\left(=\frac{q_{o2}}{q_{o1}}\right) = \frac{a_2}{a_1} = \frac{V_2}{V_1} \quad \text{(17.2-1)}$$

(The analytical solution is rather tedious, but is aided by computer algebra software.)

The result contained in equation 17.2-1 is that the feed should be split in the same ratio as the sizes of the tanks. An implication of this result is that the space times in the two tanks should be equal:

$$\tau_1 = V_1/q_{o1} = V_2/q_{o2} = \tau_2 \quad \text{(17.2-2)}$$

Furthermore, from equations 17.2-2, (A), (B), and (C)

$$c_{A1} = c_{A2} = c_A \quad \text{(17.2-3)}$$

That is, the two exit streams are at the same composition before being mixed at M.

With the feed divided as in equation (17.2-1), the two tanks act together the same as one tank of volume $V = V_1 + V_2$. There is thus no inherent performance advantage of multiple CSTRs in parallel, but there may be increased operating flexibility.

17.2.2 CSTRs in Series: RTD

For CSTRs in *parallel* with the feed split as for optimal performance, the fact that two (or more) reactors behave the same as one CSTR of the same total volume means that the RTD is also the same in each case. Here, we consider the RTD for CSTRs in *series*, as in a multistage CSTR (Section 14.4). In the following example, the RTD is obtained for two tanks in series. The general case of N tanks in series is considered in Chapter

EXAMPLE 17-4

(a) Derive the equation for the RTD function $E(\theta)$ for constant-density flow of a fluid through two stirred tanks in series; that is, the flow in each tank is BMF. The volume of each tank is $V/2$, and the steady-state flow rate is q.
(b) On the same plot, graphically compare the result from (a) (for $N = 2$), with that for a single stirred tank ($N = 1$).
(c) What is $F(\theta)$ for two stirred tanks in series?

SOLUTION

(a) The flow diagram and symbols are shown in Figure 17.3.

To develop $E(\theta)$ for two CSTRs in series, we use a slightly different, but equivalent, method from that used for a single CSTR in Section 13.4.1.1. Thus, consider a small amount (moles) of tracer M, $n_{Mo} \equiv F_t dt$, where F_t is the total steady-state molar flow rate, added to the first vessel at time 0. The initial concentration of M is $c_{Mo} \equiv n_{Mo}/(V/2)$. We develop a material balance for M around each tank to determine the time-dependent outlet concentration of M from the second vessel, $c_{M2}(t)$.

For vessel 1, there is no *subsequent* input of tracer, the outlet flow rate is qc_{M1}, and the accumulation rate is $(V/2)dc_{M1}/dt$. Thus, at any subsequent time t,

$$0 - qc_{M1} = (V/2)dc_{M1}/dt \quad \text{(A)}$$

or

$$\int_{c_{Mo}}^{c_{M1}} \frac{dc_{M1}}{c_{M1}} = -\int_0^t \frac{2q}{V} dt$$

which integrates to

$$c_{M1} = c_{Mo} e^{-2qt/V} \quad \text{(B)}$$

For vessel 2, the input rate is qc_{M1}, the output rate is qc_{M2}, and the accumulation rate is $(V/2)dc_{M2}/dt$. Therefore,

$$qc_{M1} - qc_{M2} = (V/2)dc_{M2}/dt$$

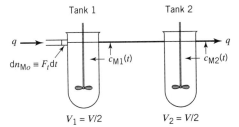

Figure 17.3 Arrangement of two tanks in series for Example 17-4

Substituting for c_{M1} from (B), and rearranging in the form of a first-order differential equation, we obtain

$$\frac{dc_{M2}}{dt} + \frac{2q}{V} c_{M2} = \frac{2q}{V} c_{Mo} e^{-2qt/V} \tag{C}$$

Integration of (C) from $c_{M2} = 0$ at $t = 0$ to c_{M2} at t results in the tracer concentration in vessel 2 as a function of time:

$$c_{M2} = (2qt/V) c_{Mo} e^{-2qt/V} \tag{D}$$

To obtain $E(t)$ for the two-vessel configuration, consider fluid leaving vessel 2 between t and $t + dt$. The total amount is $F_t dt \equiv n_{Mo}$, and the amount of tracer is $c_M q dt$. Thus,

$$\text{fraction tracer} = \frac{\text{amount of tracer}}{\text{amount of flow}} = \frac{c_M q dt}{n_{Mo}} = \frac{(2qt/V) c_{Mo} e^{-2qt/V} q dt}{n_{Mo}}$$

$$= \frac{(2qt/V)[n_{Mo}/(V/2)] e^{-2qt/V} q dt}{n_{Mo}} = 4(q/V)^2 t e^{-2qt/V} dt \tag{E}$$

This is also the fraction of the stream leaving vessel 2 that is of age between t and $t + dt$, since only the amount n_{Mo} entered at $t = 0$. But this fraction is also $E(t)dt$, from the definition of $E(t)$. Thus, with $V/q = \bar{t}$, the mean residence time for fluid in the *two* tanks (of total volume V), from (E),

$$E(t) = (4t/\bar{t}^2) \exp(-2t/\bar{t}) \quad (\text{BMF}, N = 2) \tag{17.2-4}$$

In terms of dimensionless time $\theta = t/\bar{t}$, since $E(\theta) = \bar{t} E(t)$, from equation 13.3-7, equation 17.2-4 becomes

$$E(\theta) = 4\theta \exp(-2\theta) \quad (\text{BMF}, N = 2) \tag{17.2-5}$$

(b) Figure 17.4 compares the RTD for $N = 2$, given by equation 17.2-5, with that for $N = 1$, $E(\theta) = \exp(-\theta)$, given by equation 13.4-3. The behavior differs in two main respects:
(1) At $\theta = 0$, $E(\theta) = 1$ for $N = 1$, but $E(\theta) = 0$ for $N = 2$.
(2) For $N = 2$, $E(\theta)$ goes through a maximum value of 0.736 at $\theta = 0.5$ (obtained by setting $dE(\theta)/d\theta = 0$, from equation 17.2-5).
(c) From equations 13.3-7, -9, and -10, and 17.2-5,

$$F(\theta) = \int_0^\theta E(\theta) d\theta = \int_0^\theta 4\theta \exp(-2\theta) d\theta$$

Integration by parts yields the result

$$F(\theta) = 1 - e^{-2\theta}(1 + 2\theta) \quad (\text{BMF}, N = 2) \tag{17.2-6}$$

Comparison may be made of the reactor performance of tanks in series and tanks in parallel, in light of the different RTDs shown in Figure 17.4. If the two tanks shown in Figure 17.3 are arranged in parallel, and the flow, q, is split evenly between them, they act together

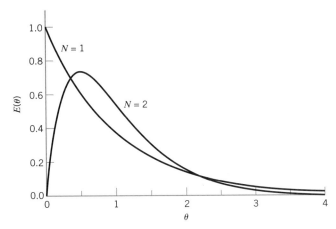

Figure 17.4 Comparison of $E(\theta)$ for two tanks in series (BMF, $N = 2$) with that for a single tank (BMF, $N = 1$) of the same total volume

as a single tank of volume $V = V_1 + V_2$ (Section 17.2.1), with the same performance, and the same RTD shown in Figure 17.4 for $N = 1$. For the same two tanks arranged in series (Figure 17.3), the RTD is different (Figure 17.4 for $N = 2$), and we expect a different performance. Thus, as discussed in Section 14-4, for positive, power-law (normal) kinetics, the series arrangement gives a better performance (e.g., f_A (series) $>$ f_A (parallel)), since reaction takes place with a higher rate at the higher reactant concentration in the first tank (the opposite conclusion applies for the case of abnormal kinetics, which is much less common).

17.2.3 PFR and CSTR Combinations in Series

17.2.3.1 Illustration of Effect of Earliness of Mixing

To this point it has been shown, mainly by illustration, that the performance of a reactor depends on three factors:

(1) The kinetics of reaction, as described by the rate law; thus, the fractional conversion depends on the order of reaction.
(2) The characteristics of flow through the vessel as described by the RTD; thus, for given kinetics, the fractional conversion in a reactor with PF (a PFR) is different from that in one with BMF (a CSTR).
(3) The nature of mixing of fluid elements during flow through the vessel as characterized, so far, by the degree of segregation (Section 13.5); thus, in a CSTR (i.e., a given RTD) with given kinetics, as shown in Table 14.1, the fractional conversion is different for the extremes of nonsegregated flow and completely segregated flow, *except for* a first-order reaction, which is a linear process. Reactions of other orders are examples of nonlinear processes.

A fourth aspect also has to do with the nature of mixing of fluid elements of different ages: the point or region in a vessel in which mixing takes place. Vessels with BMF and PF can be used to illustrate extremes described by "early" mixing and "late" (or no) mixing. In BMF, mixing of fluid elements of all ages takes place "early"—at the point of entry. In PF, mixing takes place "late"—in fact, not at all (in the axial direction). We may then use a combination of two vessels in series, one with BMF and one with

414 Chapter 17: Comparisons and Combinations of Ideal Reactors

PF, to model the effect of earliness or lateness of mixing, depending on the sequence, on the performance of a single-vessel reactor. The following two examples explore the consequences of such series arrangements—first, for the RTD of an equivalent single vessel, and second, for the fractional conversion. The results are obtained by methods already described, and are not presented in detail.

EXAMPLE 17-5

Consider a reactor system made up to two vessels in series: a PFR of volume V_{PF} and a CSTR of volume V_{ST}, as shown in Figure 17.5. In Figure 17.5(a), the PFR is followed by the CSTR, and in Figure 17.5(b), the sequence is reversed. Derive $E(\theta)$ for case (a) and for case (b). Assume constant-density isothermal behavior.

SOLUTION

(a) The material-balance method for obtaining $E(t)$ or $E(\theta)$, involving introduction of a "virtual" tracer as a pulse, is illustrated in Section 13.4.1.1 and in Example 17-4. We use it here to obtain $E(t)$, and hence $E(\theta)$.

Consider the entry of a small amount of fluid as "tracer" into the PFR at time $t = 0$. No tracer leaves the PFR until $t = V_{PF}/q_o = \bar{t}_{PF}$, the mean residence time in the PFR, and hence no tracer leaves the two-vessel system, at the exit from the CSTR, during the period $0 \leq t < \bar{t}_{PF}$. As a result,

$$E(t) = 0 \, ; \, 0 \leq t < \bar{t}_{PF} \qquad (17.2\text{-}7)$$

All tracer leaves the PFR at $\bar{t}_{PF}$, and enters the CSTR at this time as a pulse. Once the tracer leaves the PFR, its concentration in the CSTR immediately increases to $c_{Mo} \equiv c_{MST}$ at $\bar{t}_{PF}$. We then perform a material balance for tracer around the CSTR, for $t > \bar{t}_{PF}$. The input rate is 0, the output rate is qc_{MST}, and the accumulation rate is $V_{ST}dc_{MST}/dt$. Thus,

$$0 - qc_{MST} = V_{ST}dc_{MST}/dt$$

or

$$\int_{c_{Mo}}^{c_{MST}} \frac{dc_{MST}}{c_{MST}} = -\int_{\bar{t}_{PF}}^{t} \frac{q}{V_{ST}} dt$$

which integrates to

$$c_{MST} = c_{Mo} e^{-(q/V_{ST})(t-\bar{t}_{PF})}$$

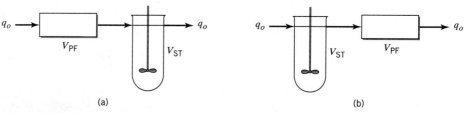

Figure 17.5 Flow diagrams for Example 17-5: (a) PFR followed by CSTR; (b) CSTR followed by PFR

Forming the fraction of tracer in the outlet stream from vessel 2 as in Example 17-4, and identifying this with $E(t)dt$, we obtain eventually

$$E(t) = \frac{1}{\bar{t}_{ST}} \exp\left(-\frac{t - \bar{t}_{PF}}{\bar{t}_{ST}}\right); \; t \geq \bar{t}_{PF} \quad (17.2\text{-}8)$$

where $\bar{t}_{ST} = V_{ST}/q_o$, the mean residence time in the CSTR. This result can be visualized as the same as that in equation 13.4-2 for BMF in a single vessel with a time delay of $\bar{t}_{PF}$.

The results in equations 17.2-7 and -8 together constitute $E(t)$ for the two-vessel configuration in Figure 17.5(a). To convert $E(t)$ to $E(\theta)$, where $\theta = t/\bar{t}$, and $\bar{t} = \bar{t}_{PF} + \bar{t}_{ST}$, we use $E(\theta) = \bar{t}E(t)$, equation 13.3-7. The results are

$$E(\theta) = 0; \; 0 \leq \theta < \bar{t}_{PF}/\bar{t} = V_{PF}/V \quad (17.2\text{-}9)$$

where $V = V_{PF} + V_{ST}$; and

$$E(\theta) = \frac{\bar{t}}{\bar{t}_{ST}} \exp\left(\frac{\bar{t}_{PF} - \bar{t}\theta}{\bar{t}_{ST}}\right); \; \theta \geq \bar{t}_{PF}/\bar{t}$$

$$= \frac{V}{V_{ST}} \exp\left(\frac{V_{PF} - V\theta}{V_{ST}}\right); \; \theta \geq V_{PF}/V \quad (17.2\text{-}10)$$

(b) The method used in part (a) applied to the configuration in Figure 17.5(b) leads to the *same* results as in equations 17.2-7 to -10 for $E(t)$ and $E(\theta)$.

Example 17-5 shows that differences in earliness or lateness of mixing need not result in differences in RTD. Part (a) involves relatively late mixing and part (b) early mixing, but both arrangements lead to the same RTD. The next example, using the same configuration, explores whether earliness or lateness of mixing affects reactor performance, through the fractional conversion.

EXAMPLE 17-6

Consider a liquid-phase reaction, A → products, taking place in the reactor systems shown in Figure 17.5(a) and (b), with $V_{ST}/V_{PF} = 4$.

(a) Calculate f_A, if the reaction is first-order, with $k_A \bar{t}_{PF} = 1$, for (i) the arrangement in Figure 17.5(a), and (ii) that in Figure 17.5(b).
(b) Repeat part (a) above for a second-order reaction with $k_A c_{Ao} \bar{t}_{PF} = 1$.

SOLUTION

(a) (i) If f_{A1} is the fractional conversion at the exit of vessel 1 (PFR), from equation 17.1-14,

$$f_{A1} = 1 - e^{-k_A \bar{t}_{PF}} = 0.632$$

If f_{A2} is the fractional conversion at the exit of vessel 2 (CSTR), from equation 14.4-1,

$$V_{ST} = F_{Ao}(f_{A2} - f_{A1})/(-r_A)_2 \quad (A)$$

On substitution of the rate law, we obtain

$$V_{ST} = \frac{F_{Ao}(f_{A2} - f_{A1})}{k_A(F_{Ao}/q)(1 - f_{A2})}$$

and on substitution of $q = V_{ST}\bar{t}_{ST}$, this leads to

$$f_{A2} = \frac{f_{A1} + k_A\bar{t}_{ST}}{1 + k_A\bar{t}_{ST}} = \frac{0.632 + 4}{1 + 4} = 0.926$$

since $k_A\bar{t}_{ST} = k_A(4\bar{t}_{PF}) = 4(1)$.

(ii) If f_{A1} is now the fractional conversion at the exit of the CSTR as vessel 1, from equation 14.3-21 (Table 14.1, nonsegregated flow, for $n = 1$),

$$f_{A1} = \frac{k_A\bar{t}_{ST}}{1 + k_A\bar{t}_{ST}} = \frac{4}{1 + 4} = 0.8$$

For the PFR, since

$$V_{PF} = F_{Ao}\int_{f_{A1}}^{f_{A2}} \frac{df_A}{(-r_A)}$$

on integration and rearrangement, we have

$$f_{A2} = 1 - (1 - f_{A1})e^{-k_A\bar{t}_{PF}} = 0.926$$

The performance results for the two configurations are the same for a *first-order*, or linear, process, even though the earliness and lateness of mixing differ.

(b) (i) Using equation 17.1-15 for the PFR, to calculate f_{A1}, we obtain

$$f_{A1} = k_A c_{Ao}\bar{t}_{PF}/(1 + k_A c_{Ao}\bar{t}_{PF}) = 0.500$$

From equation (A), corresponding to the treatment in (a)(i), but for $n = 2$,

$$f_{A2} = 0.750$$

(ii) Using the same procedure as in part (a)(ii) for the CSTR (but with $n = 2$ and equation 14.3-24) and for the PFR in turn, we obtain

$$f_{A1} = 0.610 \text{ and } f_{A2} = 0.719$$

In this case, the results for f_{A2} are different, and this is typical for nonlinear processes. The lower result is for case (ii) in which earlier mixing occurs; that is, when the CSTR precedes the PFR. Thus, although the kinetics, RTD, and degree of segregation (microfluid) are the same, a different mixing pattern in terms of location of mixing may result in a different performance.

17.2.3.2 Use for Autocatalytic Reaction

In the discussion following Example 15-10 about reactor volumes for an autocatalytic reaction, it is suggested that a further possibility exists for an even smaller volume than that obtained with a recycle PFR.

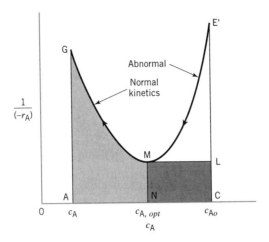

Figure 17.6 Graphical illustration of basis for using CSTR + PFR series combination for minimum volume for autocatalytic reaction

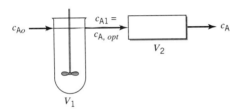

Figure 17.7 Flow diagram for CSTR + PFR series combination for minimum volume for autocatalytic reaction

The basis for this suggestion is shown in Figure 17.6, which is the same reciprocal-rate plot as in Figure 15.8. It involves taking advantage of the different consequences of normal and abnormal kinetics. In Figure 17.6, E'M represents abnormal kinetics (($-r_A$) increases as c_A decreases), and MG represents normal kinetics; M is the minimum point on the reciprocal-rate curve, or the maximum point on the rate curve (Figure 8.6). For normal kinetics, the comparisons in Sections 17.1-3 and -4 emphasize that $V_{PF} < V_{ST}$ for the same fractional conversion. The opposite is true for abnormal kinetics. Thus, in Figure 17.6, the dark-shaded rectangular area MLCN corresponds to the volume of a CSTR to effect the concentration change ($c_{Ao} - c_{A,opt}$). The volume of a PFR for the same change corresponds to the larger area ME'CN. To achieve a further change ($c_{A,opt} - c_A$), the light-shaded area MNAG corresponds to the volume of a PFR. This latter volume is smaller than the volume of a CSTR for the same change (not indicated in Figure 17.6, but equal to a rectangle of area AG × AN). The conclusion is that a reactor configuration consisting of a series arrangement of a CSTR followed by a PFR provides the smallest possible (total) volume for an autocatalytic reaction. This arrangement is shown in Figure 17.7. The outlet concentration from the CSTR is $c_{A,opt}$. The conclusion is illustrated in the following example.

EXAMPLE 17-7

For the autocatalytic reaction described in Example 15-10 and the data given there, calculate the volume of a combined CSTR + PFR reactor arranged as in Figure 17.7.

SOLUTION

For a minimum total volume, the CSTR operates at $c_{A1} = c_{A,opt}$, which, as shown in Example 8-1, is given by $c_{A,opt} = (c_{Ao} + c_{Bo})/2 = (1.5 + 0)/2 = 0.75$ mol L^{-1}, from equations 8.3-7 and -3a. The rate of reaction at this concentration is $(-r_{A,max}) = k_A(c_{Ao} + c_{Bo})^2/4 = 0.002(1.5 + 0)^2/4 = 0.001125$ mol L^{-1} s^{-1}, from equations 8.3-8 and -3a. Then, from the design equation, 14.3-5, for a CSTR,

$$V_{ST} = V_1 = F_{Ao}f_{A1}/(-r_{A,max}) = c_{Ao}q_o[1 - (c_{A1}/c_{Ao})]/(-r_{A,max})$$
$$= 1.5(0.5)[1 - (0.75/1.5)]/0.001125 = 333 \text{ L}$$

The process stream then enters the PFR at $c_{A1} = 0.75$ mol L^{-1}, and leaves at $c_A = c_{Ao}(1 - f_A) = 1.5(1 - 0.90) = 0.15$ mol L^{-1}. Thus, the volume of the second stage, the PFR, is, from equation 15.2-17,

$$V_{PF} = V_2 = -q_o \int_{c_{A1}}^{c_A} \frac{dc_A}{(-r_A)} = -\frac{q_o}{k_A}\int_{c_{A1}}^{c_A} \frac{dc_A}{c_A(c_{Ao} - c_A)}$$
$$= \frac{q_o}{k_A c_{Ao}} \ln\left[\frac{c_{A1}(c_{Ao} - c_A)}{c_A(c_{Ao} - c_{A1})}\right]$$
$$= \frac{0.5}{0.002(1.5)} \ln\left[\frac{0.75(1.5 - 0.15)}{0.15(1.5 - 0.75)}\right] = 367 \text{ L}$$

(The integral can be evaluated by means of the substitution $c_A = 1/x$.)

The total volume for the two-vessel reactor is $V_{ST} + V_{PF} = 333 + 367 = 700$ L. This volume is smaller than that required for a PFR with optimal recycle, 760 L (Example 15-10). Determining whether this difference is significant from an economic point of view would require a cost analysis taking equipment size and other factors into account.

Alternatively, this example may be solved using the E-Z Solve software (file ex17-7.msp), which allows the integral to be evaluated numerically.

17.3 PROBLEMS FOR CHAPTER 17

17-1 Determine the type of reactor with the smallest volume for the second-order liquid-phase reaction A → products, where $(-r_A) = k_A c_A^2$, and the desired fractional conversion is 0.60. Calculate the volume required.

Data: $k_A = 0.75$ L mol^{-1} h^{-1}; $c_{Ao} = 0.25$ mol L^{-1}; $q_o = 0.05$ m^3 h^{-1}.

17-2 For the first-order gas-phase reaction A → B + C, determine the reactor type and volume required, as in problem 17-1, to achieve 95% conversion of A, given $q_o = 40$ L s^{-1}; $k_A = 0.80$ s^{-1}; $c_{Ao} = 0.50$ mol L^{-1}.

17-3 Determine the reactor type and volume required, as in problem 17-1, for 75% conversion of A, if the reaction A → products is zero-order.

Data: $q_o = 25$ L min^{-1}; $k_A = 0.75$ mol L^{-1} min^{-1}; $c_{Ao} = 0.20$ mol L^{-1}. Assume constant-density conditions.

17-4 Determine the reactor configuration and minimum total volume for the reaction A + B → products with the rate law

$$(-r_A) = \frac{k_1 c_A}{K_2 + c_A + c_B}$$

The desired fractional conversion is 0.85.

Data:

$k_1 = 0.25$ mol L^{-1} min^{-1} $K_2 = 0.50$ mol L^{-1}
$c_{Ao} = c_{Bo} = 0.25$ mol L^{-1} $q_o = 5.0$ L min^{-1}

17-5 The rate law for the autocatalytic liquid-phase reaction A → B + C is $(-r_A) = k_A c_A c_B$.
(a) Calculate the volume of a single CSTR which gives 95% conversion of A.
(b) Calculate the minimum total volume of a CSTR + PFR combination which gives the same conversion of A (95%).
Assume steady-state operation. $c_{Ao} = 0.350$ mol L^{-1}; $c_{Bo} = 0.010$ mol L^{-1}; $k_A = 0.005$ L mol^{-1} s^{-1}; $q_o = 25$ L min^{-1}.

17-6 An autocatalytic reaction, A → B, with $(-r_A) = k_A c_A c_B$, is to be conducted in a reactor network consisting of a PFR and CSTR.
Determine the minimum total volume and optimum reactor sequence for 90% conversion of A, if
(a) $c_{Bo} = 0$
(b) $c_{Bo} = 0.25$ mol L^{-1}
Other data: $q_o = 10$ L min^{-1}; $k_A = 0.08$ L mol^{-1} min^{-1}; $c_{Ao} = 0.50$ mol L^{-1}.

17-7 For a liquid-phase reaction represented by A → products carried out (separately) in a BR and in a CSTR, show, for the same rate of production, f_A, and T in the two reactors,
(a) that the ratio V_{BR}/V_{ST} decreases as order of reaction (n) increases for a given BR downtime ratio $a = t_d/t$; to do this by illustration, calculate the ratio for $f_A = 0.8$, $a = 1$ and $n = 1/2$, 1 and 2;
(b) that the ratio in (a) becomes $1 + a$ at low conversion (f_A), independent of order of reaction.

17-8 For a gas-phase reaction represented by A → B + C carried out (separately) isothermally in a constant-volume BR, and isothermally and isobarically in a PFR, show that t_{BR} and t_{PF}, for a feed of pure A,
(a) are equal for a first-order reaction;
(b) are unequal for other than a first-order reaction (illustrate by obtaining the results for second-order kinetics).

17-9 (a) For the second-order, liquid-phase reaction A + bB → products, with $(-r_A) = k_A c_A c_B$, show that the ratio of the volume of a CSTR to that of a PFR to achieve (steady-state) conversion f_A in each case at a given T and for a given throughput is

$$\frac{V_{ST}}{V_{PF}} = \frac{f_A(C-b)}{(1-f_A)(C-bf_A)} \left[\ln \frac{C-bf_A}{C(1-f_A)}\right]^{-1}$$

where $C = c_{Bo}/c_{Ao}$ ($C > b$; i.e., A is the limiting reactant).
(b) Comment on the result for $C \to \infty$,
(c) Comment on the result for $C \to b$.

17-10 (a) For a constant-density, first-order reaction, A → products, show that the ratio of the volume of a CSTR to that of a batch reactor (V_{ST}/V_{BR}) to achieve the same rate of production depends on the conversion of A (f_A) and the ratio of down-time to reaction time in the batch reactor, $a = t_d/t$, according to

$$\frac{V_{ST}}{V_{BR}} = \frac{f_A}{(1+a)[(f_A - 1)\ln(1 - f_A)]}$$

The batch reactor operates isothermally at the same temperature as the CSTR. The CSTR operates at steady-state.
(b) Illustrate the basis for the result in (a) by calculating the following for the hydrolysis of acetic anhydride (A) in *dilute* aqueous solution, $(CH_3CO)_2O(A) + H_2O \to 2CH_3COOH$, at 20°C ($k_A = 0.111$ min^{-1}) with a feed concentration (c_{Ao}) of 0.3 mol L^{-1} and a conversion of 80%:
  (i) the rate of production of acetic acid (mol min^{-1}) from a batch reactor of 10,000-L capacity, if the down-time is the same as the reaction time;

(ii) the rate of production of acetic acid from a CSTR if the feed rate is 345 L min^{-1}; calculate also the volume of the CSTR.

(c) Compare the ratio V_{ST}/V_{BR} from (b) with that calculated from (a).

17-11 (a) A PFR of volume V_1 and a CSTR of volume V_2 ($= 4V_1$) are operated at steady-state in parallel to carry out a first-order, liquid-phase reaction, A → products, characterized by $k_A V_1/q_o = k_A \tau_1 = 1$. If the feed (rate q_o or F_{Ao}) is split so that 40% passes through the PFR and 60% passes through the CSTR, calculate the fractional conversion of A (f_A), when the exit streams from the two vessels are combined.

(b) For the arrangement of vessels in (a), how should be feed be split between the two vessels for the most efficient operation?

(c) Would it be more efficient to arrange the vessels in series?

(In (a), (b), and (c), assume that the PFR operates isothermally at the same T as in the CSTR.)

(d) Derive an expression for $E(\theta)$ for the situation in part (a).

17-12 Three CSTRs (each of volume $= V$) and a single CSTR (volume $= 3V$) are available for a liquid-phase, first-order reaction, A → B, where $k_A = 0.075$ min^{-1}, and $c_{Ao} = 0.25$ mol L^{-1}.

(a) Would the single CSTR or the three CSTRs arranged in series give a higher fractional conversion of A? Justify numerically.

(b) If all four reactors are used, how should they be arranged to maximize the overall fractional conversion?

17-13 A first-order, liquid-phase reaction (A → products) is being carried out isothermally and in steady-state in a PFR of volume V. The volumetric feed rate is q, in consistent units, and the conversion obtained (f_A) is 0.93. How many CSTRs, each of the same volume as above (V), would be required to obtain the same conversion, or close to it, with the same feed rate, if the tanks are connected (a) in parallel; (b) in series?

17-14 Verify the numerical results obtained in Example 17-6(b).

17-15 For the configuration in Figure 17.8, show that $E(t)$ becomes the same as that for BMF in a single tank of volume $V = V_1 + V_2$, if the flow is split at S such that $\alpha = V_2/(V_1 + V_2)$.

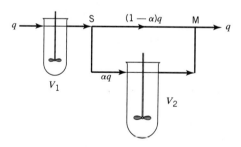

Figure 17.8 Vessel configuration for Problem 17-15

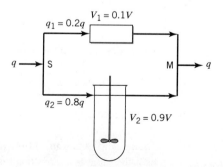

Figure 17.9 Vessel configuration for Problem 17-16

Assume that flow in each tank is BMF, and that the residence time in the transfer lines (pipes) between the inlet to the first tank and the mixing point M is negligible.

17-16 (a) Derive $E(\theta)$ for flow through a vessel of volume V, in which it is assumed that 80% of the total flow (q) occurs as BMF in 90% of the vessel, and the rest bypasses in PF in parallel (Figure 17.9).

(b) Sketch $E(\theta)$ versus θ to show essential features.

(c) Check the results for (a) by determining (analytically) the area under the $E(\theta)$ curve from $\theta = 0$ to $\theta = \infty$.

17-17 Verify the results given in equations 17.1-12 to -17, starting from the bases indicated.

Chapter 18

Complex Reactions in Ideal Reactors

In the design and analysis of reactors for carrying out complex reactions, two important considerations are reactor size and product distribution. In general, we wish to choose a reactor type, together with a mode of operation and operating conditions, to obtain a favorable product distribution for a minimum reactor size. These two objectives may or may not be compatible. When they are not, it is usually the product distribution that is more important in affecting the process economics, because of downstream processing steps required to manage byproducts and to generate a desired quantity and purity of product(s). This situation is in contrast to that in previous chapters for a simple system in which product distribution is fixed, and reactor type and minimum size are important in relation to increasing the conversion of a reactant. For a complex system, high conversion may or may not be desirable. Optimization of reactor performance is thus more complicated for a complex system, since the type and size of reactor and the operating variables (T, P, feed composition) can influence the product distribution.

In this chapter, we develop some guidelines regarding choice of reactor and operating conditions for reaction networks of the types introduced in Chapter 5. These involve features of reversible, parallel, and series reactions. We first consider these features separately in turn, and then in some combinations. The necessary aspects of reaction kinetics for these systems are developed in Chapter 5, together with stoichiometric analysis and variables, such as yield and fractional yield or selectivity, describing product distribution. We continue to consider only ideal reactor models and homogeneous or pseudohomogeneous systems.

18.1 REVERSIBLE REACTIONS

Reversible reactions pose interesting problems for reactor design, since the reactions tend toward equilibrium, short of completion, provided that the residence time is sufficiently long. Several approaches (based upon Le Chatelier's principle) may, however, be used to influence the process to favor the formation of the desired product. For example, selective removal of products or addition of reactants between stages of a cascade of CSTRs or PFRs leads to increased product formation (Figure 18.1). Similarly, the operating temperature may be used to influence the equilibrium position. Generally, if the reaction is exothermic, operating at low temperatures favors equilibrium product

18.1 Reversible Reactions

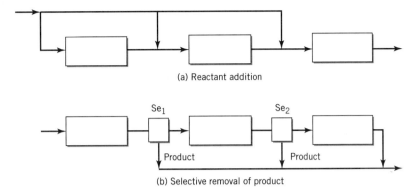

Figure 18.1 Using a series of reactors to enhance product yield in a reversible reaction by (a) reactant addition and (b) selective product removal by means of separators Se_1 and Se_2.

formation. Conversely, high temperatures are preferred if the reaction is endothermic. The concept of an optimal temperature for maximum rate for reversible, exothermic reactions is developed in Section 5.3.4. However, there is no comparable maximum for reversible endothermic reactions. If there is a change in density during a gas-phase reaction, the pressure may be adjusted to affect product formation favorably. Example 18-1 illustrates the effectiveness of selective product removal and reactant addition for enhancing product formation.

EXAMPLE 18-1

Two configurations of stirred-tank reactors are to be considered for carrying out the reversible hydrolysis of methyl acetate (A) to produce methanol (B) and acetic acid (C) at a particular temperature. Determine which of the following configurations results in the greater steady-state rate of production of methanol:

(a) a single 15-L CSTR;
(b) three 5-L CSTRs in series, with 75% of the product species B and C selectively removed between stages 1 and 2 and between stages 2 and 3, with appropriate adjustment in flow rate; a flow diagram and notation are shown in Figure 18.2.

The forward reaction is pseudo-first-order with respect to A ($k_f = 1.82 \times 10^{-4}$ s^{-1}), and the reverse reaction is second-order ($k_r = 4.49 \times 10^{-4}$ L mol^{-1} s^{-1}). The feed is a dilute aqueous solution of methyl acetate ($c_{Ao} = 0.25$ mol L^{-1}) at a rate of 0.25 L h^{-1} (q_o).

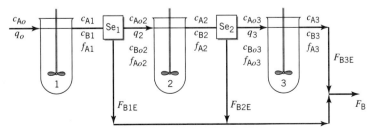

Figure 18.2 Flow diagram and notation for three tanks in series in Example 18-1(b)

424 Chapter 18: Complex Reactions in Ideal Reactors

SOLUTION

(a) The material balance for A (or for B or C) may be written in terms of the single variable f_A. Thus, for A, initially in terms of c_A, c_B, and c_C:

$$c_{Ao}q_o + k_r c_B c_C V - c_A q - k_f c_A V = 0 \qquad \text{(A)}$$

Since

$$c_A = c_{Ao}(1 - f_A)$$

and, since $c_{Bo} = c_{Co} = 0$,

$$c_B = c_C = c_{Ao} f_A$$

Equation (A) can be written, after rearrangement, as a quadratic equation in f_A:

$$f_A^2 + (1 + \frac{k_f V}{q})\frac{q}{k_r c_{Ao} V} f_A - \frac{k_f}{k_r c_{Ao}} = 0 \qquad \text{(B)}$$

or, after substitution of numerical values,

$$f_A^2 + 1.663 f_A - 1.621 = 0 \qquad \text{(C)}$$

from which $f_A = 0.689$.

As a result, $c_B = 0.25(0.689) = 0.172$ mol L^{-1} and the molar rate of production of methanol is

$$F_B = qc_B = 0.25(0.172) = 0.043 \text{ mol h}^{-1}$$

(b) For the first stage, the treatment is as in (a) except that $V = V_1 = 5$ L, and we denote the fractional conversion by f_{A1}. The equation corresponding to (C) is

$$f_{A1}^2 + 1.745 f_{A1} - 1.621 = 0 \qquad \text{(D)}$$

from which $f_{A1} = 0.671$. (Note that, in comparison with the result in part (a), tripling V to 15 L only increases f_A from 0.671 to 0.689.)

$$c_{A1} = c_{Ao}(1 - f_{A1}) = 0.082 \text{ mol L}^{-1}$$
$$c_{B1} = c_{C1} = f_{A1} c_{Ao} = 0.168 \text{ mol L}^{-1}$$

The molar rate of production of methanol from stage 1 is

$$F_{B1} = qc_{B1} = qc_{Ao} f_{A1} = 0.25(0.25)0.671 = 0.042 \text{ mol h}^{-1}$$

Seventy-five percent of this is selectively removed as final product at the reactor exit E:

$$F_{B1E} = 0.75(0.042) = 0.0315 \text{ mol h}^{-1}$$

For the second stage, the volumetric flow rate q_2 and the three inlet concentrations, c_{Ao2}, c_{Bo2}, and c_{Co2}, change by virtue of the selective removal of B and C and a proportional decrease in the flow rate. That is,

$$q_1 = q_{A1} + q_{B1} + q_{C1}$$

and

$$q_2 = q_{A1} + 0.25(q_{B1} + q_{C1})$$

(since 75% of B and C is removed). Since q_i is proportional to F_i,

$$\frac{q_1}{q_2} = \frac{q_{A1} + q_{B1} + q_{C1}}{q_{A1} + 0.25(q_{B1} + q_{C1})} = \frac{F_{A1} + F_{B1} + F_{C1}}{F_{A1} + 0.25(F_{B1} + F_{C1})} = 2.515 \quad \text{(E)}$$

so that

$$q_2 = 0.0994 \text{ L h}^{-1}$$

From material balances for A, B, and C, respectively, around the interstage separator Se_1, we obtain

$$c_{Ao2} = (q_1/q_2)c_{A1} = 0.207 \text{ mol L}^{-1} \quad \text{(F)}$$

$$c_{Bo2} = (q_1/q_2)c_{B1} = 0.106 \text{ mol L}^{-1} \quad \text{(G)}$$

$$c_{Co2} = c_{Bo2} = 0.106 \text{ mol L}^{-1} \quad \text{(H)}$$

The reactor (design) equation for stage 2, corresponding to equation (A), is

$$c_{Ao2}q_2 + k_r c_{B2} c_{C2} V_2 - c_{A2} q_2 - k_f c_{A2} V_2 = 0 \quad \text{(I)}$$

To relate c_{A2}, c_{B2}, and c_{C2} to the outlet fractional conversion, f_{A2}, we have, by definition,

$$f_{A2} = \frac{F_{Ao} - F_{A2}}{F_{Ao}} = \frac{c_{Ao}q_o - c_{A2}q_2}{c_{Ao}q_o} = 1 - \frac{c_{A2}q_2}{c_{Ao}q_o} \quad \text{(J)}$$

From (J),

$$c_{A2} = c_{Ao}(q_1/q_2)(1 - f_{A2}) \quad \text{(K)}$$

Also,

$$c_{B2} = c_{Bo2} + c_{Ao2} - c_{A2} = c_{Bo2} + c_{Ao2} - c_{Ao}(q_1/q_2)(1 - f_{A2}) = c_{C2} \quad \text{(L)}$$

Substituting equations (K) and (L) in (I), and using the known numerical values for all quantities except f_{A2}, we obtain, after rearrangement, the quadratic equation:

$$f_{A2}^2 - 0.3413 f_{A2} - 0.4051 = 0 \quad \text{(M)}$$

from which

$$f_{A2} = 0.830$$

and

$$c_{A2} = 0.107 \text{ mol L}^{-1}$$
$$c_{B2} = c_{C2} = 0.206 \text{ mol L}^{-1}$$

The rate of removal of B in the product stream from stage 2 is

$$F_{B2E} = 0.75 q_2 c_{B2} = 0.0153 \text{mol h}^{-1}$$

For the third stage, we proceed in a similar manner as in stage 2, with advancement of the subscripts, $1 \to 2$ and $2 \to 3$.

The results are:

$$q_2/q_3 = 2.471$$
$$q_3 = 0.0402 \text{ L h}^{-1}$$
$$c_{Ao3} = 0.264 \text{ mol L}^{-1}$$
$$c_{Bo3} = c_{Co3} = 0.127 \text{ mol L}^{-1}$$
$$f_{A3}^2 - 1.233 f_{A3} + 0.297 = 0$$
$$f_{A3} = 0.905$$
$$c_{A3} = 0.148 \text{ mol L}^{-1}$$
$$c_{B3} = c_{C3} = 0.243 \text{ mol L}^{-1}$$
$$F_{B3E} = q_3 c_{B3} = 0.0097 \text{ mol h}^{-1}$$

The total rate of production of B is

$$F_B = F_{B1E} + F_{B2E} + F_{B3E} = 0.0565 \text{ mol h}^{-1}$$

Thus, selective product removal between stages as described has increased the rate of production from 0.043, part (a), to 0.0565 mol h^{-1}, part (b), and the conversion from 0.689 to 0.905. Of course, this has been achieved at the expense of whatever additional equipment is required for the selective product removal. Whether this operating mode is feasible in a particular situation requires further analysis, including optimization with respect to the extent of selective removal of products and a cost analysis. This example is for illustration of the operating mode in principle.

Example 18-1 can also be solved by means of the E-Z Solve software (file ex18-1.msp). In this case, the design equations for each stage are written for species A and B in terms of c_A, c_B, and c_C, together with $c_B = c_C$. The resulting nonlinear equation set is solved by the software. In this approach, there is no need to introduce f_A, nor to relate the concentrations to f_A, although f_A can be calculated at the end, if desired.

18.2 PARALLEL REACTIONS

Parallel reactions involve simultaneous coupling of two or more reactions to form different sets of products from a common set of reactants. If one of these product sets is desirable and the others are not, we need to examine conditions that tend to promote the desired reaction and suppress the others. This may be viewed through the relative rates of the parallel processes and the means of increasing a desired rate relative to others.

Three factors that may influence relative rates are (1) choice of catalyst (if applicable); (2) temperature; and (3) concentration of reactant(s). If a reacting system is subject to catalytic activity, this is usually the dominant factor (relative to the noncatalyzed reaction), and the choice of catalyst, although difficult to quantify, can greatly influence which of two (or more) parallel reactions predominates. The next most important factor may be temperature; although all the rates (in parallel) increase with increasing T, we should determine if any feature allows one rate to increase or decrease preferentially. A similar consideration applies to concentration(s) of reactant(s).

If the concentration of reactant (c_A, say) is a significant discriminating factor, this has implications for the type of reactor to be chosen and its mode of operation. If it is desired to maintain c_A at a relatively high level in a flow system, then a BR or PFR or multistage CSTR should be chosen, rather than a single-stage CSTR; the reactor chosen should be operated at relatively low conversion (f_A). Conversely, if low c_A is desired, a single-stage CSTR or a PFR with recycle should be chosen, and operated at relatively high f_A.

An example of a parallel-reaction network is the decomposition of cyclohexane, which may undergo dehydrogenation to form benzene and isomerization to form methylcyclopentane, as follows:

$$C_6H_{12} \rightarrow C_6H_6 + 3H_2$$
$$C_6H_{12} \rightarrow C_6H_{12} \text{ (methylcyclopentane)}$$

Example 18-2 illustrates the effect of two process variables (concentration and temperature) on the instantaneous fractional yield for two parallel reactions.

EXAMPLE 18-2

Consider the following reaction network and associated rate laws:

$$A \xrightarrow{k_{A1}} B; \quad r_B = k_{A1}(T)c_A^{a_1} = A_1 \exp(-E_{A1}/RT)c_A^{a_1} \quad (1)$$

$$A \xrightarrow{k_{A2}} C; \quad r_C = k_{A2}(T)c_A^{a_2} = A_2 \exp(-E_{A2}/RT)c_A^{a_2} \quad (2)$$

B is a desired product and C is not.

(a) Should the reaction be carried out at low or high c_A?
(b) Should the reaction be carried out at low or high T?
(c) State general conclusions arising from (a) and (b).
(d) What are the implications for size and type of reactor?

SOLUTION

(a) We can maximize the formation of B relative to C by maximizing the relative rates r_B/r_C, or, what amounts to the same thing, the instantaneous fractional yield of B or of B relative to C. Thus, from equation 5.2-8,

$$\frac{\hat{s}_{B/A}}{\hat{s}_{C/A}} = \frac{r_B/(-r_A)}{r_C/(-r_A)} = \frac{r_B}{r_C} = \frac{A_1}{A_2}\exp[(E_{A2}-E_{A1})/RT]c_A^{(a_1-a_2)} \quad (3)$$

From (3)

$$\frac{\partial(r_B/r_C)}{\partial c_A} = (\text{a constant})(a_1 - a_2)c_A^{(a_1-a_2-1)}$$

Three cases arise:

(i) If $a_1 > a_2$, r_B/r_C increases as c_A increases.
(ii) If $a_1 = a_2$, r_B/r_C is independent of c_A.
(iii) If $a_1 < a_2$, r_B/r_C decreases as c_A increases.

To favor formation of B, the reaction should be carried out at high c_A in case (i) and at low c_A in case (iii); changing c_A has no effect in case (ii).

(b) From (3)

$$\frac{\partial (r_B/r_C)}{\partial T} = \text{(a constant)} \frac{E_{A1} - E_{A2}}{RT^2} \exp[(E_{A2} - E_{A1})/RT]$$

Again, three cases arise:

(i) If $E_{A1} > E_{A2}$, r_B/r_C increases as T increases.
(ii) If $E_{A1} = E_{A2}$, r_B/r_C is independent of T.
(iii) If $E_{A1} < E_{A2}$, r_B/r_C decreases as T increases.

To favor formation of B, the reaction should be carried out at high T in case (i) and at low T in case (iii); changing T has no effect in case (ii).

(c) For parallel reactions which are irreversible, and follow a power-law reactant-concentration dependence and an Arrhenius temperature dependence,

(i) the reaction of higher (or highest) order with respect to a particular reactant is favored at high concentration of that reactant;
(ii) the reaction of higher (or highest) activation energy is favored at high temperature.

(d) Effect of reaction order on choice of reactor and operating conditions:

(i) If $a_1 > a_2$, high c_A favors both formation of B and relatively small reactor size. Thus, these two desirable objectives are compatible. For a flow system, a PFR should be chosen, and operated at relatively low f_A. To conserve feedstock, the reactants and products should be separated downstream and unreacted A returned to the reactor inlet. Alternatively, if a stirred-tank reactor is used, it should be staged. A BR would be comparable to a PFR, for a nonflow system. For a gas-phase reaction in a PFR, the pressure should be relatively high, if feasible, without inert species.
(ii) If $a_2 > a_1$, low c_A favors formation of B, but does not favor a small reactor size. Since a favorable product distribution and a favorable reactor size are not compatible in this case, it may be necessary to consider balancing the respective costs involved to arrive at an optimal result. A single-stage CSTR should be chosen and operated at relatively high f_A, or a PFR with recycle should be used. Alternatively, for a PFR or a staged CSTR, the reactant could be added in stages, as shown in Figure 18.1(a)
(iii) If $a_1 = a_2$, there is no effect of c_A on product distribution, and reactor design is governed by size rather than selectivity considerations, as for a simple system.

If there is more than one reactant (e.g., A + B → products), the concentration of one reactant, say c_A, may be kept low relative to the other, c_B, by adding A in stages, as also indicated in Figure 18.1(a) for a PFR divided into stages.

18.3 SERIES REACTIONS

Series reactions involve products formed in consecutive steps. The simplest reaction network for this is given by reaction 5.5-1a:

$$A \xrightarrow{k_1} B \xrightarrow{k_2} C \tag{5.5-1a}$$

Some aspects of reactor behavior are developed in Chapter 5, particularly concentration–time profiles in a BR in connection with the determination of values of k_1 and k_2 from experimental data. It is shown (see Figure 5.4) that the concentration of the intermediate, c_B, goes through a maximum, whereas c_A and c_C continuously decrease and increase, respectively. We extend the treatment here to other considerations and other types of ideal reactors. For simplicity, we assume constant density and isothermal operation. The former means that the results for a BR and a PFR are equivalent. For flow reactors, we further assume steady-state operation.

18.3.1 Series Reactions in a BR or PFR

For first-order reaction steps in 5.5-1a in a BR or PFR, the concentration profiles $c_A(t)$, $c_B(t)$, and $c_C(t)$ are given by equations 3.4-10 and 5.5-6 and -7, respectively. These may be converted into expressions relating concentrations to f_A:

For a first-order reaction (equation 3.4-10), and from the definition of f_A,

$$c_A/c_{Ao} = \exp(-k_1 t) = 1 - f_A \tag{18.3-1}$$

Substitution of 18.3-1 into 5.5-6 gives equation 18.3-2:

$$\frac{c_B}{c_{Ao}} = Y_{B/A} = \frac{k_1}{k_2 - k_1}(e^{-k_1 t} - e^{-k_2 t}) \tag{5.5-6}$$

$$= \frac{1}{K - 1}[(1 - f_A) - (1 - f_A)^K] \tag{18.3-2}$$

where

$$K = k_2/k_1 \tag{18.3-2a}$$

From equation 5.2-2,

$$c_C/c_{Ao} = Y_{C/A} = f_A - Y_{B/A} \tag{18.3-3}$$

The selectivity of B, $\hat{S}_{B/A}$, may be obtained from equation 5.2-5:

$$\hat{S}_{B/A} = Y_{B/A}/f_A \tag{5.2-5}$$

$$= \frac{k_1(e^{-k_1 t} - e^{-k_2 t})}{(k_2 - k_1)(1 - e^{-k_1 t})} = \frac{(1 - f_A) - (1 - f_A)^K}{(K - 1)f_A} \tag{18.3-4}$$

from equations 18.3-1 and -2.

$\hat{S}_{C/A}$ may be obtained from equation 5.2-6:

$$\hat{S}_{C/A} = 1 - \hat{S}_{B/A} \tag{5.2-6}$$

430 Chapter 18: Complex Reactions in Ideal Reactors

EXAMPLE 18-3

For the liquid-phase oxidation of anthracene (AN) described in problem 5-17, Rodriguez and Tijero (1989) obtained the following values for the rate constants at 95°C in the two-step series reaction network:

$$C_{14}H_{10}(AN) \xrightarrow[k_1]{NO_2} C_{14}H_9O(ANT) \xrightarrow[k_2]{NO_2} C_{14}H_8O_2(AQ)$$

$k_1 = 7.47 \times 10^{-4}$ s^{-1}, and $k_2 = 3.36 \times 10^{-4}$ s^{-1}.

(a) Calculate the time (t_{max}) required to reach the highest concentration of the intermediate product in a semibatch reactor operated isothermally at 95°C.
(b) For the situation in (a), what are f_{AN}, $Y_{ANT/AN}$, and $\hat{S}_{ANT/AN}$?

SOLUTION

(a) The derivation of equation 5.5-8 for t_{max} by differentiation of equation 5.5-6 is shown in Example 5-7. Thus, from equation 5.5-8, $c_{ANT,max}$ is reached at

$$t_{max} = \ln(k_2/k_1)/(k_2 - k_1) = 10^4 \ln(3.36/7.47)/(3.36 - 7.47) = 1944 \text{ s}$$

(b) $f_{AN} = 1 - (c_A/c_{Ao}) = 1 - \exp(-k_1 t_{max}) = 0.766$

$$K = k_2/k_1 = 0.450$$

From equation 18.3-2,

$$Y_{ANT/AN} = [1/(0.450 - 1)][(1 - 0.766) - (1 - 0.766)^{0.450}] = 0.520$$

The same result can be obtained from 5.5-6.
From equation 5.2-5,

$$\hat{S}_{ANT/AN} = Y_{ANT/AN}/f_{AN} = 0.520/0.766 = 0.679$$

18.3.2 Series Reactions in a CSTR

For series reactions in steady-state operation of a CSTR, it is a matter of determining the stationary-state concentrations of species (product distribution) in or leaving each stage.

Consider again the first-order network in reaction 5.5-1a. From a material balance on A around a single-stage CSTR,

$$c_A/c_{Ao} = 1/(1 + k_1 \tau) = 1 - f_A \quad (18.3\text{-}5)$$

where τ is the space time, V/q_o, equal to V/q for constant-density operation.

Similarly, from a material balance on B, with $c_{Bo} = 0$, and input rate = output rate,

$$k_1 c_A V = k_2 c_B V + c_B q$$

and, on substitution for c_A from 18.3-5,

$$\frac{c_B}{c_{Ao}} = \frac{k_1 \tau}{(1 + k_1 \tau)(1 + k_2 \tau)} = \frac{(1 - f_A)f_A}{1 + (K - 1)f_A} = Y_{B/A} \quad (18.3\text{-}6)$$

Furthermore, from a material balance on C, with $c_{Co} = 0$,

$$k_2 c_B V = q c_C$$

and

$$\frac{c_C}{c_{Ao}} = \frac{k_1 k_2 \tau^2}{(1 + k_1 \tau)(1 + k_2 \tau)} = \frac{K f_A^2}{1 + (K-1) f_A} = Y_{C/A} \quad (18.3\text{-}7)$$

EXAMPLE 18-4

Repeat Example 18-3 for a single-stage CSTR, and compare the results.

SOLUTION

(a) To obtain τ_{max}, we first differentiate equation 18.3-6 to obtain $dc_B/d\tau$, set the derivative to zero, and solve for τ_{max}. The result is

$$\tau_{max} = 1/(k_1 k_2)^{1/2} \quad (18.3\text{-}8)$$

(In comparison with t_{max} in equation 5.5-8, which is the reciprocal of the logarithmic mean of k_1 and k_2, τ_{max} is the reciprocal of the geometric mean.) In this case,

$$\tau_{max} = 10^4/(7.47 \times 3.36)^{1/2} = 1996 \text{ s}$$

(b) From equation 18.3-5,

$$f_{AN} = k_1 \tau_{max}/(1 + k_1 \tau_{max}) = 0.599$$

From equation 18.3-6, with $K = 0.450$,

$$Y_{ANT/AN} = \frac{c_{ANT}}{c_{AN,o}} = \frac{(1 - f_{AN}) f_{AN}}{1 + (K-1) f_{AN}} = 0.358$$

The same result may be obtained from the other form of 18.3-6, in terms of k_1, k_2, and τ. From equation 5.2-5,

$$\hat{S}_{ANT/AN} = Y_{ANT/AN}/f_{AN} = 0.358/0.599 = 0.598$$

These results, in comparison with those from Example 18-3, illustrate that at $(Y_{ANT/AN})_{max} \equiv c_{B,max}$,

$$\tau_{max}(\text{CSTR}) > t_{max}(\text{BR or PFR})$$
$$f_A(\text{CSTR}) < f_A(\text{BR or PFR})$$
$$c_{B,max}(\text{CSTR}) < c_{B,max}(\text{BR or PFR})$$
$$\hat{S}_{B/A}(\text{CSTR}) < \hat{S}_{B/A}(\text{BR or PFR})$$

The performance of a CSTR can be brought closer to that of a PFR, if the CSTR is staged. This is considered in Chapter 20 in connection with the tanks-in-series model.

EXAMPLE 18-5

For the series-reaction network

$$A \to D \text{ (desired)}; \quad (-r_A) = k_1 c_A^\alpha = A_1 e^{-E_{A1}/RT} c_A^\alpha \quad \textbf{(A)}$$

$$D \to U \text{ (undesired)}: \quad r_U = k_2 c_D^\beta = A_2 e^{-E_{A2}/RT} c_D^\beta \quad \textbf{(B)}$$

under what circumstances is a relatively high value for the ratio r_D/r_U favored?

SOLUTION

The ratio under consideration is the ratio of instantaneous fractional yields of D and U. Thus, from equations 5.2-8, and (A) and (B),

$$\frac{r_D}{r_U} = \frac{r_D/(-r_A)}{r_U/(-r_A)} = \frac{\hat{s}_{D/A}}{\hat{s}_{U/A}} = \frac{k_1 c_A^\alpha - k_2 c_D^\beta}{k_2 c_D^\beta}$$

$$= \frac{k_1 c_A^\alpha}{k_2 c_D^\beta} - 1 = \frac{A_1 e^{-E_{A1}/RT}}{A_2 e^{-E_{A2}/RT}} \frac{c_A^\alpha}{c_D^\beta} - 1$$

$$= \frac{A_1}{A_2} \frac{c_A^\alpha}{c_D^\beta} e^{(E_{A2} - E_{A1})/RT} - 1 \quad \textbf{(C)}$$

This ratio is a function of T and the concentrations c_A and c_D.

Consider the effect of T. From (C),

$$\frac{\partial (r_D/r_U)}{\partial T} = \frac{A_1}{A_2} \frac{c_A^\alpha}{c_D^\beta} e^{(E_{A2} - E_{A1})/RT} \frac{E_{A2} - E_{A1}}{R} \left(-\frac{1}{T^2}\right) \quad \textbf{(D)}$$

From (D),

If $E_{A1} > E_{A2}$, $\partial (r_D/r_U)/\partial T > 0$ and r_D/r_U increases with increasing T.
If $E_{A1} < E_{A2}$, r_D/r_U decreases with increasing T.

Thus, a relatively high value of r_D/r_U is favored by relatively high T. This is another illustration of the conclusion (made also for parallel reactions, Example 18-2) that, of two reactions under consideration, higher T favors the reaction with the higher activation energy.

The ratio r_D/r_U also depends on, and is proportional to, the ratio c_A^α/c_D^β (equation (C)) at a given T. This is not a controllable variable in a single vessel, unlike T, since it naturally decreases as reaction proceeds (increasing f_A). The ratio c_A^α/c_D^β is highest at low conversion and decreases monotonically as f_A increases. This is another indication that the desired product D is favored at low conversion. If the reactor consists of several stages, rather than a single stage, however, the ratio c_A^α/c_D^β can be increased between stages by selective removal of products, similar to the situation for reversible reactions illustrated in Example 18-1.

18.4 CHOICE OF REACTOR AND DESIGN CONSIDERATIONS

The choice of reactor type and its design for a particular reaction network may require examination of trade-offs involving reactor size and mode of operation, product distribution (selectivity), and production rate. If, as is often the case, selectivity is

18.4 Choice of Reactor and Design Considerations

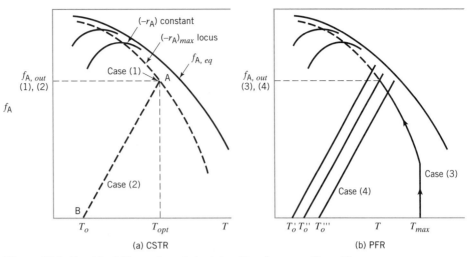

Figure 18.3 Graphical illustration of obtaining V_{min} for cases (1) to (4)

the main consideration, it may be advantageous to use combinations of reactor vessels, since greater control over process variables is possible. This may be the case even within a given reactor type: for example, use of a series arrangement of PFRs or CSTRs rather than a single PFR or CSTR. In this section, we consider illustrative situations for reversible reactions, parallel-reaction networks, series-reaction networks, and series-parallel networks.

18.4.1 Reactors for Reversible Reactions

For exothermic, reversible reactions, the existence of a locus of maximum rates, as shown in Section 5.3.4, and illustrated in Figures 5.2(a) and 18.3, introduces the opportunity to optimize (minimize) the reactor volume or mean residence time for a specified throughput and fractional conversion of reactant. This is done by choice of an appropriate T (for a CSTR) or T profile (for a PFR) so that the rate is a maximum at each point. The mode of operation (e.g., adiabatic operation for a PFR) may not allow a faithful interpretation of this requirement. For illustration, we consider the optimization of both a CSTR and a PFR for the model reaction

$$A \underset{k_r}{\overset{k_f}{\rightleftharpoons}} D \quad \text{(18.4-1)}$$

In 18.4-1, k_f and k_r are first-order rate constants with known Arrhenius parameters (A_f, E_{Af}) and (A_r, E_{Ar}), respectively. This situation, with different kinetics, arises in cases of gas-phase catalytic reactions and is further treated from this point of view in Chapter 21. Here, we consider four cases, and assume the reaction is noncatalytic.

Case (1): CSTR: minimum V or $\bar{t}$ for specified f_A, F_{Ao}, and q

To obtain V_{min} for a CSTR, the operating point is chosen so that $(-r_A)$ is a maximum at the specified value of f_A [$f_{A,out}$ and point A in Figure 18.3(a)], and V_{min} (or $\bar{t}_{min}$) is calculated from the design equation for a CSTR. The maximum rate, $(-r_A)_{max}$, is calculated at T_{opt}, which is given by equation 5.3-27, rewritten as

$$T_{opt} = \frac{(E_{Ar} - E_{Af})/R}{\ln\left(\frac{f_A}{1-f_A}\frac{A_r E_{Ar}}{A_f E_{Af}}\right)} \quad \text{(18.4-2)}$$

The rate constants k_f and k_r are then calculated at T_{opt} from the Arrhenius equation, and the maximum rate is obtained from equation 5.3-15 in the form

$$(-r_A)_{max} = c_{Ao}[k_f - (k_f + k_r)f_A] \quad (T = T_{opt}) \tag{18.4-3}$$

Finally, V_{min} is obtained from equation 14.3-5, the material balance or design equation for a CSTR:

$$V_{min} = F_{Ao}f_A/(-r_A)_{max} \tag{18.4-4}$$

From V_{min}, we may calculate $\bar{t}_{min} = V_{min}/q$.

Case (2): CSTR: T_o for V_{min} with adiabatic operation and specified f_A, F_{Ao}, and q
The result given by 18.4-4 requires only that the operating temperature within the CSTR be T_{opt}, and implies nothing about the mode of operation to obtain this, that is, nothing about the feed temperature (T_o), or heat transfer either within the reactor or upstream of it. If the CSTR is operated adiabatically without internal heat transfer, T_o must be adjusted accordingly to a value obtained from the energy balance, which, in its simplest integrated form, is, from equation 14.3-10,

$$(-\Delta H_{RA})F_{Ao}f_A = \dot{m}c_P(T_{opt} - T_o) \tag{18.4-5}$$

This case is indicated by the dashed line BA in Figure 18.3(a), which has a slope $(df_A/dT) = \dot{m}c_P/(-\Delta H_{RA})F_{Ao}$, given by equation 18.4-5.

Case (3): PFR: minimum V or $\bar{t}$ for specified f_A, F_{Ao}, and q
To obtain V_{min} for a PFR, the T profile is chosen so that the reaction path follows the locus of maximum rates as f_A increases; that is, the rate is $(-r_A)_{max}$ at each value of f_A in the design equation (from equation 15.2-2):

$$V_{min} = F_{Ao}\int_0^{f_A} \frac{df_A}{(-r_A)_{max}} \tag{18.4-6}$$

This involves solution of equation 18.4-6 together with equations 18.4-2 and -3. A stepwise algorithm is as follows, with T_{max} indicated as the highest allowable value of T for whatever reason (heat transfer considerations, product stability, etc.):

(i) Set f_A equal to a small, nonzero value (to avoid $T_{opt} \to 0$ at $f_A = 0$).
(ii) Calculate T_{opt} from 18.4-2.
(iii) If $T_{opt} \leq T_{max}$, calculate k_f, k_r at T_{opt}.
   If $T_{opt} \geq T_{max}$, calculate k_f, k_r at T_{max}.
(iv) Calculate $(-r_A)_{max}$ from 18.4-3.
(v) Repeat steps (i) to (iv) for a set of increasing values of f_A.
(vi) Calculate V_{min} by numerical integration of 18.4-6.

The "operational path" for this case is shown as case (3) in Figure 18.3(b). Achievement of this path, in a practical sense, is very difficult, since it requires a complex pattern of heat transfer for the point-by-point adjustment of T. A literal implementation is thus not feasible. A more practical situation is adiabatic operation of a PFR, considered next, which has its own optimal situation, but one different from case (3).

Case (4): PFR: V_{min} for adiabatic operation and specified f_A, F_{Ao}, and q
For a PFR operated adiabatically, V depends on the feed temperature, T_o, and is a minimum value, V_{min}, for a particular value of T_o. This is shown by the following set of

18.4 Choice of Reactor and Design Considerations 435

equations, which must be solved simultaneously to obtain $V(T_o)$. From equation 15.2-2, the material balance,

$$V = F_{Ao} \int_0^{f_A} \frac{df_A}{[-r_A(T, f_A)]} \tag{18.4-7}$$

From equation 15.2-9, the energy equation, the simplest integrated form relating f_A and T is similar to 18.4-5:

$$(-\Delta H_{RA}) F_{Ao} f_A = \dot{m} c_P (T - T_o) \tag{18.4-8}$$

The rate law is

$$[-r_A(T, f_A)] = c_{Ao}\{k_f(T) - [k_f(T) + k_r(T)] f_A\} \tag{18.4-9}$$

where

$$k_f(T) = A_f \exp(-E_{Af}/RT) \tag{18.4-10}$$

and

$$k_r(T) = A_r \exp(-E_{Ar}/RT) \tag{18.4-11}$$

This set of five equations is such that there is one degree of freedom to calculate V, and this can conveniently be chosen as T_o. The steps in the solution are then:

(i) Choose a value of T_o.
(ii) For a set of values of f_A, $0 \leq f_A \leq f_A$ (specified at exit), calculate a corresponding set of values of T from 18.4-8.
(iii) Calculate values of $[-r_A(T, f_A)]$ from equations 18.4-9 to -11 corresponding to the values of f_A and T in (ii).
(iv) Calculate $V(T_o)$ from 18.4-7.
(v) Repeat steps (i) to (iv) to obtain V_{min} (e.g., from a graph of $V(T_o)$ versus T_o).

For illustration, operating lines obtained from 18.4-8 for three values of T_o (T_o', T_o'', and T_o''') are shown in Figure 18.3(b)
Cases (1), (2), and (4) above are examined in problems 18-4 and -5.

18.4.2 Reactors for Parallel-Reaction Networks

EXAMPLE 18-6

A liquid-phase reaction involving the following two parallel steps is to be carried out in an isothermal PFR.

$$A + B \rightarrow D \text{ (desired)}; \quad r_D = k_1 c_A^2 c_B$$
$$A + B \rightarrow U \text{ (undesired)}; \quad r_U = k_2 c_A c_B^2$$

Assuming that the feed contains only A and B, compare (i) the overall fractional yield, $\hat{S}_{D/A}$, at the outlet, and (ii) the rate of production of D from a PFR consisting of (a) a single 30-L vessel (Figure 18.4a), and (b) three 10-L vessels in series, with the feed of B split

436 Chapter 18: Complex Reactions in Ideal Reactors

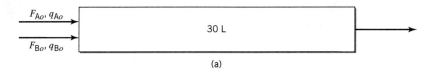

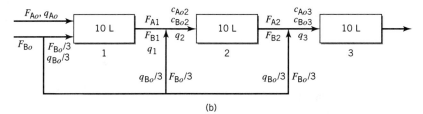

Figure 18.4 Vessel configuration for parallel reactions in Example 18-6: (a) single PFR; (b) three PFRs of same total volume in series with F_{Bo} split evenly among them

evenly among the three vessels (Figure 18.4b). The following data are available:

$$k_1 = 0.05 \text{ L}^2 \text{ mol}^{-2} \text{ min}^{-1}; \quad k_2 = 0.01 \text{ L}^2 \text{ mol}^{-2} \text{ min}^{-1}$$

$$q_{Ao} = 0.60 \text{ L min}^{-1}; \quad q_{Bo} = 0.30 \text{ L min}^{-1}$$

$$F_{Ao} = 1.2 \text{ mol min}^{-1}; \quad F_{Bo} = 0.60 \text{ mol min}^{-1}$$

SOLUTION

For both cases (a) and (b), the material balance (design) equations may be written generically as

$$\frac{dF_A}{dV} = \frac{dF_B}{dV} = -r_D - r_U = -k_1 c_A^2 c_B - k_2 c_A c_B^2 \quad \text{(A)}$$

$$\frac{dF_D}{dV} = r_D = k_1 c_A^2 c_B \quad \text{(B)}$$

$$\frac{dF_U}{dV} = r_U = k_2 c_A c_B^2 \quad \text{(C)}$$

Equations (A), (B), and (C) are not all independent but may be integrated simultaneously using the E-Z Solve software (file ex18-6.msp) to obtain concentration–volume profiles for A, B, D, and U, as well as $\hat{S}_{D/A}$.

Furthermore, by definition, at any point in the system, including the outlet,

$$\hat{S}_{D/A} = \frac{F_D - F_{Do}}{F_{Ao} - F_A} \quad \text{(from 5.2-4c)}$$

and the rate of production of D is F_D.

(a) For the single PFR system, the inlet conditions are

$$q_o = q_{Ao} + q_{Bo} = 0.90 \text{ L min}^{-1}$$

$$c_{Ao} = F_{Ao}/q_o = 1.33 \text{ mol L}^{-1}$$

$$c_{Bo} = F_{Bo}/q_o = 0.667 \text{ mol L}^{-1}$$

Integration of (A), (B), and (C) gives the following results (see file ex18-6.msp):

$$F_A = 0.720, \ F_D = 0.451, \ \text{and} \ F_U = 0.033 \text{ mol min}^{-1}$$

$$\hat{S}_{D/A} = \frac{0.451 - 0}{1.2 - 0.72} = 0.93$$

(b) For the system of 3 PFRs in series, the conditions at the inlet of each stage are as follows (see Figure 18.4b for notation):

$$q_1 = q_{Ao} + q_{Bo}/3 = 0.70 \text{ L min}^{-1}$$
$$q_2 = q_1 + q_{Bo}/3 = 0.80 \text{ L min}^{-1}$$
$$q_3 = q_2 + q_{Bo}/3 = 0.90 \text{ L min}^{-1}$$
$$c_{Ao1} = F_{Ao}/q_1 = 1.2/0.7 = 1.714 \text{ mol L}^{-1}$$
$$c_{Bo1} = F_{Bo1}/q_1 = 0.2/0.7 = 0.286 \text{ mol L}^{-1}$$
$$c_{Ao2} = F_{A1}/q_2$$
$$c_{Bo2} = (F_{B1} + 0.20)/q_2$$
$$c_{Ao3} = F_{A2}/q_3$$
$$c_{Bo3} = (F_{B2} + 0.20)/q_3$$

Again, equations (A) to (C) are integrated numerically (file ex18-6.msp) for each stage, taking into account the appropriate initial conditions. The results at the outlet of the third stage are:

$$F_A = 0.779, \ F_D = 0.407, \ \text{and} \ F_U = 0.014 \text{ mol L}^{-1}$$

$$\hat{S}_{D/A} = \frac{0.407 - 0}{1.2 - 0.779} = 0.97$$

Comparison of parts (a) and (b) shows that splitting the feed of B equally among three PFRs in series with the same total volume as a single PFR results in an increase in outlet selectivity, $\hat{S}_{D/A}$, from 0.93 to 0.97 but a decreased rate of production, F_D, of the desired product to 0.407 from 0.451 mol min^{-1}. Any gain from enhanced selectivity is offset by increased cost of vessel configuration and by a lower production rate.

Figure 18.5 shows a complete comparison of $\hat{S}_{D/A}$ as a function of reactor volume for the two cases, (a) and (b). Note that, in each case, a value of $\hat{S}_{D/A}$ is not shown at $V = 0$, since in this limit it becomes indeterminate (0/0), as inferred from defining equation 5.2-4c.

18.4.3 Reactors for Series-Reaction Networks

For the series-reaction network $A \xrightarrow{k_1} B \xrightarrow{k_2} C$, for illustration, we assume that choice of reactor type and operating conditions hinge on achieving high selectivity for whichever product species, B or C, is desired. We proceed here on the basis that B is the desired product, and that the choice is between a CSTR and a PFR operating isothermally with a constant-density system for first-order reaction steps. As far as operating T is concerned, we know that relatively high T favors formation of B, if $E_{A1} > E_{A2}$, and conversely for low T if $E_{A1} < E_{A2}$ (from Example 18-5; this conclusion applies for an isothermal PFR as well). We then consider how $\hat{S}_{B/A}$ depends on f_A and the relative magnitudes of the rate constants, $K = k_2/k_1$, separately for a PFR and a CSTR, and subsequently by comparison of the two reactor types.

438 Chapter 18: Complex Reactions in Ideal Reactors

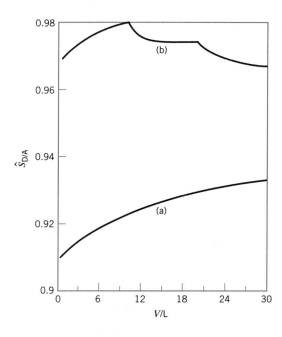

Figure 18.5 Example 18-6: comparison of $\hat{S}_{D/A}$ for (a) single-stage PFR and for (b) three-stage PFR with split feed of B

For a PFR, from equation 18.3-4,

$$\hat{S}_{B/A}(\text{PFR}) = \frac{(1 - f_A) - (1 - f_A)^K}{(K - 1)f_A} \quad (18.3\text{-}4)$$

Thus, $\hat{S}_{B/A}$ depends only on f_A and K. This behavior is shown as the full lines in Figure 18.6 for values of $K = 10$ (relatively large), 1, and 0.1 (relatively small). In all cases, $\hat{S}_{B/A}$ decreases from 1 at $f_A = 0$ to 0 at $f_A = 1$. Figure 18.6 shows that, for small values of $K(k_1 \gg k_2)$, a PFR can be designed for high f_A and still achieve high $\hat{S}_{B/A}$. Conversely, for high values of $K(k_2 \gg k_1)$, $\hat{S}_{B/A}$ decreases rapidly with increase in f_A,

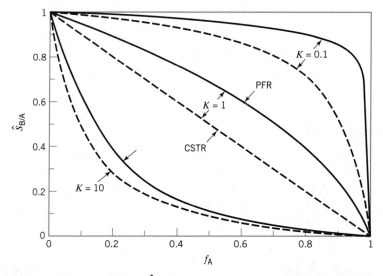

Figure 18.6 Dependence of $\hat{S}_{B/A}$ on f_A and $K(= k_2/k_1)$ for PFR (equation 18.3-4) and CSTR (equation 18.4-12) for reaction network $A \xrightarrow{k_1} B \xrightarrow{k_2} C$

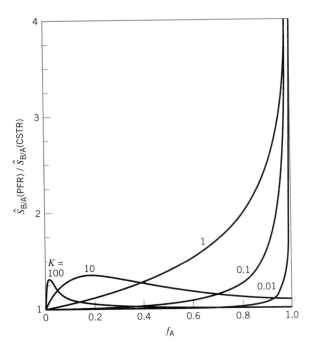

Figure 18.7 Dependence of ratio $\hat{S}_{B/A}$ (PFR)/$\hat{S}_{B/A}$ (CSTR) on f_A and $K(= k_2/k_1)$, equation 18.4-13, for reaction network $A \xrightarrow{k_1} B \xrightarrow{k_2} C$

so that a PFR should be designed for low conversion, with downstream separation of unreacted A for recycle to the reactor feed.

For a CSTR, from equations 18.3-6 and 5.2-5,

$$\hat{S}_{B/A} (\text{CSTR}) = \frac{Y_{B/A}}{f_A} = \frac{1 - f_A}{1 + (K - 1)f_A} \qquad (18.4\text{-}12)$$

Again, as for a PFR, $\hat{S}_{B/A}$ depends only on f_A and K. The relation is also shown in Figure 18.6, as the dashed lines, for three values of K. The behavior is similar to that for a PFR, but in all cases, $\hat{S}_{B/A}$ (CSTR) < $\hat{S}_{B/A}$ (PFR). Nevertheless, the same conclusions drawn for a PFR above may also be drawn for a CSTR.

The comparison between $\hat{S}_{B/A}$ (PFR) and $\hat{S}_{B/A}$ (CSTR) can be made more explicit by examination of the behavior of the ratio. From equations 18.3-4 and 18.4-12,

$$\frac{\hat{S}_{B/A} (\text{PFR})}{\hat{S}_{B/A} (\text{CSTR})} = [1 - (1 - f_A)^{K-1}]\left[1 + \frac{1}{(K-1)f_A}\right] \qquad (18.4\text{-}13)$$

This ratio is shown in Figure 18.7 as a function of f_A for values of $K = 0.01, 0.1, 1, 10,$ and 100.

If we denote the ratio in 18.4-13 by $R_{\hat{S}}$, the following conclusions may be drawn:

as $f_A \rightarrow 0$, $R_{\hat{S}} \rightarrow 1$ for all K;

as $f_A \rightarrow 1$, $R_{\hat{S}} \rightarrow K/(K - 1)$ for $K \geq 1$;
$\rightarrow \infty$ for $0 < K < 1$

440 Chapter 18: Complex Reactions in Ideal Reactors

for $K = 1$, $R_{\hat{S}} = -\ln(1 - f_A)/f_A \to \infty$ for $f_A = 1$

$\to 1$ for $f_A = 0$

as $K \to 0$, $R_{\hat{S}} \to 1$ for all f_A

as $K \to \infty$, $R_{\hat{S}} \to 1$ for all f_A

A more specific comparison is between choice of a PFR and a CSTR to achieve the maximum concentration of B, $c_{B,max}$, at the reactor outlet, and the residence time, t_{max}, required.

For a PFR, from Section 5.5, the residence time required to achieve $c_{B,max}$ is

$$\bar{t}_{max}(\text{PFR}) = \frac{\ln(k_2/k_1)}{k_2 - k_1} \tag{5.5-8}$$

and, at $\bar{t}_{max}$,

$$\frac{c_{B,max}}{c_{Ao}} = \left(\frac{k_2}{k_1}\right)^{\frac{k_2}{k_1 - k_2}} \tag{5.5-9}$$

For a CSTR, from Section 18.3.2, the mean residence time (equal to the space time τ_{max} for constant density) is

$$\bar{t}_{max}(\text{CSTR}) = (k_1 k_2)^{-1/2} \tag{18.3-8}$$

Substitution of $\bar{t}_{max}$ from 18.3-8 for $\bar{t}(=\tau)$ in 18.3-6,

$$\frac{c_B}{c_{Ao}} = \frac{k_1 \bar{t}}{(1 + k_1 \bar{t})(1 + k_2 \bar{t})} \tag{18.3-6}$$

results in

$$\frac{c_{B,max}}{c_{Ao}} = \frac{1}{[1 + (k_2/k_1)^{1/2}]^2} \tag{18.4-14}$$

We compare the performance of a PFR and a CSTR from two points of view: (1) the relative time required to achieve $c_{B,max}$, which is a measure of relative reactor size for a specified throughput, and (2) the relative value of $c_{B,max}$.

(1) Comparison of time required to achieve $c_{B,max}$; from equation 5.5-8 and 18.3-8,

$$R_{tmax} = \frac{\bar{t}_{max}(\text{PFR})}{\bar{t}_{max}(\text{CSTR})} = \frac{(k_1 k_2)^{1/2} \ln(k_2/k_1)}{k_2 - k_1} = \frac{K^{1/2} \ln K}{K - 1} \leq 1 \tag{18.4-15}$$

where

$$K = k_2/k_1 \tag{18.3-2a}$$

A plot of R_{tmax} against $\log_{10} K$ is shown in Figure 18.8. The following conclusions can be drawn:

(i) $R_{tmax} < 1$ for $K \neq 1$; i.e., $\bar{t}_{max}(\text{PFR}) < \bar{t}_{max}(\text{CSTR})$.
(ii) $R_{tmax} = 1$ for $K = 1$; i.e., $\bar{t}_{max}(\text{PFR}) = \bar{t}_{max}(\text{CSTR})$ at this singular point.
(iii) $R_{tmax} \to 0$ as $K \to 0$ or ∞.
(iv) The function is symmetrical; i.e., $R_{tmax}(K) = R_{tmax}(1/K)$.

18.4 Choice of Reactor and Design Considerations

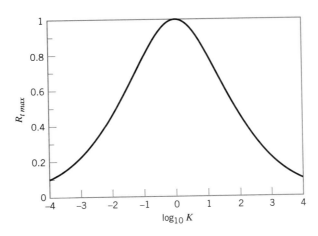

Figure 18.8 Comparison of PFR and CSTR for reaction network $A \xrightarrow{k_1} B \xrightarrow{k_2} C$: R_{tmax} as a function of $K (= k_2/k_1)$, equation 18.4-15

Table 18.1 Comparison of PFR and CSTR for series-reaction network $A \xrightarrow{k_1} B \xrightarrow{k_2} C$ (isothermal, constant-density system; $K = k_2/k_1$)

	PFR		CSTR		Comparison	
	$\bar{t}_{max}$	$\frac{c_{B,max}}{c_{Ao}}$	$\bar{t}_{max}$	$\frac{c_{B,max}}{c_{Ao}}$	$R_{tmax} = \frac{\bar{t}_{max}(\text{PFR})}{\bar{t}_{max}(\text{CSTR})}$	$R_{c_{B,max}} = \frac{c_{B,max}(\text{PFR})}{c_{B,max}(\text{CSTR})}$
General result (equation)	$\frac{\ln K}{k_1(K-1)}$ (5.5-8)	$K^{K/(1-K)}$ (5.5-9)	$\frac{1}{k_1 K^{1/2}}$ (18.3-8)	$(1 + K^{1/2})^{-2}$ (18.4-14)	$\frac{K^{1/2} \ln K}{K-1}$ (18.4-15)	$(1 + K^{1/2})^2 K^{K/(1-K)}$ (18.4-16)
$K \to 0$	∞	1	∞	1	0	1
$K = 1$	$\frac{1}{k_1} = \frac{1}{k_2}$	$e^{-1} = 0.3679$	$\frac{1}{k_1} = \frac{1}{k_2}$	0.25	1	$4/e = 1.472$
$K \to \infty$	0	0	0	0	0	1

In Table 18.1, values of R_{tmax} are given for $K \to 0$, $K = 1$, and $K \to \infty$.

(2) Comparison of $c_{B,max}$; from equations 5.5-9 and 18.4-14, with K replacing k_2/k_1,

$$R_{c_{B,max}} = \frac{c_{B,max}(\text{PFR})}{c_{B,max}(\text{CSTR})} = (1 + K^{1/2})^2 K^{\frac{K}{1-K}} \geq 1 \qquad (18.4\text{-}16)$$

A plot of $R_{c_{B,max}}$ against $\log_{10} K$ is shown in Figure 18.9. The following conclusions can be drawn:

(i) $R_{c_{B,max}} > 1$ for $K \neq 0$ or ∞; i.e., $c_{B,max}(\text{PFR}) > c_{B,max}(\text{CSTR})$.
(ii) $R_{c_{B,max}} \to 1$ as $K \to 0$ or ∞.
(iii) $R_{c_{B,max}}$ has a maximum value of $4/e = 1.472$ at $K = 1$.
(iv) The function is symmetrical; i.e., $R_{c_{B,max}}(K) = R_{c_{B,max}}(1/K)$.

In Table 18.1, values of $R_{c_{B,max}}$ are given for $K \to 0$, $K = 1$ and $K \to \infty$.

For this type of network, with normal kinetics, the performance of a CSTR cannot surpass that of a PFR; that is, use of a PFR leads to a greater $c_{B,max}$ at a smaller t_{max}. However, if the CSTR is staged, its performance can be closer to that of a PFR.

18.4.4 Reactors for Series-Parallel Reaction Networks

It is more difficult to develop general guidelines regarding the selection and design of a reactor for a series-parallel reaction network than for a parallel-reaction or a series-reaction network separately. It is still necessary to take into account the relative

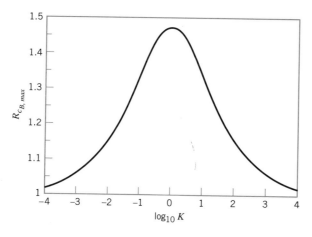

Figure 18.9 Comparison of PFR and CSTR for reaction network $A \xrightarrow{k_1} B \xrightarrow{k_2} C$: $R_{c_{B,max}}$ as a function of $K(= k_2/k_1)$, equation 18.4-16

importance of yield, selectivity, and production rate of a desired product in association with the type, size, mode of operation, and configuration of reactor. It is usually necessary to carry out a complete analysis for the particular situation under consideration based on the requirements specified and the data available.

The analysis includes various steps to assess the interrelationships of the various factors to be taken into account. This is the same type of analysis carried out in reactor design to this point, but it is reiterated here because of the greater complexity involved, which usually requires numerical methods for solution of the working equations developed, rather than analytical methods. A list of steps is as follows:

(1) A kinetics analysis, if not available *a priori*, leading to a reaction network with specified associated rate laws, together with values of the rate parameters, including their dependence on T.

(2) A stoichiometric analysis based on the species expected to be present as reactants and products to determine, among other things, the maximum number of independent material balance (continuity) equations and kinetics rate laws required, and the means to take into account change of density, if appropriate. (A stoichiometric table or spreadsheet may be a useful aid to relate chosen process variables (F_i, n_i, c_i, etc.) to a minimum set of variables as determined by stoichiometry.)

(3) A material balance analysis taking into account inputs and outputs by flow and reaction, and accumulation, as appropriate. This results in a proper number of continuity equations expressing, for example, molar flow rates of species in terms of process parameters (volumetric flow rate, rate constants, volume, initial concentrations, etc.). These are differential equations or algebraic equations.

(4) An energy analysis, if required, usually resulting in a differential or algebraic equation relating temperature and extent of reaction (see note and Example 18-8 below).

(5) Solution of the system of equations in (3) and (4), subject to any constraints imposed by (2). The solution in some cases may be possible analytically, but, more generally numerical methods are required. The solution may give, for example, F_i as a function of axial position or volume, or c_i as a function of time.

(6) Repetition of step (5) to examine sensitivity of the design to the chosen process parameters.

(7) Calculation of selectivity, yield, or production rate in relation to size and type/configuration of reactor.

An illustration of a series-parallel network is provided by the step-change polymerization kinetics model of Section 7.3.2. The following example continues the application of this model to steady-state operation of a CSTR.

18.4 Choice of Reactor and Design Considerations 443

EXAMPLE 18-7

For step-change polymerization at steady-state in a CSTR represented by the kinetics model in Section 7.3.2, derive equations giving the weight fraction of r-mer, w_r, on a monomer-free basis, as a function of r, the number of monomer units in polymer P_r at the reactor outlet. This is a measure of the distribution of polymers in the product leaving the reactor.

SOLUTION

We define w_r as

$$w_r = \frac{\text{weight of } r\text{-mer in reactor}}{\text{weight of all polymers in reactor}} = \frac{M_M r c_{P_r} V}{M_M (c_{Mo} - c_M) V} = \frac{r c_{P_r}}{c_{Mo} - c_M} \quad (18.4\text{-}17)$$

where M_M is the molar mass of monomer, c_{P_r} is the concentration of r-mer in the reactor (equal to the outlet concentration), V is the reactor volume, c_{Mo} is the concentration of monomer in the feed, and c_M is the concentration in the reactor. From equation 7.3-15, c_{P_r} is given by

$$c_{P_r} = c_M [1 + (k c_M \tau)^{-1}]^{1-r} \quad (7.3\text{-}15)$$

where k is the second-order step rate constant, and τ is the space time in the reactor. Combining 18.4-17 and 7.3-15, we obtain

$$w_r = \frac{r c_M [1 + (k c_M \tau)^{-1}]^{1-r}}{c_{Mo} - c_M} \quad (18.4\text{-}18)$$

The fraction of monomer converted is

$$f_M = \frac{c_{Mo} - c_M}{c_{Mo}} = 1 - \frac{c_M}{c_{Mo}} \quad (18.4\text{-}19)$$

The dimensionless second-order reaction number, from equation 4.3-4, with $\tau \equiv \bar{t}$ (constant density), is

$$M_2 = k c_{Mo} \tau \quad (18.4\text{-}20)$$

Substitution of 18.4-19 and -20 in 18.4-18 results in:

$$w_r(r) = \frac{r(1 - f_M)}{f_M} \{1 + [M_2(1 - f_M)]^{-1}\}^{1-r} \quad (18.4\text{-}21)$$

Equation 18.4-21 expresses $w_r(r)$ in terms of two reaction parameters f_M and M_2. These two parameters are not independent. We may relate them by means of equation 7.3-11a:

$$c_{Mo} - c_M = k c_M \tau (2 c_M + k c_M^2 \tau) \quad (7.3\text{-}11a)$$

We again use equations 18.4-19 and -20 to introduce f_M and M_2, and obtain from 7.3-11a:

$$f_M = M_2 (1 - f_M)^2 [2 + M_2 (1 - f_M)] \quad (18.4\text{-}22)$$

444 Chapter 18: Complex Reactions in Ideal Reactors

The solution of 18.4-22 for f_M in terms of M_2 is the cubic equation:

$$f_M^3 - \left(\frac{3M_2 + 2}{M_2}\right)f_M^2 + \left[\frac{(3M_2 + 1)(M_2 + 1)}{M_2^2}\right]f_M - \left(\frac{M_2 + 2}{M_2}\right) = 0 \quad (18.4\text{-}23)$$

Equation 18.4-23 may be solved by trial or by the E-Z Solve software.
The solution of 18.4-22 for M_2 in terms of f_M is a quadratic equation:

$$M_2^2 + \left(\frac{2}{1 - f_M}\right)M_2 - \frac{f_M}{(1 - f_M)^3} = 0 \quad (18.4\text{-}24)$$

from which

$$M_2 = \frac{[1 + f_M/(1 - f_M)]^{1/2} - 1}{1 - f_M} = \frac{1 - (1 - f_M)^{1/2}}{(1 - f_M)^{3/2}} \quad (18.4\text{-}25)$$

Calculation of the distribution $w_r(r)$ depends on whether f_M or M_2 is specified. If M_2 is specified, f_M is calculated from 18.4-23, and $w_r(r)$ is then calculated from 18.4-21. If f_M is specified, M_2 is obtained from 18.4-25. The use of these equations is illustrated in problem 18-26.

For nonisothermal operation, the energy analysis, point (4) above, requires that the energy balance be developed for a *complex* system. The energy (enthalpy) balance previously developed for a BR, or CSTR, or PFR applies to a *simple* system (see equations 12.3-16, 14.3-9, and 15.2-9). For a complex system, each reaction (i) in a specified network contributes to the energy balance (as $(-\Delta H_{Ri})r_i$), and, thus, each must be accounted for in the equation. We illustrate this in the following example.

EXAMPLE 18-8

Develop the energy equation as an enthalpy balance for the partial oxidation of methane to formaldehyde, occurring nonisothermally in a PFR, according to the reaction network of Spencer and Pereira (1987) in Example 5-8.

SOLUTION

From Example 5-8, the reaction network is

(1) $CH_4 + O_2 \xrightarrow{k_1(T)} HCHO + H_2O; \ r_1(T); \ \Delta H_{R1}(T)$
(2) $HCHO + \frac{1}{2}O_2 \xrightarrow{k_2(T)} CO + H_2O; \ r_2(T); \ \Delta H_{R2}(T)$
(3) $CH_4 + 2O_2 \xrightarrow{k_3(T)} CO_2 + 2H_2O; \ r_3(T); \ \Delta H_{R3}(T)$

This network is a series-parallel network: series with respect to HCHO in steps (1) and (2), parallel with respect to CH_4 in steps (1) and (3), and parallel with respect to O_2 and H_2O in all steps. The rate constants k_1, k_2, and k_3 are *step* rate constants (like k in equation 4.1-3a), the rates r_1, r_2, and r_3 are *step* rates (like r in equation 1.4-8 or 4.1-3), and ΔH_{R1},

ΔH_{R2}, and ΔH_{R3} are enthalpies of reaction for the three steps; as indicated in the reaction network, all these quantities depend on T.

The energy balance for this complex reaction in a PFR can be developed as a modification of the balance for a simple system, such as A + ... → products. In this latter case, equation 15.2-5, combined with equation 15.2-8, may be written as

$$\dot{m}c_P dT = (4U/D)(T_S - T)dV + (-\Delta H_{RA})(-r_A)dV = 0 \qquad (18.4\text{-}26)$$

For the reaction network in this example, the product $(-\Delta H_{RA})(-r_A)$ is replaced by the sum of the products:

$$-[\Delta H_{R1}r_1 + \Delta H_{R2}r_2 + \Delta H_{R3}r_3] \qquad (A)$$

Note that each r in this sum is positive, but, in general, each ΔH_R may be positive or negative (in this example, each ΔH_R is negative). As a standard enthalpy of reaction, each ΔH_R^o is obtained, at a particular T, in the usual way from standard enthalpies of formation (ΔH_f^o) or of combustion (ΔH_c^o) of the species, such as are provided by Stull et al. (1969) or by JANAF (1986):

$$\Delta H_{Rj}^o = \sum_{i=1}^{N} \nu_{ij} \Delta H_{fi}^o = -\sum_{i=1}^{N} \nu_{ij} \Delta H_{ci}^o; \qquad j = 1, 2, 3 \qquad (18.4\text{-}27)$$

The effect of T on ΔH_{Rj}^o is taken into account by the Kirchhoff equation (Denbigh, 1981, p. 145):

$$\frac{d\Delta H_{Rj}^o}{dT} = \sum_{i=1}^{N} \nu_{ij} C_{Pi}(T); \qquad j = 1, 2, 3 \qquad (18.4\text{-}28)$$

Finally, the effect of T on r_j is taken into account by the Arrhenius equation:

$$r_j = A_j \exp(-E_{A_j}/RT); \qquad j = 1, 2, 3 \qquad (18.4\text{-}29)$$

where A_j is like A in equation 4.1-4a.

The solution of this set of equations, 18.4-26 (with expression (A) incorporated) to -29, must be coupled with the set of three independent material-balance or continuity equations to determine the concentration profiles of three independent species, and the temperature profile, for either a specified size (V) of reactor or a specified amount of reaction. A numerical solution of the coupled differential equations and property relations is required. Equations (A), (B), and (C) in Example 18-6 illustrate forms of the continuity equation.

18.5 PROBLEMS FOR CHAPTER 18

18-1 The reaction between ethyl alcohol and formic acid in acid solution to give ethyl formate and water, $C_2H_5OH + HCOOH \rightleftarrows HCOOC_2H_5 + H_2O$, is first-order with respect to formic acid in the forward direction and first-order with respect to ethyl formate in the reverse direction, when the alcohol and water are present in such large amounts that their concentrations do not change appreciably. At 25°C, the rate constants are $k_f = 1.85 \times 10^{-3}$ min^{-1} and $k_r = 1.76 \times 10^{-3}$ min^{-1}. If the initial concentration of formic acid is 0.07 mol L^{-1} (no formate present initially), calculate the time required for the reaction to reach 90% of the equilibrium concentration of formate in a batch reactor.

18-2 The hydrolysis of methyl acetate (A) in dilute aqueous solution to form methanol (B) and acetic acid (C) is to take place in a batch reactor operating isothermally. The reaction is reversible, pseudo-first-order with respect to acetate in the forward direction ($k_f = 1.82 \times 10^{-4}$ s^{-1}), and first-order with respect to each product species in the reverse direction ($k_r = 4.49 \times 10^{-4}$ L mol^{-1} s^{-1}). The feed contains only A in water, at a concentration of 0.050 mol L^{-1}. Determine the size of the reactor required, if the rate of product formation is to be 100 mol h^{-1} on a continuing basis, the down-time per batch is 30 min, and the optimal fractional conversion (i.e., that which maximizes production) is obtained in each cycle.

18-3 The reaction in problem 18-2 is to be conducted in an isothermal 60-L PFR, with a feed rate (F_{Ao}) of 0.75 mol min^{-1}. All other data (c_{Ao}, k_r, k_f) remain the same.
(a) Calculate the fractional conversion f_A at the outlet of the reactor.
(b) What is the maximum possible value of f_A, regardless of reactor type or size?

18-4 For the exothermic, liquid-phase, reversible isomerization reaction, $A \underset{k_r}{\overset{k_f}{\rightleftharpoons}} B$, taking place in a CSTR, with a feed rate (F_{Ao}) of 500 mol min^{-1}, a feed concentration (c_{Ao}) of 5 mol L^{-1} ($c_{Bo} = 0$), and a fractional conversion (f_A) of 0.70, calculate
(a) the optimal temperature (T/K) in the reactor for minimum volume;
(b) the minimum volume required for the reactor;
(c) the feed temperature (T_o) for adiabatic operation at the optimal T.
Additional information:

$(-r_A) = r_B = k_f c_A - k_r c_B$ mol L^{-1} min^{-1}
$k_f = \exp(17.2 - 5840/T)$ min^{-1}
$k_r = \exp(41.9 - 14900/T)$ min^{-1}
$\Delta H_{RA} = -75{,}300$ J mol^{-1}; c_P (stream) $= 4.2$ J g^{-1} K^{-1}
ρ (stream) $= 1000$ g L^{-1}

(Parameter values taken from Levenspiel, 1972, p. 232.)

18-5 For the reaction, conditions, and data given in problem 18-4, calculate the minimum volume required for a PFR operated adiabatically.

18-6 Carbon dioxide undergoes parallel reactions in aqueous solution in the presence of ammonia and sodium hydroxide, leading to the formation of different products, P1, and P2 (Miyawaki, et al., 1975):

$$CO_2 + NH_3 \rightarrow P1$$
$$CO_2 + NaOH \rightarrow P2$$

At 298 K, the rate constants for these reactions are 0.40 and 9.30 m^3 mol^{-1} s^{-1}, respectively. What is the product distribution (i.e., liquid-phase concentrations of CO_2, NaOH, NH_3, P1, and P2) after 100 s in a constant-volume batch reactor? The initial concentrations for each species are as follows: CO_2: 1.0×10^{-4} mol L^{-1}; NH_3: 3.0×10^{-5} mol L^{-1}; NaOH: 5.0×10^{-5} mol L^{-1}.

18-7 Consider three reactions, of the same order, in parallel:

$$A \rightarrow R; \quad k_1 = A_1 \exp(-E_1/RT)$$
$$A \rightarrow S; \quad k_2 = A_2 \exp(-E_2/RT)$$
$$A \rightarrow T; \quad k_3 = A_3 \exp(-E_3/RT)$$

and the (instantaneous) fractional yield of R, the desired product, $\hat{s}_{R/A} = r_R/(-r_A)$. Of all the possible cases for differing relative magnitudes for E_1, E_2, and E_3, choose one case (there are two cases) for which an optimal temperature (T) exists to maximize $\hat{s}_{R/A}$ (and hence the rate of production of R for a given $\bar{t}$) with respect to T, and obtain an expression for T_{opt} in terms of the Arrhenius parameters for this case.

18-8 A (desired) liquid-phase dimerization $2A \to A_2$, which is second-order in $A (r_{A_2} = k_2 c_A^2)$, is accompanied by an (undesired) isomerization of A to B, which is first-order in $A (r_B = k_1 c_A)$. Reaction is to take place isothermally in an inert solvent with an initial concentration $c_{Ao} = 5$ mol L^{-1}, and a feed rate (q) of 10 L s^{-1} (assume no density change on reaction). Fractional conversion (f_A) is 0.80.
  (a) What is, in principle, the proper type of reactor to use (PFR or CSTR) for favorable product distribution and/or reactor size? Explain (without calculations).
  (b) For the reactor type chosen in (a), calculate the product distribution (c_A, c_{A_2}, c_B), and the volume of reactor required, if $k_2 = 10^{-2}$ L mol^{-1} s^{-1} and $k_1 = 10^{-2}$ s^{-1}.

18-9 A substance A undergoes isomerization (desired) and dimerization (not desired) simultaneously in a liquid-phase reaction according to

$$A \to B; \quad r_B = k_1 c_A; \quad k_1 = 0.01 \text{ s}^{-1} \text{ at } T_1 \quad \textbf{(desired)}$$

$$2A \to A_2; \quad r_{A2} = k_2 c_A^2; \quad k_2 = 0.01 \text{ L mol}^{-1} \text{ s}^{-1} \text{ at } T_1 \quad \textbf{(undesired)}$$

A reactor is to be designed for 90% conversion of A (f_A) with a feed concentration (c_{Ao}) of 10 mol L^{-1} and a volumetric throughput (q) of 10 L s^{-1}.
  (a) For a single CSTR, calculate (i) the product distribution (concentrations), and overall fractional yields of B and A_2, and (ii) the size of reactor for operation at T_1.
  (b) Repeat (a) if two equal-sized CSTRs are to be used in series, each operated at T_1.
  (c) Repeat (a) for a plug-flow reactor operated isothermally at T_1.
  (d) Comment on the results of (a), (b), and (c), and suggest generalizations that are illustrated.

18-10 For the kinetics scheme $A \xrightarrow{k_1} B \xrightarrow{k_2} C$, calculate (i) the maximum concentration of B, $c_{B,max}$, and (ii) the time, t_{max} or τ_{max}, required to achieve this concentration, for reaction carried out at constant density in (a) a batch reactor (BR), isothermally, and (b) in a CSTR. Only A is present initially, at a concentration, c_{Ao}, of 1.5 mol L^{-1}, and $k_1 = 0.004$ s^{-1}, and $k_2 = 0.02$ s^{-1}. For convenience, calculate (i) and (ii) for the BR as part (a) and for the CSTR as part (b). Finally, (c) comment on any difference in results.

18-11 In pulp and paper processing, anthraquinone accelerates delignification of wood and improves liquor selectivity (see problem 5-17 and Examples 18-3 and -4). Anthraquinone (AQ) can be produced by oxidation of anthracene (AN) in the presence of NO_2, using acetic acid as a solvent (Rodriguez and Tijero, 1989). The reaction, which proceeds via an intermediate compound, anthrone (ANT), may be written:

$$AN \xrightarrow{k_1} ANT \xrightarrow{k_2} AQ$$

Using an initial anthracene concentration of 0.025 mol L^{-1}, determine the product distribution in an isothermal, constant-volume batch reactor, when the concentration of ANT is at a maximum at 95°C. $k_1 = 0.0424$ min^{-1}, and $k_2 = 0.0224$ min^{-1}.

18-12 The reaction described in problem 18-11 is to be conducted in a 1.0 m^3 CSTR operating at 95°C. The flow rate through the reactor is 1.5 m^3 h^{-1}.
  (a) Determine the product distribution obtained, if the feed contains anthracene (AN) at a concentration of 0.025 mol L^{-1}.
  (b) Calculate the fractional conversion of AN, and the overall fractional yield of AQ.
  (c) Repeat parts (a) and (b), if the reactor consists of two CSTRs in series, each of volume 0.5 m^3.

18-13 For the kinetics scheme in problem 18-11, and from the information given there, compare the following features of a CSTR and a PFR operated isothermally.
(a) Calculate the size of reactor that maximizes the yield of the intermediate, ANT.
(b) Calculate the minimum reactor volume required to ensure that the fractional yield of AQ is at least 0.80.
The volumetric flow rate through the reactor is 1.25 m³ h⁻¹.

18-14 The reaction of A to form C occurs through the formation of an intermediate species B, with both steps being first-order: A $\xrightarrow{k_1}$ B $\xrightarrow{k_2}$ C. The rate constants at 300 K are $k_1 = 5 \times 10^{-3}$ s⁻¹ and $k_2 = 1 \times 10^{-3}$ s⁻¹. For a reactor operating at 300 K, the feed concentration (c_{Ao}) is 5 mol L⁻¹, and the feed rate (q) is 400 L min⁻¹.
(a) Which type of reactor (PFR or CSTR) should be chosen to favor formation of B? Explain (without calculations).
(b) Calculate the size (L) of the reactor chosen in (a) to maximize the yield of B (for this particular type of reactor).
(c) For the reactor size obtained in (b), calculate (i) the fractional conversion of A (f_A), and (ii) the overall fractional yield of B ($\hat{S}_{B/A}$).

18-15 Reaction in a certain liquid-phase system can be represented kinetically by:

$$A + B \rightarrow C; \quad (-r_A) = k_1 c_A c_B; \quad k_1 = A_1 \exp(-E_1/RT)$$
$$C + B \rightarrow D; \quad r_D = k_2 c_C c_B; \quad k_2 = A_2 \exp(-E_2/RT)$$

Reaction is carried out in a CSTR with a large excess of B, and it is desired to achieve a maximum concentration of C in the exit stream. For certain (nonkinetics) reasons, it is desirable to operate either at 31°C or at 77°C. Assume $E_1/R = 10,000$ K, $E_2/R = 10,700$ K, and $A_1 = A_2$.
(a) Which temperature should be used to obtain the largest concentration of C? Justify your answer without doing detailed calculations.
(b) For operation at 31°C (not necessarily the better value), calculate the maximum yield of C that can be obtained.
(c) If $c_{Bo} = 5$ mol L⁻¹, $q = 7$ L s⁻¹, $k_1 = 0.025$ s⁻¹ at 31°C, calculate (i) the size (V/L) of reactor required and (ii) the fractional conversion of A at 31°C.

18-16 (a) Suppose species A undergoes (liquid-phase) reaction according to the following kinetics scheme:

Reaction	Desired?	Rate Law	Activation Energy
A → B	no	$r_B = k_1 c_A$	E_{A1}
2A → A₂	yes	$r_{A2} = k_2 c_A^2$	$E_{A2} < E_{A1}$

(i) Does a favorable product distribution result from relatively high or relatively low c_A? Explain briefly.
(ii) Does a favorable product distribution result from relatively high or relatively low T? Explain briefly.
(iii) Does a favorable (i.e., small) reactor size result from high or low c_A? Explain briefly.
(iv) Does a favorable reactor size result from high or low T? Explain briefly.
(v) Which type of reactor, PFR or single-stage CSTR, should be chosen for this situation, and should it be operated at high, low, or optimal T to achieve favorable product distribution consistent with small size of reactor? Explain briefly.

(b) For the kinetics scheme in (a), and regardless of the proper choice of reactor, suppose a CSTR of volume 10 m³ is available to process 0.5 m³ min⁻¹ of a feed stream with a concentration (c_{Ao}) of 3000 mol m⁻³. If the temperature is such that $k_1 = 0.001$ s⁻¹, and $k_2 = 5 \times 10^{-5}$ m^3 mol^{-1} s^{-1}, calculate the fractional conversion of A (f_A) and the product distribution (c_A, c_B, c_{A2}) in the exit stream at steady-state.

18-17 (a) The following parallel, liquid-phase reactions take place in a batch reactor operating isothermally at 320 K.

$$A \xrightarrow{k_1} B + C; \qquad r_B = k_1 c_A; \qquad k_1 = A_1 e^{-E_1/RT}$$

$$A \xrightarrow{k_2} D + E; \qquad r_D = k_2 c_A; \qquad k_2 = A_2 e^{-E_2/RT}$$

Calculate, for this reacting system, the half-life of A ($t_{1/2}$) and the concentration of B and of D at time $t_{1/2}$.

Data: $A_1 = 1.5 \times 10^{11}$ s^{-1}; $A_2 = 9.0 \times 10^{13}$ s^{-1}; $E_1 = 89{,}800$ J mol^{-1}; $E_2 = 107{,}000$ J mol^{-1}; $c_{Ao} = 2.0$ mol L^{-1}; $c_{Bo} = c_{Do} = 0$.

(b) Repeat the calculations for isothermal operation at (i) 280 K and at (ii) 360 K. Assume all else is unchanged.

(c) Make a generalization about the effect of temperature on the relative yields (Y_B/Y_D) of products of parallel reactions which have different energies of activation.

18-18 A network of ideal-gas-phase reactions is to be conducted in an isothermal, isobaric PFR:

$$A + B \rightarrow R + 2S; \qquad (-r_A) = k_1 c_A c_B^{0.5}$$

$$R + B \rightarrow U; \qquad r_U = k_2 c_B c_R^{1.5}$$

R is the desired product. Determine the reactor volume that:
(a) maximizes the outlet concentration of R;
(b) maximizes the yield of R with respect to limiting reactant A.

Data:

$$F_{Ao} = 2 \text{ mol min}^{-1}; \qquad q_o = 45 \text{ L min}^{-1}$$

$$F_{Bo} = 3 \text{ mol min}^{-1}; \qquad k_1 = 50 \text{ L}^{0.5} \text{mol}^{-0.5} \text{min}^{-1}$$

$$F_{Ro} = F_{So} = F_{Uo} = 0; \qquad k_2 = 500 \text{ L}^{1.5} \text{mol}^{-1.5} \text{min}^{-1}$$

18-19 For a liquid-phase reaction A $\underset{k_r}{\overset{k_f}{\rightleftharpoons}}$ S, the rate of reaction of A is given by $(-r_A) = k_f c_A - k_r c_S$, where $k_f = 5 \times 10^6 \exp(-5000/T)$, and $k_r = 1 \times 10^{11} \exp(-11000/T)$; k_f and k_r have units of h^{-1}. The reaction is studied in a CSTR with a residence time of 0.4 h. Although the reactor is insulated, a heat loss of 1.6 kW is observed.

Data: density = 0.95 g cm^{-3} (constant); $c_P = 3.25$ J g^{-1} K^{-1} (constant); $\Delta H_{RA} = -35$ kJ mol^{-1}; $V = 4000$ L

(a) From the design equation for steady-state operation and the rate law, develop an expression for f_A in terms of k_f, k_r, c_{Ao}, c_{So}, and the mean residence time. Develop a graph of $f_A(T)$ based upon the rate law, with $c_{Ao} = 10.0$ mol L^{-1}, and $c_{So} = 0.50$ mol L^{-1}.

(b) Using the graph developed in (a), determine f_A and T at steady-state, if (i) $T_o = 280$ K; (ii) $T_o = 288$ K; (iii) $T_o = 295$ K. Comment on your findings.

(c) Discuss, without performing any calculations, what would happen to f_A and T from case (iii), if some of the insulation fell off the reactor, increasing the heat loss.

18-20 Supercritical-water oxidation has been proposed as a method for removal of phenolics from industrial waste streams. Ding et al. (1995) studied the degradation of phenol (C$_6$H$_5$OH) in a PFR, using both catalytic and noncatalytic reactions. The primary objective is to convert phenol to CO$_2$. The degradation of phenol may be reduced to the following scheme of pathways (producing 2 moles of C$_3$ oxygenates):

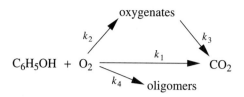

Assume the formation of oligomers (k_4) is first-order with respect to phenol, and the formation of the other products (k_1, k_2, k_3) is first-order with respect to each of phenol and O_2. The following rate constants were obtained from noncatalytic studies and from experiments using a MnO_2/CeO catalyst:

Rate constant	Noncatalytic	Catalytic
k_1/L mol^{-1} s^{-1}	0.39	5.52
k_2/L mol^{-1} s^{-1}	0.067	0.27
k_3/L mol^{-1} s^{-1}	0.38	4.00
k_4/s^{-1}	0.19	0.54

In the noncatalytic studies, the space time τ through the reactor was 15.3 s, with an inlet phenol concentration of 0.0023 mol L^{-1}, and 400% excess oxygen in the feed (based on complete oxidation of phenol to CO_2 and H_2O). In the catalytic studies, the space time was 5.3 s, with an inlet phenol concentration of 0.0066 mol L^{-1}, and 400% excess oxygen in the feed.

(a) Compare the product distribution obtained in the catalytic and noncatalytic cases, and comment on how the catalyst changes the relative importance of the pathways.

(b) To increase the yield of carbon dioxide, the inlet oxygen concentration can be adjusted. For a range of oxygen concentrations between 300% and 900% above the stoichiometric requirement, determine:
  (i) the fractional conversion of phenol;
  (ii) the yield of CO_2, defined on a carbon basis, $Y_{CO_2} = c_{CO_2}/(6c_{o,phenol})$;
  (iii) the outlet concentration of the partial oxidation products.

Use the same space time and inlet phenol concentrations specified for part (a). Comment on any differences in the effects of oxygen on the yield and selectivity in the noncatalytic and catalytic processes.

18-21 The conversion of methanol to olefins is an intermediate step in the conversion of methanol to gasoline. Schoenfelder et al. (1994) described the reaction network using a lumped kinetics analysis, as follows:

$$2A \xrightarrow{k_1} B + 2W$$
$$A + B \xrightarrow{k_2} C + W$$
$$A + C \xrightarrow{k_3} D + W$$
$$3B \xrightarrow{k_4} 2E$$
$$C \xrightarrow{k_4} E$$
$$3D \xrightarrow{k_4} 4E$$
$$2A \xrightarrow{k_5} 3F + W$$

where A = oxygenates (methanol and dimethylether), B = ethene, C = propene, D = butene, E = paraffins, F = dissociation products of the oxygenates (methane, hydrogen, carbon monoxide), and W = steam. It is assumed that the same rate constant applies to the conversion of each alkene into paraffins. The rate constants at 400°C and 500°C are as follows:

Rate constant	400°C	500°C
$k_1/\text{m}^3\ \text{h}^{-1}\ \text{kg}^{-1}$	4.61	24.3
$k_2/\text{m}^6\ \text{h}^{-1}\ \text{kg}^{-1}\ \text{mol}^{-1}$	2.09	12.3
$k_3/\text{m}^6\ \text{h}^{-1}\ \text{kg}^{-1}\ \text{mol}^{-1}$	0.56	1.8
$k_4/\text{m}^3\ \text{h}^{-1}\ \text{kg}^{-1}$	0.27	0.075
$k_5/\text{m}^3\ \text{h}^{-1}\ \text{kg}^{-1}$	0.22	0.99

Consider the conversion of methanol in a 50-L reactor (volume of catalyst) similar to that shown in Figure 1.2 (which operates like a CSTR). The reactor contains 800 g of catalyst (zeolite H-ZSM5), and the space time through the reactor is 0.1 h. The methanol feed rate is 1.3 kg h^{-1}. For each reaction temperature, determine the yield and selectivity to each olefin, and comment on your results.

18-22 For the liquid-phase oxidation of anthracene (AN), a two-step series reaction network (see problem 18-11), Rodriguez and Tijero (1989) obtained the following Arrhenius expressions for the two rate constants over the range 60–95°C:

$$k_1 = 2.42 \times 10^4 \exp(-52{,}730/RT)$$
$$k_2 = 3.98 \times 10^7 \exp(-77{,}724/RT)$$

where k_1 and k_2 are in s^{-1} and T is in K.

If the reaction takes place in a constant-volume (for the liquid phase) BR operated isothermally, calculate (i) f_{AN}; (ii) $Y_{ANT/AN}$; (iii) $Y_{AQ/AN}$; (iv) $\hat{S}_{ANT/AN}$; and (v) $\hat{S}_{AQ/AN}$ (where ANT is anthrone, the intermediate product, and AQ is anthraquinone, the final product) at $t = 60$ min, and (a) 60°C; (b) 75°C; and (c) 90°C. (d) Comment on the results.

18-23 For the decomposition of NH$_4$HCO$_3$ (A) in aqueous solution, as described in problem 4-20, Nowak and Skrzypek (1989) obtained the rate law:

$$(-r_A) = 3.5 \times 10^6 \exp(-7700/T) c_A^2 \text{ mol L}^{-1}\ \text{s}^{-1}$$

Separately, for the decomposition of (NH$_4$)$_2$CO$_3$ (B), they obtained:

$$(-r_B) = 1.3 \times 10^{11} \exp(-11{,}900/T) c_B^2 \text{ mol L}^{-1}\ \text{s}^{-1}$$

In the following, state any assumptions made.
(a) If a solution of NH$_4$HCO$_3$ ($c_{Ao} = 1$ mol L^{-1}) is stored in a well-ventilated tank at 40°C, what is the highest concentration reached by (NH$_4$)$_2$CO$_3$, and at what time (t)?
(b) Without doing any further calculations, state what effect a lower T would have on the results in (a). Explain briefly.
(c) If the solution in (a) were passed through a heat exchanger, in which BMF occurred, to raise T from 20 to 80°C, what would the space time (τ) be to reach the highest concentration of (NH$_4$)$_2$CO$_3$?
(d) Repeat (c), if the heating were done in two stages, each with BMF and the same value of τ.

18-24 Confirm the entries in Table 18.1 for $K \to 0$, $K = 1$, and $K \to \infty$.

18-25 The liver, known as the chemical reactor of the body, serves to eliminate drugs and other compounds from the body. In one such process, lidocaine (a drug given to heart attack patients) may be converted either into MEGX by deethylation, or into hydroxylidocaine (OHLID) by hydroxylation. Assuming that the liver may be represented by a CSTR, determine the product distribution which is obtained, if lidocaine (2.5 mmol m^{-3}) is supplied at a rate of 30×10^{-6} m^3 min^{-1}. The following reactions and rate expressions apply:

$$\text{lidocaine } (c_1) \rightarrow \text{MEGX}; \qquad r_{\text{MEGX}} = k_1 c_1; \qquad k_1 = 2.6 \text{ min}^{-1}$$
$$\text{lidocaine} \rightarrow \text{hydroxylidocaine}; \qquad r_{\text{OHLID}} = k_2 c_1; \qquad k_2 = 14 \text{ min}^{-1}$$

The volume of the liver, for the species studied, is approximately 1.5×10^{-5} m^3. (Simplified from Saville et al., 1986.)

18-26 This problem continues Example 18-7 for use of the step-change polymerization kinetics model of Chapter 7 in a CSTR at steady-state.

(a) If $f_M = 0.98$, $c_{Mo} = 9000$ mol m^{-3}, and $\tau = 1000$ s, calculate the value of the rate constant k, and specify its units.

(b) Using the data in (a), plot the $w_r(r)$ distribution for the polymer stream leaving the reactor, from $r = 2$ to a value that makes w_r very small.

(c) Derive an equation relating $r_{w_r,max}$ to f_M (only), where $r_{w_r,max}$ is the number of monomer units in the r-mer with the largest weight fraction in the outlet polymer stream, for a given value of M_2. Use the equation derived to confirm the result obtained in (b).

(d) Show what the effect (increase or decrease) of increasing c_{Mo} is on f_M.

18-27 Based on the reaction network in Example 18-8, calculate and plot the temperature (T)–volume (V) profile and the concentration (c_i)–volume profiles for a set of independent species in a PFR operated adiabatically. Consult the paper by Spencer and Pereira (1987) for appropriate choice of feed conditions and for kinetics data. For thermochemical data, consult the compilation of Stull et al. (1969), or an equivalent one.

Chapter 19

Nonideal Flow

In this chapter, we consider nonideal flow, as distinct from ideal flow (Chapter 13), of which BMF, PF, and LF are examples. By its nature, nonideal flow cannot be described exactly, but the statistical methods introduced in Chapter 13, particularly for residence time distribution (RTD), provide useful approximations both to characterize the flow and ultimately to help assess the performance of a reactor. We focus on the former here, and defer the latter to Chapter 20. However, even at this stage, it is important to realize that ignorance of the details of nonideal flow and inability to predict accurately its effect on reactor performance are major reasons for having to do physical scale-up (bench → pilot plant → semi-works → commercial scale) in the design of a new reactor. This is in contrast to most other types of process equipment.

We first describe features of nonideal flow qualitatively, and then in terms of mixing aspects. For the rest of the chapter, we concentrate on its characterization in terms of RTD. This involves (1) description of the experimental measurement of RTD functions (E, F, W), and development of techniques for characterizing nonideal flow; and (2) introduction of two simple models for nonideal flow that can account for departures from ideal flow.

19.1 GENERAL FEATURES OF NONIDEAL FLOW

The flow of a fluid through a vessel usually falls between the extremes of BMF (complete mixing) and PF (no axial mixing). In a stirred tank, it is difficult to achieve complete instantaneous mixing at the inlet, and there may be "dead zones" arising from stagnation near a baffle at the wall, in which there is little or no exchange of fluid with the "active zone," the central portion of the vessel (shown schematically in Figure 19.1(a)). Similarly, PF may be difficult to achieve, particularly if frictional losses at the wall and the relative effects of diffusion and convection on the transport of material within the vessel are considered; these may result in significant axial mixing or dispersion. PF is approached if the flow is fully turbulent, if frictional effects are small, and if the length-to-diameter ratio is large ($L/D \gg 1$); if any one of these conditions is not met, deviations from PF may occur. In a packed-bed (shown schematically in Figure 19.1(b)), deviations from PF may show as channeling or bypassing, arising from uneven distribution of fluid across the bed, or spatial differences in the resistance to flow through the bed. Bypassing or short-circuiting from inlet to outlet (and internal recycling) may occur in general in a stirred tank; they occur as well in BMF, but in accountable ways so as to obtain a predictable spread in residence times; in nonideal flow, they are not "accountable" *a priori*.

The effects on the RTD function E of stagnation, dispersion (or partial backmixing), and channeling or excessive bypassing are shown schematically in Figures 19.2(a), (b), and (c), respectively.

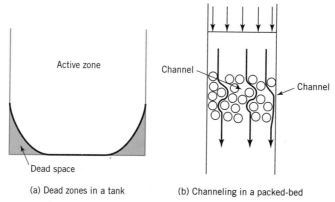

Figure 19.1 Examples of nonideal flow in stirred-tank and packed-bed vessels

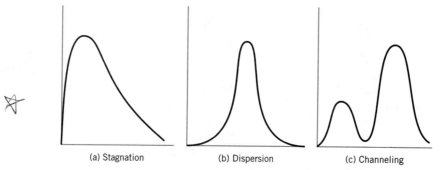

Figure 19.2 Effect of some features of nonideal flow on $E(t)$

19.2 MIXING: MACROMIXING AND MICROMIXING

As discussed in Section 17.2.3.1, reactor performance in general depends on (1) the kinetics of reaction, (2) the flow pattern as represented by the RTD, and (3) mixing characteristics within the vessel. The performance predicted by ideal reactor models (CSTR, PFR, and LFR) is determined entirely by (1) and (2), and they do not take (3) into account.

Mixing within a vessel can be a very complex process, which can be analyzed on both a macroscopic and microscopic ("molecular") scale. Nonideal flow results in irregularities relating to the mixing of fluid elements of differing ages at the microscopic level. This is not taken into account by the RTD, which provides a measure of macroscopic mixing. That is, the RTD is unable to account for how fluid elements at the microscopic level interact with each other. Such interactions can be very important, especially in a reactor, if reaction requires two different chemical species to come together.

Ideal flow models contain inherent assumptions about mixing behavior. In BMF, it is assumed that all fluid elements interact and mix completely at both the macroscopic and microscopic levels. In PF, microscopic interactions occur completely in any plane perpendicular to the direction of flow, but not at all in the axial direction. Fluid elements at different axial positions retain their identities as they progress through the vessel, such that a fluid element at one axial position never interacts with a fluid element at another position.

Following the discussion on segregated flow in Section 13.5, we may consider two extremes of micromixing behavior:

(1) Complete segregation: any fluid element is isolated from all other fluid elements and retains its identity throughout the entire vessel. No micromixing occurs, but macromixing may occur.

(2) Complete dispersion: fluid elements interact and mix completely at the microscopic level.

Micromixing between these two extremes (partial segregation) is possible, but not considered here. A model for (1) is the segregated-flow model (SFM) and for (2) is the maximum-mixedness model (MMM) (Zwietering, 1959). We use these in reactor models in Chapter 20.

Besides the degree of segregation of fluid elements (level of subdivision at which mixing takes place), another aspect of mixing is the relative time in passage through a vessel at which fluid elements mix ("earliness" of mixing). A simple model illustrating the effect of this on RTD and performance is given in Examples 17-5 and -6.

19.3 CHARACTERIZATION OF NONIDEAL FLOW IN TERMS OF RTD

19.3.1 Applications of RTD Measurements

RTD measurements may be used for the following:

(1) as a diagnostic tool for detecting and characterizing flow behavior (Section 19.3.2);
(2) the estimation of values of parameters for nonideal flow models (Section 19.4);
(3) assessment of performance of a vessel as a reactor (Chapter 20).

19.3.2 Experimental Measurement of RTD

19.3.2.1 Stimulus-Response Technique

The experimental measurement of the RTD for flow of a fluid through a vessel is usually carried out by the stimulus-response technique. In this, a tracer material that is experimentally distinguishable from the fluid being studied is injected at the inlet, and its concentration is monitored at the outlet. The injected tracer is the stimulus or signal, and the result of the monitoring is the response. Ideally, the signal should be well defined and introduced so as not to disturb the flow pattern. In principle, any input stimulus may be used, provided that the response is linear (i.e., directly proportional) with respect to the tracer; this condition is generally met if the change in tracer concentration is not too large. Two types of stimulus that can be made relatively well defined are the pulse or Dirac signal and the step-change signal. A pulse input involves the rapid injection of a small, known amount of tracer, as a physician might use a syringe to inject medication. A step-change input involves an instantaneous change in inlet concentration of tracer from an initial steady-state level (which may be zero) to a second level. We describe each of these in turn here, and their use in more detail in Section 19.3.2.3 below, after discussion of tracer characteristics. An extensive discussion of tracer methods is given by Duduković (1985).

A pulse input of tracer A at a vessel inlet ($c_{A,in}$ at t_o) and a response to it at the vessel outlet ($c_{A,out}$ at t subsequent to t_o) are shown in Figures 19.3(a) and (b), respectively.

A *unit* pulse (at $t = 0 \equiv t_o$) is represented by the Dirac delta function $\delta(t-0) \equiv \delta(t)$, such that the area of the pulse, $\int_0^\infty \delta(t)dt$, is unity (see equations 13.4-6 and -7). A pulse of arbitrary amount, m_o kg (or n_o moles), as indicated in Figure 19.3(a), is represented

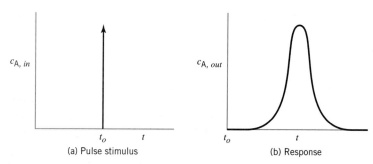

Figure 19.3 Pulse stimulus (a)–response (b) experiment with tracer A

by $(m_o/q_o)\delta(t), \equiv c_{A,in}$, where q_o is the (total) steady-state flow rate of fluid. Then the area of the pulse is such that

$$\int_0^\infty c_{A,in} dt = \int_0^\infty (m_o/q_o)\delta(t) dt = m_o/q_o \qquad (19.3\text{-}1)$$

This provides an important check on the accuracy of the pulse-tracer experiment, since the area under the response curve represents the same quantity, if the tracer is completely accounted for by a material balance. Thus,

$$m_o/q_o = \int_0^\infty c_{A,out}(t) dt \qquad (19.3\text{-}2)$$

is a condition to be satisfied. If m_o is specific mass, c_A is a specific-mass concentration. If m_o is replaced by number of moles, n_o, c_A is a molar concentration. Other consistent sets of units may be used. Another requirement is that the time to inject the pulse must be much less than $\bar{t}$, the mean residence time of A in the vessel.

Advantages of using a pulse input include:

(1) requiring only a small amount of tracer (relatively small cost or hazardous conditions, if these are factors);
(2) involving usually only a small impact on process operation (tracer study does not require shutdown).

Disadvantages include:

(1) difficulty in achieving a perfect pulse (may complicate interpretation in relating response to input);
(2) difficulty in achieving accurate material balance on tracer by means of equation 19.3-2 (frequent sampling may be required, particularly to capture a "peak" concentration).

A *step* input of tracer may itself be one of two types: a step increase from one steady-state value to another ($c_{A,in}$), or a step decrease. Usually, as illustrated in Figure 19.4, the step increase is from a zero value, and, as illustrated in Figure 19.5, the step decrease is to a zero value; in the latter case it is called a washout. Figures 19.4 and .5 also show responses ($c_{A,out}$) to a step increase and to a washout-step decrease. In Figure 19.4, note that $c_{A,out} \rightarrow c_{A,in}$ as $t \rightarrow \infty$, and, similarly in Figure 19.5, $c_{A,out} \rightarrow 0$.

A tracer study may use a step increase followed at a later time by a step decrease; the transient responses in the two cases are then checked for consistency. When considered separately, the washout technique has advantages: less tracer is required, and it avoids having to maintain a steady-state value of $c_{A,in}$ for a lengthy period.

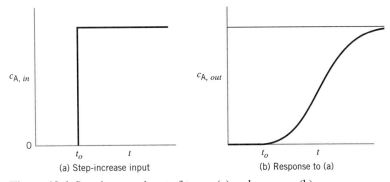

Figure 19.4 Step-increase input of tracer (a) and response (b)

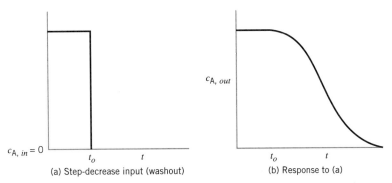

Figure 19.5 Step-decrease (washout) input of tracer (a) and response (b)

In comparison with a pulse input, the step input has the following advantages:

(1) A step change is usually easier to achieve.
(2) A material balance is usually easier to achieve.

The disadvantages are:

(1) Continuous delivery requires a greater amount of tracer (relatively large cost and hazard impact, if these are factors).
(2) It may have a significant impact on process operation, forcing a shutdown.

19.3.2.2 *Selection of a Tracer*

Accurate determination of RTD in a vessel requires proper selection and introduction of a tracer. Here, we consider characteristics and examples of tracers.

Ideally, a tracer should have the following characteristics:

(1) The tracer should be stable and conserved, so that it can be accounted for by a material balance relating the response to the input; if the tracer decays (e.g., a radio-labeled tracer), its half-life should be such that $t_{1/2}$ (tracer) $> 25\bar{t}$ (fluid).
(2) The analysis for tracer should be convenient, sensitive, and reproducible.
(3) The tracer should be inexpensive and easy to handle; this is particularly important for a step input, in which relatively large quantities of tracer may be required.
(4) The tracer should not be adsorbed on or react with the surface of the vessel. Alternatively, the tracer should be chemically and physically similar to the fluid flowing, so that any adsorption (or diffusion) behavior may be replicated.

Examples of tracers are:

(1) gas-phase tracers such as He, Ne, and Ar used with thermal conductivity detectors;
(2) pH indicators such as phenol red and methylene blue;
(3) electrolytes such as K^+ and Na^+ used with electrical conductivity detectors or specific-ion electrodes;
(4) dyes (e.g., India ink) used with color intensity;
(5) radioactive isotopes such as ^3H, ^{14}C, ^{18}O; ^{51}Cr-labeled red-blood cells used to investigate hepatic blood flow; isotopes of iodine, thallium, and technetium used to investigate cardiac blood flow;
(6) stereoisomers and structural analogs used for diffusion-limited processes (e.g., ℓ- versus d-glucose in biological systems).

19.3.2.3 RTD Analysis from Pulse Input

The response to a pulse input of tracer may be monitored continuously or by discrete measurements in which samples are analyzed at successive intervals. If discrete measurements are used, m_o/q_o in equation 19.3-2 is approximated by

$$m_o/q_o \simeq \sum_i c_i(t)\Delta t_i \qquad (19.3\text{-}3)$$

where $c_i(t)$ is the measured concentration corresponding to the ith interval Δt_i.

Since tracer experiments are used to obtain RTD functions, we wish to establish that the response to a pulse-tracer input is related to $E(t)$ or $E(\theta)$. For this purpose, $c(t)$ must be normalized appropriately. We call $c(t)$, in arbitrary units, the nonnormalized response, and define a normalized response $\mathbf{C}(t)$ by

$$\mathbf{C}(t) = \frac{c(t)}{m_o/q_o} = \frac{c(t)}{\int_0^\infty c(t)dt} \simeq \frac{c_i(t)}{\sum_i c_i(t)\Delta t_i} \qquad (19.3\text{-}4)$$

It is understood that all concentrations refer to those of the tracer A at the vessel outlet.
From equations 19.3-2 and -4, with $c(t) \equiv c_{A,out}(t)$,

$$\int_0^\infty \frac{c(t)}{m_o/q_o}dt \equiv \int_0^\infty \mathbf{C}(t)dt = 1 \qquad (19.3\text{-}5)$$

From equation 13.3-1, $E(t)$ is such that

$$\int_0^\infty E(t)dt = 1 \qquad (13.3\text{-}1)$$

Therefore, we conclude that

$$E(t) = \mathbf{C}(t) \qquad (19.3\text{-}6)$$

and use one of the three forms of equation 19.3-4 to calculate $E(t)$ from the experimental tracer data. Once $E(t)$ is determined, the mean-residence time, $\bar{t}$, and the variance

19.3 Characterization of Nonideal Flow in Terms of RTD

of the distribution, σ_t^2, may be calculated based on equations 13.3-14 and -16a for moments:

$$\bar{t} = \int_0^\infty tE(t)dt \simeq \sum_i t_i E_i(t)\Delta t_i \qquad (19.3\text{-}7)$$

$$\sigma_t^2 = \left[\int_0^\infty t^2 E(t)dt\right] - \bar{t}^2 \simeq \left[\sum_i t_i^2 E_i(t)\Delta t_i\right] - \bar{t}^2 \qquad (19.3\text{-}8)$$

As in the case of $C(t)$ or $E(t)$, the integral form in each equation is used for a continuous response, and the summation form for discrete response data. The result for $\bar{t}$ from equation 19.3-7 serves either as a second check on the accuracy of the tracer study, since $\bar{t} = V/q_o$, for constant density, or as a means of determining $\bar{t}$, if the true value of V is unknown.

Equations corresponding to those above may also be used in terms of dimensionless time, $\theta = t/\bar{t}$. Thus, from equation 13.3-7 and the results above,

$$E(\theta) = \bar{t}E(t) = \bar{t}C(t) = \frac{V}{q_o}\frac{c(t)}{m_o/q_o} = \frac{c(t)}{m_o/V} = \frac{c(t)}{c(0)} \equiv C(\theta) \qquad [\text{ by definition of } C(\theta)]$$

That is,

$$E(\theta) = C(\theta) = c(t)/c(0) \qquad (19.3\text{-}9)$$

where $c(0)$ is the initial concentration of tracer in the vessel, m_o/V. Alternatively, if we use equation 19.3-2 and retain $\bar{t}$,

$$E(\theta) = C(\theta) = \bar{t}c(t)\bigg/\int_0^\infty c(t)dt \qquad (19.3\text{-}10)$$

The moments corresponding to equations 19.3-7 and -8 are

$$\bar{\theta} = 1 \qquad (13.3\text{-}17)$$

$$\sigma_\theta^2 = \sigma_t^2/\bar{t}^2 \qquad (13.3\text{-}18)$$

For numerical work with tracer data directly, we use the dimensional quantities $E(t)$, $\bar{t}$, and σ_t^2 as illustrated in the examples in this section. We reserve use of the dimensionless quantities $E(\theta)$, $\bar{\theta}$, and σ_θ^2 for use with the models in Section 19.4 below.

Obtaining $E(t)$, $\bar{t}$, and σ_t^2 from experimental tracer data involves determining areas under curves defined continuously or by discrete data. The most sophisticated approach involves the use of E-Z Solve or equivalent software to estimate parameters by nonlinear regression. In this case, standard techniques are required to transform experimental concentration versus time data into $E(t)$ or $F(t)$ data; the subsequent parameter estimation is based on nonlinear regression of these data using known expressions for $E(t)$ and $F(t)$ (developed in Section 19.4). In the least sophisticated approach, discrete data, generated directly from experiment or obtained from a continuous response curve, are treated using the approximate (summation) forms of equations 19.3-4 (with 19.3-6), -7, and -8. The latter treatment lends itself to spreadsheet analysis, which we illustrate in the examples to follow.

We describe two simple ways in which discrete data may be treated to obtain the required areas: (1) use of the data in histogram form, and (2) use of the trapezoid rule. These are illustrated in Figures 19.6 and .7, respectively, in which 10 data points are plotted to represent a response curve. The curve drawn in each case is unnecessary for the calculations, but is included to indicate features of the approximations used.

In the histogram method, Figure 19.6, the area under the response curve is the sum of the rectangular areas whose heights are the n_c individual measured responses $c_i(t)$:

$$\text{area} = \sum_{i=1}^{n_c} c_i(t)\Delta t_i \qquad (19.3\text{-}11)$$

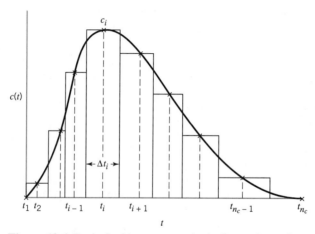

Figure 19.6 Basis for histogram method of area determination, by equations 19.3-11 to -14, based on points marked x (curve unnecessary—drawn for illustration only)

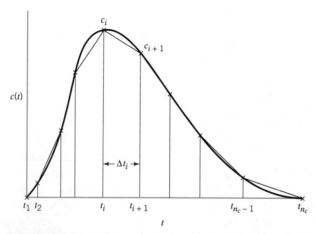

Figure 19.7 Basis for trapezoidal method of area determination, by equation 19.3-15, based on points marked x (curve unnecessary—drawn for illustration only)

19.3 Characterization of Nonideal Flow in Terms of RTD

where Δt_i is a time interval (the width of the ith rectangle) defined by

$$\Delta t_1 = \frac{t_1 + t_2}{2} \quad (i = 1) \quad (19.3\text{-}12)$$

$$\Delta t_i = \frac{(t_{i+1} - t_i) + (t_i - t_{i-1})}{2} = \frac{t_{i+1} - t_{i-1}}{2} \quad (i = 2 \text{ to } n_c - 1) \quad (19.3\text{-}13)$$

$$\Delta t_{n_c} = t_{n_c} - t_{n_c - 1} \quad (i = n_c) \quad (19.3\text{-}14)$$

$\bar{t}$ and σ_t^2 are obtained by means of the discrete forms of equations 19.3-7 and -8, respectively (see Example 19-3, below).

In the use of the trapezoid rule, Figure 19.7, the area under the response curve is given by

$$\text{area} = \sum_{i=1}^{n_c - 1} c_i(t)_{av} \Delta t_i = \sum_{i=1}^{n_c - 1} \left(\frac{c_i + c_{i+1}}{2} \right)(t_{i+1} - t_i) \quad (19.3\text{-}15)$$

Similarly, if the trapezoid rule is applied to evaluate $\bar{t}$ and σ_t^2, equations 19.3-7 and -8 become

$$\bar{t} = \sum_{i=1}^{n_c - 1} [t_i E_i(t)]_{av} \Delta t_i \quad (19.3\text{-}7a)$$

$$\sigma_t^2 = \left\{ \sum_{i=1}^{n_c - 1} [t_i^2 E_i(t)]_{av} \Delta t_i \right\} - \bar{t}^2 \quad (19.3\text{-}8a)$$

where the "av" quantities are defined as in equation 19.3-15.

EXAMPLE 19-1

In an effort to determine the cause of low yields from a reactor, a tracer study was conducted. An amount $m_o = 3.80$ kg of an inert tracer A was injected into the feed port of the 1.9-m³ reactor. The volumetric flow rate was constant at $q_o = 3.1$ L s^{-1}. The following tracer-response data were acquired:

t/min:	0.0	0.2	0.4	2.0	4.0	10.0	20.0	40.0	60.0	80.0
c_A/kg m^{-3}:	2.00	1.96	1.93	1.642	1.344	0.736	0.268	0.034	0.004	0.000

Calculate values of $E(t)$, $\bar{t}$, and σ_t^2 for flow through the vessel.

SOLUTION

Since the response data are discrete, we may use either the histogram method or the trapezoid rule. We use the trapezoid rule for illustration, but both methods give the same results.

Table 19.1 shows the calculations, in spreadsheet form, for $E(t)$ (from summation form of equation 19.3-4, with $C(t) = E(t)$, and using equation 19.3-15), $\bar{t}$ (equation 19.3-7a), and σ_t^2 (equation 19.3-8a). (Subscript i has been omitted from the headings for brevity.)

The sum from column (6) may be used to check the material balance on the tracer from equation 19.3-3:

$$m_o/q_o = 3.80/[3.1(60)/1000] = 20.4 \text{ kg min m}^{-3}$$

Table 19.1 Spreadsheet calculations for Example 19-1, using the trapezoid rule

(1)	(2)	(3)	(4)	(5)	(6)	(7)	(8)	(9)	(10)	(11)
i	$t/$ min	$c/$ kg m^{-3}	$c_{av}/$	$\Delta t/$ min	$c_{av}\Delta t/$ kg min m^{-3}	$E(t)/$ min^{-1} $= (3)/\sum(6)$	$[tE(t)]_{av}$	$(8)\Delta t/$ min	$[t^2 E(t)]_{av}$	$(10)\Delta t/$ min^2
1	0.0	2.000				0.0938				
			1.980	0.2	0.396		0.00919	0.002	0.002	0.00
2	0.2	1.960				0.0919				
			1.945	0.2	0.389		0.02729	0.005	0.009	0.00
3	0.4	1.930				0.0905				
			1.786	1.6	2.858		0.0951	0.152	0.161	0.26
4	2.0	1.642				0.0770				
			1.493	2.0	2.986		0.203	0.406	0.658	1.32
5	4.0	1.344				0.0630				
			1.040	6.0	6.240		0.2985	1.791	2.229	13.38
6	10.0	0.736				0.0345				
			0.502	10.0	5.020		0.2982	2.982	4.245	42.38
7	20.0	0.268				0.0126				
			0.151	20.0	3.020		0.1575	3.151	3.792	75.77
8	40.0	0.034				0.0016				
			0.019	20.0	0.380		0.0375	0.750	1.614	32.26
9	60.0	0.004				0.0002				
			0.002	20.0	0.040		0.0057	0.113	0.342	6.75
10	80.0	0.000				0.0000				
Σ					21.33			9.35		172.1

from column (6),

$$\sum_i c_i \Delta t_i = 21.3 \text{ kg min m}^{-3}$$

The difference is 4.4%, which may be acceptable.

Column (7) gives the value of $E(t)$ calculated at each of the ten points.

The sum of column (9) gives $\bar{t} = 9.35$ min. This may be compared with $V/q_o = 1.9/[3.1(60)/1000] = 10.21$ min. The difference of 8.3% may indicate some stagnation within the vessel to account for a lower-than-expected yield. Put another way, the effective volume is less than 1.9 m^3. The difference may also indicate that more frequent sampling is needed.

The sum of column (11) gives the first term in equation 19.3-8. Therefore,

$$\sigma_t^2 = 172.1 - 9.35^2 = 84.7 \text{ min}^2$$

By comparison, estimation of $\bar{t}$ and σ_t^2 by nonlinear regression (file ex19-1.msp) leads to the following values: $\bar{t} = 9.9$ min, and $\sigma_t^2 = 97.7$ min^2. The only way to determine which parameter set is more "correct" is to predict the experimental concentrations using these parameters in an appropriate mixing model. This procedure is explained in Section 19.4.

19.3.2.4 RTD Analysis from Step Input

As discussed in Section 19.3.2.1, a step change (increase or decrease) in inlet concentration of a tracer to a vessel causes the outlet concentration to change over time from an initial steady-state level (say, c_{A1}) to a second steady-state level (c_{A2}). As shown in Figure 19.4(a), c_{A1} may be zero for a step increase, or, as shown in Figure 19.5(a), c_{A2} is zero for a washout step decrease. It is assumed that the transition in inlet concentration from the initial to the final steady state is instantaneous, as in the opening or closing of a valve. In this section we consider only a step increase, but a washout may be treated correspondingly.

For a step change, a material-balance criterion, analogous to equation 19.3-2 for a pulse input, is that the *steady-state* inlet and outlet tracer concentrations must be equal, both before and after the step change. Then, it may be concluded that the response of the system is linear with respect to the tracer, and that there is no loss of tracer because of reaction or adsorption.

Unlike the response to a pulse input, which is related to $E(t)$, the response to a step increase is related to $F(t)$. The normalized response, which is equal to $F(t)$, is obtained as follows. Consider, for simplicity, a step-change in tracer A from $c_{A,in} = 0$ to $c_{A,in} = c_o$. Then

(1) Fraction of outlet stream that is tracer $= c(t)/c_o$ where $c(t)$ is the response in units consistent with those of c_o. This result is consistent with fraction $= 0$ for no tracer stream in outlet, and fraction $= 1$ for the outlet, being all tracer stream;

Also, from the definition of $F(t)$,

(2) Fraction of outlet stream of age $\leq t = F(t)$.
(3) Only material from the tracer stream is of age $\leq t$ (no tracer has been in the vessel longer than t).
(4) It follows that

$$F(t) = c(t)/c_o = \mathbf{C}_F(t) \tag{19.3-16}$$

where $\mathbf{C}_F(t)$ is the normalized response analogous to $\mathbf{C}(t)$ for a pulse input. Since $F(t) = F(\theta)$, equation 13.3-9,

$$F(\theta) = c(t)/c_o = \mathbf{C}_F(\theta) \tag{19.3-17}$$

More generally, if the increase in tracer concentration is from c_1 to c_2,

$$F(t) = \frac{c(t) - c_1}{c_2 - c_1} \tag{19.3-18}$$

The normalized response data may be converted to $E(t)$, if desired, since

$$E(t) = dF(t)/dt \tag{13.3-11}$$

The results are sensitive to the differencing technique used to approximate $dF(t)/dt$. The two most common methods are backward differencing:

$$E(t) = \frac{dF(t)}{dt} = \frac{F_i - F_{i-1}}{t_i - t_{i-1}}; \quad i = 2, 3, \ldots \tag{19.3-19}$$

and central differencing:

$$E(t) = \frac{dF(t)}{dt} = \frac{F_{i+1} - F_{i-1}}{t_{i+1} - t_{i-1}}; \quad i = 2, 3, \ldots \tag{19.3-20}$$

464 Chapter 19: Nonideal Flow

Differences between equations 19.3-19 and 19.3-20 are most significant if samples are collected infrequently. Ultimately, if they lead to substantially different estimates of $\bar{t}$ and σ_t^2, it is necessary to verify the results using an appropriate mixing model. Example 19-2 illustrates the method for evaluating $\bar{t}$ and σ_t^2 from a step response, using both central and backward differencing.

EXAMPLE 19-2

A step increase in the concentration of helium (tracer A), from 1.0 to 2.0 mmol L^{-1}, was used to determine the mixing pattern in a fluidized-bed reactor. The response data were as follows:

t/min:	0	5	10	15	20	30	45	60	90	120	150
$c_{A,out}$/mmol L^{-1}:	1.00	1.005	1.02	1.06	1.20	1.41	1.61	1.77	1.92	1.96	2.00

Determine $F(t)$, $E(t)$, $\bar{t}$, and σ_t^2 for flow through the vessel, using both backward and forward differencing for $E(t)$, and calculating $\bar{t}$ and σ_t^2 from $E(t)$

SOLUTION

The results for backward differencing are given in Table 19.2 in the form of a spreadsheet analysis, analogous to that in Example 19-1. Column (4) gives $F(t)$ calculated from

Table 19.2 Spreadsheet analysis for Example 19-2 using backward differencing

(1)	(2)	(3)	(4)	(5)	(6)	(7)	(8)	(9)
i	t/ min	c/ mmol L^{-1}	$F(t)$ 19.3-18	$E(t)$/ min^{-1} 19.3-19	$\Delta F(t)$	Δt/ min	$[tE(t)]_{av}\Delta t$/ min	$[t^2 E(t)]_{av}\Delta t$/ min^2
1	0	1.000	0.000	—				
					0.005	5	0.01	0.1
2	5	1.005	0.005	0.0010				
					0.015	5	0.09	0.8
3	10	1.020	0.020	0.0030				
					0.040	5	0.38	5.3
4	15	1.060	0.060	0.0080				
					0.140	5	1.70	32.5
5	20	1.200	0.200	0.0280				
					0.210	10	5.95	150.5
6	30	1.410	0.410	0.0210				
					0.200	15	9.23	344.2
7	45	1.610	0.610	0.0133				
					0.160	15	9.30	490.5
8	60	1.770	0.770	0.0107				
					0.150	30	16.35	1183.5
9	90	1.920	0.920	0.0050				
					0.040	30	9.15	895.5
10	120	1.960	0.960	0.0013				
					0.040	30	5.40	738.0
11	150	2.000	1.000	0.0013				
Σ					1.000		57.6	3840.9

19.3 Characterization of Nonideal Flow in Terms of RTD

Table 19.3 Spreadsheet analysis for Example 19-2 using central differencing (columns (2), (3), (4), (6), and (7) are the same as in Table 19.2)

(1)	(5)	(8)	(9)
i	$E(t)/$ min^{-1} 19.3-20	$[tE(t)]_{av}\Delta t/$ min	$[t^2 E(t)]_{av}\Delta t/$ min^2
1	—		
		0.03	0.1
2	0.0020		
		0.16	1.5
3	0.0055		
		0.81	11.5
4	0.0180		
		1.84	33.5
5	0.0233		
		4.79	120.5
6	0.0164		
		7.74	293.0
7	0.0120		
		7.15	368.3
8	0.0069		
		10.48	756.7
9	0.0032		
		6.67	672.7
10	0.0013		
		2.50	303.3
11	—		
Σ		42.18	2561.1

equation 19.3-18, and column (5) gives $E(t)$ from equation 19.3-19. The sum of Column (6) merely confirms that the material balance on the tracer is correct. Columns (8) and (9) are used to calculate $\bar{t}$ and σ_t^2, respectively, by means of the trapezoid rule, as in Example 19-1; only the final quantity required is tabulated. From Table 19.2,

$$\bar{t} = 57.6 \text{ min}$$
$$\sigma_t^2 = 3841 - (57.6)^2 = 529 \text{ min}^2$$

The results for central differencing are given similarly in Table 19.3. Only the columns which are different from those in Table 19.2 are shown. Thus, $F(t)$ is the same as in Table 19.2, but $E(t)$, $\bar{t}$, and σ_t^2 are different. From Table 19.3,

$$\bar{t} = 42.2 \text{ min}$$
$$\sigma_t^2 = 2561 - (42.2)^2 = 780 \text{ min}^2$$

Estimation of $\bar{t}$ and σ_t^2 by nonlinear regression (file ex19-2.msp) leads to the following values: $\bar{t} = 43.1$ min, and $\sigma_t^2 = 715$ min^2. The significant variation between the estimates of $\bar{t}$ and σ_t^2 from the central differencing and backward differencing methods is mainly due to infrequent sampling. The sampling frequency is less likely to affect the results obtained from nonlinear regression, provided that enough samples have been collected to

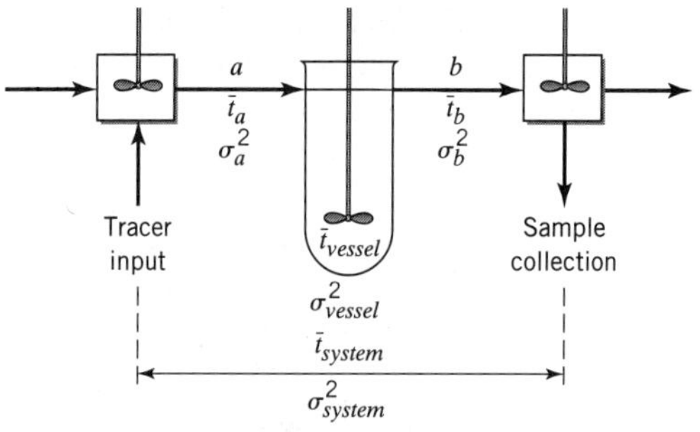

466 Chapter 19: Nonideal Flow

Figure 19.8 Time delays in a vessel system

characterize properly the tracer profile. To determine which method provides the "correct" estimate of mixing behavior (and $\bar{t}$), it is necessary to compare the experimental data with tracer curves predicted from the values of $\bar{t}$ and σ_t^2, using one of the RTD models discussed in Section 19.4 (see Example 19-9 below).

19.3.2.5 Time Delays; Nonideal Tracer Behavior

Occasionally, either because of problems in achieving a perfect pulse or step input, or because of restrictions on where the tracer may be injected or samples collected, the outlet concentrations must be adjusted prior to data analysis and calculation of $E(t)$. Problems with tracer studies often arise because of imperfect tracer delivery, or because of time delays in the pipe between the vessel and the tracer injection and sampling ports. Ideally, the distances denoted by a and b in Figure 19.8 are small, implying that the time spent by fluid in these sections is negligibly small compared with the time spent in the vessel. If this assumption does not hold, the experimental data must be corrected such that they represent the behavior of the vessel alone. If plug flow within the pipes surrounding the vessel is assumed, the sampling times (which represent the pipes and vessel) need to be adjusted by subtracting the time delay in the pipe, since

$$\bar{t}_{system} = \bar{t}_{vessel} + \bar{t}_a + \bar{t}_b$$

If it is suspected (or known) that the plug flow assumption does not hold, a separate tracer study is needed to characterize the flow distribution within the pipes. These data are then used to adjust both the concentration and the sampling time, as required. If the nature of flow and mixing in the vessel is independent of the flow characteristics in the pipes, then the σ^2 curve for the vessel may be calculated from

$$\sigma_{system}^2 = \sigma_{vessel}^2 + \sigma_a^2 + \sigma_b^2$$

19.3.2.5.1 Reactive Tracer.
If the tracer is reactive, the measured concentrations reflect both mixing characteristics and decay of the tracer. Therefore, the data must be adjusted, such that the residence time distribution within the reactor can be obtained. Example 19-3 illustrates how to adjust the data following a step input of a reactive tracer.

EXAMPLE 19-3

A pulse study was performed by adding 16.0 mg of a reactive tracer, A, which has a half-life of 25 s, to an inlet stream flowing at 0.625 L s^{-1}. Determine $E(t)$, $\bar{t}$, and σ_t^2 from the following response data:

t/s:	0	1	5	10	15	20	25	30	35	40	50	60	80
c_A/mg L^{-1}:	0	0.25	0.38	0.55	0.58	0.51	0.39	0.22	0.11	0.05	0.02	0.01	0.00

19.3 Characterization of Nonideal Flow in Terms of RTD

Table 19.4 Spreadsheet calculations for Example 19-3, using the histogram method

(1)	(2)	(3)	(4)	(5)	(6)	(7)	(8)	(9)
i	$t/$ s	$c_A/$ mg L^{-1}	$c_{adj}/$ mg L^{-1} eq. (A)	$\Delta t/$ s	$c_{adj}\Delta t/$ mg L^{-1} s	$E(t)/$ s^{-1} = (4)/$\sum$(6)	$tE(t)\Delta t/$ s	$t^2 E(t)\Delta t/$ s^2
1	0	0.00	0.000	0.5	0.00	0.0000	0.00	0.0
2	1	0.25	0.257	2.5	0.64	0.0100	0.03	0.0
3	5	0.38	0.436	4.5	1.96	0.0170	0.38	1.9
4	10	0.55	0.726	5.0	3.63	0.0283	1.42	14.2
5	15	0.58	0.879	5.0	4.40	0.0342	2.57	54.7
6	20	0.51	0.888	5.0	4.44	0.0346	3.46	69.2
7	25	0.39	0.780	5.0	3.90	0.0304	3.80	95.0
8	30	0.22	0.505	5.0	2.53	0.0197	2.96	88.8
9	35	0.11	0.290	5.0	1.45	0.0113	1.98	69.3
10	40	0.05	0.152	7.5	1.14	0.0059	1.77	70.8
11	50	0.02	0.080	10.0	0.80	0.0031	1.55	77.5
12	60	0.01	0.053	15.0	0.80	0.0021	1.89	113.4
13	80	0.00	0.000	20.0	0.00	0.0000	0.00	0.0
$\sum$					25.69		21.81	654.8

SOLUTION

Since the tracer reacts as it moves through the vessel, the measured concentration of tracer at the outlet is less than it otherwise would be. The adjusted concentration, c_{adj}, is obtained by taking the first-order-decay into account at each time: $(-r_A) = k_A c_A$. From equation 3.4-14,

$$k_A = \ln 2 / t_{1/2} = \ln 2 / 25 = 0.0277 \text{ s}^{-1}$$

Then, from equation 3.4-10,

$$c_A = c_{adj} e^{-k_A t}$$

or

$$c_{adj} = c_A e^{k_A t} = c_A e^{0.0277 t} \quad \text{(A)}$$

Table 19.4 gives the calculations of $E(t)$, $\bar{t}$, and σ_t^2 based on the histogram method. Column (4) lists the values of c_{adj} calculated from (A). Column (5) gives Δt, required for the calculations in subsequent columns, from equations 19.3-12 to -14. Column (6) gives values for the tracer material-balance (see below). Column (7) gives values of $E(t)$ from equation 19.3-4 with $C(t) = E(t)$. Columns (8) and (9) give values required for the calculation of $\bar{t}$ in equation 19.3-7, and of σ_t^2 in 19.3-8, respectively.

For the tracer material balance from equation 19.3-3:

$$m_o/q_o = 16.0/0.625 = 25.6 \text{ mg s L}^{-1}$$

From the sum of column (6):

$$\sum c_{adj} \Delta t = 25.7 \text{ mg s L}^{-1}$$

From the sum of column (8):

$$\bar{t} = 21.8 \text{ s}$$

From this and the sum of column (9):

$$\sigma_t^2 = 654.8 - (21.8)^2 = 180 \text{ s}^2$$

Note that, if the data were not adjusted to account for decay of the tracer, the material balance criterion would not be met, and $E(t)$, $\bar{t}$, and σ_t^2 would be estimated incorrectly.

Values of $\bar{t}$ and σ_t^2 obtained using the trapezoidal rule are 21.9 s and 161.4 s^2, respectively. These values are used in Examples 19-8 and -10 below.

Estimation of $\bar{t}$ and σ_t^2 by nonlinear regression (file ex19-3.msp) leads to the following values: $\bar{t} = 22.2$ s, and $\sigma_t^2 = 165$ s^2. These results are quite close to the values predicted by the trapezoidal rule approximation; the histogram technique leads to a slightly larger estimate of σ_t^2.

19.3.2.5.2 *Tailing.* In tracer studies, the measured outlet concentrations may produce a long "tail" at times significantly greater than the mean residence time (Figure 19.9a). This long tail usually indicates a region of the vessel that is poorly accessible to the tracer, or may reflect slow release of tracer adsorbed on vessel components. In many cases, these "tail" concentrations are very low, often approaching the detection limit of the assay technique. Nevertheless, they cannot be arbitrarily eliminated from the data analysis. Alternatively, in some cases, the tracer experiment may be terminated before the tracer concentration reaches zero. Fortunately, in many cases, data in the tail region decay exponentially, which permits the development of an analytical expression to describe tracer concentrations during this time period. As a check on the assumption of exponential decay, the response data may be plotted as $\ln(c_{A,out})$ versus t (Figure 19.9b); the presence of linearity indicates a tail region may be represented by an exponential function. In Figure 19.9, the tail region is shown to begin at time t_T.

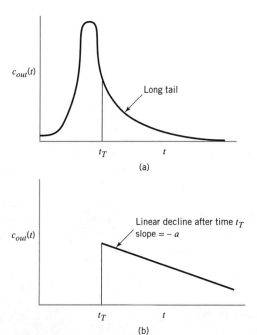

Figure 19.9 Tailing of tracer response: (a) linear coordinates; (b) semi-log coordinates

19.3 Characterization of Nonideal Flow in Terms of RTD 469

The calculation procedures described above to evaluate integrals from discrete experimental response data to a pulse input may be modified to take tailing into account. This is done by dividing the data into time periods before and after t_T. The data for $t < t_T$ are treated as already described, and the data for $t > t_T$ are treated analytically based on an exponential relation. We consider various quantities and the corresponding equations and integrals in turn.

For the tracer material-balance relation, equation 19.3-3 is written as

$$m_o/q_o = \sum_{i=1}^{n_c} c_i(t)\Delta t_i = \sum_{i=1}^{n_T} c_i(t)\Delta t_i + \int_{t_T}^{\infty} c(t)dt$$

$$= \sum_{i=1}^{n_T} c_i(t)\Delta t_i + c(t_T)\int_{t_T}^{\infty} e^{-a(t-t_T)}dt$$

$$= \sum_{i=1}^{n_T} c_i(t)\Delta t_i + \frac{c(t_T)}{a} \tag{19.3-21}$$

In equation 19.3-21, n_c is the total number of data points, and n_T is the number of data points up to t_T; the final result comes from using as the equation for the "tail" in Figure 19.9(b) the form $c(t) = c(t_T)e^{-a(t-t_T)}$, where $-a$ is the slope of the semilogarithmic linear relation shown.

For the calculation of $\bar{t}$, equation 19.3-7 is similarly modified to

$$\bar{t} = \sum_{i=1}^{n_c} t_i E_i(t)\Delta t_i = \sum_{i=1}^{n_T} t_i E_i(t)\Delta t_i + \int_{t_T}^{\infty} tE(t)dt \tag{19.3-22}$$

Here, $E(t)$ comes from the response data normalized according to equation 19.3-4 (with $C(t) = E(t)$), and using the result of equation 19.3-21. The "tail" in this case refers to the normalized response data in the form of $E(t)$, but the form and slope $-a$ remain the same. The integral in equation 19.3-22 may be evaluated analytically using integration by parts:

$$\int_{t_T}^{\infty} tE(t)dt = (t_T)\int_{t_T}^{\infty} te^{-a(t-t_T)}dt = \frac{E(t_T)}{a^2}(at_T + 1) \tag{19.3-23}$$

Substitution of equation 19.3-23 in 19.3-22 gives

$$\bar{t} = \sum_{i=1}^{n_T} t_i E_i(t)\Delta t_i + \frac{E(t_T)}{a^2}(at_T + 1) \tag{19.3-24}$$

For the calculation of σ_t^2, equation 19.3-8 is similarly modified, and the final result after integration by parts is

$$\sigma_t^2 = \sum_{i=1}^{n_T} t_i^2 E_i(t)\Delta t_i + \frac{E(t_T)}{a^3}(a^2 t_T^2 + 2at_T + 2) - \bar{t}^2 \tag{19.3-25}$$

Example 19-4 illustrates the method for analyzing data with long tails.

EXAMPLE 19-4

Using the data from Example 19-1, recalculate values of $E(t)$, $\bar{t}$, and σ_t^2 by assuming that significant "tailing" begins at $t_T = 10$ min.

470 Chapter 19: Nonideal Flow

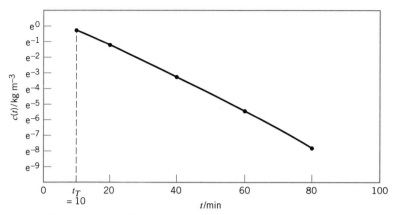

Figure 19.10 Semilogarithmic plot for Example 19-4

SOLUTION

In Figure 19.10, the experimental response data for $t \geq 10$ min are plotted as indicated in Figure 19.9(b). The semilogarithmic plot is linear with a slope $-a = -0.104$ min^{-1} (i.e., $a = 0.104$). Therefore, since $c(t_T) = c(10) = 0.736$ kg m^{-3} (from data given in Example 19-1), the tail region is represented by $c(t) = 0.736 \exp[-0.104(t - 10)]$ kg m^{-3}.

The revised calculations are given in Table 19.5, which may be compared with Table 19.1. The calculations for points 1 to 6 are the same as before, but note that the normalizing

Table 19.5 Calculations for Example 19-4

(1)	(2)	(3)	(4)	(5)	(6)	(7)	(8)	(9)	(10)	(11)
	$t/$	$c/$	$c_{av}/$	$\Delta t/$	$c_{av}\Delta t/$ kg m^{-3}	$E(t)/$ min^{-1} $= (3)/\Sigma(6)$	$[tE(t)]_{av}$	$(8)\Delta t/$	$[t^2E(t)]_{av}/$	$(10)\Delta t/$
i	min	kg m^{-3}		min	min			min	min	min^2
1	0.0	2.000				0.1000				
			1.980	0.2	0.396		0.0098	0.00	0.0020	0.00
2	0.2	1.960				0.0980				
			1.945	0.2	0.389		0.0291	0.01	0.0097	0.00
3	0.4	1.930				0.0965				
			1.786	1.6	2.858		0.1014	0.16	0.1719	0.27
4	2.0	1.642				0.0821				
			1.493	2.0	2.986		0.2165	0.43	0.7018	1.41
5	4.0	1.344				0.0672				
			1.040	6.0	6.240		0.3184	1.91	2.3776	14.30
6	10.0	0.736				0.0368				
above totals					12.87			2.51		15.98
7	20.0	0.268			from^a eqn. 19.3-21 area = 7.08	0.0134		from^b eqn. 19.3-24 area = 6.94		from^c eqn. 19.3-25 area = 168.9
8	40.0	0.034				0.0017				
9	60.0	0.004				0.0002				
10	80.0	0.000				0.0000				
Σ					20.0			9.45		184.8

$^a c(t_T)/a = 0.736/0.104 = 7.08$.
$^b E(t_T)(at_T + 1)/a^2 = 0.0368[0.104(10) + 1]/0.104^2 = 6.94$.
$^c E(t_T)(a^2t_T^2 + 2at_T + 2)/a^3 = 168.9$.

19.4 One-Parameter Models for Nonideal Flow

factor to convert $c(t)$ to $E(t)$ is 20.0 instead of 21.3. The contributions to the areas required for calculation of $E(t)$, $\bar{t}$, and σ_t^2 in columns (6), (9), and (11), respectively are from the treatment of the "tail" region by means of the equations indicated in the table.

For the tracer material balance corresponding to equation 19.3-3:

$$m_o/q_o = 20.4 \text{ kg min m}^{-3} \text{ (from Example 19-1)}$$

$$\int_0^\infty c(t)dt = 20.0 \text{ (from column(6))}$$

The difference is 2.0%, which is less than the value of 4.4% obtained in Example 19-1.

From Table 19.5, column (9), $\bar{t} = 9.45$ min, which is slightly greater than the value in Example 19-1 (9.35 min), suggesting slightly less dead space.

From column (11) in Table 19.5 and equation 19.3-25,

$$\sigma_t^2 = 184.8 - (9.45)^2 = 95.5 \text{ min}^2$$

This is somewhat different from the value in Example 19-1 (84.7 min²). However, these results for $\bar{t}$ and σ_t^2 are closer to the results obtained by nonlinear regression using E-Z Solve (9.9 min, 97.7 min²; file ex19-1.msp). One advantage of nonlinear regression is that it automatically accounts for the "long tail," without requiring additional treatment of the data.

Apart from the material balance on the tracer, we cannot tell, from the data provided, which set of values for $E(t)$, $\bar{t}$, and σ_t^2 is more correct.

19.4 ONE-PARAMETER MODELS FOR NONIDEAL FLOW

A mathematical model for nonideal flow in a vessel provides a characterization of the mixing and flow behavior. Although it may appear to be an independent alternative to the experimental measurement of RTD, the latter may be required to determine the parameter(s) of the model. The ultimate importance of such a model for our purpose is that it may be used to assess the performance of the vessel as a reactor (Chapter 20).

In this section, we develop two simple models, each of which has one adjustable parameter: the tanks-in-series (TIS) model and the axial-dispersion or dispersed-plug-flow (DPF) model. We focus on the description of flow in terms of RTD functions and related quantities. In principle, each of the two models is capable of representing flow in a single vessel between the two extremes of BMF and PF.

19.4.1 Tanks-in-Series (TIS) Model

19.4.1.1 E for TIS Model

The tanks-in-series (TIS) model for arbitrary flow through a vessel assumes that it can be represented by flow through a series of N equal-sized stirred tanks with BMF in each tank (Figure 19.11). N is the adjustable parameter of the model. Its value describes the degree of mixing in the vessel. As N increases from 1 to ∞, the flow represented in the vessel varies from BMF to PF; in fact, PF is virtually achieved for $N > 30$. The volume of the vessel modeled, V, is equal to the sum of the volumes of the N tanks:

$$V = NV_i; \quad i = 1, 2, \ldots, N \qquad (19.4\text{-}1)$$

472 Chapter 19: Nonideal Flow

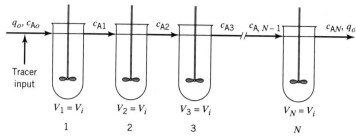

Figure 19.11 Tanks-in-series (TIS) Model

which also implies that

$$V_1 = V_2 = \ldots = V_i = \ldots = V_N \quad (19.4\text{-}2)$$

The RTDs for the first two members of the TIS model are derived in Sections 13.4.1 ($N = 1$) and 17.2.2 ($N = 2$). Thus,

$$E_1(\theta) = e^{-\theta} \quad (N = 1) \quad (13.4\text{-}3)$$

$$E_2(\theta) = 4\theta e^{-2\theta} \quad (N = 2) \quad (17.2\text{-}5)$$

(See Figure 17.4 for a graphical comparison of these two expressions.)

We could obtain a general expression for $E_N(\theta)$ by induction by using the same method used to obtain the two equations above, and increasing N incrementally. Instead, we proceed by using the Laplace transform in a more compact procedure.

Consider the steady flow of a constant-density fluid at q_o m³ s⁻¹ through the N stirred tanks (Figure 19.11). At $t = 0$, a quantity of n_o moles of a nonreacting tracer A is introduced into the first tank as a pulse or Dirac input, $\delta(t - 0) \equiv \delta(t)$. At any subsequent time t, a material balance for tracer around the i th tank is:

$$q_o c_{A,i-1} - q_o c_{Ai} = \frac{dn_{Ai}}{dt} = V_i \frac{dc_{Ai}}{dt}$$

or

$$c_{A,i-1} - c_{Ai} = \bar{t}_i \frac{dc_{Ai}}{dt} \quad (19.4\text{-}3)$$

where

$$c_{Ai} = n_{Ai}/V_i \quad (i = 1, 2, \ldots, N) \quad (19.4\text{-}4)$$

and

$$\bar{t}_i = V_i/q_o \quad (i = 1, 2, \ldots, N) \quad (19.4\text{-}5)$$

n_{Ai} is the number of moles of tracer in the i th tank, and $\bar{t}_i$ is the mean residence time of fluid in the i th tank ($\bar{t}_i$ is the same for all tanks). In terms of dimensionless time θ, equation 19.4-3 becomes

$$c_{A,i-1} - c_{Ai} = \frac{1}{N} \frac{dc_{Ai}}{d\theta} \quad (19.4\text{-}6)$$

19.4 One-Parameter Models for Nonideal Flow

since

$$\theta = t/\bar{t} = t/(V/q_o) = t/(NV_i/q_o) = t/N\bar{t}_i \quad \text{(19.4-7)}$$

and dt in 19.4-3 is replaced by

$$dt = N\bar{t}_i d\theta \quad \text{(19.4-8)}$$

The Laplace transformation of equation 19.4-6 is

$$L(c_{A,i-1}) - L(c_{Ai}) = \frac{1}{N}[sL(c_{Ai}) - c_{Ai}(0)] = \frac{s}{N}L(c_{Ai}) \quad \text{(19.4-9)}$$

where L is the Laplace transform operator, s is the Laplace parameter, and $c_{Ai}(0) = 0$. Rearrangement of 19.4-9 results in

$$\frac{L(c_{Ai})}{L(c_{A,i-1})} = \frac{1}{1 + s/N} \quad \text{(19.4-10)}$$

Since

$$\frac{L(c_{AN})}{L(c_{Ao})} = \frac{L(c_{A1})}{L(c_{Ao})}\frac{L(c_{A2})}{L(c_{A1})} \cdots \frac{L(c_{AN})}{L(c_{A,N-1})}$$

equation 19-4-10 can be written as

$$\frac{L(c_{AN})}{L(c_{Ao})} = \frac{1}{(1 + s/N)^N} \quad \text{(19.4-11)}$$

In signal-response terms, $c_{Ao} \equiv \delta(\theta)$, the Dirac input, and $c_{AN} \equiv \mathbf{C}_{AN}$, the normalized response, $\mathbf{C}(\theta)$. For the Dirac delta function,

$$L[\delta(\theta)] = \int_0^\infty e^{-s\theta}\delta(\theta)d\theta \quad \text{(by definition of } L[\delta(\theta)])$$

$$= e^{-s\theta}\Big]_{\theta=0} \quad \text{(by equation 13.4-10)} = 1$$

Thus, equation 19.4-11 becomes

$$L(\mathbf{C}_{AN}) = 1/(1 + s/N)^N \quad \text{(19.4-12)}$$

From a table of Laplace transforms (Jenson and Jeffreys, 1963, p. 532),

$$L\left[\frac{t^{n-1}e^{at}}{(n-1)!}\right] = \frac{1}{(s-a)^n} \quad (n = 1, 2, \ldots) \quad \text{(19.4-13)}$$

If we let $n = N$, $t = \theta$, and $a = -n = -N$, equation 19.4-13 can be written as

$$L\left[\frac{\theta^{N-1}e^{-N\theta}}{(N-1)!}\right] = \frac{1}{(s+N)^N} = \frac{1}{N^N}\left(\frac{1}{1+s/N}\right)^N$$

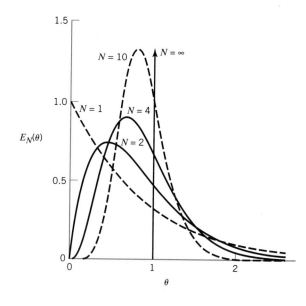

Figure 19.12 $E_N(\theta)$ for various values of N in the TIS model

Therefore,

$$\frac{1}{(1 + s/N)^N} = L(\mathbf{C}_{AN}) = L\left[\frac{N^N \theta^{N-1} e^{-N\theta}}{(N-1)!}\right]$$

and, since $\mathbf{C}_{AN} = E_N(\theta)$, from equation 19.3-9,

$$E_N(\theta) = \frac{N^N \theta^{N-1} e^{-N\theta}}{(N-1)!} \quad (N = 1, 2, \ldots, N) \tag{19.4-14}$$

The behavior of $E_N(\theta)$ as a function of N and θ is shown in Figure 19.12. As N increases from 1, the spread of the distribution decreases, and, for $N > 2$, the height of the peak increases, until for $N = \infty$, the behavior is that for PF, with zero spread and infinite peak height (the proof of this last is left to problem 19-19). The location and value of $E_N(\theta)_{max}$, and the location and spread of $E_N(\theta)$ are established quantitatively in the next two examples.

EXAMPLE 19-5

In the TIS model, for a given value of N, what is the maximum value of $E_N(\theta)$, and at what value θ_{max} does it occur?

SOLUTION

From equation 19.4-14,

$$\frac{\partial E_N(\theta)}{\partial \theta} = \frac{N^N}{(N-1)!}\left[(N-1)\theta^{N-2} e^{-N\theta} - N\theta^{N-1} e^{-N\theta}\right] = 0$$

for $E_N(\theta)_{max}$. From this,

$$(N-1)\theta_{max}^{N-2} = N\theta_{max}^{N-1}$$

19.4 One-Parameter Models for Nonideal Flow

or

$$\theta_{max} = (N - 1)/N \qquad (19.4\text{-}15)$$

Substituting this value into equation 19.4-14, we obtain, after simplification,

$$E_N(\theta)_{max} = \frac{N(N - 1)^{N-1}}{(N - 1)!} e^{(1-N)} \qquad (19.4\text{-}16)$$

EXAMPLE 19-6

For the TIS model, what are the mean and the variance of $E_N(\theta)$?

SOLUTION

This involves obtaining the mean-residence time, $\bar{\theta}$, and the variance, σ_θ^2, of the distribution represented by equation 19.4-14. Since, in general, these are related to the first and second moments, respectively, of the distribution, it is convenient to connect the determination of moments in the time domain to that in the Laplace domain. By definition of a Laplace transform,

$$L[E_N(\theta)] = \int_0^\infty e^{-s\theta} E_N(\theta) d\theta \qquad (19.4\text{-}17)$$

and

$$\frac{dL[E_N(\theta)]}{ds} = -\int_0^\infty \theta e^{-s\theta} E_N(\theta) d\theta \qquad (19.4\text{-}18)$$

$$\lim_{s \to 0}(-1)\frac{dL[E_N(\theta)]}{ds} = \int_0^\infty \theta E_N(\theta) d\theta = \mu_1 = \bar{\theta} \qquad (19.4\text{-}19)$$

from equation 13.3-14, where μ_1 is the first moment about the origin, which, in turn, is the location expressed as the mean residence time $\bar{\theta}$.

From equation 19.4-12, with $\mathbf{C}_{AN}$ replaced by its equivalent $E_N(\theta)$,

$$L[E_N(\theta)] = \frac{1}{(1 + s/N)^N} \qquad (19.4\text{-}20)$$

$$\frac{dL[E_N(\theta)]}{ds} = \frac{-1}{(1 + s/N)^{N+1}} \qquad (19.4\text{-}21)$$

Therefore,

$$\bar{\theta} = \lim_{s \to 0} \frac{(-1)(-1)}{(1 + s/N)^{N+1}} = 1 \qquad (19.4\text{-}22)$$

From equation 19.4-17,

$$\frac{d^2 L[E_N(\theta)]}{ds^2} = \int_0^\infty \theta^2 e^{-s\theta} E_N(\theta)\, d\theta \qquad (19.4\text{-}23)$$

$$\lim_{s \to 0} \frac{d^2 L[E_N(\theta)]}{ds^2} = \int_0^\infty \theta^2 E(\theta)\, d\theta = \mu_2 \qquad (19.4\text{-}24)$$

from equation 13.3-12, where μ_2 is the second moment *about the origin*. From equation 13.3-16a, the variance, which is the second moment *about the mean*, is related to μ_1 and μ_2 by

$$\sigma_\theta^2 = \mu_2 - \mu_1^2 \qquad (19.4\text{-}25)$$

From equations 19.4-20, -22, -24, and -25,

$$\sigma_\theta^2 = \lim_{s \to 0} \frac{d^2 L[E_N(\theta)]}{ds^2} - 1$$

$$= \lim_{s \to 0} \frac{N+1}{N} \frac{1}{(1+s/N)^{N+2}} - 1 = \frac{N+1}{N} - 1 = \frac{1}{N}$$

Thus,

$$\sigma_\theta^2 = 1/N \quad (N = 1, 2, \ldots, N) \qquad (19.4\text{-}26)$$

Note that the determination of moments in the Laplace domain in this case is much easier than in the time domain, since tedious integration of a complicated function is replaced by the easier operation of differentiation. The result for $\bar{\theta}$ we could have anticipated, but not the very simple result for σ_θ^2 in equation 19.4-26.

19.4.1.2 TIS Model for Nonintegral Values of N

In equation 19.4-14, $E_N(\theta)$ applies to integral values of N, and is thus a discrete function of N. This is a disadvantage, since it leaves "gaps" between values of N. The TIS model can be generalized (Buffham and Gibilaro, 1968) to be continuous with respect to N by using the generalized factorial for $(N - 1)!$. This is the gamma (Γ) function defined by

$$\Gamma(N) = \int_0^\infty x^{N-1} e^{-x}\, dx \qquad (N > 0) \qquad (19.4\text{-}27)$$

where x is a dummy variable. The definition involves an improper integral that must be evaluated numerically. By using equation 19.4-27 for $\Gamma(N + 1)$ and integrating by parts, we can show that

$$\Gamma(N+1) = N\Gamma(N) \qquad (19.4\text{-}28)$$

This provides a recursion formula to obtain $N!$:

$$\Gamma(N+1) = N\Gamma(N) = N(N-1)\Gamma(N-1) = \ldots$$
$$= N(N-1)(N-2)\ldots(2)(1)\Gamma(1) = N! \qquad (19.4\text{-}29)$$

since $\Gamma(1) = 1$, from integration of equation 19.4-27.

19.4 One-Parameter Models for Nonideal Flow

For the calculation of $N!$ for nonintegral N, values of the Γ function can be generated from equation 19.4-27 by a computer program such as provided by the E-Z Solve software. (In theory, the gamma function must be integrated from $x = 0$ to $x = \infty$. However, a practical upper limit is reached when $x = 50$ or 60, since the exponential term goes to zero for large values of x.) Alternatively, for hand calculation, equations 19.4-28 and -29 can be used, once values of $\Gamma(N)$ have been evaluated numerically over the range $1 < N < 2$ from equation 19.4-27 (such values are available in mathematical tables).

EXAMPLE 19-7

Calculate $5.575!$ given that $\Gamma(1.575) = 0.8909$.

SOLUTION

From equation 19.4-29,

$$5.575! = \Gamma(6.575) = (5.575)(4.575)(3.575)(2.575)(1.575)\Gamma(1.575) = 329.46$$

(The same result is obtained from the E-Z Solve software; file ex19-7.msp.)

Since, from equation 19.4-29, $(N-1)! = \Gamma(N)$, equation 19.4-14 can be generalized as a continuous function of N for any value[1] $N \geq 1$:

$$E_N(\theta) = \frac{N^N \theta^{N-1} e^{-N\theta}}{\Gamma(N)} \qquad (19.4\text{-}30)$$

Equation 19.4-30 may be adapted for use in conjunction with the nonlinear regression feature in E-Z Solve to estimate $\bar{t}$ and N for a pulse input. The experimental concentration versus time data must be transformed into $E(t)$ data, using conventional methods; these data may then be regressed with equation (19.4-30a) to obtain $\bar{t}$ and N:

$$\bar{t}E(t) = \frac{N^N (t/\bar{t})^{N-1} e^{-Nt/\bar{t}}}{\Gamma(N)} \qquad (19.4\text{-}30a)$$

19.4.1.3 Determining N from Tracer Data

We describe two approximate methods of determining the value of N in the TIS model from pulse-tracer experiments. One is based on the first moment or mean $\bar{\theta}$, and the other on the variance σ_θ^2 as determined from the tracer data.

For the first method, using the value of $E_N(\theta)$ from the $C(\theta)$ response curve at $\bar{\theta} = 1$, we have, from equations 19.4-30 and -29,

$$E_N(\bar{\theta}) = \frac{N^N e^{-N}}{\Gamma(N)} = \frac{N^N e^{-N}}{(N-1)!} = \frac{N^{N+1} e^{-N}}{N!} \qquad (19.4\text{-}31)$$

---

[1] Equation 19.4-30 actually allows any value $N \geq 0$. See Stokes and Nauman (1970) for discussion of this and a physical interpretation of nonintegral values of N for the TIS model; the case of $N < 1$ corresponds to bypassing of some entering fluid (see also Nauman and Buffham, 1983, pp. 61–2).

We may use Stirling's approximation for $N!$:

$$N! = N^N e^{-N} (2\pi N)^{1/2} \quad (19.4\text{-}32)$$

(accurate to within 1% for $N \geq 10$, to within 2% for $N \geq 6$, and to within 5% for $N \geq 2$; also applicable to both integral and nonintegral values). Combining equations 19.4-31 and -32, we obtain an expression for N:

$$N = 2\pi [E_N(\bar{\theta})]^2 \quad (19.4\text{-}33)$$

The second method uses the result for the variance obtained in Example 19-6:

$$N = 1/\sigma_\theta^2 \quad (19.4\text{-}26)$$

The first method is based on only a single point, and assumes that the model and the experimental data have the same mean. The second method is based on all the tracer data, and assumes the model and the data have the same variance.

EXAMPLE 19-8

Using the data in Example 19-3, estimate the value of N for the TIS model (a) based on the mean value, and (b) based on the variance.

SOLUTION

(a) From the results (trapezoidal rule) of Example 19-3, $\bar{t} = 21.9$ s and $E(\bar{t}) \approx 0.034^{-1}$. Thus, $E(\bar{\theta}) = \bar{t} E(\bar{t}) = 21.9(0.034) = 0.745$, and substitution in equation 19.4-33 yields

$$N = 2\pi (0.745)^2 = 3.5$$

(b) From Example 19-3, $\sigma_t^2 = 161.4$ s². Thus, from equation 13.3-18, $\sigma_\theta^2 = \sigma_t^2/\bar{t}^2 = 161.4/21.9^2 = 0.337$, and substitution in equation 19.4-26 yields

$$N = 1/0.337 = 3.0$$

The value of N obtained if the histogram method is used is 2.6. By comparison, nonlinear regression with E-Z Solve (file ex19-3.msp) predicts that $N = 3.0$.

19.4.1.4 F and W for TIS Model

The cumulative RTD function $F(\theta)$ in the TIS model is the fraction of the outlet stream from the Nth tank that is of age 0 to θ, or the probability that a fluid element that entered the first tank at $\theta = 0$ has left the Nth tank by the time θ. It can be obtained from $E_N(\theta)$ by means of the relation given in Table 13.1:

$$F(\theta) = \int_0^\theta E(\theta) \, d\theta \quad (19.4\text{-}34)$$

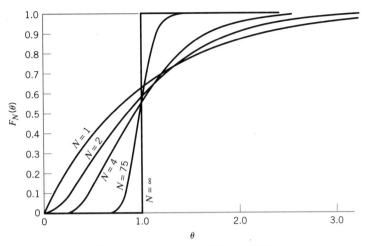

Figure 19.13 $F_N(\theta)$ for various values of N in the TIS model

From equation 19.4-14 for $E_N(\theta)$, for integral values of N,

$$F_N(\theta) = \int_0^\theta \frac{N^N \theta^{N-1} e^{-N\theta}}{(N-1)!} d\theta \tag{19.4-35}$$

The integral in equation 19.4-35 can be evaluated analytically using a recursion formula to give:

$$F_N(\theta) = 1 - e^{-N\theta} \sum_{i=0}^{N-1} \frac{N^i \theta^i}{i!} \tag{19.4-36}$$

The result is shown graphically in Figure 19.13 for several values of N. For $N = 1$, the result is that for BMF in a single vessel, as shown in Figure 13.5. As $N \to \infty$, the shape becomes that of the step function, corresponding to PF, as shown in Figures 13.7 and 19.13. The other values of N show behavior between these two extremes, corresponding to varying degrees of backmixing or dispersion. $F_N(\theta)$ in equation 19.4-36 has the same dimensionless variance as $E_N(\theta)$ (Stokes and Nauman, 1970):

$$\sigma_\theta^2[F_N(\theta)] = 1/N \tag{19.4-37}$$

For nonintegral values of N, from equation 19.4-30,

$$F_N(\theta) = \frac{1}{\Gamma(N)} \int_0^{N\theta} x^{N-1} e^{-x} dx \tag{19.4-38}$$

(Stokes and Nauman, 1970), where the integral is an *incomplete* gamma function (the upper limit is finite) that can be evaluated by the E-Z Solve software (file ex19-7.msp). It has the same variance as given in 19.4-37. Equation 19.4-38 can also be used to estimate $\bar{t}$ and N from tracer data obtained by a step input, using the nonlinear regression capabilities of E-Z Solve (see Example 19-9 for further discussion of this technique).

The washout function, $W_N(\theta)$, can be obtained from $F_N(\theta)$ in equation 19.4-36 and -38 by means of the relation given in Table 13.1:

$$W_N(\theta) = 1 - F_N(\theta) \quad (19.4\text{-}39)$$

$W_N(\theta)$ in the TIS model is the fraction of the outlet stream from the Nth tank that is of age $> \theta$, or the probability that an element of fluid that entered the first tank at $\theta = 0$ has *not* left the Nth tank by the time θ.

19.4.1.5 Concentration Profiles from the TIS Model

Occasionally, various methods for evaluating tracer data and for estimating the mixing parameter in the TIS model lead to different estimates for $\bar{t}$ and N. In these cases, the accuracy of $\bar{t}$ and N must be verified by comparing the concentration-versus-time profiles predicted from the model with the experimental data. In general, the predicted profile can be determined by numerically integrating N simultaneous ordinary differential equations of the form:

$$q_o(c_{Ai-1} - c_{Ai}) = V_i \, dc_{Ai}/dt \quad (19.4\text{-}40)$$

The value of c_{Ao} depends upon the input function (whether step or pulse), and the initial condition ($c_{Ai}(0)$) for each reactor must be specified. For a pulse input or step increase from zero concentration, $c_{Ai}(0)$ is zero for each reactor. For a washout study, $c_{Ai}(0)$ is nonzero (Figure 19.5b), and c_{Ao} must equal zero. For integer values of N, a general recursion formula may be used to develop an analytical expression which describes the concentration transient *following a step change*. The following expressions are developed based upon a step increase from a zero inlet concentration, but the resulting equations are applicable to all types of step inputs.

Setting $\bar{t} = NV_i/q_o$, and $c_{Ai}(0) = 0$ (initial condition), and solving 19.4-40, we obtain:

$$c_{Ai}(t) = e^{-Nt/\bar{t}} \int_0^t N c_{A,i-1}(t) \bar{t}^{-1} e^{-Nt/\bar{t}} \, dt \quad (19.4\text{-}41)$$

This integral may be evaluated for each tank in the network. For the first tank, where the inlet concentration is c_{Ao}:

$$\frac{c_{A1}(t)}{c_{Ao}} = 1 - e^{-Nt/\bar{t}} \quad (19.4\text{-}42)$$

By combining 19.4-41 and 19.4-42, we may relate the concentration in the second tank ($c_{A2}(t)$) to the inlet concentration:

$$c_{A2}(t) = \frac{2c_{Ao} e^{-2t/\bar{t}}}{\bar{t}} \int_0^t (1 - e^{-2t/\bar{t}}) e^{-2t/\bar{t}} \, dt \quad (19.4\text{-}43)$$

which, upon integration produces:

$$\frac{c_{A2}(t)}{c_{Ao}} = 1 - e^{-2t/\bar{t}}(1 + 2t/\bar{t}) \quad (19.4\text{-}44)$$

For N equal-sized tanks in series, the concentration at any time within tank i may be determined from:

$$\frac{c_{Ai}(t)}{c_{Ao}} = 1 - e^{-\alpha}[1 + \alpha + \alpha^2/2! + \alpha^3/3! + \ldots + \alpha^{i-1}/(i-1)!] \quad (19.4\text{-}45)$$

where $\alpha = Nt/\bar{t}$.

19.4 One-Parameter Models for Nonideal Flow

Note that if $i = N$, equation 19.4-45 generates the response profile for the entire network. Furthermore, from the definition of $F(t)$ for a step change:

$$F(t) = \frac{(c_{out}(t) - c_1)}{(c_2 - c_1)} = \frac{c_{AN}(t)}{c_{Ao}} \qquad (19.3\text{-}18)$$

since the initial steady-state concentration (c_1) is zero. Equation 19.4-45 may therefore be recast as a general equation for $F(t)$, applicable to any type of step change:

$$F(t) = 1 - e^{-\alpha}\left[1 + \alpha + \frac{\alpha^2}{2!} + \frac{\alpha^3}{3!} + \ldots + \frac{\alpha^{N-1}}{(N-1)!}\right] \qquad (19.4\text{-}46)$$

which is the same as equation 19.4-36.

EXAMPLE 19-9

Using the data from Example 19-2, estimate values of N based on both central and backward differencing, and determine which differencing technique best describes the tracer outflow concentrations.

SOLUTION

For central differencing: $N = \bar{t}^2/\sigma_t^2 = (42.2 \text{ min})^2/780 \text{ min}^2 = 2.3$
For backward differencing: $N = \bar{t}^2/\sigma_t^2 = (57.6 \text{ min})^2/529 \text{ min}^2 = 6.3$

For simplicity of comparison, we round these values to 2 and 6, respectively.
For $N = 2$ and $\bar{t} = 42.2$ min, $F(t)$ can be calculated at each time using equation 19.4-46:

$$\alpha = Nt/\bar{t} = 2t/42.2 \text{ min}$$
$$F(t) = 1 - e^{-\alpha}(1 + \alpha) \qquad (19.4\text{-}46a)$$

For $N = 6$ and $\bar{t} = 57.6$ min, $F(t)$ can similarly be calculated at each time:

$$\alpha = Nt/\bar{t} = 6t/57.6 \text{ min}$$
$$F(t) = 1 - e^{-\alpha}\left[1 + \alpha + \frac{\alpha^2}{2!} + \frac{\alpha^3}{3!} + \frac{\alpha^4}{4!} + \frac{\alpha^5}{(5)!}\right] \qquad (19.4\text{-}46b)$$

In each case, $c_{out}(t)$ may be calculated from $F(t)$:

$$F(t) = \frac{c_{out}(t) - c_1}{(c_2 - c_1)}$$

where $c_2 = 2.00$ mmol L^{-1} and $c_1 = 1.00$ mmol L^{-1}. Therefore,

$$c_{out}(t) = c_1 + F(t)(c_2 - c_1) = 1.00 \text{ mmol L}^{-1} + F(t)$$

As an alternative, equation 19.4-38 may be solved using the E-Z Solve software to obtain the concentration profiles. The gamma function can be evaluated by numerical integration using the user-defined functions *gamma*, *f_rateqn*, and *rkint* provided within the software.

482 Chapter 19: Nonideal Flow

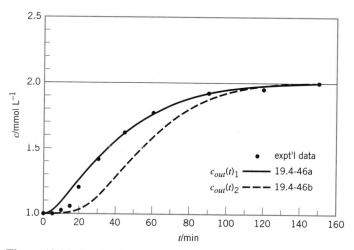

Figure 19.14 Graphical comparison for Example 19-9

Table 19.6 Results for Example 19-9

Point	t/ min	conc/ mmol L^{-1}	$F(t)_1{}^a$	$F(t)_2{}^b$	$c_{out}(t)_1$/ mmol L^{-1}	$c_{out}(t)_2$/ mmol L^{-1}	$c_{out}(t)_3$/^{c} mmol L^{-1}
1	0	1.000	0.000	0.000	1.000	1.000	1.000
2	5	1.005	0.024	0.000	1.024	1.000	1.010
3	10	1.020	0.082	0.001	1.082	1.001	1.047
4	15	1.060	0.160	0.005	1.160	1.005	1.110
5	20	1.200	0.245	0.020	1.245	1.020	1.190
6	30	1.410	0.416	0.097	1.416	1.097	1.368
7	45	1.610	0.629	0.329	1.629	1.329	1.609
8	60	1.770	0.776	0.594	1.776	1.594	1.779
9	90	1.920	0.926	0.905	1.926	1.905	1.939
10	120	1.960	0.977	0.985	1.977	1.985	1.985
11	150	2.000	0.993	0.998	1.993	1.998	1.997

^{a}From equation 19.4-46a (central).
^{b}From equation 19.4-46b (backward).
^{c}From equation 19.4-38, with $N = 2.6$ and $\bar{t} = 43.1$ min.

Figure 19.14 shows the values of $c_{out}(t)$ predicted using 19.4-46a and 19.4-46b. The central differencing technique $(c_{out}(t)_1)$ provides a superior representation of the experimental data. The individual calculated results are given in columns 6 and 7 of Table 19.6, together with those calculated using equation 19.4-38 $(c_{out}(t)_3,$ column 8) in conjunction with the best estimates of N and $\bar{t}$ obtained by nonlinear regression (file ex19-2.msp). These last provide the best representation of the three, partly because the values of N were rounded for the central-differencing and back-differencing calculations.

19.4.1.6 Comments on Methods for Parameter Estimation

Estimation of $\bar{t}$ and N by nonlinear regression has several advantages over conventional methods. First, for pulse data, uncertainties associated with use of histogram versus trapezoidal integration of discrete data are avoided. For step changes, uncertainties associated with the use of backward versus central differencing when transforming $F(t)$ to $E(t)$ data are avoided, since this transformation is not required when the nonlinear regression technique is used. Use of nonlinear regression is especially advantageous

19.4 One-Parameter Models for Nonideal Flow

if the data are "noisy," since differencing would lead to large variations in $E(t)$. Consequently, the nonlinear regression technique, using equation 19.4-30 (for $E_N(\theta)$) or 19.4-38 (for $F_N(\theta)$), has the potential to provide better estimates of $\bar{t}$ and N than are obtained by conventional techniques.

An operational issue associated with nonlinear regression is the need to evaluate the gamma function. In theory, the gamma function must be integrated from $x = 0$ to $x = $ infinity. However, a practical upper limit is reached when $x = 50$ or 60, since the exponential term is essentially zero for large values of x. The E-Z Solve software can readily evaluate the gamma function by numerical integration using the user-defined functions *gamma, f_rateeq,* and *rkint* provided within the software.

19.4.2 Axial Dispersion or Dispersed Plug Flow (DPF) Model

19.4.2.1 *Continuity Equation for DPF*

In PF, the transport of material through a vessel is by convective or bulk flow. All elements of fluid, at a particular axial position in the direction of flow, have the same concentration and axial velocity (no radial variation). We can imagine this ideal flow being "blurred" or dispersed by backmixing of material as a result of local disturbances (eddies, vortices, etc.). This can be treated as a diffusive flow superimposed on the convective flow. If the disturbances are essentially axial in direction and not radial, we refer to this as axial dispersion, and the flow as dispersed plug flow (DPF). (Radial dispersion may also be significant, but we consider only axial dispersion here.)

In considering axial dispersion as a diffusive flow, we assume that Fick's first law applies, with the diffusion or effective diffusion coefficient (equation 8.5-4) replaced by an axial dispersion coefficient, D_L. Thus, for unsteady-state behavior with respect to a species A (e.g., a tracer), the molar flux (N_A) of A at an axial position x is

$$N_A = -D_L \frac{\partial c_A(x, t)}{\partial x} \qquad (19.4\text{-}47)$$

where $c_A(x, t)$ is the concentration of A.

This diffusive flow must be taken into account in the derivation of the material-balance or continuity equation in terms of A. The result is the axial dispersion or dispersed plug flow (DPF) model for nonideal flow. It is a single-parameter model, the parameter being D_L or its equivalent as a dimensionless parameter. It was originally developed to describe relatively small departures from PF in pipes and packed beds, that is, for relatively small amounts of backmixing, but, in principle, can be used for any degree of backmixing.

Consider a material balance for A around the differential control volume shown in Figure 19.15. We assume steady flow overall, but not with respect to (tracer) A; there is no reaction taking place. In words, we write this as:

$$\begin{pmatrix} \text{rate of input} \\ \text{of A by} \\ \text{convective flow} \end{pmatrix} + \begin{pmatrix} \text{rate of input} \\ \text{of A by} \\ \text{dispersion} \end{pmatrix} - \begin{pmatrix} \text{rate of output} \\ \text{of A by} \\ \text{convective flow} \end{pmatrix} - \begin{pmatrix} \text{rate of output} \\ \text{of A by} \\ \text{dispersion} \end{pmatrix}$$

$$= \begin{pmatrix} \text{rate of accumulation} \\ \text{of A in volume} \\ dV = A_c\, dx \end{pmatrix}$$

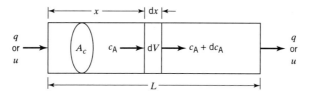

Figure 19.15 Control volume for continuity equation for axial dispersion model

That is, with $c_A \equiv c_A(x, t)$, and A_c and D_L constant,

$$A_c u c_A - D_L A_c \frac{\partial c_A}{\partial x} - A_c \left[u c_A + \frac{\partial(u c_A)}{\partial x} \right] + D_L A_c \left[\frac{\partial c_A}{\partial x} + \frac{\partial}{\partial x}\left(\frac{\partial c_A}{\partial x}\right) dx \right] = A_c \frac{\partial c_A}{\partial t} dx \quad (19.4\text{-}48)$$

If, in addition, the system has a constant density (u is constant), equation 19.4-48 simplifies to

$$D_L \frac{\partial^2 c_A}{\partial x^2} - u \frac{\partial c_A}{\partial x} = \frac{\partial c_A}{\partial t} \quad (19.4\text{-}49)$$

We may put 19.4-49 in nondimensional form with respect to t and x by letting $\theta = t/\bar{t} = t/(L/u) = (u/L)t$ and $z = x/L$. Then the continuity equation becomes

$$\frac{1}{\text{Pe}_L} \frac{\partial^2 c_A}{\partial z^2} - \frac{\partial c_A}{\partial z} = \frac{\partial c_A}{\partial \theta} \quad (19.4\text{-}50)$$

where Pe_L is a dimensionless Peclet number based on vessel length, defined by[2]

$$\text{Pe}_L = uL/D_L \quad (19.4\text{-}51)$$

Pe_L can be used in place of D_L as the single parameter of the axial dispersion model. The physical interpretation of Pe_L is that it represents the ratio of the convective flux to the diffusive (dispersed) flux:

$$\text{Pe}_L = \frac{uL}{D_L} \equiv \frac{u c_A}{D_L(c_A/L)} \equiv \frac{\text{convective flux}}{\text{dispersed flux}} \quad (19.4\text{-}52)$$

The extreme values for Pe_L represent PF and BMF:

$$\lim_{D_L \to 0} \text{Pe}_L = \lim_{D_L \to 0} \frac{uL}{D_L} = \infty \quad \text{(for negligible dispersion (PF))} \quad (19.4\text{-}53)$$

$$\lim_{D_L \to \infty} \text{Pe}_L = \lim_{D_L \to \infty} \frac{uL}{D_L} = 0 \quad \text{(for very large dispersion)} \quad (19.4\text{-}54)$$

In equation 19.4-50, c_A may be replaced by $\mathbf{C}(\theta)$, the normalized response to a Dirac delta (pulse) tracer input at the vessel outlet ($z = 1$); the normalizing factor to convert

---

[2] There is controversy over the naming of the dimensionless group uL/D_L (or its reciprocal), and, in particular, over naming it a Peclet number (see discussion by Weller, 1994).

c_A to $C(\theta)$ cancels term-by-term. In turn, $C(\theta)$ may be replaced by $E(\theta)$, from equation 19.3-9. Thus, any solution of this equation generates $E(\theta)$ for the vessel.

The method and results of solution of equation 19.4-50 depend on the choice of boundary conditions at inlet ($z = 0$) and outlet ($z = 1$). There are several possibilities with respect to "closed" and "open" vessels (Section 13.2.4), which do not necessarily satisfy actual conditions. A closed boundary condition at a point (inlet or outlet) implies that the vessel is isolated (in the sense of communication) from connecting piping at that point. An open boundary condition, conversely, presumes that the same flow distribution and mixing behavior occur in both the vessel and the piping at that point. It also implies that any tracer injected at the inlet can appear upstream of the inlet. Since, in many cases, PF occurs in the process piping, differences between the two types of boundary conditions are most prominent if there is significant backmixing within the vessel. In the next section, we consider solutions in terms of $E(\theta)$ arising from these different boundary conditions, after introducing a solution for a relatively small amount of dispersion that is independent of these boundary conditions.

19.4.2.2 Solutions of Continuity Equation for DPF

19.4.2.2.1 Small Amount of Dispersion.
For a relatively small amount of dispersion or relatively large value of Pe_L, we may assume $E(\theta)$ is approximately symmetrical or "normal" in the Gaussian sense, as depicted in Figure 19.16.

The Gaussian or "normal" solution to equation 19.4-50 with c_A replaced by $E(\theta)$ is (Levenspiel, 1972, pp. 273–5):

$$E(\theta) = E(\theta)_{max} \exp[-Pe_L(1 - \theta)^2/4] \qquad (19.4\text{-}55)$$

where $E(\theta)_{max}$ is the response peak height given by

$$E(\theta)_{max} = (1/2)(Pe_L/\pi)^{1/2} \qquad (19.4\text{-}56)$$

For this solution, moments are

$$\bar{\theta} = 1 \qquad (19.4\text{-}57)$$
$$\sigma_\theta^2 = 2/Pe_L \qquad (19.4\text{-}58)$$

Equations 19.4-56 and -58 provide two methods of estimating Pe_L for small dispersion from tracer data.

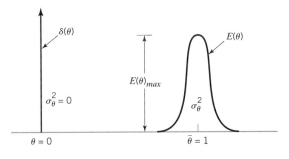

Figure 19.16 Assumed Gaussian response to pulse input for small dispersion

486   Chapter 19: Nonideal Flow

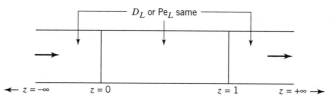

Figure 19.17 Flow conditions for open-open vessel

19.4.2.2.2 Solution for "Open" Vessel. For an "open" vessel (i.e., open-open, at both inlet and outlet), the flow conditions are illustrated schematically in Figure 19.17. The inlet is at $z = 0$, and the outlet at $z = 1$; upstream positions (in connecting piping) are represented by $z < 0$, and downstream positions by $z > 1$. It is assumed that the dispersion, as represented by D_L or Pe_L, is the same from just before the inlet to just beyond the outlet.

The boundary conditions for equation 19.4-50 with c_A replaced by the normalized response $\mathbf{C}(\theta)$ (at $z = 1$) are:

$z \to -\infty$:	$\partial \mathbf{C}(\theta)/\partial z = 0$	(19.4-59)
$z \to +\infty$:	$\partial \mathbf{C}(\theta)/\partial z = 0$	(19.4-60)
$\theta \le 0$:	$\mathbf{C}(\theta) = 0$	(19.4-61)

The solution of equation 19.4-50 is (Levenspiel and Smith, 1957):

$$\mathbf{C}(\theta) \equiv E(\theta) = \frac{1}{2}\left(\frac{Pe_L}{\pi\theta}\right)^{1/2} \exp\left[-\frac{Pe_L}{4\theta}(1-\theta)^2\right] \quad (19.4\text{-}62)$$

The moments of the solution are:

$$\bar{\theta} = 1 + 2/Pe_L \quad (19.4\text{-}63)$$

$$\sigma_\theta^2 = \frac{2}{Pe_L}\left(1 + \frac{4}{Pe_L}\right) \quad (19.4\text{-}64)$$

19.4.2.2.3 Solution for "Closed Vessel." For a "closed" vessel (i.e., closed-closed, at both inlet and outlet), the flow conditions are illustrated schematically in Figure 19.18. The flow upstream of the vessel inlet ($z < 0$) is PF characterized by $D_L = 0$ or $Pe_L = \infty$. Since flow inside the vessel is dispersed, and the pulse injection is at the inlet ($z = 0$), there is a discontinuity in c_A across the inlet. There is assumed to be continuity in c_A across the outlet ($z = 1$), even though downstream flow ($z > 1$) is PF.

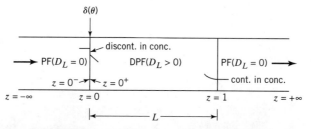

Figure 19.18 Flow conditions for closed-closed vessel

19.4 One-Parameter Models for Nonideal Flow

The boundary conditions for equation 19.4-50 with c_A replaced by the normalized response (at $z = 1$) are:

At $z = 0$: pulse input (at $z = 0^-$) = tracer output (at $z = 0^+$) by bulk flow and dispersed flow

that is,

$$n_o \delta(t) = u A_c c_A - D_L A_c \frac{\partial c_A}{\partial x} \tag{19.4-65}$$

where n_o is the amount of the pulse (in moles). In dimensionless form, with $\delta(\theta) = \bar{t}\delta(t) = (L/u)\delta(t)$, $\mathbf{C}(\theta) = c_A/(n_o/V) = c_A/(n_o/LA_c)$, and $z = x/L$, we have

$$n_o \frac{u}{L} \delta(\theta) = \frac{n_o}{LA_c} A_c \mathbf{C}(\theta) - D_L A_c \frac{n_o/LA_c}{L} \frac{\partial \mathbf{C}(\theta)}{\partial z}$$

or

$$\delta(\theta) = \mathbf{C}(\theta) - \frac{1}{\mathrm{Pe}_L} \frac{\partial \mathbf{C}(\theta)}{\partial z} \tag{19.4-66}$$

At $z = 1$: we assume there is no flow by dispersion across the outlet, consistent with

$$\frac{\partial \mathbf{C}(\theta)}{\partial z} = 0 \tag{19.4-67}$$

$\theta \leq 0$: initial condition:

$$\mathbf{C}(\theta) = 0 \tag{19.4-68}$$

An analytical solution to equation 19.4-50 (with c_A replaced by $\mathbf{C}(\theta)$) with these boundary conditions is given by Otake and Kunigita (1958), but is rather complex, and we omit it here. However, moments are relatively simple and are given by

$$\bar{\theta} = 1 \tag{19.4-69}$$

$$\sigma_\theta^2 = \frac{2}{\mathrm{Pe}_L^2} (\mathrm{Pe}_L - 1 + e^{-\mathrm{Pe}_L}) \tag{19.4-70}$$

19.4.2.2.4 Comparison of Solutions of Continuity Equation for DPF. The results for the three cases discussed above, together with those for the case with closed-open boundary conditions, are summarized in Table 19.7. Included in Table 19.7 are the boundary conditions, the expression for $\mathbf{C}(\theta)$ at $z = 1$, and the mean and variance of $\mathbf{C}(\theta)$. For large values of Pe_L, the solutions are not very different, but for small values, the results differ considerably.

19.4.2.2.5 Determining Pe_L from Tracer Data. As noted in Section 19.4.2.2.1, values of Pe_L, the single parameter in the axial dispersion model, may be obtained from the characteristics of the pulse-tracer response curve, $\mathbf{C}(\theta) \equiv E(\theta)$.

For the "open" vessel boundary conditions, Pe_L may be estimated from the value of $\mathbf{C}(\theta)$ or $E(\theta)$ at $\theta = 1$. Thus, substituting $\theta = 1$ into equation 19.4-62 and rearranging,

Table 19.7 Comparison of solutions of continuity equation for DPF

Case	Boundary conditions	$C(\theta)$ at $z = 1$	Moments $\bar{\theta}$	σ_θ^2
(1) Gauss[a]	N/A	$\dfrac{1}{2}\left(\dfrac{Pe_L}{\pi}\right)^{1/2} \exp\left[-\dfrac{Pe_L}{4}(1-\theta)^2\right]$	1	$\dfrac{2}{Pe_L}$
(2) open-open[b]	$z \to -\infty$: $\partial C(\theta)/\partial z = 0$ $z \to +\infty$: $\partial C(\theta)/\partial z = 0$ $\theta \leq 0$: $C(\theta) = 0$	$\dfrac{1}{2}\left(\dfrac{Pe_L}{\pi}\right)^{1/2} \exp\left[-\dfrac{Pe_L}{4\theta}(1-\theta)^2\right]$	$1 + \dfrac{2}{Pe_L}$	$\dfrac{2}{Pe_L}\left(1 + \dfrac{4}{Pe_L}\right)$
(3) closed-closed[c]	$z = 0$: $\delta(\theta) = C(\theta) - \dfrac{1}{Pe_L}\dfrac{\partial C(\theta)}{\partial z}$ $z = 1$: $\partial C(\theta)/\partial z = 0$ $\theta \leq 0$: $C(\theta) = 0$	complex expression (omitted here)	1	$\dfrac{2}{Pe_L^2}(Pe_L - 1 + e^{-Pe_L})$
(4) closed-open[d]	$z = 0$: as for (3) $z \to \infty$: $C(\theta) \to 0$ $\theta \leq 0$: $C(\theta) = 0$	complex expression (omitted here)	$1 + \dfrac{1}{Pe_L}$ (19.4-71)	$\dfrac{2}{Pe_L}\left(1 + \dfrac{1.5}{Pe_L}\right)$ (19.4-72)

[a] Levenspiel (1972, p. 273).
[b] Levenspiel and Smith (1957).
[c] Otake and Kunigita (1958).
[d] Villermaux and van Swaaij (1969).

we obtain:

$$Pe_L = 4\pi[E(\theta = 1)]^2 \qquad (19.4\text{-}73)$$

The measured variance σ_θ^2 may also be used to estimate Pe_L in conjunction with equations 19.4-64 (open-open), 19.4-70 (closed-closed), and 19.4-72 (closed-open, Table 19.7). For large Pe_L values the results are nearly the same, but for small Pe_L values, they differ significantly. For these three equations, and also the Gauss solution in equation 19.4-58, as $Pe_L \to \infty$, $\sigma_\theta^2 \to 0$, consistent with PF behavior. These results are illustrated in Table 19.8, which (conversely) gives values of σ_θ^2 calculated from the four equations for specified values of Pe_L. Note that σ_θ^2 becomes ill-behaved for small values of Pe_L for the Gauss solution, and for open-open and closed-open conditions. This is most apparent for $Pe_L = 0$ for these cases, in which $\sigma_\theta^2 \to \infty$. The expected result, $\sigma_\theta^2 = 1$, is given by equation 19.4-70 for closed-closed conditions. This is the expected result because $Pe_L = 0$ corresponds to BMF, which in turn corresponds to $N = 1$ in the TIS model, and for this, $\sigma_\theta^2 = 1$, from equation 19.4-26.

One conclusion from these results is that the axial diffusion model begins to fail as $Pe_L \to$ small, when an open boundary condition is used at the outlet. The case $Pe_L \to$ small means increasing backmixing, or that the diffusive flux becomes increasingly significant compared with the convective flux. For an open boundary condition, it is also questionable whether the actual response $C(\theta)$ can be identified with $E(\theta)$. Furthermore, regardless of the boundary conditions chosen, it is difficult to envisage that c_A

19.4 One-Parameter Models for Nonideal Flow 489

Table 19.8 Comparison of variances (σ_θ^2) in $E(\theta)$ calculated from different solutions of the continuity equation for DPF (19.4-50)

	σ_θ^2			
Pe_L	Gauss (19.4-58)	Open-open (19.4-64)	Closed-closed (19.4-70)	Closed-open (19.4-72)
∞ (PF)	0	0	0	0
"large"	$2/Pe_L$	$2/Pe_L$	$2/Pe_L$	$2/Pe_L$
500	0.004	0.00403	0.00399	0.00401
40	0.05	0.055	0.04875	0.05188
5	0.4	0.72	0.3205	0.52
2	1	3	0.5677	1.75
1	2	10	0.736	5
"small"	$2/Pe_L$	$8/Pe_L^2$	1	$3/Pe_L^2$
0 (BMF)	∞	∞	1	∞

is independent of radial position for large extents of backmixing, an assumption that is implicit if radial dispersion is ignored.

EXAMPLE 19-10

Using the results from Example 19-3 for $\bar{t}$ and σ_θ^2, estimate values for Pe_L for both open and closed boundary conditions, as far as possible, from (a) $E(\theta)$ and (b) the variance.

SOLUTION

(a) From Example 19-3, $\bar{t} = 21.9$ s (trapezoidal rule) and $E(\bar{t}) \simeq 0.034$ s^{-1}. Thus, $E(\bar{\theta}) = \bar{t}E(\bar{t}) = 21.9(0.034) = 0.745$. From equation 19.4-73 for open-open conditions,

$$Pe_L = 4\pi(0.745)^2 = 6.97 \simeq 7$$

The same result is obtained from the Gauss distribution, equation 19.4-56, but we haven't given the basis here for using the closed-closed and closed-open conditions.

(b) From Example 19-3, $\sigma_t^2 = 161.4$ s^2. Thus, $\sigma_\theta^2 = \sigma_t^2/\bar{t}^2 = 161.4/21.9^2 = 0.337$. Using the Gauss, open-open, closed-closed, and closed-open solutions, respectively, we obtain the following results (file ex19-10.msp):

Solution	Equation	Pe_L
Gauss	19.4-58	5.9
open-open	19.4-64	8.7
closed-closed	19.4-70	4.7
closed-open	19.4-72 (Table 19.7)	7.2

Different estimates of Pe_L are obtained if values of $\bar{t}$ (21.8 s) and σ_t^2(180s^2) from the histogram method are used.

The results again demonstrate that, since there is significant backmixing within the vessel (Pe_L relatively small), the different boundary conditions and method of solution lead to markedly different values of Pe_L.

19.4.3 Comparison of DPF and TIS Models

The TIS model is relatively simple mathematically and hence easy to use. The DPF or axial dispersion model is mathematically more complex and yields significantly different results for different choices of boundary conditions, if the extent of backmixing is large (small Pe_L). On this basis, the TIS model may be favored.

Nevertheless, we may compare the two models more quantitatively by means of the two methods already used for estimating values of N and Pe_L from tracer data: (1) matching $E(\bar{\theta})$ for the two models; and (2) matching variances, σ_θ^2.

(1) In matching $E(\bar{\theta})$ from tracer data, if we equate $E_N(\bar{\theta})$ in equation 19.4-33 (TIS model) with $E(\theta = 1)$ in 19.4-73 (DPF model), it follows that N and Pe_L are related by

$$N = Pe_L/2 \qquad (19.4\text{-}74)$$

(2) In matching σ_θ^2 from tracer data, we have four choices from above based on the use of the Gauss solution and the three different boundary conditions for the DPF model on the one hand, and equation 19.4-26, $N = 1/\sigma_\theta^2$, for the TIS model on the other hand. In summary, if we equate the right sides of equations 19.4-58, -64, -70, and -72 (Table 19.7) with $1/N$ from equation 19.4-26, in turn, we may collect the results in the form

$$N = \frac{Pe_L}{2}\left(\frac{1}{1 + X/Pe_L}\right) \qquad (19.4\text{-}75)$$

where $X = 0, 4, e^{-Pe_L} - 1$, and 1.5 for the Gauss solution, and open-open, closed-closed, and closed-open boundary conditions, respectively. For $Pe_L \to$ large (the only situation in which the Gauss solution should be used), the results from equation 19.4-75 are all the same, and the same as that in equation 19.4-74:

$$N = Pe_L/2 \quad (Pe_L \to \text{large}) \qquad (19.4\text{-}76)$$

For $Pe_L \to$ small, the results of the matching to obtain N from Pe_L give increasingly different values, as already illustrated for σ_θ^2 in Table 19.8.

The consequences of the differences in these and other mixing models for reactor performance are explored in Chapter 20.

19.5 PROBLEMS FOR CHAPTER 19

19-1 Stimulus-response experiments were used to evaluate the operation of a 1.465-L laboratory stirred-tank reactor as a CSTR. The response curves were obtained (a) by using acetic acid (A, $c_A = 0.85$ mol L^{-1}) as a tracer "chasing" water in step-change experiments, and (b) by using a small pulse of glacial acetic acid (density = 1.05 g cm^{-3}, $M = 60$ g mol^{-1}) on a flow of water.

(a) In one experiment to determine the F curve, the volumetric flow rate (q) was 47 cm^3 min^{-1}, and the concentration of acid leaving the tank (c_A) was 0.377 mol L^{-1} 21 minutes after the acid flow was begun. Calculate the coordinates ($\mathbf{C}(\theta)$ and θ) for this point.

(b) Similarly, in one experiment to determine the $C(\theta)$ curve, q was 23 cm^3 min^{-1}, the pulse was 0.5 cm^3 of glacial acetic acid, and c_A was 0.0044 mol L^{-1} 15 minutes after the introduction of the pulse. Calculate the coordinates ($C(\theta)$ and θ) for this point.

19-2 The accompanying table gives values of trace height at various times as read from a recorder for a pulse-tracer experiment (Thurier, 1977). In this experiment 1.5 cm^3 of N$_2$ was injected into a stream of He flowing steadily at 150 cm^3 s^{-1} through a stirred-tank reactor of volume 605 cm^3. A thermal conductivity detector was used to compare the outlet stream (N$_2$ + He) with the He feed stream, and the output from this as a trace on a recorder is a measure of the concentration of N$_2$ in the outlet stream.

(a) Construct a plot of $C(\theta)$ or $E(\theta)$ against θ and compare the (plotted) points with a plot of the $C(\theta)$ curve for completely back-mixed flow.

t/s	trace height	t/s	trace height
0.3	62.2	9.5	6.7
0.9	54	10.5	5.1
1.5	46	11.5	4.1
2.5	36	12.5	3.1
3.5	28.2	13.5	2.5
4.5	22.1	14.5	2.0
5.5	17.2	15.5	1.7
6.5	13.5	16.5	1.2
7.5	10.7	17.5	1.1
8.5	8.4		

(b) Compare the mean residence time ($\bar{t}$) obtained from the data in the table with that obtained from the volumetric quantities.

(c) Determine the variance of the distribution defined by the data in the table.

19-3 In order to obtain a residence-time distribution model, $E(\theta)$, for flow through a vessel of volume V, suppose the vessel is considered to be made up of two regions: a region of complete back-mixing of volume V_1, and a stagnant region of volume V_2, such that $V = V_1 + V_2$ and there is no exchange of material between the two regions. The steady-state rate of flow is q, and the flow is at constant density.

(a) Derive $E(\theta)$ for this model.
(b) What is $F(\theta)$ for this model?
(c) If $V = 10$ m^3, $V_1 = 7$ m^3, and $q = 0.4$ m^3 min^{-1}, what fraction of the exit stream
  (i) is of age less than 15 min?
  (ii) has been in the vessel longer than 40 min?

19-4 Consider steady flow of a fluid down a cylindrical vessel of volume (V) 10 m^3, at a rate (q) of 2 m^3 min^{-1}, to occur through two regions (1 and 2) in parallel. Flow through region 1 results from channeling; the flow area for this is 20% of the total area, and 1/3 of the total flow occurs in this part. Flow through each part is by dispersed PF with a relatively large value of Pe$_L$ (say > 100). Use subscripts 1 and 2 to denote quantities (such as q, V, and $\bar{t}$) for regions 1 and 2, respectively, where appropriate. Sketch $E(t)$ versus t for flow through the vessel, and show numerical values on the sketch (with supporting calculations) for the following:

(a) the location (mean) of any curve drawn;
(b) the maximum height of any curve drawn;
(c) the variance of any curve drawn;
(d) the area under any curve drawn;
(e) the value of $\bar{t} = V/q$ for total flow through the vessel.
Clearly state any assumptions made.

492 Chapter 19: Nonideal Flow

19-5 (a) Suppose flow in a "closed" vessel can be represented by the dispersion model with $Pe_L = uL/D = 2.56$. What would be the value of N (number of tanks), if the tanks-in-series model were used for this vessel with the same variance?

(b) In (a), if the solution for the dispersion model for an "open" vessel in terms of σ_θ^2 is used, what is the value of N for the tanks-in-series model? Explain briefly whether or not this value is allowed in the TIS model. Repeat similarly for "closed-open" boundary conditions. Comment on the appropriateness of these sets of boundary conditions for such a low Pe_L (i.e., with relatively large backmixing) in the dispersion model.

19-6 Consider $E_N(\theta)$ for the TIS model with $N = 5$. Calculate each of the following:
(a) the mean $\bar{\theta}$;
(b) the variance σ_θ^2;
(c) the coordinates, θ_{max} and E_{max}, of the peak.

19-7 Levenspiel and Smith (1957) conducted an experiment with a 2.85-cm diameter (internal) tube. 16.2 cm³ of a solution of KMnO₄ was rapidly injected into a water stream which flowed through the tube at a velocity of 0.357 m s⁻¹. A photoelectric cell positioned 2.75 m downstream from the injection point was used to monitor the effluent concentration (c_{KMnO_4}) from the tube. Determine, using the data given below,
(a) the mean residence time of the tracer in the tube;
(b) the number of tanks required for the tanks-in-series model.

t/s:	0	2	4	6	8	10	12	14	16	18	20	22	24	26	28	30	32	34	36	38	40
c_{KMnO_4}/arbitrary units:	0.0	11	53	64	58	48	39	29	22	16	11	9	7	5	4	3.2	2.5	2	1.5	1.3	1

19-8 A pulse input of a tracer to a 420-L vessel yielded the following tracer concentrations measured at the reactor outlet:

t/min	0	1	2.5	3	4	5	6	7.5	9	10	12	15
c/g m⁻³	0.0	1.2	7.0	8.6	9.5	8.0	6.2	3.7	2.0	1.2	0.4	0.0

(a) Determine the dispersion parameter (Pe_L), assuming both completely open and completely closed boundary conditions.
(b) Given that the volumetric flow rate is 60 L min⁻¹, determine the fraction of the reactor volume which may be classified as "dead" or inactive.

19-9 A tracer experiment was conducted in a certain vessel. The following data were obtained from a washout experiment:

t/s:	0	5	10	15	20	30	45	60	90	120	240
tracer concentration/arbitrary units:	1.000	0.928	0.843	0.751	0.593	0.376	0.199	0.094	0.021	0.007	0.000

(a) Determine the mean residence time of the tracer in this vessel.
(b) Determine the variance of the tracer curve.

19-10 It was discovered that the tracer used in problem 19-9 was subject to decay, and the measured concentrations could reflect both nonideal mixing and tracer decay. Therefore, a study of the kinetics of the tracer decay was conducted, and the following data were acquired:

t/s:	0	20	35	50	75	100
tracer activity:	21.50	15.93	12.72	10.16	6.98	4.80

Repeat parts (a) and (b) of problem 19-9, accounting for the instability of the tracer.

19-11 The Central Bank is responsible for maintaining a steady supply of currency in circulation (it just never seems to find a way into our pockets). Suppose the targeted quota for the number of $20 bills in circulation is 0.75 billion. Bills must be removed from circulation when in poor condition, but this is independent of their age (e.g., that new $20 bill in your jeans that

19.5 Problems for Chapter 19

just went through the laundry is probably slated for replacement). Assuming that $20 bills are put into circulation at a constant rate of 400 million y^{-1}, and are randomly removed from circulation and that there is no change in the total number of bills in circulation, determine:

(a) The age distribution of all $20 bills currently in circulation.
(b) The average "life" of a $20 bill.
(c) The fraction of $20 bills that are used for four or more years.
(d) On one extremely productive working (?) day, a gang of counterfeiters put 500,000 $20 bills into circulation. After five years (assuming that the bills have not been detected), how many of these bills remain in circulation? State any additional assumptions you must make to solve this problem.

19-12 In studying pyrolysis kinetics, Liliedahl et al. (1991) determined the RTD of their apparatus by pulse-injection of a gaseous hydrocarbon tracer.

(a) Given the following data, determine the mean-residence time and variance of gaseous elements in the apparatus.
(b) Find N and Pe_L for the TIS and DPF models, respectively.

t/ms:	0	68.7	103.1	152.7	206.1	290.2	404.7	652.7	984.7	1488.5
10^3 detector response:	0	0.49	1.50	3.00	3.81	2.49	1.00	0.19	0.03	0

19-13 Waldie (1992) collected RTD data by using a washout technique to determine the separation and residence time of larger particles in a spout-fluid bed.

(a) Given the following data, determine the mean residence time and variance of larger particles in the spout-fluid bed.
(b) Find N and Pe_L for the TIS and DPF models respectively.
(c) Investigate the accuracy of the TIS model for predicting the experimentally measured $F(t)$ data.

t/s:	2.057	4.000	5.710	9.486	17.257	25.370	38.011
$F(t)$:	0	0.166	0.357	0.695	0.924	0.985	1

19-14 Shetty et al. (1992) studied gas-phase backmixing for the air-water system in bubble-column reactors by measuring RTDs of pulse-injected helium tracer.

(a) Given the following data, determine the normalized variance of gaseous elements in the bubble column reactor.
(b) Find N and Pe_L for the TIS and DPF models respectively.
(c) Verify the accuracy of the TIS model for predicting the experimentally measured $E(\theta)$ data.

normalized t/dimensionless	0.1	0.2	0.4	0.6	0.7	0.9	1.3	1.9	2.8	3.4
normalized conc./dimensionless	0	0.01	0.56	0.97	1.00	0.88	0.48	0.17	0.02	0

19-15 Rhodes et al. (1991) conducted RTD measurements to study longitudinal solids mixing in a circulating fluidized-bed riser by pulse-injection of a sodium chloride tracer.

(a) Given the following RTD data, calculate the mean residence time and variance of solid particles in the circulating fluidized-bed riser. Shorten your calculations by assuming first-order decay in the tail section and verify.
(b) Find N and Pe_L for the TIS and DPF models respectively.

t/s:	0.4	0.6	1.1	1.8	2.4	2.8	3.5	4.3	5.3	6.4	7.7	9.0	10.4	11.9
tracer conc./ppt:	0	0.1	0.5	2.0	2.5	2.9	3.0	2.7	1.9	1.0	0.5	0.3	0.1	0

19-16 In studying a liquid-liquid extraction column, Dongaonkar et al. (1991) determined the column RTD by pulse-injection of tetrazine (Acid Yellow 23) in water. Given the following *incomplete* RTD data, calculate the mean-residence time and variance of a fluid element, and find N for the TIS model. Assume the tail section undergoes first-order decay.

t/min:	0	0.94	1.63	2.94	4.93	7.94	11.89
tracer conc./arbitrary units:	0	1.09	2.44	3.37	2.60	1.26	0.413

19-17 Asif et al. (1991) studied distributor effects in liquid-fluidized beds of low-density particles by measuring RTDs of the system by pulse injection of methylene blue. If PF leads into and follows the fluidized bed with a total time delay of 10 s, use the following data to calculate the mean-residence time and variance of a fluid element, and find N for the TIS model.

t/s:	29.4	32.8	37.3	42.4	47.0	51.8	55.2	64.8	69.6
normalized tracer conc./ dimensionless:	0	0.0149	0.0420	0.0528	0.0461	0.0331	0.0190	0.0024	0

19-18 Pudjiono and Tavare (1993) used the pulse-response technique to study the residence time distribution in a continuous Couette flow device with rotating inner cylinder and stationary outer cylinder. If the tracer dye decomposed with a half-life of 500 s, use the following data to determine the variance of a liquid element and N for the TIS model. The mean residence time was 760 s.

normalized time/ dimensionless:	0.30	0.40	0.45	0.50	0.60	0.81	1.00	1.25	1.55	1.72	2.00	2.60
tracer conc./ arbitrary units:	0	0.0656	0.1811	0.3112	0.5022	0.5188	0.3581	0.1680	0.0568	0.0289	0.0088	0

19-19 For the TIS model, show that, as $N \to \infty$, $E_N(\theta)$ in equation 19.4-14 becomes the RTD function for PF, $E(\theta) = \delta(\theta - 1)$, as indicated in Figure 19.12.

Chapter 20

Reactor Performance with Nonideal Flow

The TIS and DPF models, introduced in Chapter 19 to describe the residence time distribution (RTD) for nonideal flow, can be adapted as reactor models, once the single parameters of the models, N and Pe_L (or D_L), respectively, are known. As such, these are macromixing models and are unable to account for nonideal mixing behavior at the microscopic level. For example, the TIS model is based on the assumption that complete backmixing occurs within each tank. If this is not the case, as, perhaps, in a polymerization reaction that produces a viscous product, the model is incomplete.

In addition to these two macromixing reactor models, in this chapter, we also consider two micromixing reactor models for evaluating the performance of a reactor: the segregated flow model (SFM), introduced in Chapters 13 to 16, and the maximum-mixedness model (MMM). These latter two models also require knowledge of the kinetics and of the global or macromixing behavior, as reflected in the RTD.

As a preliminary consideration for these two micromixing models, we may associate three time quantities with each element of fluid at any point in the reactor (Zwietering, 1959): its residence time, t, its age, t_a, and its life expectancy in the reactor (i.e., time to reach the exit), t_e:

$$t = t_a + t_e \qquad (20\text{-}1)$$

At the inlet of a reactor, $t_a = 0$, and $t_e = t$; at the outlet, $t_e = 0$, and $t_a = t$. In the SFM, fluid elements are grouped by age, and mixing of elements of various ages does not occur within the vessel, but only at the outlet; $t_a \equiv t$. In the MMM Model, on the other hand, fluid elements are grouped by life expectancy, such that $t_e \equiv t$. This important consideration is reflected in the detailed development of each model.

In the following sections, we first develop the two macromixing models, TIS and DPF, and then the two micromixing models, SFM and MMM.

20.1 TANKS-IN-SERIES (TIS) REACTOR MODEL

The tanks-in-series (TIS) model for a reactor with nonideal flow uses the TIS flow model described in Section 19.4.1 and illustrated in Figure 19.11. The substance A is now a reacting species (e.g., A → products) instead of a tracer.

A material balance for A around the ith tank of volume V_i in the N-tank series (all tanks of equal size), in the case of unsteady-state behavior of a variable-density system, is

$$c_{A,i-1}q_{i-1} - c_{Ai}q_i - (-r_{Ai})V_i = dn_{Ai}/dt \quad (i = 1, 2, \ldots, N) \quad \textbf{(20.1-1)}$$

where n_{Ai} is the number of moles of A in the ith tank. In the special case of a constant-density system in steady-state, equation 20.1-1 becomes,

$$c_{A,i-1}q_o - c_{Ai}q_o - (-r_{Ai})V_i = 0 \quad (i = 1, 2, \ldots, N) \quad \textbf{(20.1-2)}$$

where q_o is the steady-state flow rate. This may be written in terms of the mean-residence time in the reactor system,

$$\bar{t} = V/q_o = NV_i/q_o = N\bar{t}_i \quad \textbf{(20.1-3)}$$

where $\bar{t}_i$ is the mean-residence time in the ith tank, equation 19.4-5. Equation 20.1-2 then becomes

$$c_{A,i-1} - c_{Ai} - (\bar{t}/N)(-r_{Ai}) = 0 \quad (i = 1, 2, \ldots, N) \quad \textbf{(20.1-4)}$$

More specific forms of 20.1-4 depend on the form of r_{Ai}.

For unsteady-state operation, equation 20.1-1 constitutes a set of N ordinary differential equations that must be solved simultaneously (usually numerically) to obtain the time-dependent concentration within each tank. For a constant-density system, dn_{Ai}/dt is replaced by $V_i \, dc_{Ai}/dt$. We focus on steady-state operation in this chapter.

EXAMPLE 20-1

The first-order, liquid-phase reaction of lidocaine (A) to monoethylglycinexylidide (MEGX) is conducted in a "reactor" (the liver) with arbitrary flow conditions that is to be modeled by the TIS reactor model with N tanks. Derive an expression for c_{AN}, the steady-state outlet concentration from the Nth tank, in terms of system parameters.

SOLUTION

The system parameters are the feed concentration, c_{Ao}, the rate constant in the rate law, $(-r_A) = k_A c_A$, the mean-residence time, $\bar{t}$, and N. Equation 20.1-4 for this case is

$$c_{A,i-1} - c_{Ai} - (k_A \bar{t}/N)c_{Ai} = 0$$

From this,

$$\frac{c_{Ai}}{c_{A,i-1}} = \frac{1}{1 + k_A \bar{t}/N}; \quad i = 1, 2, \ldots, N \quad \textbf{(20.1-5)}$$

Equation 20.1-5 applies to each of the N tanks. Since

$$\frac{c_{AN}}{c_{Ao}} \equiv \frac{c_{A1}}{c_{Ao}} \frac{c_{A2}}{c_{A1}} \frac{c_{A3}}{c_{A2}} \cdots \frac{c_{Ai}}{c_{A,i-1}} \cdots \frac{c_{AN}}{c_{A,N-1}} \quad \textbf{(20.1-6)}$$

$$\frac{c_{AN}}{c_{Ao}} = \frac{1}{(1 + k_A \bar{t}/N)^N} \qquad (20.1\text{-}7)$$

As $N \to \infty$, this result becomes identical with that for a PFR

$$\lim_{N \to \infty}\left(\frac{c_{AN}}{c_{Ao}}\right) = e^{-k_A \bar{t}} \qquad (20.1\text{-}8)$$

The proof of this is left to problem 20-6.

Equation 20.1-7 may also be used to calculate performance in terms of f_A, since, for a constant-density system, $f_{AN} = 1 - c_{AN}/c_{Ao}$.

For a second-order reaction, c_{AN} cannot be expressed in explicit form as in 20.1-7, but a recursion formula can be developed that makes the calculation straightforward.

EXAMPLE 20-2

The second-order, liquid-phase reaction A $\to$ products takes place in a 15-m^3 vessel. Calculate f_A at the reactor outlet, if the reactor is modeled by the TIS model with $N = 5$.
Data: $c_{Ao} = 3000$ mol m^{-3}; $q = 0.5$ m^3 min^{-1}; $k_A = 5 \times 10^{-5}$ m^3 mol^{-1} min^{-1}.

SOLUTION:

Substitution of the rate law $(-r_A) = k_A c_A^2$ in equation 20.1-4 results in

$$c_{A,i-1} - c_{Ai} - (k_A \bar{t}/N)c_{Ai}^2 = 0 \qquad (20.1\text{-}9)$$

This is a quadratic equation in c_{Ai}, the solution of which can be written in terms of $c_{A,i-1}$:

$$c_{Ai} = \frac{[1 + 4(k_A \bar{t}/N)c_{A,i-1}]^{1/2} - 1}{2k_A \bar{t}/N} \qquad (20.1\text{-}10)$$

or as

$$\frac{c_{Ai}}{c_{Ao}} = \frac{[1 + 4(k_A c_{Ao}\bar{t}/N)(c_{A,i-1}/c_{Ao})]^{1/2} - 1}{2k_A c_{Ao}\bar{t}/N}$$
$$= \frac{[1 + (4M_{A2}/N)(c_{A,i-1}/c_{Ao})]^{1/2} - 1}{2M_{A2}/N} \qquad (20.1\text{-}11)$$

where $M_{A2} = k_A c_{Ao}\bar{t}$, the dimensionless reaction number for a second-order reaction, as defined by equation 4.3-4. This is a recursion formula that can be solved for any value of N by successively setting $i = 1, 2, \ldots, N$.

In this case, $N = 5$, $M_{A2} = 5 \times 10^{-5}(3000)(15/0.5) = 4.5$, and equation 20.1-11 becomes

$$\frac{c_{Ai}}{c_{Ao}} = \frac{[1 + 3.6(c_{A,i-1}/c_{Ao})]^{1/2} - 1}{1.8}$$

If we apply this in turn for $i = 1, 2, 3, 4, 5$, we obtain the following results:

i:	1	2	3	4	5
c_{Ai}/c_{Ao}:	0.636	0.452	0.345	0.276	0.229

498 Chapter 20: Reactor Performance with Nonideal Flow

From the last entry, $c_{A5}/c_{Ao} = 0.229$, and $f_A = 1 - 0.229 = 0.771$. The five equations for c_{Ai}/c_{Ao} can also be solved simultaneously using the E-Z Solve software (file ex20-2.msp).

For a more complex kinetics scheme, a combination of the explicit and recursion-formula approaches may be required.

EXAMPLE 20-3

(a) The reaction of A to form C occurs through the formation of an intermediate species B, with both steps being first-order:

$$A \xrightarrow{k_1} B \xrightarrow{k_2} C$$

If A, with concentration (c_{Ao}) of 1.5 mol L^{-1}, enters a reactor through which the flow is nonideal, and the reactor operates at 300 K, calculate the outlet product distribution (c_A, c_B, c_C) using the TIS model with $N = 3$ to represent the reactor and the following data:

$$k_1 = 2 \times 10^{-4} \text{ s}^{-1}; k_2 = 4 \times 10^{-4} \text{ s}^{-1} \text{ (at 300 K)}$$
$$V = 100 \text{ L}; q_o \text{ (feed rate)} = 2 \text{ L min}^{-1}$$

(b) Calculate the fractional conversion of A, and the fractional yield of C.

SOLUTION

(a) Since reaction is represented by two chemical steps, we have two independent material-balance equations to develop, say, for A and for B.

For the first-order (irreversible) reaction of A, the material balance stems from equation 20.1-4, and the solution for c_A at the outlet is given by equation 20.1-7, with $N = 3$, $c_{Ao} = 1.5$ mol L^{-1}, and $k_A \bar{t} \equiv k_1 \bar{t} = (2 \times 10^{-4} \text{ s}^{-1})[100 \text{ L}/(2/60) \text{ L s}^{-1}] = 0.60 (\equiv M_{A1})$.

$$c_{A3} = \frac{c_{Ao}}{(1 + k_1 \bar{t}/N)^N} = \frac{1.5}{(1 + 0.6/3)^3} = 0.868 \text{ mol L}^{-1} = c_A$$

We develop the material balance for B in general first, and then simplify using $N = 3$. For the ith tank in an N-tank series, we have

$$c_{B,i-1} q_o - c_{Bi} q_o + k_1 c_{Ai} V_i - k_2 c_{Bi} V_i = 0 \quad \textbf{(20.1-12)}$$

From this,

$$c_{Bi} = \frac{c_{B,i-1} + k_1(V_i/q_o) c_{Ai}}{1 + k_2(V_i/q_o)}$$

$$= \frac{c_{B,i-1} + (k_1 \bar{t}/N) c_{Ai}}{1 + k_2 \bar{t}/N} \quad \textbf{(from equation 20.1-3)}$$

$$= \frac{c_{B,i-1} + (k_1 \bar{t}/N) c_{Ao}(1 + k_1 \bar{t}/N)^{-i}}{1 + k_2 \bar{t}/N} \quad \textbf{(20.1-13)}$$

(from equations 20.1-5 to -7). Equation 20.1-13 can be used as a recursion formula by setting $i = 1, 2, 3$ in turn, with the following results:

i:	1	2	3
c_{Bi}:	0.1786	0.2764	0.3214

Thus, $c_{B3} = 0.321$ mol L$^{-1} = c_B$.

From an overall material balance,

$$c_C = c_{Ao} - c_A - c_B = 0.311 \text{ mol L}^{-1}$$

(b)

$$f_A = 1 - c_A/c_{Ao} = 0.421$$

$$\hat{S}_{C/A} = \frac{c_C}{c_{Ao} - c_A} = 0.492$$

Equivalent results are obtained if the system of equations is solved with the E-Z Solve software (file ex20-3.msp).

20.2 AXIAL DISPERSION REACTOR MODEL

In this section, we apply the axial dispersion flow model (or DPF model) of Section 19.4.2 to design or assess the performance of a reactor with nonideal flow. We consider, for example, the effect of axial dispersion on the concentration profile of a species, or its fractional conversion at the reactor outlet. For simplicity, we assume steady-state, isothermal operation for a simple system of constant density reacting according to A $\rightarrow$ products.

The derivation of the material-balance or continuity equation for reactant A is similar to that of equations 19.4-48 and -49 for nonreacting tracer A, except that steady state replaces unsteady state (c_A at a point is not a function of t), and a reaction term must be added. Thus, using the control volume in Figure 19.15, we obtain the equivalent of equation 19.4-48 as:

$$A_c u c_A - D_L A_c \frac{dc_A}{dx} - A_c u \left(c_A + \frac{dc_A}{dx} dx\right) + D_L A_c \left[\frac{dc_A}{dx} + \frac{d}{dx}\left(\frac{dc_A}{dx}\right) dx\right] - A_c(-r_A) dx = 0 \quad (20.2\text{-}1)$$

where the term in $(-r_A)$ is the reaction-rate term, and the right side vanishes, since there is no accumulation of A. Equation 20.2-1 simplifies to

$$D_L \frac{d^2 c_A}{dx^2} - u \frac{dc_A}{dx} - (-r_A) = 0 \quad (20.2\text{-}2)$$

which corresponds to equation 19.4-49. In equation 20.2-2, the effect of dispersion is in the first term on the left; otherwise, if $D_L = 0$, the result is that for a PFR.

In partial nondimensional form, with $z = x/L$ and $Pe_L = uL/D_L$, equation 20.2-2 becomes

$$\frac{d^2 c_A}{dz^2} - Pe_L \frac{dc_A}{dz} - Pe_L \bar{t}(-r_A) = 0 \quad (20.2\text{-}3)$$

In general, equation 20.2-3 must be solved numerically for nonlinear kinetics. However, an analytical solution is available for first-order kinetics and a "closed" vessel.

For first-order kinetics, $(-r_A) = k_A c_A$, and equation 20.2-3 becomes

$$\frac{d^2 c_A}{dz^2} - Pe_L \frac{dc_A}{dz} - Pe_L \bar{t} k_A c_A = 0 \quad (20.2\text{-}4)$$

The boundary conditions for a closed-vessel reactor are analogous to those for a tracer in a closed vessel without reaction, equations 19.4-66 and -67, except that we are assuming steady-state operation here. These are called the Danckwerts boundary conditions (Danckwerts, 1953).[1] With reference to Figure 19.18,

At $z = 0$, we assume continuity of the flux of A; that is, flux at $z = 0^-$ by convective flow only = flux at $z = 0^+$ by convective and diffusive flow:

$$uc_{Ao} = uc_A - D_L \left(\frac{dc_A}{dz}\right)_{z=0^+} \quad (20.2\text{-}5)$$

This implies a discontinuous decrease in c_A across the inlet plane at $z = 0$, as shown in Figure 19.18 (the discontinuity at the inlet of a CSTR is the most extreme example of this, if there is PF on one side and BMF on the other). In partial nondimensional form, equation 20.2-5 is

$$c_{Ao} - c_A + \frac{1}{\text{Pe}_L}\frac{dc_A}{dz} = 0 \; (z = 0) \quad (20.2\text{-}6)$$

At $z = 1$, the continuity-of-flux argument is invalid, since it implies (incorrectly) that $c_A(z = 1^+) > c_A(z = 1^-)$; instead, we assume continuity in c_A itself (an assumption that is also analogous to that made for a CSTR); this is consistent with

$$\frac{dc_A}{dz} = 0 \quad (z = 1) \quad (20.2\text{-}7)$$

The solution of equation 20.2-4 with equations 20.2-6 and -7 as boundary conditions, although tedious, can be done by conventional means; the result (Danckwerts, 1953; Wehner and Wilhelm, 1956) is:

$$\frac{c_A}{c_{Ao}} = 1 - f_A = \frac{4\beta \exp(\text{Pe}_L/2)}{(1+\beta)^2 \exp(\beta \text{Pe}_L/2) - (1-\beta)^2 \exp(-\beta \text{Pe}_L/2)} \quad (20.2\text{-}8)$$

where

$$\beta = (1 + 4k_A \bar{t}/\text{Pe}_L)^{1/2} \quad (20.2\text{-}9)$$

For $\text{Pe}_L \to \infty$, equation 20.2-8 reduces to the result for a PFR, and for $\text{Pe}_L \to 0$, it becomes that for a CSTR. For large, but not infinite, values of Pe_L, it provides a "correction" for small deviations from PF (Danckwerts, 1953):

$$\frac{c_A}{c_{Ao}} = 1 - f_A = \exp\left(-k_A \bar{t} + \frac{k_A^2 \bar{t}^2}{\text{Pe}_L}\right) = \left(1 + \frac{k_A^2 \bar{t}^2}{\text{Pe}_L}\right)\exp(-k_A \bar{t}) \; (\text{large Pe}_L) \quad (20.2\text{-}10)$$

on expansion of $\exp(1 + k_A^2 \bar{t}^2/\text{Pe}_L)$.

In comparing the TIS and DPF reactor models, we note that the former is generally easier to use for analysis of reactor performance, particularly for nonlinear kinetics and unsteady-state operation.

---

[1] These boundary conditions and the derivation of equation 20.2-2 were first obtained by Langmuir (1908). See discussion by Weller (1994).

20.3 SEGREGATED-FLOW REACTOR MODEL (SFM)

The segregated-flow reactor model (SFM) represents the micromixing condition of complete segregation (no mixing) of fluid elements. As noted in Section 19.2, this is one extreme model of micromixing, the maximum-mixedness model being the other.

The SFM model for a reactor is developed in Section 13.5, and results in the following expression for f_A or c_A at the reactor outlet:

$$1 - f_A = \frac{c_A}{c_{Ao}} = \int_0^\infty \left[\frac{c_A(t)}{c_{Ao}}\right]_{BR} E(t)\,dt \tag{13.5-2}$$

The SFM requires knowledge of the reaction kinetics and the macroscopic RTD. The factor $[c_A(t)/c_{Ao}]_{BR}$ is determined by the kinetics (the designation BR indicates that the result to be inserted is that for a batch reactor, but it is the same for a PFR with a constant-density system). For a rate law of the form

$$(-r_A) = k_A c_A^n \tag{3.4-1}$$

$$\left[\frac{c_A(t)}{c_{Ao}}\right]_{BR} = e^{-k_A t} \quad (n = 1) \tag{3.4-10}$$

or

$$\left[\frac{c_A(t)}{c_{Ao}}\right]_{BR} = \left[\frac{1}{1 + (n-1)k_A c_{Ao}^{n-1} t}\right]^{\frac{1}{n-1}} \quad (n \neq 1) \tag{14.3-20}$$

However, any kinetics may be inserted as required. The RTD function $E(t)$ may be known either from a flow model (ideal or nonideal) or from experimental tracer data.

The SFM is applied to a (single-stage) CSTR in Chapter 14, to a PFR in Chapter 15, and to an LFR in Chapter 16. In these cases, $E(t)$ is known in exact analytical form. It is shown that the SFM gives equivalent results for a PFR and an LFR for any kinetics. For a CSTR, however, it gives an equivalent result only for first-order (i.e., linear) kinetics. This raises the question as to the usefulness of the SFM both for arbitrary kinetics and for arbitrary flow through a vessel. We first consider two methods of using equation 13.5-2 in conjunction with discrete experimental tracer data from a pulse input.

The first method uses the (normalized) response data, **C**, directly in a spreadsheet analysis similar to that used in Chapter 19:

$$\frac{c_A}{c_{Ao}} \approx \sum_i \left[\frac{c_A(t_i)}{c_{Ao}}\right]_{BR} \mathbf{C}(t_i)\,\Delta t_i \tag{20.3-1}$$

where $\mathbf{C}(t_i) \equiv E(t_i)$.

The second method uses the TIS model fitted to the tracer data. The value of N is obtained from σ_θ^2 (evaluated from the tracer data as described in Chapter 19):

$$N = 1/\sigma_\theta^2 \tag{19.4-26}$$

This value is used with $E_N(\theta)$ from equation 19.4-30 to obtain c_A/c_{Ao} from equation 13.5-2. For use in the latter, we convert $E_N(\theta)$ to $E_N(t)$; since $\theta = t/\bar{t}$ and $E(\theta) = \bar{t}E(t)$,

$$E_N(t) = \left(\frac{N}{\bar{t}}\right)^N \frac{t^{N-1}e^{-Nt/\bar{t}}}{\Gamma(N)} \tag{20.3-2}$$

and equation 13.5-2 for the TIS model is

$$\frac{c_A}{c_{Ao}} = \int_0^\infty \left[\frac{c_A(t)}{c_{Ao}}\right]_{BR} \left(\frac{N}{\bar{t}}\right)^N \frac{t^{N-1}e^{-Nt/\bar{t}}}{\Gamma(N)} \, dt \tag{20.3-3}$$

The integral in this equation may be evaluated analytically or numerically (e.g., using the E-Z Solve software).

EXAMPLE 20-4

Using the SFM and the data from Example 19-8(b), calculate f_A for the first-order, liquid-phase, isothermal reaction A → products, if $k_A = 0.05$ s^{-1}. For comparison, calculate f_A for the reactor as a PFR and as a CSTR.

SOLUTION

From Example 19-8(b), $\bar{t} = 21.9$ s, and $N = 3$ (rounded for simplicity). With the numerical values inserted, together with equation 3.4-10 for the kinetics, equation 20.3-3 for the TIS/SFM becomes

$$f_A = 1 - \frac{c_A}{c_{Ao}} = 1 - \left(\frac{3}{21.9}\right)^3 \frac{1}{(3-1)!} \int_0^\infty e^{-0.05t} t^2 e^{-3t/21.9} \, dt$$

$$= 1 - 1.303 \times 10^{-3} \int_0^\infty t^2 e^{-0.187t} \, dt = 0.607$$

using the E-Z Solve software to evaluate the integral (file ex20-4.msp). Since the reaction is first-order, the same result is obtained from the TIS model itself, equation 20.1-7.

For a PFR, from equation 3.4-10,

$$f_A = 1 - c_A/c_{Ao} = 1 - e^{-k_A t} = 1 - e^{-0.05(21.9)} = 0.666$$

For a CSTR, from equation 14.3-21,

$$f_A = \frac{k_A \bar{t}}{1 + k_A \bar{t}} = \frac{0.05(21.9)}{1 + 0.05(21.9)} = 0.523$$

20.4 MAXIMUM-MIXEDNESS REACTOR MODEL (MMM)

The maximum-mixedness model (MMM) for a reactor represents the micromixing condition of complete dispersion, where fluid elements mix completely at the molecular level. The model is represented as a PFR with fluid (feed) entering continuously incrementally along the length of the reactor, as illustrated in Figure 20.1 (after Zwietering, 1959). The introduction of feed incrementally in a PFR implies complete mixing

20.4 Maximum-Mixedness Reactor Model (MMM)

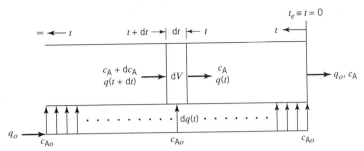

Figure 20.1 Maximum-mixedness model (MMM) (after Zwietering, 1959)

(radially) at each point. The location of each inlet is a function of t_e, the time required to exit the reactor. A particular fraction of the fluid appears immediately in the outlet stream and its inlet is therefore at $t_e = 0$. At the other extreme, another fraction has a very long residence time, and therefore enters near the beginning of the PFR, where $t_e \to \infty$. The time t_e is effectively the residence time t, and we use the latter symbol in the development to follow.

Consider a material balance for reactant A around the control volume dV in Figure 20.1. The fraction of fluid entering the vessel between t and $t + dt$ is $E(t)\,dt$; that is, if the actual flow through the vessel is q_o,

$$dq(t)/q_o = E(t)\,dt$$

or

$$dq(t) = q_o E(t)\,dt \qquad (20.4\text{-}1)$$

The total flow from the reactor inlet ($t \equiv t_e \to \infty$) to a point at t is, from equation 20.4-1,

$$q(t) = \int dq(t) = q_o \int_\infty^t E(t)\,dt = q_o W(t) \qquad (20.4\text{-}2)$$

from equation 13.3-3 and the definition of the washout function $W(t)$ in Section 13.3.3 (or see Table 13.1). The control volume dV is related to $W(t)$ by

$$dV = q(t)\,dt = q_o W(t)\,dt \qquad (20.4\text{-}3)$$

Thus, for the steady-state material balance for A:

$$\begin{pmatrix}\text{input of A}\\ \text{by flow in}\\ \text{vessel at}\\ t + dt\end{pmatrix} + \begin{pmatrix}\text{input of A}\\ \text{by flow}\\ \text{from feed}\\ \text{at } t\end{pmatrix} - \begin{pmatrix}\text{output of A}\\ \text{by flow in}\\ \text{vessel at } t\end{pmatrix} - \begin{pmatrix}\text{output of A}\\ \text{by reaction}\\ \text{in } dV\end{pmatrix} = 0$$

or,

$$(c_A + dc_A)[q(t) + dq(t)] + c_{Ao}\,dq(t) - c_A q(t) - (-r_A)\,dV = 0 \qquad (20.4\text{-}4)$$

or, from equations 20.4-1 to -3,

$$(c_A + dc_A)q_o[W(t) + dW(t)] + c_{Ao}q_o E(t)\,dt - c_A q_o W(t) - (-r_A)q_o W(t)\,dt = 0 \qquad (20.4\text{-}5)$$

which, on simplification, including division by q_o, $W(t)$ and dt, and realization that $E(t) = -dW(t)/dt$ (from Table 13.1), becomes eventually

$$\frac{dc_A}{dt} + \frac{E(t)}{W(t)}(c_{Ao} - c_A) - (-r_A) = 0 \qquad (20.4\text{-}6)$$

Equation (20.4-6) is the continuity equation for the MMM. Its solution for c_A at $t \equiv t_e = 0$ at the outlet is subject to the boundary condition:

$$\text{at } t \equiv t_e = \infty: \quad dc_A/dt = 0 \qquad (20.4\text{-}7)$$

An alternative (and much simpler) way of solving 20.4-6 is to covert it from a boundary-value problem to an initial-value problem, which may then be numerically integrated. This may be accomplished by recognizing that $E(t) \to 0$ for times much longer than the mean residence time. Practically speaking, this usually occurs whenever t is 5 to 10 times the mean residence time. Hence, for an irreversible reaction, $c_A(t = 10\bar{t}) = 0$, and for a reversible reaction, $c_A(t = 10\bar{t}) = c_{Aeq}$. Either of these *initial* conditions consequently represents c_A at the inlet of the reactor, and permits the solution of 20.4-6.

20.5 PERFORMANCE CHARACTERISTICS FOR MICROMIXING MODELS

An important aspect of the micromixing models is that they define the maximum and minimum conversion possible for a given reaction and RTD. Zwietering (1959) showed that, for the reaction A $\to$ products, with power-law kinetics, $(-r_A) = k_A c_A^n$,

(1) For $n > 1$, the segregated flow model provides the upper bound on conversion, and the maximum-mixedness model defines the lower bound.
(2) For $n < 1$, the maximum-mixedness model sets the upper bound, while the lower bound is determined by the segregated-flow model.

The bounds for a complex reaction scheme in which the rate goes through a maximum or minimum, with respect to concentration, depend on the nature of the rate law and the desired fractional conversion.

Micromixing may also have a major impact upon the yield and selectivity of complex reaction networks. Consider, for example, the following parallel reaction network, where both a desired product (D) and an undesired product (U) may be formed:

$$A \to D; \quad r_D = k_1 c_A^\alpha$$
$$A \to (1/2)U; \quad r_U = k_2 c_A^\beta$$

If both reactions are first order ($\alpha = \beta = 1$), then micromixing is irrelevant; yield, selectivity, and fractional conversion depend solely on the RTD. If, however, either α or β is not equal to 1, then the degree of micromixing can have a significant impact upon performance, as illustrated in the following example.

EXAMPLE 20-5

Consider the following liquid-phase reactions, which occur in parallel:

$$A \to D; \quad r_D = k_1 c_A^2$$
$$A \to (1/2)U; \quad r_U = k_2 c_A$$

20.5 Performance Characteristics for Micromixing Models

For a feed of pure A, determine the outlet concentrations of A and D, and f_A, $Y_{D/A}$, and $\hat{S}_{D/A}$, for the following models and flow characteristics:

(a) CSTR;
(b) PFR;
(c) SFM, with $E(t)$ equal to that for both a CSTR ($N = 1$), and for two CSTRs in series ($N = 2$);
(d) MMM, with $E(t)$ as in (c).

Data: $c_{Ao} = 1.5$ mol L^{-1}; $k_1 = 2.50$ L mol^{-1} s^{-1}; $k_2 = 0.50$ s^{-1}; $\bar{t} = 1.00$ s

SOLUTION

Define parameters (with $c_{Do} = 0$):

$$f_A = \frac{c_{Ao} - c_A}{c_{Ao}} \tag{14.3-12}$$

$$Y_{D/A} = \frac{c_D}{c_{Ao}} \tag{5.2-1d}$$

$$\hat{S}_{D/A} = \frac{c_D}{c_{Ao} - c_A} \tag{5.2-4d}$$

(a) CSTR:

$$(-r_A) = \frac{F_{Ao} - F_A}{V} = \frac{q(c_{Ao} - c_A)}{V} = \frac{c_{Ao} - c_A}{\bar{t}} = k_1 c_A^2 + 2k_2 c_A \tag{i}$$

which may be rearranged to solve for c_A:

$$c_A = \frac{-1 - 2k_2\bar{t} \sqrt{(2k_2\bar{t} + 1)^2 + 4k_1\bar{t}c_{Ao}}}{2k_1\bar{t}} = 0.47 \text{ mol L}^{-1} \tag{ii}$$

Solving for c_D, we first obtain

$$r_D = \frac{F_D - F_{Do}}{V} = \frac{q_o c_D}{V} = \frac{c_D}{\bar{t}} = k_1 c_A^2 \tag{iii}$$

Thus,

$$c_D = k_1 \bar{t} c_A^2 = 0.56 \text{ mol L}^{-1} \tag{iv}$$
$$f_A = (1.5 - 0.47)/1.5 = 0.69$$
$$Y_{D/A} = 0.56/1.5 = 0.37$$
$$\hat{S}_{D/A} = 0.56/(1.5 - 0.47) = 0.54$$

(b) PFR: The design equations for A and D must be solved. Thus, for A and constant density:

$$(-r_A) = \frac{-dF_A}{dV} = \frac{-dc_A}{dt} = k_1 c_A^2 + 2k_2 c_A$$

and

$$t \equiv \bar{t} = -\int_{c_{Ao}}^{c_A} \frac{dc_A}{k_1 c_A^2 + 2k_2 c_A} = -\int_{c_{Ao}}^{c_A} \frac{dc_A}{c_A(k_1 c_A + 2k_2)} \tag{v}$$

Chapter 20: Reactor Performance with Nonideal Flow

The integral in (v) may be solved using partial fractions:

$$\bar{t} = -\int_{c_{Ao}}^{c_A} \frac{dc_A}{2k_2 c_A} + \int_{c_{Ao}}^{c_A} \frac{k_1 dc_A}{2k_2(k_1 c_A + 2k_2)}$$

which integrates to

$$\ln\left[\frac{c_{Ao}(k_1 c_A + 2k_2)}{c_A(k_1 c_{Ao} + 2k_2)}\right] = 2k_2 \bar{t} \quad \text{(vi)}$$

Solution for c_A gives

$$c_A = \frac{2c_{Ao}k_2 e^{-2k_2 \bar{t}}}{2k_2 + k_1 c_{Ao}(1 - e^{-2k_2 \bar{t}})} \quad \text{(vii)}$$

Similarly, the design equation may be used to solve for $c_D(t)$. In this case, it is simpler to use the implicit relationship between $c_A(t)$ and $c_D(t)$, as follows:

$$\frac{(-r_A)}{r_D} = \frac{-dF_A/dV}{dF_D/dV} = \frac{-dF_A}{dF_D} = \frac{-dc_A}{dc_D} \quad \text{(viii)}$$

Thus,

$$dc_D = -\int \frac{r_D}{(-r_A)} dc_A = -\int_{c_{Ao}}^{c_A} \frac{k_1 c_A^2}{k_1 c_A^2 + 2k_2 c_A} dc_A = -\int_{c_{Ao}}^{c_A} \frac{k_1 c_A}{k_1 c_A + 2k_2} dc_A \quad \text{(ix)}$$

Integration leads to:

$$c_D - c_{Do} = \left[-\frac{1}{k_1}(k_1 c_A + 2k_2 - 2k_2 \ln(k_1 c_A + 2k_2))\right]$$

$$+ \left[\frac{1}{k_1}(k_1 c_{Ao} + 2k_2 - 2k_2 \ln(k_1 c_{Ao} + 2k_2))\right]$$

which may be rearranged to (with $c_{Do} = 0$):

$$c_D = c_{Ao} - c_A + \frac{2k_2}{k_1} \ln\left[\frac{k_1 c_A + 2k_2}{k_1 c_{Ao} + 2k_2}\right] \quad \text{(x)}$$

Equations (vii) and (x) may be solved by substituting known values of k_1, k_2, c_{Ao}, and $\bar{t}$, which leads to the following outlet concentrations of A and D:

$$c_A = 0.16 \text{ mol L}^{-1}; \qquad c_D = 0.85 \text{ mol L}^{-1}$$

Therefore, $f_A = 0.89$, $Y_{D/A} = 0.57$, and $\hat{S}_{D/A} = 0.63$.

(c) Segregated-flow model: This case may be solved by applying equation 13.5-2, with $c_A(t)$ and $c_D(t)$ determined from equations (vii) and (x), developed in (b), which are equivalent to those for a BR (constant density):

$$c_A = \int_0^\infty [c_A(t)]_{BR} E(t) \, dt \quad \text{(xi, from 13.5-2)}$$

Table 20.1 Results for Example 20-5(c)(SFM)

Case	c_A/mol L^{-1}	f_A	c_D/mol L^{-1}	$Y_{D/A}$	$\hat{S}_{D/A}$
1 CSTR	0.39	0.74	0.80	0.53	0.72
2 CSTRs	0.28	0.81	0.84	0.56	0.69

Table 20.2 Results for Example 20-5(d)(MMM)

Case	c_A/mol L^{-1}	f_A	c_D/mol L^{-1}	$Y_{D/A}$	$\hat{S}_{D/A}$
1 CSTR	0.47	0.69	0.56	0.37	0.54
2 CSTRs	0.35	0.77	0.65	0.43	0.57

As indicated in the problem statement, $E(t)$ is based upon two cases: a single CSTR, and two CSTRs in series. The respective $E(t)$ expressions are:

$$E_1(t) = \frac{1}{\bar{t}}e^{-t/\bar{t}} \tag{13.4-2}$$

$$E_2(t) = \frac{4t}{\bar{t}^2}e^{-2t/\bar{t}} \tag{17.2-4}$$

Equation (xi) must be numerically integrated, using either $E_1(t)$ or $E_2(t)$, and the appropriate expressions for $c_A(t)$ and $c_D(t)$ (see E-Z Solve file ex20-5.msp). Table 20.1 gives the outlet concentration, conversion, yield, and selectivity obtained for each of the two cases.
(d) Maximum-mixedness model: For the maximum-mixedness model, the rate laws for A and D are substituted into Equation 20.4-6, and the two resulting ordinary differential equations (in dc_A/dt and dc_D/dt) must be numerically integrated. The respective equations are:

$$\frac{dc_A(t)}{dt} = k_1 c_A^2 + 2k_2 c_A + \frac{E(t)}{W(t)}[c_A(t) - c_{Ao}] \tag{xii}$$

$$\frac{dc_D(t)}{dt} = -k_1 c_A^2 + \frac{E(t)}{W(t)}[c_D(t) - c_{Do}] \tag{xiii}$$

Equations (xii) and (xiii) must be numerically integrated from $t = 10\bar{t}$ to $t = 0$. The following initial conditions for $c_A(t)$ and $c_D(t)$ may be applied:

$$c_A(t = 10\bar{t}) = \frac{2c_{Ao}k_2 e^{-k_2(10\bar{t})}}{2k_2 + k_1 c_{Ao}[1 - e^{-2k_2(10\bar{t})}]} = 1.43 \times 10^{-5} \quad \text{[from (vii)]}$$

$$c_D(t = 10\bar{t}) = c_{Ao} - c_A + 2\frac{k_2}{k_1}\ln\left[\frac{k_1 c_A + 2k_2}{k_1 c_{Ao} + 2k_2}\right] = 0.877 \quad \text{[from (x)]}$$

Table 20.2 gives the results obtained for the two cases, with $E(t)$ based upon a single CSTR, and upon two CSTRs in series.

To obtain the results in Table 20.2, equations (xii) and (xiii) are solved numerically using E-Z Solve (file ex20-5.msp) and the initial conditions above for c_A and c_D. For $N = 1$, equation 13.4-2 is used for $E_1(t)$ and is integrated to obtain $W_1(t)$, and similarly for $N = 2$ and equation 17.2-4. Note that most software for numerical integration cannot directly handle the negative step sizes required to solve the maximum-mixedness model

differential equation(s). This restriction can be overcome through the use of dummy variables, as follows:
Let $t' = 10\bar{t} - t$ and specify $E(t)$ by

$$E(t') = \frac{N^N t'^{N-1} e^{-Nt'/\bar{t}}}{\bar{t}^N \Gamma(N)} \qquad (20.5\text{-}1)$$

The integration may now be performed with respect to t', from $t' = 0$ to $t' = 10\bar{t}$. This is equivalent to numerical integration with respect to t from $t = 10\bar{t}$ to $t = 0$.

The results from this example illustrate several points.

(1) Identical performance is obtained from a CSTR and the maximum mixedness model, with $E(t)$ based upon BMF.
(2) Comparison of the CSTR and PFR models shows that the latter gives better performance.
(3) Comparison of the segregated-flow and maximum-mixedness models, with identical RTD functions, shows that the former gives better performance. This is consistent with the observations of Zwietering (1959), who showed that for power-law kinetics of order $n > 1$, the segregated-flow model produces the highest conversion.
(4) Changing $E(t)$ from that for $N = 1$ to that for $N = 2$ in the TIS model has only a small effect on performance parameters, regardless of the micromixing model used (Tables 20.1 and 20.2). In contrast, as noted in point (3), for a particular RTD function, the micromixing behavior has a relatively large effect upon performance.

20.6 PROBLEMS FOR CHAPTER 20

20-1 A tracer study was conducted on a flow reactor, and it was found that the RTD was consistent with the tanks-in-series model, with $N = 5$. The mean residence time in the system was 7.8 minutes.
  (a) For a first-order reaction, A $\rightarrow$ products, with $k_A = 0.25$ min^{-1}, determine the exit fractional conversions predicted by the TIS, segregated-flow and maximum-mixedness models.
  (b) Repeat part (a) for a second-order reaction, with $c_{Ao} = 0.25$ mol L^{-1}, and $k_A = 0.05$ L mol^{-1} min^{-1}.

20-2 For a particular reactor system, involving three tanks in series, develop a graph which shows the change in c_A as a function of time, where $(-r_A) = k_A c_A$, $c_{Ao} = 0.25$ mol L^{-1}, $k_A = 0.125$ min^{-1}, and $\bar{t}$ for the *system* is 7.5 min; $c_A(0) = 0$.

20-3 Show that, for a first-order reaction $[(-r_A) = k_A c_A]$, the outlet concentration predicted by the maximum-mixedness model is

$$c_A = c_{Ao} \int_0^\infty E(t) e^{-k_A t} dt$$

Compare this result with that predicted by the segregated-flow model.

20-4 For arbitrary kinetics, with $E(t) = (1/\bar{t}) e^{-t/\bar{t}}$, develop an expression for c_A based on the inlet concentration, c_{Ao}, and the reaction rate, $(-r_A)$, using the maximum-mixedness model.

20-5 For the following gas-phase, series-parallel reaction network, determine the effect of micromixing on the yield and selectivity to the desired species, CH$_3$OH. The RTD function for the reactor may be represented by two equal-sized tanks in series.

$$CH_4 + \frac{1}{2}O_2 \xrightarrow{k_1} CH_3OH \tag{1}$$

$$CH_3OH + \frac{3}{2}O_2 \xrightarrow{k_2} CO_2 + 2H_2O \tag{2}$$

Data:

$T = 500$ K (constant); $q_o = 0.40$ L min^{-1}
$r_1 = k_1 c_{CH_4} c_{O_2}$, where $k_1 = 0.05$ L mol^{-1} min^{-1}
$r_2 = k_2 c_{CH_3OH} c_{O_2}$, where $k_2 = 0.25$ L mol^{-1} min^{-1}

The feed contains 60 mol % CH_4 and 40 % O_2. The total reactor volume is 30 L.

20-6 Consider a first-order, liquid-phase reaction (A $\rightarrow$ products; $(-r_A) = k_A c_A$) taking place in a series of N equal-sized stirred tanks of total volume V, each tank being at the same T, with feed concentration c_{Ao} and flow rate q (all quantities in consistent units). In Example 20-1, it is shown that c_A in the exit stream from the Nth tank is given by

$$c_{AN}/c_{Ao} = (1 + k_A \bar{t}/N)^{-N} \tag{20.1-7}$$

where $\bar{t} = V/q$, or that, on rearrangement,

$$\bar{t} = (N/k_A)[(c_{Ao}/c_{AN})^{1/N} - 1]. \tag{20.1-7a}$$

(a) From 20.1-7, show that, for a given $\bar{t}$, $c_{AN} \rightarrow c_{A,PFR}$ as $N \rightarrow \infty$.
(b) From 20.1-7a, similarly show that, for a given c_A, $\bar{t} \rightarrow \bar{t}_{PFR} (= t_{PFR} = t_{BR})$.
(c) Show that, for fixed $\bar{t}$, c_{AN} decreases (or f_A increases) as N increases (subject to the limit given in (a)).
(d) Show that, for fixed c_{AN} (or f_A), $\bar{t}$ decreases as N increases (subject to the limit given in (b)).

20-7 Apply the tanks-in-series model to the following kinetics scheme involving reactions in parallel:

$$A \rightarrow B + C; \quad r_B = k_1 c_A$$
$$A \rightarrow D; \quad r_D = k_2 c_A$$

Assume constant density and temperature, and steady-state operation.
(a) By considering the jth tank in a set of N tanks, obtain a recursion formula for calculating c_{Bj} (in terms of $c_{B,j-1}$ and various parameters).
(b) For a mean-residence time ($\bar{t}$) of 100 min, calculate c_A, c_B, c_D, and c_C, from the outlet of the second tank, using the results of (a) and the following data: $k_1 = 0.0215$ min^{-1}, $k_2 = 0.0143$ min^{-1}, and $c_{Ao} = 3$ mol L^{-1}.
(c) In the model used in (a) and (b), can N be a noninteger? (If necessary, attempt the calculations for $N = 1.5$.)

20-8 A pulse-tracer experiment for a particular flow rate in a certain vessel (assume "closed") shows that the first ($\bar{t}$) and the second (σ_θ^2) moments of the RTD are 15.4 min and 0.670, respectively. If a pseudo-first-order reaction, A + B $\rightarrow$ products, with $k_A = 0.299$ min^{-1} at a particular temperature T, takes place in the vessel, isothermally at T, and at the same flow rate, calculate the fractional conversion of A, f_A, according to
(a) the plug-flow model (i.e., a PFR);
(b) the axial-dispersion model;
(c) the back-mix-flow model in a single tank (i.e., a CSTR);
(d) the tanks-in-series model.

20-9 (a) Zoulalian and Villermaux (1970) measured $E(t)$ by means of a pulse-tracer injection for a certain reactor used for liquid-phase reactions. For a certain (total) flow rate, smoothed results (as read from a graph) are as follows:

t/s:	≤150	175	200	225	250	275	300	325	350	375	400	≥425
$10^3 E(t)/s^{-1}$:	0	0.9	4.1	8.3	9.8	8.0	4.9	2.3	1.0	0.4	0.2	0

Calculate the mean ($\bar{t}/s$) and variance (σ_θ^2) of the distribution.

(b) If the hydrolysis of t-isobutyl chloride (A) is carried out in the same vessel at 25.1°C ($k_A = 0.0115$ s^{-1}) at the same (total) flow rate, calculate the following:

(i) the outlet fractional conversion of A(f_A), based on the segregated-flow model;
(ii) the number of tanks (N) in the TIS model required to represent $E(t)$, stating the assumption(s) on which this is based;
(iii) the outlet value of f_A predicted by the TIS model.

20-10 For a liquid-phase reaction, A → products, consider an experimental reactor 20 cm in diameter and 50 cm long that is packed with nonporous pellets (to improve the approach to plug flow), such that the bed voidage (ϵ_B) is 0.4. The rate constant for the reaction is 0.02 s^{-1} at a particular temperature. For isothermal operation at this temperature, if the feed rate (q) is 0.05 L s^{-1} of pure A, what is the fractional conversion of A(f_A) at the outlet of the reactor

(a) for plug flow?
(b) if the Peclet number is 50?
(c) Compare the results with more extreme conditions (departure from PF) with Pe$_L$ = 5 and = 0 (BMF as in a CSTR).

20-11 For the vessel in problem 19-7, use N obtained in part (b) to determine the steady-state value of f_A at the outlet for a second-order reaction, with $(-r_A) = k_A c_A^2$, $c_{Ao} = 2.0$ mol L^{-1}, and $k_A = 0.25$ L mol^{-1} s^{-1}. Compare this result with the results for f_A predicted by the PFR and CSTR models.

20-12 (a) Based on the results of problem 19-8, for a first-order isomerization reaction, A → B, with $k_A = 0.18$ min^{-1}, calculate the steady-state value of f_A predicted by the axial dispersion model with closed boundary conditions.

(b) Compare the results from part (a) with values of f_A predicted by the PFR, CSTR, and TIS models.

20-13 For a second-order, constant-density reaction, A → products, carried out in the vessel in problem 19-5, what fractional conversion (f_A) does the tanks-in-series model predict, if, for the same reaction carried out in a plug-flow reactor, the conversion is 5/6?

20-14 Suppose Figure 20.2 gives the experimental $E(t)$ for a reactor in which a liquid-phase, second-order reaction (A → products) is taking place isothermally at steady-state. The

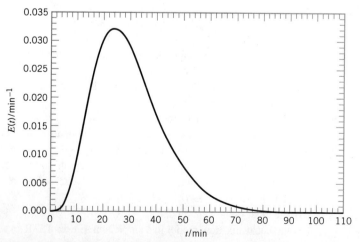

Figure 20.2 $E(t)$ for problem 20-14

volume (V) of the reactor is 15 m^3, the feed rate (q) is 0.5 m^3 min^{-1}, the feed concentration (c_{Ao}) is 3000 mol m^{-3}, and the rate constant (k_A) is 5×10^{-5} m^3 mol^{-1} min^{-1}. Based on the segregated-flow model, what fractional conversion of A (f_A) would be obtained at the outlet of the reactor? Comment on whether the result is correctly calculated in this way, and compare with the result in Example 20-2.

20-15 For a relatively small amount of dispersion, what value of Pe_L would result in a 10% increase in volume (V) relative to that of a PFR (V_{PF}) for the same conversion (f_A) and throughput (q)? Assume: the reaction, A $\rightarrow$ products, is first-order; and isothermal, steady-state, constant-density operation; and the reaction number, $M_{A1} = k_A \bar{t}$, is 2.5. For this purpose, first show, using equation 20.2-10, for the axial-dispersion model with relatively large Pe_L, that the % increase $\equiv 100(V - V_{PF})/V_{PF} = 100 M_{A1}/Pe_L$.

20-16 (a) Consider two equal-sized stirred tanks in series, with each tank acting as a CSTR. For such a configuration, $E(t) = (4t/\bar{t}^2)\exp(-2t/\bar{t})$, where $\bar{t} = V/q$, $V = V_1 + V_2$, $V_1 = V_2 = V/2$, and q is the steady-state-flow rate.
For a first-order, single-phase reaction (A $\rightarrow$ products, $(-r_A) = k_A c_A$) taking place in the tanks, calculate the fractional conversion of A(f_A) leaving the second tank based on the SFM, if $q = 0.5$ m^3 min^{-1} (constant-density flow), $V = 10$ m^3, and $k_A = 0.1$ min^{-1}.
(b) Repeat the calculation in (a) using the TIS model, and comment on the result in comparison with that in (a).
(c) If the reaction in (a) and (b) were second-order, would the comparison of the results from (a) and (b) be the same as above? Explain without doing any further calculations.

Chapter 21

Fixed-Bed Catalytic Reactors for Fluid-Solid Reactions

This chapter is devoted to fixed-bed catalytic reactors (FBCR), and is the first of four chapters on reactors for multiphase reactions. The importance of catalytic reactors in general stems from the fact that, in the chemical industry, catalysis is the rule rather than the exception. Subsequent chapters deal with reactors for noncatalytic fluid-solid reactions, fluidized- and other moving-particle reactors (both catalytic and noncatalytic), and reactors for fluid-fluid reactions.

In a fixed-bed catalytic reactor for a fluid-solid reaction, the solid catalyst is present as a bed of relatively small individual particles, randomly oriented and fixed in position. The fluid moves by convective flow through the spaces between the particles. There may also be diffusive flow or transport within the particles, as described in Chapter 8. The relevant kinetics of such reactions are treated in Section 8.5. The fluid may be either a gas or liquid, but we concentrate primarily on catalyzed gas-phase reactions, more common in this situation. We also focus on steady-state operation, thus ignoring any implications of catalyst deactivation with time (Section 8.6). The importance of fixed-bed catalytic reactors can be appreciated from their use in the manufacture of such large-tonnage products as sulfuric acid, ammonia, and methanol (see Figures 1.4, 11.5, and 11.6, respectively).

In this chapter, we first cite examples of catalyzed two-phase reactions. We then consider types of reactors from the point of view of modes of operation and general design considerations. Following introduction of general aspects of reactor models, we focus on the simplest of these for pseudohomogeneous and heterogeneous reactor models, and conclude with a brief discussion of one-dimensional and two-dimensional models.

21.1 EXAMPLES OF REACTIONS

Examples of reactions carried out in fixed-bed catalytic reactors of various types are as follows:

(1) The oxidation of sulfur dioxide to sulfur trioxide is a step in the production of sulfuric acid:

$$SO_2 + \frac{1}{2}O_2 \rightleftharpoons SO_3 \qquad (A)$$

This is a reversible, exothermic reaction carried out adiabatically in a multistage, fixed-bed reactor with axial flow of fluid and interstage heat transfer for temperature adjustment; see Figure 1.4. The catalyst is promoted V_2O_5.

(2) The synthesis of ammonia:

$$N_2 + 3H_2 \rightleftarrows 2NH_3 \qquad \textbf{(B)}$$

This is also a reversible, exothermic reaction carried out in various types of fixed-bed reactors, involving different arrangements for flow (axial or radial), and temperature adjustment; see Figure 11.5. The traditional catalyst is promoted Fe, but more active Ru-based catalysts are finding use, despite the added cost.

(3) The synthesis of methanol:

$$CO + 2H_2 \rightleftarrows CH_3OH \qquad \textbf{(C)}$$

This is also a reversible, exothermic reaction. Some reactors used for this reaction are similar to those used for ammonia synthesis; see Figure 11.6. The standard catalyst is $Cu/ZnO/Al_2O_3$.

(4) The production of styrene monomer by dehydrogenation of ethylbenzene:

$$C_8H_{10} \rightleftarrows C_8H_8 + H_2 \qquad \textbf{(D)}$$

This is a reversible, endothermic reaction carried out adiabatically in a multistage reactor with either axial or radial flow, and interstage heat transfer for temperature adjustment.

(5) The alkylation of benzene to ethylbenzene:

$$C_6H_6 + C_2H_4 \rightarrow C_8H_{10} \qquad \textbf{(E)}$$

This reaction precedes reaction (D) in the production of styrene.

(6) The production of formaldehyde by partial oxidation of methanol:

$$CH_3OH + \frac{1}{2}O_2 \rightarrow HCHO + H_2O \qquad \textbf{(F)}$$

Catalysts used are based on Ag or ferric molybdate.

Reaction (F) represents one of the uses of methanol (reaction (C)), and is also an example in which reaction selectivity is an important issue. The reaction cannot be allowed to go to ultimate completion, since the complete oxidation of CH_3OH would lead to CO_2 and H_2O as products. Similarly, in reaction (D), benzene and other (unwanted) products are produced by dealkylation reactions.

Many of these processes involve reversible reactions (Section 5.3), but may otherwise be considered to be simple in the sense of requiring only one chemical equation. In general, reversibility implies that the equilibrium constraint on fractional conversion or yield must be taken into account in the design of the reactor. In turn, this has implications for operating conditions: the choice of P; the effect of T, adjustment of T, and the means of achieving the adjustment; all this is in order to minimize any adverse effects of the equilibrium constraint, so as to increase yield. The following example illustrates elementary consideration of choice of P for a gas-phase reaction; design aspects relating to the effect and adjustment of T are developed throughout this chapter.

EXAMPLE 21-1

(a) In the production of CH_3OH by reaction (C), what is the implication of reversibility, with a potential equilibrium constraint, for the choice of reactor operating pressure (P)?
(b) Repeat (a) for the production of styrene monomer by reaction (D).
(c) For a reaction such as that in (b), if an inert gas is added to the feed at constant P, what is the effect on equilibrium yield?

SOLUTION

(a) Consider the system to be in a state of equilibrium at particular conditions, and then consider the effect of change of P on the position of equilibrium in terms of the yield of CH_3OH. The effect on equilibrium is contained in the Principle of Le Chatelier (1884): for a change in P (at fixed values of other variables affecting equilibrium), the position of equilibrium changes in a direction to counteract the change in P imposed. In the case of reaction (C), if P is increased, formation of CH_3OH results in a decrease of total moles, which is in a direction to offset the increase in P. Thus, an increase in P increases the equilibrium yield of CH_3OH. Reactors for the synthesis of CH_3OH operate at relatively high P, 50 to 100 bar.
(b) Similar reasoning applied to reaction (D) for production of styrene monomer leads to the opposite conclusion: the equilibrium yield of C_8H_8 increases as P decreases. Reactors for this reaction operate at low P, essentially at ambient pressure of 1 bar.
(c) Adding an inert gas at constant P lowers the partial pressure of the reacting system. It is the same as if P itself were lowered: the equilibrium yield increases as the initial ratio of inert gas to reactant(s) increases at fixed P (and T). This is typical for a gas-phase reaction accompanied by an increase in moles; the opposite occurs for a reaction with a decrease in moles. In reaction (D), steam (H_2O) is added to the reactor feed partly for this purpose.

21.2 TYPES OF REACTORS AND MODES OF OPERATION

21.2.1 Reactors for Two-Phase Reactions

In this section, we consider arrangements for flow of fluid, adjustment of T, and configuring beds of solid for fixed-bed catalytic reactors. These are summarized in Figure 21.1.

21.2.2 Flow Arrangement

In Figure 21.1, the first division is with respect to flow arrangement. Traditionally, most fixed-bed reactors are operated with axial flow of fluid *down* the bed of solid particles. A more recent trend is to use radial flow, either outward, as depicted in Figure 21.1, or inward. In the case of styrene monomer production (reaction (D) in Section 21.1), the purpose is to reduce the pressure drop ($-\Delta P$) by increasing the flow area for a given bed volume. We restrict attention to axial flow in this chapter.

21.2.3 Thermal and Bed Arrangement

In Figure 21.1, for axial flow of fluid, the main division for thermal considerations is between adiabatic and nonadiabatic operation.

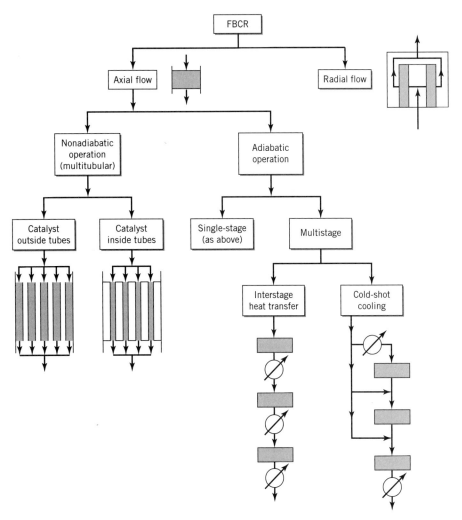

Figure 21.1 Fixed-bed catalytic reactors (FBCR) for two-phase reactions: modes of operation (each rectangle or shaded area represents a bed of catalyst; each circle represents a heat exchanger)

21.2.3.1 *Adiabatic Operation*

In adiabatic operation, no attempt is made to adjust T *within* the bed by means of heat transfer. The temperature thus increases for an exothermic reaction and decreases for an endothermic reaction. For a reactor consisting of one bed of catalyst (a single-stage reactor), this defines the situation thermally. However, if the catalyst is divided into two or more beds arranged in series (a multistage reactor), there is an opportunity to adjust T between stages, even if each stage is operated adiabatically. This may be done in one of two ways, as shown in Figure 21.1. The first involves interstage heat transfer by means of heat exchangers; this method may be used for either exothermic or endothermic reactions. The second, called cold-shot cooling, can be used for exothermic reactions. It involves splitting the original cold feed such that part enters the first stage, and part is added to the outlet stream from each stage (entering the subsequent stage), except the last. Thus, T of the outlet stream from any stage (except the last) can be lowered by mixing with cold feed, without the use of a heat exchanger. The extent of reduction of T depends on the distribution of feed among the stages.

Regardless of how achieved, the purpose of adjustment of T is twofold: (1) to shift an equilibrium limit, so as to increase fractional conversion or yield, and (2) to maintain a relatively high rate of reaction, to decrease the amount of catalyst and size of vessel required.

21.2.3.2 *Nonadiabatic Operation*

In nonadiabatic operation, heat transfer for control of T is accomplished within the bed itself. This means that the reactor is essentially a shell-and-tube exchanger, with catalyst particles either inside or outside the tubes, and with a heating or cooling fluid flowing in the shell or in the tubes accordingly.

21.3 DESIGN CONSIDERATIONS

In addition to flow, thermal, and bed arrangements, an important design consideration is the amount of catalyst required (W), and its possible distribution over two or more stages. This is a measure of the size of the reactor. The depth (L) and diameter (D) of each stage must also be determined. In addition to the usual tools provided by kinetics, and material and energy balances, we must take into account matters peculiar to individual particles, collections of particles, and fluid-particle interactions, as well as any matters peculiar to the nature of the reaction, such as reversibility. Process design aspects of catalytic reactors are described by Lywood (1996).

21.3.1 Considerations of Particle and Bed Characteristics

Characteristics of a catalyst particle include its chemical composition, which primarily determines its catalytic activity, and its physical properties, such as size, shape, density, and porosity or voidage, which determine its diffusion characteristics. We do not consider in this book the design of catalyst particles as such, but we need to know these characteristics to establish rate of reaction at the surface and particle levels (corresponding to levels (1) and (2) in Section 1.3). This is treated in Section 8.5 for catalyst particles. Equations 8.5-1 to -3 relate particle density ρ_p and *intraparticle* voidage or porosity ϵ_p.

The shape of a particle may be one of many that can be formed by extrusion or tabletting. As in Chapter 8, we restrict attention to three shapes: (solid) cylinder, sphere, and flat plate. The size for use in a packed bed is relatively small: usually about a few mm.

At the bed level (corresponding to level (3) in Section 1.3), we also use characteristics of density and voidage. The volume of the bed for a cylindrical vessel is

$$V = \pi D^2 L/4 \tag{15.4-1}$$

The bulk density, ρ_B, of the bed is the ratio of the total mass W to the total volume V:

$$\rho_B = W/V \tag{21.3-1}$$

The bed voidage, ϵ_B, is the ratio of the *interparticle* void space to the total volume V:

$$\epsilon_B = \frac{V - \text{volume of particles}}{V} = \frac{V - V\rho_B/\rho_p}{V}$$
$$= 1 - \rho_B/\rho_p \tag{21.3-2}$$

where ρ_p is the density of the particle, equation 8.5-1. From equations 21.3-2 and 8.5-3,

$$\rho_B = \rho_p(1 - \epsilon_B) = \rho_s(1 - \epsilon_p)(1 - \epsilon_B) \quad (21.3\text{-}3)$$

where ρ_s is the true density of the solid (no intraparticle voidage).

21.3.2 Fluid-Particle Interaction; Pressure Drop $(-\Delta P)$

When a fluid flows through a bed of particles, interactions between fluid and particles lead to a frictional pressure drop, $(-\Delta P)$. Calculation of $(-\Delta P)$ enables determination of both L and D, for a given W (or V). This calculation is done by means of the momentum balance, which results in the pressure gradient given by

$$\frac{dP}{dx} + \frac{fu^2\rho_f}{d'_p} = 0 \quad (21.3\text{-}4)$$

where x is the bed-depth coordinate in the direction of flow, f is the friction factor, u is the superficial linear velocity, ρ_f is the density of the fluid, and d'_p is an effective particle diameter. Integration of equation 21.3-4 on the assumption that the second term on the left is constant results in the pressure drop equation:

$$\frac{(-\Delta P)}{L} = \frac{fu^2\rho_f}{d'_p} \quad (21.3\text{-}5)$$

where L is the depth of the bed ($L = x$ at the outlet); note that $\Delta P = P(\text{outlet}) - P(\text{inlet})$ is negative. In equation 21.3-5, d'_p is the effective particle diameter, which accounts for shape, defined as

$$d'_p = 6 \times \text{volume of particle/external surface area of particle} \quad (21.3\text{-}6)$$

For a spherical particle of diameter d_p, $d'_p = d_p$, and for a solid cylindrical particle, $d'_p = 3d_p/(2 + d_p/L_p)$ or $1.5d_p$, if $d_p/L_p \ll 2$, where L_p is the length of the particle.

For the friction factor, we use the correlation of Ergun (1952):

$$f = [1.75 + 150(1 - \epsilon_B)/\text{Re}'](1 - \epsilon_B)/\epsilon_B^3 \quad (21.3\text{-}7)$$

where ϵ_B is the bed voidage, and Re′ is a Reynolds number defined by

$$\text{Re}' = d'_p u \rho_f / \mu_f = d'_p G / \mu_f \quad (21.3\text{-}8)$$

where μ_f is the viscosity of the fluid, and G is the mass velocity:

$$G = \dot{m}/A_c = 4\dot{m}/\pi D^2 \quad (21.3\text{-}9)$$

where $\dot{m}$ is the mass flow rate, and A_c is the cross-sectional area of the bed.

If D or L is specified (for a given W), $(-\Delta P)$ may be calculated from equations 21.3-5 and -7 to -9, together with equations 15.4-1 and 21.3-1.

Alternatively, we may use these equations to determine D (and L) for a specified allowable value of $(-\Delta P)$. The equations may be solved explicitly for D (or L):

$$D^6 - \beta \kappa D^2 - \kappa = 0 \qquad (21.3\text{-}10)$$

where

$$\beta = \frac{150}{4(1.75)} \frac{\pi \mu_f (1-\epsilon_B)}{d'_p \dot{m}} = \frac{67.32 \mu_f (1-\epsilon_B)}{d'_p \dot{m}} \qquad (21.3\text{-}11)$$

$$\kappa = \frac{64 \alpha \dot{m}^2 W}{\pi^3 \rho_f d'_p (-\Delta P) \rho_B} \qquad (21.3\text{-}12)$$

and

$$\alpha = 1.75(1-\epsilon_B)/\epsilon_B^3 \qquad (21.3\text{-}13)$$

Equation 21.3-10 provides the value of the bed diameter D for a given allowable pressure drop, $(-\Delta P)$, a value of W calculated as described in Section 21.5, and known values of the other quantities. Since α, β, and κ are all positive, from the Descartes rule of signs (Section 14.3.3), there is only one positive real root of equation 21.3-10. If the equations are solved for L instead of D, a cubic equation results.

In the choice of a value for the allowable $(-\Delta P)$ on the one hand, or for D (or L) on the other (given W), there is a trade-off between the cost of the vessel and the cost of pumping or compressing the fluid. The smaller D, the greater the L/D ratio and the greater $(-\Delta P)$; thus, the cost of the vessel is less, but the cost of pumping is greater, and conversely.

EXAMPLE 21-2

The feed to the first stage of a sulfur dioxide converter is at 100 kPa and 700 K, and contains 9.5 mol % SO_2, 11.5% O_2, and 79% N_2. The feed rate of SO_2 is 7.25 kg s^{-1}. The mass of catalyst (W) is 6000 kg, the bed voidage (ϵ_B) is 0.45, the bulk density of the bed (ρ_B) is 500 kg m^{-3}, and the effective particle diameter (d'_p) is 15 mm; the fluid viscosity (μ_f) is 3.8 x 10^{-5} kg m^{-1} s^{-1}. The allowable pressure drop $(-\Delta P)$ is 7.5 kPa.

(a) Calculate the bed diameter (D) and the bed depth (L) in m, assuming that the fluid density and viscosity are constant.
(b) How sensitive are these dimensions to the allowable $(-\Delta P)$? Consider values of $(-\Delta P)$ from 2.5 to 15 kPa.

SOLUTION

(a) To use equations 21.3-10 to -13, we need to determine the total mass flow rate, $\dot{m}$, and the fluid density, ρ_f.

$$\dot{m} = \dot{m}_{SO_2} + \dot{m}_{O_2} + \dot{m}_{N_2} = \dot{m}_{SO_2} + x_{O_2} F_t M_{O_2} + x_{N_2} F_t M_{N_2}$$

where x is mole fraction, F_t is total molar flow rate, and M is molar mass.

$$F_t = \dot{m}_{SO_2}/x_{SO_2} M_{SO_2} = 7.25/[0.095(64.1)] = 1.191 \text{ kmol s}^{-1}$$

$$\dot{m} = 7.25 + 0.115(1.191)32.0 + 0.79(1.191)28.0 = 38.0 \text{ kg s}^{-1}$$

Assuming ideal-gas behavior, we have

$$\rho_f = PM_{av}/RT = P(\dot{m}/F_t)/RT$$
$$= 100000(38.0)/1191(8.3145)700 = 0.548 \text{ kg m}^{-3}$$

We can now calculate α, β, and κ from equations 21.3-13, -11, and -12, respectively:

$$\alpha = 1.75(1 - \epsilon_B)/\epsilon_B^3 = 1.75(1 - 0.45)/0.45^3 = 10.56$$

$$\beta = \frac{67.32 \mu_f (1 - \epsilon_B)}{d_p' \dot{m}} = \frac{67.32(3.8 \times 10^{-5})(1 - 0.45)}{0.015(38.0)} = 2.47 \times 10^{-3} \text{ m}^{-2}$$

$$\kappa = \frac{64 \alpha \dot{m}^2 W}{\pi^3 \rho_f d_p'(-\Delta P)\rho_B} = \frac{64(10.56)(38.0)^2 6000}{(3.142)^3 0.548(0.015)7500(500)} = 6135 \text{ m}^6$$

Equation 21.3-10 becomes

$$D^6 - 15.13 D^2 - 6135 = 0$$

Solving for D by trial, or using a root-finding technique as in the E-Z Solve software (see file ex21-2.msp), we obtain

$$D = 4.31 \text{ m}$$

The bed depth (L) can be calculated from equations 15.4-1 and 21.3-1:

$$L = \frac{4W}{\rho_B \pi D^2} = \frac{4(6000)}{500(3.142)4.31^2} = 0.82 \text{ m}$$

(b) The procedure described in (a) is repeated for values of $(-\Delta P)$ in increments of 2.5 kPa within the range 2.5 to 15 kPa, with results for D and L given in the following table:

$(-\Delta P)$/kPa	D/m	L/m
2.5	5.19	0.566
5.0	4.62	0.717
7.5	4.31	0.822
10.0	4.11	0.906
12.5	3.95	0.977
15.0	3.84	1.039

As expected, D decreases and L increases as $(-\Delta P)$ increases. For a given amount of catalyst, a reduced pressure drop (and operating power cost) can be obtained by reducing the bed depth at the expense of increasing the bed diameter (and vessel cost).

21.3.3 Considerations Relating to a Reversible Reaction

Since many important gas-phase catalytic reactions are reversible, we focus on the implications of this characteristic for reactor design. These stem from both equilibrium and kinetics considerations.

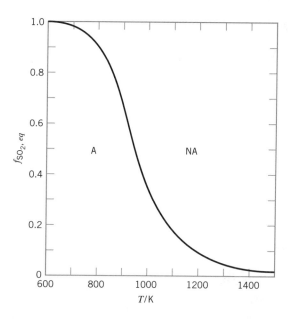

Figure 21.2 Dependence of $f_{SO_2,eq}$ on T for $SO_2 + (1/2)O_2 \rightleftharpoons SO_3$; $P = 0.101$ MPa; feed is 9.5 mole %SO_2, 11.5% O_2, 79% N_2; A = accessible, NA = nonaccessible region (Smith and Missen, 1991, p. 176)

21.3.3.1 Equilibrium Considerations

In Example 21-1 above, the effects on equilibrium of changing P (at constant T and r, initial molar ratio of inert species to limiting reactant) and changing r (at constant P and T) are examined. Here, we focus on the effect of T (at constant P, r) through the dependence of fractional conversion on T. For the reaction $A + \ldots \rightleftharpoons$ products (at equilibrium), we examine the behavior of $f_{A,eq}(T)$, where $f_{A,eq}$ represents f_A at equilibrium.

There is an important difference in this behavior between an exothermic reaction and an endothermic reaction. From equation 3.1-5, the van't Hoff equation, the equilibrium constant (K_{eq}) decreases with increasing T for an exothermic reaction, and increases for an endothermic reaction. The behavior of $f_{A,eq}(T)$ corresponds to this.

As an example for an exothermic reaction, the oxidation of SO_2 (reaction (A) in section 21.1), Figure 21.2 shows $f_{SO_2,eq}$ for a specified pressure and initial composition (Smith and Missen, 1991, p. 176); these conditions are approximately those for some commercial operations. $f_{SO_2,eq}$ decreases in a typically sigmoidal manner from virtually 1 at 600 K to nearly 0 at 1500 K. For the conditions stated, the only nonequilibrium states allowed are in the region A (for accessible); the region NA is not accessible. That is, for the initial conditions given, an actual reactor can be operated in region A, but not in region NA. The equilibrium curve is the boundary between these two regions, and represents an upper limit for f_{SO_2}. As noted in Example 21-1, the position of the equilibrium curve can be altered by change of P or feed composition (see Problem 21-5).

As an example for an endothermic reaction, we use the dehydrogenation of ethylbenzene, reaction (D) in Section 21.1. This is developed in more detail in the following example.

EXAMPLE 21-3

For the dehydrogenation of ethylbenzene at equilibrium, C_8H_{10} (EB) $\rightleftharpoons C_8H_8$ (S) + H_2, calculate and plot $f_{EB,eq}(T)$, at $P = 0.14$ MPa, with an initial molar ratio of inert gas (steam, H_2O) to EB of $r = 15$ (these conditions are also indicative of commercial operations). Assume ideal-gas behavior, with $K_p = 8.2 \times 10^5 \exp(-15,200/T)$ MPa.

SOLUTION

We first develop an expression for $f_{EB,eq}(T)$, and then plot values calculated from this expression. The expression comes from the form of K_p, which involves partial pressures, and which may be related to $f_{EB,eq}$ by means of the following stoichiometric table:

Species	Initial moles	Change	Final moles
C_8H_{10}(EB)	1	$-f_{EB,eq}$	$1 - f_{EB,eq}$
C_8H_8(S)	0	$f_{EB,eq}$	$f_{EB,eq}$
H_2	0	$f_{EB,eq}$	$f_{EB,eq}$
H_2O (inert)	r	0	r
total	$1+r$		$1+r+f_{EB,eq}$

Applying the definition of partial pressure for each species, $p_i = x_i P$, we have

$$K_p = \frac{p_S p_{H_2}}{p_{EB}} = \frac{f_{EB,eq}^2 P^2}{(1+r+f_{EB,eq})^2} \frac{(1+r+f_{EB,eq})}{(1-f_{EB,eq})P}$$

On solution for $f_{EB,eq}$, this becomes

$$f_{EB,eq} = \frac{r}{2(1+P/K_p)}\left[\left(1 + \frac{4(1+r)(1+P/K_p)}{r^2}\right)^{1/2} - 1\right]$$

For $P = 0.14$ MPa and $r = 15$,

$$f_{EB,eq} = \frac{7.5}{1+0.14/K_p}[(1+(0.2844)(1+0.14/K_p)^{1/2} - 1]$$

which, together with,

$$K_p = 8.2 \times 10^5 \exp(-15,200/T)$$

is used to obtain $f_{EB,eq}(T)$. These equations may be solved for $f_{EB,eq}$ as a function of T using E-Z Solve (file ex21-3.msp). The results are plotted in Figure 21.3. This shows a typical sigmoidal increase in $f_{EB,eq}$ with increase in T, in this case, from virtually 0 at 400 K to virtually 1 at 1200 K. The regions A and NA again represent accessible and nonaccessible regions, respectively.

21.3.3.2 Kinetics Considerations

In Section 5.3 for reversible reactions, it is shown that the rate of an exothermic, reversible reaction goes through a maximum with respect to T at constant fractional conversion f, but decreases with respect to increasing f at constant T. (These conclusions apply whether the reaction is catalytic or noncatalytic.) Both features are illustrated graphically in Figure 21.4 for the oxidation of SO_2, based on the rate law of Eklund (1956):

$$(-r_{SO_2}) = k_{SO_2}\left(\frac{p_{SO_2}}{p_{SO_3}}\right)^{1/2}\left[p_{O_2} - \left(\frac{p_{SO_3}}{p_{SO_2}K_p}\right)^2\right] \quad (21.3\text{-}14)$$

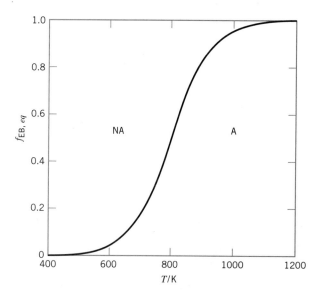

Figure 21.3 $f_{EB,eq}$ versus T for C_8H_{10} (EB) $\rightleftharpoons C_8H_8 + H_2$; $P = 0.14$ MPa; $r = 15$ (based on data from Stull et al., 1969)

In Figure 21.4, f_{SO_2} is plotted against $T/°C$ with $(-r_{SO_2})$ as a parameter; $f_{SO_2,eq}$, from Figure 21.2, is shown in part as the uppermost curve. Several curves of constant $(-r_{SO_2})$ are shown for the region of commercial operation in the accessible region (A) of Figure 21.2. However, each point in this region in Figure 21.4 represents a value of $(-r_{SO_2})$ at $\{f_{SO_2}, T\}$. It is assumed that the maximum allowable temperature (T_{max}) for the catalyst is 615°C (to avoid catalyst deactivation).

In Figure 21.4, each curve for constant $(-r_{SO_2})$ exhibits a maximum at a particular value of $\{f_{SO_2}, T\}$. For a given f_{SO_2}, this maximum is the highest reaction rate at that conversion. The dashed curve drawn through these points is therefore called the locus of maximum rates. For a given T, there is no comparable behavior: $(-r_{SO_2})$ decreases monotonically as f_{SO_2} increases.

For an endothermic, reversible reaction (such as the dehydrogenation of ethylbenzene), as also shown in Section 5.3 and illustrated in Figure 5.2(b), the rate does *not* exhibit a maximum with respect to T at constant f, but increases monotonically with increasing T. The rate also decreases with increasing f at constant T, as does the rate of an exothermic reaction.

Care must be taken to specify properly the basis for the reaction rate. The most useful basis for design of an FBCR is the catalyst mass, that is, the rate is $(-r_A)_m$, in units of, say, mol (kg cat)$^{-1}$ s^{-1}. $(-r_A)_m$ is related to the rate per unit reactor (or bed) volume, $(-r_A)_v$ through the bed density:

$$(-r_A)_v = \rho_B(-r_A)_m \qquad (21.3\text{-}15)$$

It is also convenient to express the rate on a particle volume basis, $(-r_A)_p$, which is related to $(-r_A)_m$ through the particle density:

$$(-r_A)_p = \rho_p(-r_A)_m \qquad (21.3\text{-}16)$$

Equation 21.3-16 may be used to convert rate constants from a mass to a particle volume basis for calculation of the Thiele modulus (e.g., equation 8.5-20b). In this chapter, $(-r_A)$ without further designation means $(-r_A)_m$.

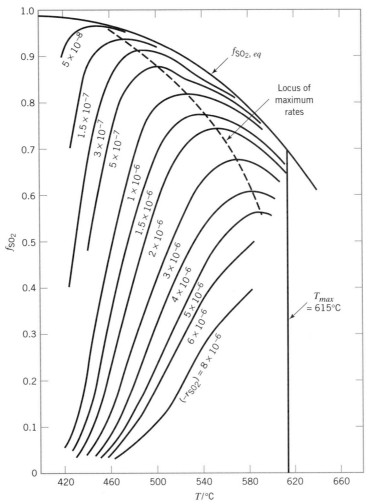

Figure 21.4 f_{SO_2} versus T with $(-r_{SO_2})$ as parameter for $SO_2 + (1/2)O_2 \rightleftharpoons SO_3$, according to equation 21.3-14 (Eklund, 1956); $(-r_{SO_2})$ in mol SO_2 (g cat)$^{-1}$ s^{-1}; $P = 0.101$ MPa; feed: 9.5 mol % SO_2, 11.5% O_2, 79% N_2; 8 × 25 mm cylindrical catalyst particles

21.4 A CLASSIFICATION OF REACTOR MODELS

The process design of an FBCR involves exploiting the continuity (material-balance) and energy equations to determine, among other things, the amount of a specified catalyst required for a given feed composition, fractional conversion, and throughput; concentration and temperature profiles; and the thermal mode of operation to achieve the objectives. The appropriate forms of these equations, together with rate equations for reaction and heat transfer, constitute the main working equations of a reactor model. Before using any particular model for calculations, we describe a classification as a basis for consideration of models in general for an FBCR.

The basis for a classification is shown in Figure 21.5, and control volumes are shown in Figure 21.6 for axial flow. These diagrams could apply to catalyst packed inside a vessel or inside a tube in a multitubular arrangement (but not outside the tubes without modification—see Section 21.5.4.2).

524 Chapter 21: Fixed-Bed Catalytic Reactors for Fluid-Solid Reactions

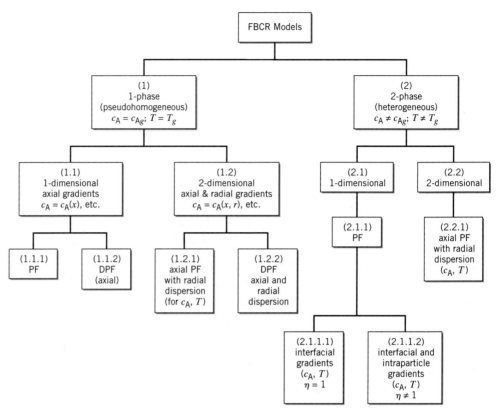

Figure 21.5 A classification of FBCR models

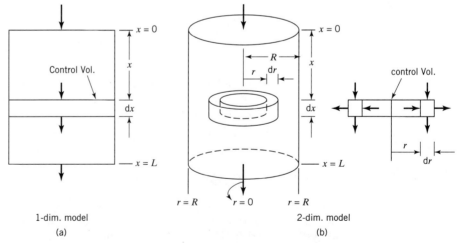

Figure 21.6 Control volumes for (a) one- and (b) two-dimensional models

As indicated in Figure 21.5, there are three levels of bifurcation:

(1) "one-phase" (pseudohomogeneous) versus two-phase (heterogeneous) behavior;

(2) one-dimensional versus two-dimensional gradients of concentration and temperature;

(3) plug flow (PF) versus dispersed plug flow (DPF).

Points (1) and (2) require further explanation, and for this we must distinguish between gradients at the local or particle level, as considered in Chapter 8, and gradients at the bed level (both axial and radial). (In point (3), the axial flow can be described by means of either the dispersed plug-flow model or the tanks-in-series model (Chapter 19).)

(1) In a "one-phase" or pseudohomogeneous model, intraparticle gradients are ignored, so that everywhere in the catalyst bed, the value of concentration (c_A) or temperature (T) is the same as the local value for bulk fluid (c_{Ag} or T_g). The actual two-phase system (fluid and catalyst) is treated as though it were just one phase. In a two-phase or heterogeneous model, intraparticle gradients are allowed, so that, locally, within the particle, $c_A \neq c_{Ag}$ and $T \neq T_g$. The effects of these gradients are reflected in the particle effectiveness factor, η (Section 8.5.4), or an overall effectiveness factor, η_o (Section 8.5.6). If the reactor operates nearly isothermally, a single value of η or η_o may be sufficient to describe thermal and concentration gradients. However, if operation is nonisothermal, η and/or η_o may vary along the length of the vessel, and it may be necessary to account explicitly for this behavior within the reactor model. A further discussion of the effects of interparticle and intraparticle gradients is presented in Section 21.6.

(2) In a one-dimensional model, gradients of c_A and T at the bed level are allowed only in the axial direction of bulk flow. In a two-dimensional model, gradients at the bed level in both the axial and radial directions are taken into account.

The following example illustrates the derivation of the continuity and energy equations for model 1.2.2 in Figure 21.5, a pseudohomogeneous, two-dimensional model with DPF.

EXAMPLE 21-4

Derive the continuity and energy equations for an FBCR model based on pseudohomogeneous, two-dimensional, DPF considerations. State any assumptions made, and include the boundary conditions for the equations.

SOLUTION

The notation and bed control volume used are as in Figure 21.6(b). Assumptions are as follows:

(1) The reaction is $A(g) + \ldots \xrightarrow{cat(s)}$ products.
(2) Operation is at steady state.
(3) Fluid density is not necessarily constant.
(4) $c_A = c_A(x, r)$; $T = T(x, r)$.
(5) The linear velocity, $u = u(x)$; that is, the mass velocity G, D_{er} and k_{er} are independent of r, where D_{er} and k_{er} are the effective radial dispersion coefficient and thermal conductivity, respectively.
(6) c_P (fluid) and $(-\Delta H_{RA})$ are constant.

Continuity equation: from a material balance for A around the control volume,

$$\begin{pmatrix} \text{input rate of A by} \\ \text{bulk flow} \\ + \text{axial dispersion} \\ + \text{radial dispersion} \end{pmatrix} = \begin{pmatrix} \text{output rate of A by} \\ \text{bulk flow} \\ + \text{axial dispersion} \\ + \text{radial dispersion} \\ + \text{chemical reaction} \end{pmatrix}$$

that is,

$$uc_A 2\pi r dr - D_{ex} 2\pi r dr \frac{\partial c_A}{\partial x} - D_{er} 2\pi r dx \frac{\partial c_A}{\partial r} =$$

$$2\pi r dr \left[uc_A + \frac{\partial (uc_A)}{\partial x} dx \right] - D_{ex} 2\pi r dr \left[\frac{\partial c_A}{\partial x} + \frac{\partial}{\partial x} \left(\frac{\partial c_A}{\partial x} \right) dx \right]$$

$$- D_{er} 2\pi dx \left[r \frac{\partial c_A}{\partial r} + \frac{\partial}{\partial r} \left(r \frac{\partial c_A}{\partial r} \right) dr \right] + \rho_B(-r_A) 2\pi r dr dx$$

which becomes a second-order partial differential equation for c_A on simplification:

$$D_{ex} \frac{\partial^2 c_A}{\partial x^2} + D_{er} \left(\frac{\partial^2 c_A}{\partial r^2} + \frac{1}{r} \frac{\partial c_A}{\partial r} \right) - \frac{\partial (uc_A)}{\partial x} - \rho_B(-r_A) = 0 \quad (21.4\text{-}1)$$

where D_{ex} is the effective axial dispersion coefficient, and all quantities are in consistent units, mol A m^{-3} (bed or vessel) s^{-1}. In this connection, note that u is *not* equal to the superficial linear velocity u_s, as it would be in an empty vessel, but takes the bed voidage into account; that is, $u = q/A_c$ m^3 (fluid) s^{-1} m^{-2} (vessel), where q is the volumetric flow rate of fluid through the interparticle bed voidage, m^3 (fluid) s^{-1}. Similarly, D_{ex} and D_{er} take voidage into account, and their apparent units, m^2 s^{-1}, are in reality m^3 (fluid) m^{-1} (vessel) s^{-1} (see Section 8.5.3).

Energy equation: from an enthalpy balance around the control volume,

$$\begin{pmatrix} \text{enthalpy input rate by} \\ \text{bulk flow} \\ +\text{axial conduction} \\ +\text{radial conduction} \\ +\text{chemical reaction} \end{pmatrix} = \begin{pmatrix} \text{enthalpy output rate by} \\ \text{bulk flow} \\ +\text{axial conduction} \\ +\text{radial conduction} \end{pmatrix}$$

that is,

$$G 2\pi r dr c_P (T - T_{ref}) - k_{ex} 2\pi r dr \frac{\partial T}{\partial x} - k_{er} 2\pi r dx \frac{\partial T}{\partial r}$$

$$+ \rho_B(-r_A)(-\Delta H_{RA}) 2\pi r dr dx = G 2\pi r dr c_P \left[(T - T_{ref}) + \frac{\partial T}{\partial x} dx \right]$$

$$- k_{ex} 2\pi r dr \left[\frac{\partial T}{\partial x} + \frac{\partial}{\partial x} \left(\frac{\partial T}{\partial x} \right) dx \right]$$

$$- k_{er} 2\pi dx \left[r \frac{\partial T}{\partial r} + \frac{\partial}{\partial r} \left(r \frac{\partial T}{\partial r} \right) dr \right]$$

which becomes a second-order partial differential equation for T on simplification:

$$k_{ex} \frac{\partial^2 T}{\partial x^2} + k_{er} \left(\frac{\partial^2 T}{\partial r^2} + \frac{1}{r} \frac{\partial T}{\partial r} \right) - G c_P \frac{\partial T}{\partial x} + \rho_B(-r_A)(-\Delta H_{RA}) = 0 \quad (21.4\text{-}2)$$

where k_{ex} is the effective axial thermal conductivity. Note the similar forms of equations 21.4-1 and -2.

21.5 Pseudohomogeneous, One-Dimensional, Plug-Flow Model

The boundary conditions may be chosen as (for a "closed" vessel):

at $x = 0$,
$$u(c_{Ao} - c_A) = -D_{ex}\frac{dc_A}{dx} \quad (21.4\text{-}3)$$

$$Gc_P(T_o - T) = -k_{ex}\frac{dT}{dx} \quad (21.4\text{-}4)$$

(from Danckwerts' boundary conditions, Chapter 20)

at $x = L$ (outlet),
$$\frac{dc_A}{dx} = \frac{dT}{dx} = 0 \quad (21.4\text{-}5)$$

(no flow by dispersion across outlet plane)

at $r = 0$,
$$\frac{\partial c_A}{\partial r} = \frac{\partial T}{\partial r} = 0 \quad (21.4\text{-}6)$$

at $r = R$,
$$\frac{\partial c_A}{\partial r} = 0 \quad (21.4\text{-}7)$$

$$-k_{er}\left(\frac{\partial T}{\partial r}\right)_{r=R} = U(T_{r=R} - T_s) \quad (21.4\text{-}8)$$

where T_s is the temperature of the surroundings.

The continuity and energy equations for the simpler pseudohomogeneous models 1.1.1, 1.1.2, and 1.2.1 in Figure 21.5 may be obtained from equations 21.4-1 and -2, respectively. Model 1.1.1 is used, for simplicity, in Section 21.5. For this model (and for the other one-dimensional model, 1.1.2), we must consider what the radial conduction terms become, both for conduction through the wall, and for adiabatic operation. In Section 21.6, we extend the treatment briefly to the corresponding heterogeneous model (2.1.1.2 in Figure 21.5). In Section 21.7, we give a short discussion on one-dimensional versus two-dimensional models.

21.5 PSEUDOHOMOGENEOUS, ONE-DIMENSIONAL, PLUG-FLOW MODEL

21.5.1 Continuity Equation

For the simplest model in Figure 21.5, model 1.1.1, pseudohomogeneous, one-dimensional plug flow, equation 21.4-1 becomes, with $D_{ex} = D_{er} = 0$ and $c_A = c_A(x)$,

$$\frac{d(uc_A)}{dx} + \rho_B(-r_A) = 0 \quad (21.5\text{-}1)$$

In terms of fractional conversion f_A, since

$$uc_A = qc_A/\pi r^2 = F_A/\pi r^2 = F_{Ao}(1 - f_A)/\pi r^2$$

then,

$$d(uc_A) = -(F_{Ao}/\pi r^2)df_A$$

and equation 21.5-1 becomes

$$\frac{df_A}{dx} - \frac{\pi r^2 \rho_B (-r_A)}{F_{Ao}} = 0 \qquad (21.5\text{-}2)$$

The most useful form of the continuity equation (or material balance) is that which enables calculation of the amount of catalyst (W) required in terms of f_A (at the bed depth x). Since

$$W = \rho_B V = \rho_B \pi r^2 x$$

$$dx = dW/\rho_B \pi r^2$$

and equation 21.5-2 becomes

$$\frac{dW}{df_A} - \frac{F_{Ao}}{(-r_A)} = 0 \qquad (21.5\text{-}3)$$

or

$$W = F_{Ao} \int \frac{df_A}{(-r_A)} \qquad (21.5\text{-}4)$$

Equation 21.5-4 is a design equation for an FBCR of the same form as equation 15.2-2 for the volume V of a PFR. An important difference relates to the units, and hence numerical value, of $(-r_A)$ in the two equations. In equation 21.5-4, $(-r_A)$ is in mass units, say, mol A (kg cat)$^{-1}$ s^{-1}, and in equation 15.2-2, $(-r_A)$ is in volume units, say, mol A m^{-3} s^{-1}. As noted in Section 21.3.3.2, they are related by

$$\rho_B (-r_A)_m = (-r_A)_v \qquad (21.3\text{-}15)$$

Equation 21.3-15 reconciles the various forms of the continuity equation for pseudo-homogeneous systems in this section with those in Chapter 15 for actual homogeneous reactions.

An alternative form of the continuity equation, in terms of molar flow rates, is

$$\frac{dF_A}{dx} + \pi r^2 \rho_B (-r_A) = 0 \qquad (21.5\text{-}5)$$

since $F_A = \pi r^2 u c_A$. This form is useful for variable-density operation, and for complex-reaction systems, where f_A cannot be used exclusively for design purposes.

Since $(-r_A)$ depends on T as well as on f_A, we need to use the energy equation in addition to the continuity equation to obtain W. The form of the energy equation depends on the thermal configuration used (Figure 21.1), and we consider various cases in the following sections. We neglect any effect of pressure drop in obtaining W, because $(-\Delta P)$ is usually relatively small; however, $(-\Delta P)$ is important in determining vessel diameter and bed depth from W, as described in Section 21.3.2.

21.5.2 Optimal Single-Stage Operation

From equation 21.5-4, the amount of catalyst is a minimum, W_{min}, if $(-r_A)$ is the maximum rate at f_A. For an exothermic, reversible reaction, this means operating

21.5 Pseudohomogeneous, One-Dimensional, Plug-Flow Model

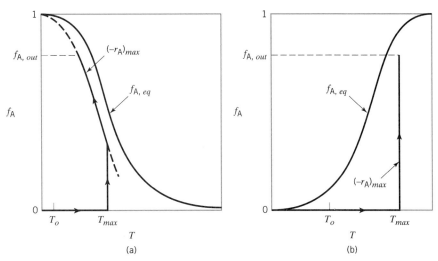

Figure 21.7 Path for W_{min} in FBCR: (a) exothermic reaction (nonadiabatic, nonisothermal operation), (b) endothermic reaction (isothermal)

nonadiabatically and nonisothermally on the locus of maximum rates, subject to any limitation imposed by T_{max}. For an endothermic, reversible reaction, it means operating isothermally at the highest feasible value of T. The reaction paths (f_A versus T) for the two cases are shown schematically in Figure 21.7(a) for an exothermic reaction and 21.7(b) for an endothermic reaction. In Figure 21.7, T_{max} represents the highest feasible value of T, usually dictated by catalyst stability.

In Figure 21.7(a), it is assumed that $f_{Ao} = 0$, and $T_o < T_{max}$. T_o is first increased to T_{max} in a preheater, and operation in the FBCR is then isothermal until T_{max} intersects with $(-r_A)_{max}$, after which it follows $(-r_A)_{max}$ until a specified final value of fractional conversion, $f_{A,out}$, is reached. This last part requires appropriate adjustment of T at each point, and is thus nonadiabatic and nonisothermal. Such precise and continuous adjustment is impractical, but any actual design path attempts to approximate the essence of this.

In Figure 21.7(b), the path in the FBCR itself is isothermal at T_{max}, since this represents the locus of maximum rates in this case. Even strict adherence to isothermal operation is usually impractical, but it represents an ideal case for minimizing W.

21.5.3 Adiabatic Operation

21.5.3.1 *Multistage Operation with Interstage Heat Transfer*

For one-dimensional plug flow, with $k_{ex} = k_{er} = 0$ and $T = T(x)$, equation 21.4-2 reduces to

$$Gc_P \frac{dT}{dx} - \rho_B(-r_A)(-\Delta H_{RA}) = 0 \quad \text{(21.5-6)}$$

The one required boundary condition can be chosen as

$$T = T_o \quad \text{at } x = 0 \quad \text{(21.5-7)}$$

Equation 21.5-6 is in a form to determine the axial temperature profile through the catalyst bed. To determine W from equation 21.5-4, however, we transform 21.5-6 to relate f_A and T, corresponding to equation (D) in Example 15-7 for adiabatic operation of a PFR. Since

$$G = \dot{m}/\pi r^2 \quad (21.3\text{-}9)$$

and

$$dx = dV/\pi r^2 = dW/\rho_B \pi r^2 = F_{Ao} df_A/(-r_A)\rho_B \pi r^2$$

from equations 15.2-7, 21.3-1 and 21.5-3, substituting for G and dx in equation 21.5-6 and rearranging, we obtain

$$df_A = \frac{\dot{m}c_P}{(-\Delta H_{RA})F_{Ao}} dT \quad (21.5\text{-}8)$$

On integration, with $f_A = 0$ at T_o, and the coefficient of dT constant, this becomes

$$f_A = \frac{\dot{m}c_P}{(-\Delta H_{RA})F_{Ao}}(T - T_o) \quad (21.5\text{-}9)$$

Integration of equation 21.5-8 from the inlet to the outlet of the ith stage of a multistage arrangement, again with the coefficient of dT constant, results in

$$\Delta f_{Ai} = f_{Ai} - f_{A,i-1} = \frac{\dot{m}c_P}{(-\Delta H_{RA})F_{Ao}}(T_i - T_{i-1}) \quad (21.5\text{-}9a)$$

The product $F_t C_P$, where F_t is the *total* molar flow rate and C_P is the molar heat capacity of the flowing stream, may replace $\dot{m}c_P$ in these equations, if the stream is an ideal solution. Integration of equation 21.5-8, or its equivalent, may need to take into account the dependence of $\dot{m}c_P$ and $F_t C_P$ on T and/or f_A, and of $(-\Delta H_{RA})$ on T (see Example 15-7). However, compared to the effect of T on k_A, the effect of T on $(-\Delta H_{RA})$ and c_P is usually small.

For adiabatic operation, with constant c_P and $(-\Delta H_{RA})$, equation 21.5-9 is a linear relation for $f_A(T)$, with a positive slope for an exothermic reaction and a negative slope for an endothermic reaction. It may be regarded as an operating line, since each point (f_A, T) actually exists at some position x in the reactor.

Figure 21.8 shows schematically the linear energy equation or operating line added to lines shown in Figure 21.7, both for an exothermic reaction, Figure 21.8(a), and for an endothermic reaction, Figure 21.8(b). Both cases illustrate the limits imposed by the equilibrium constraint for reversible reactions.

In Figure 21.8(a), the operating line AB represents equation 21.5-9; the line cannot extend past the equilibrium curve, but the question remains about an optimal "distance" of point B from it. Figure 21.8(a) also illustrates how the equilibrium constraint can be overcome by using multistage operation. If the line AB represents operation in a first stage, the line CD represents operation in a second stage, in which f_A (at D) > f_A (at B). Between the two stages, the system is cooled in a heat exchanger, the change represented by the line BC. A second question for optimal operation in this fashion

21.5 Pseudohomogeneous, One-Dimensional, Plug-Flow Model

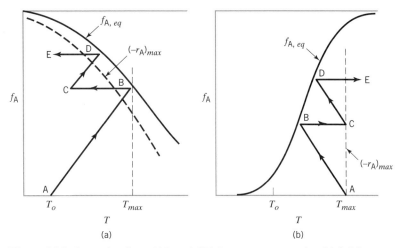

Figure 21.8 Operating lines (AB and CD) from energy equation 21.5-8 for two-stage (adiabatic) FBCR: (a) exothermic reversible reaction; (b) endothermic reversible reaction

relates to the position of C. Additional stages (not shown) may be added to achieve still higher conversion, but a law of diminishing returns sets in, since large amounts of catalyst are required in the high-conversion region, where reaction rates are low. Note that the operating lines, both for the catalyst beds and for the heat exchanger(s), tend to straddle the locus of maximum rates in order to reduce W.

Figure 21.8(b) shows that staging can also be used to "climb" the equilibrium barrier in adiabatic operation for an endothermic reaction. The differences between endothermic and exothermic systems are:

(1) The energy equation (operating lines AB and CD for the two-stages shown) has a negative slope.
(2) There is no locus of maximum rates.
(3) The interstage heat transfer is heating rather than cooling.
(4) The feed to each stage, in principle, is at T_{max}.

The following example illustrates the simultaneous use of equations 21.5-4 and -9 to calculate the amount of catalyst, W, for an FBCR.

EXAMPLE 21-5

Sheel and Crowe (1969) carried out a simulation of an ethylbenzene (EB) dehydrogenation reactor in Sarnia, Ontario, as it existed at that time. In this example, we use some of the data they provided. They considered a complex system to allow for the formation of byproducts, but, for simplicity, we consider a simple system involving only the dehydrogenation reaction ((D) in Section 21.1), and use the kinetics data of Wenner and Dybdal (1948). They used the pseudohomogeneous, one-dimensional PF model with adiabatic operation for a single-stage reactor. (Reactor design for this system has been discussed in greater detail by Elnashaie and Elshishini, 1993, pp. 364–379, based on the work of Sheel and Crowe, with corrections.)

From the data given below, calculate (a) the amount of catalyst, W, for $f_{EB} = 0.40$, and (b) the bed diameter D and bed depth L.

Data:

$F_{EBo} = 11$ mol s^{-1}; $T_o = 922$ K; $P_o = 0.24$ MPa; allowable $(-\Delta P) = 8.1$ kPa;
$F_{H_2O} = 165$ mol s^{-1}; $\Delta H_{REB} = 126$ kJ mol^{-1};
$c_P = 2.4$ J g^{-1} K^{-1}; $\epsilon_B = 0.50$; $\mu_f = 3 \times 10^{-5}$ Pa s;
Assume $\rho_B = 500$ kg m^{-3} and the particles are cylindrical with $d_p = 4.7$ mm;
Rate law: $(-r_{EB}) = k_{EB}(p_{EB} - p_S p_{H_2}/K_p)$;
$k_{EB} = 3.46 \times 10^4 \exp(-10,980/T)$ mol (kg cat)$^{-1}$ s^{-1} MPa^{-1} with T in K;
$K_p = 8.2 \times 10^5 \exp(-15200/T)$ MPa

SOLUTION

(a) Calculation of the amount of catalyst, W, is done by simultaneous (numerical) solution of the material and energy balances given by equations 21.5-4 and -9, respectively:

$$W = \int_0^{0.4} F_{EBo} df_{EB}/(-r_{EB}) \qquad \text{(A)}$$

$$T = T_o + [(-\Delta H_{REB})F_{EBo}/\dot{m}c_P]f_{EB} \qquad \text{(B)}$$

The latter corresponds to the operating line AB in Figure 21.8(b). To evaluate the integral in (A), we need the rate law:

$$(-r_{EB}) = k_{EB}(p_{EB} - p_S p_{H_2}/K_p) \qquad \text{(C)}$$

where

$$k_{EB} = 3.46 \times 10^4 \exp(-10,980/T) \qquad \text{(D)}$$

$$K_p = 8.2 \times 10^5 \exp(-15,200/T) \qquad \text{(E)}$$

For use in (C), the partial pressures can be determined using a stoichiometric table as in Example 21-3, in conjunction with $C_8H_{10}(EB) = C_8H_8(S) + H_2$:

Species	Initial moles	Change	Final moles
EB	F_{EBo}	$-F_{EBo}f_{EB}$	$F_{EBo}(1 - f_{EB})$
S	0	$+F_{EBo}f_{EB}$	$F_{EBo}f_{EB}$
H$_2$	0	$+F_{EBo}f_{EB}$	$F_{EBo}f_{EB}$
H$_2$O	$15F_{EBo}$	0	$15F_{EBo}$
total			$F_{EBo}(16 + f_{EB})$

Therefore,

$$p_{H_2} = p_S = [f_{EB}/(16 + f_{EB})]P_o \qquad \text{(F)}$$

$$p_{EB} = [(1 - f_{EB})/(16 + f_{EB})]P_o \qquad \text{(F')}$$

where P_o is the inlet pressure, and the small pressure drop is ignored for this purpose, since it is only 6% of P_o.

These equations, (A) to (F'), may be solved using the E-Z Solve software or the trapezoidal rule for evaluation of the integral in (A). In the latter case, the following algorithm may be used:

(1) Choose a value of f_{EB}.
(2) Calculate T from (B).
(3) Calculate k_{EB} and K_p from (D) and (E) respectively.
(4) Calculate p_{H_2}, p_S, and p_{EB} from (F) and (F').
(5) Calculate $(-r_{EB})$ from (C).
(6) Calculate $G = F_{EBo}/(-r_{EB})$.
(7) Repeat steps (1) to (6) for values of f_{EB} between 0 and 0.40.
(8) Calculate $W = \sum G^* \Delta f_{EB}$, where G^* is the average of two successive values of G, and Δf_{EB} is the increment or step-size for f_{EB}.

Results are given in the following table for a step-size of 0.1. The estimated amount of catalyst is $W = 2768$ kg.

f_{EB}	T/K	$10^3(-r_{EB})/$ mol (kg cat)$^{-1}$ s^{-1}	G/kg	G^*/kg	$G^*\Delta f_{EB}$/kg
0	922	3.49	3149		
				3697	370
0.1	908	2.59	4244		
				5077	508
0.2	894	1.86	5911		
				7313	730
0.3	880	1.26	8715		
				11596	1160
0.4	866	0.76	14476		
				total	2768

The amount of catalyst required for each 0.10 increment in f_{EB} increases considerably as f_{EB} increases. This is due to the significant decrease in $(-r_{EB})$ between the inlet and the outlet of the bed. This, in turn, is due to the decrease in T, which decreases k_{EB}, and due to the accumulation of products, which enhances the reverse reaction.

If the E-Z Solve software is used (file ex21-5.msp), $W = 2704$ kg, as a more accurate result. However, the trapezoidal-rule approximation provides a good estimate for the relatively large step-size used.

(b) The calculation of D and L is similar to that in Example 21-2. In this case,

$$d'_p = 1.5 d_p = 0.0071 \text{ m} \quad \text{(from 21.3-6)}$$

$$\dot{m} = [11(106.17) + 165(18.02)]/1000 = 4.136 \text{ kg s}^{-1}$$

$$\rho_f \text{ (at inlet conditions)} = P_o(\dot{m}/F_{to})/RT_o$$
$$= 0.24(4.14)/176(8.3145)922 = 0.429 \text{ kg m}^{-3}$$

$$\alpha = 1.75(1 - \epsilon_B)/\epsilon_B^3 = 1.75(1 - 0.50)/0.50^3 = 7.0 \quad \text{(from 21.3-13)}$$

$$\beta = 67.32 \mu_f (1 - \epsilon_B)/d'_p \dot{m} \quad \text{(from 21.3-11)}$$
$$= 67.32(3 \times 10^{-5})(1 - 0.50)/0.0071(4.136) = 0.0346 \text{ m}^{-2}$$

$$\kappa = 64\alpha \dot{m}^2 W/\pi^3 \rho_f d'_p(-\Delta P)\rho_B = 64(7.0)(4.136)^2 2704/\pi^3(0.429)0.0071(8100)500$$
$$= 54.5 \text{ m}^6 \quad \text{(from 21.3-12)}$$

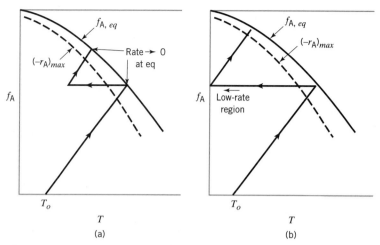

Figure 21.9 Extreme cases for FBCR, both requiring relatively large W(catalyst): (a) approach to equilibrium; (b) extent of interstage cooling

Equation 21.3-10 becomes

$$D^6 - 1.89D^2 - 54.5 = 0$$

from which (file ex21-5.msp)

$$D = 1.99 \text{ m}$$

From equations 15.4-1 and 21.3-1,

$$L = 4W/\rho_B \pi D^2 = 4(2704)/500\pi(1.99)^2 = 1.74 \text{ m}$$

21.5.3.2 Optimal Multistage Operation with Interstage Cooling

In this section, we consider one type of optimization for adiabatic multistage operation with intercooling for a single, reversible, exothermic reaction: the minimum amount of catalyst, W_{min}, required for a specified outlet conversion. The existence of an optimum is indicated by the two questions raised in Section 21.5.3.1 about the degree of approach to equilibrium conversion (f_{eq}) in a particular stage, and the extent of cooling between stages. A close approach to equilibrium results in a relatively small number of stages (N), but a relatively large W per stage, since reaction rate goes to zero at equilibrium; conversely, a more "distant" approach leads to a smaller W per stage, since operation is closer to the locus of maximum rates, but a larger N. Similarly, a large extent of cooling (lower T at the inlet to a stage) results in a smaller N, but a larger W per stage, since operation is farther from the locus of maximum rates, and conversely. The extremes are illustrated in Figure 21.9(a) for approach to equilibrium, and in Figure 21.9(b) for cooling.

We outline two situations for determining W_{min}. The method, based on the technique of dynamic programming, is described in more detail by Aris (1965, pp. 236–247), and by Froment and Bischoff (1990, pp. 416–423); see also Levenspiel (1972, pp. 510–511). Optimization has been considered by Chartrand and Crowe (1969) for an SO_2 converter in a plant in Hamilton, Ontario, as it existed then.

(1) W_{min} for specified N, $f_{N,out}$

For an N-stage reactor, there are $2N - 1$ decisions to make to determine W_{min}: N values of $T_{i,in}$ and $N - 1$ values of $f_{i,out}$, where sub i refers to the ith stage. Two

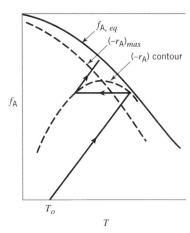

Figure 21.10 Graphical illustration of criterion 21.5-10 and its consequences for determination of W_{min}.

criteria provided (Konocki, 1956; Horn, 1961) for these are:

$$(-r_A)_{i,out} = (-r_A)_{i+1,in} \qquad (21.5\text{-}10)$$

and

$$\left[\int_{f_{A,in}}^{f_{A,out}} \frac{1}{(-r_A)} \frac{\partial(-r_A)}{\partial T} df_A\right]_i = 0 \qquad (21.5\text{-}11)$$

The first of these states that the extent of cooling should be such that the rate of reaction entering a given stage should be the same as that leaving the previous stage. The second is applied over the adiabatic operation of a particular stage to determine $f_{A,out}$ for that stage. The first is relatively easy to implement, and is illustrated graphically in Figure 21.10. The second is much more difficult to implement. A program for implementation of both is given by Lombardo (1985). The procedure requires a choice of feed temperature ($T_o \equiv T_{1,in}$) in a one-dimensional search.

(2) W_{min} for specified f_{out}

A more general case than (1) is that in which f_{out} is specified but N is not. This amounts to a two-dimensional search in which the procedure and criteria in case (1) constitute an inner loop in an outer-loop search for the appropriate value of N. Since N is a small integer, this usually entails only a small number of outer-loop iterations.

21.5.3.3 Multistage Operation with Cold-Shot Cooling

An alternative way to adjust the temperature between stages is through "cold-shot" (or "quench") cooling. In adiabatic operation of a multistage FBCR for an exothermic, reversible reaction with cold-shot cooling (Figure 21.1), T is reduced by the mixing of cold feed with the stream leaving each stage (except the last). This requires that the original feed be divided into appropriate portions. Interstage heat exchangers are not used, but a preheater and an aftercooler may be required.

A flow diagram indicating notation is shown in Figure 21.11 for a three-stage FBCR in which the reaction A $\rightleftarrows$ products takes place. The feed enters at T_o and $\dot{m}$ kg s^{-1}, or, in terms of A, at F_{Ao} and $f_{Ao} = 0$. The feed is split at S_1 so that a fraction r_1 enters stage 1 after passing through the preheater E1, where the temperature is raised from T_o to T_{o1}. A subsequent split occurs at S_2 so that a feed fraction r_2 mixes with the effluent

536 Chapter 21: Fixed-Bed Catalytic Reactors for Fluid-Solid Reactions

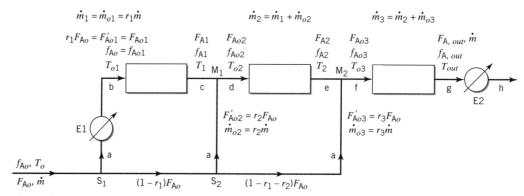

Figure 21.11 Flow diagram and notation for three-stage FBCR with cold-shot cooling

from stage 1 at M_1, and the resulting stream enters stage 2. The remainder of the original feed mixes with the effluent from stage 2 at M_2, and the resulting stream enters stage 3. The fraction of the original feed entering any stage i is defined by

$$r_i = \dot{m}_{oi}/\dot{m} = F'_{Aoi}/F_{Ao} \tag{21.5-12}$$

where $\dot{m}_{oi}$ and F'_{Aoi} are the portions of the feed, in specific mass and molar terms, respectively, entering stage i, such that

$$\sum_i \dot{m}_{oi} = \dot{m} \tag{21.5-13}$$

and

$$\sum_i F'_{Aoi} = F_{Ao} \tag{21.5-13a}$$

It follows that

$$\sum_i r_i = 1 \tag{21.5-14}$$

For the purpose of determining the size of the reactor from equation 21.5-4, the inlet composition and flow rate are required for each stage. The former is in terms of f_{Aoi} in Figure 21.11. f_{Aoi} can be related to $f_{A,i-1}$, the fractional conversion leaving the previous stage and r_k ($k = 1, 2, \ldots, i$) by a material balance for A around the corresponding mixing point, M_{i-1}. Thus, for $i = 2$, around M_1,

$$F_{Ao2} = F_{A1} + F'_{Ao2} = F_{A1} + r_2 F_{Ao} \tag{21.5-15}$$

Since

$$f_{Ao2} = \frac{F'_{Ao1} + F'_{Ao2} - F_{Ao2}}{F'_{Ao1} + F'_{Ao2}} = \frac{(r_1 + r_2)F_{Ao} - F_{Ao2}}{(r_1 + r_2)F_{Ao}} \tag{21.5-16}$$

and

$$f_{A1} = \frac{F'_{Ao1} - F_{A1}}{F'_{Ao1}} = \frac{r_1 F_{Ao} - F_{A1}}{r_1 F_{Ao}} \tag{21.5-17}$$

21.5 Pseudohomogeneous, One-Dimensional, Plug-Flow Model

substitution of 21.5-16 and -17 in 21.5-15 to eliminate F_{Ao2} and F_{A1}, respectively, results in

$$(r_1 + r_2)F_{Ao}(1 - f_{Ao2}) = r_1 F_{Ao}(1 - f_{A1}) + r_2 F_{Ao}$$

from which

$$f_{Ao2} = \frac{r_1}{r_1 + r_2} f_{A1} \qquad (21.5\text{-}18)$$

Similarly, for $i = 3$, around M_2,

$$f_{Ao3} = \frac{r_1 + r_2}{r_1 + r_2 + r_3} f_{A2} = (r_1 + r_2) f_{A2} \qquad (21.5\text{-}18a)$$

since $r_1 + r_2 + r_3 = 1$ for a three-stage reactor.

In general, for the ith stage (beyond the first, for which $f_{Ao1} = f_{Ao}$) of an N-stage reactor,

$$f_{Aoi} = \left(\sum_{k=1}^{i-1} r_k \Big/ \sum_{k=1}^{i} r_k \right) f_{A,i-1} \qquad (i = 2, \ldots, N) \qquad (21.5\text{-}19)$$

Note that $f_{Aoi} < f_{A,i-1}$. That is, the mixing of fresh feed with the effluent from stage $i-1$ results in a decrease in f_A as well as in T. This is a major difference between cold-shot cooling and interstage cooling, in which f_A does not change between stages.

The temperature at the inlet to stage i, T_{oi}, may similarly be related to T_{i-1}, the temperature at the outlet of the previous stage, by means of an energy balance around M_{i-1}. If we assume c_P is constant for the relatively small temperature changes involved on mixing (and ignore any compositional effect), an enthalpy balance around M_1 is

$$\dot{m}_1 c_P (T_1 - T_{ref}) + \dot{m}_{o2} c_P (T_o - T_{ref}) = (\dot{m}_1 + \dot{m}_{o2}) c_P (T_{o2} - T_{ref})$$

Setting the reference temperature, T_{ref}, equal to T_o, and substituting for $\dot{m}_1 = \dot{m}_{o1}$ and $\dot{m}_{o2}$ from equation 21.5-12, we obtain, after cancelling c_P,

$$r_1 \dot{m}(T_1 - T_o) = (r_1 + r_2) \dot{m}(T_{o2} - T_o)$$

from which

$$T_{o2} = T_o + \frac{r_1}{r_1 + r_2}(T_1 - T_o) \qquad (21.5\text{-}20)$$

In general, for stage i (beyond the first) in an N-stage reactor,

$$T_{oi} = T_o + \left(\sum_{k=1}^{i-1} r_k \Big/ \sum_{k=1}^{i} r_k \right)(T_{i-1} - T_o) \qquad (i = 2, 3, \ldots, N) \qquad (21.5\text{-}21)$$

The operating lines for an FBCR with cold-shot cooling are shown schematically and graphically on a plot of f_A versus T in Figure 21.12, which corresponds to Figure

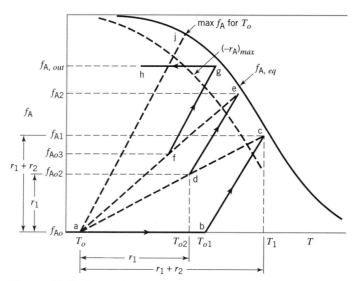

Figure 21.12 Conversion-temperature diagram showing operating lines for cold-shot cooling operation of FBCR for exothermic, reversible reaction

21.8(a) for multistage adiabatic operation with interstage cooling. The letters a, b, c, etc. are states corresponding to the points marked a, b, c, etc. in Figure 21.11. The line ab represents heating the feed in E1 from T_o to T_{o1}. The line bc is the operating line for the first stage, and is obtained from the energy equation 21.5-8 (assumed to be linear as in equation 21.5-9). The point d on the line joining a to c represents the feed to the second stage, with fractional conversion f_{Ao2} given by equation 21.5-18. This may be located graphically, given r_1 and r_2, on the line ac from the geometry shown and in accordance with equation 21.5-18 (with $f_{Ao} = 0$):

$$\frac{ac}{ad} = \frac{f_{A1}}{f_{Ao2}} = \frac{r_1 + r_2}{r_1} \tag{21.5-22}$$

An alternative, corresponding construction to locate point d can be based on equation 21.5-20 in terms of temperature. The temperature rise in stage 1 is $T_1 - T_{o1}$, and the temperature decrease around M_1 is $T_{o2} - T_1$. The line de is the operating line for the second stage. The point f on ae is the feed point for the third stage, and may be located in a manner similar to the location of d. The line fg is the operating line for the third stage, and gh represents cooling of effluent in the aftercooler E2. Note that the various operating lines are represented by *full* lines, since each point (as a state) on such a line actually exists somewhere in the system. Other lines, such as ac and ae, are not operating lines, and are represented by *dashed* lines. For example, on ac, only the points a, d, and c represent actual states (they may be called operating points); other points on ac have no meaning for existing states. The slopes of the operating lines are all the same and are given by equation 21.5-8 (see problem 21-18).

Within each stage, the amount of catalyst, W, may be calculated from equation 21.5-4, together with an appropriate rate law, and the energy equation 21.5-8. Optimization problems relating to minimizing W may be considered in terms of choice of values of r_i and f_{Ai}.

A number of factors may influence the choice of cold-shot cooling rather than interstage cooling in adiabatic operation of an FBCR. (Some of these are explored in

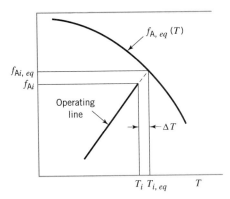

Figure 21.13 Interpretation of ΔT in equation 21.5-23 for exothermic, reversible reaction with cold-shot cooling, ith stage

problem 21-9.) An important difference is the limiting value of fractional conversion, f_A, that can be attained in principle. In operation with interstage cooling, as illustrated in Figure 21.8(a), it is possible to "step-up" f_A indefinitely to avoid the equilibrium limitation; this can be done with an increasing number of stages (and amount of catalyst), but the economics of diminishing returns eventually sets in (marginal increase in f_A at the expense of a large increase in amount of catalyst, as the rate diminishes). In operation with cold-shot cooling, however, the upper limit in f_A, for a given T_o, is $f_{A,eq}$ at the intersection with the equilibrium curve of an operating line (aj) drawn from point a in Figure 21.12; aj has the same slope as the other operating lines (bc, de, fg) shown. Consequently, to obtain the same conversion, T_o for an FBCR with cold-shot cooling (Figure 21.12) must be much lower than T_o (first-stage inlet temperature, Figure 21.8a) for an FBCR with interstage cooling. Since reaction rate is relatively small at low temperature, the feed to the first stage of a cold-shot-cooled reactor is preheated (from T_o to T_{o1} in Figure 21.12).

21.5.3.4 Calculations for a FBCR with Cold-Shot Cooling

The calculations for an N-stage FBCR with cold-shot cooling for a reversible, exothermic reaction may involve several types of problems: the design problem of determining N and the amount and distribution of catalyst (W_i, $i = 1, 2, \ldots, N$) for a specified feed rate and composition and fractional conversion (f_A), or the converse performance problem of determining f_A for specified N and W_i, and include variations of these for optimal design and performance. Here, we describe an algorithm for the first-named design problem, but it may be suitably revised for other problems.

In general, for an N-stage reactor, there are $2N$ degrees of freedom or free parameters from among r_i and T_i (or f_{Ai}). This number may be reduced to $N + 1$ if a criterion such as a constant degree of approach to equilibrium, ΔT, is used for each stage, where

$$\Delta T = T_{i,eq}(f_{Ai,eq}) - T_i \qquad (21.5\text{-}23)$$

That is, the outlet temperature from stage i, T_i, is determined by calculating the equilibrium temperature $T_{i,eq}$ from the intersection of $f_{A,eq}(T)$ with the energy balance operating line, equation 21.5-9, and then subtracting ΔT. The interpretation of ΔT is illustrated in Figure 21.13. The free parameters may then be chosen as r_i ($i = 1, 2, \ldots, N - 1$), ΔT and T_{oi} (choice of ΔT replaces choice of $T_i(1, 2, \ldots, N)$). In addition to the feed rate and composition ($\dot{m}$, F_{A_o}), we assume knowledge of a rate law, $(-r_A)$, and $f_{A,eq}(T)$, together with the thermal characteristics c_P (of the flowing stream) and $(-\Delta H_{RA})$, which are treated as constants.

The steps in an algorithm are as follows:

(1) Calculate the operating line slope from equation 21.5-8: $df_A/dT = \dot{m}c_P/(-\Delta H_{RA})F_{Ao}$.
(2) Choose ΔT.
(3) Calculate T_o from an integrated form of 21.5-8:

$$f_{A,out} \equiv f_{AN} = f_{Ao} + \frac{\dot{m}c_P}{(-\Delta H_{RA})F_{Ao}}(T_{out} - T_o) \qquad (21.5\text{-}24)$$

where

$$T_{out} \equiv T_N = T_{N,eq}(f_{AN,eq}) - \Delta T \qquad (21.5\text{-}25)$$

$$f_{AN,eq} = f_{AN} + (df_A/dT)\Delta T \qquad (21.5\text{-}26)$$

and $f_{Ao} = 0$, usually.

(4) Choose r_1 and T_{o1}.
(5) Calculate W_1 by simultaneous solution of equation 21.5-4, with the rate law incorporated, and -8. In equation 21.5-4, F_{Ao} is replaced by $r_1 F_{Ao}$, and the limits of integration are f_{Ao} and f_{A1}, where f_{A1} is

$$f_{A1} = f_{A1,eq} - (df_A/dT)\Delta T \qquad (21.5\text{-}26a)$$

and $f_{A1,eq}$ is obtained from the intersection of the operating line and $f_{A,eq}(T)$, that is, by the simultaneous solution of

$$f_A = f_{Ao} + (df_A/dT)(T - T_{o1}) \qquad (21.5\text{-}27)$$

and

$$f_A = f_{A,eq}(T) \qquad (21.5\text{-}28)$$

(6) Calculate T_1, corresponding to f_{A1}, from equation 21.5-27, with $f_A = f_{A1}$, and $T = T_1$.
(7) Choose r_2.
(8) Calculate f_{Ao2} from equation 21.5-19.
(9) Calculate T_{o2} from equation 21.5-21.
(10) Calculate W_2 as in step (5) for W_1. The inlet conditions are $F_{A,in} = (r_1 + r_2)(1 - f_{Ao2})F_{Ao}$, f_{Ao2} and T_{o2}. The outlet conditions are f_{A2} and T_2, which are calculated as in steps (5) and (6) for f_{A1} and T_1, respectively, with subscript 1 replaced by 2, and subscript 0 by 1.
(11) Repeat steps (7) to (10) by advancing the subscript to N until $f_{AN} \geq$ specified $f_{A,out}$. It may be appropriate to adjust r_i so that $f_{AN} \simeq f_{A,out}$.

EXAMPLE 21-6

For an FBCR operated with cold-shot cooling for the reaction A $\rightleftarrows$ products, determine, from the information given below,

(a) the maximum possible fractional conversion (f_A);
(b) the fractional conversion at the outlet of a three-stage reactor.

21.5 Pseudohomogeneous, One-Dimensional, Plug-Flow Model

The feed is split such that 40% enters stage 1 and 30% enters stage 2. The feed entering stage 1 is preheated from 375°C (T_o) to 450°C (T_{o1}). The equilibrium temperature-fractional conversion relation is

$$T_{eq} = \frac{11835}{\ln\left[\frac{7.825 \times 10^4 f_{A,eq}}{1-f_{A,eq}}\left(\frac{1-0.0475 f_{A,eq}}{0.115-0.0475 f_{A,eq}}\right)^{1/2}\right]} \quad (A)$$

For each stage, the outlet temperature, T_i, is to be 25°C lower than the equilibrium temperature (i.e., in equation 21.5-23, $\Delta T = 25°C$).

Other data: $\dot{m} = 10$ kg s^{-1}; $F_{Ao} = 62$ mol s^{-1}; $c_P = 1.1$ J g^{-1} K^{-1}; $\Delta H_{RA} = -85$ kJ mol^{-1}.

SOLUTION

Numerical results are obtained by means of the E-Z Solve software (file ex21-6.msp).

(a) The maximum possible conversion is obtained by applying the criterion for degree of approach to equilibrium ($\Delta T = 25°C$) to the intersection of the operating line aj (Figure 21.12) drawn from (f_{Ao}, T_o) with

$$\text{slope} = \frac{\dot{m}c_P}{(-\Delta H_{RA} F_{Ao})} = \frac{10(1.1)}{85(62)} = 2.09 \times 10^{-3} \text{ K}^{-1}$$

Simultaneous solution of equation (A) and the equation for the operating line with this slope gives the coordinates of the intersection at point j:

$$f_{A,eq} = 0.579; \quad T_{eq} = 925 \text{ K}$$

Thus, at the outlet point h (Figure 21.12), $T_{out} = (925 - 25) = 900$ K, and

$$f_{A,out} = (\text{slope})(T_{out} - T_o) = 2.09 \times 10^{-3}(900 - 648) = 0.527$$

That is, the maximum possible fractional conversion for these conditions, regardless of the number of stages, is 0.527.

(b) We proceed by treating the three stages in order to obtain f_{A1}, f_{A2}, and f_{A3}. The procedure is described in detail for stage 1, and the results are summarized in Table 21.1.

For stage 1, the equation of the operating line bc (Figure 21.12) through b (f_{Ao1}, T_{o1}) with the slope calculated in (a) is

$$f_A = f_{Ao1} + 2.09 \times 10^{-3}(T - T_{o1}) \quad (B)$$

where $f_{Ao1} = 0$ and $T_{o1} = 723$ K. Solving equations (A) and (B) simultaneously for the intersection of the operating line and the equilibrium line, we obtain

$$f_{A1,eq} = 0.485 \text{ and } T_{1,eq} = 955 \text{ K}$$

Thus,

$$T_1 = 955 - 25 = 930 \text{ K}$$

Table 21.1 Results for Example 21-6(b)

Stage i	Operating line Figure 21.12	r_i	T_{oi}/K	$T_{i,eq}/K$	T_i/K	f_{Aoi}	$f_{Ai,eq}$	f_{Ai}
1	bc	0.40	723	955	930	0	0.485	0.433
2	de	0.30	809	942	917	0.247	0.525	0.473
3	fg	0.30	837	937	912	0.331	0.541	0.489

Substitution for $T = T_1 = 930$, in (B) gives

$$f_A = f_{A1} = 0.433$$

For stage 2, we calculate f_{Ao2} from equation 21.5-18,

$$f_{Ao2} = [r_1/(r_1 + r_2)]f_{A1} = [0.4/(0.4 + 0.3)]0.433 = 0.247$$

and T_{o2} from equation 21.5-20,

$$T_{o2} = T_o + [r_1/(r_1 + r_2)](T_1 - T_o)$$
$$= 648 + [0.4/(0.4 + 0.3)](930 - 648) = 809 \text{ K}$$

We then proceed as for stage 1, writing the equation for the operating line de (Figure 21.12), equivalent to (B), and continuing from there. The results obtained for all three stages (see file ex21-6.msp) are given in Table 21.1. The final conversion is $f_{A3} = 0.489$, which is close to the maximum possible (with many stages) value of 0.527 from part (a). The amount of catalyst to achieve this is not required in this example, but may be calculated as described in the algorithm in Example 21-5, provided an appropriate rate law for $(-r_A)$ is available.

21.5.4 Nonadiabatic Operation

21.5.4.1 Multitubular Reactor; Catalyst Inside Tubes

The arrangement for a multitubular reactor is depicted in Figure 15.1(c), with catalyst packed either inside or outside the tubes; here, we consider the former case (Figure 11.3(c)). We assume all tubes behave in the same way as a set of reactors in parallel, and apply the continuity and energy equations to a single tube. The number of tubes, N_t, must be determined as part of the design, to establish the diameter, D, of the vessel.

For a single tube, the continuity equation, 21.5-4, may be written

$$W' = F'_{Ao} \int df_A / [-r_A(T, f_A)] \qquad (21.5\text{-}29)$$

where $W' = W/N_t$, the amount of catalyst per tube, and $F'_{Ao} = F_{Ao}/N_t$, the feed rate per tube. Equation 21.5-29 must be solved in conjunction with an appropriate rate law and the energy equation to relate f_A and T.

The appropriate form of the energy equation may be obtained from equation 21.4-2 with $k_{ex} = k_{er} = 0$, and boundary condition 21.4-8 incorporated in the equation to describe the rate of heat transfer through the wall: $\delta \dot{Q} = U(T_s - T)dA'_p/dV'$, as in Sec-

21.5 Pseudohomogeneous, One-Dimensional, Plug-Flow Model

tion 15.2.1.2, where A'_p is the (peripheral) heat transfer surface area per tube, and V' is the volume enclosed per tube (the rate of heat transfer is referred to unit volume, through dV', to correspond to the units of the other terms in equation 21.4-2).

Since

$$dA'_p = \pi d_t dx$$

$$dV' = \pi d_t^2 dx/4 = dW'/\rho_B$$

$$G' = G/N_t = 4\dot{m}'/\pi d_t^2$$

and

$$(-r_A)dW = F'_{Ao}df_A$$

equation 21.4-2 may be written, with rearrangement, as:

$$\dot{m}'c_P dT = \left[(-\Delta H_{RA}) + \frac{4U(T_S - T)}{d_t \rho_B(-r_A)}\right] F'_{Ao} df_A \qquad (21.5\text{-}30)$$

Equation 21.5-30 for flow through a tube of diameter d_t corresponds to equation 15.2-9 for flow through a PFR of diameter D (the inclusion of ρ_B in the heat transfer term stems from equation 21.3-15).

The solution of equations 21.5-29 and -30 must be done by a suitable numerical procedure, such as one analogous to that used in Example 15-4. A typical problem may involve determining W, N_t, and L_t (length of tubes), given information about $\dot{m}$, F_{Ao}, $(-r_A)$, d_t, T_o, allowable $(-\Delta P)$, and the other parameters in the two equations. On the assumption that the heat transfer characteristics depend on the arrangement of tubes, this should be done on a single-tube basis, with iterations in terms of N_t. A procedure or algorithm such as the following could be used:

(1) Choose a value of N_t.
(2) Calculate $\dot{m}'$ and F'_{Ao}.
(3) Calculate W' from equations 21.5-29 and -30.
(4) Calculate $L_t = 4W'/\rho_B \pi d_t^2$.
(5) Calculate $(-\Delta P)$ from equation 21.3-5, and compare with the allowable $(-\Delta P)$.
(6) Adjust N_t based on the result in (5), and repeat steps (2) to (5) until the $(-\Delta P)$ criterion is satisfied.

The value of N_t, together with d_t and standard triangular or square pitch for tubes in a shell-and-tube arrangement, determines the diameter, D, of the vessel (shell).

21.5.4.2 Multitubular Reactor; Catalyst Outside Tubes

Instead of being placed inside the tubes (with a heat-transfer or process fluid on the shell side), the catalyst may be placed outside the tubes (Figure 11.5(a)). The result is to have a fixed bed of diameter D, say, with N_t holes, each of diameter d_t, in it. The amount of catalyst in the entire bed may be calculated from equation 21.5-4 in conjunction with equation 21.5-30 modified to accommodate the changed geometry. In this case, $dA_p = N_t \pi d_t dx$, $dV = (\pi/4)(D^2 - N_t d_t^2)dx = dW/\rho_B$. These, together with $(-r_A)dW = F_{Ao}df_A$, and incorporation of boundary condition 21.4-8, as in the development leading to equation 21.5-30, and rearrangement, allow equation 21.4-2 to be written as:

$$\dot{m}c_P dT = \left[(-\Delta H_{RA}) + \frac{4N_t d_t U(T_S - T)}{\rho_B(D^2 - N_t d_t^2)(-r_A)}\right] F_{Ao} df_A \quad (21.5\text{-}31)$$

The typical problem outlined in the previous section may be solved in this case in a similar manner:

(1) Choose a value of N_t, which implies a value of D.
(2) Calculate W by numerical solution of equations 21.5-4 and -31.
(3) Calculate value of L (i.e., L_t).
(4) Calculate $(-\Delta P)$ and compare with allowable $(-\Delta P)$.
(5) Adjust value of N_t from result in (4), and repeat steps (2) to (4) until the $(-\Delta P)$ criterion in satisfied.

21.6 HETEROGENEOUS, ONE-DIMENSIONAL, PLUG-FLOW MODEL

In Section 21.5, the treatment is based on the pseudohomogeneous assumption for the catalyst + fluid system (Section 21.4). In this section, we consider the local gradients in concentration and temperature that may exist both within a catalyst particle and in the surrounding gas film. The system is then "heterogeneous." We retain the assumptions of one-dimensional, plug-flow behavior, and a simple reaction of the form $A(g) + \ldots \xrightarrow{\text{cat}(s)}$ products.

The starting points for the continuity and energy equations are again 21.5-1 and 21.5-6 (adiabatic operation), respectively, but the rate quantity $(-r_A)$ must be properly interpreted. In 21.5-1 and 21.5-6, the implication is that the rate is the intrinsic surface reaction rate, $(-r_A)_{int}$. For a heterogeneous model, we interpret it as an overall observed rate, $(-r_A)_{obs}$, incorporating the transport effects responsible for the gradients in concentration and temperature. As developed in Section 8.5, these effects are lumped into a particle effectiveness factor, η, or an overall effectiveness factor, η_o. Thus, equations 21.5-1 and 21.5-6 are rewritten as

$$\frac{d(uc_A)}{dx} + \rho_B(-r_A)_{obs} = 0 \quad (21.6\text{-}1)$$

and

$$Gc_P \frac{dT}{dx} - \rho_B(-r_A)_{obs}(-\Delta H_{RA}) = 0 \quad (21.6\text{-}2)$$

where, in terms of η,

$$(-r_A)_{obs} = \eta(-r_A)_{int} \quad (21.6\text{-}3)$$

η is a function of the Thiele modulus ϕ'', defined in equation 8.5-20b for power-law kinetics, and more generally in equation 8.5-22; see Figure 8.11 and Tables 8.1 to 8.3 for the $\eta(\phi'')$ relationship for various shapes of catalyst particles and orders of reaction.[1]

---

[1] Note that if $(-r_A)$ or a corresponding rate constant k_A is specified on a mass basis, k_A must be multiplied by the particle density to obtain the form of the rate constant to be used in the Thiele modulus; that is, $k_A \equiv (k_A)_p = \rho_p(k_A)_m$, which follows from equation 21.3-16.

21.6 Heterogeneous, One-Dimensional, Plug-Flow Model

For a simple system, the continuity equation 21.6-1 may be put in forms analogous to 21.5-2 and 21.5-4 for the axial profile of fractional conversion, f_A, and amount of catalyst, W, respectively:

$$\frac{df_A}{dx} - \frac{\pi r^2 \rho_B \eta (-r_A)_{int}}{F_{Ao}} = 0 \qquad (21.6\text{-}4)$$

and

$$W = F_{Ao} \int \frac{df_A}{\eta(-r_A)_{int}} \qquad (21.6\text{-}5)$$

For a complex system, the continuity equation 21.6-1 must be written for more than one species, and equations 21.6-4 and -5 are not sufficient by themselves. Thus, instead of 21.6-4, it is more appropriate to develop profiles for the proper number of concentrations (c_i) or molar flow rates (F_i); see problem 21-21. However, we restrict attention to simple systems in this section.

The simplest case to utilize η is that of an isothermal situation with no axial gradient in T. In this case, a constant, average value of η may describe the situation reasonably well, and equation 21.6-5 becomes

$$W = \frac{F_{Ao}}{\eta} \int \frac{df_A}{(-r_A)_{int}} \qquad (\eta \text{ constant}) \qquad (21.6\text{-}6)$$

This approach is analytically correct for isothermal reactors and first-order rate laws, since concentration does not appear in the expression for the Thiele modulus. For other (nonlinear) rate laws, concentration changes along the reactor affect the Thiele modulus, and hence produce changes in the local effectiveness factor, even if the reaction is isothermal. Problem 21-15 uses an average effectiveness factor as an approximation.

A more realistic case is a nonlinear reaction rate or a nonisothermal situation in which axial variation in η must be allowed for. Within a fixed bed of catalyst particles, the rate of heat transfer to or from a particle is usually controlled by the resistance of the gas film surrounding the particle; this is because of the high thermal conductivity of most solid catalysts. Conversely, mass transport is usually limited by pore diffusion, external gas-film resistance being relatively small. Thus, concentration gradients are primarily confined to the region inside the particle, and temperature gradients, if present, are primarily confined to the surrounding gas film (Minhas and Carberry, 1969). For a given particle, the process is essentially isothermal, but each layer of particles in the bed may be at a different temperature.

To calculate W from equation 21.6-5, ϕ'' and η must be calculated at a series of axial positions or steps, since each depends on T and c_A; $(-r_A)_{obs}$ is then calculated from η and $(-r_A)_{int}$ at each step. For adiabatic operation, an algorithm for this purpose (analogous to that in Example 21-5) is as follows:

(1) Choose a value of f_A.
(2) Calculate T from an integrated form of 21.6-2, such as 21.5-9.
(3) Calculate $(-r_A)_{int}$ at f_A and T from a given rate law.
(4) Calculate ϕ'', e.g., from equation 8.5-20b; if necessary, use ρ_p to convert k_A (mass basis) to k_A (volume basis).
(5) Calculate η from ϕ'' (Section 8.5).
(6) Calculate $(-r_A)_{obs}$ from equation 21.6-3.
(7) Repeat steps (1) to (6) for values of f_A between $f_{A,in}$ and $f_{A,out}$.

(8) Evaluate the integral in equation 21.6-5 by means of the E-Z Solve software or an approximation such as the trapezoidal rule, as in Example 21-5.
(9) Calculate W from equation 21.6-5.

Problem 21-20 illustrates a situation in which the design of an FBCR must take into account variation in η.

As an alternative to the use of η, an empirical determination of $(-r_A)_{obs}$ may be obtained in laboratory experiments with different reactors or in pilot plant runs. In these, the same catalyst (size, shape, etc.) and ranges of operating conditions expected in the full-scale reactor are used. An example of such an empirical rate law is equation 21.3-14 for SO_2 oxidation (Eklund, 1956), with values of the rate constant, k_{SO_2}, given in problem 8-19 (see also Figure 21.4). The data for the relatively large cylindrical particles indicate both strong-pore-diffusion-resistance, and reaction-rate-control regimes for lower temperature and higher temperature ranges, respectively. Such a rate law can be incorporated directly into the continuity equation. Several problems at the end of this chapter are to be done using this approach, which is actually a modification of the pseudohomogeneous model used in Section 21.5. A disadvantage of the use of an empirical rate law is that it cannot be generalized to different particle sizes (except under special circumstances: see problem 21-11(c) and (d)) or to different catalyst site densities; furthermore, it cannot be extrapolated with confidence outside the range of conditions used to obtain it.

21.7 ONE-DIMENSIONAL VERSUS TWO-DIMENSIONAL MODELS

Radial gradients in both temperature and concentration may develop for extremely exothermic reactions. This results from the positive interaction between reaction rate and temperature. The temperature rises more rapidly at the center ($r = 0$) of the bed, if there is heat transfer at the wall. The gradient in concentration may be further affected by radial dispersion. Quantitative modeling of radial gradients (in addition to axial gradients) requires a two-dimensional model. Reactions with small thermal effects, either from dilution of the reactants or from small enthalpies of reaction, do not develop significant radial gradients. For these reactions, a one-dimensional model usually suffices for process-simulation and reactor modeling.

Systems that require two-dimensional treatment are sensitive to the parameters in the model, and, as a result, the transport coefficients (k_e and D_e) must be well known. Consequently, a one-dimensional model is usually used for preliminary process design, and a two-dimensional model is developed subsequently. More sophisticated "one-dimensional" models for systems with thermal gradients use appropriate averaging over an assumed radial temperature profile (usually parabolic). These models yield the best temperature for inclusion in a one-dimensional model and for calculating reaction rate. They reproduce thermal profiles reasonably well, provided the temperature gradients are not large, but generally fail for reactions which are highly exothermic and have large activation energies (Carberry and White, 1969). It is desirable to achieve a final design that avoids situations in which extreme sensitivity to conditions exists, from both safety and economic considerations.

21.8 PROBLEMS FOR CHAPTER 21

21-1 (a) What is the mean residence time ($\bar{t}$) of gas flowing through a fixed bed of particles, if the bed voidage is 0.38, the depth of the bed is 1.5 m, and the superficial linear velocity of the gas is 0.2 m s^{-1}?

(b) What is the bulk density of a bed of catalyst, if the bed voidage is 0.4 and the particle density is 1750 kg m^{-3} (particle)?

(c) What is the mass (kg) of catalyst contained in a 100-m^3 bed, if the catalyst particles are made up of a solid with an *intrinsic* density of 2500 kg m^{-3}, the bed voidage is 0.4, and the particle voidage is 0.3?

21-2 In "cold-shot-cooling" or "quench" operation of a three-stage fixed-bed catalytic reactor for an exothermic reaction (see Figure 21.11), temperature is controlled by splitting the feed to the various stages. For the reaction A → products, suppose the feed of pure A is divided so that 45% enters the first stage and 30% enters the second stage, that is, joining the outlet from the first stage to form the feed to the second. No reaction occurs between stages, and f_A is based on original feed.

(a) If the fraction of A converted in the stream leaving the first stage (f_{A1}) is 0.40, what is the fraction of A converted (f_{Ao2}) in the feed entering the second stage?

(b) If $f_{A2} = 0.60$, what is f_{Ao3}?

21-3 Consider a two-stage fixed-bed catalytic reactor, with axial flow and temperature adjustment between stages, for the dehydrogenation of ethylbenzene (A) to styrene (S) (monomer), reaction (D) in Section 21.1. From the data given below, for adiabatic operation, if the fractional conversion of A at the outlet of the second stage (f_{A2}) is 0.70, calculate the upper and lower limits (i.e., the range of allowable values) for the fractional conversion at the outlet of the first stage (f_{A1}).

Data:

Feed:

$$T_o = 925\text{K}; \quad P_o = 1.4 \text{ bar};$$

$$F_{Ao} = 100 \text{ mol s}^{-1}; \quad F_{H_2O,o}(\text{inert}) = 1200 \text{ mol s}^{-1}$$

Equilibrium:

$$K_p/\text{MPa} = 8.2 \times 10^5 \exp(-15{,}200/T)$$

Thermochemical:

$$\Delta H_{RA} = 126{,}000 \text{ J mol}^{-1} \text{ (constant)}$$

$$c_P = 2.4 \text{ J g}^{-1}\text{K}^{-1} \text{ (constant)}$$

The maximum allowable temperature is 925 K. See Figure 21.8(b) for a schematic representation of one particular situation between the limits. The slopes of the operating lines AB and CD are equal and constant. The points B, C, and D are not fixed. No reaction occurs during temperature adjustment between stages, or after the second-stage outlet.

21-4 (a) For the dehydrogenation of ethylbenzene (A), reaction (D) in Section 21.1, a reversible reaction, on a single plot of *equilibrium* conversion f_A versus T, sketch each of the following three cases to show relative positions of the equilibrium curves:

(i) given constant total pressure P_1, and ratio of inert gas to A in the feed (r_1); explain the direction of $f_{A,eq}$ with respect to T;

(ii) given constant total pressure $P_2 > P_1$, and r_1; explain the position of the curve relative to that in (i);

(iii) given constant total pressure P_1 and ratio $r_2 > r_1$; explain the position of the curve relative to that in (i).

(b) State, with explanations, two reasons for the use of an inert gas (steam) in the feed. (The reaction is carried out adiabatically.)

21-5 For the oxidation of SO_2(A), reaction (A) in Section 21.1, a reversible reaction, on a single plot of *equilibrium* conversion f_A versus T (assume A to be limiting reactant), sketch each of the following three cases to show relative positions of the equilibrium curves:
(a) Given constant total pressure P_1, with air for oxidation (N_2 inert); explain the direction of $f_{A,eq}$ with respect to T.
(b) Given constant total pressure P_1, with pure O_2 for oxidation; explain the position of the curve relative to that in (a).
(c) Given constant total pressure $P_2 < P_1$, with air; explain as in (b).

21-6 Consider a fixed-bed catalytic reactor (FBCR), with axial flow, for the dehydrogenation of ethylbenzene (A) to styrene (S) (monomer). From the information given below, calculate the temperature (T/K) in the first-stage bed of the reactor,
(a) at the outlet of the bed (i.e., at L_1); and
(b) $L = 0.38\,L_1$.

Assume steady-state, adiabatic operation, and use the pseudohomogeneous, one-dimensional plug-flow model.

Data:

Feed: $T_o = 925$ K; $P_o = 2.4$ bar; $F_{Ao} = 85$ mol s^{-1}; F_{H_2O}(inert) $= 1020$ mol s^{-1}
Fractional conversion at outlet: $f_{A1} = 0.45$
Rate law: $(-r_A) = k_A(p_A - p_S p_{H_2}/K_p)$
k_A/mol (kg cat)$^{-1}$ s^{-1} bar^{-1} $= 3.46 \times 10^3 \exp(-10{,}980/T)$
K_p/bar $= 8.2 \times 10^6 \exp(-15{,}200/T)$
(T in K for k_A and K_p)
$\Delta H_{RA} = 126{,}000$ J mol^{-1}; $c_P = 2.4$ J g^{-1} K^{-1}
$W_1 = 25{,}400$ kg (of catalyst)

21-7 Consider a two-stage fixed-bed catalytic reactor (FBCR), with axial flow, for the dehydrogenation of ethylbenzene (A) to styrene (S) (monomer). From the data given below, for adiabatic operation, calculate the amount of catalyst required in the first stage, W_1/kg.

Feed: $T_o = 925$ K; $P_o = 2.4$ bar; $F_{Ao} = 100$ mol s^{-1}; F_{H_2O}(inert) $= 1200$ mol s^{-1}
Fractional conversion at outlet: $f_{A1} = 0.4$; use the model and other data as in problem 21-6.

21-8 Consider a fixed-bed catalytic reactor (FBCR), with axial flow, for the dehydrogenation of ethylbenzene (A) to styrene (S) (monomer). From the data given below, calculate the amount of catalyst required, W/kg, to reach a fractional conversion (f_A) of 0.2 in the (adiabatic) first stage (of a two-stage reactor). What is the temperature at this point?

Feed: $T_o = 925$ K; $P_o = 2.4$ bar; $F_{Ao} = 120$ mol s^{-1}; F_{H_2O}(inert) $= 1200$ mol s^{-1}; use the model and other data as in problem 21-6.

21-9 Consider a reversible, exothermic, gas-phase reaction (A $\rightleftarrows$ products) occurring in a three-stage, fixed-bed catalytic reactor, with either (1) adiabatic operation within each stage and interstage cooling with heat exchangers, or (2) adiabatic operation and "cold-shot" cooling between stages.
(a) Sketch a conversion (f_A)–temperature (T) diagram showing operating lines for each of the two modes of operation, for say, 90% (final) conversion of A; include a sketch of the equilibrium fractional conversion (f_A) in each case.
(b) Indicate if each of the following factors affects the slope of the adiabatic operating line; if there is an effect, indicate the direction of it:
  (i) the enthalpy of reaction;
  (ii) the feed temperature;
  (iii) whether the feed is pure A or relatively dilute in A (i.e., with inert gas present).
(c) Which one of the three factors in (b) most influences the choice of mode of operation, (1) or (2) above? Explain.

21-10 For the SO_2 converter in a 1000-tonne day^{-1} H_2SO_4 plant (100% H_2SO_4 basis), calculate the following:

(a) The amount (kg) of catalyst (V_2O_5) required for the first stage of a four-stage adiabatic reactor, if the feed temperature (T_o) is 430°C, and the (first-stage) outlet fractional conversion of SO_2 (f_{SO_2}) is 0.687; the feed composition is 9.5 mol % SO_2, 11.5% O_2, and 79 % N_2.

(b) The depth (L/m) and diameter (D/m) of the first stage.

Data:

Use the Eklund rate law (equation 21.3-14), with data for k_{SO_2} from problem 8-19 (B particles); assume $K_p/MPa^{-1/2} = 7.97 \times 10^{-5} \exp(12{,}100/T)$, with T in K;
For bed of catalyst: $\rho_B = 500$ kg m^{-3}; $\epsilon_B = 0.40$;
For gas: $\mu_f = 4 \times 10^{-5}$ kg m^{-1} s^{-1}; $c_P = 0.94$ J g^{-1} K^{-1};
f_{SO_2} is 0.98 over four stages; $P_o \simeq 101$ kPa; $\Delta H_R = -100$ kJ (mol SO_2)$^{-1}$;
Allowable $(-\Delta P)$ for first stage is 2.5 kPa.

21-11 Suppose the operating conditions for a four-stage SO_2 converter, with adiabatic operation within each stage and interstage cooling, are as given in the table below. The other conditions are: $P = 1$ bar; feed is 9.5 mol % SO_2, 11.5% O_2, 79% N_2.

Stage	T_{in}/°C	T_{out}/°C	$f_{SO_2,out}$	$(-r_{SO_2,out})$/mol g^{-1} s^{-1}
1	430	614	68.7	5 x 10^{-7}
2	456	515	90.7	1 x 10^{-7}
3	436	452	97	3 x 10^{-8}
4	430	433	98	~ 10^{-8}

(a) In order to extract more energy in a "waste-heat" boiler upstream, if it is proposed that the feed temperature be lowered from 430 to 410°C, what would be the effect on the conversion in the first stage (no other changes made)? Explain, without doing any calculations, but with the aid of a sketch, whether this change should be made.

(b) In order to take advantage of higher rates of reaction, if it is proposed that the feed temperature be increased from 430 to 450°C, what would be the main consequence? Explain, without doing any calculations, whether this change should be made.

(c) Based on the results of problem 8-19, for which stages of the converter would you expect the effectiveness factor (η) to be close to 1 and for which parts < 1 (with what consequences for translating Eklund's data to other particle sizes)? Explain.

(d) If the amount of catalyst required for the second stage is calculated to be 44,000 kg (by the method in 21-10(a) and for the same catalyst particles, rate of production, and feed conditions), what would the amount be if the diameter of the particles were changed from 8 mm to 5 mm? Explain.

21-12 (a) Write, without proof, the equations, together with boundary conditions, that describe a steady-state (reactor) model for a fixed-bed catalytic reactor (FBCR) and that allow for the following: axial convective (bulk) flow of mass and energy, radial dispersion/conduction of mass/energy, chemical reaction (e.g., A $\rightarrow$ products), and energy (heat) transfer between reactor (tube) and surroundings. Write the equations in terms of c_A and T. Define the meaning of each symbol used. Assume catalyst is packed inside tubes in a multitubular arrangement.

(b) Classify the model constructed in terms of the usual descriptive phrases (three considerations).

550 Chapter 21: Fixed-Bed Catalytic Reactors for Fluid-Solid Reactions

(c) Repeat (a) and (b), if radial dispersion/conduction of mass/energy is replaced by axial dispersion/conduction of mass/energy.
(d) Repeat (a) and (b), if axial dispersion/conduction of mass/energy is included.
(e) Under what conditions may the axial dispersion and conduction terms be (separately) ignored?

21-13 Outline a procedure, with an algorithm and illustrative sketches, to determine W_{min} for an FBCR, with f_{out} specified (analogous to that described in Section 21.5.3.2) for each of the following cases:

(a) a reversible, endothermic reaction, subject to T_{max};
(b) an irreversible, exothermic reaction, subject to T_{max}.

21-14 Using the equation for K_p given in problem 21-10 for the oxidation of SO_2, determine the outlet conversion from the third stage of a three-stage FBCR, given that, for each stage, the outlet temperature is 20 K from the $f_{eq}(T)$ curve. The inlet temperature to the first stage is 700 K. Between stages 1 and 2, the temperature is reduced by 120 K, and between stages 2 and 3, the temperature is reduced by 85 K. The feed contains 19.7 mol % SO_2(A), 23.8 mol % O_2, 56.5 mol % N_2, and a trace (say 0.0001 mol %) of SO_3.

Data: $F_{Ao} = 220$ mol s^{-1}; $\dot{m} = 40.3$ kg s^{-1}; $\Delta H_{RA} = -100$ kJ mol^{-1}; $c_P = 0.94$ J g^{-1} K^{-1}; $P_o = 101$ kPa.

21-15 Determine the mass of catalyst required for stage 1 of the three-stage FBCR described in problem 21-14. Assume that the effectiveness factor is 0.90. Use the Eklund equation, 21.3-14, with k_{SO_2} from problem 8-19 (B particles).

21-16 A multistage FBCR is to be designed for the production of H_2 from CO and steam (the so-called "shift reaction"). Our objective is to design the reactor such that the outlet stream contains less than 0.6 mol % CO (dry basis). The outlet temperature from each stage is 10 K less than the maximum equilibrium temperature for that stage. Interstage coolers are used to reduce the temperature between stages: by 130 K between stages 1 and 2, and by 110 K between any stages beyond the second. The catalyst undergoes significant deactivation if the temperature exceeds 475°C. Determine the number of stages required, and the outlet conditions from each stage, given the following data:

Feed rate (dry basis) = 3.5 mol s^{-1}
Steam-to-CO mole ratio in feed = 6:1
$c_P = 2.12$ J g^{-1} K^{-1}; $\Delta H_R = -39.9$ kJ mol^{-1}; $T_o = 300°C$
$K_{eq} = 0.008915 e^{4800/T}$; $k = 0.779 e^{-4900/T}$ (mol CO) s^{-1} (kg cat)$^{-1}$ bar^{-2}
Feed composition (dry basis):

CO: 13.0 mol % CO_2: 7.0 mol % H_2: 56.8 mol % N_2: 23.2 mol %

21-17 Determine the catalyst requirement for each stage of the multistage FBCR in problem 21-16, given the following additional data:

$P_o = 2760$ kPa; $(-\Delta P)_{max} = 70$ kPa
$\rho_B = 1100$ kg m^{-3}; $\rho_p = 2000$ kg m^{-3}
$(-r_{CO}) = k(p_{CO} p_{H_2O} - p_{CO_2} p_{H_2}/K_{eq})$

The catalyst pellets are 6 mm × 6 mm semi-infinite slabs, with all faces permeable. (Problems 21-16 and 21-17 are based on data provided by Rase, 1977, pp. 44–60.)

21-18 In cold-shot-cooling operation of a multistage FBCR for the exothermic reaction A $\rightleftarrows$ products, if c_P (for the flowing stream) and $(-\Delta H_{RA})$ are constant, show that the slopes of the operating lines (Figure 21.12) are all given by equation 21.5-8 (that is, they are all the same).

21-19 (a) Based on the information given in problem 5-19(e) for the partial oxidation of benzene (B, C_6H_6) to maleic anhydride (M, $C_4H_2O_3$), calculate the amount of catalyst (W/kg) required for a PFR operating isothermally at 350°C and isobarically at 2 bar to achieve a maximum yield ($Y_{M/B}$) of $C_4H_2O_3$ at the outlet. Assume the molar feed ratio of air (79 mol % N_2, 21% O_2) to C_6H_6 is 100/1, and the total feed rate is 500 mol h^{-1}. State any assumptions made.

(b) What is the overall fractional yield or selectivity, $\hat{S}_{M/B}$, at the outlet?

21-20 An exothermic first-order reaction $A \xrightarrow{k_1} B$ is conducted in an FBCR, operating adiabatically and isobarically. The bed has a radius of 1.25 m and is 4 m long. The feed contains pure A at a concentration of 2.0 mol m^{-3}, and flowing at $q = 39.3$ m^3 s^{-1}. The reaction may be diffusion limited; assume that the relationship between η and ϕ'' is $\eta = (\tanh \phi'')/\phi''$. The diffusivity is proportional to $T^{1/2}$, and L_e for the particles is 0.50 mm. Determine the fractional conversion of A and the temperature at the bed outlet. How would your answer change, if (a) diffusion limitations were ignored, and (b) a constant effectiveness factor, based on inlet conditions, was assumed.

Data:

$T_o = 495°C$; $D_A = 0.10$ cm^2 s^{-1} at 495°C
$k_1 = 4.0 \times 10^{-3}$ m^3 (kg cat)$^{-1}$ s^{-1} at 495°C; $E_A = 125.6$ kJ mol^{-1}
$\Delta H_{RA} = -85$ kJ mol^{-1}; $\dot{m} = 26$ kg s^{-1}
$\rho_B = 700$ kg m^{-3}; $\rho_p = 2500$ kg m^{-3}; $c_P = 2.4$ J g^{-1} K^{-1}

21-21 It was discovered that product "B" in problem 21-20 was subject to further reaction (i.e., $A \xrightarrow{k_1} B \xrightarrow{k_2} C$). Determine the outlet concentrations of A, B, and C, and the selectivity to the desired product B. Compare your results to those in which diffusional effects are ignored.

Data:

$k_2 = 4.0 \times 10^{-4}$ m^3 (kg cat)$^{-1}$ s^{-1} at 495°C; $E_{A2} = 94.2$ kJ mol^{-1}
$\Delta H_{R2} = -125.6$ kJ mol^{-1}; $D_B = 0.10$ cm^2 s^{-1} at 495°C

Chapter 22

Reactors for Fluid-Solid (Noncatalytic) Reactions

In this chapter, we develop matters relating to the process design or analysis of reactors for fluid-solid noncatalytic reactions; that is, for reactions in which the solid is a reactant. To construct reactor models, we make use of:

(1) the shrinking-core kinetics models from Chapter 9;
(2) the residence-time distribution (RTD) functions from Chapter 13;
(3) the segregated-flow model (SFM) from Chapter 13.

We restrict attention to relatively simple models involving solid particles in fixed-bed or ideal-flow situations. More-complex flow situations are considered in Chapter 23, for both catalytic and noncatalytic reactions.

22.1 REACTIONS AND REACTION KINETICS MODELS

The types of reactions involving fluids and solids include combustion of solid fuel, coal gasification and liquefaction, calcination in a lime kiln, ore processing, iron production in a blast furnace, and regeneration of spent catalysts. Some examples are given in Sections 8.6.5 and 9.1.1.

The sulfation reaction, (D), in Section 9.1.1, is the second step of a process to remove SO_2 from the flue gas of a coal-fired utility furnace. The first step is the calcination of limestone ($CaCO_3$) particles, injected either dry or in an aqueous slurry, to produce CaO:

$$CaCO_3(s) \rightleftarrows CaO(s) + CO_2(g)$$

$$2CaO(s) + 2SO_2(g) + O_2(g) \rightleftarrows 2CaSO_4(s)$$

The first step is endothermic, and the second is exothermic. The (single-particle) kinetics of the overall reaction are relatively complex (Shen, 1996).

Chemical vapor deposition (CVD) reactions in electronics materials processing provide further examples. One is the deposition of Si from silane:

$$SiH_4(g) \rightarrow Si(s) + 2H_2(g)$$

The regeneration of coked catalysts (Section 8.6.5) can be represented by "coke burning" with air:

$$C(s) + O_2 \rightarrow CO_2(g)$$

Two models developed in Chapter 9 to describe the kinetics of such reactions are the shrinking-core model (SCM) and the shrinking-particle model (SPM). The SCM applies to particles of constant size during reaction, and we use it for illustrative purposes in this chapter. The results for three shapes of single solid particle are summarized in Table 9.1 in the form of the integrated time (t)–conversion (f_B) relation, where B is the solid reactant in model reaction 9.1-1:

$$A(g) + bB(s) \rightarrow \text{products}[(s), (g)] \tag{9.1-1}$$

These results take into account three possible processes in series: mass transfer of fluid reactant A from bulk fluid to particle surface, diffusion of A through a reacted product layer to the unreacted (outer) core surface, and reaction with B at the core surface; any one or two of these three processes may be rate-controlling. The SPM applies to particles of diminishing size, and is summarized similarly in equation 9.1-40 for a spherical particle. Because of the disappearance of the product into the fluid phase, the diffusion process present in the SCM does not occur in the SPM.

The results given in Table 9.1 are all for first-order surface-reaction kinetics. If it is necessary to take other kinetics into account (other orders, LH kinetics, etc.), it may be necessary to develop numerical, rather than analytical, methods of solution.

22.2 REACTOR MODELS

22.2.1 Factors Affecting Reactor Performance

The performance of a reactor for a fluid (A) + solid (B) reaction may be characterized by f_B obtained for a given feed rate (F_{B_o}) and size of reactor; the latter is related to the holdup (W_B) of solid in the reactor, whether the process is continuous or batch (fixed bed) with respect to B.

In addition to the single-particle kinetics of reaction, the performance depends on the flow pattern of each phase, the contacting pattern of the two phases, and the particle-size characteristics of the solid. If the process is continuous, we assume ideal flow for the solid particles, as described by the appropriate RTD. For the fluid, we assume that the rate of flow is sufficiently large that there is no appreciable change in its composition through the vessel; that is, the fluid is assumed to be uniform in composition. Consequently, both the flow pattern for the fluid and the contacting pattern for the two phases (e.g., cocurrent or countercurrent flow) are irrelevant. With respect to particle sizes, the solid is characterized by a particle-size distribution (PSD), which is usually in discrete rather than continuous form. Particle size and its distribution are important factors, since small particles reach a given value of f_B faster than large particles (although the intrinsic surface reaction rate is the same).

Aspects of the various factors affecting performance in the types of reactions and reactors considered in Chapter 9 and in this chapter are summarized in Figure 22.1.

22.2.2 Semicontinuous Reactors

A semicontinuous reactor for a fluid-solid reaction involves the axial flow of fluid downward through a fixed bed of solid particles, the same arrangement as for a fixed-bed catalytic reactor (see Figure 15.1(b)). The process is thus continuous with respect to the fluid and batch with respect to the solid (Section 12.4).

Reactor calculations could relate to the fractional conversion of solid and the time for replacement of a given amount of solid (bed weight) to ensure that the outlet gas composition is within specifications. Such calculations are complicated by the unsteady-

554 Chapter 22: Reactors for Fluid-Solid (Noncatalytic) Reactions

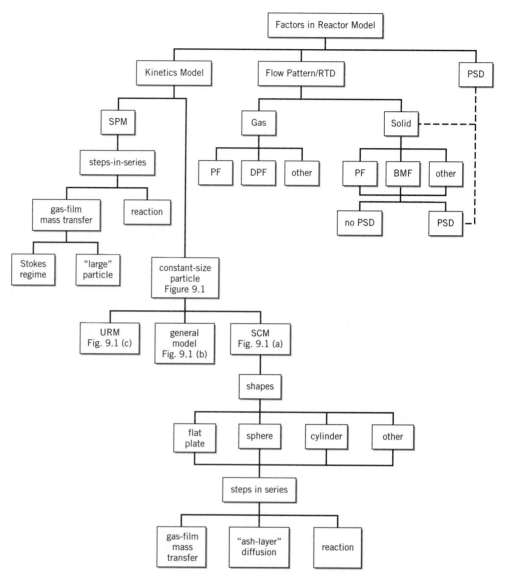

Figure 22.1 Factors affecting reactor performance for a fluid (A) + solid (B) reaction, A + bB → products

state nature of the process for the solid particles. Even if we assume one-dimensional PF for the gas, the composition of particles depends on both axial position in the bed and on time. Two distinct reaction fronts exist: a reaction front that penetrates a particle and a reaction front that moves down the bed. The calculations may instead be based on a lumped "experience" or "loading" factor that takes into account behavior at both the particle level and the bed level (see problem 22-1). Because of the complexity and the fact that semicontinuous, fixed-bed reactors for gas-solid reactions may be replaced by continuous reactors (e.g., fluidized-bed reactors, Chapter 23), we do not consider the detailed fundamental calculations further.

22.2.3 Continuous Reactors

Continuous reactors for fluid-solid reactions involve continuous flow for both fluid and solid phases. With the assumptions made in Section 22.2.1 about the fluid, we focus only on the flow pattern for the solid particles. For these, one feature is that the flow is completely segregated. That is, the particles may mix at the particle or macroscopic level, but do not exchange material; they constitute a macrophase. On the other hand, elements of fluid may mix at the microscopic level; the fluid is a microphase. As a result, we may apply the segregated-flow model (SFM) of Chapter 13 to the reacting system as an exact representation, rather than one that is not necessarily exact for a single-fluid system (see Sections 14.3.2.3, 15.2.2.3, and 16.2.6 for CSTR, PFR, and LFR, respectively). However, the SFM, which averages performance over residence time, is by itself not sufficient, if the particles are not all of the same size. In this case, we must also average performance over particle size, in what may be called an extended SFM. We first treat these two averaging aspects separately in turn, and then together in a general way, and subsequently apply the results to specific cases of ideal flow.

22.2.3.1 *Performance Averaged with Respect to Residence Time t: SFM*

If the particles are all of the same size, and each particle acts as a batch system, the fraction of solid reactant B unconverted at the reactor outlet, $1 - f_B(t)$, is the same for a particular residence time t. The contribution of particles with this residence time to the (average) fraction unconverted at the outlet, $1 - \bar{f}_B$, is the product of $1 - f_B(t)$ and the fraction of particles with residence time t to $t + dt$, which is $E(t)dt$. By integrating over t, we obtain the following form of the SFM, which is equivalent to that in equation 13.5-2:

$$1 - \bar{f}_B = \int_0^\infty [1 - f_B(t)] E(t) dt$$
$$= \int_0^{t_1} [1 - f_B(t)] E(t) dt + \int_{t_1}^\infty [1 - f_B(t)] E(t) dt$$

That is,

$$1 - \bar{f}_B = \int_0^{t_1} [1 - f_B(t)] E(t) dt \qquad (22.2\text{-}1)$$

since particles for which $t > t_1$ (time for complete reaction) do not contribute to $1 - \bar{f}_B$. In equation 22.2-1, the factor $1 - f_B(t)$ is obtained from single-particle kinetics (summarized for the SCM in Table 9.1), and $E(t)$ is obtained from the RTD for the solid particles.

22.2.3.2 *Performance Averaged with Respect to Particle Size R*

If the particles all have the same residence time, the contribution to $1 - \bar{f}_B$ depends on particle size. For spherical or cylindrical particles of radius R, the contribution is the product of $1 - f_B(t, R)$ for that size and the fraction of particles of that size. By integrating over all sizes, we obtain a form analogous to that in equation 22.2-1:

$$1 - \bar{f}_B = \int_{R_{min}}^{R_{max}} [1 - f_B(t, R)] P(R) dR \qquad (22.2\text{-}2)$$

where $P(R)dR$ is the fraction of particles with radius R to $R+dR$, and $P(R)$ is the particle-size distribution (PSD) analogous to the RTD. In equation 22.2-2, R_{max} is unambiguously the size of the largest particle(s), but the interpretation of R_{min} is not so clear cut. Depending on the flow pattern for the solid particles, R_{min} may be the smallest size of particle(s), or it may be the smallest size for which $t_1 > t$, the residence time; if $t_1 < t$, the particle is completely converted and does not contribute to $1 - \bar{\bar{f}}_B$.

22.2.3.3 Performance Averaged with Respect to Both t and R: Extended SFM

If the particles do not have the same residence time and are not all of the same size, we must average with respect to both t and R, to obtain the fraction converted at the reactor outlet. This double averaging is represented by $1 - \bar{\bar{f}}_B$, and may be obtained by combining equations 22.2-1 and -2:

$$1 - \bar{\bar{f}}_B = \int_{R_{min}}^{R_{max}} \left\{ \int_0^{t_1(R)} [1 - \bar{f}_B(t, R)]E(t)dt \right\} P(R)dR \qquad (22.2\text{-}3)$$

In the remainder of this chapter, for simplicity, we use a discrete form for the PSD given by the fraction of a particular size R in the solid feed. For identical particles of the same density, this is $F_{Bo}(R)/F_{Bo}$, the ratio of molar, or mass, flow rates. The outer integral in equation 22.2-3 then becomes a summation, and the equation may be rewritten as:

$$1 - \bar{\bar{f}}_B = \sum_{R_{min}}^{R_{max}} \left\{ \int_0^{t_1(R)} [1 - \bar{f}_B(t, R)]E(t)dt \right\} \frac{F_{Bo}(R)}{F_{Bo}} \qquad (22.2\text{-}4)$$

22.2.4 Examples of Continuous Reactor Models

22.2.4.1 Reactor Models Based on Solid Particles in PF

A reactor model based on solid particles in PF may be used for situations in which there is little or no mixing of particles in the direction of flow. An example is that of a conveyor belt carrying the particles; others that approximate this include a rotating cylindrical vessel slightly inclined to allow axial movement of particles by gravity flow.

In such a case, the particles all have the same residence time, and the RTD is represented by the δ function:

$$E(t) = \delta(t - \bar{t}) \qquad (13.4\text{-}8)$$

where $\bar{t}$ is the mean-residence time. Furthermore, based on the property of the δ function given by equation 13.4-10, equation 22.2-1 becomes

$$\begin{aligned} 1 - \bar{f}_B &= 1 - \bar{f}_B(\bar{t}); & 0 < \bar{t} < t_1 \\ &= 0; & \bar{t} > t_1 \end{aligned} \qquad (22.2\text{-}5)$$

for particles of a given size.

If *all* particles are of the same size, equation 22.2-5 is also the model for the reactor. This result can also be obtained from equation 22.2-3, since then

$$P(R) = \delta(R - \bar{R}) \tag{22.2-6}$$

where $\bar{R}$ is the mean-particle size. It follows from similar application of equation 13.4-10 that

$$1 - \bar{f}_B = 1 - \bar{f}_B(\bar{t}, \bar{R}) \tag{22.2-7}$$

In this case, the performance of the reactor is governed entirely by single-particle kinetics (e.g., as given in Table 9.1).

EXAMPLE 22-1

For the gas-solid reaction $A(g) + bB(s) \rightarrow$ products $[(s), (g)]$, calculate the solid holdup (W_B) required in a reactor for $f_B = 0.80$, if $F_{Bo} = 50$ kg min^{-1} and $t_1 = 1.5$ h. Assume the following apply: the solid particles are in PF, the gas is uniform in composition, the particles are spherical and all of one size, the reaction is a first-order surface reaction, and the SCM applies, with ash-layer-diffusion control.

SOLUTION

The holdup is calculated from the definition of mean residence time of solid particles;

$$\bar{t}_B = W_B/F_{Bo} \tag{22.2-8}$$

where F_{Bo} is given and $\bar{t}_B$ is obtained from the performance equation, 22.2-5 or -7, together with the given value of f_B and the kinetics model. Thus, from Table 9.1, for spherical particles and diffusion control, with $t_1 = \rho_{Bm}R^2/6bc_{Ag}D_e$,

$$t(f_B) \equiv \bar{t}_B = t_1[1 - 3(1 - \bar{f}_B)^{2/3} + 2(1 - \bar{f}_B)] = 0.561 \text{ h}$$

for $t_1 = 1.5$ h and $f_B = 0.80$. Then

$$W_B = \bar{t}_B F_{Bo} = 0.561(50)60 = 1680 \text{ kg}$$

The size of reactor is chosen to accommodate this holdup. The diameter is determined from the *gas* flow rate, q, together with, for example, an allowable superficial linear gas velocity, u (in lieu of an allowable $(-\Delta P)$): $D = (4q/\pi u)^{1/2}$ for a cylindrical vessel. The volume could be determined from an appropriate "bed" density, together with an overhead space for disengagement of solid and gas phases (we assume no carryover of solid in the gas exit stream).

If the particles in PF are not all of the same size, the performance equation, from equation 22.2-2 in the summation form of equation 22.2-4, is

$$1 - \bar{f}_B = \sum_{R(t_1 = \bar{t}_B)}^{R_{max}} [1 - f_B(t, R)] \frac{F_{Bo}(R)}{F_{Bo}} \qquad (22.2\text{-}9)$$

where $R(t_1 = \bar{t}_B)$ is the smallest size of particle for which the time for complete conversion (t_1) is greater than the mean-residence time of solid particles ($\bar{t}_B$); particles smaller than this are completely converted in the time $\bar{t}_B$, and do not contribute to $1 - \bar{f}_B$.

EXAMPLE 22-2

For the reaction and assumptions in Example 22-1, except that reaction-rate control replaces ash-layer-diffusion control, suppose the feed contains 25% of particles of size R for which $t_1 = 1.5$ h, 35% of particles of size $2R$, and 40% of particles of size $3R$. What residence time of solid particles, $\bar{t}_B$, is required for $\bar{f}_B = 0.80$?

SOLUTION

If we denote the three sizes in increasing order by subscripts 1, 2, and 3, equation 22.2-9 becomes

$$1 - \bar{f}_B = [1 - f_{B1}(t)] \frac{F_{Bo}(R_1)}{F_{Bo}} + [1 - f_{B2}(t)] \frac{F_{Bo}(R_2)}{F_{Bo}} + [1 - f_{B3}(t)] \frac{F_{Bo}(R_3)}{F_{Bo}} \quad \text{(A)}$$

although at this stage, we do not know if all three terms contribute to $1 - \bar{f}_B$. From Table 9.1, for spherical particles with reaction-rate control,

$$t = t_1[1 - (1 - f_B)^{1/3}]$$

or

$$1 - f_B = (1 - t/t_1)^3$$

where $t_1 = \rho_{Bm} R/bc_{Ag} k_{As}$. Thus, for the three particle sizes, we write

$$1 - f_{B1} = (1 - t/t_{11})^3 \qquad \textbf{(B)}$$

$$1 - f_{B2} = (1 - t/t_{12})^3 \qquad \textbf{(C)}$$

$$1 - f_{B3} = (1 - t/t_{13})^3 \qquad \textbf{(D)}$$

where $t_{11} = 1.5$ h, and, since $t_1 \propto R$, $t_{12} = 3.0$ h, and $t_{13} = 4.5$ h. With these values known, together with those of the fractional amounts of the different sizes in the feed, and the value of $\bar{f}_B(0.8)$, equations (A) to (D) are four equations which contain the four unknown quantities t, f_{B1}, f_{B2}, and f_{B3}. They may be solved by selecting values of t, and calculating f_{B1}, f_{B2}, and f_{B3} from equations (B) to (D) until equation (A) is satisfied. This results in the following:

$$t = 1.33 \text{ h}$$

$$f_{B1} = 0.999; \; f_{B2} = 0.828; \; f_{B3} = 0.651$$

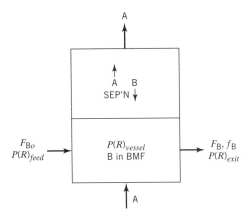

Figure 22.2 Schematic representation of fluid (A) + solid (B) reacting system with solid in BMF and fluid uniform

Equations (A) to (D) may also be solved simultaneously using the E-Z Solve software (file ex22-2.msp).

Alternatively, equation (A) may be solved with the user-defined function $fbsr(t, t_{1i})$ to estimate the conversion for each group of particles, where t_{1i} is the value of t_1 for the ith group of particles. This function for spherical particles with reaction control is included in the E-Z Solve software.

22.2.4.2 Reactor Models Based on Particles in BMF

A reactor model based on solid particles in BMF may be used for situations in which there is deliberate mixing of the reacting system. An example is that of a fluid-solid system in a well-stirred tank (i.e., a CSTR)—usually referred to as a "slurry" reactor, since the fluid is normally a liquid (but may also include a gas phase); the system may be semibatch with respect to the solid phase, or may be continuous with respect to all phases (as considered here). Another example involves mixing of solid particles by virtue of the flow of fluid through them; an important case is that of a "fluidized" bed, in which upward flow of fluid through the particles brings about a particular type of behavior. The treatment here is a crude approximation to this case; the actual flow pattern and resulting performance in a fluidized bed are more complicated, and are dealt with further in Chapter 23.

A schematic representation of this reactor model is shown in Figure 22.2. Particles of solid reactant B are in BMF, and fluid reactant A is uniform in composition, regardless of its flow pattern. The solid product, consisting of reacted and/or partially reacted particles of B, leaves in only one exit stream as indicated. That is, we assume that no solid particles leave in the exit fluid stream (no elutriation or entrainment of solid). This assumption, together with the assumption, as in the SCM, that particle size does not change with reaction, has an important implication for any particle-size distribution, represented by $P(R)$. The implication is that $P(R)$ must be the same for both the solid feed and the solid exit stream, since there is no accumulation in the vessel in continuous operation. Furthermore, in BMF, the exit-stream properties are the same as those in the vessel. Thus, $P(R)$ is the same throughout the system:

$$P(R)_{feed} = P(R)_{vessel} = P(R)_{exit} \quad (22.2\text{-}10)$$

or, with a discrete form of $P(R)$ as in equation 22.2-4,

$$\frac{F_{Bo}(R)}{F_{Bo}} = \frac{W_B(R)}{W_B} = \left[\frac{F_B(R)}{F_B}\right]_{exit} \quad (22.2\text{-}11)$$

where $W_B(R)$ is the holdup of particles of size R in the vessel.

A further consequence of the assumptions made, and of equation 22.2-11, is that the mean residence time of solid particles is independent of particle size:

$$\bar{t}_B(R) = \frac{W_B(R)}{F_{Bo}(R)} = \frac{W_B}{F_{Bo}} = \bar{t}_B \quad (22.2\text{-}12)$$

If the solid particles are in BMF, $E(t)$ in equation 22.2-1 (or -3, or -4) is given by

$$E(t) = (1/\bar{t})e^{-t/\bar{t}} \quad (13.4\text{-}2)$$

To describe the performance of the reactor, we first assume the particles are all of the same size, and then allow for a distribution of sizes.

For particles all the same size, equation 22.2-1, with equation 13.4-2 incorporated, becomes

$$1 - \bar{f}_B = \frac{1}{\bar{t}} \int_0^{t_1} [1 - f_B(t)] e^{-t/\bar{t}} dt \quad (22.2\text{-}13)$$

In equation 22.2-13, $[1 - f_B(t)]$ comes from single-particle kinetics, such as the SCM, for which results for three shapes are summarized in Table 9.1. The following example illustrates the use of the SCM model with equation 22.2-13.

EXAMPLE 22-3

The performance of a reactor for a gas-solid reaction ($A(g) + bB(s) \rightarrow$ products) is to be analyzed based on the following model: solids in BMF, uniform gas composition, and no overhead loss of solid as a result of entrainment. Calculate the fractional conversion of B ($\bar{f}_B$) based on the following information and assumptions: $T = 800$ K; $p_A = 2$ bar; the particles are cylindrical with a radius of 0.5 mm; from a batch-reactor study, the time for 100% conversion of 2-mm particles is 40 min at 600 K and $p_A = 1$ bar. Compare results for $\bar{f}_B$ assuming (a) gas-film (mass-transfer) control; (b) surface-reaction control; and (c) ash-layer diffusion control. The solid flow rate is 1000 kg min^{-1}, and the solid holdup (W_B) in the reactor is 20,000 kg. Assume also that the SCM is valid, and the surface reaction is first-order with respect to A.

SOLUTION

A flow diagram for the reactor is shown schematically in Figure 22.3, together with the given data and assumptions.

Since the particles are all of one size and are in BMF, the reactor model is given by equation 22.2-13. The expression for $[1 - f_B(t)]$ in the integral depends on the relevant kinetics (for a single particle), and is obtained from Table 9.1.

(a) From Table 9.1, for a cylindrical particle with gas-film control,

$$t(f_B) = t_1 f_B \quad (22.2\text{-}14)$$

or

$$1 - f_B(t) = 1 - t/t_1 \quad (22.2\text{-}14a)$$

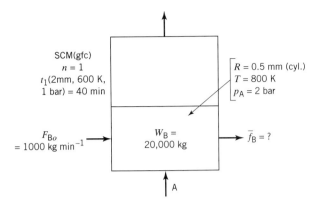

Figure 22.3 Flow diagram, data, and assumptions for Example 22-3

where

$$t_1 = \rho_{Bm} R / 2 b k_{Ag} c_{Ag} \quad (22.2\text{-}15)$$

With equation 22.2-14a substituted in equation 22.2-13, the reactor model is

$$1 - \bar{f}_B = \frac{1}{\bar{t}} \int_0^{t_1} \left(1 - \frac{t}{t_1}\right) e^{-t/\bar{t}} dt = 1 - (\bar{t}/t_1)(1 - e^{-t_1/\bar{t}}); \quad (22.2\text{-}16)$$

(cylindrical particles BMF, SCM, gas-film control)

on integration and insertion of the limits. (The two terms in the integrand, after multiplication, can be integrated separately, but the second requires integration by parts.) Equation 22.2-16 may be written as

$$\bar{f}_B = (\bar{t}/t_1)(1 - e^{-t_1/\bar{t}}) \quad (22.2\text{-}16a)$$

Information is given directly for $\bar{t}$, but indirectly for t_1.

$$\bar{t} = W_B / F_{Bo} = 20{,}000/1000 = 20 \text{ min}$$

To obtain t_1 at flow-reactor conditions from t_{1B} obtained in the batch reactor, we use equation 22.2-15, together with ideal-gas behavior for reactant A:

$$c_{Ag} = p_A / RT \quad (4.2\text{-}3a)$$

Denoting BR conditions by subscript 1 for radius R, T and p_A, and flow-reactor conditions by subscript 2, from 22.2-15, for fixed values of the other parameters, we have

$$\frac{t_1}{t_{1B}} = \frac{R_2/c_{Ag2}}{R_1/c_{Ag1}} = \frac{R_2 p_{A1} T_2}{R_1 p_{A2} T_1} = \frac{0.5(1)800}{2(2)600} = \frac{1}{6}$$

Thus,

$$t_1 = 40/6 = 6.67 \text{ min}$$

and from equation 22.2-16a,

$$\bar{f}_B \text{ (gas-film control)} = 0.850$$

(b) From Table 9.1, for a cylindrical particle with surface-reaction control,

$$t(f_B) = t_1[1 - (1 - f_B)^{1/2}] \qquad (22.2\text{-}17)$$

or

$$1 - f_B(t) = (1 - t/t_1)^2 \qquad (22.2\text{-}17a)$$

where

$$t_1 = \rho_{Bm} R / b c_{Ag} k_{As} \qquad (9.1\text{-}31a)$$

With equation 22.2-17a substituted into 22.2-13, the reactor model becomes, after integration,

$$\bar{f}_B = (2\bar{t}/t_1)[1 - (\bar{t}/t_1)(1 - e^{-t_1/\bar{t}})] \qquad (22.2\text{-}18)$$

(cylindrical particles, BMF, SCM, $n = 1$, reaction control)

Equation 22.2-18 corresponds to equation 22.2-16a for gas-film control.

The values of both $\bar{t}$ and t_1 are the same as in part (a), 20 and 6.67 min, respectively. The latter is because, for reaction and gas-film control, the forms of the expressions for t_1 are the same with respect to the parameters that change.

Thus, with these values used in 22.2-18,

$$\bar{f}_B \text{ (reaction control)} = 0.898$$

This value of $\bar{f}_B$ may also be obtained, by means of the E-Z Solve software, by simultaneous solution of equation 22.2-17 and numerical integration of 22.2-13 (with user-defined function $fbcr(t, t_1)$, for cylindrical particles with reaction control; see file ex22-3.msp). This avoids the need for analytical integration leading to equation 22.2-18.

(c) From Table 9.1, for a cylindrical particle with ash-layer diffusion control,

$$t(f_B) = t_1[f_B + (1 - f_B)\ln(1 - f_B)] \qquad (22.2\text{-}19)$$

where

$$t_1 = \rho_{Bm} R^2 / 4 b c_{Ag} D_e \qquad (22.2\text{-}20)$$

Unlike the situation in (a) and (b), equation 22.2-19 cannot be inverted functionally to obtain $1 - f_B$ or f_B explicitly in terms of t, for ultimate use in equation 22.2-13. Rather than an analytical expression corresponding to 22.2-16a or -18, we *must* use the alternative numerical procedure described in part (b). Furthermore, for ash-layer diffusion control, since $t_1 \propto R^2$, $t_1 = 40/24 = 1.667$ min. Using this value for t_1 and $\bar{t} = 20$ min, as before, we obtain (see file ex22-3.msp):

$$\bar{f}_B \text{ (ash-layer control)} = 0.980$$

Note that in the numerical solution of equation 22.2-19, f_B must be less than 1 to avoid $\ln(0)$; in the use of E-Z Solve, an initial guess of 0.5 is entered, and the upper and lower limits may be set to 0 and 0.9999, respectively. Since upper and lower limits must be specified, a user-defined function is not applicable to this case.

As shown in Example 22-3, for solid particles of the same size in BMF, the form of the reactor model resulting from equation 22.2-13 depends on the kinetics model used for a single particle. For the SCM, this, in turn, depends on particle shape and the relative magnitudes of gas-film mass transfer resistance, ash-layer diffusion resistance and surface reaction rate. In some cases, as illustrated for cylindrical particles in Example 22-3(a) and (b), the reactor model can be expressed in explicit analytical form; additional results are given for spherical particles by Levenspiel (1972, pp. 384–5). In other cases, it is convenient or even necessary, as in Example 22-3(c), to use a numerical procedure. As also illustrated in the example, the E-Z Solve software is well-suited for this purpose (file ex22-3.msp).

If the solid particles in BMF are not all of the same size, the result obtained for $\bar{f}_B$ from equation 22.2-13 must be combined with the PSD as in equation 22.2-3 or -4 to obtain $\bar{\bar{f}}_B$. We emphasize again that (1) the mean residence time ($\bar{t}$) is independent of particle size (R), provided there is only one exit stream for the solid particles, and (2) t_1 in equation 22.2-13 depends on R. Furthermore, the interpretation of R_{min} in equation 22.2-3 or -4 for particles in BMF is different from that for particles in PF (in equation 22.2-9). For BMF, the averaging to obtain $\bar{\bar{f}}_B$ must be done over *all* particle sizes; that is, R_{min} is the smallest size in the feed. Thus, equation 22.2-4 is written:

$$1 - \bar{\bar{f}}_B = \sum_{\text{all sizes}} \left\{ \frac{1}{\bar{t}} \int_0^{t_1(R)} [1 - f_B(t, R)] e^{-t/\bar{t}} dt \right\} \frac{F_{Bo}(R)}{F_{Bo}} \quad (22.2\text{-}21)$$

The calculation of $\bar{\bar{f}}_B$ with a range of particle sizes is straightforward using the E-Z Solve software with user-defined functions, such as $fbcr(t, t_1)$ and $fbsr(t, t_1)$ included in the software.

22.2.5 Extension to More Complex Cases

We use two examples to illustrate (1) the dependence of $\bar{f}_B$ on $\bar{t}$ and the sensitivity to the type of rate-controlling process, and (2) the situation in which more than one rate process contributes resistance. We use particles of the same size in BMF in both examples, but PF could be considered, as could a particle-size distribution.

22.2.5.1 *Dependence of $\bar{f}_B$ on $\bar{t}$ and Sensitivity to Rate-Controlling Process*

EXAMPLE 22-4

Extend Example 22-3 to examine the dependence of $\bar{f}_B$ on $\bar{t}$ and the sensitivity to the type of rate-controlling process. For this purpose, using the data of Example 22-3, calculate and plot $\bar{f}_B$ as a function of $\bar{t}$ over the range 0 to 25 min for each of the three cases in Example 22-3(a), (b), and (c).

SOLUTION

For gas-film mass transfer control, we use equation 22.2-16a; for reaction control, we use equation 22.2-18; and for ash-layer diffusion control, we integrate equation 22.2-13 numerically in conjunction with 22.2-19, as described in Example 22-3(c). The results generated by the E-Z Solve software (file ex22-4.msp) are shown in Figure 22.4.

For the data and conditions of Example 22-3, $\bar{f}_B$ depends strongly on $\bar{t}$ up to about 15 min, and is also very sensitive to the type of rate-controlling process. For example, for

564 Chapter 22: Reactors for Fluid-Solid (Noncatalytic) Reactions

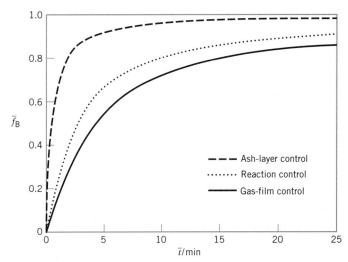

Figure 22.4 Example 22-4: Dependence of mean fractional conversion ($\bar{f}_B$) on mean residence time ($\bar{t}$)–effect of rate-limiting process; data of Example 22-3 (cylindrical particles of all same size in BMF; SCM; $n = 1$)

$\bar{f}_B = 0.80$, $\bar{t}$, which is a measure of the size of reactor, is about 1.7 min for ash-layer control, 9.5 min for reaction control, and 14.5 min for gas-film control. The relatively favorable behavior for ash-layer diffusion control in this example reflects primarily the low value of t_1 (1.67 min versus 6.67 min for the other two cases) imposed.

Although it relates to a specific set of conditions, Figure 22.4 nevertheless illustrates the importance of knowing the underlying process (or processes) controlling the kinetics. If the results cited above are put another way, from the point of view of determining $\bar{f}_B$ for a given $\bar{t}$ (or reactor size), a mean residence time of 1.7 min gives $\bar{f}_B = 0.80$ for ash-layer control, as noted, but only 0.37 for reaction control, and only 0.23 for gas-film control.

A figure such as Figure 22.4 can also be used in general to determine the size of reactor for a given throughput, F_{Bo}, and specified conversion, $\bar{f}_B$, in terms of solid holdup, since $W_B = F_{Bo}\bar{t}$, from equation 22.2-8. The determination of $\bar{t}$ for a given value of $\bar{f}_B$ with equations such as 22.2-4, -9, and -21 is awkward. The converse problem of calculating $\bar{f}_B$ from $\bar{t}$ is straightforward, as illustrated in Example 22-4. Once generated, a figure corresponding to Figure 22.4 can then be used as a design chart for various possible operating requirements.

22.2.5.2 Cases Where More Than One Rate Process Is Important

The cases considered thus far have all been based upon the premise that one process, ash-layer diffusion, surface reaction, or gas-film mass transfer, is rate controlling. However, in some cases, more than one process affects the overall kinetics for the conversion of the solid. This has two implications:

(1) Values for k_{As}, k_{Ag}, and D_e must be known explicitly; the kinetics experiments must be designed to provide this information.
(2) It is no longer possible to obtain an explicit relationship between $\bar{f}_B$ and $\bar{t}$ from the expressions in Table 9.1.

The following example illustrates a numerical method used to assess reactor performance when more than one process affects the overall kinetics of the reaction.

EXAMPLE 22-5

A kinetics study of the reaction $A(g) + 2B(s) \rightarrow$ products (s, g) (problem 9-21) yielded the following values of parameters for 2-cm (diameter) spherical particles at 700 K and $p_A = 200$ kPa:

$t_1 = 4.92 \times 10^5$ s; $k_{Ag} = 9.2 \times 10^{-5}$ m s^{-1}
$k_{As} = 8.85 \times 10^{-5}$ m s^{-1}; $D_e = 8.37 \times 10^{-7}$ m^2 s^{-1}

For the value of t_1 obtained, determine, using the SCM model, the mean fractional conversions ($\bar{f}_B$) for mean residence times ($\bar{t}$) of (i) 50 h, and (ii) 100 h, assuming

(a) reaction rate-control;
(b) ash-layer diffusion control;
(c) gas-film mass transfer control;
(d) resistance contributed by all three processes.

SOLUTION

Although it is possible to obtain analytical expressions for the reactor model for spherical particles for parts (a), (b), and (c) using the approach in Example 22-3(a) and (b) for cylindrical particles (expressions for $\bar{f}_B$ for spherical particles are given by Levenspiel, 1972, pp. 384-5), we use the numerical procedure described in Example 22-3 (c) for all four parts of this example, (a) to (d), since this procedure must be used for (d).

For numerical solution with the E-Z Solve software (file ex22-5.msp), a procedure equivalent to that in Example 22-3(c) is used. First, equation 22.2-13 is expressed as a differential equation:

$$\frac{d\bar{f}_B}{dt} = -\frac{1}{\bar{t}} e^{-t/\bar{t}} [1 - f_B(t)] \quad (22.2\text{-}22)$$

Equation 22.2-22 is then numerically integrated simultaneously with solution of the algebraic equation $t = t(f_B)$ given below for each of the cases (a) to (d). The integration is from $t = 0$ to $t = t_1$, subject to the initial condition that $\bar{f}_B = 0$ at $t = 0$. The expressions for $t(f_B)$ are taken from Table 9.1, as follows.

(a) $$t = t_1[1 - (1 - f_B)^{1/3}]$$

where $$t_1 = \rho_{Bm} R / b c_{Ag} k_{As}$$

(b) $$t = t_1[1 - 3(1 - f_B)^{2/3} + 2(1 - f_B)]$$

where $$t_1 = \rho_{Bm} R^2 / 6 b c_{Ag} D_e$$

(c) $$t = t_1 f_B$$

where $$t_1 = \rho_{Bm} R / 3 b c_{Ag} k_{Ag}$$

(d) $$t = \frac{\rho_{Bm} R}{b c_{Ag}} \left\{ \frac{f_B}{3 k_{Ag}} + \frac{R}{6 D_e}[1 - 3(1 - f_B)^{2/3} + 2(1 - f_B)] + \frac{1}{k_{As}}[1 - (1 - f_B)^{1/3}] \right\}$$

(22.2-23)

Table 22.1 Results for Example 22-5

$\bar{t}$/h	$\bar{f}_B$			
	(a)	(b)	(c)	(d)
50	0.569	0.669	0.342	0.515
100	0.734	0.793	0.545	0.692

File ex22-5.msp illustrates the implementation of the solution in each case. Note that, to avoid evaluation of zero to a fractional power, f_B is bounded between zero and 0.9999. Note also that t_1 is not used for case (d). The results for $\bar{f}_B$ for all cases, (a) to (d), are given in Table 22.1. Because of the way the problem is stated, with t_1 the same for each case, the result for case (d), all three rate processes involved, is an average of some sort of the results for cases (a) to (c), each of which involves only one rate process. The assumption of a single rate-controlling process introduces significant error in each of cases (a) to (c), relative to case (d).

22.3 PROBLEMS FOR CHAPTER 22

22-1 Desulfurization of natural gas (e.g., removal of H_2S from the feedstock for a methanol plant) can be carried out by allowing the gas to flow downward through a fixed bed of ZnO particles so that the reaction $H_2S(g) + ZnO(s) \rightarrow ZnS(s) + H_2O(g)$ occurs. In such a case, the desulfurizer may be sized by means of a "loading factor" (as an experience kinetics parameter). Based on the information given below, calculate the following:
(a) the amount (m³) of ZnO required for 1 year's operation, if the plant operates 8000 h year^{-1};
(b) the fraction of ZnO reacted in the bed at the end of 1 year;
(c) the space velocity (reciprocal of space time) for the gas in h^{-1};
(d) the diameter (D) and depth (L) of the bed, and the pressure drop (ΔP) through the bed.

Data:

20 ppmv H_2S in feed gas and 0.2 ppmv at outlet;
$T = 344°C$ (isothermal); $P = 18$ bar; gas feed rate = 2500 kmol h^{-1};
Loading factor = 0.20 kg S reacted (cumulative) (kg ZnO initially)$^{-1}$;
$\rho_{ZnO} = 1120$ kg m^{-3} (bulk density); $\epsilon_B = 0.35$ (bed voidage);
$z = 1$ (gas compressibility factor); $\mu_f = 2.2 \times 10^{-5}$ N s m^{-2};
$d_p = 4.8$ mm (spherical particles of ZnO); $u_s = 0.18$ m s^{-1} (superficial gas velocity).

22-2 (a) In equation 22.2-9 for PF, averaging is done over the fraction of B unconverted, $1 - f_B$. If averaging is done over f_B, what changes, if any, are required?
(b) Repeat (a) for equations 22.2-13 and -21 for BMF.

22-3 Consider the fractional conversion (f_B) of solid (B) in a gas-solid reaction taking place in a reactor, with mean-residence time of solid $\bar{t}_B$, and time for complete reaction of solid particles $t_1(R)$. The particles are not all of the same size.
(a) If the solids are in BMF and $\bar{t} > t_1(R_{max})$, is $f_B = 1$ or < 1? Explain briefly. Is it advantageous to operate with $\bar{t} > t_1(R_{max})$?
(b) Repeat (a) for the case of the solids in PF.

22-4 Consider the reduction of 3-mm (radius) spherical particles of Fe_3O_4 by H_2 at 1000 K in a reactor in which the solid is in PF and the gas is of uniform composition, with $p_{H_2} = 1$ bar. What size of reactor (in terms of holdup of solid, W_B/kg) is required so that the particles just completely react for a feed rate (F_{Bo}) of 500 kg min^{-1}? Assume the particles are of constant size and the SCM applies.

Data: $\rho_{Bm} = 20,000$ mol m^{-3}; k_{As} (first-order rate constant) = $1930 \exp(-12,100/T)$ m s^{-1}; and $D_e = 3 \times 10^{-6}$ m^2 s^{-1} for diffusion through the "ash" layer. Clearly state any other assumption(s) made (data from Levenspiel, 1972, p. 401).

22-5 (a) A steady-state solid feed consisting of 2-mm (radius) spherical particles enters a rotating kiln-type reactor (assume PF), and reacts with gas of uniform composition at particular T and P. If the SCM is applicable and the process is ash-layer-diffusion controlled, calculate the residence time t required for 80% conversion of the solid feed to solid product, if the time required for complete conversion (t_1) of 4-mm particles is 4 h at the same T and P.

(b) Calculate the holdup (W_B), if the feed rate (F_{Bo}) is 10,000 kg h^{-1}.

22-6 Suppose cylindrical particles of ZnS (B) of radius 1.5 mm are roasted (oxidized) in air (21 mol% O$_2$ (A)) at 900°C and 1 bar to form ZnO (and SO$_2$).

(a) Calculate the time, t/h, required to achieve a fractional conversion, f_B, of 0.90 in a single particle.

(b) If the feed rate (F_{Bo}, for the reaction as in (a)) is 15,000 kg h^{-1}, what is the solid holdup (W_B/kg) in the reactor?

Additional information:

Assume the SCM is valid, the reaction is first-order, the gas phase is uniform, the solid particles are in PF, and that gas-film resistance is negligible.

$\rho_{Bm} = 0.043$ mol cm^{-3}; $D_e = 0.08$ cm^2 s^{-1}; $k_{As} = 2$ cm s^{-1} (Data from Levenspiel, 1972, p. 402).

22-7 Consider a gas-solid reaction, $A(g) + bB(s) \rightarrow$ products, occurring in a vessel in which the solid is in PF and the gas composition is uniform. The solid is made up of 2-mm (radius) cylindrical particles (35 wt %), 3-mm particles (45%), and 4-mm particles (20%). Assume the shrinking-core model (SCM) applies with reaction-rate control. For cylindrical particles of 1-mm radius, the time for complete reaction, $t_1(1)$ is 0.25 h at $p_{Ag} = 1$ bar and a particular temperature, T. The reactor operates at $p_{Ag} = 0.5$ bar at the same T.

(a) If the reactor operates at steady-state so that the 2-mm particles are (just) completely reacted at the outlet [$f_B(2) = 1$], calculate
  (i) the fraction converted of 3-mm particles at the outlet, $f_B(3)$;
  (ii) $f_B(4)$;
  (iii) the overall fraction converted, $\bar{f}_B$.

(b) If the total feed rate (F_{Bo}) is 10,000 kg h^{-1}, what is the (total) holdup of solid (W_B/kg) in the reactor, and what is the distribution of the particle sizes?

22-8 The performance of a reactor for a gas-solid reaction ($A(g) + bB(s) \rightarrow$ products) is to be analyzed based on the model: solids in BMF, uniform gas composition, and no overhead product as a result of entrainment. Calculate (a) the fractional conversion of B obtained ($\bar{f}_B$), and (b) the diameter of the reactor (D/m), based on the following information and assumptions: $T = 800$ K; $P = 1.5$ bar; superficial linear gas velocity (based on the empty vessel) is $u_s = 0.1$ m s^{-1}; total molar gas flow rate, including any inert gas, = 50 mol s^{-1}; the particles are spherical with a radius of 0.3 mm; from a batch-reactor study, time for 100% conversion of 1-mm particles is 20 min; assume rate is reaction-controlled; solid flow rate (F_{Bo}) is 1000 kg min^{-1}, and the solid holdup (W_B) in the reactor is 15,000 kg. Assume also that the SCM is valid, and the reaction is first-order with respect to A.

22-9 Repeat problem 22-8(a), if the rate is controlled by ash-layer diffusion.

22-10 Repeat problem 22-8(a), if the rate is controlled by gas-film mass transfer.

22-11 The performance of a reactor for a gas-solid reaction ($A(g) + bB(s) \rightarrow$ products) is to be analyzed based on the following model: solids in BMF, uniform gas composition, and no overhead product as a result of elutriation. Calculate the mean fractional conversion of B obtained ($\bar{f}_B$), based on the following information and assumptions: particle sizes and relative amounts: 30% by weight of 0.2-mm (radius) spherical particles, and 70% 0.3-mm; from a batch-reactor study, for the operating conditions used, time for 100% conversion of 1-mm particles is 20 min; assume rate is reaction-controlled; solid feed rate (F_{Bo}) is 1000 kg min^{-1}, and the solid holdup (W_B) in the reactor is 15,000 kg. Use the SCM, and assume the reaction is first-order in A.

22-12 Repeat problem 22-11, if the rate is controlled by ash-layer diffusion.

22-13 Repeat problem 22-11, if the rate is controlled by gas-film mass transfer.

22-14 If cylindrical particles of two different sizes ($R_2 = 2R_1$) of species B undergoing reaction with species A (A(g) + bB(s) → products) are in BMF in a vessel with uniform gas concentration, what is the mean fractional conversion of B ($\bar{f}_B$) for particles leaving the vessel, if the feed is 60% of the smaller size, and the mean residence time ($\bar{t}$) of particles is equal to t_1 (time for complete conversion) for the smaller size (1 h)? Assume the SCM with gas-film rate control applies for a first-order surface reaction.

22-15 Repeat problem 22-14, if the rate is reaction controlled.

22-16 Repeat problem 22-14, if the rate is controlled by ash-layer diffusion.

Chapter 23

Fluidized-Bed and Other Moving-Particle Reactors for Fluid-Solid Reactions

In this chapter, we consider reactors for fluid-solid reactions in which the solid particles are in motion (relative to the wall of the vessel) in an arbitrary pattern brought about by upward flow of the fluid. Thus, the solid particles are neither in ideal flow, as in the treatment in Chapter 22, nor fixed in position, as in Chapter 21. We focus mainly on the fluidized-bed reactor as an important type of moving-particle reactor. Books dealing with fluidization and fluidized-bed reactors include those by Kunii and Levenspiel (1991), Yates (1983), and Davidson and Harrison (1963).

The first commercial use of a fluidized-bed reactor, in the 1920s, was for the gasification of coal to supply CO and H_2 for the production of synthetic chemicals. The main impetus, however, came from the need for large quantities of aviation fuel in the early 1940s in World War II by the catalytic cracking of petroleum fractions. Since the catalyst is rapidly deactivated by coking, it was desired to replace intermittent operation of fixed-bed reactors with continuous operation for both the cracking process and the regeneration process. This led to the widespread use of "fluid catalytic cracking" for the production of gasoline. The technology has been developed also for a number of other processes, including other catalytic reactions for the production of chemicals and synthetic gasoline by the Fischer-Tropsch process, and noncatalytic processes for the roasting of sulfide ores, and the incineration of solid waste. Fluidized-bed and similar technology has also been used for nonchemical operations in heat exchange, drying of solids, and coating processes analogous to chemical vapor deposition.

After introducing some types of moving-particle reactors, their advantages and disadvantages, and examples of reactions conducted in them, we consider particular design features. These relate to fluid-particle interactions (extension of the treatment in Chapter 21) and to the complex flow pattern of fluid and solid particles. The latter requires development of a hydrodynamic model as a precursor to a reactor model. We describe these in detail only for particular types of fluidized-bed reactors.

23.1 MOVING-PARTICLE REACTORS

23.1.1 Some Types

23.1.1.1 Fluidized-Bed and Related Types

Consider a bed of solid particles initially fixed in a vessel, and the upward flow through the bed of a fluid introduced at many points below the bed, as indicated schematically in Figure 23.1. The rate of flow of fluid is characterized by the superficial linear velocity u_s, that is, the velocity calculated as though the vessel were empty. At relatively low values of u_s, the bed remains a fixed bed of particles, although it expands somewhat as u_s increases. As u_s increases further, a point is reached at which the bed begins to "lift." Beyond this point, over a range of increasing u_s, the action of the multiphase fluid + solid region (bed) may resemble that of a vigorously boiling liquid with an apparent upper surface from which rising fluid disengages. This action is referred to in general as fluidization of the bed (in the case above, a "bubbling fluidized bed"). As u_s increases still further, solid particles become elutriated from the fluidized region by entrainment in the rising fluid. At sufficiently high values of u_s, the entire bed of particles may be entrained with the fluid and carried overhead out of the vessel. Other types of intermediate behavior occur, usually undesirable, and we have simplified the picture considerably.

When a chemical reaction occurs in the system, each of these types of behavior gives rise to a corresponding type of reactor. These range from a fixed-bed reactor (Chapter 21—not a moving-particle reactor), to a fluidized-bed reactor *without* significant carryover of solid particles, to a fast-fluidized-bed reactor *with* significant carryover of particles, and ultimately a pneumatic-transport or transport-riser reactor in which solid particles are completely entrained in the rising fluid. The reactors are usually operated commercially with continuous flow of both fluid and solid phases. Kunii and Levenspiel (1991, Chapter 2) illustrate many industrial applications of fluidized beds.

Figure 23.2 depicts some of the essential features of (a) fluidized-bed, (b) fast-fluidized-bed, and (c) pneumatic-transport reactors (after Yates, 1983, p. 35).

Figure 23.2(a) shows a fluidized-bed reactor, essentially the same as that in Figure 23.1.

Figure 23.2(b) shows a fast-fluidized-bed reactor, together with external equipment, such as cyclones, for separation of fluid and solid particles carried out of the reactor, and subsequent recirculation to the reactor. In a fast-fluidized bed, the fluidization velocity

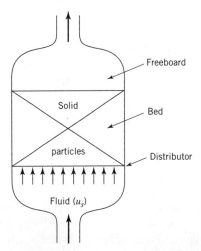

Figure 23.1 Schematic representation of (incipient) particle movement brought about by upward flow of a fluid, leading to fluidization

23.1 Moving-Particle Reactors

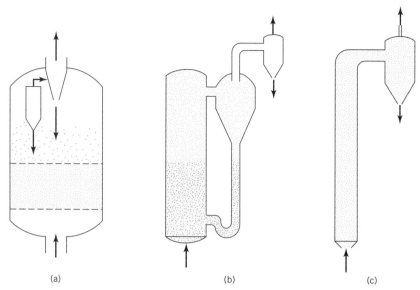

Figure 23.2 Some features of (a) a fluidized-bed reactor; (b) a fast-fluidized-bed reactor; and (c) a pneumatic-transport reactor

is very high, resulting in significant entrainment of solid particles. Consequently, there is a concentration gradient of solid particles through both the bed and freeboard regions. Continuous addition of fresh solid particles may be required for some operations (e.g., coal gasification). The performance is typically described using both a fluidized-bed model and a freeboard-reaction model. Applications of fast-fluidized beds are in fluidized-bed combustion and Fischer-Tropsch synthesis of hydrocarbons from CO and H_2.

Figure 23.2(c) shows a pneumatic-transport reactor. In this type, fluid velocities are considerably greater than the terminal velocities of the particles, so that virtually all of the particles are entrained. The vessel may be extremely tall, with no solid recirculation (e.g., coal combustion), or it may provide for solid recirculation with external cyclones. The process stream is extremely dilute in solid particles because of the high volume of gas passing through the "bed." Fluid-catalyzed cracking of gasoil is an important example of pneumatic transport with external recirculation (and regeneration) of catalyst pellets. A major design issue is the configuration of the recirculation system, which must carry out heat transfer, catalyst regeneration, solid recovery, and recirculation.

23.1.1.2 Spouted Bed

If the fluid enters the vessel at one central point, as indicated in Figure 23.3, rather than at many points spaced across a circular distributor, as in Figure 23.1, the action is different as u_s increases: a *spouted* bed results rather than a fluidized bed. A spouted bed is characterized by a high-velocity spout of gas moving up the center of the bed, carrying particles to the top. This action induces particle circulation, with particle motion toward the wall and downward around the spout and toward the center. The particles in a spouted bed are relatively large and uniformly sized.

Spouted beds have been used for drying and low-temperature chemical-treatment operations. Examples are the low-temperature roasting of agricultural products, and particle-coating and crystal-growth operations.

Further consideration of spouted beds is outside our scope, but see the work by Mathur and Epstein (1974).

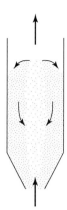

Figure 23.3 Schematic representation of a spouted bed

23.1.2 Examples of Reactions

Reactions in moving-particle reactors in general, and in fluidized-bed reactors in particular, may be catalytic or noncatalytic. That is, the particles may be catalyst particles or reactant particles. Examples are as follows:

(1) Catalytic cracking of gasoil: an impetus for the development of fluidized-bed reactors over 50 years ago was the desire to make the catalytic cracking of gasoil (to gasoline) a continuous process, in spite of the rapid deactivation of the catalyst particles by coke and tarry deposits. Originally, both the catalytic-cracking reactor ("cracker") itself and the catalyst regenerator were fluidized-bed reactors, with solid particles moving continuously between the two in an overall continuous process, but more recently the cracker is made a pneumatic-transport reactor.

(2) Production of acrylonitrile by ammoxidation of propylene (SOHIO process):

$$2\,NH_3 + 3\,O_2 + 2\,C_3H_6 \rightarrow 2\,C_3H_3N + 6\,H_2O$$

The fluidized-bed process for this reaction has several advantages over a fixed-bed process. First, the process is highly exothermic, and the selectivity to C_3H_3N is temperature dependent. The improved temperature control of the fluidized-bed operation enhances the selectivity to acrylonitrile, and substantially extends the life of the catalyst, which readily sinters at temperatures in excess of 800 K. Furthermore, since both the reactants and products are flammable in air, the use of a fluidized bed enables the moving particles to act to quench flames, preventing combustion and ensuring safe operation.

(3) Oxidation of napthalene to produce phthalic anhydride:

$$C_{10}H_{10} + O_2 \rightarrow C_8H_4O_3 \rightarrow CO_2 + H_2$$

The reaction may proceed directly to phthalic anhydride, or it may proceed via naphthaquinone as an intermediate. Phthalic anhydride may also undergo subsequent conversion to CO_2 and H_2O. Thus, the selectivity to phthalic anhydride is a crucial aspect of the design. Proper control of temperature is required to limit napthaquinone production and avoid the runaway (and possibly explosive) reaction which leads to the production of CO_2 and H_2O. A fluidized-bed is thus preferred over a fixed-bed process.

(4) Production of synthetic gasoline by the Fischer-Tropsch process:

$$nCO + 2nH_2 \rightarrow (CH_2)_n + nH_2O$$

This is another example of a highly exothermic process which requires strict temperature control to ensure appropriate selectivity to gasoline, while limiting the production of lighter hydrocarbons. Again, the enhanced temperature control provided by a fluidized-bed system greatly improves the feasibility of this process.

(5) Noncatalytic roasting of ores such as zinc and copper concentrates:

$$2\,ZnS + 3\,O_2 \rightarrow 2\,ZnO + 2\,SO_2$$

The fluidized-bed process replaced rotary kilns and hearths; its primary advantages are its higher capacity and its lower air requirement, which leads to a product gas richer in SO_2 for use in a sulfuric acid plant.

(6) Noncatalytic complete or partial combustion of coal or coke in fluidized-bed combustors:

$$C + O_2 \rightarrow CO_2$$

$$C + \frac{1}{2}O_2 \rightarrow CO$$

These reactions may serve as a means of regeneration of coked catalysts. Both reactions are exothermic, and the improved temperature control provided by a fluidized bed is critical for regeneration of catalysts prone to sintering.

This process (usually with addition of steam) can also be used to generate gas mixtures from partial oxidation of coal for synthetic gasoline production (see (4), above). Fluidized beds offer convenience for continuous handling of the feed solids.

23.1.3 Advantages and Disadvantages

The advantages and disadvantages of moving-particle reactors may be considered relative to the characteristics and operating conditions of fixed-bed reactor (Chapter 21). Fixed-bed reactors are best suited for processes in which the catalyst is very stable. To avoid large pressure drop, the catalyst particles are relatively large; this, however, may lead to concentration gradients through the particle (large L_e). Furthermore, if the enthalpy of reaction is large, significant temperature gradients may be present. The suitability of a fixed-bed reactor for these types of reactions depends upon the sensitivity of the process to temperature, and the ease of design of the required heat-transfer equipment. By comparison, the contacting scheme in moving-particle systems leads to significant operational advantages, although it introduces other problems which need to be considered. The following comments relate primarily to bubbling fluidized-bed reactors.

(1) **Advantages**
  (i) *Mode of operation:* operation can be made continuous with respect to both the processing fluid and the solid; this allows, for example, for the continuous regeneration of a deactivating catalyst.
  (ii) *Thermal:* there is near-uniformity of T throughout the bed, which allows for better control of T and avoidance of hot spots in highly exothermic reactions; the uniformity of T is due to such things as the high degree of turbulence (resulting in relatively high heat transfer coefficients), and the large interfacial area between fluid and small particles.
  (iii) *Chemical performance:* the use of relatively small particles (e.g., 0.1 to 0.3 mm) can result in lower pore-diffusion resistance in solid particles and an

effectiveness factor (η) much closer to 1; by itself, this, in turn, results in a smaller catalyst holdup.

(2) Disadvantages
  (i) *Mechanical:* abrasion causes erosion of pipes and internal parts (e.g., heat transfer surface); attrition of particles leads to greater entrainment and elutriation, requiring equipment (cyclones) for recovery; these mechanical features lead to higher operating and maintenance costs, as well as greater complexity.
  (ii) *Fluid-mechanical:* There is a larger ($-\Delta P$), requiring greater energy consumption; the complex flow and contacting patterns are difficult to treat rationally, and create difficulties of scale-up from small-diameter, shallow beds to large-diameter, deep beds.
  (iii) *Chemical performance:* in fluidized-beds, there is a "bypassing effect" which leads to inefficient contacting; fluid in large bubbles tends to avoid contact with solid particles; this leads to a larger catalyst holdup and/or lower conversion, which may even be lower than that predicted on the basis of BMF, which in turn is lower than that based on PF; the turbulence and resulting backmixing may result in adverse effects on selectivity, for example, depressing the intermediate in a series reaction (Chapter 18); staging may be required to obtain sufficiently high conversion.

If larger particles are used in a moving-bed reactor, there is some sacrifice over temperature control and fluid-solid exchange. However, the pressure drop is much less than in bubbling fluidized beds, and erosion by particles is largely avoided. Furthermore, the fluid-solid contacting is close to ideal, and so performance is enhanced.

In fast-fluidized beds, which use extremely small catalyst particles, the flow of particles and fluid is cocurrent, and nearly PF, leading to high conversion and selectivity. However, severe erosion of equipment is common, and thus, the design of the solids recovery and recirculation system is important.

23.1.4 Design Considerations

For moving-particle reactors, in addition to the usual reactor process design considerations, there are special features that need to be taken into account. Many of these features, particularly those that relate to fluid-particle interactions, can only be described empirically. Typical design requirements include calculations of catalyst or reactant solid holdup for a given fractional conversion and production rate (or vice versa), the bed depth, the vessel diameter and height, and heat transfer requirements. The reactor model may also need to account for conversion in regions of the vessel above ("freeboard" region) and below ("distributor" region) the bed, if there is a significant fraction of the solid in these regions, and/or the reaction is very rapid. The overall design must consider special features related to the superficial velocity, and the flow characteristics of the solid and fluid phases within the vessel. A reactor model which incorporates all of these features together with a kinetics model can be rather complicated.

23.2 FLUID-PARTICLE INTERACTIONS

This section is a continuation of Section 21.3.2 dealing with pressure drop ($-\Delta P$) for flow through a fixed bed of solid particles. Here, we make further use of the Ergun equation for estimating the minimum superficial fluidization velocity, u_{mf}. In addition, by analogous treatment for free fall of a single particle, we develop a means for estimating terminal velocity, u_t, as a quantity related to elutriation and entrainment.

23.2 Fluid-Particle Interactions

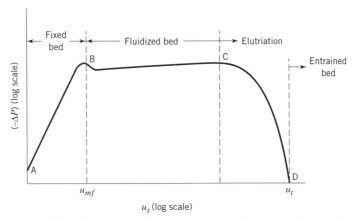

Figure 23.4 Dependence of pressure drop $(-\Delta P)$ on fluid velocity (u_s) for upward flow of fluid through bed of particles illustrating different conditions of the bed (schematic)

23.2.1 Upward Flow of Fluid Through Solid Particles: $(-\Delta P)$ regimes

The dependence of $(-\Delta P)$ on the superficial velocity of fluid, u_s, flowing upward through a bed of particles is shown schematically in Figure 23.4 (after Froment and Bischoff, 1990, p. 570). The plot is $\log(-\Delta P)$ versus $\log u_s$, for given fluid and particle properties.

The following regimes of $(-\Delta P)$, somewhat idealized for simplification, corresponding to different conditions of the bed may be distinguished:

AB: $(-\Delta P)$ increases with increasing u_s (slope in Figure 23.4 is about 2), as the bed retains the character of a fixed bed, with ϵ_B increasing somewhat.
BC: $(-\Delta P)$ is relatively constant, as the bed becomes fluidized; B is the point of incipient fluidization at u_{mf}.
CD: $(-\Delta P)$ decreases with increasing u_s, from C, at which elutriation begins, to D, at which point the particles may be said to be entrained (at u_t).

23.2.2 Minimum Fluidization Velocity (u_{mf})

The minimum fluidization velocity (u_{mf}) can be estimated by means of the Ergun equation (Chapter 21) for pressure drop, $(-\Delta P)$, for flow of fluid through a bed of particles (Bin, 1986). In this case, the flow is upward through the bed. At incipient fluidization, the bed is on the point of "lifting." This condition is characterized by the equality of the frictional force, corresponding to $(-\Delta P)$, acting upward, and the gravity force on the bed, acting downward:

$$(-\Delta P)_{mf} A_c = gW = g\rho_{B,app} V$$
$$= g(\rho_p - \rho_f)(1 - \epsilon_{mf})L_{mf} A_c$$

from which

$$(-\Delta P)_{mf} = g(\rho_p - \rho_f)(1 - \epsilon_{mf})L_{mf} \qquad (23.2\text{-}1)$$

where subscript mf refers to the *bed* at minimum-fluidization conditions. In equation 23.2-1, $\rho_{B,app}$ is the apparent density of the bed, which is $(\rho_p - \rho_f)$ on allowing for the buoyancy of the fluid; equations corresponding to 21.3-1 to -3 are used to obtain the final form of 23.2-1.

Rewriting the definition of the friction factor f from equation 21.3-5, and the Ergun correlation for f given by equation 21.3-7, both at mf, we obtain

$$f = \frac{d'_p(-\Delta P)_{mf}}{L_{mf} u^2_{mf} \rho_f} \qquad (23.2\text{-}2)$$

and

$$f = \frac{1 - \epsilon_{mf}}{\epsilon^3_{mf}} \left(1.75 + 150 \frac{1 - \epsilon_{mf}}{\text{Re}'_{mf}} \right) \qquad (23.2\text{-}3)$$

where

$$\text{Re}'_{mf} = d'_p u_{mf} \rho_f / \mu_f \qquad (23.2\text{-}4)$$

and d'_p is the effective particle diameter given by equation 21.3-6.

Eliminating $(-\Delta P)_{mf}$, f, and Re'_{mf} by means of equations 23.2-1 to -4, we obtain a quadratic equation for u_{mf} in terms of parameters for the fluid, solid, and bed:

$$u^2_{mf} + \frac{150(1 - \epsilon_{mf})\mu_f}{1.75 \rho_f d'_p} u_{mf} - \frac{g(\rho_p - \rho_f)\epsilon^3_{mf} d'_p}{1.75 \rho_f} = 0 \qquad (23.2\text{-}5)$$

If the fluid is a gas, values of u_{mf} range between about 0.005 and 0.5 m s^{-1}. Note that u_{mf} is independent of bed depth.

Two special or limiting cases of equation 23.2-5 arise, depending on the relative magnitudes of the two terms for f in equation 23.2-3. This is shown in the following example.

EXAMPLE 23-1

Obtain the special forms of equation 23.2-5 for (a) relatively small particles, and (b) relatively large particles.

SOLUTION

(a) For relatively small particles, Re is relatively small, and, in equation 23.2-3, we assume that

$$1.75 \ll 150(1 - \epsilon_{mf})/\text{Re}'_{mf}$$

This is equivalent to ignoring the first term (u^2_{mf}) in equation 23.2-5, which then may be written as:

$$u_{mf} = \frac{g(\rho_p - \rho_f)(d'_p)^2}{K \mu_f} \quad (\text{Re}'_{mf} \text{ small}) \qquad (23.2\text{-}6)$$

where

$$K = 150(1 - \epsilon_{mf})/\epsilon^3_{mf} \qquad (23.2\text{-}7)$$

23.2 Fluid-Particle Interactions

This is a commonly used form, with $K = 1650$, which corresponds to $\epsilon_{mf} = 0.383$.

(b) For relatively large particles, Re is relatively large, and, in equation 23.2-3, we assume that

$$1.75 \gg 150(1 - \epsilon_{mf})/\text{Re}'_{mf}$$

This is equivalent to ignoring the second (linear) term in equation 23.2-5, which then becomes

$$u_{mf} = \left[\frac{g(\rho_p - \rho_f)\epsilon_{mf}^3 d'_p}{1.75\rho_f}\right]^{1/2} \quad (\text{Re}'_{mf} \text{ large}) \quad (23.2\text{-}8)$$

EXAMPLE 23-2

Calculate u_{mf} for particles of ZnS fluidized by air at 1200 K and 200 kPa. Assume $d'_p = 4 \times 10^{-4}$ m, $\epsilon_{mf} = 0.5$, $\rho_p = 3500$ kg m^{-3}, and $\mu_f = 4.6 \times 10^{-5}$ N s m^{-2}; $g = 9.81$ m s^{-2}.

SOLUTION

At the (T,P) conditions given, the density of air (ρ_f), assumed to be an ideal gas ($z = 1$) with $M_{av} = 28.8$, is

$$\rho_f = \frac{m}{V} = \frac{PM_{av}}{RT} = \frac{200(1000)28.8}{8.314(1200)1000} = 0.577 \text{ kg m}^{-3}$$

Since the particles are relatively small, we compare the results obtained from equations 23.2-5 and -6. With given values of parameters inserted, the former becomes

$$u_{mf}^2 + 8.542 u_{mf} - 1.700 = 0$$

from which $u_{mf} = 0.195$ m s^{-1} (the units of each term should be confirmed for consistency).

From equation 23.2-6,

$$u_{mf} = \frac{9.81(3500)4^2(10^{-4})^2(1 - 0.5)^3}{150(1 - 0.5)4.6(10^{-5})} = 0.199 \text{ m s}^{-1}$$

which is within 2% of the more accurate value above. (See also file ex23-2.msp.)

23.2.3 Elutriation and Terminal Velocity (u_t)

At sufficiently high velocity of fluid upward through a bed of particles, the particles become entrained and do not settle; that is, the particles are carried up with the fluid. Elutriation is the selective removal of particles by entrainment, on the basis of size. The elutriation velocity (of the fluid) is the velocity at which particles of a given size are entrained and carried overhead.

The minimum elutriation velocity for particles of a given size is the velocity at incipient entrainment, and is assumed to be equal to the terminal velocity (u_t) or free-falling velocity of a particle in the fluid. This is calculated, in a manner analogous to that used for u_{mf}, by equating the frictional drag force F_d (upward) on the particle with the gravity force on the particle (downward):

$$F_d = g(\rho_p - \rho_f)V_p$$
$$= \pi g(\rho_p - \rho_f)d_p^3/6 \tag{23.2-9}$$

for a spherical particle of diameter d_p.

The dimensionless drag coefficient C_d, analogous to the friction factor, is defined by

$$C_d = 2F_d/A_{proj}\rho_f u^2$$
$$= 8F_d/\pi d_p^2 \rho_f u^2 \tag{23.2-10}$$

for a spherical particle at terminal velocity, where A_{proj} is the projected area of the particle in the direction of motion ($\pi d_p^2/4$ for a sphere). C_d depends on Re and shape of the particle. Correlations have been given by Haider and Levenspiel (1989). For small spherical particles at low Re (<0.1), these reduce to the result for the Stokes' regime:

$$C_d = 24/\text{Re} \tag{23.2-11}$$

where, at u_t,

$$\text{Re} \equiv \text{Re}_t = d_p u_t \rho_f/\mu_f \tag{23.2-12}$$

To obtain an expression for u_t, we eliminate F_d, C_d, and Re_t from equations 23.2-9 to -12:

$$u_t = g(\rho_p - \rho_f)d_p^2/18\mu_f$$
(spherical particles, small Re_t) \hfill (23.2-13)

Equation 23.2-13 for u_t corresponds to 23.2-6 for u_{mf}.

Correlations for u_t for larger values of Re and other shapes are also given by Haider and Levenspiel (1989).

23.2.4 Comparison of u_{mf} and u_t

To obtain proper fluidization, the actual fluid velocity, u_{fl}, must be considerably greater than the minimum fluidization velocity, u_{mf}. However, to avoid excessive entrainment, u_{fl} should be less than the terminal velocity, u_t. Thus, the ratio u_t/u_{mf} is a guide to selection of the value of u_{fl}.

Since relatively small particles are used in a fluidized bed, corresponding to relatively small Re, we use equations 23.2-6 and -13 for comparison of spherical particles ($d_p' = d_p$). The result is

$$\frac{u_t}{u_{mf}} = \frac{150(1-\epsilon_{mf})}{18\epsilon_{mf}^3} \quad \text{(small, spherical particles)} \tag{23.2-14}$$

This ratio is very sensitive to the value of ϵ_{mf}, ranging from 15 at $\epsilon_{mf} = 0.60$ to 92 at $\epsilon_{mf} = 0.383$. In practice, values of u_{fl} are 30 to 50 times the value of u_{mf}.

23.3 HYDRODYNAMIC MODELS OF FLUIDIZATION

A hydrodynamic model of fluidization attempts to account for several essential features of fluidization: mixing and distribution of solids and fluid in a so-called "emulsion region," the formation and motion of bubbles through the bed (the "bubble region"), the nature of the bubbles (including their size) and how they affect particle motion/distribution, and the exchange of material between the bubbles (with little solid content) and the predominantly solid emulsion. Models fall into one of three classes (Yates, 1983, pp. 74–78):

(1) two-region models, which take into account a bubble region and an emulsion region, with very little variation in properties within each region;
(2) bubble models, which are based upon a mean bubble size; all system properties are functions of this bubble size;
(3) bubble-growth models, which also endeavor to account for bubble coalescence and bubble splitting.

The first two classes of models are simplest, but may require substantial experimental information to predict rates of exchange between the bubble and emulsion regions. Class (1) models are too simplistic to be of practical use, while class (3) models tend to be relatively complicated. Yates (1983, Chapter 2) gives an excellent discussion of the various types of models and their assumptions. The most important model assumptions, in terms of their effect on performance, are those which describe interphase mass transfer and mixing of gas in the emulsion region. A typical class (1) model is that of van Deemter (1961), which assumes that the bed contains a bubble region and an emulsion region, that the emulsion region is in plug flow both upward and downward, that there is no solid within the bubbles, and that all reaction occurs in the emulsion region. The bubbling-bed model of Kunii and Levenspiel (1991) is a class (2) model. In addition to the two regions proposed by van Deemter, Kunii and Levenspiel add a "cloud and wake" region surrounding the bubble. Furthermore, more complex flow (including down-flow) within the emulsion phase is allowed. Other models include those of Davidson and Harrison (1963), Partridge and Rowe (1968), and Kato and Wen (1969).

Overall, the hydrodynamic behavior of a fluidized bed depends upon the nature of the particles used, and the ease of fluidization. Spherical solid particles that are not "sticky" fluidize easily; "sticky" particles, conversely, do not fluidize well, since they tend to agglomerate, leading to uneven distribution of solid through the bed, and nonuniform circulation of solid and fluid. A more detailed description of types of particles and their effect upon fluidization is provided by Geldart (1973, 1978) and by Grace (1986). The fluid may be a liquid or a gas. If the fluid is a liquid, the bed tends to expand uniformly with increasing fluidization velocity u_{fl}, and bubbles are generally not formed; this is called particulate fluidization. If the fluid is a gas, bubbles are usually formed at the inlet distributor; these bubbles travel upward through the bed, and may drag solid particles along with them as a "wake"; bubbles may coalesce and/or split, depending upon local conditions; in this "bubbling fluidization," the fluidized bed may resemble a boiling liquid, as bubbles burst upon reaching the upper "surface" of the bed. We focus primarily on this latter type. Ultimately, the hydrodynamic model which describes these phenomena must be coupled with a kinetics model to develop an overall model of a fluidized-bed reactor.

23.3.1 Two-Region Model (Class (1))

The discussion above suggests a hydrodynamic flow model based on two distinct regions in the fluidized bed: a "bubble" region made up mostly of gas, but also containing solid

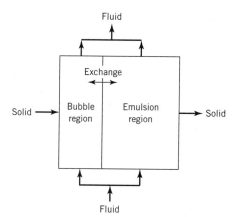

Figure 23.5 Schematic representation of two-region model for fluidized bed

particles, and a fluid + solid ("emulsion") region, resembling the bed at mf conditions. This is illustrated schematically in Figure 23.5; the two regions are actually interspersed.

In Figure 23.5, the fluid entering is depicted as being split between the two regions; most fluid flows through in the bubble region, and there is provision for exchange ("mass transfer") between the two regions characterized by an exchange coefficient K_{be}. The solid entering (in a continuous-flow situation) is also depicted as split between the two regions, but most solid is in the emulsion region. The simplest flow pattern for the bubble region is PF and for the emulsion region is BMF. In general, however, each region could be characterized by a Peclet number (Pe, Chapter 19; Pe_b and Pe_e for the bubble and emulsion regions, respectively) to represent the degree of dispersion/backmixing.

This model can have as many as six parameters for its characterization: K_{be}, Pe_b, Pe_e, and ratios of volumes of regions, of solid in the regions, and of fluid in the regions. The number can be reduced by assumptions such as PF for the bubble region ($Pe_b \rightarrow \infty$), all solid in the emulsion, and all fluid entering in the bubble region. Even with the reduction to three parameters, the model remains essentially empirical, and doesn't take more detailed knowledge of fluidized-bed behavior into account.

23.3.2 Kunii-Levenspiel (KL) Bubbling-Bed Model (Class (2))

A one-parameter model, termed the bubbling-bed model, is described by Kunii and Levenspiel (1991, pp. 144–149, 156–159). The one parameter is the size of bubbles. This model endeavors to account for different bubble velocities and the different flow patterns of fluid and solid that result. Compared with the two-region model, the Kunii-Levenspiel (KL) model introduces two additional regions. The model establishes expressions for the distribution of the fluidized bed and of the solid particles in the various regions. These, together with expressions for coefficients for the exchange of gas between pairs of regions, form the hydrodynamic + mass transfer basis for a reactor model.

The assumptions are as follows (Levenspiel, 1972, pp. 310–311):

(1) Bubbles are all the same size, and are distributed evenly throughout the bed, rising through it.
(2) Gas within a bubble essentially remains in the bubble, but recirculates internally, and penetrates slightly into the emulsion to form a transitional cloud region around the bubble; all parameters involved are functions of the size of bubble (Davidson and Harrison, 1963).

(3) Each bubble drags a wake of solid particles up with it (Rowe and Partridge, 1965). This forms an additional region, and the movement creates recirculation of particles in the bed: upward behind the bubbles and downward elsewhere in the emulsion region.

(4) The emulsion is at mf conditions.

There are thus two other regions, the cloud and wake regions, in addition to the bubble and emulsion regions. These are illustrated in Figure 23.6: Figure 23.6(a) depicts an idealized bubble, with wake, cloud and recirculation patterns indicated (after Levenspiel, 1972, p. 311); Figure 23.6(b) is a schematic representation of the various regions, similar to Figure 23.5, with the wake and cloud regions combined.

The main model parameter, the mean bubble diameter, d_b, can be estimated using various correlations. It depends on the type of particle and the nature of the inlet distributor. For small, sand-like particles that are easily fluidized, an expression is given for d_b as a function of bed height x by Werther (Kunii and Levenspiel, 1991, p. 146):

$$d_b/\text{cm} = 0.853[1 + 0.272(u_{fl} - u_{mf})]^{1/3}(1 + 0.0684x)^{1.21} \quad \textbf{(23.3-1)}$$

where u_{fl} and u_{mf} are in cm s^{-1} and x is in cm.

The rise velocity of bubbles is another important parameter in fluidized-bed models, but it can be related to bubble size (and bed diameter, D). For a single bubble, the rise velocity, u_{br}, relative to emulsion solids is (Kunii and Levenspiel, 1991, p. 116):

$$u_{br} = 0.711(gd_b)^{1/2} \quad [\text{small bubbles}, (d_b/D) < 0.125] \quad \textbf{(23.3-2)}$$

$$u_{br} = [0.711(gd_b)^{1/2}]1.2e^{-1.49d_b/D} \quad [0.125 < (d_b/D) < 0.6] \quad \textbf{(23.3-3)}$$

(For $(d_b/D) > 0.6$, the bed is not a bubbling bed; slugging occurs.)

Another measure of bubble velocity is the absolute rise velocity of bubbles in the bed, u_b; this can be taken in the first instance as the sum of u_{br} and the apparent rise velocity of the bed ahead of the bubbles, $u_{fl} - u_{mf}$:

$$u_b = u_{br} + u_{fl} - u_{mf} \quad \textbf{(23.3-4)}$$

We next consider the volume fractions of the various regions in the fluidized bed. The volume fraction of bubbles, f_b, m^3 bubbles (m^3 bed)$^{-1}$, can be assessed from the point of view of either voidage or velocity. In terms of voidage, if we assume the void fraction

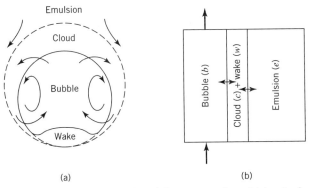

Figure 23.6 Bubbling-bed model representation of (a) a single bubble and (b) regions of a fluidized bed (schematic)

in the bubbles is 1 and the remainder of the bed is at mf conditions with voidage ϵ_{mf}, the volume-average voidage in the fluidized bed is

$$\epsilon_{fl} = f_b(1) + (1 - f_b)\epsilon_{mf}$$

from which

$$f_b \simeq \frac{\epsilon_{fl} - \epsilon_{mf}}{1 - \epsilon_{mf}} \tag{23.3-5}$$

In terms of velocity, if we assume, for a vigorously bubbling bed with $u_{fl} \gg u_{mf}$, that gas flows *through* the bed only in the bubble region ($q \simeq q_b$, or $u_{fl}A_c \simeq f_b u_b A_c$, where A_c is the cross-sectional area of the bed),

$$f_b \simeq u_{fl}/u_b \tag{23.3-6}$$

Equation 23.3-6 may need to be modified to take into account the relative magnitude of u_b (Kunii and Levenspiel, 1991, pp. 156–157):

(1) For slowly rising bubbles, $u_b < u_{mf}/\epsilon_{mf}$,

$$f_b = (u_{fl} - u_{mf})/(u_b + 2u_{mf}) \tag{23.3-6a}$$

(2) For the intermediate case, $u_{mf}/\epsilon_{mf} < u_b < 5u_{mf}/\epsilon_{mf}$,

$$f_b = (u_{fl} - u_{mf})/(u_b + u_{mf}); \text{ for } u_b \simeq u_{mf}/\epsilon_{mf} \tag{23.3-6b}$$

$$f_b = (u_{fl} - u_{mf})/u_b; \text{ for } u_b \simeq 5u_{mf}/\epsilon_{mf} \tag{23.3-6c}$$

(3) For fast bubbles, $u_b > 5u_{mf}/\epsilon_{mf}$,

$$f_b = (u_{fl} - u_{mf})/(u_b - u_{mf}) \tag{23.3-6d}$$

The ratio of cloud volume to bubble volume is given by Kunii and Levenspiel (1991, p. 157), and from this we obtain the volume fraction in the cloud region[1]

$$f_c = \frac{3u_{mf}}{\epsilon_{mf}u_{br} - u_{mf}} f_b \tag{23.3-7}$$

The ratio of wake volume to bubble volume is difficult to assess, and is given by Kunii and Levenspiel (1991, p. 124) in the form of graphical correlations with particle size for various types of particles. From this, we assume that the bed fraction in the wakes[1] is

$$f_w \simeq \alpha f_b \quad (\alpha = 0.2 \text{ to } 0.6) \tag{23.3-8}$$

The bed fraction in the emulsion, f_e, is obtained by difference, since

$$f_b + f_c + f_w + f_e = 1 \tag{23.3-9}$$

---

[1] The notation used here for bed fraction is used for ratio with respect to bubble volume by Kunii and Levenspiel.

23.3 Hydrodynamic Models of Fluidization

For the distribution of solid particles in the various regions, we define the following ratios:

$$\gamma_b = (m^3 \text{ solid in bubbles})(m^3 \text{ bubbles})^{-1}$$

$$\gamma_{cw} = (m^3 \text{ solid in cloud} + \text{wakes})(m^3 \text{ bubbles})^{-1}$$

$$\gamma_e = (m^3 \text{ solid in emulsion})(m^3 \text{ bubbles})^{-1}$$

The sum of these can be related to ϵ_{mf} and f_b:

$$\gamma_b + \gamma_{cw} + \gamma_e \equiv \frac{m^3 \text{ total solid}}{m^3 \text{ bubbles}} \equiv \left(\frac{m^3 \text{ total solid}}{m^3 \text{ bed at } mf}\right)\left(\frac{m^3(c+w+e)}{m^3 \ fl \ bed}\right)\left(\frac{m^3 \ fl \ bed}{m^3 \text{ bubbles}}\right)$$

$$= (1 - \epsilon_{mf})(f_c + f_w + f_e)/f_b$$

$$= (1 - \epsilon_{mf})(1 - f_b)/f_b \quad (23.3\text{-}10)$$

To obtain equation 23.3-10, it is assumed that the volume of (cloud + wakes + emulsion) in the fluidized bed is equal to the volume of the bed at mf conditions. The first of these quantities, γ_b, is relatively small, but its value is uncertain. From a range of experimental data ($\gamma_b = 0.01$ to 0.001), it is usually taken as:

$$\gamma_b = 0.005 \quad (23.3\text{-}11)$$

The second quantity, γ_{cw}, can also be related to ϵ_{mf} and bed-fraction quantities:

$$\gamma_{cw} \equiv \frac{m^3 \text{ solid in }(c+w)}{m^3 \text{ bubbles}} \equiv \frac{m^3 \text{ solid in } c}{m^3 \text{ bubbles}} + \frac{m^3 \text{ solid in } w}{m^3 \text{ bubbles}}$$

$$\equiv \left(\frac{m^3 \text{ total solid}}{m^3 \text{ bed at } mf}\right)\left(\frac{m^3 \text{ bed in } c}{m^3 \ fl \ bed}\right)\left(\frac{m^3 \ fl \ bed}{m^3 \text{ bubbles}}\right) + \left(\frac{m^3 \text{ total solid}}{m^3 \text{ bed at } mf}\right)\left(\frac{m^3 \text{ bed in } w}{m^3 \ fl \ bed}\right)\left(\frac{m^3 \ fl \ bed}{m^3 \text{ bubbles}}\right)$$

$$= (1 - \epsilon_{mf})(f_c + f_w)/f_b \quad (23.3\text{-}12)$$

The third quantity, γ_e, is obtained by difference from equations 23.3-10 to -12.

Finally, we extend the hydrodynamic model to include exchange of gas between pairs of regions, analogous to mass transfer. Figure 23.6(b) implies schematically that convective flow of gas occurs primarily through the bubble region, and that there is exchange between the bubble region and the cloud (+ wake) region, and between the cloud (+ wake) region and the emulsion. These are characterized by exchange coefficients K_{bc} and K_{ce}, respectively, representing the volumetric rate of exchange of gas per volume of bubble, with units of time^{-1}. Those coefficients are calculated by the following semi-empirical relations (Kunii and Levenspiel, 1991, pp. 251–2):

$$K_{bc}/s^{-1} = 4.5\left(\frac{u_{mf}}{d_b}\right) + 5.85\left(\frac{D_m^{1/2} g^{1/4}}{d_b^{5/4}}\right) \quad (23.3\text{-}13)$$

where D_m is the molecular diffusion coefficient of the gas; the first term in 23.3-13 relates to bulk or convective flow and the second term to diffusive transport across the bubble-cloud boundary.

$$K_{ce}/s^{-1} = 6.77\left(\frac{D_m \epsilon_{mf} u_{br}}{d_b^3}\right)^{1/2} \quad (23.3\text{-}14)$$

584 Chapter 23: Fluidized-Bed and Other Moving-Particle Reactors for Fluid-Solid Reactions

23.4 FLUIDIZED-BED REACTOR MODELS

A fluidized-bed reactor consists of three main sections (Figure 23.1): (1) the fluidizing gas entry or distributor section at the bottom, essentially a perforated metal plate that allows entry of the gas through a number of holes; (2) the fluidized-bed itself, which, unless the operation is adiabatic, includes heat transfer surface to control T; (3) the freeboard section above the bed, essentially empty space to allow disengagement of entrained solid particles from the rising exit gas stream; this section may be provided internally (at the top) or externally with cyclones to aid in the gas-solid separation. A reactor model, as discussed here, is concerned primarily with the bed itself, in order to determine, for example, the required holdup of solid particles for a specified rate of production. The solid may be a catalyst or a reactant, but we assume the former for the purpose of the development.

A model of a fluidized-bed reactor combines a hydrodynamic model of bubble and emulsion flow and interphase mass transfer with a kinetics model. As discussed in Section 23.3, various hydrodynamic models exist; their suitability as reactor models depends upon the actual flow and mixing conditions within the bed. If the reaction is very slow, or the residence time through the bed is very short, then the choice of the hydrodynamic model is not important. However, for very fast reactions, or if the contact time is very long, the details of the interphase mass transfer, the location of the solid, and the nature of mixing and flow within each region become important. Furthermore, the hydrodynamic models only describe phenomena occurring within the bed. If there is significant elutriation of particles, conversion within the freeboard region may also need to be accounted for. If the reaction is very fast, it may be controlled largely by the nature of fluid flow through the inlet distributors. In this case, conversion in the distributor region must also be accounted for.

Another important issue affecting the performance of a fluidized-bed reactor is the size of the particles, which, in turn, has a large effect upon the ease of fluidization, mixing of fluid and solid within the emulsion phase, and mass transfer between the emulsion and the bubbles. Small, nonsticky particles fluidize easily, and promote efficient contact between the bubble and emulsion phases. However, if the particles are too small, entrainment may be significant, and reaction may occur in the freeboard. Larger particles minimize entrainment, but are more difficult to fluidize, and fluid-particle interactions are often inefficient and nonuniform. In some systems, the particle size changes during the process, as in coal gasification. In this case, there is a particle-size distribution, and the entrainment observed with smaller particles is an inevitable consequence of the solid reaction.

In the following sections, we discuss reactor models for fine, intermediate, and large particles, based upon the Kunii-Levenspiel (KL) bubbling-bed model, restricting ourselves primarily to first-order kinetics. Performance for both simple and complex reactions is considered. Although the primary focus is on reactions within the bed, we conclude with a brief discussion of the consequences of reaction in the freeboard region and near the distributor.

23.4.1 KL Model for Fine Particles

The following assumptions are made in addition to those in Section 23.3.2:

(1) The reaction is $A(g) + \ldots \rightarrow$ products, catalyzed by solid particles that are fluidized by a gas stream containing A and, perhaps, other reactants and inert species.

(2) The reactor operates isothermally at constant density and at steady-state.

23.4 Fluidized-Bed Reactor Models

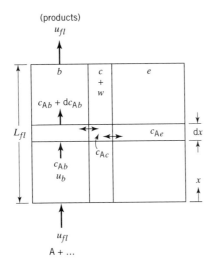

Figure 23.7 Schematic representation of control volume for material balance for bubbling-bed reactor model

(3) The fluidizing (reactant) gas is in convective flow through the bed *only* via the bubble-gas region (with associated clouds and wakes); that is, there is no convective flow of gas through the emulsion region.

(4) The bubble region is in PF (upward through the bed).

(5) Gas exchange occurs (i) between bubbles and clouds, characterized by exchange coefficient K_{bc} (equation 23.3-13), and (ii) between clouds and emulsion, characterized by K_{ce} (equation 23.3-14).

The performance equation for the model is obtained from the continuity (material-balance) equations for A over the three main regions (bubble, cloud + wake, and emulsion), as illustrated schematically in Figure 23.7. Since the bed is isothermal, we need use only the continuity equation, which is then uncoupled from the energy equation. The latter is required only to establish the heat transfer aspects (internally and externally) to achieve the desired value of T.

In Figure 23.7, the bubble, cloud, and emulsion regions are represented by b, $c + w$, and e, respectively. The control volume is a thin horizontal strip of height dx through the vessel. The overall depth of the bed is L_{fl}, which is related to the holdup of catalyst, W_{cat}. The performance equation may be used to determine W_{cat} for a given conversion f_A (and production rate), or the converse.

The continuity or material-balance equations for A stem from the flow/kinetics scheme shown in Figure 23.8, which corresponds to the representation in Figure 23.7.

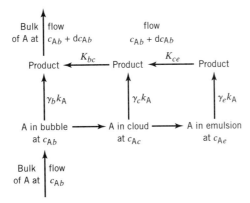

Figure 23.8 Flow/kinetics scheme for bubbling-bed reactor model for reaction $A(g) + \cdots \rightarrow$ product(s)

The continuity equations for the three main regions lead eventually to the performance equation for the reactor model.

Continuity equation for the bubble region:

$$-u_b \frac{dc_{Ab}}{dx} \equiv -\frac{dc_{Ab}}{dt} = \gamma_b k_A c_{Ab} + K_{bc}(c_{Ab} - c_{Ac}) \tag{23.4-1}$$

which states that the rate of disappearance of A from the bubble region is equal to the rate of reaction in the bubble region + the rate of transfer to the cloud region; note that γ_b serves as a weighting factor for the intrinsic rate constant k_A.

Continuity equation for the cloud + wake region:

$$K_{bc}(c_{Ab} - c_{Ac}) = \gamma_{cw} k_A c_{Ac} + K_{ce}(c_{Ac} - c_{Ae}) \tag{23.4-2}$$

Continuity equation for the emulsion region:

$$K_{ce}(c_{Ac} - c_{Ae}) = \gamma_e k_A c_{Ae} \tag{23.4-3}$$

Eliminating c_{Ac} and c_{Ae} from equations 23.4-1 to -3, and dropping the subscript b from c_{Ab}, we obtain

$$-\frac{dc_A}{dt} = k_{overall} c_A \tag{23.4-4}$$

where

$$k_{overall} = \gamma_b k_A + \cfrac{1}{\cfrac{1}{K_{bc}} + \cfrac{1}{\gamma_{cw} k_A + \cfrac{1}{\cfrac{1}{K_{ce}} + \cfrac{1}{\gamma_e k_A}}}} \tag{23.4-5}$$

Integrating equation 23.4-4 from the bed inlet ($c_A = c_{Ao}, t = x = 0$) to the bed outlet ($c_A = c_A, t = L_{fl}/u_b$), we have

$$1 - f_A = c_A/c_{Ao} = \exp(-k_{overall} L_{fl}/u_b) \tag{23.4-6}$$

The fluidized-bed depth, L_{fl}, is calculated from the fixed-bed packed depth, L_{pa}, as follows. Since, from a balance for bed solid,

$$(1 - \epsilon_{fl}) L_{fl} = (1 - \epsilon_{pa}) L_{pa} \tag{23.4-7}$$

and, from equation 23.3-5,

$$1 - \epsilon_{fl} = (1 - \epsilon_{mf})(1 - f_b) \tag{23.4-7a}$$

we have, on elimination of $1 - \epsilon_{fl}$ from 23.4-7,

$$L_{fl} = \frac{(1 - \epsilon_{pa}) L_{pa}}{(1 - \epsilon_{mf})(1 - f_b)} \tag{23.4-7b}$$

23.4 Fluidized-Bed Reactor Models

Equation 23.4-6 is one form of the performance equation for the bubbling-bed reactor model. It can be transformed to determine the amount of solid (e.g., catalyst) holdup to achieve a specified f_A or c_A:

$$W_{cat} = \frac{\rho_p q (1 - \epsilon_{mf}) u_{br}}{k_{overall} u_{fl}} \ln\left(\frac{c_{Ao}}{c_A}\right) \qquad (23.4\text{-}8)$$

(The confirmation of equations 23.4-4, -5, and -8 is left to problem 23-9.)

In this reactor model, the only free parameter (or degree of freedom) is d_b, the bubble diameter. The following example illustrates use of the model.

EXAMPLE 23-3

(a) Estimate the amount of catalyst (W_{cat}/kg) required for a fluidized-bed reactor, according to the Kunii-Levenspiel bubbling-bed model, for the production of 60,000 Mg year^{-1} of acrylonitrile by the ammoxidation of propylene with air.

Data and assumptions:

- The heat transfer configuration within the bed (for the exothermic reaction) and other internal features are ignored.
- Only C_3H_3N is formed (with water).
- The feed contains C_3H_6 and NH_3 in the stoichiometric ratio and 20% excess air (79 mole % N_2, 21% O_2); there is no water in the feed.
- Conversion based on C_3H_6(A) is 70%.
- $T = 400°C$; $P = 2$ bar.
- The annual stream service factor (fraction of time in operation) is 0.94.
- $d_b = 0.1$ m; $d_p = 0.05$ mm; $\rho_p = 2500$ kg m^{-3}; $\mu_f = 1.44$ kg h^{-1} m^{-1}.
- $u_{mf} = 0.002$ m s^{-1}; $\alpha = 0.6$; $D_m = 0.14$ m^2 h^{-1} at 400°C; $\epsilon_{pa} = 0.5$.
- $\epsilon_{mf} = 0.6$; $k_A = 1.0$ s^{-1}; $u_{fl} = 720$ m h^{-1}; $\gamma_b = 0.004$.

(b) Calculate the vessel diameter and the bed depth (fluidized) in m.
(c) For comparison, calculate W_{cat} for the two cases (assume constant density for both):
  (i) The reactor is a PFR.
  (ii) The reactor is a CSTR.

SOLUTION

(a) To calculate W_{cat}, we use equation 23.4-8 in conjunction with 23.4-5 and -6 (file ex23-3.msp). For this purpose, we also need to calculate other quantities, as indicated below.

The reaction is

$$C_3H_6(A) + NH_3 + \frac{3}{2}O_2 \rightarrow C_3H_3N(B) + 3H_2O$$

Because of the small change in moles on reaction, and the dilution with N_2 (inert), together with an anticipated (but not calculated) relatively small ($-\Delta P$), the assumption of constant density is a reasonable one. It can be calculated that for every mole of A entering, the total moles entering is 10.57 and the total leaving is 10.92, an increase of only 3.3%. The actual molar feed rates of A (F_{Ao}) and total gas (F_{to}) are:

$$F_{Ao} = \left(60{,}000(1000)\frac{\text{kg B}}{\text{yr}}\right)\left(\frac{1}{53}\frac{\text{kmol B}}{\text{kg B}}\right)\left(1\frac{\text{kmol A(reacted)}}{\text{kmol B(formed)}}\right)\left(\frac{1}{0.70}\frac{\text{kmol A(in feed)}}{\text{kmol A(reacted)}}\right)$$

$$\times \frac{1}{365(24)3600(0.94)}\frac{\text{yr}}{\text{s}} = 0.0546 \text{ kmol s}^{-1}$$

$$F_{to} = 10.57(0.0546) = 0.577 \text{ kmol s}^{-1}$$

The total volumetric feed rate is

$$q_o = \frac{F_{to}RT}{P} = \frac{0.577(1000)8.314(673)}{2(10^5)} = 16.1 \text{ m}^3 \text{ s}^{-1}$$

To calculate $k_{overall}$ in equation 23.4-5, we require K_{bc}, K_{ce}, γ_{cw}, and γ_e; these, in turn, require calculations of u_{br}, u_b, f_b, and $(f_c + f_w)$, as follows (with each equation indicated):

$$u_{br} = 0.711(gd_b)^{1/2} = 0.711[9.81(0.1)]^{1/2} = 0.704 \text{ m s}^{-1} \qquad \textbf{(from 23.3-2)}$$

$$u_b = u_{fl} - u_{mf} + u_{br} = (720/3600) - 0.002 + 0.704 = 0.902 \text{ m s}^{-1} \qquad \textbf{(from 23.3-4)}$$

$$K_{bc} = 4.5(u_{mf}/d_b) + 5.85(D_m^{1/2}g^{1/4}/d_b^{5/4})$$
$$= 4.5(0.002/0.1) + 5.85[(0.14/3600)^{1/2}(9.81)^{1/4}(0.1)^{5/4}] = 1.24 \text{ s}^{-1} \qquad \textbf{(from 23.3-13)}$$

$$K_{ce} = 6.77(\epsilon_{mf}D_m u_{br}/d_b^3)^{1/2} = 6.77[0.6(0.14/3600)0.704/(0.1)^3]^{1/2} = 0.868 \text{ s}^{-1}$$
$$\textbf{(from 23.3-14)}$$

$$f_b = u_{fl}/u_b = 0.2/0.902 = 0.222 \qquad \textbf{(from 23.3-6)}$$

$$f_c = \frac{3u_{mf}f_b}{\epsilon_{mf}u_{br} - u_{mf}} = \frac{3(0.002)0.222}{(0.60)(0.704) - 0.002} = 0.0032 \qquad (23.3\text{-}7)$$

$$f_w = \alpha f_b = (0.60)(0.222) = 0.133 \qquad (23.3\text{-}8)$$

$$f_e = 1 - 0.222 - 0.0032 - 0.133 = 0.642 \qquad (23.3\text{-}9)$$

$$\gamma_{cw} = (1 - \epsilon_{mf})(f_c + f_w)/f_b = (1 - 0.60)(0.0032 + 0.133)/0.222 = 0.245 \qquad (23.3\text{-}12)$$

$$\gamma_e = \frac{(1 - \epsilon_{mf})(1 - f_b)}{f_b} - \gamma_b - \gamma_{cw} \qquad (23.3\text{-}10)$$

$$= \frac{(1 - 0.60)(1 - 0.222)}{0.222} - 0.004 - 0.245 = 1.153$$

$$k_{overall} = \gamma_b k_A + \cfrac{1}{\cfrac{1}{K_{bc}} + \cfrac{1}{\gamma_{cw}k_A + \cfrac{1}{\cfrac{1}{K_{ce}} + \cfrac{1}{\gamma_e k_A}}}} \qquad (23.4\text{-}5)$$

$$= (0.004)(1.0) + \cfrac{1}{\cfrac{1}{1.24} + \cfrac{1}{(0.245)(1.0) + \cfrac{1}{\cfrac{1}{0.868} + \cfrac{1}{(1.153)(1.0)}}}} = 0.467 \text{ s}^{-1}$$

The bed depth, L_{fl} can now be determined using 23.4-6:

$$L_{fl} = \frac{-[\ln(1 - f_A)]u_b}{k_{overall}} = \frac{-[\ln(1 - 0.70)](0.902 \text{ m s}^{-1})}{0.467 \text{ s}^{-1}} = 2.33 \text{ m}$$

The catalyst requirement can be determined from 23.4-8:

$$W_{cat} = \frac{2500(16.1)(1 - 0.60)0.704[-\ln(1 - 0.70)]}{0.467(720/3600)} = 1.46 \times 10^5 \text{ kg}$$

(b) From $q = u_{fl}A_c = u_{fl}\pi D^2/4$,

$$D = (4q/\pi u_{fl})^{1/2} = [4(16.1)/\pi(720/3600)]^{1/2} = 10.1 \text{ m}$$

From (a), $L_{fl} = 2.33$ m

(c) **(i)** For a PFR,

$$W_{cat} = \rho_p V_{cat} = (\rho_p q/k_A)\int_0^{f_A} df_A/(1 - f_A)$$
$$= [2500(16.1)/1.0][-\ln(1 - 0.70)] = 0.48 \times 10^5 \text{ kg}$$

(ii) For a CSTR,

$$W_{cat} = \rho_p V_{cat} = (\rho_p q/k_A)[f_A/(1 - f_A)]$$
$$= [2500(16.1)/1.0][0.70/(1 - 0.70)] = 0.94 \times 10^5 \text{ kg}$$

The amount of catalyst required in (a) is even greater than that required for a CSTR, which may be accounted for by the "bypassing effect" (Section 23.1.3).

23.4.1.1 KL Model: Special Cases of First-Order Reaction

For an extremely fast reaction, with k_A relatively large, very little A reaches the emulsion and 23.4-5 reduces to:

$$k_{overall} = \gamma_b k_A + K_{bc} \tag{23.4-9}$$

If the reaction is intrinsically slow, with $k_A \ll K_{ce}$ and K_{bc}, equation 23.4-5 reduces to:

$$k_{overall} = (\gamma_b + \gamma_e + \gamma_{cw})k_A \tag{23.4-10}$$

In both cases, f_A is then determined using equation 23.4-6.

23.4.1.2 KL Model: Extension to First-Order Complex Reactions

Extension of the Kunii-Levenspiel bubbling-bed model for first-order reactions to complex systems is of practical significance, since most of the processes conducted in fluidized-bed reactors involve such systems. Thus, the yield or selectivity to a desired product is a primary design issue which should be considered. As described in Chapter 5, reactions may occur in series or parallel, or a combination of both. Specific examples include the production of acrylonitrile from propylene, in which other nitriles may be formed, oxidation of butadiene and butene to produce maleic anhydride and other oxidation products, and the production of phthalic anhydride from naphthalene, in which phthalic anhydride may undergo further oxidation.

590 Chapter 23: Fluidized-Bed and Other Moving-Particle Reactors for Fluid-Solid Reactions

Kunii and Levenspiel (1991, pp. 294–298) extend the bubbling-bed model to networks of first-order reactions and generate rather complex algebraic relations for the net reaction rates along various pathways. As an alternative, we focus on the development of the basic design equations, which can also be adapted for nonlinear kinetics, and numerical solution of the resulting system of algebraic and ordinary differential equations (with the E-Z Solve software). This is illustrated in Example 23-4 below.

We illustrate the development of the model equations for a network of two parallel reactions, A → B, and A → C, with k_1 and k_2 representing the rate constants for the first and second reactions, respectively. Continuity equations must be written for two of the three species. Furthermore, exchange coefficients (K_{bc} and K_{ce}) must be determined for each species chosen (here, A and B).

Continuity equation for A in the bubble region:

$$u_b \frac{dc_{Ab}}{dx} = -\gamma_b(k_1 + k_2)c_{Ab} - K_{bcA}(c_{Ab} - c_{Ac}) \tag{23.4-11}$$

Continuity equation for A in the cloud + wake region:

$$K_{bcA}(c_{Ab} - c_{Ac}) = \gamma_{cw}(k_1 + k_2)c_{Ac} + K_{ceA}(c_{Ac} - c_{Ae}) \tag{23.4-12}$$

Continuity equation for A in the emulsion region:

$$K_{ceA}(c_{Ac} - c_{Ae}) = \gamma_e(k_1 + k_2)c_{Ae} \tag{23.4-13}$$

Continuity equation for B in the bubble region:

$$u_b \frac{dc_{Bb}}{dx} = \gamma_b k_1 c_{Ab} - K_{bcB}(c_{Bb} - c_{Bc}) \tag{23.4-14}$$

Continuity equation for B in the cloud + wake region:

$$K_{bcB}(c_{Bb} - c_{Bc}) = -\gamma_{cw} k_1 c_{Ac} + K_{ceB}(c_{Bc} - c_{Be}) \tag{23.4-15}$$

Continuity equation for B in the emulsion region:

$$K_{ceB}(c_{Bc} - c_{Be}) = -\gamma_e k_1 c_{Ae} \tag{23.4-16}$$

Note that the continuity equations for product B reflect the fact that B, formed in the cloud + wake and emulsion regions, transfers *to* the bubble region. This is in contrast to reactant A, which transfers *from* the bubble region to the other regions.

For other reaction networks, a similar set of equations may be developed, with the kinetics terms adapted to account for each reaction occurring. To determine the conversion and selectivity for a given bed depth, L_{fl}, equations 23.4-11 and -14 are numerically integrated from $x = 0$ to $x = L_{fl}$, with simultaneous solution of the algebraic expressions in 23.4-12, -13, -15, and -16. The following example illustrates the approach for a *series* network.

EXAMPLE 23-4

(Kunii and Levenspiel, 1991, p. 298) Phthalic anhydride is produced in the following process:

$$\text{naphthalene (A)} \xrightarrow{k_1} \text{phthalic anhydride (B)} \xrightarrow{k_2} CO_2 + H_2O$$

The reaction occurs in a fluidized-bed reactor, with sufficient heat exchange to ensure isothermal operation. The bed (before fluidization) is 5 m deep (L_{pa}), with a voidage (ϵ_{pa}) of 0.52. The reaction rate constants for the two steps are $k_1 = 1.5$ m^3 gas (m^3 cat)$^{-1}$ s^{-1} and $k_2 = 0.010$ m^3 gas (m^3 cat)$^{-1}$ s^{-1}. Additional data are:

$$u_{mf} = 0.005 \text{ m s}^{-1} \quad d_b = 0.05 \text{ m} \quad D_A = 8.1 \times 10^{-6} \text{ m}^2 \text{ s}^{-1}$$
$$\epsilon_{mf} = 0.57 \quad u_b = 1.5 \text{ m s}^{-1} \quad D_B = 8.4 \times 10^{-6} \text{ m}^2 \text{ s}^{-1}$$
$$u_{fl} = 0.45 \text{ m s}^{-1} \quad \gamma_b = 0.005 \quad f_w/f_b = 0.6$$

Determine the overall fractional conversion of naphthalene, and the selectivity to phthalic anhydride.

SOLUTION

To solve the problem, continuity equations are written for A and B in all three regions of the fluidized bed. First, however, the exchange parameters (K_{bc} and K_{ce}) for A and B are determined, along with u_{br}, f_b, f_c, γ_{cw}, and γ_e.

$$u_{br} = 0.711 (g d_b)^{0.5} = 0.711[(9.8)(0.05)]^{0.5} = 0.498 \text{ m s}^{-1} \quad (23.3\text{-}2)$$

Since $u_{fl} \gg u_{mf}$, equation 23.3-6 is used to determine f_b:

$$f_b = u_{fl}/u_b = 0.45/1.5 = 0.30 \quad (23.3\text{-}6)$$

From equation 23.3-7,

$$f_c/f_b = 3u_{mf}/(\epsilon_{mf} u_{br} - u_{mf}) = 3(0.005)/[(0.57)(0.498) - 0.005] = 0.0538$$

The catalyst fractions in the cloud + wake and emulsion regions are determined using equations 23.3-12 and 23.3-10, respectively:

$$\gamma_{cw} = (1 - \epsilon_{mf})(f_c + f_w)/f_b = (1 - 0.57)(0.0538 + 0.60) = 0.281$$

$$\gamma_e = (1 - \epsilon_{mf})\frac{1 - f_b}{f_b} - \gamma_{cw} - \gamma_b = (1 - 0.57)\frac{1 - 0.30}{0.30} - 0.281 - 0.005 = 0.717$$

K_{bc} and K_{ce} are determined for naphthalene (A) and phthalic anhydride (B) using equations 23.3-13 and -14. The resulting values are: $K_{bcA} = 1.70$ s^{-1}; $K_{bcB} = 1.72$ s^{-1}; $K_{ceA} = 0.92$ s^{-1}; and $K_{ceB} = 0.93$ s^{-1}.

Continuity equations for A and B are written similar to equations 23.4-11 to -16 (in this case we have a series network rather than a parallel network):

Continuity equation for A in the bubble region:

$$u_b \frac{dc_{Ab}}{dx} = -\gamma_b k_1 c_{Ab} - K_{bcA}(c_{Ab} - c_{Ac}) \quad (A)$$

Continuity equation for A in the cloud + wake region:

$$K_{bcA}(c_{Ab} - c_{Ac}) = \gamma_{cw} k_1 c_{Ac} + K_{ceA}(c_{Ac} - c_{Ae}) \quad (B)$$

Continuity equation for A in the emulsion region:

$$K_{ceA}(c_{Ac} - c_{Ae}) = \gamma_e k_1 c_{Ae} \quad (C)$$

Continuity equation for B in the bubble region:

$$u_b \frac{dc_{Bb}}{dx} = \gamma_b k_1 c_{Ab} - \gamma_b k_2 c_{Bb} - K_{bcB}(c_{Bb} - c_{Bc}) \quad \textbf{(D)}$$

Continuity equation for B in the cloud + wake region:

$$K_{bcB}(c_{Bb} - c_{Bc}) = -\gamma_{cw} k_1 c_{Ac} + \gamma_{cw} k_2 c_{Bc} + K_{ceB}(c_{Bc} - c_{Be}) \quad \textbf{(E)}$$

Continuity equation for B in the emulsion region:

$$K_{ceB}(c_{Bc} - c_{Be}) = -\gamma_e k_1 c_{Ae} + \gamma_e k_2 c_{Be} \quad \textbf{(F)}$$

Equations (A) and (D) are numerically integrated from $x = 0$ to $x = L_{fl}$, with simultaneous solution of equations (B), (C), (E), and (F) to obtain c_{Ab} and c_{Bb} (see E-Z Solve file ex23-4.msp). From equation 23.4-7b,

$$L_{fl} = \frac{1 - \epsilon_{pa}}{(1 - \epsilon_{mf})(1 - f_b)} L_{pa} = \frac{1 - 0.52}{(1 - 0.57)(1 - 0.30)}(5\ m) = 7.97\ m$$

The resulting conversion is 0.96, with a selectivity to phthalic anhydride of 0.95.

23.4.2 KL Model for Intermediate-Size Particles

When intermediate-sized particles are used, the behavior of bubbles within the fluidized bed falls between the fast-bubble and very-slow-bubble regimes (Section 23.3.2). In this case, bubbles rise somewhat faster than gas in the emulsion region, but not so fast that the emulsion gas can be ignored (as in the previous section). Furthermore, the clouds surrounding the bubbles tend to be very large, and may overlap, possibly growing to the point where the overlapping clouds make up the entire emulsion region. Under these conditions, the bed can be viewed as consisting of only a bubble region and an emulsion region. A single exchange coefficient is required to describe mass transfer between the two regions. Unlike the model for fine particles in the previous section, which could be based upon a bubble rise velocity, u_b, we must now consider the rise velocity of the bubble plus the gas surrounding the bubble (u_b^*), such that

$$u_b^* = u_b + 3u_{mf} \quad \textbf{(23.4-17)}$$

Figure 23.9 illustrates the model and kinetics scheme for these conditions. We confine our analysis to a single first-order reaction, based on the development of Kunii and Levenspiel (1990; 1991, pp. 300–302). However, extension to other reaction orders is straightforward.

Continuity equation in the bubble region (for reactant A):

$$f_b u_b^* \frac{dc_{Ab}}{dx} = -f_b \gamma_b k_A c_{Ab} - f_b K_{be}(c_{Ab} - c_{Ae}) \quad \textbf{(23.4-18)}$$

Continuity equation in the emulsion region:

$$(1 - f_b)u_{mf} \frac{dc_{Ae}}{dx} = f_b K_{be}(c_{Ab} - c_{Ae}) - (1 - f_b)(1 - \epsilon_{mf})k_A c_{Ae} \quad \textbf{(23.4-19)}$$

The exchange coefficient, K_{be}, can be approximated by:

$$K_{be} \simeq 4.5 u_{mf}/d_b \quad \textbf{(23.4-20)}$$

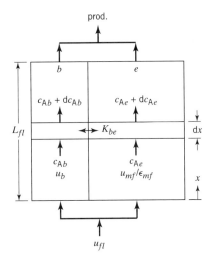

Figure 23.9 Schematic representation of regions of bubbling-bed model for intermediate-sized particles

That is, the second term in equation 23.3-13 can be neglected, since transport is primarily by convection.

The fraction of bed in the bubble phase, f_b, is determined from:

$$f_b = \frac{u_{fl} - u_{mf}}{u_b} \quad \text{(for } u_b \simeq 5 u_{mf}/\epsilon_{mf}) \tag{23.3-6c}$$

or

$$f_b = \frac{u_{fl} - u_{mf}}{u_b + u_{mf}} \quad \text{(for } u_b \simeq u_{mf}/\epsilon_{mf}) \tag{23.3-6b}$$

The fractional conversion across the bed is given in terms of the fractions of reactant in the bubble and emulsion regions:

$$f_A = 1 - \frac{f_b u_b^* c_{Ab} + (1 - f_b) u_{mf} c_{Ae}}{u_{fl} c_{Ao}} \tag{23.4-21}$$

where c_{Ao} is the concentration of A at the inlet of the bed, and c_{Ae} and c_{Ab} represent the concentrations of A in the emulsion region and at the outlet of the bubble region, respectively. To determine the performance of the reactor, the differential equations represented by 23.4-18 and -19 are numerically integrated from $x = 0$ to $x = L_{fl}$, the depth of the fluidized bed. An analytical solution for f_A for first-order kinetics is given by Kunii and Levenspiel (1991, pp. 300–302); however, the resulting equations are quite cumbersome. Use of a computer-aided numerical solution is a convenient alternative, and has the added advantage that the model equations can be extended for other reaction orders.

EXAMPLE 23-5

A first-order reaction, A → products, is conducted in a fluidized-bed reactor. Before fluidization, the bed is 1.8 m deep (L_{pa}), with a voidage (ϵ_{pa}) of 0.42. The feed gas ($\rho_f = 0.45$ kg m^{-3}; $\mu_f = 4.2 \times 10^{-5}$ Pa s) flows through the bed at a velocity three times the minimum fluidization velocity. The following data are also available:

$$\epsilon_{mf} = 0.48 \qquad \rho_p = 3100 \text{ kg m}^{-3} \qquad \gamma_b = 0$$
$$d_p = 0.45 \text{ mm} \qquad d_b = 0.09 \text{ m} \qquad k_A = 1.9 \text{ m}^3 \text{ gas (m}^3 \text{ cat)}^{-1} \text{ s}^{-1}$$

Determine the fractional conversion of A (f_A).

SOLUTION

First, the minimum fluidization velocity (u_{mf}) is calculated from 23.2-5. After substitution of the particle and fluid properties provided, the resulting quadratic equation is:

$$u_{mf}^2 + 9.24 u_{mf} - 1.922 = 0$$

which leads to a value of u_{mf} of 0.203 m s^{-1}. Thus, $u_{fl} = 3u_{mf} = 0.609$ m s^{-1}.
Next, the bubble velocities are calculated to determine the flow regime:

$$u_{br} = 0.711(gd_b)^{0.5} = 0.711[(9.81)(0.09)]^{0.5} = 0.668 \text{ m s}^{-1} \quad \text{(23.3-2)}$$

$$u_b = u_{fl} - u_{mf} + u_{br} = 0.609 - 0.203 + 0.668 = 1.074 \text{ m s}^{-1} \quad \text{(23.3-4)}$$

The flow regime is determined by comparing u_b with u_{mf}/ϵ_{mf}:

$$u_b/(u_{mf}/\epsilon_{mf}) = 1.704/(0.203/0.48) = 2.54$$

Thus, the bubble velocity is within the intermediate-bubble regime, and equation 23.3-6b is used for f_b. Parameters u_b^*, K_{be}, f_b, and L_{fl} are calculated as follows:

$$u_b^* = u_b + 3u_{mf} = 1.074 + 3(0.203) = 1.683 \text{ m s}^{-1} \quad \text{(23.4-17)}$$

$$K_{be} = 4.5(u_{mf}/d_b) = 4.5(0.203/0.09) = 10.16 \text{ s}^{-1} \quad \text{(23.4-20)}$$

$$f_b = (u_{fl} - u_{mf})/(u_b + u_{mf}) = (0.609 - 0.203)/(1.074 + 0.203) = 0.318 \quad \text{(23.3-6b)}$$

$$L_{fl} = \frac{L_{pa}(1 - \epsilon_{pa})}{(1 - \epsilon_{mf})(1 - f_b)} = \frac{(1.8)(1 - 0.42)}{(1 - 0.48)(1 - 0.318)} = 2.94 \text{ m} \quad \text{(23.4-7b)}$$

The continuity equations 23.4-18 and -19 are applicable. Thus;
Continuity equation for A in the bubble region:

$$f_b u_b^* \frac{dc_{Ab}}{dx} = -f_b \gamma_b k_A c_{Ab} - f_b K_{be}(c_{Ab} - c_{Ae}) \quad \text{(23.4-18)}$$

Continuity equation for A in the emulsion region:

$$(1 - f_b)u_{mf} \frac{dc_{Ae}}{dx} = f_b K_{be}(c_{Ab} - c_{Ae}) - (1 - f_b)(1 - \epsilon_{mf})k_A c_{Ae} \quad \text{(23.4-19)}$$

These equations are numerically integrated simultaneously from $x = 0$ to $x = L_{fl} = 2.94$ m to obtain c_{Ab} and c_{Ae}. Using an initial (inlet) condition of $c_{Ao} = c_{Ab}(0) = c_{Ae}(0) = 1.0$, we obtain (file ex23-5.msp): $c_{Ab} = 0.077$ mol L^{-1} and $c_{Ae} = 0.066$ mol L^{-1}.

These values are substituted into equation 23.4-21 to determine f_A:

$$f_A = 1 - \frac{f_b u_b^* c_{Ab} + (1 - f_b)u_{mf} c_{Ae}}{u_{fl} c_{Ao}}$$

$$= 1 - \frac{(0.318)(1.683)(0.077) + (1 - 0.318)(0.203)(0.066)}{(0.609)(1.0)} = 0.917$$

Thus, the fractional conversion for this system, which operates in the intermediate-bubble and -particle regime, is 0.92.

23.4.3 Model for Large Particles

In bubbling beds of large particles, flow of gas through the emulsion region is significant, and is usually faster than the rise velocity of the bubbles (i.e., $u_b < u_{mf}/\epsilon_{mf}$). Under these conditions, the gas in the emulsion phase shortcuts through the bed by way of the gas bubbles overtaken along the path from the bottom to the top of the bed. Thus, the reactant gas sees a series of nearly homogeneous segments within the vessel, starting with an emulsion region, followed by a bubble region, then another emulsion region, then another bubble phase, etc. The resulting flow of gas is essentially PF. Kunii and Levenspiel (1990; 1991, pp. 303–305) developed a model which describes reactor performance under these conditions.

The flow of gas through the bed (of cross-sectional area A_c) is the sum of the flows through the emulsion and bubble regions:

$$u_{fl}A_c = (1 - f_b)u_{mf}A_c + f_b(u_b + 3u_{mf})A_c \qquad (23.4\text{-}22)$$

The fraction of gas which passes through the emulsion region is given by:

$$f_{ge} = \frac{(1 - f_b)u_{mf}A_c}{u_{fl}A_c} = (1 - f_b)u_{mf}/u_{fl} \qquad (23.4\text{-}23)$$

The fraction of bed which is bubbles is, from 23.3-6a,

$$f_b = \frac{u_{fl} - u_{mf}}{u_b + 2u_{mf}} \qquad (23.3\text{-}6a)$$

Since most (if not all) of the solid is in the emulsion region, equation 23.4-23 may be used in a plug-flow model to represent the contact efficiency of the gas and solid. The resulting expression to determine the conversion is, for a first-order reaction,

$$f_A = 1 - e^{-k_A W_{cat} f_{ge}/\rho_p} = 1 - e^{-k_A W_{cat} u_{mf}(1-f_b)/\rho_p u_{fl}} \qquad (23.4\text{-}24)$$

where W_{cat} represents the catalyst mass, and ρ_p is the particle density.

23.4.4 Reaction in Freeboard and Distributor Regions

As gas bubbles reach the upper surface of a fluidized bed, they burst, releasing gas and ejecting particles into the freeboard region above the bed. The solid particles thrown into the freeboard originate from the bubble wakes, and cover the entire range of particle sizes present within the bed. However, based on their terminal velocity, u_t, larger particles fall back to the bed, while smaller particles continue to rise, possibly being carried out of the bed. This leads to a concentration gradient of particles in the freeboard region. Furthermore, unreacted gas leaving the bed can react with the particles, increasing conversion and possibly altering the yield and selectivity in a reaction network. Confirmation of conversion in the freeboard region is usually by measurement of temperature—if the temperature in the freeboard is significantly different from the temperature at the outlet of the bed, and if there is a temperature gradient within the freeboard, reaction in the freeboard is likely occurring.

Several models have been proposed to account for reaction in the freeboard. Yates and Rowe (1977) developed a simple model based upon complete mixing of particles in the freeboard, coupled with either BMF or PF of the freeboard gas. Two model parameters are the rate of particle ejection from the bed, and the fraction of wake particles ejected. Kunii and Levenspiel (1990; 1991, pp. 305–307) proposed a model of freeboard reaction which accounts for the contact efficiency of the gas with the solid, and the fraction of solid in the freeboard. A comprehensive freeboard entrainment model is used to describe this fraction and the concentration gradient of solid particles in the freeboard.

Generally, it is advantageous to avoid reaction in the freeboard, as much as possible, since the temperature control and near-isothermal conditions observed in the fluidized bed are nearly impossible to achieve in the freeboard region. This is particularly problematic for a complex reaction, since the selectivity is often temperature-dependent. Experiments have shown that the following design features influence the extent of particle entrainment, and, by extension, the likelihood of reaction in the freeboard region:

(1) Vertical internals do not affect entrainment in small-particle beds, but may increase entrainment in large-particle beds.
(2) Horizontal louvers placed near the bed surface can significantly decrease entrainment in large-particle systems.
(3) Entrainment increases markedly at high pressure.
(4) Stirrers or bed internals which reduce the size of the bubbles bursting at the surface can significantly reduce entrainment (reducing the number of particles ejected into the freeboard).

Near the inlet distributor, where development of gas bubbles begins, interphase mass transfer appears to be very fast, and the efficiency of gas-solid contact may also be very high. Consequently, for very fast reactions, significant conversion may occur near the inlet distributor. Behie and Kehoe (1973) proposed a model based on mass transfer and reaction near gas jets assumed to be formed at the inlet to the vessel. This model was subsequently extended by Grace and de Lasa (1978) to cover more complex mixing regimes near the distributor. However, there is conflicting evidence as to whether or not the gas jets exist, and, although other theories of bubble formation have been proposed, models have not yet been developed which incorporate this information into a quantitative measure of conversion in the distributor region.

23.5 PROBLEMS FOR CHAPTER 23

23-1 (a) Is u_{mf} for a fluidized-bed reactor greater if the fluid is a liquid than if it is a gas? Justify quantitatively, assuming that only fluid properties change, and without doing calculations.

(b) For a given fluid, how does u_{mf} change (increase or decrease) as the solid characteristics change (e.g., for a switch from one catalyst to another)? Justify quantitatively without doing calculations.

23-2 How does u_t change with changes in (a) fluid properties, and (b) solid properties? Justify quantitatively without doing calculations.

23-3 (a) Calculate the minimum-fluidization velocity, u_{mf}/m s^{-1}, for the fluidization of a bed of sand with air at 100°C and 103 kPa, if the density (ρ_p) of the sand is 2600 kg m^{-3}, the particles are all 0.45 mm in diameter (d_p), the viscosity (μ_f) of air at the fluidizing conditions is 2.1×10^{-5} N s m^{-2}, and the bed voidage, ϵ_{mf}, is 0.38. Assume the particles are spherical.

(b) Calculate u_t for the particles in (a).

23-4 Consider a fluidized-bed combustor with air as the fluid. Conditions and properties are as follows: $T = 900°C$, $P = 1000$ kPa, $\rho_p = 2960$ kg m^{-3}, $d_p = 3$ mm, $\mu_f = 4.6 \times 10^{-5}$ N s m^{-2}, and $\epsilon_{mf} = 0.4$. For air, $M_{av} = 28.8$. $g = 9.81$ m s^{-2}.

(a) Calculate u_{mf} in m s^{-1}.
(b) Calculate u_t in m s^{-1}, *if* equation 23.2-13 is valid.

23-5 Calculate the pressure drop ($-\Delta P$/kPa) for flow of fluid through a fluidized bed of solid particles based on the following data: the difference in density between solid particles and fluid is 2500 kg m^{-3}; at mf conditions, the bed voidage (ϵ_{mf}) is 0.4, and the bed depth (L_{mf}) is 1.2 m. State any assumption(s) made.

23.5 Problems for Chapter 23

23-6 A fluidized-bed reactor for a solid-catalyzed gas-phase reaction (of A) operates with relatively small catalyst particles. Based on the information given below, calculate the following:
  (a) the terminal velocity of the particles (u_t/m s^{-1});
  (b) the depth of the bed under fluidized conditions (L_{fl}/m), if the superficial linear velocity (u_s) is 80% of the value calculated in (a);
  (c) the diameter of the bed (D/m).
  Data: $u_{mf} = 0.005$ m s^{-1}; $\epsilon_{mf} = 0.4$; feed-gas flow rate at reactor conditions, $q_t = 4$ m^3 s^{-1}; catalyst holdup, $W_{cat} = 10{,}000$ kg; catalyst density (particle), $\rho_p = 2500$ kg m^{-3}; $L_{fl} = 1.6 L_{mf}$.

23-7 (a) Using the Kunii-Levenspiel bubbling-bed model of Section 23.4.1, calculate the fraction (f_A) of cumene (A) decomposed in a laboratory fluidized-bed reactor based on the following (cf. Iwasaki et al., 1965):

$\epsilon_{pa} = 0.45$; $\epsilon_{mf} = 0.50$; $u_{mf} = 0.002$ m s^{-1}; $u_{fl} = 0.1$ m s^{-1}; $d_b = 0.02$ m; $\gamma_b = 0.005$; $\gamma_{cw} = 0.25$; $D_e = 1.86 \times 10^{-4}$ m^2 s^{-1}; $k_A = 22.2$ s^{-1}; $L_{pa} = 0.08$ m (depth of packed bed at rest).

  (b) Compare the result in (a) with that obtained if the reactor behaved as a CSTR, and comment on the result.

23-8 Using the Kunii-Levenspiel bubbling-bed model of Section 23.4.1 for the fluidized-bed reactor in the SOHIO process for the production of acrylonitrile (C$_3$H$_3$N) by the ammoxidation of propylene with air, and the information given below, calculate the following:
  (a) the rate of production of C$_3$H$_3$N (kg h^{-1});
  (b) the depth of the bed when fluidized, L_{fl} (m);
  (c) the mean residence time of gas in the fluidized bed.

  Information: $W_{cat} = 1 \times 10^5$ kg; feed rate of propylene = 40 mol s^{-1}; feed contains C$_3$H$_6$ and NH$_3$ in the stoichiometric ratio and 25% excess air (79 mol% N$_2$); $T = 700$ K; $P = 2$ bar; D(bed) = 7 m; $\rho_p = 2500$ kg m^{-3}; $\epsilon_{mf} = 0.4$; $d_b = 12$ cm; $k_{overall} = 0.5$ s^{-1}.
  State any assumptions made.

23-9 Confirm equations 23.4-4 (with -5) and -8.

23-10 Si$_3$N$_4$ is an excellent high-temperature material (Morooka, et al., 1989). However, it must be made electroconductive to permit electrospark machining. Chemical vapor deposition (CVD) of TiN onto Si$_3$N$_4$ in a fluidized bed is one means of producing electroconductive Si$_3$N$_4$, since a thin layer of TiN is bound to the particle. TiN is formed by the following reaction, using a feed containing 3.5 mol% TiCl$_4$ and the balance NH$_3$:

$$\text{TiCl}_4 + (4/3)\text{NH}_3 \rightarrow \text{TiN} + 4\text{HCl} + (1/6)\text{N}_2$$

One feature of the process, as in other crystal-growth operations, is that the fluidization velocity must be adjusted as the particles grow. Using the following data, develop a profile of minimum fluidization velocity (u_{mf}) versus particle size (d_p), if, as the reaction proceeds, the particle size increases from 100 μm to 700 μm. Assume ideal-gas behavior.
  $P = 0.3$ bar; $T = 973$ K; $\rho_p = 2910$ kg m^{-3}; $\mu_f = 5.2 \times 10^{-5}$ kg m^{-1} s^{-1}; $\epsilon_{mf} = 0.45$.

23-11 Silicon carbide is a ceramic material which can be produced from silicon dioxide and carbon. The resulting impure SiC contains residual SiO$_2$ and C, which must be removed. Air fed to a fluidized-bed reactor may be used to oxidize the residual carbon in the impure SiC, leaving a refined SiC which is suitable for further use. The following data (Kato, et al., 1989) were obtained from studies at 950°C in which impure SiC was decarbonized in a fluidized-bed reactor. Assuming that the Kunii-Levenspiel model for fine particles is valid, determine the overall reaction-rate constant for the process, and estimate the fraction of C removed:

$k_A = 5.64$ s^{-1} (A $\equiv$ O$_2$); $\rho_p = 2600$ kg m^{-3}; $\epsilon_{mf} = 0.51$; $L_{fl} = 1.0$ m; $u_{mf} = 3.15$ cm s^{-1}; $u_{fl} = 40$ cm s^{-1}; $d_p = 310$ μm.

Further to these data, assume that the bubble diameter, d_b, is 4.5 cm, that the fraction of catalyst in the bubble region, γ_b, is 0.005, and that $f_w/f_b = 0.6$. The diffusivity of air at 950°C is 8.2×10^{-5} m² s⁻¹. The reaction rate is assumed to be first-order with respect to the oxygen concentration. Although the rate would also depend upon the surface area (or mass) of exposed carbon, assume that the reaction rate is independent of the amount of carbon present.

23-12 Various operational parameters can influence the yield and selectivity of complex reactions conducted in a fluidized-bed reactor. For the production of phthalic anhydride from naphthalene in Example 23-4, determine the sensitivity of the fractional conversion and selectivity to the fluidization velocity, u_{fl}, and to the bubble diameter, d_b.

23-13 Catalytic reactions in fluidized-bed reactors often involve the use of catalyst particles of different sizes, which may be characterized by a particle-size distribution. Consider the reaction $A(g) \rightarrow$ products, for which the following data are available:

$P = 200$ kPa $\qquad T = 673$ K $\qquad \alpha = 0.5$ $\qquad k_A = 0.25$ s⁻¹
$u_{fl} = 0.20$ m s⁻¹ $\qquad \epsilon_{pa} = 0.50$ $\qquad \epsilon_{mf} = 0.50$ $\qquad \mu_f = 4.0 \times 10^{-4}$ kg m⁻¹ s⁻¹
$L_{fl} = 2.0$ m $\qquad c_{Ao} = 1.0$ mol m⁻³ $\qquad D = 0.5$ m $\qquad D_m = 1.94 \times 10^{-5}$ m² s⁻¹
$\rho_p = 2500$ kg m⁻³ $\qquad W_{cat} = 10{,}000$ kg $\qquad \gamma_b = 0.004$ $\qquad M_{av} = 28.8$ g mol⁻¹

A mixture of 0.1 mm-, 0.55 mm-, and 1.0 mm-diameter spherical particles was considered, with the three following particle-size distributions:

Case	%0.10 mm	%0.55 mm	%1.0 mm
1	0	100	0
2	50	0	50
3	30	40	30

Determine the fractional conversion of A based on
(a) a weighted-average particle size, that is,

$$\bar{d}_p = \sum_i x_i d_{pi}$$

where d_{pi} is the diameter of catalyst fraction i, and x_i represents the fraction of total particles in catalyst fraction i. The performance is calculated using the average particle size.

(b) a weighted-average f_A, that is,

$$\bar{f}_A = \sum_i x_i f_{Ai}$$

where f_{Ai} is the fractional conversion obtained with catalyst fraction i. This latter method may thus require up to three calculations of f_{Ai} (using fine-, intermediate-, and large-particle models, as required) in order to determine the average conversion (Partaatmadja, 1998).

Chapter 24

Reactors for Fluid-Fluid Reactions

In this chapter, we consider process design aspects of reactors for multiphase reactions in which each phase is a fluid. These include gas-liquid and liquid-liquid reactions, although we focus primarily on the former. We draw on the results in Section 9.2, which treats the kinetics of gas-liquid reactions based on the two-film model. More detailed descriptions are given in the books by Danckwerts (1970), by Kaštánek et al. (1993), and by Froment and Bischoff (1990, Chapter 14).

We first present further examples of the types of reactions involved in two main classifications, and then a preliminary discussion of various types of reactors used. Following an examination of some factors affecting the choice of reactor, we develop design equations for some reactor types, and illustrate their use with examples. The chapter concludes with a brief introduction to trickle-bed reactors for three-phase gas-liquid-solid (catalyst) reactions.

24.1 TYPES OF REACTIONS

As indicated in Section 9.2, a fluid-fluid reaction may be considered from one of two points of view depending on its involvement in a separation process or in a reaction process. We give further examples of each of these in turn.

24.1.1 Separation-Process Point of View

From a separation-process point of view, a fluid-fluid reaction is intended to enhance separation (e.g., preparation of feed for a subsequent process step, product purification, or effluent control for environmental protection). Examples include the use of ethanolamines for the removal of H_2S and CO_2 (reactions (A) and (B) in Section 9.2), the removal of SO_2 by an aqueous stream of a hydroxide, and absorption of O_2 by blood or desorption of CO_2 from blood. A solid catalyst may be involved as a third phase, as in hydrodesulfurization in a trickle-bed reactor.

24.1.2 Reaction-Process Point of View

From a reaction-process point of view, a fluid-fluid reaction is used for the formation of a desired product or set of products. In addition to the three examples given in Section 9.2, we cite the following:

(1) Production of the ethanolamines mentioned above, from ethylene oxide and ammonia:

$$C_2H_4O + NH_3 \rightarrow HOCH_2CH_2NH_2 \quad (MEA)$$
$$C_2H_4O + MEA \rightarrow (HOCH_2CH_2)_2NH \quad (DEA)$$
$$C_2H_4O + DEA \rightarrow (HOCH_2CH_2)_3N \quad (TEA)$$

(2) Hydrogenation of benzene to cyclohexane:

$$C_6H_6(\ell) + 3H_2(g) \rightarrow C_6H_{12}(\ell)$$

(3) $CO_2(g) + 2NH_3(aq) + H_2O(\ell) \rightarrow (NH_4)_2CO_3(aq)$

24.2 TYPES OF REACTORS

The types of reactors used for fluid-fluid reactions may be divided into two main types: (1) tower or column reactors, and (2) tank reactors. We consider some general features of these in this section and in Section 24.3. In Sections 24.4 and 24.5, we treat some process design aspects more quantitatively.

24.2.1 Tower or Column Reactors

Tower or column reactors, without mechanical agitation, are used primarily for gas-liquid reactions. If used for a liquid-liquid reaction, the arrangement involves vertically stacked compartments, each of which is mechanically agitated. In either case, the flow is countercurrent, with the less dense fluid entering at the bottom, and the more dense fluid at the top. In the case of a gas-liquid reaction without mechanical agitation, both interphase contact and separation occur under the influence of gravity. In a liquid-liquid reaction, mechanical agitation greatly enhances the contact of the two phases. We consider here primarily the case of gas-liquid reactions.

As its name implies, a tower reactor typically has a height-to-diameter (h/D) ratio considerably greater than 1. Types of tower or column reactors (the words *tower* and *column* may be used interchangeably) go by descriptive names, each of which indicates a particular feature, such as the means of creating gas-liquid contact or the way in which one phase is introduced or distributed. The flow pattern for one phase or for both phases may be close to ideal (PF or BMF), or may be highly nonideal. In the quantitative development in Section 24.4 below, we assume the flow to be ideal, but more elaborate models are available for nonideal flow (Chapter 19; see also Kaštánek et al., 1993, Chapter 5). Examples of types of tower reactors are illustrated schematically in Figure 24.1, and are discussed more fully below. An important consideration for the efficiency of gas-liquid contact is whether one phase (gas or liquid) is dispersed in the other as a continuous phase, or whether both phases are continuous. This is related to, and may be determined by, features of the overall reaction kinetics, such as rate-determining characteristics of mass transfer and intrinsic reaction.

(1) *Packed tower.* A packed tower (Figure 24.1(a)) contains solid shapes such as ceramic rings or saddles to ensure appropriate flow and mixing of the fluids. The flow is usually countercurrent, with the less dense fluid entering at the bottom of the tower. Both phases are considered to be continuous and ideally in PF. Gas-liquid interfacial area is enhanced by contact of gas rising through the void space between particles of packing with a liquid film flowing down over the packing

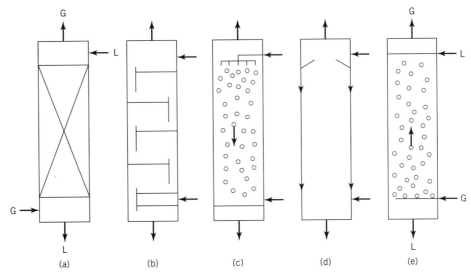

Figure 24.1 Types of tower or column reactors for gas-liquid reactions: (a) packed tower; (b) plate tower; (c) spray tower; (d) falling-film tower; (e) bubble column

surface. The packing may be arranged in (vertical) stages to decrease maldistribution of the fluids, or to facilitate adjustment of temperature by heat exchange between stages. In some cases, the packing consists of catalyst particles, as in a trickle-bed reactor, in which case the flow may be either cocurrent (downward, usual situation) or countercurrent.

(2) *Plate tower.* A plate tower (Figure 24.1(b)) contains, for example, bubble-cap or sieve plates at intervals along its height. The flow of gas and liquid is countercurrent, and liquid may be assumed to be distributed uniformly radially on each plate. On each plate or tray, gas is dispersed within the continuous liquid phase. The gas-liquid interfacial area is relatively large, and the gas-liquid contact time is typically greater than that in a packed tower.

(3) *Spray tower.* A spray tower (Figure 24.1(c)) is an "empty" vessel with liquid sprayed (as in a "shower") from the top as droplets to contact an upward-flowing gas stream. The liquid phase is dispersed within the continuous gas phase, and ideal flow for each phase is PF. The gas-liquid interfacial area is relatively large, but the contact time is small.

(4) *Falling-film column.* A falling-film column (Figure 24.1(d)) is also an "empty" vessel, with liquid, introduced at the top, flowing down the wall as a film to contact an upward-flowing gas stream. Ideal flow for each phase is PF. Since neither liquid nor gas is dispersed, the interfacial area developed is relatively small, and gas-liquid contact is relatively inefficient. This type is used primarily in the experimental determination of mass transfer characteristics, since the interfacial area is well defined.

(5) *Bubble column.* A bubble column (Figure 24.1(e)) is also an "empty" vessel with gas bubbles, developed in a sparger (see below) rising through a downward-flowing liquid stream. The gas phase is dispersed, and the liquid phase is continuous; the assumed ideal flow pattern is PF for the gas and BMF for the liquid. Performance as a reactor may be affected by the relative difficulty of controlling axial and radial mixing. As in the case of a packed tower, it may also be used for catalytic systems, with solid catalyst particles suspended in the liquid phase.

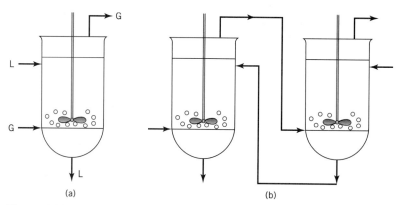

Figure 24.2 Agitated tank reactors for gas-liquid reactions: (a) single-stage; (b) two-(or multi-) stage

24.2.2 Tank Reactors

Tank reactors usually employ mechanical agitation to bring about more intimate contact of the phases, with one phase being dispersed in the other as the continuous phase. The gas phase may be introduced through a "sparger" located at the bottom of the tank; this is a circular ring of closed-end pipe provided with a number of holes along its length allowing multiple entry points for the gas.

Tank reactors are well suited for a reaction requiring a large liquid holdup or a long liquid-phase residence time. The operation may be continuous with respect to both phases, or it may be semicontinuous (batch with respect to the liquid). The simplest flow pattern for each phase is BMF. The reactor may be single-stage or multistage (Figure 24.2), and the flow for the latter may be cocurrent or countercurrent. In the case of a liquid-liquid reaction, each stage typically consists of a mixer (with agitation, for reaction) and a settler (without agitation, for separation by gravity).

Tank reactors equipped with agitators (stirrers, impellers, turbines, etc.) are used extensively for gas-liquid reactions, both in the traditional chemical process industries and in biotechnology. They are also used for three-phase gas-liquid-solid reactions, in which the solid phase may be catalyst particles; in this case, they are usually referred to as "slurry" reactors. In comparison with nonagitated tank reactors equipped only with spargers, mechanically agitated tank reactors have the advantage of providing a greater interfacial area for more efficient mass transfer. This reduces the likelihood of mass transfer limitations for a process, an advantage particularly for a viscous system. Adjustment of the stirrer rate can influence the rate of mass transfer. The enhanced mixing also ensures a nearly uniform temperature within the vessel, an advantage for processing temperature-sensitive materials, and for control of product yield and selectivity in complex systems. A major disadvantage is the cost of the energy required for agitation. Furthermore, since the mass transfer characteristics are related to agitation characteristics (e.g., stirrer rate), there are difficulties of scale-up. Thus, pilot plant studies are often necessary to obtain a basis for design of a full-scale reactor. For biological systems, shear sensitivity of the cells or enzyme may limit the range of stirrer rates that can be used.

24.3 CHOICE OF TOWER OR TANK REACTOR

The choice between a tower-type and a tank-type reactor for a fluid-fluid reaction is determined in part by the kinetics of the reaction. As described by the two-film model

Table 24.1 Typical values of gas-liquid interfacial area (a_i and a_i') for various types of vessels

Characteristic	Tower (column) packed, plate, spray	Tank agitated	Tank sparger
a_i (m² m⁻³)	~1000	~200	~20
a_i' (m² m⁻³)	~100	~180	~20
V_ℓ/V_R	~0.1	~0.9	~0.98

for gas-liquid reactions (Section 9.2), the rate of the overall process is influenced by relative rates of mass transfer and of intrinsic reaction. The two extremes (see Figure 9.9), for a nonvolatile liquid-phase reactant, are virtually instantaneous reaction in the liquid-film, which is controlled by interphase mass transfer (equation 9.2-22), and "very slow" reaction, which is controlled by reaction itself in the bulk liquid (equation 9.2-18 without the mass transfer contribution). The former is facilitated by a large interfacial area; for the latter, the extent of interfacial area is relatively unimportant.

Table 24.1 gives typical values of gas–liquid interfacial area for various reactor types. In Table 24.1,

a_i interfacial area based on unit volume of liquid phase, m² m⁻³ (liquid)
a_i' interfacial area based on unit volume of vessel (occupied by fluids), m² m⁻³ (vessel)

The two quantities a_i and a_i' are related by

$$a_i' = (V_\ell/V_R)a_i = (1 - \epsilon_g)a_i \qquad (24.3\text{-}1)$$

where V_ℓ = volume of liquid in the vessel, V_R = volume of reactor (vessel) occupied by fluid, and ϵ_g is the gas holdup, m³ gas (m³ reactor)⁻¹.

Values of the ratio V_ℓ/V_R given in Table 24.1 emphasize that most of the volume in a tower reactor (apart from a bubble column, data for which would be similar to a sparger-equipped tank) is occupied by the gas phase, and conversely for a tank reactor. This means that $a_i \gg a_i'$ in a tower and $a_i \simeq a_i'$ in a tank. For mass transfer-controlled situations, a_i is the more important quantity, and is much greater in a tower. For reaction-controlled situations, in which neither a_i nor a_i' is important, a sparger-equipped tank reactor, the cheapest arrangement, is sufficient.

24.4 TOWER REACTORS

24.4.1 Packed-Tower Reactors

We consider the problem of determining the height, h, of a tower (i.e., of the packing in the tower) and its diameter, D, for a reaction of the model type:

$$A(g) + bB(\ell) \to \text{products} \qquad (9.2\text{-}1)$$

in which A transfers from the gas phase to react with nonvolatile B in the liquid phase. The height h is determined by means of appropriate material balances or forms of the continuity equation. The tower may be treated as a physical continuum, and the material balance done over a control volume of differential height dh. A flow diagram indicating notation and control volume is shown in Figure 24.3.

604 Chapter 24: Reactors for Fluid-Fluid Reactions

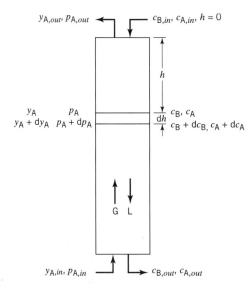

Figure 24.3 Flow diagram and notation for packed-tower reactor for reaction $A(g) + bB(\ell) \rightarrow$ products

For simplification, we make the following assumptions:

(1) The gas and liquid flow rates are constant throughout the column; that is, the transfer of material from one phase to the other does not affect the rate of flow of either phase (the "dilute-system" assumption). This assumption is not always applicable; in hydrogenation reactions, much of the gas may be consumed by reaction.
(2) Each phase is in PF.
(3) T is constant.
(4) P is constant.
(5) The operation is at steady state.
(6) The two-film model is applicable (Section 9.2).

In Figure 24.3, the other symbols are interpreted as follows:

G = total molar mass flow rate of gas, mol m^{-2} s^{-1}
L = total liquid volumetric flow rate, m^3 m^{-2} s^{-1}
(both G and L are related to unit cross-sectional area A_c of the unpacked column)
c_A = liquid-phase concentration of A, mol m^{-3}
c_B = liquid-phase concentration of B, mol m^{-3}
y_A = mole fraction of A in gas
p_A = partial pressure of A in gas = $y_A P$

Note that h is measured from the top of the column.

In a typical situation, as illustrated in Figure 24.3, the composition and flow rate of each feed stream (gas at the bottom and liquid at the top) are specified, directly or indirectly; this enables evaluation of the quantities $p_{A,in}$, $c_{A,in}$, $c_{B,in}$, L, and G. The unknown quantities to be determined, in addition to h (or V, the packed volume), are $p_{A,out}$ and $c_{A,out}$. The determination involves use of the rate law developed in Section 9.2 for an appropriate kinetics regime: (1) reaction in bulk liquid only (relatively slow intrinsic rate of reaction), or (2) in liquid film only (relatively fast reaction), or (3) in both bulk liquid and liquid film. For case (2), $c_A = 0$ throughout the bulk liquid, and the equations developed below for the more general case (3), $c_A \neq 0$, are simplified accordingly.

24.4 Tower Reactors

The continuity equations to be solved for the unknown quantities are as follows (with reference to the control volume in Figure 24.3):

Continuity equation for A in the gas phase (PF):

$$\begin{pmatrix} \text{rate of input} \\ \text{of A} \\ \text{by bulk flow} \end{pmatrix} = \begin{pmatrix} \text{rate of output} \\ \text{of A} \\ \text{by bulk flow} \end{pmatrix} + \begin{pmatrix} \text{rate of transfer} \\ \text{of A} \\ \text{to liquid film} \end{pmatrix}$$

The second term on the right is the flux of A at the gas-liquid interface, $N_A(z = 0)$. Thus, the continuity equation may be written as

$$(y_A + dy_A)GA_c = y_A G A_c + N_A(z = 0)a'_i A_c dh \quad (24.4\text{-}1)$$

which becomes, with $y_A = p_A/P$,

$$\frac{G}{P} \frac{dp_A}{dh} = N_A(z = 0)a'_i \quad (24.4\text{-}2)$$

An example of the form of $N_A(z = 0)$ is equation 9.2-45.

Continuity equation for A in the bulk liquid phase (PF):

For A in the bulk liquid, with reference to the control volume in Figure 24.3, in which the input of A is at the bottom,

$$\begin{pmatrix} \text{rate of input} \\ \text{of A} \\ \text{by bulk flow} \end{pmatrix} + \begin{pmatrix} \text{rate of transfer} \\ \text{of A} \\ \text{from L film} \end{pmatrix} = \begin{pmatrix} \text{rate of output} \\ \text{of A} \\ \text{by bulk flow} \end{pmatrix} + \begin{pmatrix} \text{rate of reaction} \\ \text{of A} \\ \text{in bulk liquid} \end{pmatrix}$$

The second term on the left is the flux of A at the fictitious liquid film–bulk liquid interface, $N_A (z = 1)$. That is,

$$(c_A + dc_A)LA_c + N_A(z = 1)a'_i A_c dh = c_A LA_c + (-r_A)_{int}(a'_i/a_i)A_c dh \quad (24.4\text{-}3)$$

where $(-r_A)_{int}$, in mol m^{-3} (liquid) s^{-1}, is the intrinsic rate of reaction of A in the liquid phase, as given by a rate law for a homogeneous reaction. Equation 24.4-3 becomes

$$L\frac{dc_A}{dh} = (-r_A)_{int}(a'_i/a_i) - N_A(z = 1)a'_i \quad (24.4\text{-}4)$$

An example of the form of $N_A(z = 1)$ is equation 9.2-45a.

Continuity equation for B in the bulk liquid phase (PF):

With reference to the control volume in Figure 24.3, in which the input of B is at the top,

$$\begin{pmatrix} \text{rate of input} \\ \text{of B} \\ \text{by bulk flow} \end{pmatrix} = \begin{pmatrix} \text{rate of output} \\ \text{of B} \\ \text{by bulk flow} \end{pmatrix} + \begin{pmatrix} \text{rate of reaction} \\ \text{of B} \\ \text{in bulk liquid} \end{pmatrix} + \begin{pmatrix} \text{rate of transfer} \\ \text{of B} \\ \text{to liquid film} \end{pmatrix}$$

That is, since the rate of diffusion of B in the liquid film is $N_B = -bN_A$ for counter-diffusion,

$$c_B LA_c = (c_B + dc_B)LA_c + b(-r_A)_{int}(a'_i/a_i)A_c dh - bN_A(z = 1)a'_i A_c dh \quad (24.4\text{-}5)$$

or

$$-L\frac{dc_B}{dh} = b[(-r_A)_{int}(a_i'/a_i) - N_A(z=1)a_i'] = bL\frac{dc_A}{dh}$$

That is,

$$-\frac{dc_B}{dh} = b\frac{dc_A}{dh} \qquad (24.4\text{-}6)$$

which is in accordance with the overall stoichiometry. The two continuity equations for the liquid phase are thus not independent relations; either one, but not both, may be chosen for convenience.

Correlations for parameters involved in equations 24.4-2 and -4, such as ϵ_g, $k_{A\ell}$, and a_i', are given by Froment and Bischoff (1990, pp. 610–613).

The continuity equations 24.4-2 and -4 are developed for the case of reaction in both liquid film and bulk liquid ($0.1 < \text{Ha} < 3$, say). In the case of reaction in liquid film only ($\text{Ha} > 3$, say), the equations simplify, since $c_A = 0$ in the bulk liquid, and equation 24.4-4 does not apply; an example of $N_A(z=0) \equiv (-r_A)$ is given by equation 9.2-48; $N_A(z=1) \to 0$. In the case of reaction in bulk liquid only ($\text{Ha} < 0.1$, say), $N_A(z=0) = N_A(z=1) \equiv (-r_A)$; an example is given by equation 9.2-18.

Overall material balance around column:

A further (independent) material balance around the entire column enables a relation to be established between, for example, $p_{A,out}$ and $c_{A,out}$.

- For A: rate of moles entering in gas + rate of moles entering in liquid = rate of moles leaving in gas + rate of moles leaving in liquid + rate of moles lost by reaction:

$$A_c\frac{G}{P}p_{A,in} + A_cLc_{A,in} = A_c\frac{G}{P}p_{A,out} + A_cLc_{A,out} + A_c\text{``}r_A\text{''}$$

which can be written

$$\text{``}r_A\text{''} = \frac{G}{P}(p_{A,in} - p_{A,out}) + L(c_{A,in} - c_{A,out}) \qquad (24.4\text{-}7)$$

where "r_A" is the total rate of consumption of A (in liquid film and bulk liquid) over the entire column.

- Similarly, for B:

$$\text{``}r_B\text{''} = L(c_{B,in} - c_{B,out}) = b\text{``}r_A\text{''} \qquad (24.4\text{-}8)$$

Combining 24.4-7 and -8, we obtain

$$\frac{G}{P}(p_{A,in} - p_{A,out}) = \frac{L}{b}(c_{B,in} - c_{B,out}) + L(c_{A,out} - c_{A,in}) \qquad (24.4\text{-}9)$$

Equation 24.4-9 may also be transformed to relate p_A, c_B, and c_A at any point in the column either to $p_{A,in}$, $c_{B,out}$, and $c_{A,out}$ (material balance around the bottom of the column from the point in question) or to $p_{A,out}$, $c_{A,in}$, and $c_{B,in}$ (around the top of the column). Thus, around the top,

$$\frac{G}{P}(p_A - p_{A,out}) = \frac{L}{b}(c_{B,in} - c_B) + L(c_A - c_{A,in}) \quad (24.4\text{-}9a)$$

and around the bottom,

$$\frac{G}{P}(p_{A,in} - p_A) = \frac{L}{b}(c_B - c_{B,out}) + L(c_{A,out} - c_A) \quad (24.4\text{-}9b)$$

Determination of the tower diameter D depends on what is specified for the system. Thus, the cross-sectional area is

$$A_c = \frac{\pi D^2}{4} = \frac{q_g}{u_{sg}} = \frac{F_{tg}}{G} = \frac{q_\ell}{L} \quad (24.4\text{-}10)$$

where q_g, u_{sg}, and F_{tg} are the volumetric flow rate, superficial linear velocity, and molar flow rate of gas, respectively, and q_ℓ is the volumetric flow rate of liquid. The gas flow rate quantities are further interrelated by an equation of state. Thus, for an ideal gas,

$$q_g = F_{tg}RT/P \quad (24.4\text{-}11)$$

Another aspect of the process design is the relative rates of flow of gas and liquid, G and L, respectively. This is considered in the following example.

EXAMPLE 24-1

If, for the situation depicted in Figure 24.3, the partial pressure of A in the gas phase is to be reduced from $p_{A,in}$ to $p_{A,out}$ at a specified gas flow rate G and total pressure P, what is the minimum liquid flow rate, L_{min}, in terms of G, P, and the partial pressures/concentrations of A and B? Assume that there is no A in the liquid feed.

SOLUTION

The criterion for $L \to L_{min}$ is that $c_{B,out} \to 0$. That is, there is just enough input of B to react with A to lower its partial pressure to $p_{A,out}$ and to allow for an outlet liquid-phase concentration of $c_{A,out}$. From equation 24.4-9, with $c_{B,out} = c_{A,in} = 0$ and $L = L_{min}$,

$$L_{min} = \frac{bG(p_{A,in} - p_{A,out})}{P(c_{B,in} + bc_{A,out})} \quad (24.4\text{-}12)$$

For reaction in the liquid film only, $c_{A,out} = 0$, and equation 24.4-12 reduces to

$$L_{min} = \frac{bG(p_{A,in} - p_{A,out})}{P(c_{B,in})} \quad (24.4\text{-}13)$$

Then,

$$L = \alpha L_{min} \quad (\alpha > 1) \quad (24.4\text{-}14)$$

24.4.2 Bubble-Column Reactors

In a bubble-column reactor for a gas-liquid reaction, Figure 24.1(e), gas enters the bottom of the vessel, is dispersed as bubbles, and flows upward, countercurrent to the flow of liquid. We assume the gas bubbles are in PF and the liquid is in BMF, although nonideal flow models (Chapter 19) may be used as required. The fluids are not mechanically agitated. The design of the reactor for a specified performance requires, among other things, determination of the height and diameter.

24.4.2.1 Continuity Equations for Bubble-Column Reactors

Continuity equation for A *in the gas phase* (PF):

Equation 24.4-2 again represents the continuity equation for A in the gas phase, which is in PF.

$$\frac{G}{P}\frac{dp_A}{dh} = N_A(z=0)a_i' \tag{24.4-2}$$

Continuity equation for A *in the bulk liquid phase* (BMF):

The continuity equation for A in the liquid phase, corresponding to the development of equation 24.4-3, is

$$c_{A,in}q_\ell + a_i'\int_0^V N_A(z=1)dV = c_{A,out}q_\ell + (1-\epsilon_g)(-r_A)_{int}V \tag{24.4-15}$$

The integral on the left side of equation 24.4-15 is required, since, although $c_A(= c_{A,out})$ is constant throughout the bulk liquid from top to bottom (BMF for liquid), p_A decreases continuously from bottom to top. These quantities are both included in $N_A(z=1)$ (see Example 24-2, below).

Overall material balance around column:

The overall material balance around the column is the same as that for a packed tower, equation 24.4-9:

$$\frac{G}{P}(p_{A,in} - p_{A,out}) = \frac{L}{b}(c_{B,in} - c_{B,out}) + L(c_{A,out} - c_{A,in}) \tag{24.4-9}$$

24.4.2.2 Correlations for Design Parameters for Bubble-Column Reactors

Before we can apply the continuity or design equations for bubble-column reactors developed in Section 24.4.2.1, we must have the means of determining values of the parameters involved. These include gas holdup, ϵ_g, mass transfer coefficients, $k_{A\ell}$ and k_{Ag}, and gas-liquid interfacial area, a_i or a_i'. For most of these, we take advantage of a review by Shah et al. (1982), although the choice among several possibilities given there in each case is somewhat arbitrary. As realized from problem 24-6, the correlation used can have a significant effect on the design or operating conditions. It may be necessary to consult the literature to obtain a specific correlation appropriate to the system under consideration.

Gas holdup, ϵ_g:
For a nonelectrolyte liquid phase, the correlation of Hikita et al. (1980) is

$$\epsilon_g = 0.672 \left(\frac{u_{sg}\mu_\ell}{\sigma}\right)^{0.578} \left(\frac{\mu_\ell^4 g}{\rho_\ell \sigma^3}\right)^{-0.131} \left(\frac{\rho_g}{\rho_\ell}\right)^{0.062} \left(\frac{\mu_g}{\mu_\ell}\right)^{0.107} \quad \textbf{(24.4-16)}$$

where
- u_{sg} superficial linear gas velocity, m s^{-1}
- g gravitational acceleration, 9.81 m s^{-2}
- μ_g, μ_ℓ viscosity of gas, liquid, Pa s
- ρ_g, ρ_ℓ density of gas, liquid, kg m^{-3}
- σ interfacial tension, N m^{-1}

Each factor in 24.4-16 is dimensionless, and so consistent units must be used, such as those given above.

Mass transfer coefficient, $k_{A\ell}$:
The liquid-film mass transfer coefficient may be given as a correlation for $k_{A\ell}$ (k_{il} in general for species i, or often denoted simply by k_L), or for $k_{A\ell}a'_i$, the product of the mass transfer coefficient and the interfacial area based on vessel volume (often denoted simply as $k_L a$).

For $k_{A\ell}$, the correlation of Calderbank and Moo-Young (1961) for small bubbles is

$$k_{A\ell} = 0.0031 \left[\frac{(\rho_\ell - \rho_g)\mu_\ell g}{\rho_\ell^2}\right]^{1/3} \left(\frac{\mu_\ell}{\rho_\ell D_A}\right)^{-1/3} \quad \textbf{(24.4-17)}$$

where D_A = molecular diffusivity of A in the liquid phase, m^2 s^{-1}

With units given above, $k_{A\ell}$ is in m s^{-1}, as derived from the second factor in 24.4-17; the third factor is dimensionless.

For $k_{A\ell}a'_i$, the correlation of Hikita et al. (1981) is

$$k_{A\ell}a'_i = 14.9 \left(\frac{g}{u_{sg}}\right) \left(\frac{u_{sg}\mu_\ell}{\sigma}\right)^{1.76} \left(\frac{\mu_\ell^4 g}{\rho_\ell \sigma^3}\right)^{-0.248} \left(\frac{\mu_g}{\mu_\ell}\right)^{0.243} \left(\frac{\mu_\ell}{\rho_\ell D_A}\right)^{-0.604} \quad \textbf{(24.4-18)}$$

With units given above, $k_{A\ell}a'_i$ is in s^{-1} as derived from the factor g/u_{sg}, since the other factors are dimensionless.

Interfacial area, a'_i:
If $k_{A\ell}$ is to be calculated from $k_{A\ell}a'_i$, a correlation for a'_i (or a_i) is required, but the result may be subject to significant error. An expression for a'_i given by Froment and Bischoff (1990, p. 637) may be written

$$a'_i = 2u_{sg}(\rho_\ell^3 g/\sigma^3)^{1/4} \quad \textbf{(24.4-19)}$$

With units given above, a'_i is in m^{-1} (i.e., m^2 interfacial area (m^3 reactor)$^{-1}$).

Mass transfer coefficient, k_{Ag}:
Shah et al. (1982) made no recommendation for the determination of k_{Ag}; in particular, no correlation for k_{Ag} in a bubble column had been reported up to that time. If the gas phase is pure reactant A, there is no gas-phase resistance, but it may be significant for a highly soluble reactant undergoing fast reaction.

24.4.2.3 Illustrative Example

The use of the design equations in Section 24.4.2.1 with the correlations in Section 24.4.2.2 for the parameters involved is illustrated in the following example.

EXAMPLE 24-2

It is required to determine the height (h) of a bubble-column reactor and the outlet partial pressure of oxygen (A, $p_{A,out}$) for the liquid-phase oxidation of o-xylene to o-methylbenzoic acid; the column diameter (D) is 1 m. The reaction is

$$3O_2(g) + 2C_8H_{10}(\ell) \rightarrow 2C_8H_8O_2(\ell) + 2H_2O(\ell)$$

or

$$A(g) + \frac{2}{3}B(\ell) \rightarrow \frac{2}{3}C + \frac{2}{3}H_2O \tag{A}$$

The production rate of acid is 1.584×10^4 kg d^{-1}. The following data are given by Froment and Bischoff (1990, p. 642):

The intrinsic liquid-phase rate of reaction is pseudo-first-order with respect to O_2 (A):

$$(-r_B)_{int} = k_B c_A = (40 \text{ min}^{-1}) c_A = b(-r_A)_{int}$$

$f_B = 0.16$
$H_A = 1.266 \times 10^4$ kPa L mol^{-1}
$D_{A\ell} = 1.444 \times 10^{-9}$ m^2 s^{-1}
$\rho_\ell = 750$ g L^{-1}
$\mu_\ell = 2.3 \times 10^{-4}$ kg m^{-1} s^{-1}; $\mu_g = 2.48 \times 10^{-5}$ kg m^{-1} s^{-1}
$\sigma = 1.65 \times 10^{-2}$ N m^{-1}

The gas feed is 25% excess air for the extent of reaction specified ($f_B = 0.16$) at 1.38 MPa and 160°C. Assume negligible gas-phase resistance.

SOLUTION

From the stoichiometry of reaction (A) and the specified fractional conversion, f_B, the molar feed rate of xylene (B, $M_B = 106.2$ g mol^{-1}) is obtained from the molar rate of production, F_C, of acid (C, $M_C = 136.2$ g mol^{-1}):

$$F_{B,in} = \frac{F_C}{f_B} = \frac{1.584 \times 10^4 (1000)}{136(3600)24(0.16)} = 8.425 \text{ mol s}^{-1}$$

The volumetric liquid feed rate of xylene is

$$q_\ell = \frac{F_{B,in} M_B}{\rho_B} = \frac{8.425(106.2)}{750} = 1.193 \text{ L s}^{-1}$$

and the liquid mass flow rate is

$$L = \frac{q_\ell}{A_c} = \frac{1.193 \times 10^{-3}}{\pi(1)^2/4} = 1.51 \times 10^{-3} \text{ m s}^{-1}$$

24.4 Tower Reactors

The molar feed rate of O_2 (A), 25% in excess of the requirement for reaction (A), is obtained from F_{Bo}, f_B and the stoichiometry:

$$F_{A,in} = 1.25(3/2) f_B F_{B,in} = 2.53 \text{ mol s}^{-1}$$

The volumetric gas feed rate of air (assumed to be 21% O_2) is

$$q_g = \frac{F_{A,in}}{0.21} \frac{RT}{P} = \frac{2.53(8.314)433}{0.21(1380)} = 31.42 \text{ L s}^{-1}$$

and the gas mass flow rate is

$$G = \frac{F_{A,in}}{0.21 A_c} = \frac{2.53}{0.21 \pi (1)^2/4} = 15.3 \text{ mol m}^{-2} \text{ s}^{-1}$$

The superficial linear velocity of the gas is

$$u_{sg} = \frac{q_g}{A_c} = \frac{31.42 \times 10^{-3}}{\pi(1)^2/4} = 0.04 \text{ m s}^{-1}$$

The gas density is

$$\rho_g = \frac{PM_{av}}{RT} = \frac{1380[0.79(28) + 0.21(32)]}{8.314(433)} = 11.05 \text{ kg m}^{-3}$$

The liquid-phase mass transfer coefficient is calculated from 24.4-18:

$$k_{A\ell} a'_i = 14.9 \left(\frac{g}{u_{sg}}\right) \left(\frac{u_{sg} \mu_\ell}{\sigma}\right)^{1.76} \left(\frac{\mu_\ell^4 g}{\rho_\ell \sigma^3}\right)^{-0.248} \left(\frac{\mu_g}{\mu_\ell}\right)^{0.243} \left(\frac{\mu_\ell}{\rho_\ell D_A}\right)^{-0.604}$$

$$= 14.9 \left(\frac{9.81}{0.04}\right) \left[\frac{0.04(2.3 \times 10^{-4})}{0.0165}\right]^{1.76} \left[\frac{(2.3 \times 10^{-4})9.81}{750(0.0165)^3}\right]^{-0.248} \left(\frac{2.484 \times 10^{-5}}{2.3 \times 10^{-4}}\right)^{0.243}$$

$$\times \left[\frac{2.3 \times 10^{-4}}{750(1.444 \times 10^{-9})}\right]^{-0.604} = 0.0883 \text{ s}^{-1}$$

From equation 24.4-19,

$$a'_i = 2 u_{sg} g^{1/4} (\rho_\ell/\sigma)^{3/4} = 2(0.04)(9.81)^{1/4}(750/1.65 \times 10^{-2})^{3/4} = 441 \text{ m}^{-1}$$

$$k_{A\ell} = (k_{A\ell} a'_i)/a'_i = 0.0883/441 = 2.00 \times 10^{-4} \text{ m s}^{-1}$$

From equation 24.4-16,

$$\epsilon_g = 0.672 \left(\frac{u_{sg} \mu_\ell}{\sigma}\right)^{0.578} \left(\frac{\mu_\ell^4 g}{\rho_\ell \sigma^3}\right)^{-0.131} \left(\frac{\rho_g}{\rho_\ell}\right)^{0.062} \left(\frac{\mu_g}{\mu_\ell}\right)^{0.107}$$

$$= 0.672 \left[\frac{0.040(2.3 \times 10^{-4})}{1.65 \times 10^{-2}}\right]^{0.578} \left[\frac{(2.3 \times 10^{-4})^4 9.81}{750(0.0165)^3}\right]^{-0.131} \left(\frac{11.05}{750}\right)^{0.062} \left(\frac{2.484 \times 10^{-5}}{2.3 \times 10^{-4}}\right)^{0.107}$$

$$= 0.152$$

From equation 9.2-40,

$$\text{Ha} = (k_A D_{A\ell})^{1/2}/k_{A\ell} = [1(1.444 \times 10^{-9})]^{1/2}/2.00 \times 10^{-4} = 0.190$$

Now, the design equations may be set up and solved. From 24.4-2,

$$h = \frac{G}{a_i'P} \int_{p_{A,in}}^{p_{A,out}} \frac{dp_A}{N_A(z=0)} \quad \text{(i)}$$

where

$$p_{A,in} = 0.21(1380) = 290 \text{ kPa}$$

and $N_A(z=0)$ is obtained from equation 9.2-45, with $1/k_{Ag} \to 0$ and $c_{Ab} \equiv c_{A,out}$:

$$N_A(z=0) = \frac{k_{A\ell}\text{Ha}}{\tanh(\text{Ha})}\left[\frac{p_A}{H_A} - \frac{c_{A,out}}{\cosh(\text{Ha})}\right]$$

From equation 24.4-15, with $c_{A,in} = 0$, $V = A_c h$, and $q_\ell = LA_c$,

$$a_i' \int_0^h N_A(z=1)dh = c_{A,out}L + (1-\epsilon_g)(-r_A)_{int}h \quad \text{(ii)}$$

where $N_A(z=1)$ is obtained from equation 9.2-45a, with changes relating to k_{Ag} and c_{Ab} as noted above:

$$N_A(z=1) = \frac{k_{A\ell}\text{Ha}}{\tanh(\text{Ha})}\left[\frac{p_A}{H_A \cosh(\text{Ha})} - c_{A,out}\right]$$

and

$$(-r_A)_{int} = \frac{1}{b}(-r_B)_{int} = \frac{3}{2}(40 \text{ min}^{-1})c_{A,out} = (1 \text{ s}^{-1})c_{A,out}$$

From equation 24.4-9, with $c_{A,in} = 0$

$$\frac{G}{LP}(p_{A,in} - p_{A,out}) = \frac{1}{b}(c_{B,in} - c_{B,out}) + c_{A,out} \quad \text{(iii)}$$

Also, h, V, and D are related by

$$V = A_c h = (\pi D^2/4)h \quad \text{(iv)}$$

The four unknown quantities in equations (i) to (iv) are $p_{A,out}$, $c_{A,out}$, h, and V.

Since the flux N_A is given in terms of p_A, the integral on the left side of equation (ii) must be recast in terms of p_A. From equation (i) (or 24.4-2),

$$dh = Gdp_A/N_A(z=0)a_i'P$$

and, hence, on substitution for dh in the integral in equation (ii),

$$\int_0^h N_A(z=1)dh = \frac{G}{a_i'P}\int_{p_{A,in}}^{p_{A,out}} \frac{N_A(z=1)}{N_A(z=0)}dp_A$$

$$= \frac{G}{a_i'P}\int_{p_{A,in}}^{p_{A,out}} \frac{\frac{k_{A\ell}\text{Ha}}{\tanh(\text{Ha})}\left[\frac{p_A}{H_A\cosh(\text{Ha})} - c_{A,out}\right]}{\frac{k_{A\ell}\text{Ha}}{\tanh(\text{Ha})}\left[\frac{p_A}{H_A} - \frac{c_{A,out}}{\cosh(\text{Ha})}\right]}dp_A$$

$$= \frac{G}{a_i'P}\int_{p_{A,in}}^{p_{A,out}} \frac{p_A - H_A c_{A,out}\cosh(\text{Ha})}{p_A\cosh(\text{Ha}) - H_A c_{A,out}}dp_A$$

This integral can be solved analytically, if we make the substitution

$$\alpha = p_A\cosh(\text{Ha}) - H_A c_{A,out}$$

so that

$$d\alpha = \cosh(\text{Ha})dp_A$$

and

$$p_A = (\alpha + H_A c_{A,out})/\cosh(\text{Ha})$$

In terms of α, on substitution for p_A and dp_A,

$$\int N_A(z=1)dh = \frac{G}{a_i'}\int \frac{\frac{\alpha + H_A c_{A,out}}{\cosh(\text{Ha})} - H_A\cosh(\text{Ha})c_{A,out}}{\alpha}\frac{d\alpha}{\cosh(\text{Ha})}$$

$$= \frac{G}{a_i'P[\cosh(\text{Ha})]^2}\left\{\int d\alpha + H_A c_{A,out}\left[\int \frac{d\alpha}{\alpha} - [\cosh(\text{Ha})]^2\int \frac{d\alpha}{\alpha}\right]\right\}$$

$$= \frac{G}{a_i'P[\cosh(\text{Ha})]^2}\left\{\alpha + Hc_{A,out}[1 - [\cosh(\text{Ha})]^2]\ln\alpha\right\}$$

On substituting back for α in terms of p_A and evaluating the integral over the limits $p_{A,in}$ to $p_{A,out}$, we obtain

$$\int_0^h N_A(z=1)dh = \frac{G}{a_i'P[\cosh(\text{Ha})]^2}\left\{p_{A,out}\cosh(\text{Ha}) - p_{A,in}\cosh(\text{Ha})\right.$$

$$\left.+ H_A c_{A,out}[1 - [\cosh(\text{Ha})]^2]\ln\left[\frac{p_{A,out}\cosh(\text{Ha}) - H_A c_{A,out}}{p_{A,in}\cosh(\text{Ha}) - H_A c_{A,out}}\right]\right\} \quad \textbf{(B)}$$

The result (B) is substituted into equation (ii) to convert it into an algebraic equation.

Similarly, although much simpler, the integral in equation (i) can be solved analytically on substitution for $N_A(z=0)$. The result is

$$h = \frac{GH_A\tanh(\text{Ha})}{a_i'Pk_{A\ell}\text{Ha}}\ln\left[\frac{p_{A,in} - H_A c_{A,out}/\cosh(\text{Ha})}{p_{A,out} - H_A c_{A,out}/\cosh(\text{Ha})}\right] \quad \textbf{(C)}$$

The E-Z Solve software is now used to solve simultaneously (file ex24-2.msp) the four equations (i), replaced by (C), (ii), with (B) introduced, (iii), and (iv) to determine the four quantities:

$$p_{A,out} = 57.8 \text{ kPa}; \quad c_{A,out} = 1.038 \text{ mol m}^{-3}; \quad h = 2.86 \text{ m}; \quad V = 2.25 \text{ m}^3$$

As an alternative to the use of the software, an iterative procedure may be used in the following stepwise manner:

(1) Choose a value of $c_{A,out}$.
(2) Calculate $p_{A,out}$ from (iii).
(3) Calculate h from equation (C), in place of (i).
(4) Calculate $c_{A,out}$ from (ii), with equation (B) introduced.
(5) Repeat steps (2) to (4) until $c_{A,out}$ satisfies a convergence criterion.
(6) Calculate V from (iv).

24.5 TANK REACTORS

For a tank reactor (Figure 24.2), the ideal flow pattern is BMF for each phase (gas and liquid). Important considerations are the dispersion of gas bubbles within the liquid phase, and the agitator power required for this. The h/D ratio is usually about 1, but this is exceeded if the stages in a multistage arrangement are stacked vertically. Gas is usually fed to the reactor below the agitator through a distributor which may be a perforated plate or pipe ring (sparger). Such a distributor can produce relatively small bubbles for a greater interfacial area. The type of agitator also can affect bubble size, and, thus, mass transfer characteristics. The most common types are rotating flat paddle mixers, with straight or inclined blades, and a turbine, with several flat blades attached to a disc (turbine). The diameter of the blades is usually about 35% of the vessel diameter (D). Several vertical baffles, each with a width of up to 10% of D, are attached to the wall of the vessel around its circumference.

We proceed in the rest of this section, as in Section 24.4.2, by developing the requisite continuity equations, giving correlations from the literature for the parameters involved, and finally, repeating Example 24-2 for a tank reactor.

24.5.1 Continuity Equations for Tank Reactors

Continuity equation for A *in the gas phase* (BMF):

Since the gas phase is in BMF, the continuity equation corresponding to 24.4-1, and based on the entire vessel of volume $V = A_c h = (\pi D^2/4)h$, is

$$y_{A,in} G A_c = y_{A,out} G A_c + N_A(z = 0) a_i' A_c h \tag{24.5-1}$$

or, with $y_A = p_A/P$,

$$\frac{G}{P}(p_{A,in} - p_{A,out}) = N_A(z = 0) a_i' h \tag{24.5-2}$$

Continuity equation for A *in the bulk liquid phase* (BMF):

Since the liquid phase is in BMF, the continuity equation for A in the bulk liquid phase is similar to equation 24.4-15, except that the integration in the second term on the left is unnecessary because both p_A and c_A in the expression for N_A are constant ($p_A = p_{A,out}$, and $c_A = c_{A,out}$). Thus, we have

$$c_{A,in} q_\ell + a_i' N_A(z = 1) V = c_{A,out} q_\ell + (1 - \epsilon_g)(-r_A)_{int} V \tag{24.5-3}$$

Overall material balance around tank:
The overall material balance around the tank is again given by equation 24.4-9:

$$\frac{G}{P}(p_{A,in} - p_{A,out}) = \frac{L}{b}(c_{B,in} - c_{B,out}) + L(c_{A,out} - c_{A,in}) \quad (24.4\text{-}9)$$

24.5.2 Correlations for Design Parameters for Tank Reactors

In addition to the parameters involved in a bubble-column reactor, that is, ϵ_g, a_i or a_i', and mass transfer coefficients, the required power input, P_I, is a design parameter for an agitated tank.

Power input, P_I:
Michell and Miller (1962) proposed the following correlation for P_I (in kW):

$$P_I = 0.34 n_{b,imp}^{1/2} (P_{Io} \dot{n}_{imp} d_{imp}^3 / q_g)^m \quad (24.5\text{-}4)$$

- P_{Io} power input without gas flow, kW
- $n_{b,imp}$ number of impeller blades
- $\dot{n}_{imp}$ rate of rotation of impeller, Hz
- d_{imp} diameter of impeller, m
- m 0.45 for normal coalescing fluids
  0.33 for ionic solutions which suppress coalescence

In equation 24.5-4, q_g is in m³ s⁻¹.

Interfacial area, a_i':
The correlation of Calderbank (1958) for a_i' is

$$a_i' = 22.8 \left(\frac{P_I}{V_\ell}\right)^{0.4} \left(\frac{u_{sg}}{u_{br}}\right)^{0.5} \left(\frac{\rho_\ell}{\sigma^3}\right)^{0.2} \quad (24.5\text{-}5)$$

where u_{br} is the rise velocity of a bubble through a quiescent liquid (equation 23.3-2).

Mass transfer coefficient, $k_{A\ell} a_i'$:
The correlations of Meister et al. (1979) for $k_{A\ell} a_i'$ for one and two impellers per stage, respectively, are:

$$k_{A\ell} a_i' = 0.0291 (P_I/V_\ell)^{0.801} u_{sg}^{0.248} \quad \text{(1 impeller)} \quad (24.5\text{-}6)$$

$$k_{A\ell} a_i' = 0.0193 (P_I/V_\ell)^{0.707} u_{sg}^{0.305} \quad \text{(2 impellers)} \quad (24.5\text{-}7)$$

With P_I in kW, V_ℓ in m³, and u_{sg} in mm s⁻¹, $k_{A\ell} a_i'$ has units of s⁻¹.

Chandrasekharan and Calderbank (1981) proposed the following correlation, which shows a much stronger inverse dependence on vessel diameter:

$$k_{A\ell} a_i' = 0.0248 (P_I/V_\ell)^{0.55} q_g^{0.55} D^{-4.5} \quad (24.5\text{-}8)$$

It was shown to be accurate to within 7.5% over a range of vessel diameters.

Gas holdup, ϵ_g:
The correlations of Hassan and Robinson (1977) for gas holdup, ϵ_g, for both non-electrolyte and electrolyte liquid phases are:

$$\epsilon_g = 0.11 (q_g \dot{n}_{imp}^2 / \sigma)^{0.57} \quad \text{(nonelectrolyte)} \quad (24.5\text{-}9)$$

$$\epsilon_g = 0.21 (q_g \dot{n}_{imp}^2 / \sigma)^{0.44} \quad \text{(electrolyte)} \quad (24.5\text{-}10)$$

These two correlations were based on laboratory-scale and pilot-plant-scale reactors ($D < 1$ m), and do not take into account vessel and impeller geometry.

An earlier correlation for ϵ_g was developed by Calderbank (1958) for coalescing liquids under turbulent agitation.

The use of the design equations in Section 24.5.1 with these correlations for the parameters involved is illustrated in the following example.

EXAMPLE 24-3

Repeat Example 24-2 for the xylene (B) oxidation reaction carried out in an agitated tank reactor (instead of a bubble-column reactor). Use the data given in Example 24-2 as required, but assume the diameter D is unknown. Additional data are: the power input without any gas flow is 8.5 kW; the impeller rotates at 2.5 Hz; the height and diameter of the tank are the same ($h = D$); the impeller diameter is $D/3$, and the impeller contains 6 blades; assume $u_{br} = 1.25 u_{sg}$. In addition to the vessel dimensions for the conversion specified ($f_B = 0.16$), determine the power input to the agitator (P_I).

SOLUTION

The solution is complicated by the fact that many of the parameters in the design equations depend on the vessel dimensions (h or D or V). In addition to the design equations, we give expressions for these parameters in terms of D. The result is a set of nonlinear algebraic equations to be solved for the unknown quantities, including D. We solve these by means of the E-Z Solve software (file ex24-3.msp).

The three design equations, with $h = D$, $V = \pi D^2 h/4 = 0.7854\, D^3$, and $c_{A,in} = 0$, may be written as:

$$\frac{G}{P}(p_{A,in} - p_{A,out}) = N_A(z = 0) a'_i D \qquad \text{(i) (from 24.5-2)}$$

$$a'_i N_A(z = 1) 0.7854 D^3 = c_{A,out} q_\ell + (1 - \epsilon_g)(-r_A)_{int} 0.7854\, D^3 \qquad \text{(ii) (from 24.5-3)}$$

$$\frac{G}{LP}(p_{A,in} - p_{A,out}) = \frac{1}{b}(c_{B,in} - c_{B,out}) + c_{A,out} \qquad \text{(iii) (from 24.4-9)}$$

In these equations, the unknown quantities are $p_{A,out}$, $c_{A,out}$, and D. The flux of A into the liquid film, $N_A(z = 0)$, is obtained from equation 9.2-45 for the special case of no gas-film mass transfer resistance ($k_{Ag} \to$ large), and with $c_{Ab} \equiv c_{A,out}$ and $p_A \equiv p_{A,out}$:

$$N_A(z = 0) = \frac{k_{A\ell} \text{Ha}}{\tanh(\text{Ha})} \left[\frac{p_{A,out}}{H_A} - \frac{c_{A,out}}{\cosh(\text{Ha})} \right] \qquad \text{(iv)}$$

(from 9.2-45, negligible gas-film resistance)

Similarly, the flux of A from the liquid film to the bulk liquid, $N_A(z = 1)$, is obtained from equation 9.2-45a:

$$N_A(z = 1) = \frac{k_{A\ell} \text{Ha}}{\tanh(\text{Ha})} \left[\frac{p_{A,out}}{H_A \cosh(\text{Ha})} - c_{A,out} \right] \qquad \text{(v)}$$

(from 9.2-45a, negligible gas-film resistance)

24.5 Tank Reactors

The intrinsic rate of reaction in the bulk liquid, $(-r_A)_{int}$, from Example 24-2, is, with $c_A \equiv c_{A,out}$:

$$(-r_A)_{int} = (-r_B)/b = \frac{3}{2}(40 \text{ min}^{-1})c_{A,out} = (1 \text{ s}^{-1})c_{A,out} \quad (\text{i.e., } k_A = 1 \text{ s}^{-1}) \quad \text{(vi)}$$

Quantities that carry over from Example 24-2 without change are:

$$f_B = 0.16; b = 2/3; P = 1380 \text{ kPa}; T = 433 \text{ K}$$
$$H_A = 1.266 \times 10^4 \text{ kPa L mol}^{-1}$$
$$D_A = 1.444 \times 10^{-9} \text{ m}^2 \text{ s}^{-1}$$
$$\rho_\ell = 750 \text{ g L}^{-1}; \rho_g = 11.05 \text{ kg m}^{-3}$$
$$\sigma = 1.65 \times 10^{-2} \text{ N m}^{-1}$$
$$q_\ell = 1.193 \text{ L s}^{-1}; q_g = 31.42 \text{ L s}^{-1}$$
$$c_{B,in} - c_{B,out} = 1130 \text{ mol m}^{-3} \text{ (from } f_B, F_{B,in}, \text{ and } q_\ell)$$
$$p_{A,in} = 290 \text{ kPa}; F_{A,in} = 2.53 \text{ mol s}^{-1}$$

In addition,

$$\epsilon_g = 0.11(q_g \dot{n}_{imp}^2/\sigma)^{0.57} = 0.11[(31.42/1000)2.5^2/0.016]^{0.57} = 0.45 \quad \textbf{(from 24.5-9)}$$

The following quantities depend on D:

$$L(D) = q_\ell/A_c = (1.193/1000)/(\pi/4)D^2 = 1.519 \times 10^{-3}/D^2 \text{ m s}^{-1} \quad \text{(vii)}$$

$$G(D) = F_{A,in}/0.21A_c = 2.53/0.21(\pi/4)D^2 = 15.34/D^2 \text{ mol m}^{-2} \text{ s}^{-1} \quad \text{(viii)}$$

$$u_{sg}(D) = q_g/A_c = 31.42 \times 10^{-3}/(\pi/4)D^2 = 0.0400/D^2 \text{ m s}^{-1} \quad \text{(ix)}$$

$$P_I(D) = 0.34 n_{b,imp}^{1/2}(P_{Io}\dot{n}_{imp}d_{imp}^3/q_g)^{0.45} = 0.34(6)^{1/2}[8.5(2.5)(D^3/27)/(31.42 \times 10^{-3})]^{0.45}$$
$$= 3.62 \, D^{1.35} \text{ kW} \quad \textbf{(x) (from 24.5-4)}$$

$$V_\ell(D) = (1 - \epsilon_g)V = (1 - 0.55)0.7854 \, D^3 = 0.432 \, D^3 \text{ m}^3 \quad \textbf{(xi) (from 24.3-1)}$$

$$k_{A\ell}a_i'(D) = 0.0291\left[\frac{P_I(D)}{V_\ell(D)}\right]^{0.801}[1000u_{sg}(D)]^{0.248}$$
$$= 0.1616\left[\frac{P_I(D)}{V_\ell(D)}\right]^{0.801}[u_{sg}(D)]^{0.248} \text{ s}^{-1} \quad \textbf{(xii) (from 24.5-6)}$$

$$a_i'(D) = 22.8\left[\frac{P_I(D)}{V_\ell(D)}\right]^{0.4}\left(\frac{u_{sg}}{u_{br}}\right)^{0.5}\left(\frac{\rho_\ell}{\sigma^3}\right)^{0.2} = 22.8\left[\frac{P_I(D)}{V_\ell(D)}\right]^{0.4}\left(\frac{1}{1.25}\right)^{0.5}\left[\frac{750}{0.0165^3}\right]^{0.2}$$
$$= 900\left[\frac{P_I(D)}{V_\ell(D)}\right]^{0.4} \text{ m}^2 \text{ m}^{-3} \text{ (vessel)} \quad \textbf{(xiii) (from 24.5-5)}$$

$$k_{A\ell}(D) = [k_{A\ell}a_i'(D)]/a_i'(D) \text{ m s}^{-1} \quad \text{(xiv)}$$

$$\text{Ha}(D) = \frac{(k_A D_{A\ell})^{1/2}}{k_{A\ell}} = \frac{[1(1.444 \times 10^{-9})]^{1/2}}{k_{A\ell}(D)} = 3.8 \times 10^{-5}/k_{A\ell}(D) \quad \textbf{(xv) (from 9.2-40)}$$

Equations (i) to (xv) are a set of nonlinear algebraic equations. Solution of these equations simultaneously by means of the E-Z Solve software (file ex24-3.msp) gives the following results:

$$D = h = 1.60 \text{ m} \qquad p_{A,out} = 57.8 \text{ kPa}$$
$$P_I = 6.72 \text{ kW} \qquad c_{A,out} = 1.03 \text{ mol m}^{-3}$$
$$\text{Ha} = 0.349 \qquad u_{sg} = 0.0155 \text{ m s}^{-1}$$
$$G = 5.95 \text{ mol m}^{-2} \text{ s}^{-1} \qquad L = 5.9 \times 10^{-4} \text{ m}^3 \text{ m}^{-2} \text{ s}^{-1}$$
$$k_{A\ell} a'_i = 0.167 \text{ s}^{-1} \qquad a'_i = 1530 \text{ m s}^{-1}$$
$$V_\ell = 1.78 \text{ m}^3 \qquad k_{A\ell} = 1.088 \times 10^{-4} \text{ m s}^{-1}$$

In addition, we calculate $d_{imp} = D/3 = 0.54$ m, and $V = \pi D^2 h/4 = 3.2$ m^3.

Experience with the software prompts the following comments. The fact that a large number of parameters and variables are coupled (through D) poses two problems for solution of the equations:

(1) Convergence may not occur to a unique solution, or may be slow (requires many iterations).
(2) The "converged" solution may not be realistic in terms of realistic values of the parameters and variables involved.

To address the first problem, reasonable initial estimates of certain quantities are provided. By default, E-Z Solve initially sets all values to one. Instead, we use other values for the following:

$$k_{A\ell} = 1 \times 10^{-4} \qquad D = 2$$
$$a'_i = 1000 \qquad p_{A,out} = 100$$
$$L = 1 \times 10^{-3} \qquad u_{sg} = 0.01$$

Furthermore, the lower bound for each quantity is set to zero, except for three quantities for which the following (more restrictive) lower and upper bounds are specified:

$$0 < \text{Ha} < 15$$
$$k_{A\ell} > 1 \times 10^{-6}$$
$$0.1 < D < 10$$

If the maximum number of allowable iterations is reached, the software provides "intermediate" values of all quantities calculated. These intermediate values are used as initial estimates to continue the solution. However, if any one of Ha, $k_{A\ell}$, or D is at the upper or lower boundary specified above, values of $k_{A\ell}$, D, a'_i, $p_{A,out}$, L, and u_{sg} are all reset to their initial values as listed above; for the other quantities, the intermediate values are used. This latter procedure is also used to check a solution that has apparently converged with any one of the three quantities at a boundary.

24.6 TRICKLE-BED REACTOR: THREE-PHASE REACTIONS

A trickle-bed reactor is a three-phase (gas + liquid + solid) reactor in which the solid is a fixed bed of particles catalyzing a gas-liquid reaction:

$$A(g) + bB(\ell) \xrightarrow{\text{cat}(s)} \text{products} \qquad (24.6\text{-}1)$$

The limiting reactant is usually B, which may be volatile or nonvolatile.

The flow of gas and liquid is usually cocurrent down the bed. Ideally, the relative flows and properties of the gas and liquid phases are such that liquid completely wets the surface of the solid particles as a thin film in its downward flow (trickle flow). The gas moves down through the interparticle void space in the bed. Trickle flow is achieved at relatively low flow rates of gas and liquid (but as high as feasible for this purpose). If the flow rates are higher, less desirable flow behavior occurs, such as spraying, slugging, foaming, pulsing, or bubbling, the type depending on the relative flow rates and properties of the fluids. Uncertainty in the ability to predict the flow pattern on scale-up contributes to the difficulty of design for this type of reactor.

Trickle-bed reactors are used in catalytic hydrotreating (reaction with H_2) of petroleum fractions to remove sulfur (hydrodesulfurization), nitrogen (hydrodenitrogenation), and metals (hydrodemetallization), as well as in catalytic hydrocracking of petroleum fractions, and other catalytic hydrogenation and oxidation processes. An example of the first is the reaction in which a sulfur compound is represented by dibenzothiophene (Ring and Missen, 1989), and a molybdate catalyst, based, for example, on cobalt molybdate, is used:

$$(C_6H_5)_2S(\ell) + H_2(g) \rightarrow (C_6H_5)_2(\ell) + H_2S(g) \qquad (24.6\text{-}2)$$

The design of a trickle-bed reactor requires combination of appropriate hydrodynamic models for the gas and liquid phases with mass transfer characteristics of the reactants, heat transfer characteristics, and a kinetics model for the intrinsic reaction rate on the catalyst surface (mostly interior pore surface). The simplest hydrodynamic models used are PF for the gas and DPF for the liquid. Mass transfer characteristics must allow for the transfer of A(g) from the bulk gas to the bulk liquid, thence to the exterior surface of the solid particles, and eventually to the interior pore structure. If B(ℓ) is nonvolatile, corresponding mass transfer characteristics must allow for transfer of B from bulk liquid to the interior pore structure. Heat transfer characteristics for nonisothermal operation may have to allow for significant axial and radial temperature gradients, and vaporization of liquid. The intrinsic kinetics may involve a rate law of the Langmuir-Hinshelwood type (Chapter 8) or a simpler power-law form. In principle, these are all taken into account in continuity equations written for A in the gas and liquid phases, and for B in the liquid phase, and in energy equations to obtain, for example, axial concentration profiles for A and B in the reactor, and axial and radial T profiles. The various models contain a large number of parameters, such as mass and heat transfer coefficients, liquid holdup, axial dispersion, interfacial area, extent of wetting, effectiveness factor and effective thermal conductivity. Uncertainty in values of these obtained from various correlations, where available, contributes further to the difficulty of designing a trickle-bed reactor. It is beyond our scope here, although this is an important type of reactor.

24.7 PROBLEMS FOR CHAPTER 24

24-1 For a gas-liquid reaction, under what main condition would a relatively large liquid holdup be required in a reactor, and what type of reactor would be used? Explain briefly.

24-2 Consider the removal of CO_2 from a gas stream by treatment with an aqueous solution of monoethanolamine (MEA) in a countercurrent-flow, packed tower operating at 25°C and 10 bar. The reaction is

$$CO_2(A) + 2RNH_2(B) \rightarrow RNH_2COO^- + RNH_2^+$$

Calculate the required height of the tower (m), based on an assessment of the appropriate kinetics regime (for possible simplification), and on the following information:

$$
\begin{aligned}
c_{B,in} &\ 2.0 \text{ mol L}^{-1} \\
y_{A,in} &\ 2 \text{ mol \%}; \ y_{A,out} = 3 \times 10^{-3} \text{ mol \%} \\
G \text{ (assume constant)} &\ 3 \times 10^{-3} \text{ mol cm}^{-2} \text{ s}^{-1} \\
L \text{ (assume constant)} &\ 5 \ L_{min} \text{ cm}^3 \text{ cm}^{-2} \text{ s}^{-1} \\
D_{A\ell} &\ 1.4 \times 10^{-5} \text{ cm}^2 \text{ s}^{-1}; \ D_{B\ell} = 0.77 \times 10^{-5} \text{ cm}^2 \text{ s}^{-1} \\
k_{A\ell} &\ 0.02 \text{ cm s}^{-1} \\
k_{Ag} a_i' &\ 8 \times 10^{-6} \text{ mol cm}^{-3} \text{ bar}^{-1} \text{ s}^{-1} \\
a_i' &\ 140 \text{ m}^2 \text{ m}^{-3} \text{ (tower)} \\
k_A &\ 7.5 \times 10^3 \text{ L mol}^{-1} \text{ s}^{-1} \\
H_A &\ 3 \times 10^4 \text{ bar cm}^3 \text{ mol}^{-1}
\end{aligned}
$$

24-3 Consider a packed tower for the absorption of A from a gas containing inert material (in addition to A) by a liquid containing B (nonvolatile) under continuous, steady-state conditions. Absorption is accompanied by the instantaneous reaction $A(g) + bB(\ell) \rightarrow$ products. Assume the overall process is liquid-film controlled.

(a) Show, by means of a material balance on B, that under certain conditions (state the assumptions made), the height of packing required is given by

$$ h = \frac{L}{a_i'} \int_{c_{B,out}}^{c_{B,in}} \frac{dc_B}{(-r_B)} \equiv \frac{L}{ba_i'} \int_{c_{B,out}}^{c_{B,in}} \frac{dc_B}{(-r_A)} $$

(b) Show that if the enhancement factor, E, is relatively large, the equation in part (a) becomes

$$ h = \frac{L}{k_{B\ell} a_i'} \ln\left(\frac{c_{B,in}}{c_{B,out}}\right) $$

(c) Calculate the height h for the following:

$c_{B,in} = 1500 \text{ mol m}^{-3}$; $y_{A,in} = 0.08$; $y_{A,out} \ll y_{A,in}$; $b = 2$; $G = 150 \text{ mol m}^{-2} \text{ s}^{-1}$, $k_{B\ell} = 5 \times 10^{-5} \text{ m s}^{-1}$; $L = 2 L_{min}$; and $a_i' = 100 \text{ m}^2 \text{ m}^{-3}$ (packing).

24-4 A carrier gas at 20 bar contains $H_2S(A)$ which is to be absorbed at 20°C by an aqueous solution of monoethanolamine (MEA, B), with a concentration c_B in a packed column. Assume reaction is instantaneous and can be represented by

$$ H_2S + RNH_2 \rightarrow HS^- + RNH_3^+, $$

where $R = HOCH_2CH_2$. The following data are available (Danckwerts, 1970):

$$
\begin{aligned}
k_{A\ell} a_i' &\ 0.030 \text{ s}^{-1} \\
k_{Ag} a_i' &\ 6 \times 10^{-5} \text{ mol cm}^{-3} \text{ s}^{-1} \text{ bar}^{-1} \\
D_{A\ell} &\ 1.5 \times 10^{-5} \text{ cm}^2 \text{ s}^{-1} \\
D_{B\ell} &\ 1.0 \times 10^{-5} \text{ cm}^2 \text{ s}^{-1} \\
H_A &\ 0.115 \text{ bar L mol}^{-1}
\end{aligned}
$$

Complete the following table, with reference to three points in the column:

	Top	Intermed. point	Bottom
p_A/bar	0.002	0.02	0.08
c_B/mol L^{-1}	0.295	0.25	0.10
$c_{B,max}$/mol L^{-1}	?	?	?
"kinetics regime"	?	?	?
E	?	?	?

Confirm (by material balances) that the sets of (p_A, c_B) are mutually consistent. What is the ratio L/L_{min}?

24-5 Consider determining the height (h/m) of a packed column for carrying out a gas-liquid reaction [$A(g) + bB(\ell) \to$ products] in which the reaction is fast, irreversible, and second-order. The gas phase is not pure A. The approximate solution of the material-balance equation, given by equations 9.2-50 to -54, provides an expression for the enhancement factor E in the rate equation $(-r_A) = k_{A\ell} c_{Ai} E$ (9.2-26a). This equation also involves c_{Ai}, which is not known *a priori*. Construct an algorithm (i.e., a stepwise procedure) that enables determination of the height of the column by an iterative solution for $(-r_A)$ at each of several points in the column. Clearly state the assumptions involved and list the information that would be required.

24-6 The results in Example 24-2 depend on values of parameters calculated from the correlations used. In particular, the value of Ha, which determines the kinetics regime to use for the two-film model, depends on the value of $k_{A\ell}$. Example 24-2 is solved on the basis that reaction occurs in both liquid film and bulk liquid, although the value of Ha (0.136) indicates a borderline situation with reaction in bulk liquid only.
   (a) Repeat Example 24-2 assuming reaction occurs in bulk liquid only.
   (b) Repeat Example 24-2 using equation 24.4-17 to estimate $k_{A\ell}$, and assuming reaction occurs in liquid film only, if Ha > 3. Comment on the result in comparison with that in Example 24-2.

24-7 The comments in problem 24-6 apply also to Example 24-3.
   (a) Repeat Example 24-3 assuming reaction occurs in bulk liquid only.
   (b) Consider Example 24-3 assuming reaction occurs in liquid film only.

24-8 Repeat Example 24-3 using P = 3500 kPa. What are the advantages/disadvantages of operating at this higher pressure?

24-9 In Example 24-2, explore the sensitivity of results obtained to (a) vessel diameter, D, and (b) rate of production of the acid product.

Appendix A

A.1 COMMON CONVERSION FACTORS FOR NON-SI UNITS TO SI UNITS[1]

Quantity	To convert from	To	Multiply by
length	ft	m	0.3048
	Å	m	10^{-10}
area	ft^2	m^2	0.09290
volume	ft^3	m^3	0.028317
	gal (US)	m^3	0.003785
	gal (IMP)	m^3	0.004545
	barrel	m^3	0.1590
pressure	atm	Pa	101325
	psi	Pa	6894.8
	mm Hg (Torr)	Pa	133.32
energy	BTU	J	1055.1
power	BTU h^{-1}	W	0.29307
	HP	W	745.7
density	lb ft^{-3}	kg m^{-3}	16.018
viscosity	cP	Pa s	0.001
temperature	°R	K	5/9
	°F	K	multiply by 5/9, add 255.37
mass	lb	kg	0.45359
	ton (short)	kg	907.18
	ton (long)	kg	1016

[1] Perry et al. (1984).

A.2 VALUES OF PHYSICOCHEMICAL CONSTANTS

Symbol	Name	Value
c	speed of light	2.9979×10^8 m s^{-1}
e	base of natural logarithms	2.71828
h	Planck constant	6.626×10^{-34} J s
k_B	Boltzmann constant	1.3807×10^{-23} J K^{-1}
N_{Av}	Avogadro number	6.022×10^{23} mol^{-1}
R	gas constant $= k_B N_{Av}$	8.3145 J mol^{-1} K^{-1} or Pa m^3 mol^{-1} K^{-1}

A.3 STANDARD SI PREFIXES

Prefix	Abbreviation	Multiplier
giga	G	10^9
mega	M	10^6
kilo	k	10^3
deci	d	10^{-1}
centi	c	10^{-2}
milli	m	10^{-3}
micro	μ	10^{-6}
nano	n	10^{-9}

Appendix B

Bibliography

B.1 BOOKS ON CHEMICAL REACTORS

Aris, R., *Introduction to the Analysis of Chemical Reactors,* Prentice-Hall, Englewood Cliffs, NJ, 1965.

Aris, R., *Elementary Chemical Reactor Analysis,* Prentice-Hall, Englewood Cliffs, NJ, 1969.

Atkinson, B., *Biochemical Reactors,* Pion, London, 1974.

Bailey, J.E., and D.F. Ollis, *Biochemical Engineering Fundamentals,* McGraw-Hill, New York, 2nd ed., 1986.

Butt, J.B., *Reaction Kinetics and Reactor Design,* Prentice-Hall, Englewood Cliffs, NJ, 1980.

Carberry, J.J., *Chemical and Catalytic Reaction Engineering,* McGraw-Hill, New York, 1976.

Carberry, J.J., and Arvind Varma (editors), *Chemical Reaction and Reactor Engineering,* Marcel Dekker, New York, 1987.

Chen, Ning Hsing, *Process Reactor Design,* Allyn and Bacon, Boston, 1983.

Cooper, A.R., and G.V. Jeffreys, *Chemical Kinetics and Reactor Design,* Prentice-Hall, Englewood Cliffs, NJ, 1971.

Danckwerts, P.V., *Gas-Liquid Reactions,* McGraw-Hill, New York, 1970.

Denbigh, K.G., and J.C.R. Turner, *Chemical Reactor Theory,* Cambridge, 3rd ed., 1984.

Doraiswamy, L.K., and M.M. Sharma, *Heterogeneous Reactions: Analysis, Examples, and Reactor Design; Volume 1: Gas-Solid and Solid-Solid Reactions; Volume 2: Fluid-Fluid-Solid Reactions,* Wiley, New York, 1984.

Elnashaie, S.S.E.H., and S.S. Elshishini, *Modelling, Simulation and Optimization of Industrial Fixed Bed Catalytic Reactors,* Gordon and Breach, New York, 1993.

Fogler, H.S., *Elements of Chemical Reaction Engineering,* Prentice-Hall, Englewood Cliffs, NJ, 3rd ed., 1999.

Froment, G.F., and K.B. Bischoff, *Chemical Reactor Analysis and Design,* Wiley, New York, 2nd ed., 1990.

Gianetto, A., and P.L. Silveston (editors), *Multiphase Chemical Reactors: Theory, Design, Scale-up,* Hemisphere, Washington, 1986.

Hill, C.G., *An Introduction to Chemical Engineering Kinetics and Reactor Design,* Wiley, New York, 1977.

Holland, C.D., and R.G. Anthony, *Fundamentals of Chemical Reaction Engineering,* Prentice-Hall, Englewood Cliffs, NJ, 2nd ed., 1989.

Horák, J., and J. Pasek, *Design of Industrial Chemical Reactors from Laboratory Data,* Heyden, London, 1978.

Hougen, O.A., and K.M. Watson, *Chemical Process Principles, Part Three, Kinetics and Catalysis,* Wiley, New York, 1947.

Kaštánek, F., J. Zahradník, J. Kratochvíl, and J. Černák, *Chemical Reactors for Gas-Liquid Systems,* Ellis Horwood, New York, 1993.

Kramers, H., and K.R. Westerterp, *Elements of Chemical Reactor Design and Operation,* Academic Press, New York, 1963.

Lapidus, L., and N.R. Amundson (editors), *Chemical Reactor Theory: A Review,* Prentice-Hall, Englewood Cliffs, NJ, 1977.

Levenspiel, O., *Chemical Reaction Engineering,* Wiley, New York, 3rd ed., 1999.

Levenspiel, O., *The Chemical Reactor Omnibook,* distributed by OSU Book Stores, Inc., Corvallis, OR, 1979.

Levenspiel, O., *The Chemical Reactor Minibook,* distributed by OSU Book Stores, Inc., Corvallis, OR, 1979.

Nauman, E.B., *Chemical Reactor Design,* Wiley, New York, 1987.

Perlmutter, D.D., *Stability of Chemical Reactors,* Prentice-Hall, Englewood Cliffs, NJ, 1972.

Petersen, E.E., *Chemical Reaction Analysis,* Prentice-Hall, Englewood Cliffs, NJ, 1965.

Ramachandran, P.A., and R.V. Chaudhari, *Three-Phase Catalytic Reactors,* Gordon and Breach, New York, 1983.

Rase, H.F., *Chemical Reactor Design for Process Plants; Volume One: Principles and Techniques; Volume Two: Case Studies and Design Data,* Wiley-Interscience, New York, 1977.

Rase, H.F., *Fixed-Bed Reactor Design and Diagnostics,* Butterworths, Boston, 1990.

Rodrigues, A.E., J.M. Calo, and N.H. Sweed (editors), *Multiphase Chemical Reactors; Volume 1: Fundamentals; Volume 2: Design Methods,* Sijthoff and Noordhoff, Alphen aan den Rijn, The Netherlands, 1981.

Rose, L.M., *Chemical Reactor Design in Practice,* Elsevier, Amsterdam, 1981.

Schmidt, L.D., *The Engineering of Chemical Reactions,* Oxford U. Press, New York, 1998.

Shah, Y.T., *Gas-Liquid-Solid Reactor Design,* McGraw-Hill, New York, 1979.

Smith, J.M., *Chemical Engineering Kinetics,* 3rd ed., McGraw-Hill, New York, 1981.

Tarhan, M. Orhan, *Catalytic Reactor Design,* McGraw-Hill, New York, 1983.

Ulrich, G.D., *A Guide to Chemical Engineering Reactor Design and Kinetics,* Ulrich Research and Consulting, Lee, NH, 1993.

Walas, S., *Reaction Kinetics for Chemical Engineers,* McGraw-Hill, New York, 1959.

Westerterp, K.R., W.P.M. van Swaaij, and A.A.C.M. Beenackers, *Chemical Reactor Design and Operation,* Wiley, New York, 1984.

B.2 BOOKS ON CHEMICAL KINETICS AND CATALYSIS

Benson, S.W., *Thermochemical Kinetics,* Wiley, New York, 2nd ed., 1976.

Boudart, M., and G. Djéga-Mariadassou, *Kinetics of Heterogeneous Catalytic Reactions,* Princeton U. Press, Princeton, NJ, 1984.

Eyring, H., S.H. Lin, and S.M. Lin, *Basic Chemical Kinetics,* Wiley, New York, 1980.

Gates, B.C., *Catalytic Chemistry,* Wiley, New York, 1992.

Laidler, K.J., *Reaction Kinetics; Volume 1: Homogeneous Gas Reactions; Volume 2: Reactions in Solution,* Pergamon, London, 1963; *Chemical Kinetics,* McGraw-Hill, New York, 2nd ed., 1965.

Parshall, G.W., and S.D. Ittel, *Homogeneous Catalysis,* Wiley, New York, 1992.

Somorjai, G.A., *Introduction to Surface Chemistry and Catalysis,* Wiley, New York, 1994.

Steinfield, J.I., J.S. Francisco, and W.L. Hase, *Chemical Kinetics and Dynamics,* Prentice-Hall, Englewood Cliffs, NJ, 1989.

Wilkinson, F., *Chemical Kinetics and Reaction Mechanisms,* Van Nostrand Reinhold, New York, 1980.

Wojciechowski, B.W., *Chemical Kinetics for Chemical Engineers,* Sterling Swift, Austin, TX, 1975.

Appendix C

Answers to Selected Problems

CHAPTER 1

1-2 (a) 4.02 L mol^{-1} h^{-1}
(b) 0.0195 mol L^{-1} h^{-1}
(c) no

1-4 (a) $C = 4$; $R = 5$; permissible set:

$$NH_4ClO_4 + 5Cl_2 + 3N_2O = 4HCl + 7NOCl$$
$$NH_4ClO_4 + Cl_2 + N_2O = 2H_2O + 3NOCl$$
$$Cl_2 + 2N_2O = N_2 + 2NOCl$$
$$4NOCl = O_2 + 2Cl_2 + 2N_2O$$
$$8NOCl = 2ClO_2 + 3Cl_2 + 4N_2O$$

(components here are: NH_4ClO_4, Cl_2, N_2O, $NOCl$)

(b) $C = 3$; $R = 3$; permissible set:

$$2CO = CO_2 + C(gr)$$
$$Zn(g) = Zn(l)$$
$$CO + Zn(g) = ZnO(s) + C(gr)$$

(components here are: $C(gr)$, CO, $Zn(g)$)

(f) $C = 4$; $R = 2$; permissible set:

$$5ClO_2^- + 2H_3O^+ = 3ClO_3^- + Cl_2 + 3H_2O$$
$$8ClO_2^- + 8H_3O^+ = 6ClO_2 + Cl_2 + 12H_2O$$

(components here are: ClO_2^-, H_3O^+, Cl_2, H_2O)

1-6 (a) 8.26×10^{-6} mol g^{-1} s^{-1}
(b) 1.503×10^{-7} mol m^{-2} s^{-1}
(c) 9.05×10^{-3} s^{-1}

1-8 (a) 2.81×10^{-10} mol m^{-3} s^{-1}
(b) 2810 mol s^{-1}
(c) 1.01 mol person^{-1} h^{-1}
(d) 4.22×10^{-10} mol m^{-3} s^{-1}; 1.41×10^{-10} mol m^{-3} s^{-1}

Appendix C: Answers to Selected Problems

CHAPTER 2

2-2 (a) $\bar{t} = \tau = 10$ min
(b) $\tau = 0.5$ min
$\bar{t} = 0.306$ min
$\bar{t} < \tau$, since τ (calculated at inlet conditions) does not take into account acceleration of fluid (because of increase in T and in F_t) at outlet conditions; $\bar{t}$ is calculated based on outlet conditions

2-4 (i) increase in F_t or q from inlet to outlet (T, P constant)
(ii) increase in T from inlet to outlet (F_t, P constant)
(iii) decrease in P from inlet to outlet (T, F_t constant)

2-6 2.42×10^{-4} mol L^{-1} s^{-1}

2-8 as a check, $n_t = n_{to} - f_B n_{Bo}$

2-10 if CO and CO_2 are chosen as noncomponents, as a check, $n_t = n_{to} + 2\Delta n_{CO} + 2\Delta n_{CO_2}$; there are 2 independent stoichiometric variables, and not just 1

CHAPTER 3

3-2 20 kPa

3-4 equation 3.4-13 reduces to $0 = 0$; equation 3.4-8 with $n = 2$ can be used with k_A replaced by $k'_A = (c_{Bo}/c_{Ao})k_A$; linear test as in Figure 3-4 can be used if c_{Ao} known

3-6 59000 J mol^{-1}

3-9 (a) $\ln(c_A/c_B) = \ln(c_{Ao}/c_{Bo}) + (3c_{Ao} - c_{Bo})k_A t$
(b) 8.31×10^{-5} L mol^{-1} s^{-1}

3-11 (b) 2; (c) $k = 0.055$ L mol^{-1} min^{-1}

CHAPTER 4

4-2 hypothesis is closely followed, with $k_A = 4.4 \times 10^{-4}$ s^{-1}

4-5 first order; $k_A = 1.24 \times 10^{-4}$ s^{-1}

4-7 first order; $k_B = 3.6 \times 10^{-3}$ s^{-1}

4-10 $A = 2.6 \times 10^6$ L mol^{-1} s^{-1}; $E_A = 6.2 \times 10^4$ J mol^{-1} (values obtained by nonlinear regression)

4-11 $n = 1.5$; $A = 4.70 \times 10^{12}$ L$^{0.5}$ mol$^{-0.5}$ s^{-1}; $E_A = 8.38 \times 10^4$ J mol^{-1}; $k_A(25) = 9.49 \times 10^{-3}$ and $k_A(35) = 28.5 \times 10^{-3}$, both in same units as A

4-12 (a) $k_{Ap} = k_A/(RT)^2$; e.g., 10.08×10^{-4} at 377 K
(b) $A = 3030$ L^2 mol^{-2} s^{-1}; $E_A = -3610$ J mol^{-1}
(c) $A_p = 2.09 \times 10^{-5}$ kPa^{-2} s^{-1}; $E_{Ap} = -12150$ J mol^{-1} (values in (b) and (c) obtained by nonlinear regression; for comparison, values obtained by linear regression, in same order and units as above: 3200, -3360; 2.10×10^{-5}, -12100)
(d) 8.74 kJ mol^{-1} versus 9.78 kJ mol^{-1} from equation 4.2-11, with $T = T$ (average)
(e) (c)

4-15 (a) first order with respect to each reactant; $k = 0.0786$ L mol^{-1} s^{-1}; rates determined by backward differencing and correlated to average c_A over each interval
(b) order as in (a), with $k = 0.0813$ L mol^{-1} s^{-1} (results from (b) represent data better)

4-18 (a) The plot of (storage time)$^{-1}$ versus T^{-1} can be interpreted as linear
(b) 153,000 J mol^{-1} (nonlinear regression); 121,000 J mol^{-1} (linear regression)

4-20 (a) $2NH_4HCO_3 = (NH_4)_2CO_3 + CO_2 + H_2O$
$(NH_4)_2CO_3 = 2NH_3 + CO_2 + H_2O$
(b) $(-r_A) = (1.68 \times 10^{-4}/\text{L mol}^{-1}\text{ s}^{-1})c_A^2$
$(-r_B) = (3.16 \times 10^{-4}/\text{L mol}^{-1}\text{ s}^{-1})c_B^2$

CHAPTER 5

5-2 $k_f = 0.350$ h^{-1}; $k_r = 0.150$ h^{-1}

5-4 (a) (i) 0.400 mmol min^{-1}; (ii) 0.477; (iii) 1.008
(b) 0.600

(c) 0.480; 0.060; 0.060
(d) (i) 0.600 mmol (g cat)$^{-1}$ min^{-1}; 0.288

5-6 (a) $3.00 \times 10^{-4}, 7.00 \times 10^{-4}$ s^{-1}; (b) 1.96, 2.80 mol L^{-1}

5-8 (a) 0.0191, 0.00638 min^{-1}
(b) 0.100, 0.040 mol L^{-1}
(c) 0.016 mol L^{-1}; (d) 0.070 mol L^{-1}

5-10 (a) $[\ln(2 - k_2/k_1)]/(k_1 - k_2); (k_2/k_1) < 2$
(b) $1/k_1 = 1/k_2$
(d) (i) 0.2558, 0.0774; (ii) 1, 0.3679; (iii) 2.558, 0.7743
(e) both increase

5-11 (a) 0.20, 1.10 mol L^{-1}
(b) $2.96 \times 10^{-3}, 4.89 \times 10^{-3}$ L mol^{-1} s^{-1}

5-13 (a) 0.12 min^{-1}; (b) 0.016 L mol^{-1} min^{-1}
(c) 1.25 mol L^{-1}; (d) 0.5 mol L^{-1}

5-15 (b) at $t = 0, P = p_A = 46.4$ kPa
at $t = 3600$, in order, 7.7, 21.4, 19.4, 34.6 kPa
at $t = \infty$, in order, 27.0, 23.2, 38.8 kPa
(c) 4.99×10^{-4} s^{-1}

5-17 $k_1 = 7.47 \times 10^{-4}$ s^{-1}; $k_2 = 3.36 \times 10^{-4}$ s^{-1} (authors' values)

CHAPTER 6

6-2 (a) 26.5% ; (b) 75.5%

6-4 (a) $10^4 k_{uni} = 2.47, 3.25, 4.50, 5.54, 6.29, 6.90$ s^{-1}
(b) yes; $k_\infty = 8.33 \times 10^{-4}$ s^{-1}; $k_1 = 0.294$ L mol^{-1} s^{-1}; $k_{-1}/k_2 = 353$ L mol^{-1}

6-7 (a) 34; (b) $\Delta S^{\circ \ddagger} > 0$ consistent with break-up of ring structure; (c) 266

6-9 (a) (i) −148; (ii) 75; (b) $\Delta S^{\circ \ddagger} < -45$ ("expected" result for $m = 2$), consistent with likelihood that the transition state has a highly organized ring-like structure of relatively low entropy
(c) <1, since $\Delta S^{\circ \ddagger}$ (calculated) < $\Delta S^{\circ \ddagger}$ ("expected")

CHAPTER 7

7-2 $3k_1 k_2 c_{O_3}^2 / (k_{-1} c_{O_2} + k_2 c_{O_3})$; SSH applied to O$^\bullet$

7-4 $r_{NO_2} = 2k_2 K_1 c_{NO}^2 c_{O_2}$
$E_A = E_{A2} + \Delta H_1^\circ (-)$
$E_A < 0$, if $E_{A2}(+) < |\Delta H_1^\circ|$

7-6 (a) (i) $\qquad\qquad\qquad\qquad\qquad C_2H_4O = CH_4 + CO$

(ii) $\qquad\qquad\qquad\qquad\qquad C_2H_4O \xrightarrow{k_1} C_2H_3O^\bullet + H^\bullet$ (1)

(iii) $\qquad\qquad\qquad\qquad\qquad C_2H_3O^\bullet \xrightarrow{k_2} CH_3^\bullet + CO$ (2)

$\qquad\qquad\qquad\qquad\qquad C_2H_4O + CH_3^\bullet \xrightarrow{k_3} C_2H_3O^\bullet + CH_4$ (3)

(iv) $\qquad\qquad\qquad\qquad\qquad C_2H_3O^\bullet \xrightarrow{k_4}$ termination products (4)

(b) $r_{CO} = (k_1 k_2 k_3 / 2k_4)^{1/2} c_{C_2H_4O}$ assumptions: SSH; $k_1 \ll 2k_3$; yes (agrees)
(c) 224 kJ mol^{-1}

7-8 (a) CH$_3^\bullet$ and CH$_3$CO$^\bullet$
(b) step 1: chain initiation
steps 2 and 3: chain transfer
steps 4 and 5: chain propagation
step 6: chain termination
(c) $r_{CH_4} = k_4(k_1/k_6)^{1/2} c_{CH_3CHO}^{3/2}$

7-10 $(-r_{C_2H_6}) = \frac{1}{2} k_1 [1 + (1 + 4k_3 k_4 / k_1 k_5)^{1/2}] c_{C_2H_6}$

7-11 $r_{C_2H_4} = \frac{1}{2}k_1[1 + (1 + 4k_3k_4/k_1k_5)^{1/2}]c_{C_2H_6}$
$r_{CH_4} = 2k_1 c_{C_2H_6}$
other choices of (2) species can be made;
$(-r_{C_2H_6}) = r_{C_2H_4} + \frac{1}{2}r_{CH_4}$
$r_{H_2} = r_{C_2H_4} - \frac{1}{2}r_{CH_4}$

7-13 (a) $n = 1; k = 4.4 \times 10^{-5}$ m$^{3/2}$ mol$^{-1/2}$ s^{-1}
(b) 0

CHAPTER 8

8-1 1; 0.11 L mol^{-1} s^{-1}

8-6 $r_P = k_{obs}p_B/p_A; k_{obs} = kK_B/K_A$; inhibition by reactant A

8-8 (a) $r_P = \dfrac{kK_A K_B p_A p_B}{(1+K_A p_A + K_B p_B + K_P p_P)^2}$
(b) $r_P = k_{obs}p_A p_B/p_P^2; k_{obs} = kK_A k_B/K_P^2$
strong inhibition by product P

8-12 $k_A = 188$ s^{-1}; $D_e = 6.8 \times 10^{-5}$ cm^2 s^{-1}; $\phi = 0.49, 0.94$; $\eta = 0.93, 0.78$

8-13 $\eta = \dfrac{3}{\phi}\left(\dfrac{1}{\tanh\phi} - \dfrac{1}{\phi}\right)$
$\phi = R(k_A/D_e)^{1/2}$

8-14 (a) $d^2\psi/dz^2 - \phi^2 = 0$
$\phi = L(k_A/D_e c_{As})^{1/2}$
(b) $\psi(\phi, z) = 1 - \phi^2 z + \frac{1}{2}\phi^2 z^2$
(c) $\eta = 1$
(d) $\phi = 2^{1/2} = 1.414$
(e) for $\phi \geq 2^{1/2}$ and $0 \leq z \leq 2^{1/2}/\phi, \psi = 1 - 2^{1/2}\phi z + \frac{1}{2}\phi^2 z^2$
for $\phi \geq 2^{1/2}$ and $2^{1/2}/\phi \leq z \leq 1, \psi = 0$
for $\phi \geq 2^{1/2}, \eta(\text{particle}) = 2^{1/2}/\phi$

8-16 0.16; 0.10

8-18 (d) comparable; (ignore); 1; $1/(k_A/k_{Ag} + 1); \eta_o k_A c_{Ag}$

8-19 (a) ≈ 194 kJ mol^{-1}
(b) yes; $T >\approx 450°$C; ≈ 65 kJ mol^{-1} (by nonlinear regression)
(c) (i) 6.2×10^{-5}; (ii) 2.23×10^{-6}

8-22 4

CHAPTER 9

9-1 see Table 9.1

9-2 see Table 9.1

9-3 see Table 9.1

9-5 1.3 bar

9-7 (a) t_1 (ash-layer control) = 17.4 min; t_1 (reaction control) = 15.2 min
(b) t_1 (ash-layer control) = 69.6 min; t_1 (reaction control) = 30.4 min
(c) 0.07 m min^{-1}

9-9 (a) increases
(b) (i) increases; (ii) c_{Ag}

9-12 (a) $E_i \to \infty$ (E of no use in this case; see 9-16(d) below
(b) (i) and (ii) $E_i \to 1$, which corresponds to physical absorption of A without reaction

9-16 (b) (i) $(-r_A) = \dfrac{p_A}{\frac{1}{k_{Ag}} + \frac{H_A}{k_{A\ell}E}}$ (ii) $\dfrac{p_A - H_A c_A}{\frac{1}{k_{Ag}} + \frac{H_A}{k_{A\ell}}}$
(d) reaction plane is at gas-liquid interface; by increasing c_B; $(-r_A) = k_{Ag}p_A$

9-17 (a) for $\psi > 5$, say (relatively large), $E \to \psi$, and $E = \dfrac{\text{Ha}^2}{2(E_i-1)}\left\{\left[1 + \dfrac{4E_i(E_i-1)}{\text{Ha}^2}\right]^{1/2} - 1\right\}$
(b) $E = 62.0 = \psi$
(c) (i) $E \to$ Ha as $E_i \to \infty$ (Ha > 5, say)
(ii) $E \to E_i$ as Ha $\to \infty$

CHAPTER 10

10-1 2.25×10^4; A, $f_{H_2O_2}$ the same in each case

10-3 $V_{max} = 0.216$ μmol L^{-1} s^{-1}; $K_m = 0.0184$ mmol L^{-1} (from nonlinear regression)

10-6 0.20 mmol L^{-1} h^{-1}; 0.20 mmol L^{-1}; 20 h^{-1}

10-9 (a) both intercept [$(1 + c_I/K_3)/V_{max}$] and slope [$K_m(1 + c_I/K_2)/V_{max}$] increase
(b) both intercept [$(1 + c_I/K_2)/V_{max}$] and slope [$K_m(1 + c_I/K_2)/V_{max}$] increase

10-11 (a) (i) 45; 595
(ii) 32.4; 365
(b) (ii) 34.5 g L^{-1}

CHAPTER 11

11-2 1270 mol s^{-1}
11-5 0.695; 892 K
11-6 437 K

CHAPTER 12

12-2 (a) 33 s; (b) 0.46
12-4 73 s
12-6 1420 min
12-8 (a) (i) 43 min; (ii) 91 min
12-10 5.0 m^3
12-12 (a) 4130 mol h^{-1}
(b) 6058 mol h^{-1}; $t_c = 1.9$ h
12-15 0.13 min

CHAPTER 13

13-10 4.1 h
13-12 (b) 11.3 min

CHAPTER 14

14-2 1.43 and 1.00 mol L^{-1}
14-4 2860 L
14-6 (a) 0.25; (b) 26.3 L h^{-1}; (c) 7.5; (d) 12.0 s; (e) 13.7 s; (f) $\bar{t}$ is based on outlet flow rate which is less than feed rate (decrease in moles)
14-8 3.75 L min^{-1}; 135 min
14-10 (a) 400 K; 0.80; (b) 318.5 K
14-12 813 K ($T_o = 536$ K)
14-14 0.731
14-16 (a) 103 L; (b) 208 L
14-18 144 L
14-20 (a) in series, order immaterial
(b) in series, smaller vessel first
14-22 134 L min^{-1}

CHAPTER 15

15-3 (a) 4.12 m^3; (b) 0.97; 4.02 m^3
15-4 (b) 0.686
15-6 23.3 m^3
15-8 (a) 1.34 m^3; (b) 164 m
15-10 1170 L
15-12 (a) 2.42 L; (b) 3.6 L s^{-1}; (c) 7.2 L
15-14 0.16 s
15-16 (a) 0.235; (b) 0.219; (c) 0.363; (d) 0.289

CHAPTER 16

16-2 0.76 (LFR); 0.80 (PFR); 0.61 (CSTR)
16-3 0.698 (LFR); 0.90 (PFR and CSTR)

CHAPTER 17

17-1 400 L (PFR)
17-3 5.0 L (CSTR or PFR)
17-5 (a) 4.6 m^3; (b) 1.1 m^3
17-7 (a) 1.24, 0.81, 0.4
17-9 (b) reduces to equation 17.1-10 for first order
(c) reduces to equation 17.1-11 for second order
17-11 (a) 0.889 ($q_2/q_1 = 1.5$)
(b) $q_2/q_1 = 1.68$ ($f_A = 0.890$; very shallow maximum)
(c) yes; $f_A = 0.926$
17-13 (a) 5; (b) 2
17-16 (a) $E(\theta) = 0.2\delta(\theta - 1/2) + (6.4/9)\exp[-(8/9)\theta]$

CHAPTER 18

18-2 6.32 m^3 for $f_A = 0.487$ at $t = 1.04$ h
18-4 (a) 69°C; (b) 338 L; (c) 6°C
18-7 for $E_3 > E_1 > E_2$ and $E_3 < E_1 < E_2$,

$$T_{opt} = \frac{E_3 - E_2}{R \ln\left[\frac{A_3}{A_2}\left(\frac{E_3 - E_1}{E_1 - E_2}\right)\right]}$$

18-8 (a) PFR; high c_A favors A$_2$, small reactor
(b) 1, 1.68, 0.65 mol L^{-1}; 310 L
18-11 $c_{AN} = 0.0064$, $c_{ANT} = 0.0122$, $c_{AQ} = 0.0063$ mol L^{-1}
18-13 (a) $V(\text{PFR}) = 675$ L
(b) $V(\text{PFR}) = 2125$ L
18-15 (a) 31°C; lower T favors reaction with lower E_A
(b) 0.578; (c) (i) 177 L; (ii) 0.76
18-17 (a) 1090 s; 0.517, 0.483 mol L^{-1}
(b) (i) 1.9×10^5 s; 0.729, 0.271 mol L^{-1}
(ii) 17 s; 0.343, 0.657 mol L^{-1}
(c) relative yield as given is favored by lower T, since reaction with lower E_A is favored
18-20 (a) catalytic: 6.1×10^{-7} (phenol), 0.0289 (CO$_2$), 1.1×10^{-5} (oxygenates), 2.03×10^{-3} (oligomers) mol L^{-1}
noncatalytic: 8.4×10^{-5}, 1.95×10^{-3}, 1.2×10^{-4}, 1.9×10^{-3}
comment: lowers phenol concentration and raises CO$_2$ (both desirable)
18-22 (a) 0.373; 0.355, 0.018; 0.952, 0.048
(b) 0.655; 0.548, 0.107; 0.837, 0.163
(c) 0.900; 0.490, 0.408; 0.543, 0.457
(d) results are independent of c_{Ao}; T should be as high as feasible to favor AQ (reaction step with higher E_A)
18-25 $c_1 = 0.27$, c (MEGX) $= 0.35$, c (hydroxylidocaine) $= 1.88$ mmol m^{-3}

CHAPTER 19

19-2 (b) $\bar{t} = 4.02$ s; $\bar{t} = V/q = 4.03$ s
(c) 16.25 s^2
19-3 (a) $E(\theta) = \frac{V}{V_1} \exp\left(-\frac{V}{V_1}\theta\right)$
(b) $F(\theta) = 1 - \exp\left(-\frac{V}{V_1}\theta\right)$
(c) (i) 0.576; (ii) 0.102

19-6 (a) 1; (b) 0.2; (c) 0.8 and 0.977
19-8 (a) 12.3 ("open" vessel), 8.1 ("closed" vessel) by spreadsheet; 11.6, 7.4 by nonlinear regression; (b) 0.28
19-10 (a) 43.2 s; (b) 717 s^2 (by nonlinear regression)
19-13 (a) 8.06 s; 19.8 s^2
  (b) 3.3; 9.4 ("open" vessel), 5.3 ("closed" vessel) (all results obtained by nonlinear regression)
19-15 (a) 4.20 s; 3.70 s^2 (by trapezoidal rule)
  (b) $N = 4.76$; $Pe_L = 12.52$ ("open" vessel), 8.39 ("closed" vessel)
19-17 35.0 s; 48.7 s^2; $N = 25.2$

CHAPTER 20

20-1 (a) 0.807 for each model
  (b) 0.0874 (TIS); 0.0879 (SFM); 0.0856 (MMM)
20-7 (a) $c_{Bj} = c_{B,j-1} + \frac{k_1 \bar{t} c_{Ao}}{N} \left[\frac{1}{1+(\bar{t}/N)(k_1+k_2)} \right]^j$; (b) 0.385, 1.570, 1.570, 1.044 mol L^{-1}
20-9 (a) 256 s; 1771 s^2 (by nonlinear regression)
  (b) (i) 0.941; (ii) 37; (iii) 0.941
20-12 (a) 0.569; (b) 0.603; 0.480; 0.564
20-14 0.793; SFM is not exact for second-order reaction
20-16 (a) 0.75; (b) 0.75 (same result as in (a), since reaction is first-order); (c) no; SFM gives incorrect result

CHAPTER 21

21-1 (a) 2.85 s; (b) 1050 kg m^{-3} (bed); (c) 105000 kg
21-3 0.291 and 0.592
21-6 (a) 852 K (b) 878 K
21-8 6730 kg (using E-Z Solve); 888 K
21-16 use of three stages meets the objective as follows:

Stage	T_{in}/K	T_{out}/K	$f_{A,out}$	$x_{CO,out}$
1	573	639	0.780	0.0286
2	509	516	0.862	0.0179
3	406	415	0.958	0.0055

CHAPTER 22

22-1 (a) 56.7 m^3 y^{-1}; (b) 0.508; (c) 359 h^{-1}; (d) $D = 3.7$ m; $L = 5.3$ m; 6.7 kPa
22-3 (a) $f_B < 1$; regardless of $\bar{t}$ and R_i, a certain fraction of particles is in the reactor for $t < t_1(R_i)$; yes (subject to law of diminishing returns), since f_B increases as $\bar{t}$ increases
  (b) $f_B = 1$, since all particles are in the reactor for $t > t_1(R_i)$
22-5 (a) 0.37 h; (b) 3700 kg
22-7 (a) (i) 0.889; (ii) 0.75 (iii) 0.90; (b) 10000 kg; same as in feed
22-9 0.927
22-11 0.916
22-14 0.552

CHAPTER 23

23-1 (a) no; $\partial u_{mf}/\partial u_f < 0$; $\partial u_{mf}/\partial \rho_f < 0$
  (b) u_{mf} increases as d_p or ρ_s increases; $\partial u_{mf}/\partial d_p$, $\partial u_{mf}/\partial \rho_s > 0$
23-3 (a) 0.14 m s^{-1}; (b) 14 m s^{-1}

23-5 about 18 kPa (value at mf conditions)
23-8 5150 kg h^{-1}
23-11 2.19 s^{-1}; 0.926

CHAPTER 24

24-2 3.4 m; show reaction occurs in liquid film only (Ha > 20); show reaction is pseudo-first-order fast ($E \to$ Ha), rather than second-order fast or instantaneous

24-3 **(c)** 4.4 m

24-4 $c_{B,max} = 0.006, 0.06, 0.24$
"kinetics regime": gas-film control (first two); gas-film + liquid-film resistance (at bottom)
$E = \infty, \infty, 1.80$

24-6 **(b)** Ha = 5.1; $p_{A,out} = 58.0$ kPa; $c_{A,out} = 2.71 \times 10^{-3}$ mol m^{-3}; $h = 13.5$ m; $V = 10.6$ m^3

24-8 $D = h = 0.95$ m; $p_{A,out} = 146$ kPa; $c_{A,out} = 3.85$ mol m^{-3}; Ha = 0.229; $P_I = 5.05$ kW

Appendix D

Use of E-Z Solve for Equation Solving and Parameter Estimation

Although there are many numerical problems within this text, they can be classified into four categories:

1. Those requiring solution of a system of algebraic equations
2. Those requiring solution of a system of differential and algebraic equations
3. Those involving simple parameter estimation, whereby the model equation is known
4. Those involving complex parameter estimation; that is, the relationship between a variable and a parameter can only be determined by numerical integration of one or more than one ordinary differential equation

In this appendix, we illustrate how E-Z Solve may be used for each of these types of problems. We expand on the basic features of E-Z Solve described in the "Help" and "Help/tutorial" menus in the software. The file "Appendix.msp" in the E-Z Solve software contains the equations used to solve examples D-1 to D-4 which follow. The main elements of the syntax are also printed within this appendix.

EXAMPLE D-1: SOLUTION OF COUPLED ALGEBRAIC EQUATIONS

Problem Statement

At 700 K, $CO_2(g)$ and $H_2(g)$ react to form $CO(g)$ and $H_2O(g)$, with an equilibrium constant of 0.11. Initially, a reactor is filled with 1.5 mol of $CO_2(g)$ and 1.2 mol of $H_2(g)$, and then heated to 700 K and a total pressure of 3.2 bar. Find the mole fraction and partial pressure of each species at equilibrium.

Solution

The reaction may be represented by:

$$CO_2 + H_2 \rightleftharpoons CO + H_2O$$

The governing relationship is that between the equilibrium constant and the equilibrium partial pressures of all species. This may be represented by:

$$K_p = \frac{P_{H_2O} P_{CO}}{P_{CO_2} P_{H_2}}$$

The partial pressures are functions of the species mole fractions, y_i, which are, in turn, dependent upon the extent of conversion of the reactants. A stoichiometric table may be used to relate the number of moles of all species at equilibrium, with x representing the moles of H_2 consumed. The moles of each species can thus be represented as follows:

$$n_{H_2} = n_{H_2,o} - x$$
$$n_{CO_2} = n_{CO_2,o} - x$$
$$n_{H_2O} = x$$
$$n_{CO} = x$$

These equations, in terms of the single unknown variable x, may be used to solve for the mole fractions (y_i) and partial pressures (p_i), according to:

$$y_i = n_i/n_t$$
$$p_i = y_i P$$

Consequently, this system may be represented by 14 algebraic equations, one for K_p, one for n_t, and four each for n_i, y_i, and p_i. These equations, plus the known values for K_p, $n_{H_2,o}$ and $n_{CO_2,o}$ may be entered into the E-Z Solve software and solved for x. The values of n_i, y_i, and p_i are then calculated based upon the value of x. The E-Z Solve syntax is listed below.

E-Z Solve Syntax

//Data

Kp = 0.11	//equilibrium constant
nAo = 1.5	//initial moles of CO2
nBo = 1.2	//initial moles of H2
P = 3.2	//total pressure, bar

//Equations

```
/*
We let x represent the number of moles of H2 (the limiting reactant) consumed during
the process. The moles of each species may thus be related to the number of moles of
H2 consumed, by stoichiometry.
*/
```

nA = nAo - x	//moles of CO2 at equilibrium
nB = nBo - x	//moles of H2 at equilibrium
nC = x	//moles of H2O at equilibrium
nD = x	//moles of CO at equilibrium
nt = nA + nB + nC + nD	//calculation of total moles in the system

yA = nA/nt	//mole fraction of CO2
yB = nB/nt	//mole fraction of H2
yC = nC/nt	//mole fraction of H2O
yD = nD/nt	//mole fraction of CO
pA = yA*P	//partial pressure of CO2
pB = yB*P	//partial pressure of H2
pC = yC*P	//partial pressure of H2O
pD = yD*P	//partial pressure of CO
Kp = pC*pD/(pA*pB)	//equilibrium expression

Results

With the equations entered as listed above, press F5 or "solve/sweep" under the solutions menu to solve the equations. The software indicates that $x = 0.333$. From this, the following mole fractions and partial pressures are obtained:

Species	Mole fraction	Partial pressure/bar
CO_2	0.432	1.38
H_2	0.321	1.03
H_2O	0.124	0.395
CO	0.124	0.395

Comments on E-Z Solve Syntax

1. Note that algebraic equations can be entered in any order. Furthermore, there is no need to isolate variables. For example, $c_A * t = \ln(c_A/t)$ is an acceptable equation which could be solved for either c_A or t (depending upon whether c_A or t is specified).
2. The maximum number of unknown variables which can be solved for is 50.
3. Variable names are case-sensitive.
4. Comments may be delimited by a double slash ("//"), in which case only one set is required, or by a slash and star, in which case "/*" represents the beginning of a comment, and "*/" represents the end of a comment. Both forms are illustrated in the syntax for the example shown above.

EXAMPLE D-2: COUPLED DIFFERENTIAL AND ALGEBRAIC EQUATIONS

Problem Statement

A 10-m³ cylindrical tank, 0.5 m in diameter, is filled with fluid. The valve at the bottom of the tank is then opened, allowing fluid to drain. The outflow rate, q_{out}, depends upon the height of fluid in the tank, h, according to

$$q_{out} = 0.50 h^{0.5} \text{ L s}^{-1}$$

Determine the height of fluid in the tank 30 minutes after the valve is opened.

Solution

A material balance around the fluid in the tank must be performed to examine the change in fluid volume over time. Since there is no inflow, the change in fluid volume is entirely due to outflow through the valve. Thus,

$$\frac{dV}{dt} = q_{out}$$

The fluid volume and fluid height are interrelated, so that dV/dt may be expressed in terms of the fluid height, h, as follows:

$$\frac{\pi D^2}{4}\frac{dh}{dt} = \frac{dV}{dt}$$

so that

$$\frac{dh}{dt} = -\frac{4}{\pi D^2}q_{out}$$

This latter differential equation, coupled with the algebraic expression relating q_{out} to h, may be solved to determine the fluid height as a function of time. The initial fluid height is 50.93 m, based upon the specified initial fluid volume and tank diameter. This serves as the initial condition to be used for the integration. The E-Z Solve syntax is:

E-Z Solve Syntax

//Data

D = 0.50   //vessel diameter, m
Vo = 10.0   //initial fluid volume, m^3

//Differential Equations

h' = -qout / (PI*D^2/4)   //differential equation, equal to dh/dt

//Algebraic Equations

qout = 0.00050*SQRT(h)   //expression for outflow rate, m^3/s
Vo = PI*D^2/4*ho   //calculation of initial fluid height

Results

Press F5 or "solve/sweep" under the solutions menu to solve the equations. A window which prompts for entry of initial conditions will appear. Numerical integration of h' from $h = 50.93$ m at $t = 0$ to $t = 15$ min (900 s) indicates that the fluid height at the end of the process is 23.5 m.

Comments on E-Z Solve Syntax

1. When solving systems of differential and algebraic equations, you must *list the differential equations first*. If a variable is first referred to in an algebraic equation,

and then is listed as a variable in a differential equation, an error message will be returned. Thus, as general practice, list all differential expressions before listing algebraic expressions.

2. The "prime" notation is used to designate a differential equation. In the equations above, h' represents dh/dt. Note that t is the default independent variable. If you wish to track a different independent variable (e.g., dh/dx), simply write the expression for h' as before, and then include a statement to change the variable, that is,

    x = t   //change of independent variable

3. Note the expression for the calculation of the initial height of fluid, h_o (immediately following the expression for q_{out}). In this expression, we are calculating h_o from V_o and D, both of which are specified. However, it is not necessary to rearrange the equation to place h_o on the left hand side, and V_o and D on the right side. All that is required is the relationship between the variables, and E-Z Solve does the rest!

4. Note the use of the function SQRT, and PI for π. These are two of many standard and engineering functions within E-Z Solve. For a complete listing, check under the Help menu.

EXAMPLE D-3: PARAMETER ESTIMATION WITH A KNOWN MODEL EQUATION

Problem Statement

Experiments were conducted to assess the effect of temperature on the pressure of a fixed quantity of gas kept in a 2.0-L vessel. Assuming that the ideal gas law is valid, determine the slope and intercept of the line which arises from the following data:

P/bar	0.804	0.856	0.910	0.965
T/K	293	313	333	353

Solution

According to the ideal gas law, the pressure of a gas is directly proportional to the absolute temperature, that is, $P = (nR/V)T$. This linear relationship can be represented by $P = mT + b$, where m is the slope, equal to (nR/V), and b is the intercept of the line, which is zero in the absence of any imprecision in the data. The slope and intercept may thus be determined by linear regression of the four datapoints versus the model equation $P = mT + b$.

In E-Z Solve, a separate equation entry is required for each data pair. The software then determines the values of m and b which minimize the sum of squared residuals (SSR). The E-Z Solve syntax is shown below.

E-Z Solve Syntax

```
//Data
    P1 = 0.804              //pressure, 1st measurement, bar
    P2 = 0.856              //2nd measurement, bar
    P3 = 0.910              //3rd measurement, bar
    P4 = 0.965              //4th measurement, bar
```

```
T1 = 293          //temperature, 1st measurement, K
T2 = 313          //2nd measurement, K
T3 = 333          //3rd measurement, K
T4 = 353          //4th measurement, K
```

//Governing Relationships

```
P1 = m*T1 + b     //assuming linear relationship P = mT + b
P2 = m*T2 + b
P3 = m*T3 + b
P4 = m*T4 + b
```

Results

The optimum values of m and b are 0.002685 and 0.0165, respectively. These parameter values may then be used to predict the experimental data, to confirm the goodness of the fit and ensure that the model equation is consistent with the trends in the data. The following table lists the experimental and predicted values of P at each T.

T /K	P (exp) /bar	P (pred) /bar
293	0.804	0.803
313	0.856	0.857
333	0.910	0.911
353	0.965	0.964

The assumed linear relationship is valid, and the parameter estimates provide excellent predictions of the experimental results.

Comments on E-Z Solve Syntax

In any regression problem, there are more equations than unknown parameters. In this instance, there are four equations, representing the four data pairs. The software then returns the optimum values of parameters in the governing equation(s). The Solutions/Statistics menu can be consulted to determine the squared residual between each experimental data point and the corresponding predicted value obtained from the parameter estimates.

EXAMPLE D-4: PARAMETER ESTIMATION WITH AN UNKNOWN MODEL EQUATION

Problem Statement

In a batch-reactor study of the kinetics of a liquid phase reaction A → products, the following data were obtained:

t /min	0	4	8	12	16	20
c_A /mmol L^{-1}	7.5	5.8	4.6	3.8	3.1	2.6

Estimate the reaction rate constant, k_A, and reaction order, n in the rate law $(-r_A) = k_A c_A^n$.

Example D-4: Parameter Estimation with an Unknown Model Equation

Solution

The material balance for a batch reactor may be used to develop a differential equation which may be solved for the $c_A(t)$ profile (see equation 3.4-1):

$$-\frac{dc_A}{dt} = (-r_A) = k_A c_A^n \qquad \text{(A)}$$

Although equation (A) can be integrated analytically (resulting in equations 3.4-9 and 3.4-10), for the sake of illustration we presume that we must integrate equation (A) numerically. For a given k_A and n, numerical integration of Equation (A) provides a predicted $c_A(t)$ profile which may be compared against the experimental data. Values of k_A and n are adjusted until the sum of squared residuals between the experimental and predicted concentrations is minimized.

E-Z Solve Syntax

```
//Data
    cAo = 7.5            //initial concentration, mmol/L
    cA1 = 5.8            //concentration at 4 min, mmol/L
    cA2 = 4.6            //concentration at 8 min, mmol/L
    cA3 = 3.8            //concentration at 12 min, mmol/L
    cA4 = 3.1            //concentration at 16 min, mmol/L
    cA5 = 2.6            //concentration at 20 min, mmol/L
    t1 = 4               //first time point, min
    t2 = 8               //second time point, min
    t3 = 12              //third time point, min
    t4 = 16              //fourth time point, min
    t5 = 20              //fifth time point, min

//Equations
/* We make use of the user-defined function rkint, a Runge-Kutta integrator, to return
function values at the specified values t1, t2, t3, t4, and t5. The unknown parameters kA
and n are passed to the rkint function, and they are passed from within the rkint function
to a second user-defined function f_rateeq, which contains the kinetics expression. The
form of f_rateeq is as follows:

    function f_rateeq(t,cA,kA,n,c)  //kA, n, and c are unknown parameters
    f = - kA*cA^n
    return f
    end

The variable "f" within f_rateeq is simply dcA/dt.
*/

//Model Equations
    cA1 = rkint(cAo, t1, kA, n, 0)
    cA2 = rkint(cAo, t2, kA, n, 0)
    cA3 = rkint(cAo, t3, kA, n, 0)
    cA4 = rkint(cAo, t4, kA, n, 0)
    cA5 = rkint(cAo, t5, kA, n, 0)
```

```
/*
The rkint function integrates the differential equation in f_rateeq from cAo at t = 0 to
cAi at t = ti. The value of cAi is then compared with the experimental value of cA; values
of kA and n are adjusted until the sum of squared residuals between the predicted and
experimental concentrations is minimized.
*/
```

Results

The parameter estimation can be started by pressing "F5", or by clicking on "solve/sweep" under the solutions menu. With any regression, it is important to set reasonable lower and upper bounds for the parameters. Furthermore, convergence can be accelerated if reasonable initial guesses for the parameters are chosen. For the current problem, a lower bound of zero was chosen for both k_A and n. The upper bound on n was set to 5, and the initial guesses were 0.25 and 1 for k_A and n, respectively. Using these initial conditions, the optimum values for k_A and n were 0.027 and 1.457. These values were subsequently used to predict the experimental data, to confirm the goodness of fit. The following table lists the experimental and predicted values of c_A at each t.

t/min	c_A (exp) /mmol L^{-1}	c_A (pred) /mmol L^{-1}
0	7.5	7.5
4	5.8	5.81
8	4.6	4.62
12	3.8	3.75
16	3.1	3.11
20	2.6	2.61

The results indicate that the model equation and parameters are appropriate.

Comments on E-Z Solve Syntax

1. The *rkint* and *f_rateeq* functions must be used for cases in which an analytical solution to the governing equation is not available. *rkint* is a standalone Runge-Kutta integration routine which may be imported and used for other problems of this type. The function *f_rateeq* contains the expression to be integrated; it may be edited as required for the problem at hand. The form of *f_rateeq* is shown in the E-Z Solve Syntax, above.
2. User-defined functions such as *rkint* and *f_rateeq* can be accessed by clicking on "user-defined functions" under the equations menu. Once within this window, functions can be imported and edited as required. User-defined functions are specific to a particular file. Thus, they must be imported into each problem file where they are used.

Nomenclature[1]

LATIN LETTERS

a	ratio of down-time to reaction time in BR, t_d/t
a_i	interfacial area, $m^2\ m^{-3}$ (liquid)
a_i'	interfacial area, $m^2\ m^{-3}$ (vessel)
a_j	exponent in rate law for reaction j
a_{ki}	subscript to element k in molecular formula of species i
aq	aqueous (phase)
A	pre-exponential factor in Arrhenius equation, $(mol^{-1}\ m^3)^{n-1}\ s^{-1}$, equation 3.1-8; area, m^2
A_p	pre-exponential factor corresponding to k_{ip}, equation 4.2-12
A_p'	pre-exponential factor corresponding to k_{ip}', equation 4.2-10
A	reacting species; molecular species
A_i	molecular formula of species i; pre-exponential factor with respect to species i
A	system formula matrix with entries a_{ki}
A*	unit-matrix form of **A**
b	stochiometric coefficient (+); bubble region of fluidized bed
B	reacting species; molecular species
c	cloud region of fluidized bed
c_E	(free) enzyme concentration, mol L^{-1}
c_{Eo}	total enzyme concentration, mol L^{-1}
c_i	volume concentration (molarity) of species i, mol L^{-1} equation 2.2-7 or 2.3-7
$c(t)$	nonnormalized tracer response (at vessel outlet)
$c(0)$	initial concentration of pulse tracer in vessel, m_o/V
c_i'	number density of molecules of species i, m^{-3}
c_{io}'	concentration of species i at inlet to recycle reactor, equation 15.3-3, mol L^{-1}
c_i^*	(fictitious) liquid-phase concentration in equation 9.2-12, mol L^{-1}
c_o	inlet concentration of step-change tracer
c_P	specific heat capacity at constant pressure, J $kg^{-1}\ K^{-1}$
C	reacting species
C	number of components = rank (**A**)
C_d	drag coefficient, dimensionless
C_o	c_{Ao}/c_{Bo}, equation 8.3-12a
C_P	molar heat capacity at constant pressure, J $mol^{-1}\ K^{-1}$
C(t)	normalized pulse tracer response (at vessel outlet), s^{-1}, equation 19.3-4
C(θ)	normalized pulse tracer reponse (at vessel outlet), equation 19.3-9 or 10
C$_{AN}$	**C**(θ), Section 19.4.1.1

---

[1] Where typical units are given, other (dimensionally equivalent) units may be used instead (e.g., diffusivity, D, in $cm^2\ s^{-1}$ rather than $m^2\ s^{-1}$).

Nomenclature

$\mathbf{C}_F(t)$	normalized step-change tracer response (at vessel outlet), equation 19.3-16
$\mathbf{C}_F(\theta)$	normalized step-change tracer response (at vessel outlet), equation 19.3-17
CL	chain length, equation 7.1-2
d_{AB}	collision diameter of A and B molecules, m, equation 6.4-3
d_b	bubble diameter, m
d_{imp}	diameter of impeller, m
d_p	particle diameter, m
d_p'	particle diameter taking shape into account, m, equation 21.3-6
d_t	tube diameter, m
D	reacting species
D	vessel or bed diameter, m; molecular diffusivity (with species subscript), $m^2 \, s^{-1}$
D_e	particle effective diffusivity, equation 8.5-4, -4d
D_K	Knudsen diffusion coefficient, $m^2 \, s^{-1}$
D_L	axial dispersion coefficient, $m^2 \, s^{-1}$ equation 19.4-47
D_m	molecular diffusivity, $m^2 \, s^{-1}$
D_i	diffusivity of species i, $m^2 \, s^{-1}$
e	base of natural logarithms, 2.71828; emulsion region of fluidized bed
E	reacting species; enzyme
E	enhancement factor, equation 9.2-26; energy, potential energy, J; exit-age RTD function, s^{-1}
ΔE	energy of reaction, J
E^*	"necessary" energy, J, equation 6.4-11
$E^\ddagger$	energy barrier for reaction, J
$E_o^\ddagger$	energy barrier for energetically neutral reaction, J
E_A	energy of activation, $J \, mol^{-1}$, equation 3.1-6
E_{Ap}	energy of activation corresponding to k_{ip}, $J \, mol^{-1}$, equation 4.2-11
E_{Ap}'	energy of activation corresponding to k_{ip}', $J \, mol^{-1}$, equation 4.2-9
EI	enzyme–inhibitor complex
EIS	enzyme-inhibitor–substrate ternary complex
ES	enzyme–substrate binary complex
ESS	enzyme–substrate ternary complex
$E(t), E(\theta)$	exit-age RTD function, s^{-1}, Section 13.3.1
$E_N(t), E_N(\theta)$	exit-age RTD function for N tanks in series (TIS), equation 19.4-14 or 19.4-30 or -30a
f	friction factor, equation 15.2-11 or 21.3-7; fluidized bed fraction (with subscript), equations 23.3-5 to -9; efficiency (Section 7.3.1)
f_i	fractional conversion of species (reactant) i, equation 2.2-3 or 2.3-5 or 14.3-12
f_{io}'	fractional conversion of species i at inlet to recycle reactor, equation 15.3-9
f_{iok}	fractional conversion of species i at inlet to kth stage in cold-shot cooling, equation 21.5-19
F	molar rate of flow, $mol \, s^{-1}$; cumulative RTD function
F_d	drag force, N
F_i	molar flow rate of species i, $mol \, s^{-1}$
F_{io}	molar feed flow rate of species i, $mol \, s^{-1}$
F_{ioj}	molar flow rate of species i entering stage j in cold-shot cooling, $mol \, s^{-1}$
F_{io}'	molar flow rate of species i to recycle PFR reactor, equation 15.3-8
F_{ioj}'	portion of total feed rate of species $i(F_{io})$ diverted to stage j in cold-shot cooling
F_t	total molar flow rate, $mol \, s^{-1}$
$F(t), F(\theta)$	cumulative RTD function, Section 13.3.2
$F_N(\theta)$	cumulative RTD function for N tanks in series (TIS), equation 19.4-36 or -38
g	gas (phase); gravitational acceleration, $9.807 \, m \, s^{-2}$
gr	graphite
G	Gibbs energy, J; mass velocity, $kg \, m^{-2} \, s^{-1}$, equation 21.3-9; molar mass velocity of gas, $mol \, m^{-2} \, s^{-1}$ (Chapter 24)
ΔG	Gibbs energy of reaction, J
ΔG°	standard Gibbs energy of reaction, J
$\Delta G^{\circ\ddagger}$	standard molar Gibbs energy of activation (TST), $J \, mol^{-1}$ equation 6.5-7
h	height, m; heat transfer film coefficient, $W \, m^{-2} \, K^{-1}$; Planck constant, $6.626 \times 10^{-34} \, J \, s$
H	enthalpy, J; holdback, Section 13.3.5
$\dot{H}$	rate of flow of enthalpy (across control surface), $J \, s^{-1}$

Latin Letters

Ha	see Dimensionless Groups
H_i	Henry's law constant for species i, Pa m^3 mol^{-1}, equation 9.2-8
ΔH_R	enthalpy of reaction for reaction as written, J
$\Delta H°$	standard enthalpy of reaction, J
$\Delta H°_{ci}$	standard enthalpy of combustion of species i, J mol^{-1}
$\Delta H°_{fi}$	standard enthalpy of formation of species i, J mol^{-1}
$\Delta H°_{Ri}$	standard molar enthalpy of reaction with respect to species i, J mol^{-1}
ΔH^{vap}	enthalpy of vaporization, J mol^{-1}
$\Delta H°^{\ddagger}$	standard molar enthalpy of activation (TST), J mol^{-1}
I	initiator (species); inhibitor; inert species
I	moment of inertia; internal-age distribution function
$I(t)$, $I(\theta)$	internal-age distribution function, s^{-1}, Section 13.3.4
J	rotational molecular energy, J
k	species-independent rate constant; thermal conductivity, W m^{-1} K^{-1}
k_{ai}	adsorption rate constant for species i, m s^{-1}
k_{api}	adsorption rate constant in terms of partial pressure, mol m^{-2} s^{-1} kPa^{-1}
k_B	Boltzmann constant, 1.381×10^{-23} J K^{-1}
k_{di}	desorption rate constant for species i, mol m^{-2} s^{-1}
k_e	effective thermal conductivity, W m^{-1} K^{-1}
k_f, k_r	rate constant for forward, reverse reaction
k_{H^+}	hydrogen-ion catalytic rate constant, equation 8.2-3
k_i	rate constant with respect to species i
k'_i	k_i/a_i, equation 9.2-17
k''_i	$k'_i c_B$, equation 9.2.-17a
k'''_i	$k_i c_B$, equation 9.2-35a
k_{ig}	mass transfer coefficient for species i across gas film, mol Pa^{-1} m^{-2} s^{-1} (equation 9.2-3), or m s^{-1} (for analog of equation 9.2-3 in terms of c_i, equation 8.5-48 or 9.1-11)
$k_{i\ell}$	mass transfer coefficient for species i across liquid film, m s^{-1}, equations 9.2-6, -7
$k_{i\ell}a'_i$	volumetric mass transfer coefficient for species i, s^{-1}
k_{ip}	rate constant defined in equation 4.2-6
k'_{ip}	rate constant defined in equation 4.2-4
k_{is}	rate constant for surface reaction (SCM, SPM)
k_o	rate constant in acid-base catalysis, equation 8.2-3
k_{OH^-}	hydroxyl-ion catalytic rate constant, equation 8.2-3
$k_{overall}$	rate constant in KL model, equation 23.4-5
k_r	rate constant for rds (Chapter 10)
k_1, k_{-1}	forward, reverse rate constants
k_2, k_{-2}	forward, reverse rate constants
k_3, k_{-3}	forward, reverse rate constants
K	ratio of rate constants, k_2/k_1; equilibrium constant
K_a	acid dissociation constant
$K_c^{\ddagger}$	equilibrium constant for formation of transition state, appropriate units, equation 6.5-3
K_{bc}, K_{ce}	interchange ("mass transfer") coefficients, bubble-cloud and cloud-emulsion in Kunii-Levenspiel fluidized-bed reactor model, s^{-1}, equations 23.3-13 and -14
K_{eq}, $K_{c,eq}$, K_p	equilibrium constant, appropriate units
K_i, K_{ip}	k_{ai}/k_{di} m^3 mol^{-1}, k_{api}/k_{di}, kPa^{-1}
K_{ig}, K_{iL}	overall mass transfer coefficient for species i based on gas, liquid film, units as for k_{ig}, k_{iL}, respectively, equations 9.2-9 and 9.2-10
K_m	Michaelis constant (Chapter 10, k_{-1}/k_1 or equation 10.2-17)
K_w	ion-product constant of water, equation 8.2-4
K_2	k_{-2}/k_2 (Chapter 10)
K_3	k_{-3}/k_3 (Chapter 10)
ℓ	liquid (phase)
l	length of unreacted zone (core) in SCM, m
L	length, depth of bed, m; mass velocity of liquid, m^3 m^{-2} s^{-1}; Laplace transform operator
m	mass, kg; molecular mass, kg; molecularity (TST)
$\dot{m}$	rate of flow of mass (across control surface), kg s^{-1}

Nomenclature

m_o	amount of pulse (tracer), kg
$\dot{m}_{oj}$	portion of total feed ($\dot{m}_o$) entering stage j in cold-shot cooling, kg s^{-1}
M	molar mass (with subscript), kg mol^{-1}; number of elements
M	molecular species (in elementary reaction); monomer; mixing point (of streams); tracer
M_k	kth moment about the mean (of a distribution); equation 13.3-15
M_{in}	reaction number for species i and nth-order reaction, equation 4.3-4, dimensionless
M_o	$c_{Ao} + c_{Bo}$, equation 8.3-3a
n	order of reaction; quantity defined by equation 5.3-12
$n_{b,imp}$	number of impeller blades
n_c	number of measured (discrete) responses $c(t)$
n_i	amount (number of moles) of species i, mol; number of surface fragments from dissociation of species i, equation 8.4-23
$\dot{n}_{imp}$	rate of rotation of impeller, Hz
n_{Mo}	amount of tracer M in pulse, mol
n_o	amount of pulse (tracer), mol
n_t	total number of moles
N	number of molecules, tanks, stages, species or substances; surface sites m^{-2}
N_A	molar flux of A, mol m^{-2} s^{-1}
$N_A(z=0)$	molar flux of A at particle exterior surface or at gas-liquid interface, mol m^{-2} s^{-1}
$N_A(z=1)$	molar flux of A from liquid film to bulk liquid, mol m^{-2} s^{-1}
N_{Av}	Avogadro number, 6.022×10^{23} mol^{-1}
N_g	number of species in gas phase
N_i	rate of transport (flux) of species i, mol m^{-2} s^{-1}
N_t	number of tubes
p	protonic charge (as an element)
p	steric factor (SCT)
p_i	partial pressure of species i, Pa, equation 4.2-1
p_i^*	vapor pressure of species i, Pa; (fictitious) partial pressure in equation 9.2-11
P	pressure, Pa
P	product (species)
ΔP	pressure drop, Pa
Pe$_L$	see Dimensionless Groups
P_I	power input, kW
P_r	r-mer (polymer)
$Pr(i)$	production rate of species i, mol s^{-1}
$P(R)$	particle-size distribution
q	volumetric flow rate, m^3 s^{-1}
q'	volumetric amount of (pulse) tracer in vessel at time t
q_o	feed flow rate, m^3 s^{-1}
q'_o	small volumetric amount (pulse) of tracer entering vessel
Q	partition function (Chapter 6)
$\dot{Q}$	rate of heat transfer, W
$Q^\ddagger$	partition function for transition state
Q_r	product of partition functions of reactant molecules
r	species-independent intensive rate of reaction, mol s^{-1} (normalizing factor)$^{-1}$; radius, radial position (variable), interatomic distance m; number of monomer units in r-mer; ratio of flow rates; molar ratio of inert species to limiting reactant (in feed)
r_{ai}	rate of adsorption of species i
r_c	radius of unreacted core (variable, SCM), m
r_{di}	rate of desorption of species i
r_f, r_r	rate of forward, reverse reaction
r_i	(intensive) rate of formation (+) or consumption (−) of species i; fraction of feed entering ith stage in cold-shot cooling, equation 21.5-12
r_{ip}	rate of reaction in terms of pressure, Pa s^{-1}, equation 4.2-6
R	gas constant, 8.3145 J mol^{-1} K^{-1}; recycle ratio; radius (fixed), m; maximum number of linearly independent chemical equations
Re, Re$'$	see Dimensionless Groups

$R_{c_{B,max}}$	ratio in equation 18.4-16
R_i	(extensive) rate of formation (+) or consumption (−) of species i, mol s^{-1}
$R_{\hat{S}}$	ratio defined following equation 18.4-13
R_{tmax}	ratio in equation 18.4-15
s	Laplace parameter; stoichiometric number; surface site; solid (phase); slope
$\hat{s}_{P/A}$	instantaneous fractional yield of P with respect to A, equation 5.2-8
S	substrate; split point (of streams)
S	entropy, J K^{-1}
$S(t-b)$	unit step function about $t = b$
$S(\theta - b)$	unit step function about $\theta = b$
Sc	see Dimensionless Groups
Sh	see Dimensionless Groups
S_v	space velocity, s^{-1}
$\hat{S}_{P/A}$	overall fractional yield of P with respect to A, equation 5.2-4
$\Delta S^{\circ\ddagger}$	standard molar entropy of activation (TST), J mol^{-1} K^{-1}
t	time; residence time, equation 2.4-9 (PFR), s
t_a	age of element of fluid, equation 20-1
t_c	cycle time (BR)
t_d	down-time (BR)
t_e	life expectancy of element of fluid, equation 20-1
t_o	time of introduction of pulse or step change (tracer)
t_T	beginning (time) of tail region, Figure 19.9
$t_{1/2}$	half-life (of reactant), s
t_1	time for complete conversion of reacting particle (SCM), s
$\bar{t}$	mean residence time, s, equation 2.3-1 (CSTR)
T	temperature, K (occasionally °C)
T_{oj}	temperature of stream entering stage j in cold-shot cooling, equation 21.5-21
ΔT	temperature difference in cold-shot cooling defined by equation 21.5-23
u	linear velocity; molecular velocity, m s^{-1}
u_b	absolute rise velocity of bubbles in fluidized bed, equation 23.3-4
u_b^*	bubble rise velocity defined by equation 23.4-17
u_{br}	rise velocity of bubble, equation 23.3-2 or -3, m s^{-1}
u_s	superficial linear velocity, m s^{-1}
u_{sg}	superficial linear velocity of gas (Chapter 24), m s^{-1}
U	internal energy, J; overall heat transfer coefficient, W m^{-2} K^{-1}
$\Delta U^{\circ\ddagger}$	standard molar internal energy of activation (TST), J mol^{-1}
v	volume, m^3; molar volume, m^3 mol^{-1}; molecular vibrational energy, J
V	volume, m^3
V_{max}	maximum-rate parameter (Chapter 10), equation 10.2-7
w	wake region of fluidized bed
w_i	weight (mass) fraction of species i
w_r	mass of r-mer/mass of all r-mers
W	mass (holdup) of catalyst, solid, kg; washout RTD function
$W(t), W(\theta)$	washout RTD function, Section 13.3.3
x	coordinate, distance, axial position (variable), m; (dummy) variable
x_i	mole fraction of species i (in a phase), equation 4.2-2 (or equivalent)
y_i	mole fraction of species i (gas phase)
$Y_{P/A}$	yield of P with respect to A, equation 5.2-1
z	dimensionless distance, length, equation 8.5-9 or 9.2-38; compressibility factor
Z_{AA}	collision frequency of like molecules, m^{-3} s^{-1}
Z_{AB}	collision frequency, molecules of type A with molecules of type B, m^{-3} s^{-1}

GREEK LETTERS

α	order of reaction with respect to (species indicated by a subscript); volume of wake/volume of bubble (fluidized bed), equation 23.3-8; quantity defined by equation 21.3-13
β	order of reaction with respect to (species indicated by a subscript); dimensionless parameter defined by equation 8.5-40; quantity defined by equation 21.3-11

γ order of reaction with respect to (species indicated by a subscript); dimensionless parameter defined by equation 8.5-39; volume of solids/volume of bubbles (fluidized bed), with subscript for region, equations 23.3-10 to -12

Γ gamma function, equation 19.4-27

δ film thickness; film thickness (variable) at position of reaction plane for instantaneous reaction; (fluidized) bed fraction in bubbles; small change in or amount of

$\delta(t-b)$ Dirac (delta) function about $t = b$

$\delta(\theta-b)$ Dirac (delta) function about $\theta = b$

Δ change in

ϵ voidage or porosity [m^3 (void) m^{-3} (relevant quantity)]; (molecular) energy, J

ϵ_g gas holdup, m^3 gas m^{-3} reactor volume (Chapter 24)

η particle effectiveness factor, equation 8.5-5

η_o overall effectiveness factor, equation 8.5-45

θ dimensionless time, $t/\bar{t}$; fraction of surface sites or surface covered

κ transmission coefficient (TST) ($\kappa \simeq 1$); quantity defined by equation 21.3-12

λ wavelength; (molecular) mean free path; c_A/c_{Ai}, equation 9.2-37

μ viscosity, Pa s or kg m^{-1} s^{-1}; reduced mass, equation 6.4-6

μ_k kth moment about the origin (of a distribution), equation 13.3-12

ν frequency, s^{-1}; stoichiometric coefficient

ν_i stoichiometric coefficient of species i

ν_{ij} stoichiometric coefficient of species i in chemical equation j

$\nu^{\ddagger}$ frequency with which transition state is transformed into product(s), s^{-1}, equation 6.5-4

ξ extent of reaction, mol, equation 2.2-5, or 2.3-6, or 5.2-9

π 3.14159

Π continued product

ρ density, kg m^{-3}

σ standard deviation; collision cross-section, m^2; interfacial tension, kg s^{-2} or N m^{-1}

σ^2 variance, equations 13.3-16, 13.3-16a, 13.3-18

Σ continued summation

τ space time, s, equation 2.3-2; tortuosity

ϕ Thiele modulus, e.g., equation 8.5-11, 8.5-19

ϕ' Thiele modulus normalized with respect to shape, equation 8.5-17

ϕ'' Thiele modulus normalized with respect to shape and order, equation 8.5-20

ϕ_G generalized Thiele modulus defined by equation 8.5-22

Φ observable modulus (equation 8.5-25); quantum yield

ψ dimensionless concentration, equation 8.5-8

SUPERSCRIPTS

$'$ quantity per tube (Section 21.5.4.1)

$\circ$ standard state (thermodynamic)

$\bullet$ free radical

$\ddagger$ (relating to) activated complex (in TST)

$*$ excited (energy) state

SUBSCRIPTS

a (of) adsorption

adj adjusted

app apparent

av average

A (of) species A

b (of) bubble, bulk

bi bimolecular

br bubble rise

B (of) species B, batch

B (of) bed

BR batch reactor

Subscripts

c	coil, core, at core surface, cloud, cross-sectional, cycle, cylinder
$calc$	calculated
cat	(of) catalyst
C	(of) species C, (of) cylinder
d	(of) decomposition, (of) desorption, down (time)
D	(of) species D
e	effective, (of) emulsion
eff	effective
$endo$	endothermic
eq	(at) equilibrium
ex	(at) exit of vessel
exo	exothermic
exp	experimental
E	(of) enzyme or species E
f	(of) film, fluid, forward (reaction), of formation
fl	fluidized (bed), (of) fluidizing
FP	flat plate
g	(in, of) gas (phase)
gen	generated
i	(of) species i, (of) initiation, (at) interface, interfacial, interstage, instantaneous, dummy index
in	(at) inlet of vessel
$init$	initial, initiation
int	intrinsic
$isoth$	isothermal
I	(of) initiator; (of) inhibitor
j	dummy index
k	dummy index, moment index ($k = 0, 1, 2, \ldots$)
ℓ	(in, of) liquid (phase)
m	mean, molar, melting
max	maximum
mf	(at) minimum-fluidized/-fluidization
min	minimum
M	(of) monomer; of tracer M
$nonseg$	nonsegregated (flow)
N	species N
N	stage N
o	feed, at inlet, initial, center-line (velocity), at center of particle
obs	observed
opt	optimal
out	(at) outlet of vessel
ov	overall
p	(of) particle, peripheral, (of) propagation
pa	packed (bed), Chapter 23
$proj$	projected
P	at constant pressure
P	(of) product, poison
PF	plug flow (reactor)
P_r	(of) r-mer (polymer)
r	radial (direction), reverse (reaction), (of) r-mer
ref	reference
rem	removal
rot	(of) rotation, rotational
R	(of) reaction, reacting species, recycled/recycle stream
s	(at) surface, sphere, superficial
seg	segregated (flow)
S	(of) substrate
S	(of) solid, surroundings

650 Nomenclature

	SCT	simple collision theory
	ST	stirred tank (CSTR)
	t	terminal, (of) termination, total, of tube(s), (with respect to dimensional) time
	tr	(of) translation, translational
	V	at constant volume; of void space
	uni	unimolecular
	vib	(of) vibration, vibrational
	w	(of) wake
	x	axial (direction)
	θ	(with respect to dimensionless) time
	$1, 2, \ldots$	species, stage, step, order $1, 2, \ldots$
	∞	(at) infinity or infinite time

OTHER

$\bar{}$	mean or (single) average of (quantity indicated)
$=$	double average
$\|x\|$	absolute value of x
$x!$	factorial x
$\propto$	varies as or is proportional to

DIMENSIONLESS GROUPS

Ha	Hatta number, equation 9.2-40 or -54 (see also equation 9.2-55)
Pe_L	Peclet number based on vessel length, uL/D_L
Re	vessel Reynolds number, $Du\rho/\mu = DG/\mu$
Re'	particle Reynolds number, $d'_p u\rho/\mu$
Sc	Schmidt number, $\mu/\rho D_m$ (D_m is diffusivity of species transferred)
Sh	particle Sherwood number, $k_{ig}d_p/D_m$ (D_m is diffusivity of species i)

GLOSSARY OF ABBREVIATIONS

BMF	backmix flow
BR	batch reactor
CRE	chemical reaction engineering
CSTR	continuous-flow stirred-tank reactor
CVD	chemical vapor deposition
DEA	diethanolamine
DPF	dispersed plug flow
exp	exponential
FBCR	fixed-bed catalytic reactor
gfc	gas-film control
IR	infrared (spectroscopy)
KL	Kunii-Levenspiel
ℓfc	liquid-film control
LF	laminar flow
LFR	laminar-flow reactor
LH	Langmuir-Hinshelwood
MEA	monoethanolamine
MMM	maximum-mixedness model
NMR	nuclear magnetic resonance
NQ	normalizing quantity (for intensive rate of reaction)
PF	plug flow
PFR	plug-flow reactor
ppmv	parts per million by volume
PSD	particle-size distribution
QSSA	quasi-steady-state approximation
rds	rate-determining step

RTD	residence-time distribution
SCM	shrinking-core model
SCT	simple collision theory
SFA	segregated-flow assumption
SFM	segregated-flow model
SPM	shrinking-particle model
SSH	stationary-state hypothesis
TEA	triethanolamine
TF	tubular flow
TIS	tanks-in-series (model)
TST	transition state theory
URM	uniform-reaction model

References

Alberty, R.A., and R.J. Silbey (1992), *Physical Chemistry*, Wiley, New York.

Aris, R. (1965), *Introduction to the Analysis of Chemical Reactors*, Prentice-Hall, Englewood Cliffs, NJ.

Aris, R. (1968), *Arch. Rational Mech. Anal.*, **27**, 356.

Arnold, L.B., and G.B. Kistiakowsky (1933), *J. Chem. Phys.*, **1**, 166.

Arrhenius, S. (1889), *Z. physik. Chem.*, **4**, 226.

Ashmore, P.G., M.G. Burnett, and B.J. Tyler (1962), *Trans. Faraday Soc.*, **58**, 685.

Asif, M., N. Kalogerakis, and L.A. Behie (1991), *AIChE J.*, **37**, 1825.

Atkinson, B., and F. Mavituna (1983), *Biochemical Engineering and Biotechnology Handbook*, The Nature Press, New York, 1983.

Austin, L.G., and P.L. Walker, Jr. (1963), *AIChE J.*, **9**, 203.

Baciocchi, E., A. Ciana, G. Illuminati, and C. Pasini (1965), *J. Am. Chem. Soc.*, **87**, 3953.

Bailey, J.E., and D.F. Ollis (1986), *Biochemical Engineering Fundamentals*, 2nd ed., McGraw-Hill, New York.

Bamford, C.H., and C.F.H. Tipper (eds.) (1969), *Comprehensive Chemical Kinetics*, Vol. 2, p. 197, Elsevier, Amsterdam.

Behie, L.A., and P.W.K. Kehoe (1973), *AIChE J.* **19**, 1070.

Benson, S.W. (1976), *Thermochemical Kinetics*, 2nd ed., Wiley, New York.

Billmeyer, F.W., Jr. (1984), *Textbook of Polymer Science*, 3rd ed., Wiley-Interscience, New York.

Bin, A.K. (1986), *Can. J. Chem. Eng.*, **64**, 854.

Blum, E.H., and R. Luus (1964), *Chem. Eng. Sci.*, **19**, 322.

Bodenstein, M., and Frl. Wachenheim (1918), *Z. Elektrochem.*, **24**, 183.

Bodenstein, M., and Frl. Lindner (1922), *Z. physik. Chem.*, **100**, 87.

Bodenstein, M., and Herr Ramstetter (1922), *Z. physik. Chem.*, **100**, 106.

Bodenstein, M., and S.C. Lind (1907), *Z. physik. Chem.*, **57**, 168.

Boikess, R.S., and E. Edelson (1981), *Chemical Principles*, 2nd ed., Harper and Row, New York.

Briggs, G.E., and J.B.S. Haldane (1925), *Biochem. J.*, **19**, 338.

Bromberg, J.P. (1984), *Physical Chemistry*, 2nd ed., Allyn and Bacon, Boston.

Brönsted, J.N., Martin Kilpatrick, and Mary Kilpatrick (1929), *J. Am. Chem. Soc.*, **51**, 428.

Brönsted, J.N., and E.A. Guggenheim (1927), *J. Am. Chem. Soc.*, **49**, 2554.

Buffham, B.A., and L.G. Gibilaro (1968), *AIChE J.*, **14**, 805.

Calderbank, P.H. (1958), *Trans. I. Chem. E.*, **36**, 443.

Calderbank, P.H., and M.B. Moo-Young (1961), *Chem. Eng. Sci.*, **16**, 39.

Caldini, C., F. Bonomi, P.G. Pifferi, G. Lanzarini, and Y.M. Galente (1994), *Enz. Microb. Technol.*, **16**, 286.

Carberry, J.J., and D. White (1969), *Ind. Eng. Chem.*, **61**, 27.

Cassano, A.E. (1980), *Chem. Eng. Educ.*, **14**, 14.

Chandrasekharan, K., and P.H. Calderbank (1981), *Chem. Eng. Sci.*, **36**, 819.

Chartrand, G., and C.M. Crowe (1969), *Can. J. Chem. Eng.*, **47**, 296.

Chase, A.M., H.C.V. Meier, and V.J. Menna, (1962), *J. Cellular Comp. Physiol.*, **59**, 1.

Chen, Ning Hsing (1983), *Process Reactor Design*, Allyn and Bacon, Boston.

Chong, A.O., and K.B. Sharpless (1977), *J. Org. Chem.*, **42**, 1587.

Cobranchi, D.P., and E.M. Eyring (1991), *J. Chem. Educ.*, **68**, 40.

Coulson, J.M., J.F. Richardson, and D.G. Peacock (1982), *Chemical Engineering*, 2nd ed., Vol. 3, Pergamon, Oxford.

Crocco, L., I. Glassman, and I.E. Smith (1959), *J. Chem. Phys.*, **31**, 506.

Czarnowski, J. (1992), *Int. J. Chem. Kinet.*, **24**, 679.

Danckwerts, P.V. (1953), *Chem. Eng. Sci.*, **2**, 1.

Danckwerts, P.V. (1970), *Gas-Liquid Reactions*, McGraw-Hill, New York.

Daniels, F., and E.H. Johnston (1921), *J. Am. Chem. Soc.*, **43**, 53.

Davidson, J.F., and D. Harrison (1963), *Fluidised Particles*, Cambridge U. Press, London.

Denbigh, K.G. (1981), *The Principles of Chemical Equilibrium*, 4th ed., Cambridge U. Press.

Denbigh, K.G., and J.C.R. Turner (1971), *Chemical Reactor Theory*, 2nd ed., Cambridge U. Press.

Denbigh, K.G., and J.C.R. Turner (1984), *Chemical Reactor Theory*, 3rd ed., Cambridge U. Press.

Dillon, R.T. (1932), *J. Am. Chem. Soc.*, **54**, 952.

Ding, Z-Y., S.N.V.K. Aki, and M.A. Abraham (1995), *Envir. Sci. Technol.*, **29**, 2748.

Dixon, D.C. (1970), *Chem. Eng. Sci.*, **25**, 337.

Dongaonkar, K.R., H.R.C. Pratt, and G.W. Stevens (1991), *AIChE J.*, **37**, 694.

Duan, K.J., J.S. Chen, and D.C. Sheu (1994), *Enz. Microb. Technol.*, **16**, 334.

Duduković, M.P. (1985), *Tracer Methods in Chemical Reactors: Techniques and Applications*, NATO Advanced Study Institute Conference, June 2–12, U. Western Ontario, London, Ontario.

Duo, W., K. Dam-Johansen, and K. Ostergaard (1992), *Can. J. Chem. Eng.*, **70**, 1014.

Eklund, R.B. (1956), *The Rate of Oxidation of Sulfur Dioxide with a Commercial Vanadium Catalyst*, Almgvist and Wiksell, Stockholm.

Elnashaie, S.S.E.H., and S.S. Elshishini (1993), *Modelling, Simulation and Optimization of Industrial Fixed Bed Catalytic Reactors*, Gordon and Breach, New York.

England, S.M. (1982), *J. Chem. Educ.*, **59**, 766, 860.

Ergun, S. (1952), *Chem. Eng. Progr.*, **48**., 93.

Evans, G.J. (1995), personal communication.

Evans, M.G., and M. Polanyi (1938), *Trans. Faraday Soc.*, **34**, 11.

Eyring, H., and F. Daniels (1930), *J. Am. Chem. Soc.*, **52**, 1473.

Fournier, M.-C., L. Falk, and J. Villermaux (1996), *Chem. Eng. Sci.*, **51**, 5053.

Froment, G.F., and K.B. Bischoff (1990), *Chemical Reactor Analysis and Design*, 2nd ed., Wiley, New York.

Geldart, D. (1973), *Powder Technol.*, **7**, 285.

Geldart, D. (1978), *Powder Technol.*, **19**, 133.

Geldart, D. (1986), *Gas Fluidization Technology*, Wiley, New York.

Giralt, F, and R. W. Missen (1974), *Can. J. Chem. Eng.*, **52**, 81.

Glasstone, S. (1946), *Textbook of Physical Chemistry*, 2nd ed., Van Nostrand, New York.

Grace, J.R. (1986), *Can. J. Chem. Eng.*, **64**, 353.

Grace, J.R., and H.I. de Lasa (1978), *AIChE J.*, **24**, 364.

Greig, J.D., and P.G. Hall (1967), *Trans. Faraday Soc.*, **63**, 655.

Guggenheim, E.A. (1967), *Thermodynamics*, 5th ed., North-Holland, Amsterdam.

Haider, A., and O. Levenspiel (1989), *Powder Technol.*, **58**, 63.

Harper, C., and J. Heicklen (1988), *Int. J. Chem. Kinet.*, **20**, 9.

Hassan, I.T., and C.W. Robinson (1977), *AIChE J.*, **23**, 48.

Hatta, S. (1932), *Technol. Repts. Tôhoku Imp. Univ.*, **10**, 119 (Danckwerts, 1970, p. 107).

Haupfear, E.A., and L.D. Schmidt (1994), *Chem. Eng. Sci.*, **49**, 2467.

Henri, V. (1902), *Compt. Rend. Acad. Sci., Paris*, **135**, 916.

Hikita, H., S. Asai, K. Tanigawa, K. Segawa, and M. Kitao (1980), *Chem. Eng. J.*, **20**, 59.

Hikita, H., S. Asai, K. Tanigawa, K. Segawa, and M. Kitao (1981), *Chem. Eng. J.*, **22**, 61.

Hill, C.G. (1977), *An Introduction to Chemical Engineering Kinetics and Reactor Design*, Wiley, New York.

Hinshelwood, C.N., and P.J. Askey (1927), *Proc. Roy. Soc.*, **A115**, 215.

Hinshelwood, C.N., and R.E. Burk (1925), *Trans. Chem. Soc.*, **127**, 1105.

Hinshelwood, C.N., and W.K. Hutchison (1926), *Proc. Roy. Soc.*, **A111**, 380.

Hlaváček, V., and M. Kubiček (1970), *Chem. Eng. Sci.*, **25**, 1537.

Horn, F. (1961), *Chem. Eng. Sci.*, **14**, 20.

Houser, T.J., and B.M.H. Lee (1967), *J. Phys. Chem.*, **71**, 3422.

Huang, I., and L. Dauerman (1969), *Ind. Eng. Chem. Prod. Res. Dev.*, **8**, 227.

Iwasaki, M., I. Furuoya, H. Sueyoshi, T. Shirasaki, and E. Echiogoya (1965), *Kagaku Kogaku* (Japan), **29**, 892.

JANAF (1986), *JANAF Thermochemical Tables*, 3rd ed., M.W. Chase, Jr., C.A. Davies, J.R. Downey, Jr., D.J. Frurip, R.A. McDonald, A.N. Syverud. Published by the American Chemical Society (Washington, DC) and the American Institute of Physics (New York) for the National Bureau of Standards.

Jensen, W.B. (1987), *J. Chem. Educ.*, **64**, 646.

Jensen-Holm, H., and G. Lyne (1994), *Sulphur* (232) May–June, p. 47.

Jenson, V.G., and G.V. Jeffreys (1963), *Mathematical Methods in Chemical Engineering*, Academic Press, London.

Kaštánek, F., J. Zahradník, J. Kratochvíl, and J. Černák, (1993), *Chemical Reactors for Gas-Liquid Systems*, Ellis Horwood, New York.

Kato, K., and C.Y. Wen (1969), *Chem. Eng. Sci.*, **24**, 1351.

Kato, K., T. Takarada, A. Koshinuma, I. Kanazawa, and T. Sugihara (1989), in *Fluidization VI*, J.R. Grace, L.W. Shemilt, and M.A. Bergougnou (eds.), Engineering Foundation, New York, pp. 351-358.

Kay, J.M., and R.M. Nedderman (1974), *An Introduction to Fluid Mechanics and Heat Transfer*, 3rd ed., Cambridge U. Press, London.

Kilty, P.A., and W.M.H. Sachtler (1974), *Cat. Rev.-Sci. Eng.*, **10**, 1.

Kirkpatrick, E.G. (1974), *Introductory Statistics and Probability for Engineering, Science, and Technology*, Prentice-Hall, Englewood Cliffs, NJ.

Knudsen, J.G., and D.L. Katz (1958), *Fluid Dynamics and Heat Transfer*, McGraw-Hill, New York.

Konocki, K.K. (1956), *Chem. Eng. (Japan)*, **21**, 408.

Kunii, D., and O. Levenspiel (1990), *Ind. Eng. Chem. Res.*, **29**, 1226.

Kunii, D., and O. Levenspiel (1991), *Fluidization Engineering*, 2nd. ed., Butterworth-Heinemann, Boston.

Laidler, K.J. (1965), *Chemical Kinetics*, 2nd ed., McGraw-Hill, New York.

Laidler, K.J., and M.T.H. Liu (1967), *Proc. Roy. Soc.*, **A297**, 365.

Laidler, K.J., and B.W. Wojciechowski (1961), *Proc. Roy. Soc.*, **A260**, 91.

Langmuir, I. (1908), *J. Am. Chem. Soc.*, **30**, 1742.

Le Chatelier, H. (1884), *Compt. Rend.*, **99**, 786.

Levenspiel, O. (1972), *Chemical Reaction Engineering*, 2nd ed., Wiley, New York.

Levenspiel, O., and W.K. Smith (1957), *Chem. Eng. Sci.*, **6**, 227.

Lewis, W.K., and W.G. Whitman (1924), *Ind. Eng. Chem.*, **16**, 1215.

Leyva-Ramos, R., and C.J. Geankopolis (1994), *Can. J. Chem. Eng.*, **72**, 262.

Liliedahl, T., K. Sjöström, and L-P. Wiktorsson (1991), *AIChE J.*, **37**, 1415.

Lindemann, F.A. (1922), *Trans. Faraday Soc.*, **17**, 599.

Lineweaver, H., and D. Burk (1934), *J. Am. Chem. Soc.*, **56**, 658.

Lister, M., and P. Rosenblum (1963), *Can. J. Chem.*, **41**, 3013.

Liu, B., P. Dagaut, R.E. Huie, and M.J. Kurylo (1990), *Int. J. Chem. Kinet.*, **22**, 711.

Lombardo, S. (1985), B.A.Sc. Thesis, University of Toronto.

Lynch, D.T. (1986), personal communication.

Lywood, W.J. (1996), in Twigg, 1996, Chapter 2.

Mahoney, J.A. (1974), *J. Catal.*, **32**, 247.

Mars, P., and D.W. van Krevelen (1954), *Special Supplement to Chem. Eng. Sci.*, **3**, 41.

Mathur, K.B., and N. Epstein (1974), *Spouted Beds*, Academic Press, New York.

Mauti, R. (1994), Ph.D. Thesis, University of Toronto.

Meister, D., T. Post, I.J. Dunn, and J.R. Bourne (1979), *Chem. Eng. Sci.*, **34**, 1367.

Michaelis, L., and M.L. Menten (1913), *Biochem. Z.*, **49**, 333.

Michell, B.J., and S.A. Miller (1962), *AIChE J.*, **8**, 262.

Millard, E.B. (1953), *Physical Chemistry for Colleges*, 7th ed., McGraw-Hill, New York.

Mills, Ian (1988) (chairman), *Quantities, Units and Symbols in Physical Chemistry*, (IUPAC Commission), Blackwell, Oxford.

Mims, C.A., R. Mauti, A.M. Dean, and K.D. Rose (1994), *J. Phys. Chem.*, **98**, 13357.

Mims, C.A., A.J. Jacobson, R.B. Hall, and J.T. Lewandowski, Jr. (1995), *J. Catal.*, **153**, 197.

Minhas, S., and J.J. Carberry (1969), *Br. Chem. Eng.*, **14**, 799.

Missen, R.W., and W.R. Smith (1989), *J. Chem. Educ.*, **66**, 217; erratum, **66**, 534.

Miyawaki, O., H. Tsujikawa, and Y. Uragughi (1975), *J. Chem. Eng. (Japan)*, **8**(1), 63.

Moelwyn-Hughes, E.A. (1957), *Physical Chemistry*, Pergamon, London.

Moore, W.J. (1972), *Physical Chemistry*, 4th ed., Prentice-Hall, Englewood Cliffs, N.

Morooka, S., K. Kusakabe, and A. Kobota (1989), in *Fluidization VI*, J.R. Grace, L.W. Shemilt, and M.A. Bergougnou (eds.), Engineering Foundation, New York, pp. 359–366.

Mueller, K.H., and W.D. Walters (1951), *J. Am. Chem. Soc.*, **73**, 1458.

Nauman, E.B., and B.A. Buffham (1983), *Mixing in Continuous Flow Systems*, Wiley, New York.

Nikolova, P., H. Goldman, and O.P. Ward (1995), *Can. J. Chem. Eng.*, **73**, 510.

Nowak, P., and J. Skrzypek (1989), *Chem. Eng. Sci.*, **44**, 2375.

Ogg, R.A. (1953), *J. Chem. Phys.*, **21**, 2079.

Otake, T., and E. Kunigita (1958), *Kagaku Kogaku*, **22**, 144 (Westerterp, et al. (1984), p. 225).

Ouellet, L., K.J. Laidler, and M.F. Morales (1952), *Arch. Biochem. Biophys.*, **39**, 37.

Parshall, G.W., and S.D. Ittel (1992), *Homogeneous Catalysis*, Wiley, New York.

Partaatmadja, L. (1998), B.A.Sc. Thesis, University of Toronto.

Partridge, B.A., and P.N. Rowe (1968), *Trans. IChemE*, **44**, 335.

Patel, R.D. (1992), *Int. J. Chem. Kinet.*, **24**, 541.

Perry, R.H., D.W. Green, and J.O. Maloney (eds.) (1984), *Perry's Chemical Engineers' Handbook*, 6th ed., McGraw-Hill, New York.

Peters, M.S., and K.D. Timmerhaus (1991), *Plant Design and Economics for Chemical Engineers*, 4th ed, McGraw-Hill, New York.

Petersen, E.E. (1965), *Chem. Eng. Sci.*, **20**, 587.

Pialis, P. (1996), M.A.Sc. Thesis, University of Toronto.

Pialis, P., and B.A. Saville (1998), *Enz. Microb. Technol.*, **22**, 261.

Porter, S.K. (1985), *J. Chem. Educ.*, **62**, 507.

Pudjiono, P.I., and N.S. Tavare (1993), *Can. J. Chem. Eng.*, **71**, 312.

Ranz, W.E., and W.R. Marshall (1952), *Chem. Eng. Progr.*, **48**, 173.

Rase, H.F. (1977), *Chemical Reactor Design for Process Plants; Volume Two: Case Studies and Design Data*, Wiley-Interscience, New York.

Rase, H.F. (1990), *Fixed-Bed Reactor Design and Diagnostics*, Butterworths, Boston.

Reid, R.C., J.M. Prausnitz, and B.E. Poling (1987), *The Properties of Gases and Liquids*, 4th ed., McGraw-Hill, New York.

Rhodes, M.J., S. Zhou, T. Hirama, and H. Cheng (1991), *AIChE J.*, **37**, 1450.

Rice, F.O., and K.F. Herzfeld (1934), *J. Am. Chem. Soc.*, **56**, 284.

Ring, Z.E., and R.W. Missen (1989), *AIChE J.*, **35**, 1821.

Rodriguez, F., and J.F. Tijero (1989), *Can. J. Chem. Eng.*, **67**, 963.

Rowe, P.N., and B.A. Partridge (1965), *Trans. IChemE*, **43**, 157.

Rowley, D., and H. Steiner (1951), *Disc. Faraday Soc.*, **10**, 198.

de Santiago, M., and I.H. Farena (1970), *Chem. Eng. Sci.*, **25**, 744.

Satterfield, C.N. (1991), *Heterogeneous Catalysis in Industrial Practice*, 2nd ed., McGraw-Hill, New York.

Saville, B.A., M.R. Gray, and Y.K. Tam (1986), *Can. J. Chem. Eng.*, **64**, 617.

Schoenfelder, H., J. Hinderer, J. Werther, and F.J. Keil (1994), *Chem. Eng. Sci.*, **49**, 5377.

Segraves, R.O., and D. Wickersham (1991), *Chem. Eng. Progr.*, **87**, June, p. 65.

Shah, Y.T., B.G. Kelkar, S.P. Godbole, and W.-D. Deckwer (1982), *AIChE J.*, **28**, 353.

Sheel, J.G.P., and C.M. Crowe (1969), *Can. J. Chem. Eng.*, **47**, 183.

Shen, H.K.Y. (1996), Ph.D. Thesis, University of Toronto.

Shetty, S.A., M.V. Kantak, and B.G. Kelkar (1992), *AIChE J.*, **38**, 1014.

Smith, H.A. (1939), *J. Am. Chem. Soc.*, **61**, 254.

Smith, W.R., and R.W. Missen (1979), *Chem. Eng. Educ.*, **13**, 26.

Smith, W.R., and R.W. Missen (1991), *Chemical Reaction Equilibrium Analysis*, Krieger, Malabar, FL.

Smith, W.R., and R.W. Missen (1997), *J. Chem. Educ.*, **74**, 1369.

Spencer, N.D., and C.J. Pereira (1987), *AIChE J.*, **33**, 1808.

Steinfeld, J.I., J.S. Francisco, and W.L. Hase (1989), *Chemical Kinetics and Dynamics*, Prentice-Hall, Englewood Cliffs, NJ.

Stokes, R.L., and E.B. Nauman (1970), *Can. J. Chem. Eng.*, **48**, 723.

Streitwieser, A., Jr., and C.H. Heathcock (1981), *Introduction to Organic Chemistry*, 2nd ed., Macmillan, New York.

Stull, D.R., E.F. Westrum, Jr., and G.C. Sinke (1969), *The Chemical Thermodynamics of Organic Compounds*, Wiley-Interscience, New York.

Subbotin, A.I., V.N. Antonov, and V.S. Etlis (1966), *Kinetics and Catalysis*, **7**, 183.

Svirbely, W.J., and J.F. Roth (1953), *J. Am. Chem. Soc.*, **75**, 3106.

Swabb, E.A., and B.C. Gates (1972), *Ind. Eng. Chem. Fundam.*, **11**, 540.

Themelis, N.J., and G.M. Freeman (1983), paper presented at A.I.M.E. Annual Meeting, Atlanta, Georgia.

Thurier, R.T. (1977), M.A.Sc. Thesis, University of Toronto.

Trickett, A.A. (1992), *Sulphur* (219), March–April, p. 26.

Twigg, M.V. (ed.), (1996), *Catalyst Handbook*, 2nd ed., second impression with revisions, Manson, London.

Vaidyanathan, K., and L.K. Doraiswamy (1968), *Chem. Eng. Sci.*, **23**, 537.

van Deemter, J. (1961), *Chem. Eng. Sci.*, **13**, 143.

van Krevelen, D.W., and P.J. Hoftijzer (1948), *Rec. Trav. Chim. Pays-Bas*, **67**, 563.

Villermaux, J., and W.P.M. van Swaaij (1969), *Chem. Eng. Sci.*, **24**, 1097.

Waldie, B. (1992), *Can. J. Chem. Eng.*, **70**, 873.

Wasserman, A. (1936), *J. Chem. Soc.*, 1028.

Wehner, J.F., and R.H. Wilhelm (1956), *Chem. Eng. Sci.*, **6**, 89.

Weisz, P.B., and J.S. Hicks (1962), *Chem. Eng. Sci.*, **17**, 265.

Weisz, P.B., and C.D. Prater (1954), *Adv. Catal.*, **6**, 143.

Weller, S.W. (1994), *Chem. Eng. Educ.*, **28**, 262.

Wenner, R.R., and E.C. Dybdal (1948), *Chem. Eng. Progr.*, **44**, 275.

Westerterp, K.R., W.P.M. van Swaaij, and A.A.C.M. Beenackers (1984), *Chemical Reactor Design and Operation*, Wiley, New York.

Whitman, W.G. (1923), *Chem. and Met. Eng.*, **29**, 146.

Wilkinson, F. (1980), *Chemical Kinetics and Reaction Mechanisms*, Van Nostrand Reinhold, New York.

Williams, S. (1996), M.A.Sc. Thesis, University of Toronto.

Wojciechowski, B.W., and K.J. Laidler (1960), *Can. J. Chem.*, **38**, 1027.

Wylie, C.R. (1960), *Advanced Engineering Mathematics*, 2nd ed., McGraw-Hill, New York.

Yates, J.G., (1983), *Fundamentals of Fluidized-Bed Chemical Processes*, Butterworths, London.

Yates, J.G., and P.N. Rowe (1977), *Trans. IChemE*, **55**, 137.

Zenz, F.A. (1972), *Chem. Eng.*, **79**, Nov. 13, p. 120.

Zoulalian, A., and J. Villermaux (1970), *Chem. Eng. J.*, **1**(1), 76.

Zwietering, Th.N. (1959), *Chem. Eng. Sci.*, **11**, 1.

Author Index

A

Abraham, M.A., 653
Aki, S.N.V.K., 653
Alberty, R.A., 83, 652
Amundson, N.R., 625
Anthony, R.G., 625
Antonov, V.N., 655
Aris, R., 95, 406, 534, 625, 652
Arnold, L.B., 187, 652
Arrhenius, S., 44, 652
Asai, S., 653
Ashmore, P.G., 83, 652
Asif, M., 494, 652
Askey, P.J., 81, 152, 653
Atkinson, B., 261, 625, 652
Austin, L.G., 217, 652

B

Baciocchi, E., 84, 652
Bailey, J.E., 261, 625, 652
Bamford, C.H., 79, 652
Beenackers, A.A.C.M., 626, 655
Behie, L.A., 596, 652
Benson, S.W., 143, 626, 652
Bergougnou, M.A., 653, 654
Billmeyer, F.W., Jr., 166, 652
Bin, A.K., 575, 652
Bischoff, K.B., 35, 233, 314, 376, 534, 575, 599, 606, 609, 610, 625, 653
Blum, E.H., 95, 652
Bodenstein, M., 84, 160, 161, 652
Boikess, R.S., 314, 652
Bonomi, F., 652
Boudart, M., 626
Bourne, J.R., 654
Briggs, G.E., 264, 266, 652
Bromberg, J.P., 277, 652
Brönsted, J.N., 70, 71, 219, 361, 652
Buffham, B.A., 476, 477, 652, 654
Burk, D., 268, 654
Burk, R.E., 221, 653
Burnett, M.G., 652
Butt, J.B., 625

C

Calderbank, P.H., 609, 615, 616, 652
Caldini, C., 316, 652
Calo, J.M., 626

Carberry, J.J., 545, 546, 625, 652, 654
Cassano, A.E., 4, 652
Černák, J., 625, 653
Chandrasekharan, K., 615, 652
Chartrand, G., 534, 652
Chase, A.M., 268, 652
Chase, M.W., 653
Chaudhari, R.V., 625
Chen, J.S., 653
Chen, Ning Hsing, 311, 625, 652
Cheng, H., 654
Chong, A.O., 177, 652
Ciana, A., 652
Cobranchi, D.P., 23, 652
Cooper, A.R., 625
Coulson, J.M., 109, 652
Crocco, L., 171, 652
Crowe, C.M., 531, 534, 652, 655
Czarnowski, J., 84, 85, 652

D

Dagaut, P., 654
Dam-Johansen, K., 653
Danckwerts, P.V., 253, 319, 500, 599, 620, 625, 652
Daniels, F., 83, 112, 652, 653
Dauerman, L., 82, 653
Davidson, J.F., 569, 579, 580, 653
Davies, C.A., 653
Dean, A.M., 175, 654
Deckwer, W.D., 655
Denbigh, K.G., 22, 44, 95, 141, 168, 293, 391, 445, 625, 653
Dillon, R.T., 22, 62, 653
Ding, Z-Y., 449, 653
Dixon, D.C., 4, 653
Djéga-Mariadassou, G., 626
Dongaonkar, K.R., 493, 653
Doraiswamy, L.K., 113, 625, 655
Downey, J.R., Jr., 653
Duan, K.J., 278, 653
Duduković, M.P., 455, 653
Dunn, I.J., 654
Duo, W., 113, 653
Dybdal, E.C., 531, 655

E

Echiogoya, E., 653
Edelson, E., 314, 652

Eklund, R.B., 223, 521, 523, 546, 653
Elnashaie, S.S.E.H., 531, 625, 653
Elshishini, S.S., 531, 625, 653
England, S.M., 294, 295, 653
Epstein, N., 571, 654
Ergun, S., 517, 653
Etlis, V.S., 655
Evans, G.J., 23, 653
Evans, M.G., 123, 653
Eyring, E.M., 23, 652
Eyring, H., 83, 626, 653

F

Falk, L., 653
Farena, I.H., 259, 655
Fogler, H.S., 625
Fournier, M.C., 653
Francisco, J.S., 626, 655
Freeman, G.M., 290, 655
Froment, G.F., 35, 233, 314, 376, 534, 575, 599, 606, 609, 610, 625, 653
Frurip, D.J., 653
Furuoya, I., 653

G

Galente, Y.M., 652
Gates, B.C., 221, 626, 655
Geankopolis, C.J., 223, 654
Geldart, D., 579, 653
Gianetto, A., 625
Gibilaro, L.G., 476, 652
Giralt, F., 110, 653
Glassman, I., 652
Glasstone, S., 82, 653
Godbole, S.P., 655
Goldman, H., 654
Grace, J.R., 579, 596, 653, 654
Gray, M.R., 655
Green, D.W., 654
Greig, J.D., 84, 653
Guggenheim, E.A., 20, 219, 652, 653

H

Haider, A., 578, 653
Haldane, J.B.S., 264, 266, 652
Hall, P.G., 84, 653
Hall, R.B., 654
Harper, C., 80, 653
Harrison, D., 569, 579, 580, 653
Hase, W.L., 626, 655
Hassan, I.T., 615, 653
Hatta, S., 250, 653
Haupfear, E.A., 259, 653
Heathcock, C.H., 390, 655
Heicklen, J., 80, 653
Henri, V., 264, 653
Herzfeld, K.F., 116, 655
Hicks, J.S., 211, 212, 213, 222, 655

Hikita, H., 609, 653
Hill, C.G., 166, 172, 625, 653
Hinderer, J., 655
Hinshelwood, C.N., 71, 81, 152, 221, 653
Hirama, T., 654
Hlaváček, V., 212, 653
Hoftijzer, P.J., 251, 259, 655
Holland, C.D., 625
Horák, J., 625
Horn, F., 535, 653
Hougen, O.A., 625
Houser, T.J., 159, 653
Huang, I., 82, 653
Huie, R.E., 654
Hutchison, W.K., 71, 653

I

Illuminati, G., 652
Ittel, S.D., 186, 626, 654
Iwasaki, M., 597, 653

J

Jacobson, A.J., 654
Jeffreys, G.V., 473, 625, 653
Jensen, W.B., 22, 653
Jensen-Holm, H., 223, 653
Jenson, V.G., 473, 653
Johnston, E.H., 112, 652

K

Kalogerakis, N., 652
Kanazawa, I., 653
Kantak, M.V., 655
Kaštánek, F., 599, 600, 625, 653
Kato, K., 579, 597, 653
Katz, D.L., 370, 653
Kay, J.M., 37, 653
Kehoe, P.W.K., 596, 652
Keil, F.J., 655
Kelkar, B.G., 655
Kilpatrick, Martin, 652
Kilpatrick, Mary, 652
Kilty, P.A., 653
Kirkpatrick, E.G., 323, 653
Kistiakowsky, G.B., 187, 652
Kitao, M., 653
Knudsen, J.G., 370, 653
Kobota, A., 654
Konocki, K.K., 535, 654
Koshinuma, A., 653
Kramers, H., 625
Kratochvíl, J., 625, 653
Kubiček, M., 212, 653
Kunigita, E., 487, 488, 654
Kunii, D., 290, 569, 570, 579, 580, 581, 582, 583, 590, 592, 593, 595, 654
Kurylo, M.J., 654
Kusakabe, K., 654

L

Laidler, K.J., 172, 175, 626, 654, 655
Langmuir, I., 500, 654
Lanzarini, G., 652
Lapidus, L., 625
de Lasa, H.I., 596, 653
Le Chatelier, H., 514, 654
Lee, B.M.H., 159, 653
Levenspiel, O., 290, 344, 406, 446, 485, 486, 488, 492, 534, 563, 565, 567, 569, 570, 578, 579, 580, 581, 582, 583, 590, 592, 593, 595, 625, 653, 654
Lewandowski, J.T., Jr., 654
Lewis, W.K., 240, 654
Leyva-Ramos, R., 223, 654
Liliedahl, T., 493, 654
Lin, S.H., 626
Lin, S.M., 626
Lind, S.C., 160, 161, 652
Lindemann, F.A., 135, 654
Lindner, Frl., 652
Lineweaver, H., 267, 654
Lister, M., 79, 654
Liu, B., 85, 654
Liu, M.T.H., 172, 654
Lombardo, S., 535, 654
Luus, R., 95, 652
Lynch, D.T., 347, 391, 654
Lyne, G., 223, 653
Lywood, W.J., 516, 654

M

McDonald, R.A., 653
Mahoney, J.A., 7, 654
Maloney, J.O., 654
Mars, P., 197, 654
Marshall, W.R., 236, 654
Mathur, K.B., 571, 654
Mauti, R., 112, 654
Mavituna, F., 261, 652
Meier, H.C.V., 652
Meister, D., 615, 654
Menna, V.J., 652
Menten, M.L., 264, 654
Michaelis, L., 264, 654
Michell, B.J., 615, 654
Millard, E.B., 82, 654
Miller, S.A., 615, 654
Mills, Ian, 4, 654
Mims, C.A., 164, 165, 654
Minhas, S., 545, 654
Missen, R.W., 9, 10, 110, 520, 619, 653, 654, 655
Miyawaki, O., 446, 654
Moelwyn-Hughes, E.A., 127, 128, 133, 654
Moore, W.J., 79, 84, 654
Moo-Young, M.B., 609, 652
Morales, M.F., 654
Morooka, S., 597, 654
Mueller, K.H., 82, 654

N

Nauman, E.B., 477, 479, 625, 654, 655
Nedderman, R.M., 37, 653
Nikolova, P., 277, 654
Nowak, P., 86, 451, 654

O

Ogg, R.A., 155, 654
Ollis, D.F., 261, 625, 652
Ostergaard, K., 653
Otake, T., 487, 488, 654
Ouellet, L., 276, 654

P

Parshall, G.W., 186, 626, 654
Partaatmadja, L., 598, 654
Partridge, B.A., 579, 581, 654, 655
Pasek, J., 625
Pasini, C., 652
Patel, R.D., 83, 389, 654
Peacock, D.G., 652
Pereira, C.J., 90, 108, 444, 452, 655
Perlmutter, D.D., 625
Perry, R.H., 283, 284, 336, 388, 623, 654
Peters, M.S., 283, 654
Petersen, E.E., 208, 625, 654
Pialis, P., 262, 277, 654
Pifferi, P.G., 652
Polanyi, M., 123, 653
Poling, B.E., 654
Porter, S.K., 23, 654
Post, T., 654
Prater, C.D., 208, 222, 655
Pratt, H.R.C., 653
Prausnitz, J.M., 654
Pudjiono, P.I., 494, 654

R

Ramachandran, P.A., 625
Ramstetter, Herr, 652
Ranz, W.E., 236, 654
Rase, H.F., 550, 625, 654
Reid, R.C., 200, 654
Rhodes, M.J., 493, 654
Rice, F.O., 116, 655
Richardson, J.F., 652
Ring, Z.E., 619, 655
Robinson, C.W., 615, 653
Rodrigues, A.E., 626
Rodriguez, F., 113, 430, 447, 451, 655
Rose, K.D., 654
Rose, L.M., 626
Rosenblum, P., 79, 654
Roth, J.F., 82, 655

Rowe, P.N., 579, 581, 595, 654, 655
Rowley, D., 79, 80, 153, 362, 655

S

Sachtler, W.M.H., 653
de Santiago, M., 259, 655
Satterfield, C.N., 200, 655
Saville, B.A., 277, 452, 654, 655
Schmidt, L.D., 259, 626, 653
Schoenfelder, H., 450, 655
Segawa, K., 653
Segraves, R.O., 22, 655
Shah, Y.T., 608, 609, 626, 655
Sharma, M.M., 625
Sharpless, K.B., 177, 652
Sheel, J.G.P., 531, 655
Shemilt, L.W., 653, 654
Shen, H.K.Y., 552, 655
Shetty, S.A., 493, 655
Sheu, D.C., 653
Shirasaki, T., 653
Silbey, R.J., 83, 652
Silveston, P.L., 625
Sinke, G.C., 655
Sjöström, K., 654
Skrzypek, J., 86, 451, 654
Smith, H.A., 218, 219, 655
Smith, I.E., 652
Smith, J.M., 626
Smith, W.K., 486, 488, 492, 654
Smith, W.R., 9, 10, 520, 654, 655
Somorjai, G.A., 626
Spencer, N.D., 90, 108, 444, 452, 655
Steiner, H., 79, 80, 153, 362, 655
Steinfeld, J.I., 123, 132, 134, 143, 145, 626, 655
Stevens, G.W., 653
Stokes, R.L., 477, 479, 655
Streitwieser, A. Jr., 390, 655
Stull, D.R., 362, 445, 452, 522, 655
Subbotin, A.I., 391, 655
Sueyoshi, H., 653
Sugihara, T., 653
Svirbely, W.J., 82, 655
Swabb, E.A., 221, 655
Sweed, N.H., 626
Syverud, A.N., 653

T

Takarada, T., 653
Tam, Y.K., 655
Tanigawa, K., 653
Tarhan, M. Orhan, 626
Tavare, N.S., 494, 654
Themelis, N.J., 290, 655
Thurier, R.T., 40, 491, 655
Tijero, J.F., 113, 430, 447, 451, 655
Timmerhaus, K.D., 283, 654
Tipper, C.F.H., 79, 652

Trickett, A.A., 655
Tsujikawa, H., 654
Turner, J.C.R., 168, 293, 391, 625, 653
Twigg, M.V., 285, 286, 287, 288, 289, 654, 655
Tyler, B.J., 652

U

Ulrich, G.D., 626
Uragughi, Y., 654
Vaidyanathan, K., 113, 655

V

van Deemter, J., 579, 655
van Krevelen, D.W., 197, 251, 259, 654, 655
van Swaaij, W.P.M., 488, 626, 655
Varma, Arvind, 625
Villermaux, J., 488, 510, 653, 655

W

Wachenheim, Frl., 652
Walas, S., 626
Waldie, B., 493, 655
Walker, P.L., Jr., 217, 652
Walters, W.D., 82, 654
Ward, O.P., 654
Wasserman, A., 315, 655
Watson, K.M., 625
Wehner, J.F., 500, 655
Weisz, P.B., 208, 211, 212, 213, 222, 655
Weller, S.W., 484, 500, 655
Wen, C.Y., 579, 653
Wenner, R.R., 531, 655
Werther, J., 655
Westerterp, K.R., 625, 626, 655
Westrum, E.F., Jr., 655
White, D., 546, 652
Whitman, W.G., 240, 654, 655
Wickersham, D., 22, 655
Wiktorsson, L.P., 654
Wilhelm, R.H., 500, 655
Wilkinson, F., 155, 184, 186, 626, 655
Williams, S., 109, 655
Wojciechowski, B.W., 175, 626, 654, 655
Wylie, C.R., 329, 655

Y

Yates, J.G., 569, 570, 579, 595, 654, 655

Z

Zahradník, J., 625, 653
Zenz, F.A., 608, 655
Zhou, S., 654
Zoulalian, A., 510, 655
Zwietering, Th.N., 455, 495, 502, 503, 504, 508, 655

Subject Index

A

Acetaldehyde, CH_3CHO, 71, 82, 172, 220, 346-347, 361
Acid-base catalysis, 183–186, 218–219
Acrylonitrile, C_3H_3N, 572, 587–589, 597
Activation energy, 44, 45, 57, 65, 145
 and strong pore-diffusion resistance, 209–210
 in terms of partial pressure, 68–69
Activation in enzyme reactions, 269–276, 278
Activator(s), 272, 278
Adsorption, 119, 148–149, 192–194, 215, 223. *See also* Chemisorption; Desorption
Age (of an element of fluid), 318, 495
Age-distribution function(s), 21, 319
 relationships among, 322, 323t
Ammonia, NH_3, 152, 221, 292, 293, 366, 446, 512, 572, 587, 597, 600
 synthesis, 22, 100, 116, 176, 214, 286, 287, 289, 513
 reactors, 286, 287, 288
Anthracene, $C_{14}H_{10}$, 113, 430, 431, 447–448, 451
Arrhenius equation, 44, 65, 69, 79, 145, 157, 445
Arrhenius parameters, 44, 64, 79–80, 115, 153
 in terms of partial pressure, 68–69
Ash layer, 229, 230, 234, 237, 258, 260
 diffusion, 233, 236, 257, 258, 553, 564
  control, 233, 234, 257, 557, 558, 560, 562, 563, 564, 565, 567, 568
 resistance, 237, 239, 258
Autocatalysis, 78, 178, 187–191, 383, 402
Autocatalytic reaction(s), 187–191, 384, 385, 386, 404, 416–418, 419
Axial-dispersion model, *see* Dispersed plug flow (DPF) model
Axial-dispersion reactor model, 499–500, 509, 510, 511. *See also* Dispersed plug flow (DPF) model

B

Backmix flow (BMF), 25, 29, 284, 317, 318, 325, 326, 327, 332, 333, 334, 335, 453, 559–563, 574, 580, 600, 601, 602, 608, 614
 E, 325–326, 332, 333, 334
 F, 325, 326–327, 332

 H, W, 325, 332, 334
 I, 325, 332, 333, 334
Backmixing, degree of, 318, 380
Balance equation(s), 16–17, 21, 282, 295. *See also* Continuity equation(s); Element balance(s); Energy balance(s); Enthalpy; Mass balance; Material balance; Momentum balance
Batch operation, comparison with continuous operation, 295
Batch reactor(s), 6, 21, 25, 26–29, 42, 283–284, 294–309, 313, 314, 315, 407
 adiabatic operation, 304–307
 comparison with CSTR, 402–404, 419–420
 comparison with PFR, 404–405, 406, 408, 419
 cycle time, 297, 307
 design equations, 296–309
 determination of rate parameters with, 49–54, 57
 down-time, 296, 297, 308
 energy balance, 297–299, 304
 general features, 26, 294
 isothermal operation, 300–304
 material balance, 27, 28
 nonisothermal operation, 304–307
 optimal performance for maximum production rate, 307–309, 315
 and parallel reactions, 101–103, 110, 427, 428, 446, 449
 rate of energy generation or loss by reaction, 298
 rate of production, 297, 300
 and reversible reactions, 97–98, 109, 112, 445, 446
 and series reactions, 103–106, 111, 112, 429–430, 431, 447, 451
 time of reaction, 296–297, 300, 301–302, 305, 307, 313, 314, 315
 uses, 26, 294
 volume, 296, 297, 300–301, 315
Bed, 280. *See also* Reactor(s)
 density, 516, 522, 547, 557, 575
 depth, 516, 517, 566, 574
 diameter, 516, 518, 566

---

t refers to table

662 Subject Index

Bed (*continued*)
  mean residence time for flow through, 546
  voidage, 516, 517, 526, 547
Benzene, C_6H_6, 113–114, 513, 550–551, 600
Bimolecular reaction, 116, 118, 125, 129–134, 137–139, 145, 153, 196–197
Biochemical reactions, 261, 263
Boltzmann, 127–128, 132, 140
Bubble-column reactor(s), *see* Reactors for gas-liquid reactions

C

Catalysis, 5, 15, 21, 176–214, 512. *See also* Acid-base catalysis; Autocatalysis; Enzyme catalysis; Heterogeneous catalysis; Homogeneous catalysis; Molecular catalysis; Surface catalysis
Catalyst, 1, 4, 155, 176–177, 181, 427
  deactivating, 310
  deactivation and regeneration, 214–218, 512, 522, 552, 569, 573
  fouling, 214–215, 216
  maximum allowable T, 522, 529
  poisoning, 215, 217
  sintering, 215–216, 217
Catalyst particle(s), 516, 601, 602. *See also* Particle(s)
  concentration gradient in porous, 198–199
  kinetics in porous, 198–214
  temperature gradient in porous, 198–199, 210–212
Catalytic (surface) site(s), 179–180, 191, 197–198, 215
Catalytic reaction(s), 155, 176
Cell-growth kinetics, 261
Chain carrier(s), 158, 162
Chain length, 158, 160, 163
Chain reaction(s), 157–162
Chemical equation(s), 7–13, 17, 22–23, 90, 93, 103, 113
  canonical form, 11, 12, 13, 90, 93
  number of linearly independent, 10, 17, 22
  procedure for generating, 9–13
Chemical kinetics, 1, 2, 21. *See also* Kinetics; Kinetics schemes
  and thermodynamics, 14–15
  and transport processes, 15
Chemical reaction engineering, 1, 2, 15–19, 21, 279–292
Chemical reactors, *see* Reactors
Chemical vapor deposition (CVD), 138, 224, 256, 259, 310, 311, 552, 569, 597
Chemisorption, 194, 215, 216
Closed vessel, 318, 325, 335, 337, 365, 393, 527
Cold-shot cooling, *see* Fixed-bed catalytic reactor(s); Quench cooling
Collision diameter, 129, 200
Collision frequency, 129–131
Collision cross-section, 129, 130
Collision theory, 115, 128
  simple collision theory (SCT), 128–139, 145–146
    and activation energy, 133
    bimolecular combination reactions, 137–139
    bimolecular reactions, 129–134
    comparison with TST, 145–146
    pre-exponential factor, 133
    rate constant, 133, 134
    rate expression, 132, 133
    steric (orientation) factor, 131, 132, 153
    termolecular reactions, 137–139
    unimolecular reactions, 134–137
Complex reactions, *see* Complex system(s)
Complex system(s), 4, 7, 9, 13, 17, 21, 87–108, 545, 584, 602
  energy balance, 444–445
  examples, 87–89
  in a fluidized-bed reactor, 589–592, 598
  reaction stoichiometry, 90
  stoichiometric table, 93–94
Component (species), 8, 9, 11, 12, 13, 22
Computer software, 21–22, 282. *See also* E-Z Solve; Maple; Mathematica
Concentration, molar (molarity), 28, 31
  and partial pressure, 66, 561
Continuity equation(s), 202, 221, 227, 229, 230, 232, 247, 250, 255, 523, 525, 527, 528, 545, 586, 590, 591, 592, 594, 603, 605, 608, 614
Continuous operation, comparison with batch operation, 295
Control surface, 16, 212
Control volume, 16, 17, 27, 34, 227, 297, 337, 367, 394, 523, 585, 603, 605
Conversion, *see* Conversion, fractional
Conversion, fractional (of a reactant), 27, 31, 91, 107, 108, 513, 516, 574, 590, 591, 598
  average (mean), 555, 556, 565, 568
CSTR (continuous stirred-tank reactor), 25, 29–32, 284, 318, 335–361, 559. *See also* Backmix flow (BMF); Stirred tank reactor(s)
  adiabatic operation, 339, 350, 351
  for autocatalytic reaction, 416–418, 419
  autothermal operation, 350, 353, 362
  in combination with PFR, 413–418
  comparison with BR, 402–404, 419–420
  comparison with LFR, 406–408
  comparison with PFR, 404, 405–408, 419, 420
  constant-density system, 339–344
  determination of rate parameters with, 54–55, 57
  energy balance, 338–339
  $E(t)$, 343

general features, 29–31, 335
mean residence time, 30, 335, 337, 340
material balance, 31, 337, 355
multiple stationary states, 339, 347–354
multistage, 29, 30, 335, 336, 355–361
    graphical solution, 356–358
    optimal operation, 356, 358–361
    parallel arrangement, 409–410, 412–413
    series arrangement, 355–361, 410–413, 420, 423–426. *See also* Tanks-in-series (TIS) model
and parallel reactions, 109, 427, 428, 447, 448, 451–452, 505, 508
performance of, and degree of segregation, 344, 345t, 364
and polymerization reactions, 443–444, 452
and recycling, 364, 380
and reversible reactions, 109, 423–426, 433–434, 446, 449
and series reactions, 103, 111, 430–431, 437, 438, 439, 440, 441, 447, 448
and SFM, 343–344
space time, 30, 337, 340
unsteady-state operation, 341–343
uses, 29, 336
variable-density system, 344–347
volume, 337, 340, 341

D

Delta function, *see* Dirac delta function
Descartes rule of signs, the, 347, 349, 518
Design, 1, 2, 15, 16. *See also* Mechanical design; Process design
Desorption, 148, 192, 193, 194, 216
Diethanolamine (DEA), 239, 600
Diffusion, 2, 15, 17, 199–200, 202. *See also* Ash layer diffusion; Pore diffusion resistance
    coefficient, *see* Diffusivity
    Fick's law, 200, 212, 227, 240, 244
    Knudsen, 200, 209
    molecular, 200, 209, 240, 583
    surface, 200
Diffusivity, 200
    effective, 200, 227, 526
    "true" units of, 201
    molecular, 200, 236, 240, 583
    relation to mass transfer coefficients, 240, 241, 244
Dilatometer, 48
Dimension(s), 19–20. *See also* Vessel dimensions.
Dimethyl ether, 61, 81, 152, 171, 390
Dirac delta function δ, 328–329, 330, 365, 375, 556
Dispersed plug flow (DPF) model, 483–489, 495. *See also* Axial dispersion reactor model
    comparison with TIS model, 490

continuity equation, 483–489
    solutions, 485–487, 488t
        comparison, 487, 488t, 489t
        moments, 488t, 489t
    determination of Pe_L, 487–489, 492, 493
Dispersion, 484, 502, 524, 527. *See also* Dispersed plug flow (DPF) model
Dispersion coefficient, *see* Diffusivity

E

E, exit-age residence-time distribution function, 319–321, 325, 332, 333, 334, 555, 560
    for BMF, 325–326, 332
    for CSTR + PFR series combination, 414–415
    definitions of $E(t)$ and $E(\theta)$, 319, 320
    for LF, 325, 330–332
    measurement, 458–471
        normalized response, **C**, 458–459, 473, 490, 491, 501
        from pulse input, 458–462, 491
        with reactive tracer, 466–468
        from step input, 462–466
        with tailing, 468–471
        with time delay, 466
    for PF, 325, 327–329, 332
    relation between $E(t)$ and $E(\theta)$, 320
    relation between $E(t)$ and $F(t)$, 321–322
    relation to other age-distribution functions, 332t
    for TIS model, 471–476, 477
    for 2 CSTRs in series, 411–413
Earliness of mixing, *see* Mixing
Effectiveness factor, particle, 201–214, 217, 525, 544, 545, 550, 551, 574
    as a function of Thiele modulus (graph), 205
    definition, 201
    dependence on T, 210–213
    effect of order of reaction on, 207
    effect of particle shape on, 205–207
    for flat-plate geometry, 201–205
    overall, 201, 212–214, 525, 544
    for spherical particle, 221, 222
Elementary chemical reaction(s), 115, 145, 152
    definition, 116
    endoergic, 121, 123
    exoergic, 121, 123
    form of rate law, 117
    requirements for, 120
    on surfaces, 119
    types, 117–119
Element balance(s), 7, 9, 13, 14, 16
Element of fluid, 2, 16, 25, 317, 365, 375, 495, 555
Elutriation, 559, 567, 570, 575, 577, 584
Energy of activation, *see* Activation energy
Energy balance(s)/equation(s), 2, 211, 228, 282, 296, 445, 523. *See also* Batch

Energy balance(s)/equation(s) (*continued*)
reactor(s); Complex system(s); CSTR;
Enthalpy; Plug-flow reactor(s)
Energy barrier, 121, 122, 123, 124, 125, 126, 131, 140
Energy in molecules, 120–128
  electronic, 125–126, 127
  kinetic, 120, 126–128, 152
  modes, 126, 143, 144, 145
  potential, 120–126
    diagram, 120, 122, 124, 126, 128, 140
Energy transfer, molecular, 134
Enhancement factor, 246–247, 259, 260, 620, 621
  definition, 246
  for fast first-order or pseudo-first-order reaction, 251, 252, 254, 259
  for fast second-order reaction, 251, 252, 254, 259
  for instantaneous reaction, 247, 251, 252, 254, 258
  and liquid film, 255
Enthalpy, 17
  of activation, 141, 142, 153
  balance(s), 228, 298, 338, 368, 526
  change, dependence on T, 299, 445
  consumption or loss or removal, 211, 338, 339, 353
  generation, 211, 338, 339, 353, 369
  input or output by flow, 338, 339, 368
  of reaction, 44, 228, 298, 445
Entrainment, 559, 560, 567, 570, 575, 577, 578, 584
Entropy of activation, 141, 142, 143, 145
Enzyme(s), 261, 262, 263, 264, 270, 602
  activity, 263, 264, 270, 273
  binary complex, 270, 273
  coenzyme, 261, 270
  cofactor, 261, 262, 264, 270
  complex, dissociation constant, 264, 270, 274, 275
  - substrate complex/system, 270
  ternary complex, 270, 273
Enzyme catalysis, 178, 186, 187, 261–264
  external effects, 270
  substrate effects, 270–272
Enzyme-catalyzed reactions, 262, 263, 266, 269
Enzyme kinetics, 21, 261–276
  maximum rate, V_{max}, 265, 267–269, 270, 271, 277, 278
  models, 264–267
  rate law, 265, 266, 267, 269, 271, 272, 274, 275, 276
  substrate effects, 270–272
    multiple-substrate inhibition, 271–272
    single-substrate inhibition, 270–271
Equation of state, 6, 28, 296, 297, 302, 607
  ideal-gas, 36, 302, 607

Equilibrium, chemical/reaction, 1, 2, 9, 15, 136, 282, 514
  considerations, 293, 520–521, 522, 547, 548
  limitations, 16, 513, 516
Equilibrium constant:
  dependence on T, 44, 157, 520
  for formation of transition state, 139, 140, 141, 143
Ergun equation, 517, 574, 575
Ethane, C_2H_6:
  dehydrogenation, 35–36, 154, 286, 366, 376–377, 379–380
  mechanism, 116, 124–125, 137, 138–139, 158, 165, 172, 173–175
Ethyl acetate, $C_4H_8O_2$, 82, 218, 314, 364, 390
Ethyl alcohol, C_2H_5OH, 78–79, 88, 97–98, 220, 314, 445
Ethylbenzene, C_8H_{10}:
  dehydrogenation to styrene, 176, 366, 513, 522
    equilibrium considerations, 520–521, 522, 547
    reactor calculations, 531–534, 547, 548
Ethylene, C_2H_4, 152, 286, 390. *See also* Ethane
  reaction with C_4H_6, 79–80, 153, 313–314, 362, 377–379
Ethylene oxide, C_2H_4O, 40, 61, 70–71, 82, 171, 361, 600
Evans-Polanyi correlation, 123
Experimental methods in kinetics:
  differential, 49, 152
  to follow extent of reaction, 46–48
  general considerations, 45
  half-life method, 53–54
  initial-rate method, 50–51
  integral, 49, 70, 152, 190
  other quantities measured, 48
Explosion(s), 4, 22, 161–162
Exponential integral, 345, 398
Extent of reaction (parameter), 27, 31, 53, 93
E-Z Solve (computer software), 22, 61, 282, 540, 590, 592, 618, 635–642
  computer files, designation of, 22
  icon, 22
  syntax, 636–637, 638–639, 639–640, 641–642
  use in regression analysis for parameter estimation, 49, 50, 59, 98, 105, 459
  user-defined function(s), 59, 483, 559, 562, 563, 641–642

F

F, cumulative residence-time distribution function, 325, 332, 333, 334
  for BMF, 325, 326–327, 332
  definitions of $F(t)$ and $F(\theta)$, 321
  for LF, 325, 331, 332
  measurement from step input, 462–466
  for PF, 325, 329–330, 332

Subject Index   665

relation between $F(t)$ and $E(t)$, 321–322
relation to other age-distribution functions, 332t
for 2 CSTRs in series, 411, 412
Fast reaction (gas-liquid), 246, 250–252, 259
   first-order or pseudo-first-order, 250–251, 252, 253, 254, 259
   second-order, 251–252, 253, 254, 259
Fick's law, 200, 212, 227, 240, 244
First-order reaction(s), 52, 53, 56–57, 65, 69–71, 76, 77, 78, 237, 248–251, 253, 254, 398–399, 400, 405–406, 406–408, 415–416
   pseudo-first-order reaction, 70, 97, 243, 248–251, 253, 254
Fixed-bed catalytic reactor(s) (FBCR), 21, 287–290, 310, 515–546, 573. *See also* Bed; Catalyst particle(s); Gas-solid (catalyst) reactions; Particle(s)
   bed arrangement, 514, 515, 516
   bed dimensions, calculation of, 518–519, 533–534, 549
   classification of reactor models, 523–527, 549–550
   comparison of one-dimensional and two-dimensional models, 546
   examples (diagrams), 18, 286, 287, 288, 289
   flow arrangement, 513, 514, 515, 523
   heterogeneous reactor model, 512, 524, 525, 551
      one-dimensional PF model, 544–546, 550
   pressure drop, 516–519
   pseudohomogeneous, one-dimensional PF model, 527–544
      adiabatic, multistage operation, 529–542
         interstage heat transfer, 529–535, 547, 548, 549, 550
         cold-shot cooling, 515, 535–542, 547, 548, 550
      continuity equation, 527–528
      nonadiabatic operation, 528–529, 542–544
   pseudohomogeneous reactor model, 512, 524, 525, 527, 546
   pseudohomogeneous, two-dimensional DPF model, 525–527
Fluid-fluid reaction(s), 512, 599, 600, 602. *See also* Gas-liquid systems or reactions; Liquid-liquid reactions
Fluidized bed(s), 559, 570, 574–583. *See also* Fluidized-bed reactors(s)
   distributor, 570, 574, 584, 595, 596
   freeboard (region), 570, 571, 574, 584, 595–596
   hydrodynamic models, 569, 579–584
      bubbling-bed model, 580–583
         assumptions, 580–581
         bubble diameter correlation, 581
         bubble rise velocities, 581
         distribution of bed in regions, 581–582

   distribution of solid particles, 583
   exchange coefficients, 580, 583, 585, 592
   two-region model, 579–580
   minimum fluidization velocity, u_{mf}, 574, 575–577, 578, 596, 597
   pressure drop, 575, 596
   terminal velocity, 574, 577–578, 595, 596, 597
Fluidized-bed reactor(s), 21, 290–291, 310, 554, 559. *See also* Fluidized-bed(s)
   advantages and disadvantages, 572, 573–574
   fast-fluidized-bed reactor, 570–571, 574
   KL model for fine particles, 584–592, 597
      assumptions, 584–585
      bed depth, 586, 587, 589, 597
      bed diameter, 587, 589
      catalyst holdup, 587, 589
      complex reactions, 589–592, 598
      continuity equations, 585–586, 590, 591–592
   KL model for intermediate-size particles, 584, 592–594
   model for large particles, 584, 595
   pneumatic transport or transport riser reactor, 570, 571
   reaction in freeboard and distributor regions, 595–596
Fluid-particle interactions, 516–519, 569, 574–578
Flux, molar, 240, 242, 244, 249, 250
Formula matrix, *see* Matrix
Formula vector, 10
Fourier's law, 212, 228
Free-radical species, 116, 158
Fractional conversion, *see* Conversion, fractional
Fractional yield, *see* Yield, fractional
Friction factor, 370, 388, 517, 576, 577

G
Gamma function, 476–477, 483
Gas film, 229, 234, 247. *See also* Mass transfer
   mass transfer, 228, 236, 237, 257, 258, 564, 567
      control, 222, 233, 234, 236, 257, 560, 561, 562, 563, 564, 565, 567, 568
      resistance, 250, 258
Gas-liquid systems or reactions, 21, 239–255, 311, 599, 600, 602, 603. *See also* Fast reaction (gas-liquid); Reactors for gas-liquid reactions; Two-film model for gas-liquid systems
   correlations for parameters, 606, 608–609, 615–616
   examples of reacting, 239, 600
   mass transfer in, *see* Gas film; Mass transfer
   reaction in bulk liquid only, 242–243, 247, 254, 258, 603, 604, 606, 621
   reaction in liquid film and bulk liquid, 247–250, 254, 604, 606, 621

Gas-liquid systems or reactions (*continued*)
  reaction in liquid film only, 244–247, 603, 604, 606, 607, 621
  types of reactions, 599–600
Gas-solid (catalyst) reactions, 512–513, 551, 572. *See also* Fixed-bed catalytic reactor(s) (FBCR)
Gas-solid (reactant) systems, 21, 224–239. *See also* Reactors for gas-solid (noncatalytic) reactions
  examples of reacting, 224–225, 552, 573
  general kinetics model for constant-size particles, 225–229
  intrinsic kinetics, 255–257, 258
  shrinking-core model (SCM), 227, 228, 229–236, 237, 239, 257, 258, 260, 552, 553, 560, 561, 562, 563, 565, 566, 567, 568
    spherical particle, 229–234
    summary for various shapes, 234, 235t
  shrinking-particle model (SPM), 237–239, 258, 553
Gibbs energy of activation, 141, 142

H

Half-life, of a reactant, 40, 53–54, 71
Hatta number (Ha), 255, 259, 260
  as criterion for kinetics regime, 252–253
  definition, 249, 251, 254
  interpretation, 252–253
Heat capacity, 228, 298, 299
Heat transfer, 2, 15, 16, 17, 33, 37, 210, 228, 229, 286, 298, 338, 339, 341, 350, 354, 368, 369, 387, 515, 529, 542, 543, 584, 587
Heat transfer coefficients, 229, 298, 369, 573
Heaviside unit function, *see* Unit step function
Henry's law, 241, 243, 244
Heterogeneous catalysis, 178–179, 198–214, 263
Holdback H, 322, 325, 334
  and age-distribution functions, 322, 332t
Holdup:
  catalyst, 574, 584, 585, 587
  gas, 603, 608, 609, 615–616
  liquid, 603, 619
  solid, 557, 559, 560, 564, 566, 567, 574, 584
Homogeneous catalysis, 178, 180, 263, 276
Hydrogen-bromine reaction, 160–161, 171

I

I, internal-age distribution function, 325, 332, 333, 334
  definitions of $I(t)$ and $I(\theta)$, 322
  relation to other age-distribution functions, 332t
Ideal flow, 21, 25, 317–332, 552, 600, 601
  age-distribution functions, 325–332
Ideal reactor models, 21, 25, 38, 317, 333, 454
Inhibition, 78, 79, 161, 196, 269–276

Inhibitor(s), 261, 264, 270, 272, 275
Instantaneous reaction (gas-liquid), 244–246, 247, 250, 252, 253, 254, 255, 259, 603, 620
  enhancement factor, 247, 251, 253, 254, 258
  gas-film control, 245, 246, 254, 255, 258, 259
  liquid-film control, 245, 246, 254, 255, 258, 259, 620
  reaction plane, 244, 245, 258
Integrated forms of rate laws, 29, 51, 52, 53, 54, 57, 72, 74, 76, 152–153, 190, 269
Interfacial area, 573, 600, 602, 603
  gas-liquid, 600, 601, 603t, 608, 609 (correlation), 615 (correlation)

K

k_{Ag}, 195, 213, 228, 231, 236, 240, 258, 260, 564, 608, 609
$k_{A\ell}$, 241, 608, 609, 615, 621
k_{As}, rate constant for intrinsic surface reaction, 231, 237, 260
Kinetics, 2, 3, 6, 14, 15. *See also* Chemical kinetics; Kinetics scheme(s); Rate constant; Rate law(s); Rate parameters; Rate of reaction
  abnormal, 187, 189, 336, 350, 381, 383, 386, 404, 406, 417
  Langmuir-Hinshelwood, *see* Langmuir-Hinshelwood (LH) kinetics
  normal, 189, 336, 338, 350, 355, 382, 383, 386, 404, 406, 417
Kinetics scheme(s), 13, 14, 21, 101. *See also* Reaction network(s)
Kirchhoff equation, 445

L

Laminar flow (LF), 25, 36, 37, 38, 317, 318, 330–331, 332, 334, 393
  E, 325, 330–331, 332, 334, 393
  F, 325, 331, 332, 393
  H, I, W, 325, 332, 334
Laminar-flow reactor(s), 25, 36–38, 284, 318, 393–400
  comparison with CSTR and PFR, 406–408
  $E(t)$, 400, 401
  fractional conversion and order, 397–399, 400t
  material balance, 394–395
  and SFM, 400
  size determination, 397, 398–399
  uses, 393–394
Langmuir adsorption isotherm, 193, 215, 220, 223
Langmuir-Hinshelwood (LH) kinetics, 191, 192, 195–197, 208, 217, 219–221, 256, 259, 553
  beyond, 197–198
  and enzyme kinetics, 263, 269, 276
Laplace transform:
  use in deriving $E_N(\theta)$ for TIS model, 472–474
  use in deriving moments of $E_N(\theta)$, 474–476

Subject Index   667

Le Chatelier's Principle, 514
L'Hôpital's rule, applications of, 52, 62, 204, 236, 392
Lineweaver-Burk plot, 267, 268, 275, 277, 278
Liquid film, 247, 250, 255
Liquid-liquid reaction(s), 599, 600, 602

M

Macrofluid, 343
Macromixing, 343, 454, 495
Macrophase, 555
Maple, 10
Mass balance, 17, 282, 295. *See also* Material balance
Mass transfer, 2, 15, 16, 214, 225, 230, 237, 239, 242, 243, 553, 579, 584, 602, 603. *See also* Gas film
  film coefficients, *see* k_{Ag}, $k_{A\ell}$
    gas-film control, 241, 560, 561
    liquid-film control, 241
    relation to diffusivities, 240, 241, 244
  overall coefficients, for gas-liquid, 241
Material balance, *See* Batch reactor(s); Continuity equation(s); CSTR; Laminar-flow reactor(s); Plug-flow reactor(s)
Mathematica, 10, 12, 90
Matrix, 8, 10, 11, 12
  formula, 8, 11
  reduction, for generating chemical equations, 10
Maximum-mixedness model (MMM), 455, 495, 501, 502–504, 507–508
Mean residence time, 26, 30, 556, 557, 558, 560, 563, 565, 568
Mechanical design, 16, 279, 283
Mechanism, reaction, 2, 9, 13, 15, 115, 116, 154, 165
  Briggs-Haldane, 266–267
  for chain-reaction polymerization, 166–167, 172
  closed-sequence, 155, 157–162, 177
  for decomposition of acetaldehyde, 172
  for decomposition of dimethyl ether, 171
  for decomposition of ethylene iodide, 188
  for decomposition of ethylene oxide, 171
  for decomposition of ozone, 170
  derivation of rate law from, examples of, 155–157, 159–161
  Eley-Rideal, 197
  for enzyme activation, 273, 278
  for enzyme-reaction inhibition, 273, 275
  for ethane dehydrogenation, 116, 124–125, 137, 138–139, 158, 165, 172, 173–175
  for formation of HBr, 160–161, 171
  Lindemann, 135, 136, 137, 152
  Mars-van Krevelen for oxidation, 197
  for methane oxidative coupling, 164–165, 172
  for methanol synthesis, 180–181, 220
  Michaelis-Menten, 264
  for multiple-substrate inhibition, 270
  open-sequence, 155–157
  for pyrolysis of ethyl nitrate, 159–160
  for single-substrate inhibition, 270
  for step-change polymerization, 168–170
Methane, CH_4, 88
  partial oxidation to HCHO, 88, 90, 94, 108, 113, 444–445, 452
  oxidative coupling, 164–165, 172
Methyl acetate, $C_3H_6O_2$, 88, 109, 361, 423–426, 446
Methyl alcohol (methanol), CH_4O, 180–181, 218–219, 220, 221, 423, 446, 450–451, 512, 513, 514
  synthesis, 23, 96–97, 176, 196, 197, 219–220, 286, 289–290, 513
  reactors, 289, 366
Michaelis constant, 264, 265, 267–269, 271, 272, 274, 277, 278, 316
Michaelis-Menten:
  equation, 266, 267, 269, 270, 271, 272, 274, 276, 277, 278
  model, 264–266, 277
  parameters, 274, 277
Microfluid, 343
Micromixing, 343, 454–455, 495, 501, 502, 504, 508
Microphase, 555
Mixing, 2, 15, 16, 332, 333, 335, 343, 375, 393, 413, 453, 454–455, 579, 602
  axial, 318, 365, 393, 601
  earliness of, 413–416, 455
  radial, 318, 365, 393, 601
Mode(s) of operation, 21, 25, 280–281, 427, 428, 432, 512, 573. *See also* Steady state; Unsteady state
Molecular catalysis, 178, 180, 182–187. *See also* Acid-base catalysis; Organometallic catalysis
Molecular energy, *see* Energy in molecules
Molecular formula(s), 8, 9, 10
Molecularity, 116, 141, 142
  and order, 116
Molecular velocity, 128, 129, 148
Mole fraction, 66
Moments of distribution functions. *See also* Dispersed plug flow (DPF) model; Tanks-in-series (TIS) model
  about the mean, 324–325
    relation between σ_t^2 and σ_θ^2, 324–325
  about the origin, 323
  measurement:
    from pulse input, 458–462, 492, 493, 494
    with reactive tracer, 466–468
    from step input, 464–466, 492, 493

Moments of distribution functions, measurement (*continued*)
  with tailing, 468–471
  with time delay, 466
  use for determining model parameters:
    DPF, Pe$_L$, 492, 493
    TIS, N, 477–478, 492, 493, 494
Momentum balance, 366, 370, 517
Monoethanolamine (MEA), 239, 600, 619, 620
Moving-particle reactors, 512, 569, 570–574
Multiphase system(s) or reaction(s), 3, 16, 21, 224, 512, 599, 602, 618. *See also* Gas-liquid systems or reactions; Gas-solid (catalyst) reactions; Gas-solid (reactant) systems; Liquid-liquid reactions
Multiple vessel configurations, 355–361, 387–389, 408–418, 422–426, 602. *See also* CSTR; Plug-flow reactor(s); Tanks-in-series (TIS) model

N

$N_A(z = 0)$, 249, 250, 254, 605, 606, 608, 612, 614, 616
$N_A(z = 1)$, 250, 254, 605, 606, 608, 612, 614, 616
Nitric oxide, NO, 73, 84, 170
  reaction with O_2, 73, 390
    effect of T on rate, 5, 73, 83–84, 171
Nitrogen pentoxide, N_2O_5, 61, 83, 88, 112, 155–157
Natural gas, steam-reforming of, 12, 285, 286
Noncomponent (species), 8, 11, 12, 13, 17, 93
Nonideal flow, 21, 25, 26, 317, 319, 332, 333, 453–490, 495, 600. *See also* Dispersed plug flow (DPF) model; Macromixing; Maximum-mixedness model (MMM); Micromixing; Mixing; Residence time distribution (RTD); Tanks-in-series (TIS) model
  and reactor performance, 495–508
Nonsegregated flow, 335, 333, 344, 345, 364, 408, 413. *See also* Segregated flow
Nonsegregation, 332, 333, 343. *See also* Segregation

O

Open vessel, 318
Order of reaction, 42–43, 65, 75, 115
  comparison, 75–78
  and molecularity, 116
  and strong pore-diffusion resistance, 209
Organometallic catalysis, 186–187
Ozone, 23, 170, 182–183

P

Parallel reactions, 21, 87, 88, 100–103, 426–428, 435–437, 504–508
  activation energy, 110-111
  in a BR, 101–103, 110, 427, 428, 446, 449
  in a CSTR, 109, 427, 428, 447, 448, 451–452, 505, 508
  determination of rate constants, 101–102
  in a fluidized-bed reactor, 590
  in a PFR, 427, 428, 435–437, 446, 447, 448, 449–450, 505–506
  product distribution, 101, 102
  reactors for, 427, 428, 435–437
Parameter estimation, 21, 57–61, 482–483
  by regression analysis, 21
    guidelines for choices, 59–61
    linear regression, 49, 50, 51, 58, 98
    nonlinear regression, 49, 50, 51, 58, 98, 459, 462, 465, 468, 471, 477, 478, 482
    use of E-Z Solve, 49, 50, 59, 98, 105, 459
Partial pressure, 6, 66
  Arrhenius parameters in terms of, 68–69
  rate and rate constant in terms of, 67–68
Particle(s), *See also* Catalyst particle(s)
  density, 199, 226, 516, 517, 522, 547
  diameter, effective, 517, 576
  effectiveness factor, *see* Effectiveness factor
  shapes, 234, 235, 516
  size, effect on diffusion resistance, 208
  size parameters, 234, 235
  terminal velocity, 574, 577–578, 595, 596, 597
  tortuosity, 200
  voidage (porosity), 199, 200, 516, 547
Particle-size distribution (PSD), 553, 556, 559, 563, 584, 598
Partition function, 143–145, 153
Performance, reactor, 1, 2, 16, 21, 555, 556, 560, 573, 574, 601
  factors affecting, 16, 21, 413–414, 454, 455, 553, 554 (chart)
Phosphine, PH_3, 40, 220, 313
Photochemical reaction(s), 5, 118, 149–150, 163–164
Phthalic anhydride, 590–592, 598
Physicochemical constants, values, 623
Planck's constant, 118, 140
Plasmas, reactions in, 150–151
Plug flow (PF), 25, 33, 37, 317, 318, 327–330, 332, 334, 365, 388, 453, 454, 479, 483, 524, 554, 556–559, 563, 566, 567, 574, 579, 580, 585, 595, 600, 601, 604, 605, 608, 619
  dispersed, 524, 525. *See also* Dispersed plug flow (DPF) model
  E, 325, 327–329, 332
  F, 325, 329–330, 332
  H, I, W, 325, 332, 334
Plug-flow reactor(s), 25, 33–36, 284, 318, 365–389
  adiabatic operation, 369
  in combination with CSTR, 413–418, 419, 420
  comparison with BR, 404–405, 406, 408, 419
  comparison with CSTR, 405–408, 419, 420
  comparison with LFR, 406–408

configurational forms, 387–389
constant-density system, 370–375
   and batch reactor, 371, 375
     isothermal operation, 370–373
     nonisothermal operation, 373–374
determination of rate parameters with, 55–57
differential, 55–56, 108, 109
energy balance, 368–369, 444–445
$E(t)$, 374
general features, 33–34, 365
integral, 56–57, 113
material balance 34, 367–368
and parallel reactions, 427, 428, 435–437, 446, 447, 448, 449–450, 505–506, 508
pressure gradient, 370
recycle operation, 380–387, 416, 418
   constant-density system, 381–386, 390
   variable-density system, 386–387
residence time, 34, 35, 36, 365, 370, 371
and reversible reactions, 422, 433, 434–435, 446
and series reactions, 103, 429–430, 431, 437–442, 448, 449
and SFM, 374–375
space time, 35, 36, 367, 370, 371
uses, 33, 365–366
variable-density system, 376–380
Polymerization reactions, 165–170, 186
  chain-reaction, 166–167, 172
  step-change, 168–170, 173, 443–444, 452
Pore diffusion resistance:
  negligible, 204, 205, 208, 209
  significant, 205
  strong, 203, 204, 205, 207, 208, 209
    consequences, 209–210
Pressure drop, 15, 367, 370, 514, 516–519, 566, 574, 575, 596
Pre-exponential factor, 44, 57, 65, 145
  in terms of partial pressure, 68–19
Process design, 15, 279–282, 283, 294, 296, 336, 366, 394, 516, 523, 552, 599, 600, 607
Product distribution, 21, 87, 101, 102, 422, 432, 443, 446, 450, 452. See also Selectivity; Yield, fractional

Q
Quasi-steady-state approximation (QSSA), 231, 234
Quench cooling, 287, 289, 290 See also Fixed-bed catalytic reactor(s), cold shot cooling

R
Ranz-Marshall correlation for k_{Ag}, 236, 237, 258
Rate constant, 43, 44, 65, 67, 168, 444, 544
Rate-determining step (rds), 106, 136, 157, 233, 257, 259, 260, 267
Rate law(s), 5, 13, 14, 21, 42–45, 64, 66, 68, 69, 72, 76, 77, 78, 115

Rate parameters:
  experimental strategies for determining, 48–57
  methodology for estimation, 57–61
Rate of reaction, 1, 2, 3, 4, 5–6. See also Rate law(s)
  basis (mass, particle, volume), 3, 522, 528
  intrinsic, 208, 210, 224, 242, 247, 605, 617
  point, 4, 6, 54, 55
  sign, 3, 27
  species-independent, 3, 9, 65
  in terms of partial pressure, 67–68
Reaction coordinate, 121, 122, 123, 124, 140
Reaction mechanism, see Mechanism, reaction
Reaction network(s), 87, 100, 113, 114, 422, 427, 429, 444, 445, 504. See also Kinetics schemes
  compartmental (box) representation of, 89
  determination, 106–108
  for partial oxidation of CH_4 to HCHO, 90, 108
  and primary, secondary, and tertiary products, 107–108, 114
Reaction number, dimensionless, 75, 76, 343, 398, 406, 443, 498
Reactive intermediate, 116, 135, 154, 155, 197
Reactor(s), 1, 2, 5, 6, 280. See also Batch reactor(s); CSTR; Fixed-bed catalytic reactor(s); Fluidized-bed reactor(s); Laminar-flow reactor(s); Plug-flow reactor(s); Reactors for gas-liquid reactions; Reactors for gas-solid (noncatalytic) reactions; Stirred-tank reactor(s)
  dimensions, see Vessel dimensions
  examples of (diagrams), 18, 283, 284, 285, 286, 287, 288, 289, 290, 291
  moving-particle, 512, 569, 570–574
  size and product distribution, 422
  slurry, 559, 602
  trickle-bed, 599, 601, 618–619
  tubular-flow, 284, 285, 287
Reactors for gas-liquid reactions, 599–619. See also Gas-liquid systems or reactions
  tank, 602–603
    continuity equations, 614–615
    correlations for parameters, 615–616
    dimensions, 614, 616
  tower or column, 600–601, 602, 603–614
    bubble-column, 21, 601, 608–614
      continuity equations, 608
      correlations for parameters, 608–609
      dimensions, 610, 612, 613, 614, 621
    packed tower, 600, 603–608, 619, 620
      continuity equations, 605–606
      correlations for parameters, 606
      dimensions, 603, 607, 619, 620, 621
      minimum liquid flow rate, 607, 621

Reactors for gas-solid (noncatalytic) reactions, 552–566. *See also* Gas-solid (reactant) systems
   continuous reactors, 554–566
      solid particles in PF 556–559, 566, 567
      solid particles in BMF, 559–563, 566, 567
      factors affecting reactor performance, 553, 554 (chart)
   semicontinuous reactors, 553–554, 566
Reduced (molecular) mass, 130, 145
Residence time, 25, 33, 34, 35, 36, 318, 555, 567
Residence-time distribution (RTD), 16, 21, 26, 317, 318, 333, 552. *See also* Age-distribution function(s); F, cumulative residence-time distribution function; E, exit-age distribution function; W, washout residence-time distribution function
   applications, 455
   experimental measurement, 455–471
      normalized response, 458–459, 463, 501
      pulse input (signal), 455, 456, 458–462
      reactive tracer, 466–468
      step input (signal), 455, 456, 457, 462–466
      stimulus-response technique, 455–457
      tailing, 468–471
      time delay, 466
      tracer selection, 457–458
   for multiple-vessel configurations, 408, 410–413, 414–415, 420–421
Reversible (opposing) reactions, 21, 87, 94–100, 422–426, 445, 446, 449, 513
   in a BR, 97–98, 109, 112, 445, 446
   in a CSTR, 109, 423–426, 433–434, 446, 449
   equilibrium considerations, 514, 519, 520–521, 522, 547, 548
   examples, 87–88, 512–513
   locus of maximum rates, 99, 433, 522
   optimal T for exothermic, 99–100, 433, 521
   in a PFR, 422, 433, 434–435, 446
   rate behavior, for exothermic and endothermic, 100, 101, 521–522, 523
   rate law, 94–95, 97–98
      thermodynamic restrictions on, 95–97
   reactors for, 423–426, 433–435
Runaway reaction, 161–162, 368, 572

S

Second-order reaction(s), 68, 71–72, 76, 77, 78, 251–252, 253, 254, 259, 400, 405–406, 406–408, 415–416, 419
Segregated flow, 332–333, 344, 345, 364, 408, 413. *See also* Nonsegregated flow
Segregated-flow model (SFM), 317, 333, 335, 343, 374–375, 400, 455, 495, 501, 552, 555, 556. *See also* Segregated flow reactor model
Segregated-flow reactor model, 501–502, 504, 506–507, 508, 510, 511. *See also* Segregated-flow model (SFM)
Segregation, 16, 343. *See also* Nonsegregation
   degree of, 332, 343, 344, 413, 455
Selectivity, 92, 108, 109, 381, 422, 429, 432, 504, 508, 513, 551, 572, 573, 574, 589, 590, 591, 595, 598, 602. *See also* Product distribution; Yield, fractional
Semibatch reactor(s), 113, 309, 310, 311–313, 316, 559
Semicontinuous reactor(s), 309–310, 311–313, 553–554, 566, 602
Series reaction(s), 21, 87, 88, 103–106, 113, 429–432, 574
   in a BR or PFR, 103–106, 111, 112, 429–430, 431, 447, 451
   in a CSTR, 111, 430–431, 447, 448
   in a fluidized-bed reactor, 590–592, 598
   operating conditions, 432
   and TIS reactor model, 498–499
Shrinking-core model (SCM), *see* Gas-solid (reactant) system(s)
Shrinking-particle model (SPM), *see* Gas-solid (reactant) system(s)
Simple collision theory (SCT), *see* Collision theory
Simple system(s), 4, 7, 8, 9, 13, 17, 21, 25, 42, 45, 46, 64, 335, 365, 545
Sintering, 215–216
Space time, 26, 30, 114, 337, 340
   as a scaling factor, 372
Space velocity, 26, 566
Spouted bed, 571–572
Stationary-state hypothesis (SSH), 135, 155, 260, 266, 267, 272, 274, 275, 276
Steady state operation or behavior, 17, 29, 30, 31, 32, 33, 34, 37, 40, 200, 201, 225, 230, 240, 243, 249, 325, 335, 337, 338, 339, 340, 341, 353, 355, 367, 368, 393, 512
Step function, *see* Unit step function
Steric factor, 131, 132, 153
Stirred-tank reactor(s), 21, 284, 423. *See also* CSTR
Stoichiometric coefficient(s), 8, 9, 13–14, 43
Stoichiometric number, 156, 159
Stoichiometric table, 39, 56, 93–94, 302, 345–346, 346–347, 348, 377
Stoichiometry, 1, 6, 13
   chemical reaction, 6, 7, 10, 113, 114
   and reaction mechanism, 154, 156
   use in relating rate constants, 43
   use in relating rates of reaction, 9, 13–14
Styrene, C_8H_8, 513, 514. *See also* Ethylbenzene
Substrate, 262, 266, 270
Sucrose, hydrolysis of, 262, 268, 277

Subject Index 671

Sulfur burner, 293
Sulfur dioxide, SO_2, 552, 573, 599
   oxidation, 5, 7, 18, 40, 87, 100, 152, 176, 223, 292, 293, 366, 512, 513, 520
      Eklund rate law, 223, 521–522, 523 (chart), 546, 549, 550
      equilibrium considerations, 520, 548
      industrial reactor, 18–19
      reactor calculations, 518–519, 549, 550
Sulfuric acid, 5, 7, 18, 292, 293, 366, 512, 573
Surface catalysis, 178, 180, 187, 191–198, 263
Surface reaction(s), 119, 125, 148, 149, 152, 191, 192, 197–198, 214, 237, 257, 553, 557, 560, 564
   activation energy of, and strong pore-diffusion resistance, 209–210
   order of, and strong pore-diffusion resistance, 209
   - rate control, 214, 222, 233, 257, 560, 562
System, 2, 3, 15, 17, 26, 34, 280. *See also* Complex system(s); Gas-liquid systems or reactions; Gas-solid (reactant) systems; Multiphase system(s) or reaction(s); Simple system(s)

T

t_1, time required for complete conversion of particles, 233, 234, 235, 238, 239, 257, 558, 568
$t(f_B)$, for various shapes of particlet, SCM, 235
Tanks-in-series (TIS) model, 471–483, 495, 525. *See also* Tanks-in-series (TIS) reactor model
   comparison with DPF model, 490
   determination of N, 477–478, 481, 492, 493, 494
   $E_N(\theta)$, 471–478
   $F_N(\theta)$, $F(t)$, 478–479, 481–482
   moments, 475–476
   nonintegral values of N, 476–477
   $W_N(\theta)$, 479–480
Tank reactor, *see* Reactors for gas-liquid reactions; Stirred-tank reactor(s)
Tanks-in-series (TIS) reactor model, 495–499, 508, 509, 510, 511. *See also* Tanks-in-series (TIS) model
Termolecular reaction, 116, 118, 137–139, 145
Theories of reaction rate, 115, 128–152. *See also* Collision theory; Transition state theory
Thermal conductivity, effective, 211, 228, 526
Thermodynamics, 1, 14–15, 95–97, 141–143. *See also* Equilibrium, chemical/reaction; Equilibrium constant
Thiele modulus, 203, 216, 217, 221, 222, 223, 253, 522, 544, 545
   normalized for particle shape, 206
   for nth-order reaction, 207
   general form, 207–208

Tower reactor, *see* Reactors for gas-liquid reactions
Third-order reaction(s), 72–75, 83–84, 170, 171, 390
Tracer, 325, 326, 327, 328, 330, 331, 334, 411, 414, 455, 457–458, 466–468
Transition state, 121, 122, 124, 125, 132, 140, 144, 145, 153
   equilibrium constant for formation, 139, 140, 141, 143
   partition function, 143
Transition state theory (TST), 115, 139–145, 153
   and activation energy, 141
   comparison with SCT, 145–146
   pre-exponential factor, 141, 142, 145
   rate constant, 140, 141
   and statistical mechanics, 143–145
   thermodynamic formulation, 141–143
Trickle-bed reactor, 599, 601, 618–619
Tubular flow, 25, 318, 393
Tubular-flow reactors, 284, 285, 287
Two-film model for gas-liquid systems, 240–255, 258, 259, 599, 602, 604, 621. *See also* Enhancement factor; Fast reaction; Hatta number; Instantaneous reaction
   profiles, 240, 243, 244, 247, 259
   summary of rate expressions, 254 (chart), 255

U

Uniform reaction model (URM), 227
Unimolecular reaction, 116, 117, 125, 134–137, 195–196
Unit impulse function, *see* Dirac delta function
Unit(s), 19, 20, 623
Unit step function, 329, 330, 365
Unsteady state operation or behavior, 17, 26, 29, 31, 33, 34, 37, 225, 230, 325, 335, 337, 341–342, 553–554

V

van't Hoff equation, 44, 157, 520
Variance, 324
Velocity profile, 25, 37, 330
Velocity, superficial linear, 557, 566, 567, 570, 574, 607, 611
Vessel dimensions, 388, 543, 544, 557, 567, 574, 600, 603, 608, 610, 619, 620
W, washout residence-time distribution function, 325, 332, 333, 334, 503
   definitions of $W(t)$ and $W(\theta)$, 322
   relation to other age-distribution functions, 332t

W

Weisz-Prater criterion, 208–209, 222

Y

Yield of a product, 91, 422, 514, 589, 595, 598
Yield, fractional, 92, 107, 109, 513, 516. *See also* Product distribution; Selectivity
    instantaneous, 92, 427, 432
    overall, 92, 102, 435, 551

Z

Zero-order reaction(s), 65, 76, 77, 78, 187, 197, 209, 221–222, 259, 364, 397–398, 400, 401, 406–408

Directory of Examples Solved Using E-Z Solve

Example #	Text page	Filename	Topic
3-8	59	ex3-8.msp	estimation of kinetics rate parameters by nonlinear regression
4-3	70	ex4-3.msp	estimation of kinetics rate parameters by nonlinear regression
4-4	71	ex4-4.msp	estimation of kinetics rate parameters by nonlinear regression
4-8	79	ex4-8.msp	estimation of Arrhenius parameters by nonlinear regression
5-4	98	ex5-4.msp	estimation of kinetics rate parameters for a reversible reaction
9-6	253	ex9-6.msp	enhancement factor: gas-liquid kinetics
10-1	268	ex10-1.msp	estimation of kinetics rate parameters for an enzyme reaction
12-2	300	ex12-2.msp	performance of a constant-volume batch reactor
12-5	305	ex12-5.msp	adiabatic operation of a batch reactor
12-6	308	ex12-6.msp	optimization of batch-reactor operation
12-7	311	ex12-7.msp	semibatch operation
14-3	341	ex14-3.msp	transient operation of a CSTR
14-6	346	ex14-6.msp	gas-phase reaction in a CSTR
14-7	348	ex14-7.msp	multiple steady states in an isothermal CSTR
14-8	351	ex14-8.msp	multiple steady states in an adiabatic CSTR
14-9	357	ex14-9.msp	multistage CSTR
14-11	359	ex14-11.msp	economics of multistage CSTR design
15-3	372	ex15-3.msp	isothermal gas-phase reaction in a PFR
15-4	373	ex15-4.msp	nonadiabatic operation of a PFR
15-6	376	ex15-6.msp	variable-density operation of a PFR
15-7	377	ex15-7.msp	adiabatic, variable-density operation of a PFR
15-8	379	ex15-8.msp	effect of pressure drop on PFR performance
16-3	398	ex16-3.msp	sizing of an LFR
17-2	405	ex17-2.msp	comparison of CSTR and PFR performance
17-7	417	ex17-7.msp	reactor combination for autocatalytic reaction
18-1	423	ex18-1.msp	effect of reactor staging on conversion for a reversible reaction
18-6	435	ex18-6.msp	reactor design for parallel reaction network
19-1	461	ex19-1.msp	RTD analysis for a pulse input
19-2	464	ex19-2.msp	RTD analysis for a step input
19-3	466	ex19-3.msp	RTD analysis with a reactive tracer
19-4	469	ex19-1.msp	RTD analysis - pulse input with tailing
19-7	477	ex19-7.msp	calculation of gamma function
19-8	478	ex19-3.msp	parameter estimation for tanks-in-series model
19-9	481	ex19-2.msp	parameter estimation for tanks-in-series model
19-10	489	ex19-10.msp	parameter estimation for axial dispersion model
20-2	497	ex20-2.msp	2nd order reaction in 5 tanks-in-series
20-3	498	ex20-3.msp	series reaction network in 3 tanks-in-series
20-4	502	ex20-4.msp	segregated flow model for 1st order reaction
20-5	504	ex20-5.msp	segregated flow, maximum mixedness models for parallel reactions
21-2	518	ex21-2.msp	sizing bed depth and diameter in a fixed bed catalytic reactor
21-3	520	ex21-3.msp	equilibrium profile for a reversible reaction
21-5	531	ex21-5.msp	ethylbenzene dehydrogenation in a fixed bed catalytic reactor
21-6	540	ex21-6.msp	design of a fixed bed reactor with cold-shot cooling
22-2	558	ex22-2.msp	gas-solid reaction: spherical particles in PF, particle size distribution
22-3	560	ex22-3.msp	gas-solid reaction: cylindrical particles in BMF
22-4	563	ex22-4.msp	gas-solid reaction: sensitivity of design to rate-limiting process
22-5	565	ex22-5.msp	gas-solid reaction with more than one rate limiting process
23-2	577	ex23-2.msp	minimum fluidization velocity in a fluidized-bed reactor
23-3	587	ex23-3.msp	design of a fluidized-bed reactor: 1st order reaction
23-4	590	ex23-4.msp	series reactions in a fluidized-bed reactor
23-5	593	ex23-5.msp	operation of fluidized-bed reactor in intermediate particle regime
24-2	610	ex24-2.msp	gas-liquid systems: performance of a bubble-column reactor
24-3	616	ex24-3.msp	gas-liquid systems: design of a tank reactor